Marton Dénes

WILEY SERIES IN ION CHEMISTRY AND PHYSICS

The interests of ion chemists and physicists are very diverse, covering processes relevant to organic, inorganic, and analytical chemistry, and atmospheric, interstellar, plasma, and material sciences. Ion physicists and chemical physicists in general are concerned about detailed studies of simple atomic and molecular ion systems for comparison with first principle theoretical calculations, while ion chemists are usually interested in establishing trends in structures, stabilities, and reactivities of more complex molecular ion systems of chemical importance. Experimental and rigorous theoretical comparisons made in the studies of small ionic systems often reveal the fundamental nature of chemical and physical interactions involving ions. Results from reactivity and energetics measurements of more complex systems may provide guidance in the selection of chemically interesting systems for future detailed studies. Partly due to the historical developments, scientific communication between ion physicists and ion chemists has been sparse. As a result of many recent advances in experimental techniques and in theoretical methodologies, detailed studies concerning the structures, energetics, and reaction dynamics of state- or energy-selected polyatomic molecular ions can now be performed with increasing sophistication, and the artificial boundary between the ion chemistry and ion physics groups is disappearing. Over the past two decades, the community of chemists and physicists interested in the study of ions has grown significantly and they are in need of a permanent forum for scientific exchanges. This review book series, which publishes reviews of fundamental studies in ion chemistry and physics, is established partly with this goal in mind. In order to better serve the ion chemistry and physics community, we have decided that each volume of this series will be dedicated to the review a specific topic, emphasizing new experimental and theoretical developments in the study of ions. Other than serving as an up-dated reference book series, it is our hope that this series will help to stimulate new research directions and to point to future opportunities in the field of ion chemistry and physics.

SERIES EDITORS

EDITORIAL BOARD

LOW ENERGY ION–SURFACE INTERACTIONS

Wiley Series in Ion Chemistry and Physics

Edited by

J. Wayne Rabalais

Department of Chemistry, University of Houston, Houston, Texas 77204-5641, USA

JOHN WILEY & SONS

Chichester · New York · Brisbane · Toronto · Singapore

Telephone: National Chichester (0243) 779777
International +44 243 779777

Other Wiley Editorial Offices

John Wiley & Sons, Inc., 605 Third Avenue,
New York, NY 10158-0012, USA

Jacaranda Wiley Ltd, 33 Park Road, Milton,
Queensland 4064, Australia

John Wiley & Sons (Canada) Ltd, 22 Worcester Road,
Rexdale, Ontario M9W 1L1, Canada

John Wiley & Sons (SEA) Pte, 37 Jalan Pemimpin #05-04,
Block B, Union Industrial Building, Singapore 2057

Library of Congress Cataloging-in-Publication Data

Low energy ion–surface interactions / editor, J. Wayne Rabalais.
p. cm. — (Wiley series in ion chemistry and physics ; vol.)
Includes bibliographical references and index.
ISBN 0 471 93891 2
1. Ions—Scattering. 2. Surfaces (Physics) 3. Surface chemistry.
I. Rabalais, J. Wayne, 1944– II. Series.
QC702.7.S3L69 1994
530.4′27—dc20 93-21244
CIP

British Library Cataloguing in Publication Data

A catalogue record for this book is available from the British Library

ISBN 0 471 93891 2

Typeset in 10/12pt Times by Techset Composition Ltd, Salisbury, Wiltshire
Printed and bound in Great Britain by Biddles Ltd, Guildford, Surrey

CONTENTS

LIST OF CONTRIBUTORS

R. A. Baragiola, *Department of Materials Science and Engineering, University of Virginia, Charlottesville, VA 22901, USA*

E. R. Behringer, *Laboratory of Atomic and Solid State Physics, Cornell University, Ithaca, NY 14853-2501, USA*

P. G. Bertrand, *Laboratory of Physical Chemistry and Physics, Catholic University of Louvain, 1 Place Croix du Sud, B1348 Louvain-la-Neuve, Belgium*

R. Bruinsma, *Department of Physics, University of California at Los Angeles, Los Angeles, CA 90024, USA*

B. H. Cooper, *Laboratory of Atomic and Solid State Physics, Cornell University, Ithaca, NY 14853-2501, USA*

W. Heiland, *Fachberich Physik, University of Osnabrück, Barbarastrasse 7, Postfach 4469, D-4500 Osnabrück, Germany*

O. C. Hellmann, *NTT Basic Research Laboratories, 3-9-11 Midoricho, Musashino-shi, Tokyo 180, Japan*

N. Herbots, *Department of Physics and Astronomy, Arizona State University, Tempe, AZ 85287, USA*

D. C. Langreth, *Department of Physics, Rutgers University, Piscataway, New Jersey 08855-1849, USA*

D. Marton, *Department of Chemistry, University of Houston, Houston TX 77204-5641, USA*

P. Nordlander, *Department of Physics, Rice University, Houston, TX 77251-1892, USA*

J. W. Rabalais, *Department of Chemistry, University of Houston, Houston, TX 77204-5641, USA*

J. Rudnick, *Department of Physics, University of California at Los Angeles, Los Angeles, CA 90024, USA*

H. Shao, *Department of Physics, Rutgers University, Piscataway, NJ 08855-1849, USA*

O. Vancauwenberghe, *Department of Materials Science and Engineering, Massachusetts Institute of Technology, Cambridge, MA 02139, USA*

P. Varga, *Institut für Allgemeine Physik, Technische Universität Wien, Wiener Hauptstrasse 8-10/134, A-1040 Vienna, Austria*

X. Wang, *Department of Physics and Astronomy, Arizona State University, Tempe, AZ 85287-1504, USA*

R. S. Williams, *Department of Chemistry, University of California at Los Angeles, Los Angeles, CA 90024-1569, USA*

P. Yeh, *Department of Physics and Astronomy, Arizona State University, Tempe, AZ 85287-1504, USA*

PREFACE

A broad range of physical and chemical phenomena are accessible by means of low-energy ion beams. The ability of such ion beams to stimulate selected physical and chemical processes on surfaces has been the subject of intense research over the past decade. Although it is now well appreciated that there are significant technological applications for these ion–surface processes, a rigorous understanding of the fundamental physics and chemistry involved in the interactions is often lacking and, as a result, progress in developing the applications is impeded. The purpose of this volume is to attempt to provide a treatment of the state of the art of the understanding of low-energy ion–surface interactions along with some of the promising applications.

The regime that we consider as 'low energy' must be defined. A reasonable lower limit is a kinetic energy of the order of bond energies, *ca* 5 eV. The energy released upon bond formation is dissipated as phonon-like excitations and lattice distortions which perturb the local environment in a manner similar to the excess kinetic energy supplied by ion beams. At this lower limit, chemical-bonding interactions become significant, the binary-collision approximation becomes questionable, and inelastic interactions between the ions and substrate can alter the ion trajectories. A reasonable higher limit is a kinetic energy of the order of a few keV. In this range the sputtering yield becomes equivalent to or higher than the beam flux, classical ion-trajectory simulations using a binary-collision approximation provide a satisfactory description of the collision events, and the impinging ions remain in the near-surface layers. Hence, this volume focuses on the 'subplantation' regime where the energies of the impinging particles are higher than those of thermal species, which result in surface deposition, but lower than those of highly accelerated particles, which are implanted into bulk layers. Although we use the term 'ion' extensively to identify the impinging species, it is known that ions with energies at the lower limit of this regime are efficiently neutralized near surfaces. Therefore the actual particles colliding with the surface are often neutrals in various electronic states produced by the neutralizing electronic transitions.

This volume provides some of the basic principles relating to the interactions of low-energy ions with surfaces including some of the physical methods applied to these studies. Due to the large amount of research being carried out in this field, we have to be selective in the material that we consider. This selection is guided by my own interests and prejudices. Some important topics, such as sputtering and molecular-dynamics simulations, have been left out because it was felt that there were already excellent recent reviews or treatments available. The progression of topics covered by the chapters in this volume, i.e. scattering, ion–surface charge exchange, and deposition, proceeds from fundamental physics and chemistry towards applications.

Chapter 1 provides a theoretical treatment of the classical scattering and recoiling trajectories for ions impinging on well-defined crystal surfaces. These simulations are important in interpreting experimental scattering and recoiling data in terms of models of the atomic structure of the target surface.

Chapter 2 presents the experimental details of ion scattering and recoiling for surface elemental analysis and structure determinations. Surface-structure determinations from low-energy ion scattering and recoiling have been the subject of many advances in the last decade and the technique is rapidly developing into a surface crytallography.

Chapter 3 is a theoretical treatment of the subject of electronic charge transfer in atom–surface collisions. A method for describing the dynamics of the charge transfer probability in atom–surface scattering, a major unsolved problem, is described in this chapter.

Chapter 4 treats ion-stimulated electron emission from surfaces. Electron emission is a consequence of inelastic collision events and accompanies most of the processes described in this volume. The underlying theoretical concepts are presented and experimental data are interpreted in terms of these concepts.

Chapter 5 considers the dynamics of ion–surface collisions in the low and 'hyperthermal' *ca* <100 eV, energy regime. Many of the fundamental ion–surface interactions in this hyperthermal regime are poorly understood due to both experimental and theoretical complexities. This chapter presents experimental techniques and theoretical models appropriate to this energy regime.

Chapter 6 discusses the interaction of low-energy molecular ions with surfaces. Due to the many degrees of freedom existing in molecules, there are many more inelastic-energy channels available as compared to atom collisions with surfaces. Experimentally observed phenomena involving molecular ion scattering are presented, experimental systems are described, and theoretical models for interpretation are presented.

Chapter 7 focuses on the influence of the potential energy or charge state of impinging hyperthermal rare-gas ions on the sputtering of adsorbates, such as CO on Ni, and insulators, such as LiF. The charge state of the projectile is found to have a large effect on the secondary-ion yield. Experimental results demonstrating these effects are presented and the phenomena are modeled in terms of electron charge exchange processes with the surface.

Chapter 8 considers synthesis of thin films on surfaces by means of low-energy ion-induced chemical reactions. The specific cases of ion-beam oxidation and nitridation of the semiconductors Si, Ge, Si_xGe_{1-x}, and GaAs are considered.

The mechanisms of these chemical reactions, including diffusion and kinetic pathways, have been incorporated into a model for simulating the experimental observations.

Chapter 9 treats the subjects of film deposition using low-energy ions in which the elements in the film are directly deposited from ion beams. The hyperthermal nature of the ions results in subsurface growth called 'subplantation', producing films with distinctive qualities. The experimental techniques and apparatus for direct ion-beam deposition are described, experimental results for insulating, semiconducting and metallic films are presented, and models for interpretation are developed.

Chapter 10 reviews some of the most popular models of thin-film growth from various deposition processes. Emphasis is placed on the consequences of growth instabilities with regard to the roughness of the film surface, the fractal dimensionality of the bulk film, and the general morphology of the resulting surface.

Houston, Texas
April 1993

J. Wayne Rabalais

1 QUANTITATIVE INTENSITY ANALYSIS OF LOW-ENERGY SCATTERING AND RECOILING FROM CRYSTAL SURFACES

R. Stanley Williams
Department of Chemistry and Biochemistry,
University of California Los Angeles,
Los Angeles, California 90024-1569, USA

CONTENTS

Low Energy Ion–Surface Interactions
Edited by J. W. Rabalais

1.1 INTRODUCTION

This chapter presents a status report of the current state of the art in calculating the intensities of ions and/or atoms that have scattered or recoiled from the surface of a crystal. These calculations are useful for interpreting experimental scattering and recoiling data in terms of models of the atomic structure of the surface that acts as a target for a beam of incident projectiles. This is an area to which many researchers have contributed, and perhaps surprisingly, since the calculations are based on classical mechanics, it is still an area of active development. Many different approaches have evolved, and each has its champions and practitioners. The chapter is an attempt at collecting, describing and comparing these methods. It is strongly biased by the author's viewpoint and experience with low-energy scattering and recoiling. It may suffer from incompleteness or glaring misrepresentations of the work of others, but it should at least provide the reader with the proper background to understanding the various types of calculations that have been or could be performed and provide an entry into the literature on this topic.

1.1.1 BACKGROUND

Ion scattering first attracted attention as a spectroscopic technique for materials analysis over 25 years ago. Two different applications appeared nearly simultaneously in the United States. Turkevich (1961) began a series of investigations into using the large-angle scattering of energetic charged particles for the chemical analysis of surfaces. This work gained wide recognition as a result of a scattering experiment that was part of the scientific mission of Surveyor V, in which radioactive sources emitted MeV alpha particles and solid-state detectors were utilized to measure the energy of alphas that back-scattered from the lunar surface (Turkevich *et al.*, 1968). At about the same time, Smith (1967) built a rather simple analytical instrument in which beams of approximately 1 keV ions (He^+, Ne^+, or Ar^+) were directed at a solid target and an energy spectrum of the ions that scattered from the surface was collected with an electrostatic analyzer. Since the amount of energy transferred in an elastic collision depends upon the mass of the target atom, analyzing the energy of the scattered particles provided a means of identifying the atoms that scattered the ions. In both of the above experiments, elastic scattering of the incident particles from individual atoms with different masses in the respective targets into the energy-dispersive detectors produced features in the spectra at characteristic energies. Since the scattering process is mass-dispersive, the energy spectra of the scattered particles were also mass spectra of the surfaces of the targets (Chu *et al.*, 1973; Brongersma and Mul, 1973).

In the same time period, Robinson and Oen (1963) performed computer simulations that showed that ions traveling through a crystal would have dramatically different transmission probabilities depending on the direction of travel. Ions incident along a major crystallographic direction have a very high transmission probability, since they are channeled into the gaps between the rows of atoms by the action of multiple low-angle collisions with the atoms in the rows. This leads to a strong anisotropy in the angular dependence of the intensity of high-energy ions that are scattered from a single crystal, which became the basis of the channeling techniques for the study of the bulk atomic structure of crystals (Gemmel, 1974).

A very active experimental and theoretical effort centered on keV-energy ion scattering also started in the Soviet Union in the early 1960s. However, much of that work was not accessible or known to researchers outside the Soviet Union until relatively recently. Very thorough English-language reviews have now appeared in books by Mashkova and Molchanov (1985) and by Parilis *et al.* (1993), the latter of which is dedicated primarily to theoretical studies and contains over 1000 references covering the period up to 1989.

By the late 1970s, several researchers had realized that the different ion-scattering techniques, including Rutherford back-scattering spectrometry (RBS—incidence energies in the MeV range), medium energy ion scattering (MEIS—incidence energies in the 100 keV range) and low energy ion scattering (LEIS), otherwise known simply as ion-scattering spectroscopy (ISS—incidence energies from 1 to 10 keV) could also be used to determine the atomic structure of solid surfaces (Stensgaard *et al.*, 1978; Turkenburg *et al.*, 1978; Poelsema *et al.*, 1977a). These approaches utilized the local shadowing and blocking of the projectiles by the target atoms near the surface of a crystalline solid, as shown in Figure 1.1. In principle, one should be able to geometrically triangulate the positions of atoms on a surface by determining the angles of particular features, such as peaks or valleys, in the back-scattered yield of particles from the surface (Aono *et al.*, 1981; Williams, 1982).

The details of how the experiments are performed in each of the energy ranges and the case histories of data analysis are covered in numerous previous reviews (Feldman and Mayer, 1986; van der Veen, 1985; Overbury, 1987; Echstein, 1987; Rabalais, 1988; Fauster, 1988; Niehus, 1988; Taglauer *et al.*, 1988; Chang *et al.*, 1989; Aono *et al.*, 1989; Rabalais, 1990; Niehus and Spitzl, 1991), and in Chapter 2 of this monograph. The complications and variations are especially numerous for the ISS regime, because of the strong neutralization of low energy

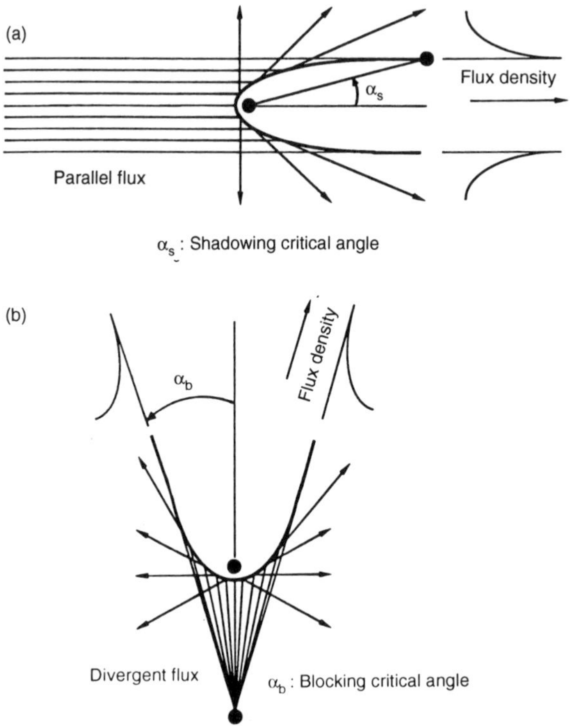

Figure 1.1 Illustration of (a) shadowing and (b) blocking cones. The edges of the cones are regions of increased flux arising from the focusing of the projectile trajectories. The critical angles are the minimum possible values for which the scattered projectiles can intercept a second atom with a small impact parameter (a) or exit to reach the detector (b)

ions that occurs near a solid surface. These phenomena have been investigated at length by Woodruff (1982) and are also reviewed in Chapter 3 of this monograph. The discussion in the present chapter will be restricted to the fundamentals of the kinematics, particle trajectories and scattering intensities for elastic collisions, ignoring all other processes. If the back-scattered yield being analyzed in an experiment is affected by some other process, the basic physics of that process must also be incorporated in any algorithm for calculating the yield from a structural model of the target.

1.1.2 SURFACE-STRUCTURE ANALYSIS IN REAL SPACE

The major advantage of ion scattering over other surface crystallographic techniques, such as thermal-energy atom diffraction, low-energy electron diffraction (LEED), or X-ray diffraction, is that the structural information is available in coordinate space, instead of reciprocal space. For incidence energies greater than 1 keV, the de Broglie wavelength $\lambda = \hbar/p$, where p is the momentum of the projectile, is less than 10^{-4} nm, which is much smaller than the lattice constant of any solid. Thus, except for the case of rainbow scattering as discussed below, quantum-mechanical effects can be ignored because there are neither interference effects nor diffraction from the lattice. The anisotropy in the ion yield of keV–MeV ions scattered from a crystalline target is caused by a sequence of classical collisions, and thus Newtonian mechanics suffices to calculate the intensity of the back-scattered yield. This makes interpretation of the data more straightforward in principle, since peaks and valleys in the angular dependence of the scattering yield can often be directly related to relative positions of atoms with respect to each other near the surface. Ion scattering techniques are primarily probes of the local structure near a particular scattering center, and thus they do not require long-range order to provide structural information.

There is a 10-dimensional space of variables experimentally available for the investigation of surface structures in scattering and recoiling experiments: (1) the identity (i.e. mass and atomic number) of the incident projectile, (2) the energy of the incident projectile, (3) the polar and (4) azimuthal angles of the projectile incidence with respect to the sample coordinate system, (5) the energy of the scattered projectiles, (6) the polar and (7) azimuthal angles of the scattered projectiles, (8) the energy of the target atoms recoiled from the surface, and (9) the polar and (10) azimuthal angles of the recoiled targets. This is an enormous parameter space to explore, and it is quite impossible to collect a data set in which all the available probe and response variables are systematically varied. In fact, experimental limitations usually fix most of the parameters, and the scattering intensity is usually measured as a function of only one or two of the variables (see the reviews listed in Section 1.1.1). However, with recent developments in detectors for scattered and recoiled particles and in laboratory computers for data acquisition and storage, it is common to collect complete energy spectra of scattered projectiles or recoiled targets over an entire range of both incident azimuthal and polar angles (Aono *et al.*, 1989; Rabalais, 1990). Such rich data sets can in principle improve the quality and uniqueness of an atomic structural determination, but they also pose major challenges to those who would attempt to quantitatively analyze them.

1.1.3 ENERGY REGIMES AND SCATTERING PHYSICS

Each of the different energy ranges used for ion scattering and recoiling has particular advantages and disadvantages, since the nature of the scattering and thus the particle trajectories are strongly energy-dependent. Traditionally, ion

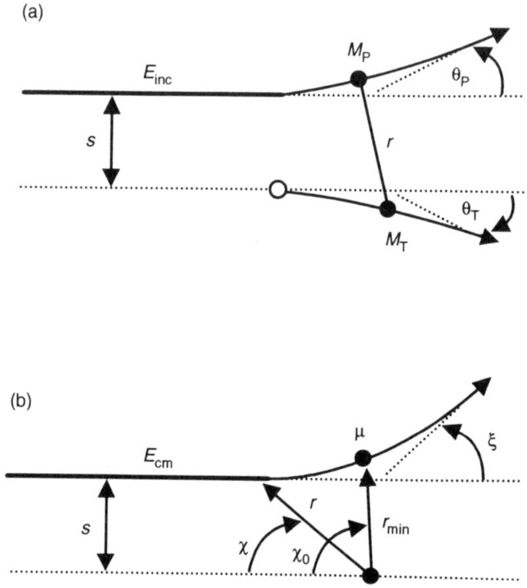

Figure 1.2 Illustration of the trajectories of the projectile and target in (a) the LAB coordinate system. The angles θ_P and θ_T are the asymptotic scattering and recoiling angles of the projectile and target, respectively, with respect to the incident trajectory. The trajectory of the μ-particle in (b) the CM coordinate system. The angle ξ is the asymptotic scattering angle of the μ-particle, whereas χ is the instantaneous angular position of the μ-particle with respect to the scattering center

scattering techniques have been divided into the three regimes defined above: RBS, MEIS and LEIS, primarily depending on the equipment utilized for the experiments. At the low end of the low-energy range, roughly less than 1000 eV, the physics of the scattering events is significantly different from that at higher energies, and thus a fourth category, the hyperthermal energy regime (Hulpke and Mann, 1983; Horn *et al.*, 1988), must also be defined.

In the RBS regime, the kinetic energy of the projectiles is so high that at the distance of closest approach, the nucleus of the projectile is entirely within the electron cloud of the target. This can be seen by equating the initial or incident kinetic energy E_{inc} for a light projectile scattering from a heavy target with the repulsive Coulomb potential of the bare nuclei and solving for the distance of closest approach, r_{ca}:

$$E_{inc} \approx \frac{Z_P Z_T e^2}{r_{ca}}, \tag{1}$$

where Z_P is the atomic number of the projectile, Z_T is the atomic number of the target atom, and e is the unit electrical charge. For the case of a 1 MeV

He^+ ion scattering from a Ag atom, r_{ca} is $\sim 1.3 \times 10^{-4}$ nm. This means that electron screening of the nuclei from each other can reasonably be neglected and scattering trajectories involving nearly head-on collisions can be calculated almost exactly from the bare Coulomb potentials of the interacting nuclei, just as was originally done by Rutherford (1911).

Equation (1) also gives a good indication of the effective size of the target atom, which can be roughly approximated as a sphere with radius r_{ca}, as sensed by the projectile. The effective cross-sectional area of the sphere presented to incoming targets is πr_{ca}^2, which varies as the inverse square of the projectile energy. Comparing this cross-sectional area to that of the unit cell of elemental Ag yields a probability for an incident 1 MeV He^+ ion to scatter off the first monolayer of Ag atoms of $\sim 5 \times 10^{-7}$. In the MeV range, the target atoms are so small that the probability of hitting any one of them is extremely small, and multiple scattering is insignificant. A projectile that has been collected by a detector has scattered from one and only one atom with a very high probability. These facts make RBS an extremely quantitative technique for determining the areal density of species that are exposed to the incident projectiles, and a great deal of surface-structure information can be obtained by performing triangulation measurements of the number of exposed atoms near a surface (Feldman and Mayer, 1986). The primary disadvantage of RBS for surface crystallography is the poor depth resolution of the technique, which is at best 50 Å.

By lowering the energy of the incident projectiles into the MEIS regime, the depth resolution is enhanced by an order of magnitude, which is sufficient for detailed analyses of surface structures. This gain is achieved with only minor disadvantages compared to RBS, i.e. a bare Coulomb potential no longer suffices for calculating the projectile trajectories, since $r_{ca} \sim 10^{-3}$ nm and the probability of back-scattering from the surface monolayer has risen to $\sim 5 \times 10^{-5}$. The scattering cross sections have increased two orders of magnitude with respect to RBS, which is large enough that multiple scattering can no longer be neglected, and in fact becomes a major part of the analysis of the scattering data (van der Veen, 1985). However, in this energy range the sensitivity of the cross sections to electron screening of the nuclei is so small that any reasonable approximation to the scattering potential gives essentially the correct result. Thus, MEIS is also very quantitative in terms of measuring absolute amounts of species present on a surface.

When projectile energies are in the LEIS regime, the surface sensitivity is enhanced even more, but the penalties for this gain are severe. Electron screening becomes extremely important, since r_{ca} can approach 0.1 nm, and the calculated cross sections depend strongly on the potential model that is employed. Also, the cross sections are large enough that multiple scattering becomes very important, since the probability of scattering from the first atomic monolayer increases into the range from 0.005 to 0.5 for projectile energies of 10 to 1 keV. These two factors alone make it extremely difficult to use LEIS to determine absolute areal densities of species on a surface. The major physical advantage of LEIS is that the scattering cross sections are so large that it is possible to determine which atomic species terminates a surface with absolute certainty (Buck *et al.*, 1980). Since this is often a major clue in choosing among a number of different models for a particular surface, this aspect of LEIS makes it very appealing. In recent years, the most often used experimental implementation of LEIS has been impact-collision ion-scattering spectroscopy (ICISS), for which the scattering angle is fixed at some large angle as close to π as the detector will allow, and the intensity of the elastic scattering from a particular feature is monitored as a function of the angle between the incident ion direction and the sample surface. The advantage of this approach is that the number of scattering events is minimized, thus making data analysis easier, as discussed in detail below.

The complexity of the scattering events in the hyperthermal energy regime ($E_{inc} < 1000$ eV) is much greater than for LEIS as it is defined here. At such low energies, the cross sections are so large that projectiles do not interact with a solid via a series of binary collisions, but rather they interact with several target atoms simultaneously along their entire trajectory (Hulpke and Mann, 1983; Horn *et al.*, 1988). Also, at these energies, attractive forces between the projectile and the target atoms are no longer negligible, and thus scattering calculations are extremely sensitive to the potentials that are used. The projectile velocities are slow enough that the changes induced in the solid by the projectile must be included in calculating the trajectory (Garrison, 1979). Given that the interactions of hyperthermal ions with surfaces are so complex, scattering of particles in this regime is not useful for determining surface structures. Rather, hyperthermal ions are scattered from surfaces with known structures to study the scattering process in this technologically important regime.

1.1.4 SIMULATION OF SCATTERED AND RECOILED INTENSITIES

When ion scattering techniques were first applied to surface crystallography, the hope was that it would be possible to interpret the data simply by examining the positions of peaks in energy spectra or angular distributions and relating them directly to atomic structures with algebraic or geometric arguments. However, true surface structures are complex, and scattering features such as shadow and blocking cones that arise from different parts of a surface often overlap in energy or angle. Thus, the key to utilizing scattering and recoiling for surface crystallography is the ability to compare the intensity of the energy or angular distributions of scattered ions obtained experimentally to those calculated for a series of structural models. Even though this is in principle very straightforward, it has proven to be surprisingly difficult to implement. The computational method that is used must provide a realistic simulation of the scattering processes and at the same time be rapid enough to allow a large number of calculations to be performed to enable the refinement of the structural model to obtain agreement with experimental data.

Many different approaches have been developed for this task, and understanding them requires a reasonable familiarity with classical scattering theory. In order to lay the foundations for understanding scattering and recoiling from solid surfaces, the next section of this chapter reviews the physics of binary scattering, i.e. the interaction of a single projectile with a single target. Section 1.3 then presents the overview of the different computational techniques that have been developed to model the scattering and recoiling from surfaces. Finally, Section 1.4 offers the conclusions of this chapter.

1.2 BINARY SCATTERING AND RECOILING

For the purposes of the discussion presented here, a binary collision involves an incident projectile that could be an ion or an atom and a stationary target that is a single atom. The projectile and target interact with each other through a potential energy $V(r)$. During the collision, the projectile loses kinetic energy and is scattered away from its original direction of flight; the target gains kinetic energy in the collision and recoils away from its initial position (see Figure 1.2). As long as no internal degrees of freedom of the projectile or target are excited, the scattering is termed 'elastic'. Detectors can be set up in the laboratory to intercept the scattered projectile, the recoiled target, or both. The data to be measured are the energies and the intensities of the scattered and recoiled particles into any particular direction that contains a detector. This section describes how to calculate this same information, given that the identities of the projectile and target and their interaction potential are known.

1.2.1 KINEMATICS

The energy and momentum conservation conditions entirely determine the kinematical relationships for an elastic collision between two particles (Marion, 1970; Goldstein, 1980). Figure 1.2(a) shows the coordinate system and defines the relevant angles for a binary collision in the laboratory frame of reference (LAB). The scattering angle (θ_P) is defined as the *asymptotic* angle through which the projectile is scattered by the target. If there is no collision, the projectile continues on a straight trajectory and has a zero scattering angle, and if it scatters back along its incident trajectory, the scattering angle is π.

The recoil angle (θ_T) is the angle between the *asymptotic* trajectory of the target atom and the incidence direction of the projectile.

Equations (2) and (3) below show the energies of the projectile (E_P) and the target (E_T) after the scattering event as functions of θ_P and θ_T, respectively:

$$E_P = E_{inc} \times \gamma^2 \left(\frac{\cos\theta_P \pm \sqrt{\gamma^{-2} - \sin^2\theta_P}}{(1+\gamma)} \right)^2, \tag{2}$$

$$E_T = E_{inc} \times \frac{4\gamma}{(1+\gamma)^2} \cos^2\theta_T, \tag{3}$$

where $\gamma = M_P/M_T$ is the ratio of the masses of the projectile M_P and target M_T. For the case in which the projectile mass is smaller than the target mass ($\gamma < 1$), the plus sign is taken in equation (2) and the range over which θ_P can vary is 0 to π, i.e. from the case of no scattering to complete back-scattering.

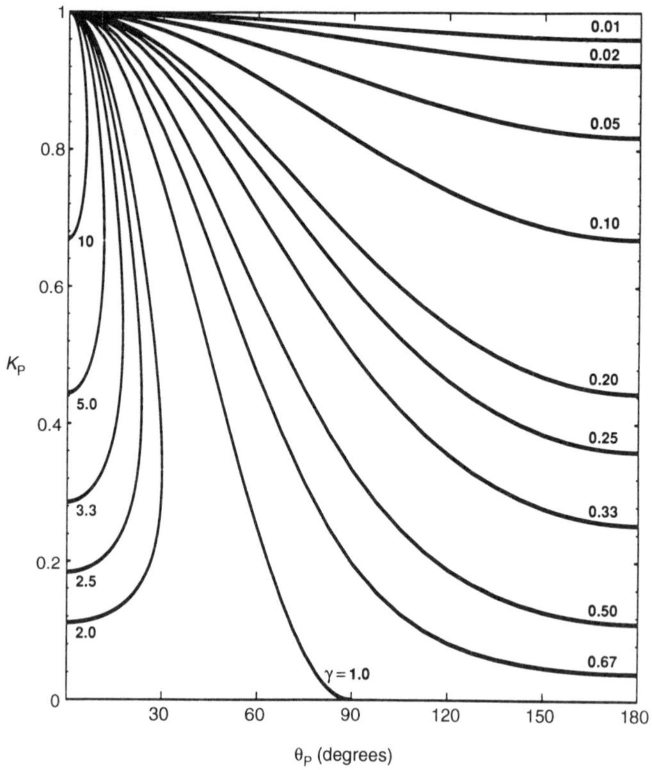

Figure 1.3 The kinematic factor $K_P(\theta_P) = E_P/E_{inc}$ for the scattered projectile in the LAB frame. The parameter γ is the ratio of the projectile to the target mass

However, if the projectile mass is larger than the target ($\gamma > 1$), then both the plus and minus signs are taken, but only forward scattering, i.e. $\theta_P < \pi/2$, is possible. There is a maximum angle that depends only on the mass ratio γ at which the projectile can scatter,

$$\theta_{P,\,max} = \arcsin \gamma^{-1}, \tag{4}$$

and that angle decreases with increasing γ. For the recoiled targets, θ_T varies from 0 to $\pi/2$ radians, independent of the ratio of the masses.

Figures 1.3 and 1.4 show plots of the kinematic factors, $K_P(\theta_P) = E_P/E_{inc}$ and $K_T(\theta_T) = E_T/E_{inc}$, as functions of the scattering and recoiling angles, respectively, for several different values of the mass ratio γ. Such plots can be used to determine the mass of a target atom that either scattered the projectile or recoiled into an energy detector that is placed at some angle with respect to the incident projectile trajectory (Mashkova and Molchanov, 1985; Chu *et al.*,

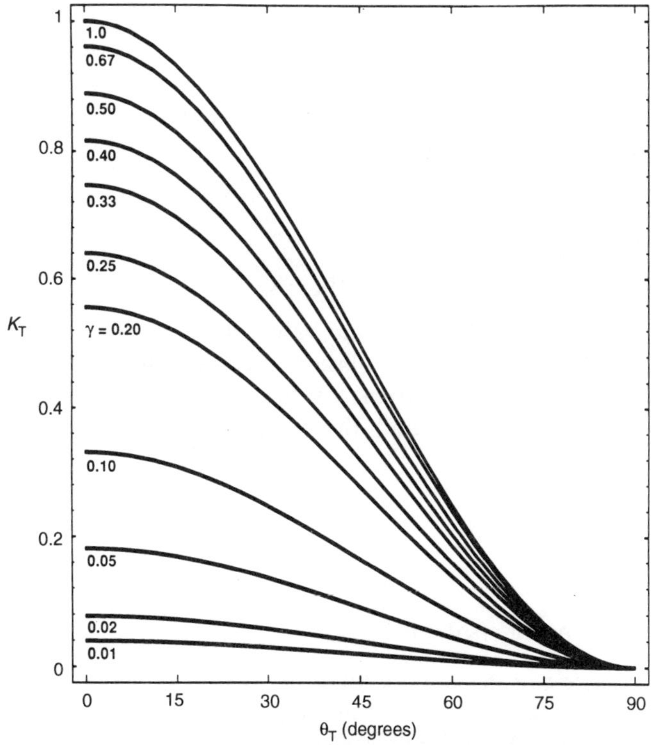

Figure 1.4 The kinematic factor $K_T(\theta_T) = E_T/E_{inc}$ for recoiled target in the LAB frame. The K_T curves for γ^{-1} are identical to those for γ

1978). For scattered projectiles, the kinematics are completely different for values of γ that are less than one, for which $K_P(\theta_P)$ is a monotonic function, from those for which γ is greater than one, for which $K_P(\theta_P)$ is double-valued. There is more symmetry present for the energies of the recoiled targets, which are the same for values of γ and γ^{-1}.

For many calculations, it is easiest to transform into the center of mass (CM) coordinate system, illustrated in Figure 1.2(b). The coupled two-particle system is replaced by two one-particle systems (Marion, 1970; Goldstein, 1980): one containing the total mass of the system, $M_P + M_T$, which is completely unaffected by the collision and translates along the center of mass, and the other the reduced mass or μ-particle, for which the motion is defined within the CM system. The μ-particle moves in the potential $V(r)$ centered at a fixed point, has a reduced mass $\mu = M_P M_T / M_P + M_T$, and has the same kinetic energy in the CM system before and after the collision:

$$E_{CM} = E_{inc} \times \frac{M_T}{M_P + M_T}. \tag{5}$$

In the discussion that follows, both LAB and CM coordinate systems will be utilized where convenient, but for comparison with experimental data the results will ultimately be expressed in LAB coordinates. The transformation from the CM scattering angle ξ to the LAB scattering and recoil angles is accomplished with the following formulae:

$$\tan \theta_P = \frac{\sin \xi}{\gamma + \cos \xi}, \tag{6}$$

$$\theta_T = \tfrac{1}{2}(\pi - \xi). \tag{7}$$

In the CM system, ξ is the *asymptotic* value of the scattering angle of the μ-particle a long time after the collision measured from the undeflected continuation of the trajectory, whereas χ is the *instantaneous* value of the angle between the scattering center and the μ-particle at any point along the trajectory.

1.2.2 DYNAMICS

1.2.2.1 Equations of motion

In order to be able to calculate the intensity or probability of scattering, one must have an accurate description of the trajectory of the scattered and/or the recoiled particles over the entire binary collision event, as illustrated in Figure 1.2. The calculations are first carried out in the CM coordinate system, and the development here has been adapted from those of Mashkova and Molchanov

(1985), Marion (1970), Goldstein (1980) and Chu *et al.* (1978). In general, the values of the scattering and recoil angles will depend strongly on the initial conditions for the system, such as the value of E_{inc}, the nature of the interaction potential $V(r)$, and the value of the impact parameter s, which is the minimum distance by which the projectile would miss the target if there were no interaction between the two and the projectile flew straight with no deflection. The equations of motion for this system come from the the conservation of angular momentum $\mathbf{L}$ and energy:

$$|\mathbf{L}| = |\mathbf{r} \times \mathbf{p}| = s\sqrt{2\mu E_{CM}} = \mu r^2 \frac{d\chi}{dt}, \tag{8}$$

$$E_{CM} = \frac{1}{2}\mu\left[\left(\frac{dr}{dt}\right)^2 + r^2\left(\frac{d\chi}{dt}\right)^2\right] + V(r), \tag{9}$$

where χ is the angular degree of freedom of the μ-particle in Figure 1.2(b). Two initial conditions are required to determine the equations of motion; most often E_{CM} and s are specified. Eliminating $d\chi/dt$ between equations (8) and (9) yields

$$E_{CM} = \frac{1}{2}\mu\left(\frac{dr}{dt}\right)^2 + \frac{L^2}{2\mu r^2} + V(r), \tag{10}$$

which can be rewritten in the following form:

$$\frac{dr}{dt} = \sqrt{\frac{2}{\mu}(E_{CM} - V(r)) - \frac{L^2}{\mu^2 R^2}}. \tag{11}$$

Integrating the above expression yields

$$t = \int \frac{dr}{\sqrt{\frac{2}{\mu}(E_{CM} - V(r)) - \frac{L^2}{\mu^2 r^2}}} + \text{const.} \tag{12}$$

This expression is known as the *time integral*, and it relates the distance between the μ-particle and the scattering center to time via the functional form $t = t(r)$, which must be inverted either analytically if possible or numerically to provide the more useful $r = r(t)$ form. In order to determine the complete temporal behaviour of the μ-particle, one also needs to know the time dependence of the angular coordinate, $\chi = \chi(t)$. This can be found by substituting the solution for $r(t)$ into equation (8) and integrating:

$$\chi(t) = L \int \frac{dt}{\mu r^2(t)} + \text{const.} \tag{13}$$

However, usually one does not need to know the explicit time dependence of the motion of the μ-particle, but rather just the trajectory $r(\chi)$. In such a case, one can make a change of variables by going back to equation (8) to define

$$\mathrm{d}t = \frac{\mu r^2}{L}\,\mathrm{d}\chi, \tag{14}$$

which can be substituted into equation (11) to yield the *scattering integral*,

$$\chi(r) = \int \frac{\frac{L}{r^2}\,\mathrm{d}r}{\sqrt{2\mu(E_{\mathrm{CM}} - V(r)) - \frac{L^2}{r^2}}} + \text{const.} \tag{15}$$

The resulting function is inverted to yield $r(\chi)$.

1.2.2.2 Interaction potentials

The behaviour of the equations of motion presented above depends upon the interaction potential $V(r)$ for the projectile and target. The equations can be solved analytically for several different forms of $V(r)$, such as a hard-sphere potential,

$$V_{\mathrm{HS}}(r) = \begin{cases} \infty \text{ for } r \le a \\ 0 \text{ for } r > a \end{cases}, \tag{16}$$

where a is the radius of the sphere. This provides a simple example that will be used often in this section to illustrate much of classical scattering theory. Another $V(r)$ that provides an important limit for ion/atom scattering problems is the Coulomb potential,

$$V_{\mathrm{C}}(r) = \frac{Z_{\mathrm{P}} Z_{\mathrm{T}} e^2}{r}, \tag{17}$$

which also yields analytical solutions for the equations of motion.

For the RBS scattering regime, the kinetic energies are so high that the interaction is almost completely determined by the Coulomb repulsions of the bare nuclei (Chu *et al.*, 1978). For the hyperthermal energy regime, the interaction potential has been evaluated self-consistently using quantum-chemical codes in order to include the effects of the electrons, e.g. screening of the nuclei, interparticle attractive interactions, and electron repulsions (Horn *et al.*, 1988). In between these two extremes, a number of different semiempirical potentials have been developed to describe the collisions of low- and medium-energy ions with target atoms.

One of the first and most popular of these potentials is the Molière (1947) approximation to the Thomas–Fermi potential, which is a screened Coulomb interaction:

$$V_M(r) = \frac{Z_P Z_T e^2}{r}\left[0.35\exp\left(-\frac{0.3r}{a_{sc}}\right) + 0.55\exp\left(-\frac{1.2r}{a_{sc}}\right) + 0.10\exp\left(-\frac{6.0r}{a_{sc}}\right)\right], \tag{18}$$

where a_{sc} is a scaling length that determines the strength of the screening. The screening length most often used is that of Firsov (1958),

$$a_F = 0.8853a_0(Z_P^{1/2} + Z_T^{1/2})^{-2/3}, \tag{19}$$

where a_0 is the Bohr radius. However, the experience of many investigators has shown that a_F is too large, i.e. the screening experienced in experiments is stronger than Firsov's theoretical estimate, and a_F must be multiplied by a factor of 0.6 to 0.8 for calculations involving $V_M(r)$ to achieve agreement with experimental results (Torrens, 1972). Thus, the Molière potential as used in practice (Poelsema *et al.*, 1977b) contains an adjustable parameter, the screening length, that is varied to optimize agreement with experimental results.

To address this discrepancy, Ziegler *et al.* (1985) have defined a universal potential (often known as the ZBL potential) that in principle does not have to be adjusted for individual cases. These authors fit a reasonable functional form to a data set containing thousands of measurements:

$$V_{ZBL}(r) = \frac{Z_P Z_T e^2}{r}\left[0.1818\exp\left(-\frac{3.2r}{a_{ZBL}}\right) + 0.5099\exp\left(-\frac{0.9423r}{a_{ZBL}}\right)\right.$$
$$\left. + 0.2802\exp\left(-\frac{0.4029r}{a_{ZBL}}\right) + 0.02817\exp\left(-\frac{0.2016r}{a_{ZBL}}\right)\right], \tag{20}$$

where the ZBL screening length is

$$a_{ZBL} = 0.88534a_0(Z_P^{0.23} + Z_T^{0.23})^{-1}. \tag{21}$$

In comparisons with LEIS data, the ZBL potential is better than the unadjusted Molière potential, but still does not agree as well as calculations in which the screening length is adjusted on a case-by-case basis (Daley and Williams, 1990 and unpublished results). Small changes in the potentials can lead to rather large differences in the trajectories evaluated from equation (15), because $r(\chi)$ is the result of an integral over the potential.

Whenever screened potentials such as $V_M(r)$ or $V_{ZBL}(r)$ are used, the solutions to the equations of motion must be obtained numerically.

1.2.2.3 Deflection functions and time integrals

For many experiments, knowing the detailed trajectory $r(\chi)$ is not important. In the case of a detector located far from the target (the far field), it is only necessary to know the asymptotic CM scattering angle ξ, which can then be transformed to the asymptotic laboratory angles θ_P and θ_T as necessary. The trajectory $r(\chi)$ is symmetric about the line connecting the scattering center to the position of closest approach, r_{min}, of the μ-particle, so that by integrating from r_{min} to $r = \infty$, one can determine the angle χ_0 from the incidence direction to r_{min}, as shown in Figure 1.2(b). Thus it follows that the center-of-mass scattering angle of the μ-particle is $\xi = \pi - 2\chi_0$, or by substituting $L = s(2\mu E_{CM})^{1/2}$ into equation (15),

$$\xi = \pi - 2s \int_{r_{min}}^{\infty} \frac{dr}{r\sqrt{(1 - V(r)/E_{CM})r^2 - s^2}} \tag{22}$$

Since the upper limit of this integral is infinity and the integrand has a singularity, it is often best to make a change of variables before evaluating equation (22), especially if the calculation is to be done numerically. The choice of variable described in most textbooks is $u = r_{min}/r$ (Marion, 1970; Goldstein, 1980), which yields the following integral defined over the interval from 0 to 1:

$$\xi = \pi - 2s \int_0^1 \frac{du}{\sqrt{\left(1 - V\left(\frac{r_{min}}{u}\right)/E_{CM}\right)r_{min}^2 - s^2u^2}} \tag{23}$$

This formula has the disadvantage that it often has a singularity within the range of integration. There have been several examinations of this problem which have led to other forms for the change of variables and the resulting integral (Robinson, 1963, 1970a; Latta and Scanlon, 1974; Mendenhall and Weller, 1990), with the primary goal being the rapid numerical evaluation of equation (22) and the avoidance of divergences. The value of r_{min} that appears in the solution of equation (22) must be determined for each value of s, and is obtained by finding the extremum of $r(t)$, i.e. the value of r for which $dr/dt = 0$. From equation (11), this condition becomes

$$E_{CM}\left(1 - \frac{s^2}{r_{min}^2}\right) = V(r_{min}), \tag{24}$$

which can be solved analytically in some cases, but often must be determined by a numerical procedure such as Newton–Raphson iteration (Robinson, 1963, 1970a; Mendenhall and Weller, 1990).

The result of evaluating the integral in equation (22) is the asymptotic CM scattering angle expressed as a function of the impact parameter, $\xi = \xi(s)$, which is called the CM deflection function. The CM deflection functions for the hard-sphere and for the Coulomb potentials can be determined analytically, and are, respectively,

$$\xi_{\mathrm{HS}}(s) = \pi - 2\arcsin\left(\frac{s}{a}\right) = 2\arccos\left(\frac{s}{a}\right), \quad \text{for } 0 \le s \le a \tag{25}$$

$$\xi_{\mathrm{C}}(s) = \pi - 2\arctan\left(\frac{2sE_{\mathrm{CM}}}{Z_{\mathrm{P}}Z_{\mathrm{T}}e^2}\right) \qquad \text{for } 0 \le s \le \infty. \tag{26}$$

The CM deflection function for the hard-sphere potential is independent of the energy of the projectile, whereas the energy acts as a scaling parameter for the Coulomb potential. Once the CM deflection function has been defined, the deflection functions for the scattered projectile and the recoiled target can be determined in the LAB system using the coordinate transformations of equations (6) and (7). As examples, the LAB deflection functions for the hard-sphere potential are found to be

$$\theta_{\mathrm{P,HS}}(s) = \arctan\left(\frac{2\frac{s}{a}\sqrt{1-\left(\frac{s}{a}\right)^2}}{2\left(\frac{s}{a}\right)^2 - 1 + \gamma}\right), \tag{27}$$

$$\theta_{\mathrm{T,HS}}(s) = \arcsin\left(\frac{s}{a}\right). \tag{28}$$

For the screened potentials of equations (18) and (20), the integral in equation (22) must be evaluated numerically for a sequence of the values of the impact parameter s to define an array that contains the deflection function. The solution of equation (22) for zero impact parameter is just $\xi(0) = \pi$, and as long as $V(r)$ is repulsive and monotonic, i.e. it contains no local minima or maxima, then the deflection function $\xi(s)$ monotonically decays to zero as s increases. Figure 1.5 shows how the impact parameter s required to yield a scattering angle in the LAB frame varies with the incident kinetic energy of the projectile for He scattering from Ag modeled with the ZBL potential.

There are also occasions in which the detector of scattered or recoiled particles can be in the near field, such as when the detector is actually another target atom located in a solid (Mashkova and Molchanov, 1985). In this case, the asymptotic angle of the trajectory given by the deflection function is not sufficient by itself to determine whether the near-field detector will intercept the exit trajectory or not, because the impact parameter for the first collision may not be negligible with respect to the spacing between the target and the detector. However, instead of determining the entire trajectory, in many cases it is possible to replace $r(\theta)$ with straight line segments that are the asymptotic limits of the incident and exit trajectories. The key to this replacement is determining the starting points for the asymptotes by finding the distances Δ_{P}

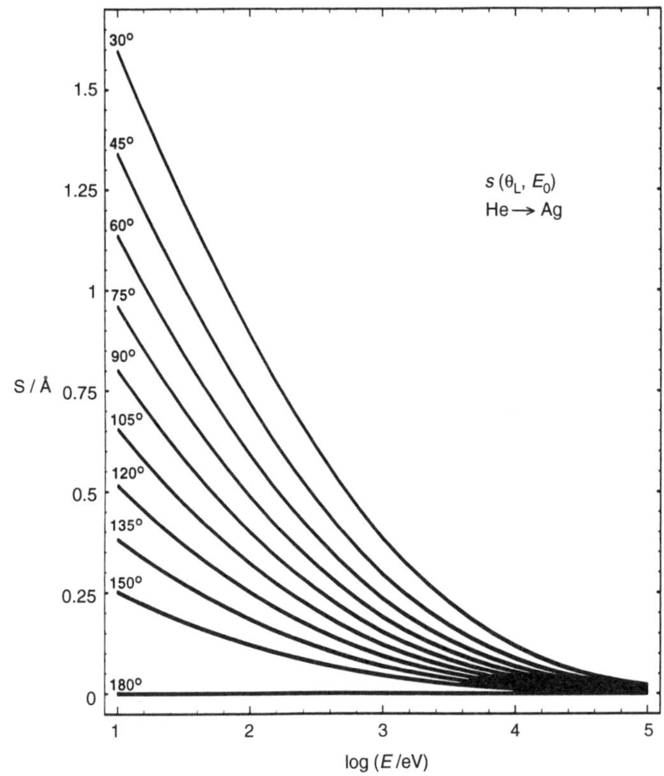

Figure 1.5 A plot of the impact parameter s versus the incident projectile kinetic energy E for several different LAB scattering angles, as calculated for He scattering from Ag with a ZBL potential. For the case of 180° back-scattering, the impact parameter is always zero. The value of s required to scatter the He projectile at any other angle depends strongly on the energy. At low energies and scattering angles, the impact parameters are large compared to the interatomic spacings of atoms in a crystal

and Δ_T as defined in Figure 1.6 (Mashkova and Molchanov, 1985). Robinson (1970b) showed that in order to determine the values of the Δ's, it was first necessary to compare the interval defined by the time integral of equation (12) for a particle to travel from infinity to r_{min} to the corresponding interval for the case of the time integral in which $V(r) = 0$ for all r. Both of these time integrals diverge, but in subtracting one from the other, a finite value is determined for most potential functions (Lehman, 1977)—excluding the Coulomb potential. The length τ is defined as the difference in the distance traveled by the μ-particle to r_{min} in the presence and absence of the scattering potential, and is calculated by multiplying the velocity of the μ-particle by the difference

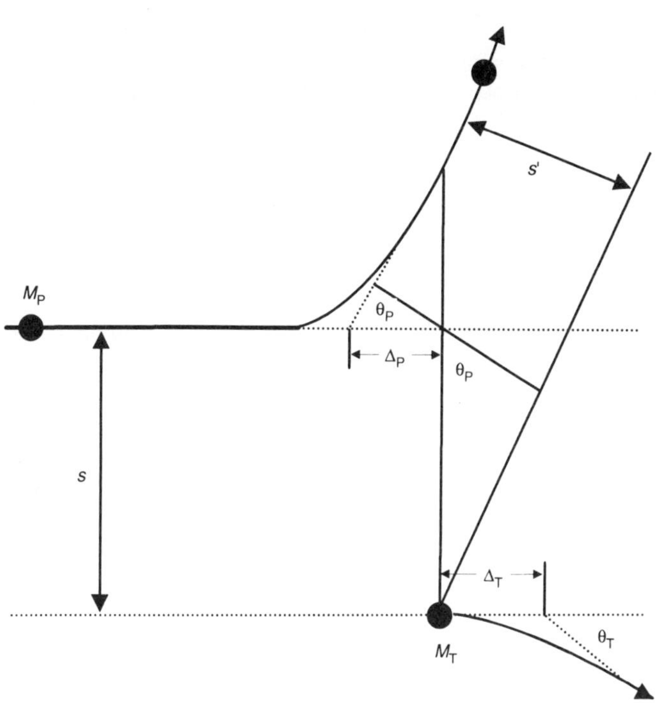

Figure 1.6 The intersection of the incident and exit trajectories defines the quantities Δ_P and Δ_T, which are related to the time integrals

of the two time integrals:

$$\tau(s) = \sqrt{r_{\min}^2 - s^2} - \int_{r_{\min}}^{\infty} \left[\frac{1}{\sqrt{1 - \dfrac{s^2}{r^2} - \dfrac{V(r)}{E_{\mathrm{CM}}}}} - \frac{1}{\sqrt{1 - \dfrac{s^2}{r^2}}} \right] \mathrm{d}r. \tag{29}$$

This length, which is also commonly called the time integral, is transformed to the LAB frame by the following (Robinson, 1970b):

$$\Delta_P(s) = \frac{2\tau(s)M_P}{M_P + M_T} + s\,\frac{M_T - M_P}{M_T + M_P}\tan\left(\frac{\xi}{2}\right), \tag{30}$$

$$\Delta_T(s) = s\tan\left(\frac{\xi}{2}\right) - \Delta_P(s). \tag{31}$$

For most cases in which the path of a projectile can be approximated as a sequence of binary collisions, everything that one needs to know about the trajectory of the particle in LAB coordinates can be obtained from the values of the deflection functions and the time integrals, which are often both tabulated numerically as functions of the impact parameter (Robinson, 1963, 1970a). These numerical arrays are then used to follow the trajectory of a particle in a stepwise fashion.

1.2.3 CROSS SECTIONS

1.2.3.1 General derivation

The relation between the incident particle flux and the intensity of scattered projectiles (see Figure 1.7) or recoiled targets at some angle in the far field is determined by the differential collision cross sections for the scattered projectiles $d\sigma_P/d\Omega$ and the recoiled targets $d\sigma_T/d\Omega$. In order to understand the material in the next section, it is essential to understand how to derive a cross section from a deflection function. The following derivation (Marion, 1970; Goldstein, 1980) is valid for either the CM or LAB frame of reference, and also for scattered

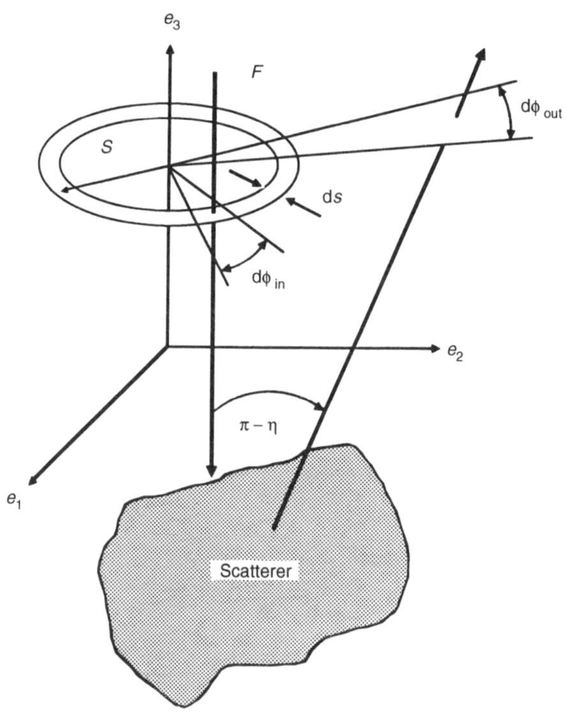

Figure 1.7 Scattering from an arbitrary target in three-dimensional space. The projectile flux is incident parallel to the e_3 axis. For an impact parameter s, the scattering angle is η

projectiles and scattered targets. Consider a uniform flux of incident projectiles, F, which has the dimension of number of particles per unit area per unit time. What one wants to know is the rate of arrival in particles per second of scattered projectiles in a detector at an angle η from the direction of incidence that accepts a solid angle $d\Omega$, i.e. $I(\eta) \sin \eta \, d\eta \, d\phi$. The detector function $I(\eta)$ has the dimensions of number of particles (either projectiles or recoils) per steradian per second, and has no explicit dependence on ϕ because the binary scattering problem has cylindrical symmetry. If the deflection function $\eta(s)$ (which can be $\xi(s)$, $\theta_P(s)$, or $\theta_T(s)$) is monotonic and well behaved, then the particles that arrive at the detector will all have been scattered from the same location in space, which is the unit of area defined by $s \, ds \, d\phi_{in}$, as shown in Figure 1.7. Since the number of particles must be conserved,

$$Fs \, ds \, d\phi_{in} = I(\eta) \sin \eta \, d\eta \, d\phi_{out}. \tag{32}$$

The solution for the detector function is formally expressed as

$$I(\eta) = F \frac{s}{\sin \eta} \left| \frac{\partial s \, \partial \phi_{in}}{\partial \eta \, \partial \phi_{out}} \right|, \tag{33}$$

where the Jacobian determinant is evaluated in the following fashion:

$$\left| \frac{\partial s \, \partial \phi_{in}}{\partial \eta \, \partial \phi_{out}} \right| = \left| \frac{\partial s \, \partial \phi_{in}}{\partial \eta \, \partial \phi_{out}} - \frac{\partial s}{\partial \phi_{out}} \frac{\partial \phi_{in}}{\partial \eta} \right| = \left| \frac{\partial s}{\partial \eta} \right| \tag{34}$$

for $\phi_{in} = \phi_{out}$. The Jacobian is bypassed completely in the usual derivations of the cross section (Marion, 1970; Goldstein, 1980), but it has been introduced here because it is essential in deriving cross sections for systems that are not cylindrically symmetric. The differential cross section is just the detector function divided by the flux, or

$$\frac{d\sigma(\eta)}{d\Omega} = \frac{I(\eta)}{F} = \frac{s}{\sin \eta} \left| \frac{d\eta}{ds} \right|^{-1}, \tag{35}$$

where $\eta(s)$ is the deflection function for the system and $d\sigma(\eta)/d\Omega$ has dimensions of area per steradian. Integrating this three-dimensional differential cross section over the total solid angle available yields the integral or total cross section σ_{tot}.

The differential cross sections for scattering or recoiling can be calculated directly in LAB coordinates by substituting $\theta_P(s)$ or $\theta_T(s)$, respectively, for $\eta(s)$ in equation (35) if one knows these deflection functions. In most textbooks, the CM differential cross section is calculated first and then transformed into LAB coordinates. The transformation begins with the conservation of the number of scattered particles in the following fashion:

$$\frac{d\sigma_{LAB}(\theta)}{d\Omega_{LAB}} \sin \theta \, d\theta \, d\phi = \frac{d\sigma_{CM}(\xi)}{d\Omega_{CM}} \sin \xi \, d\xi \, d\phi, \tag{36}$$

which can be solved to yield the desired cross section

$$\frac{\mathrm{d}\sigma_{\mathrm{LAB}}(\theta)}{\mathrm{d}\Omega_{\mathrm{LAB}}} = \frac{\sin\xi}{\sin\theta}\frac{\mathrm{d}\xi}{\mathrm{d}\theta}\frac{\mathrm{d}\sigma_{\mathrm{CM}}(\xi)}{\mathrm{d}\Omega_{\mathrm{CM}}}, \tag{37}$$

or from equations (23) and (24)

$$\frac{\mathrm{d}\sigma_{\mathrm{LAB}}(\theta_{\mathrm{P}})}{\mathrm{d}\Omega_{\mathrm{LAB}}} = \frac{[\gamma\cos\theta_{\mathrm{P}} + \sqrt{1-\gamma^2\sin^2\theta_{\mathrm{P}}}]^2}{\sqrt{1-\gamma^2\sin^2\theta_{\mathrm{P}}}}\frac{\mathrm{d}\sigma_{\mathrm{CM}}(\xi)}{\mathrm{d}\Omega_{\mathrm{CM}}}, \tag{38}$$

$$\frac{\mathrm{d}\sigma_{\mathrm{LAB}}(\theta_{\mathrm{T}})}{\mathrm{d}\Omega_{\mathrm{LAB}}} = 4\sin\frac{\xi}{2}\frac{\mathrm{d}\sigma_{\mathrm{CM}}(\xi)}{\mathrm{d}\Omega_{\mathrm{CM}}} = 4\cos\theta_{\mathrm{T}}\frac{\mathrm{d}\sigma_{\mathrm{CM}}(\xi)}{\mathrm{d}\Omega_{\mathrm{CM}}}. \tag{39}$$

1.2.3.2 Hard-sphere cross sections

By substituting the deflection function from equation (25) into the cross section formula of equation (35), one obtains the differential hard-sphere scattering cross section in CM coordinates:

$$\frac{\mathrm{d}\sigma_{\mathrm{CM}}(\xi)}{\mathrm{d}\Omega_{\mathrm{CM}}} = \frac{a^2}{4}, \tag{40}$$

which is independent of γ or s (or ξ). Integrating the hard-sphere differential cross section over 4π steradians yields the total cross section $\sigma_{\mathrm{tot}} = \pi a^2$, which is simply the area of the sphere projected onto a plane, as one should expect. This total cross section should have the same value independent of coordinate system. The relation in equation (40) can be transformed into the LAB system quite easily using equations (38) and (39) to yield:

$$\frac{\mathrm{d}\sigma_{\mathrm{LAB}}(\theta_{\mathrm{P}})}{\mathrm{d}\Omega_{\mathrm{LAB}}} = \frac{a^2}{4}\frac{[\gamma\cos\theta_{\mathrm{P}} + \sqrt{1-\gamma^2\sin^2\theta_{\mathrm{P}}}]^2}{\sqrt{1-\gamma^2\sin^2\theta_{\mathrm{P}}}}, \tag{41}$$

$$\frac{\mathrm{d}\sigma_{\mathrm{LAB}}(\theta_{\mathrm{T}})}{\mathrm{d}\Omega_{\mathrm{LAB}}} = a^2\cos\theta_{\mathrm{T}} \quad 0 \le \theta_{\mathrm{T}} \le \frac{\pi}{2}. \tag{42}$$

The scattered-projectile differential cross section now depends explicitly on both θ_{P} and γ, whereas the recoiled-target cross section depends only on θ_{T}.

There are three different cases of interest for the scattered projectile:

$$\gamma \ll 1 \quad \frac{\mathrm{d}\sigma_{\mathrm{LAB}}(\theta_{\mathrm{P}})}{\mathrm{d}\Omega_{\mathrm{LAB}}} \approx \frac{a^2}{4} \qquad 0 \le \theta_{\mathrm{P}} \le \pi, \tag{43}$$

$$\gamma = 1 \quad \frac{\mathrm{d}\sigma_{\mathrm{LAB}}(\theta_{\mathrm{P}})}{\mathrm{d}\Omega_{\mathrm{LAB}}} = a^2\cos\theta_{\mathrm{P}} \qquad 0 \le \theta_{\mathrm{P}} \le \frac{\pi}{2}, \tag{44}$$

$$\gamma \gg 1 \quad \frac{\mathrm{d}\sigma_{\mathrm{LAB}}(\theta_{\mathrm{P}})}{\mathrm{d}\Omega_{\mathrm{LAB}}} \approx \frac{a^2}{4}\frac{\gamma[1 \pm \sqrt{\gamma^{-2} - \theta_{\mathrm{P}}^2}]^2}{\sqrt{\gamma^{-2} - \theta_{\mathrm{P}}^2}} \qquad 0 \le \theta_{\mathrm{P}} < \gamma^{-1}. \tag{45}$$

The deflection functions and their respective differential cross sections from equations (43–45) are illustrated in Figure 1.8. In the first case (equation (43)), the mass of the target is so large that it essentially represents the center of mass of the system, so that $d\sigma_{LAB}(\theta_P)/d\Omega_{LAB}$ is the same as $d\sigma_{CM}(\xi)/d\Omega_{CM}$. Integrating equation (43) over 4π steradians yields $\sigma_{tot} = \pi a^2$, just as in CM coordinates. Equation (44) represents the familiar case of billiard balls, in which the masses of the target and projectile are equal. Just as the condition $\gamma = 1$ is satisfied, the maximum value that θ_P can take drops discontinuously from π to $\pi/2$, since it is not possible for the projectile to scatter back from a free target of equal mass. The total cross section $\sigma_{tot} = \pi a^2$ is again obtained by integrating equation (44) over the half-sphere, or 2π steradians, for forward scattering.

Finally, equation (45) represents the case of a massive projectile and a light target. In this case, the differential cross section appears to diverge at $\theta_P = \gamma^{-1}$. Since the inverse deflection function $s(\theta_P)$ is double-valued, there are two different branches for the relation between the differential cross section and the scattering angle, as shown in Figure 1.8. The lower branch arises from the part of the deflection function for $0 \leq s < s_{crit}$, and the upper, higher cross section

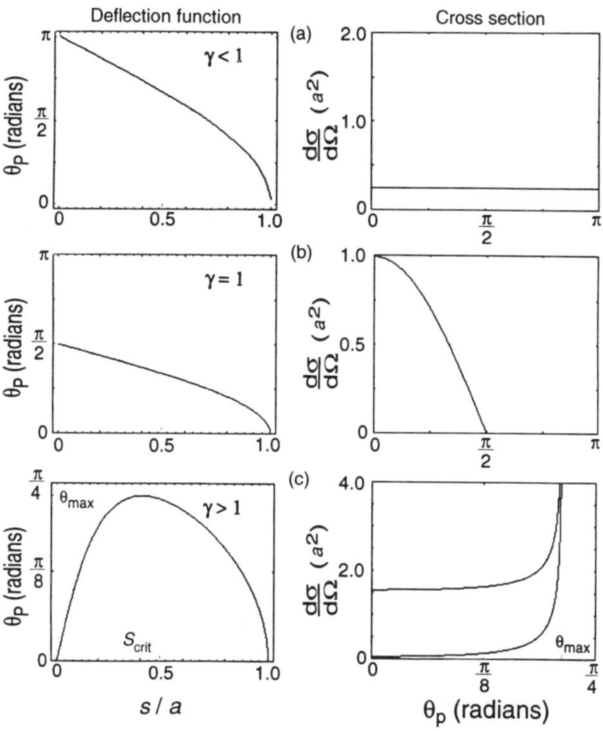

Figure 1.8 Deflection functions and differential scattering cross sections for hard-sphere scattering. In (a), the projectile is lighter than the target, in (b) they have the same mass, and in (c), the projectile is heavier than the target

branch for $a \geq s > s_{crit}$. These two branches correspond to the lower and higher kinetic energy branches, respectively, of the kinetic parameter curves $K_P(\theta_P)$ for $\gamma > 1$ in Figure 1.3. Since the differential cross section does not diverge in the CM coordinate system, there is an upper bound to the value that θ_P can obtain. The maximum value $\theta_{P,max}$ that can be assumed by θ_P is determined by setting the integral over both branches of equation (45) equal to πa^2, and solving for the upper limit of the integral that satisfies that restriction. This still yields a giant resonance in the differential cross section at the maximum scattering angle, since most of the contribution to σ_{tot} comes from values of θ_P very near $\theta_{P,max}$.

1.2.3.3 Rutherford cross section

If the hard-sphere radius a in equation (40) is replaced by the distance of closest approach for a repulsive Coulomb interaction $r_{ca} \approx Z_P Z_T e^2/E_{inc}$ from equation (1), the result is a very good approximation to the differential cross section for back-scattering at large angles of a light projectile such as He from a massive target at high energy. Thus, much of the insight gained from hard-sphere scattering can be carried over into other systems. The Coulomb potential of equation (17) is also analytically integrable and yields the famous cross-section formula of Rutherford (1911),

$$\frac{d\sigma_R(\xi)}{d\Omega_{CM}} = \left(\frac{Z_P Z_T e^2}{4E_{CM}}\right)^2 \frac{1}{\sin^4\left(\frac{\xi}{2}\right)}. \tag{46}$$

in CM coordinates. This formula can also be transformed analytically into LAB coordinates for the scattered projectile

$$\frac{d\sigma_R(\theta_P)}{d\Omega_{LAB}} = \left(\frac{Z_P Z_T e^2}{2E_{inc}}\right)^2 \frac{((1-(\gamma \sin\theta_P)^2)^{1/2} + \cos\theta_P)^2}{\sin^4\theta_P(1-(\gamma\sin\theta_P)^2)^{1/2}}, \tag{47}$$

as derived by Ziegler and Lever (1973), but in most treatments the transformation to the LAB frame is only outlined (Marion, 1970; Goldstein, 1980). For the recoiled target, the cross section is

$$\frac{d\sigma_R(\theta_T)}{d\Omega_{LAB}} = \left(\frac{Z_P Z_T e^2}{2E_{inc}}\right)^2 \frac{1}{\cos^3(\theta_T)}. \tag{48}$$

Equations (47) and (48) are the high-energy limits to the scattering cross sections that are more accurately calculated from the screened interaction potentials.

1.2.3.4 Two-dimensional approximations

An approximation that is made at times is to use a two-dimensional version of a cross section, σ_{2d}. This is often rationalized as a good approximation if the entire scattering process occurs in a single plane. For the case of binary scattering, the particle conservation condition is obtained by equating ds times the line flux f with dη times the two-dimensional detector function $I_{2d}(\eta)$:

$$f \, ds = I_{2d}(\eta) \, d\eta, \tag{49}$$

which leads to the 'two-dimensional differential cross section'

$$\frac{d\sigma_{2d}(\eta)}{d\eta} = \left| \frac{d\eta}{ds} \right|^{-1}. \tag{50}$$

This differs significantly from the cross section of equation (35), even though binary scattering always occurs within a single plane. Thus, the dimensionality of the problem is important; there is a big difference between scattering that occurs in a two-dimensional universe and that which occurs in a three-dimensional universe but is confined to a plane by symmetry. This is an important distinction that is often not appreciated when comparing calculated results to experimental data. For the two-dimensional analog of a hard-sphere potential (the hard circle), the CM 'cross section' is $d\sigma_{2d}(\xi)/d\xi = (a/2) \sin(\xi/2)$, which has completely different functional form from the three-dimensional cross section of equation (40).

1.2.3.5 Rainbow scattering

There is an interesting anomaly in determining the cross section of the scattered projectile for cases in which the potential is not monotonic (Child, 1974; Newton, 1966). In binary scattering, this occurs when there is an attractive well in the potential, and thus it is only important at very low projectile incidence energies (<100 eV or so). However, for the multicenter targets discussed in the next section, the total effective scattering potential is never monotonic. Therefore, that discussion will be anticipated here by presenting the simplest case of rainbow scattering.

Any time there is a peak or valley in the potential function, there will be a corresponding local maximum or minimum in the deflection function (Figure 1.9). This has two major consequences. First, for some values of the scattering angle ξ in the CM coordinate system, there will be two or more values of the impact parameter. Thus, calculating the differential cross section at any scattering angle will require summing over all values of the impact parameter s which yield that scattering angle ξ:

$$\frac{d\sigma(\xi)}{d\Omega} = \sum_s \frac{s}{\sin \xi} \left| \frac{d\xi}{ds} \right|^{-1}. \tag{51}$$

The differential cross section can have branches that must be summed to obtain the complete value at a given ξ. For the cases in which $\gamma < 1$, the energy of

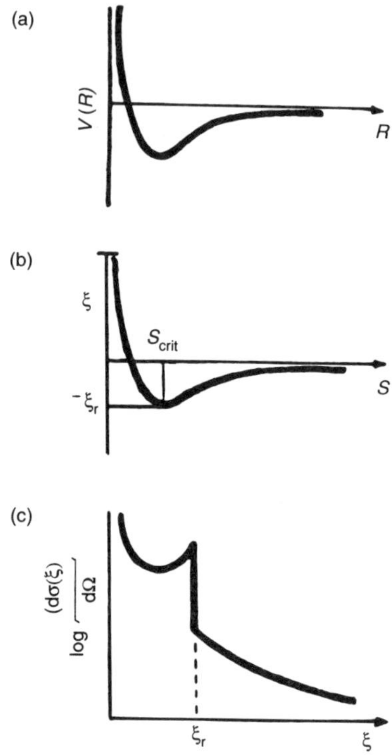

Figure 1.9 An illustration of a rainbow in the case of binary scattering. If the interaction potential has a well, as in (a), then the deflection function also displays a local minimum at a negative value of the scattering angle, as shown in (b). The differential cross section displays a sharp local maximum at the critical angle for rainbow scattering, ξ_r, as shown in (c)

the scattered projectile will be the same at a given value of θ_P on the different branches of the cross section vs. s relation even in the LAB coordinate system, which is different from the case for $\gamma > 1$ discussed above. Second, whenever the deflection function has an extremum, then $|d\xi/ds|^{-1}$ and thus the CM differential cross section diverges at the finite critical impact parameter s_{crit} that characterizes that extremum. The literal interpretation of this fact is that the number of particles scattered at that corresponding angle is infinitely large, no matter what the incident flux is. Since this violates the assumptions used in deriving the cross section, it represents a breakdown in the validity of classical mechanics. The occurrence of this type of divergence in the classical cross section is called rainbow scattering (Child, 1974; Newton, 1966), and the resolution of the apparent paradox lies in quantum mechanics.

The fact that the classical differential cross section diverges is more than just a theoretical curiosity but is a major problem when we are trying to compare calculated and experimental intensities. At extremely low energies there can be significant oscillations in the scattering cross section, but at higher particle kinetic energies the primary effect is a sharp increase in the differential cross section at the rainbow angle. The obvious solution to the divergence is to truncate the calculation of the classical cross section at an impact parameter that is smaller than s_{crit} by some value Δs so that the divergence in the calculation is avoided, just as in the case of the divergence in the LAB differential cross section for $\gamma > 1$ discussed above. However, choosing the value of Δs correctly is very important, since near s_{crit} small variations in the choice of Δs are amplified tremendously in the values of the calculated cross section, which can dramatically alter a comparison of the calculated cross section with experimental data. The choice of Δs is not arbitrary, but is determined by the details of the interaction potential and by Heisenberg's uncertainty principle (Child, 1974; Newton, 1966), since it is not possible to simultaneously specify the exact values of the impact parameter and the momentum of the incident projectile.

It is easiest to determine the value of Δs in CM coordinates. If the impact parameter is multiplied by the magnitude of the momentum of the μ-particle, the result is the magnitude of the angular momentum, as shown in equation (8). The deflection function can be expressed as a function of the angular momentum and expanded about the critical angular momentum for rainbow scattering $L_{crit} = s_{crit}(2\mu E_{CM})^{1/2} = s_{crit}p_{CM}$ in the following fashion:

$$\xi(L) = \xi_{crit} + \Gamma(L - L_{crit})^2 + \dots, \tag{52}$$

where $\Gamma = \frac{1}{2}\partial^2\xi/\partial L^2$ and the term linear in ΔL is missing because the critical angle is an extremum. Thus, the deviation of ξ from the critical value can be expressed as

$$\Delta\xi = \xi(L) - \xi_{crit} = \Gamma(L - L_{crit})^2 = \Gamma(\Delta L)^2. \tag{53}$$

This condition is combined with the uncertainty principle of the angular momentum and its conjugate coordinate the scattering angle

$$\Delta\xi\Delta L \geq \frac{\hbar}{2}, \tag{54}$$

to yield the condition for the minimum value of Δs within which classical mechanics is not valid (Child, 1974; Newton, 1966):

$$\Delta s \geq \frac{1}{p}\left(\frac{\hbar}{2\Gamma}\right)^{1/3}. \tag{55}$$

Thus, the classical formula, equation (35), is not valid over the region defined by $s_{crit} - \Delta s < s < s_{crit} + \Delta s$, and inside that region the differential cross section must be calculated quantum-mechanically or semiclassically. The actual cross section does not diverge in this region because the wave functions for the

μ-particle interfere destructively to yield a finite scattering intensity. Over the range from $s_{crit} - \Delta s$ to $s_{crit} + \Delta s$, the average of the quantum-mechanical cross section is well approximated by the classical cross section at $s_{crit} - \Delta s$. Thus, an excellent approximation to the cross section is as shown in Figure 1.9(c).

1.3 SCATTERING AND RECOILING FROM SURFACES

The complexity of the scattering problem increases dramatically when the target consists of many atoms. As long as the kinetic energies are high enough that the projectile interacts with only one atom in the target at a time, the collision sequence can be treated computationally as a series of binary collisions. For this condition to hold, the scattering cross section must be small with respect to the interatomic spacing in the target. The presence of a multi-atom target introduces several new phenomena to the scattering process and these must be considered in calculating scattering and recoiling intensities; the key to quantitative interpretation of experimental data is to be able to rapidly and reliably evaluate these multiple-collision effects. Many different approaches have been developed to treat scattering and recoiling from surfaces, and this section presents an overview of them.

1.3.1 FORESHORTENING

There is one point to consider that is common to all treatments of scattering from surfaces that has caused difficulties. That is the foreshortening effect, which affects the intensity of particles scattered from a line or plane of targets, such as the atoms on a surface. The foreshortening effect is very familiar to anyone who has driven by a line of trees or telephone poles. When observed at a right angle to the line, the spacing between the targets is large, but as the angle between the line of targets and the direction of observation gets smaller, the apparent distance between the targets gets smaller, as shown in Figure 1.10. Another way of stating this is that the apparent density of the scatterers increases as the angle of observation decreases. Thus, even in the complete absence of multiple-scattering effects, the total scattering intensity from a line (or a plane) of scatterers increases with decreasing angle as

$$I(\alpha) \sim \frac{1}{\sin \alpha}, \tag{56}$$

where α is the angle between the projectile incidence direction and the sample surface. This factor must be included in any calculation in which the angle of

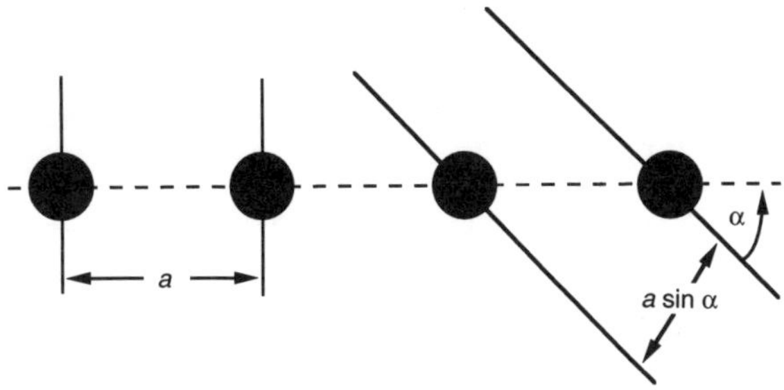

Figure 1.10 When viewed at an angle α from the line of targets, the inter-target spacing decreases as $a \sin \alpha$. Thus, it is more difficult for a projectile to pass between the targets, since their effective density has increased

incidence of the projectile beam changes with respect to the target surface in order to obtain agreement with experimental results.

1.3.2 TRAJECTORY ENSEMBLES: SHADOWING AND BLOCKING

The yield of scattered projectiles or recoiled targets from a crystal surface is highly anisotropic when examined as a function of the projectile incidence angle or the scattering angle. This anisotropy is caused by the shadowing and blocking effects, which are illustrated in Figure 1.1 with ensembles of line-segment trajectories to show the effects of forward scattering on parallel or diverging trajectories.

For the case in which $\gamma < 1$, the ensemble of possible trajectories for a parallel beam of particles incident on an isolated target atom has a region behind the target, called the shadow cone, from which trajectories are totally excluded because the scattering angle decreases as the impact parameter increases (Turkenburg *et al.*, 1978). A small fraction of the incident trajectories are back-scattered, but the vast majority scatter into the forward direction. Most of the trajectories excluded from the inside are piled up at the boundary of the shadow cone, and thus the effective flux of projectiles at the edge is greatly enhanced over the incident flux. In a solid, the shadow cone of one atom can either shield a neighbor or focus a large flux of particles on it, thus giving rise to a large anisotropy in the yield of particles scattered from the second atom as a function of the incidence angle of the ion beam. In the case where $\gamma > 1$, the shadowing is incomplete. The density of allowed trajectories behind the shadowing atom is decreased but not eliminated. However, there is still a substantial increase in the effective flux at the edge of the partial shadow cone.

The shape of the shadow cone can be analyzed quantitatively by considering the down-field point where two scattered trajectories that initially had slightly different impact parameters cross. The shadow cone radius r_{sh} at a distance d down-field from the target atom is

$$r_{sh} = s + (d + \Delta_P(s)) \tan \theta_P(s). \tag{57}$$

At the edge of the shadow cone, the derivative of r_{sh} with respect to s must vanish, $dr_{sh}/ds = 0$, since trajectories with arbitrarily close impact parameters will cross at the same point, which leads to the following formula for d given a particular value of s:

$$d = -\left(1 + \frac{d\Delta_P(s)}{ds} \tan \theta_P(s)\right) \cos^2 \theta_P(s) \left(\frac{d\theta_P(s)}{ds}\right)^{-1} - \Delta_P(s). \tag{58}$$

The shadow cone can be mapped out by first using equation (58) to find values of d for an appropriate range of s values over which the scattering is all in the forward direction, and then generating (d, r_{sh}) pairs by substituting d and s into equation (57). As discussed by Oen (1983), the shadow-cone radius for a repulsive Coulomb potential at high-incidence energies, $\gamma \ll 1$, and $d \gg s$ is

$$r_{sh,C} = 2\sqrt{\frac{dZ_P Z_T e^2}{E_{inc}}}. \tag{59}$$

This equation is obtained from equation (57) by ignoring $\Delta_P(s)$, using the Coulomb deflection function from equation (26), and taking the small-angle approximation $\tan \theta \approx \theta$. Since evaluating equations (57) and (58) can be a cumbersome process for an arbitrary potential, Oen (1983) has presented a simple formula for correcting the Coulomb value in equation (59) to yield the shadow-cone radius for the Molière potential. This formula yields a good representation of the shape and energy dependence of the shadow-cone radius for $\gamma \ll 1$, but the actual values obtained for r_{sh} must be fit to experimental data because of the inadequacy of the Firsov screening length. Figure 1.11 shows the energy dependence of the shadow cone for Li^+ scattering from Ag atoms calculated from equation (58) using a Molière potential.

The related phenomenon of blocking occurs for projectiles that originate at a point inside a solid target. This can happen if a nucleus decays by emitting an alpha particle, for instance, or if an atom below the surface layer back-scatters a projectile. In this case, the initial trajectories are not *parallel*, but *diverging*. The blocking atom near the point emitter deflects the ions further, producing an angular region in space where there are no ($\gamma < 1$) or fewer ($\gamma > 1$) allowed trajectories than there would have been in the absence of the blocking atom. This blocking cone is responsible for anisotropies in the yield of the emitted projectiles as a function of the orientation of the particle detector with respect to the emitter and the blocking atom. The total scattering angle

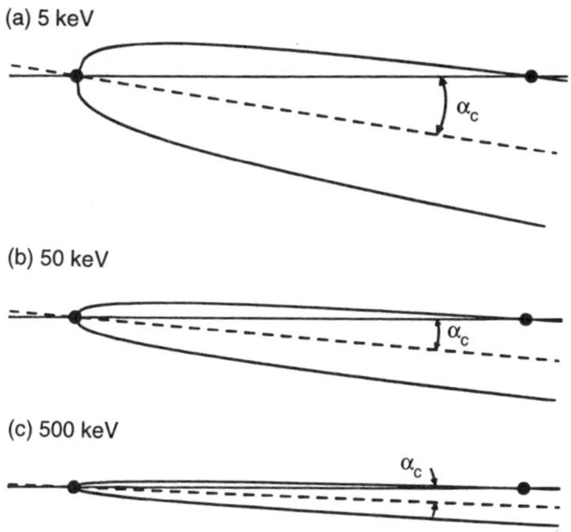

Figure 1.11 The shadow cones for Li^+ ions scattering from Ag atoms calculated in the Molière approximation. The shadow-cone width decreases dramatically with increasing kinetic energy

is the sum of the initial angle of emission of the projectile with respect to the emitter–blocker axis θ_e and the angle of scattering from the blocking atom $\theta_P(s)$,

$$\theta_{tot}(s) = \theta_e + \theta_P(s) = \arcsin\left(\frac{s}{d}\right) + \theta_P(s), \tag{60}$$

where d is the distance between the emitter and the blocker. In this case, the blocking-cone angle θ_{bl} is the minimum value of $\theta_{tot}(s)$, which is determined by setting $d\theta_{tot}(s)/ds = 0$ to yield the transcendental equation

$$\frac{1}{d}\frac{1}{\sqrt{1-\left(\frac{s_{bl}}{d}\right)^2}} = -\frac{d\theta_P(s_{bl})}{ds}. \tag{61}$$

The solution s_{bl} to equation (61) is then used to calculate θ_{bl} from equation (60).

As illustrated in Figure 1.1, shadowing (blocking) produces strong intensity variations as the angle of projectile incidence (detection) is changed with respect to the internuclear axis of the pair of atoms involved. The scattered yield rises rapidly from a minimum to a peak at the shadowing and blocking critical angles, which represent the edge of the shadow and blocking cones. In the simplest cases, experimental measurements of these critical angles can be used to triangulate the positions of atoms near a surface with respect to each other, such as in the case of ICISS (Aono *et al.*, 1981). Since the potentials for scattering in the keV range are not well known, the shadow-cone dimensions must be determined by performing experiments on known structures. However, for more complex systems, which are usually those of most interest, shadowing and blocking features can overlap and thus make a straightforward determination of critical angles impossible. In these cases, it is necessary to be able to calculate the scattering intensity, in addition to the angular positions of features, thus resolving such difficulties. Oen (1965, 1982) and Mashkova and Molchanov (1985) have shown in detail how to calculate the intensity of ion scattering in the near field, which is a quantitative measure of the effective flux created inside the target surface by shadowing and blocking.

1.3.3 TRAJECTORY TRACING OR MOLECULAR DYNAMICS

The most straightforward and reliable way to calculate the energy and/or angle distribution of scattered and recoiled particles from a surface is to simulate an experiment by following a large number of projectiles as they propagate through the target. This approach was pioneered and popularized through the use of the computer program MARLOWE introduced by Robinson and Torrens (1974), who referred to this algorithm as a Monte Carlo simulation because of the method of choosing the initial configurations of the projectile and target atoms for starting a calculation. In this chapter, any technique that actually integrates the equations of motion (12) and (13) or otherwise follows trajectories of projectiles and/or targets will be termed a molecular-dynamics (MD) simulation, which is more in line with modern terminology. (This will allow a precise distinction to be made between MD simulations and the Monte Carlo calculations to be discussed later in the next section.)

As implemented by Robinson (Robinson and Torrens, 1974; Hou and Robinson, 1976; Oen and Robinson, 1976; Heiland *et al.*, 1976) and many others (Karpuzov *et al.*, 1967; Yurasova *et al.*, 1968; Karpuzov and Yurasova, 1971; Preuss, 1978; Garrison, 1979; Buck *et al.*, 1980; Taglauer *et al.*, 1980; Jackson *et al.*, 1981; Overbury, 1981; von dem Hagen *et al.*, 1982; Helbig *et al.*, 1982; Yarmoff and Williams, 1982), a three-dimensional model of the solid target of interest is constructed, and then the computer simulation starts trajectories with the proper initial conditions at random positions over the surface and follows them until the projectile either becomes implanted in the solid, or the projectile or a recoiled target atom exits the surface with enough energy to be detected. For such events, the kinetic energy and direction of emission are written to a file for later analysis. In the most complete treatments (Robinson and Torrens, 1974; Karpuzov and Yurasova, 1971; Garrison, 1979), the positions of the projectile and all the target atoms in a cluster used to model the surface region are computed at each time step, and these positions are used to compute the forces acting on all the bodies in the simulation to determine the positions at the next time step. In this case, following only one projectile becomes a major chore that requires a sophisticated computer and program.

Such rigorous MD simulations have provided valuable insights into the mechanisms of ion–solid interactions (Robinson and Torrens, 1974; Hou and Robinson, 1976; Oen and Robinson, 1976) and they have been essential for understanding the particle and cluster emission processes in secondary ion mass spectrometry (SIMS) (Harrison *et al.*, 1978; Winograd *et al.*, 1978; Chang *et al.*, 1988). Statistically significant simulated SIMS data sets can be obtained by MD with a reasonable expenditure of CPU time because the sputter yield of particles from a surface is roughly one atom from the solid ejected for every incident projectile or greater, and thus the events accumulate rapidly. However, experimental scattering and recoil measurements often require in excess of 10^{10} incident projectiles to obtain reasonable statistics in the data, so a true simulation of these experiments is extremely difficult because of the huge amounts of CPU time that would be needed (Garrison, 1979). This is especially relevant considering that determining a surface structure requires that many different models be analyzed and compared with experimental data to find an optimum geometry. Thus, many approximations and short cuts had to be developed to apply MD to scattering and recoiling.

In an MD simulation of the interaction of keV projectiles incident normal to a surface, fewer than 1 in 10^3 of the projectiles reexits from the solid if the trajectories are started at truly random positions above the unit cell of the surface (Williams and Yarmoff, 1983). Thus, most of the computer time in such simulations is spent following trajectories that will become implanted in the solid and not contribute to the back-scattered yield. In order to reduce the number of wasted trajectories, the surface unit cell can be interrogated with a light mesh of incident trajectories to find those positions that will lead to back-scattered particles (Garrison, 1979; Williams and Yarmoff, 1983), and then the simulation is carried out with an extremely large set of initial trajectories restricted to regions over the surface most likely to yield a back-scattering event. This simple procedure can save more than an order of magnitude in computing time, but still other CPU-time-saving strategies are required.

In the frozen-lattice approximation (Poelsma *et al.*, 1976a, 1976b, 1976c, 1977a; Boers, 1977), the target atoms in the solid are fixed in a configuration in which the atoms are displaced from their equilibrium positions by random distances and directions that have moments determined by the thermal ellipsoids of the solid. These displacements are essential to a simulation; many important trajectories in a realistic model are only present if the perfect symmetry of the lattice is broken by the vibrational motion of the atoms (Williams and Yarmoff, 1983; Boers, 1977). Correlations can be built into the displacements to account for dominant phonon modes of the lattice, if necessary (Poelsema *et al.*, 1976c). This approximation is reasonably accurate as long as the time scale for the entire collision process is short compared to a single vibrational period of an atom in the lattice, which is true for projectile energies above 1 keV.

The binary-collision approximation (BCA) (Hou and Robinson, 1976; Sigmund, 1969) assumes the projectile only interacts with one target atom at a time, and allows the trajectory of the projectile to be modeled as a series of straight line segments. Instead of following the projectile during a series of equal time steps (Karpuzov and Yurasova, 1971; Garrison, 1979), the projectile is propagated through the crystal using tables of deflection functions and time integrals (Jackson *et al.*, 1981; Overbury, 1981; von dem Hagen *et al.*, 1982; Helbig *et al.*, 1982; Yarmoff and Williams, 1982, 1983). The projectile is allowed to move forward until it has an impact parameter with a target atom that is within some allowed value for scattering defined by the program—usually a deflection of 1° or more. The particle is then positioned at the intersection of the ingoing and outgoing trajectory asymptotes as calculated from equation (30) from the time integral, its new direction of motion is determined from the deflection function, and its new kinetic energy is calculated from equation (2) using the known scattering angle. This process is repeated as the projectile scatters though the crystal until it loses sufficient energy to be no longer interesting, penetrates so deeply that it will not be able to get back out, or it exits the surface. The details of the computer algorithms used for such calculations have been presented in Parilis *et al.* (1993).

These types of calculations were often run to simulate the energy spectra of projectiles scattered from a crystalline surface for specific incidence and scattering angles (Buck *et al.*, 1980; Taglauer *et al.*, 1980; Jackson *et al.*, 1981; Overbury, 1981; von dem Hagen *et al.*, 1982; Helbig *et al.*, 1982; Yarmoff and Williams, 1982, 1983). The idea behind these simulations was that if a sufficiently good agreement could be obtained between experimental data and a computer model of a surface, then the structure of the surface had been determined. An experimental data set collected by Buck *et al.* (1979) for 2.4 and 5 keV Ne^+ ions scattered from a clean Ni(001) surface served as a benchmark for several MD algorithms (Buck *et al.*, 1980; Helbig *et al.*, 1982; Yarmoff and Williams, 1983). Time-of-flight data were collected for a scattering angle of 90°, with the beam incident along the [100] and the [110] azimuths. Reasonable but not total agreement was found between the simulated and experimental energy spectra. Because of the relatively large amount of computer time required to generate a single energy spectrum, the number of trajectories that could be followed was limited to about 10^6 in the best cases (Yarmoff and Williams, 1983). The residual differences observed between the simulated and experimental data, as shown in Figure 1.12, could have been caused by statistical noise, approximations in the algorithms, or differences in the actual scattering potential or surface structure from those used in the simulation (Yarmoff and Williams, 1983). These uncertainties limit the usefulness of such comparisons for surface-structure determination, and thus ways to increase the efficiency of the MD simulations even further are required.

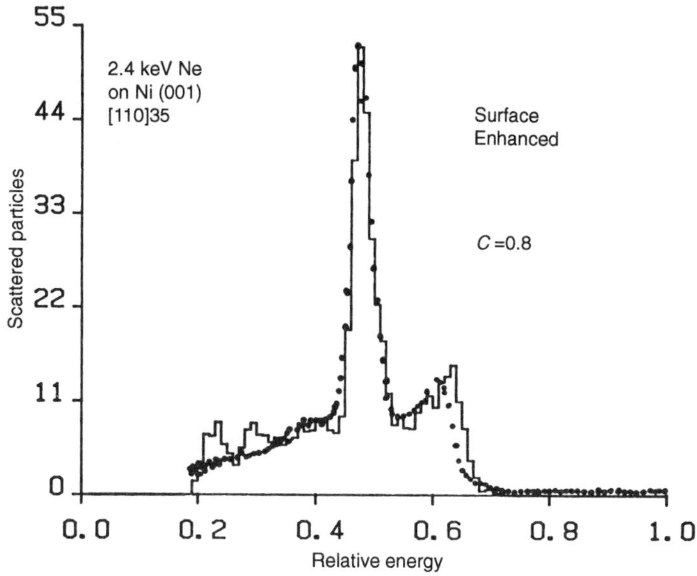

Figure 1.12 Comparison of the experimental time-of-flight ion-scattering data from Buck *et al.* (1979) for 2.4 keV Ne^+ scattering from a Ni(001) surface with MD simulations from Yarmoff and Williams (1982). The experimental data are plotted as points and the MD simulations are plotted as a solid-line histogram

The earliest MD-type simulations were used to calculate the energy spectra of particles that were incident at a fairly low angle α from the surface along a high-symmetry azimuth of the crystal and were then scattered at a small total angle θ_L into the detector (Yurasova *et al.*, 1968; Poelsema *et al.*, 1976a, 1976b, 1976c, 1977a; Boers, 1977; Kivilis *et al.*, 1967; Nizhnaya *et al.*, 1979). This geometry had two major advantages for both computational efficiency and experimental convenience. The small incidence angle ensured a high probability of scattering from a near-surface atom because of foreshortening, and the forward-scattering geometry was chosen to make the binary cross sections large. The combined effect was to significantly increase the count rates for both the simulations and the experiments. However, the large target density and cross sections produced a high probability for complex multiple-scattering involving many atoms (Boers, 1977). The energy spectra contained prominent peaks for quasi-single, quasi-double and sometimes even quasi-triple scattering (Poelsema *et al.*, 1976b), for which there were one, two, or three major scattering events accompanied by several small-angle scattering events. The positions and intensities of the multiple scattering peaks were very sensitive to the surface structure, but they were also very sensitive to nonstructural parameters, especially the scattering potential (Poelsema *et al.*, 1976b). This increase in the scattering complexity placed very stringent demands on the quality of both the simulation and the interaction potential, since any errors would be magnified through the sequence of collisions. Given the noise in the experiments and the simulations, it was very difficult to draw firm conclusions about surface structures from such simulations, especially for complex surfaces.

Other approximations were made to further increase the statistics in simulated energy spectra. One approach was to use two-dimensional simulations in which only a single plane or even just one row (the 'chain model') of atoms contributed to the scattering (Poelesma *et al.*, 1976a, 1976b, 1976c, 1977a; Boers, 1977; Winograd *et al.*, 1978). This approach was generally found to be reasonable as long as the shadow-cone radii were small enough that they did not induce a significant amount of interplanar scattering near the surface. Another was to simply use a 'detector' in the simulations that had a significantly larger solid collection angle than the detector used in the experiments (Helbig *et al.*, 1982). The problem with this approach is that since the energy of a scattered particle is strongly dependent on the scattering angle, collecting particles over a larger solid angle changes the energy spectra significantly (Yarmoff and Williams, 1983).

An alternative approach is to utilize MD simulations for situations in which the symmetry of the experimental geometry is high. A particular example was the analysis of an experimental data set (Bernheim and Slodzian, 1976) in which the scattered energy- and azimuthal-angle-dependent yield from 9.5 keV Ne^+ ions incident normal to a Cu(001) surface and scattered at 135° was measured. Yarmoff (1985) performed a full three-dimensional MD simulation of this system in which 6.125×10^6 trajectories were launched over a preselected area of the surface unit cell of the system. The back-scattered yield was summed over the eightfold rotation and reflection symmetry-equivalent azimuthal orientations in determining the simulated yields. These operations produced a yield equivalent to the launching of 2.5×10^8 projectiles. The statistics in the simulated distributions and the agreement with the experimental distributions were both very good, as shown in Figure 1.13. In this case, a single three-dimensional MD simulation was used to generate a data set that could be compared to hundreds of experimental measurements, and with this type of leverage the use of rigorous MD simulations becomes feasible.

The situation is completely reversed for the case of ICISS, in which the ion beam and the detector are fixed with respect to one another and the sample is rotated, so that a different MD simulation is required for each orientation of the sample. Since a single ICISS angular distribution usually contains measurements from 90 different sample orientations with respect to the ion beam and detector, large-scale MD simulations cannot be used to analyze experimental data. However, it is possible to apply MD simulations involving just two target atoms, one a shadowing atom and the other the back-scattering target, which is actually just a detector in the near field, or alternatively a fixed emitter that represents the back-scattering target and a blocking atom (Yarmoff *et al.*, 1986).

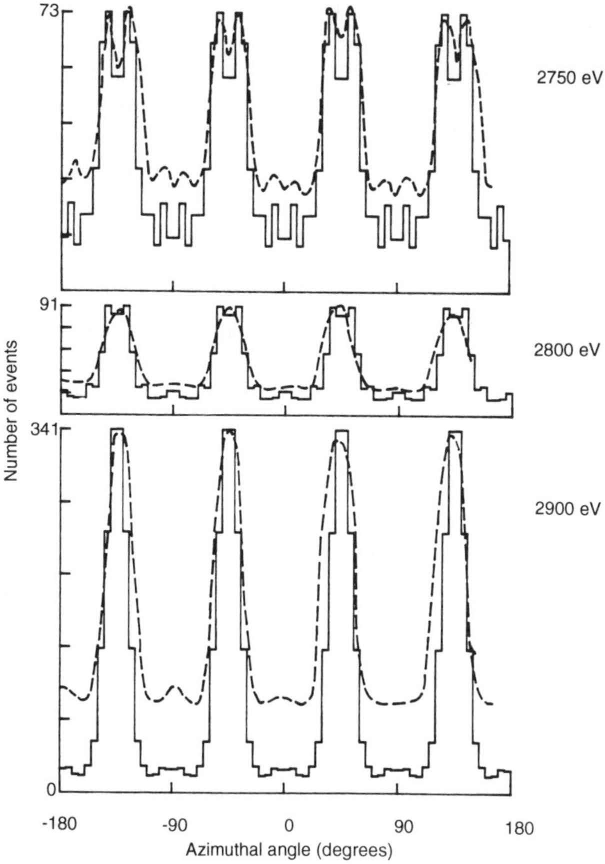

Figure 1.13 Comparison of experimental azimuthal angular distributions measured for 9.5 Ne^+ ions incident normal to a Cu(001) surface and scattered at 135° with MD simulations. The energies of the scattered ions were measured at 2750, 2800 and 2900 eV, as well as other kinetic energies. The dashed lines show the experimental measurements from Bernheim and Slodzian (1976), and the solid-line histogram the simulations from Yarmoff (1985)

Simulations are performed by running a moderately large number of trajectories to evaluate the relative probability that a projectile will be scattered into a particular direction. The effect of thermal vibrations is modeled by moving the atom a randomly chosen distance from a two-dimensional Gaussian distribution with a mean-square vibrational amplitude corresponding to the thermal ellipsoid of the appropriate atom. By evaluating such probabilities for all reasonable pairs of atoms in a solid target that can be involved in shadowing or blocking interactions, an angular distribution can be obtained by taking the product of all such distributions for equivalent scatterers, and then summing over inequivalent scatterers (Yarmoff *et al.*, 1986). This approach works for LEIS energies because when the detector is set at an angle near but not exactly 180°, the vast majority of back-scattered particles experience only one large-angle collision and usually no more than one small-angle collision, so a two-atom model is sufficient to simulate most of the scattering trajectories. The primary drawback to this approach is that there is still noise in the calculations that result from summing over a finite number of trajectories.

In general, the collective experience accumulated with MD routines has shown that the amount of computer time required to simulate a single scattering geometry is too long to enable an investigator to compare an experimental data set with a sufficiently large number of variations of structural models to be able to choose an optimized structure from a large number of possibilities. The MD simulations are very valuable for examining details of the scattering mechanisms. Once a candidate structure is determined by other techniques, MD simulations may be the best way to optimize the final structural refinements, especially if experiments are performed with a high degree of symmetry in the scattering geometry.

1.3.4 HITTING PROBABILITIES

An alternative to trajectory *counting* techniques is to determine the *relative probability* for a projectile to hit a particular target atom given the incidence direction with respect to the crystal lattice and the subsequent probability that the scattered projectile will reach the detector for a specific orientation with respect to the target. The basic idea for evaluating such probabilities was introduced by Oen (1965, 1982) for a two-atom target (of which one of the atoms is the detector or emitter of projectiles), and subsequently adopted or rediscovered and applied to an increasing number and complex geometry of atoms by many researchers (Barrett, 1971; Poelsema and Boers, 1977, 1979; Stansgaard *et al.*, 1978; van der Veen *et al.*, 1978, 1980; Tromp and van der Veen, 1983; Frenken *et al.*, 1986; van der Gon *et al.*, 1991; Sato *et al.*, 1985; Huang *et al.*, 1987; Huang and Williams, 1988a, 1988b; Porter *et al.*, 1988; Daley *et al.*, 1989; Durr *et al.*, 1990; Spitzl *et al.*, 1990). For the two-target-atom case, a hitting probability analysis is similar in concept to the single-atom MD simulations of Yarmoff *et al.* (1986). Instead of launching a large number of trajectories and then counting the number that are intercepted by a target atom in the near field of a shadowing atom or that are steered toward the detector by a blocking atom, the atomic configurations that can produce scattering in the correct direction are integrated over the thermal ellipsoid of the shadowing or the blocking atoms (Oen, 1965; van der Veen *et al.*, 1978; Sato *et al.*, 1985;

Daley *et al.*, 1989). Such an integral produces the probability that two atoms can have the proper orientation in space to scatter a projectile into the detector given their equilibrium separation, orientation, and scattering potentials.

The hitting probabilities may be evaluated for the ICISS case in which the sample is rotated and the angle between the target and detector remains fixed by considering the diagrams in Figure 1.14 (Oen, 1965; van der Veen *et al.*, 1978; Daley *et al.*, 1989). The scattering atom is considered to be fixed in space, and the shadowing or blocking atoms assume the entire vibrational amplitude, which is the sum in quadrature of the components of the vibrational amplitudes of both atoms. Consider first the shadowing case in which the projectile is initially incident at an angle α with respect to the internuclear axis D of the atoms in their equilibrium positions.

For a large angle of incidence, the existence of the shadowing atom is not a factor and the probability of hitting the target atom is the same as if the shadowing atom did not exist, which in this case is assigned the value of 1. For

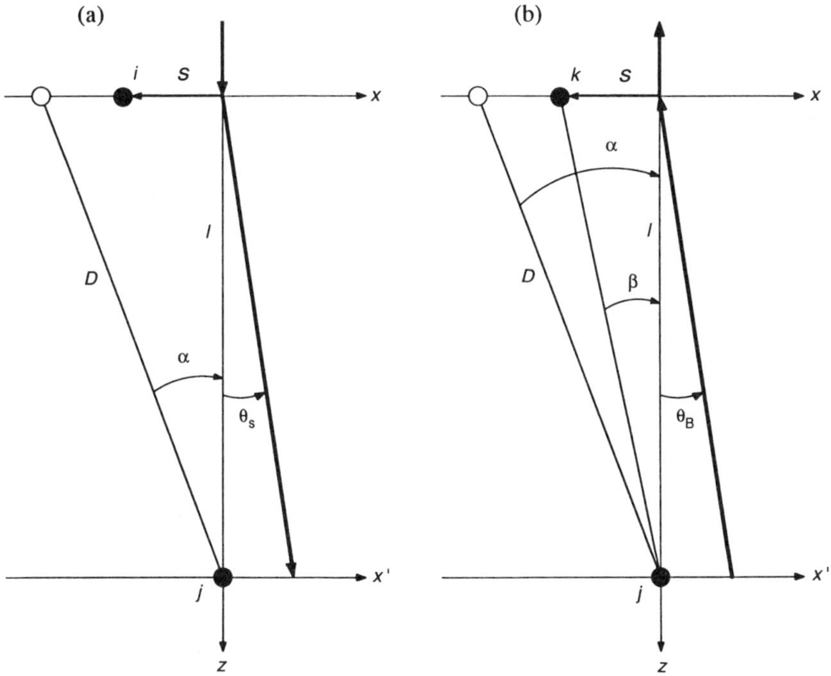

Figure 1.14 Diagrams illustrating hitting probability calculations for (a) shadowing and (b) blocking. The equilibrium positions of the shadowing and blocking atoms are indicated with open circles, and the positions required for impact parameters corresponding to a scattering angle of α in the incoming or outgoing path are shown (after Daley *et al.*, 1989)

a projectile incident with $\alpha = 0$, the shadowing atom must be moved away from its equilibrium position by an amount at least equal to the shadow-cone radius R_S at the location of the target atom in order for the target atom to be hit. Since the shadowing atom is vibrating about its equilibrium position, there is some probability for hitting the target atom. If the two-dimensional thermal ellipsoid of the shadowing atom is modeled by a Gaussian distribution with a mean value of ρ, then the probability that the target atom will be hit is the integral of this Gaussian distribution outside a circle of radius R_S. If $R_S \gg \rho$, then the hitting probability is extremely small and shadowing is effective. However, if $R_S \approx \rho$, then there is a very high (>1) *relative* probability that the target atom will be hit, since it has a high probability of sitting at the edge of the shadow cone where the effective ion flux is much greater than the incident flux. This is the general idea behind the string calculations used to determine the equivalent number of atoms per string detected in RBS channeling experiments (Barrett, 1971; Stensgaard *et al.*, 1978).

For the case of a general incidence angle α, the probability Ps_{ij} for a projectile to hit the target atom j after scattering from the shadowing atom i is determined by the following integral (Oen, 1982; van der Veen *et al.*, 1978; Daley *et al.*, 1989):

$$\mathrm{Ps}_{ij}(\alpha) = \int_0^\infty \int_0^{2\pi} \exp\left(-\frac{[(s + l\sin\theta)\cos\phi + l\sin\alpha]^2 + (s + l\sin\theta)^2\sin^2\phi}{\rho^2}\right) \times \frac{s\,\mathrm{d}s\,\mathrm{d}\phi}{\pi\rho^2}, \tag{62}$$

where s is the impact parameter of the projectile with respect to the shadowing atom, ϕ is the azimuthal angle about the z-axis, $l = D\cos\alpha$, and $\theta(s)$ is the deflection function for the projectile scattering from the shadowing atom. After analytically integrating over ϕ, this integral becomes

$$\mathrm{Ps}_{ij}(\alpha) = \frac{2}{\rho^2}\int_0^{s_{\max}} I_0\left(\frac{2l\tan\alpha[s + l\tan\theta(s)]}{\rho^2}\right) \times \exp\left(-\frac{(l\tan\alpha)^2 + [s + l\tan\theta(s)]^2}{\rho^2}\right) s\,\mathrm{d}s, \tag{63}$$

where I_0 is the modified Bessel function of the first kind. A similar expression exists (Daley *et al.*, 1989) for the blocking probability $\mathrm{Pb}_{jk}(\alpha)$, where atom k can block projectiles scattered by atom j from reaching the detector.

To approximate the angular dependence of the total yield $I(\alpha)$ from a solid target, all possible two-atom shadowing and blocking distributions involving the same back-scattering atom are multiplied together, and then the sum over all inequivalent atoms is evaluated to produce a fan-like distribution (Daley *et al.*, 1989; Durr *et al.*, 1990; Spitzl *et al.*, 1990):

$$I(\alpha) = \sum_{j} \prod_{i \neq j} \mathrm{Ps}_{ij}(\alpha) \prod_{k \neq j} \mathrm{Pb}_{jk}(\alpha). \tag{64}$$

If necessary, the expression in equation (64) can be modified to include foreshortening and/or neutralization effects. However, this expression assumes that each of the scattering events is independent. This is usually a good approximation, but in some cases the relative positions of several atoms may be involved in an important successful trajectory. Direct comparisons between hitting probability simulations for 5 keV Li^+ ions scattering from the Ni atoms of an epitaxial $NiSi_2$ film grown on Si(001) and experimental results (Huang *et al.*, 1987) are shown in Figure 1.15 for the case in which the incidence angle is fixed and the scattering angle is varied and in Figure 1.16 for the case in which the scattering angle is fixed and the incidence angle is varied. Other comparisons between experimental data for the low-energy ion-scattering regime and hitting probability calculations have been presented (Huang and Williams, 1988a, 1988b; Porter *et al.*, 1988; Daley *et al.*, 1989; Durr *et al.*, 1990; Spitzl *et al.*, 1990).

For the multiple-scattering case, the series of interactions leads to a multidimensional integral over all the thermal ellipsoids of the target atoms involved (Tromp and van der Veen, 1983; Frenken *et al.*, 1986; van der Gon *et al.*, 1991; Daley *et al.*, 1989). Since the amount of computer time required to numerically evaluate a multidimensional integral scales as q^N, where N is the number of dimensions and q is the number of steps used in evaluating each integral, simulating ion-scattering intensities in this fashion can quickly become unmanageable. For the case of scattering from two or three atoms, Poelsema and Boers (1977, 1979) placed an additional constraint on the scattering probability by explicitly considering the detected yield to be a function of both projectile incidence angle and scattered energy $I(\alpha, E_P)$. This constraint decreases the dimensionality of the integral to be evaluated by one, so that a two-atom system does not require the evaluation of an integral and a three-atom system only requires a one-dimensional integral. This procedure also has further advantages: the measurements of the scattered intensities are energy-resolved, so the comparison between simulated and experimental angular yields are direct, and energy spectra can also be calculated from the same equations. In practice, the time required to calculate an angular distribution $I(\alpha)$ at a particular scattered energy with a constrained hitting probability calculation was reduced by two to three orders of magnitude compared to that required for a similar MD chain calculation (Poelsema and Boers, 1979).

For more than three target atoms, the most efficient method for evaluating the hitting probabilities is by a Monte Carlo approach (Barrett, 1971; Stensgaard *et al.*, 1978; van der Veen *et al.*, 1980; Tromp and van der Veen, 1983; Frenken *et al.*, 1986). A large number, 10^5 or more, of trajectories are generated

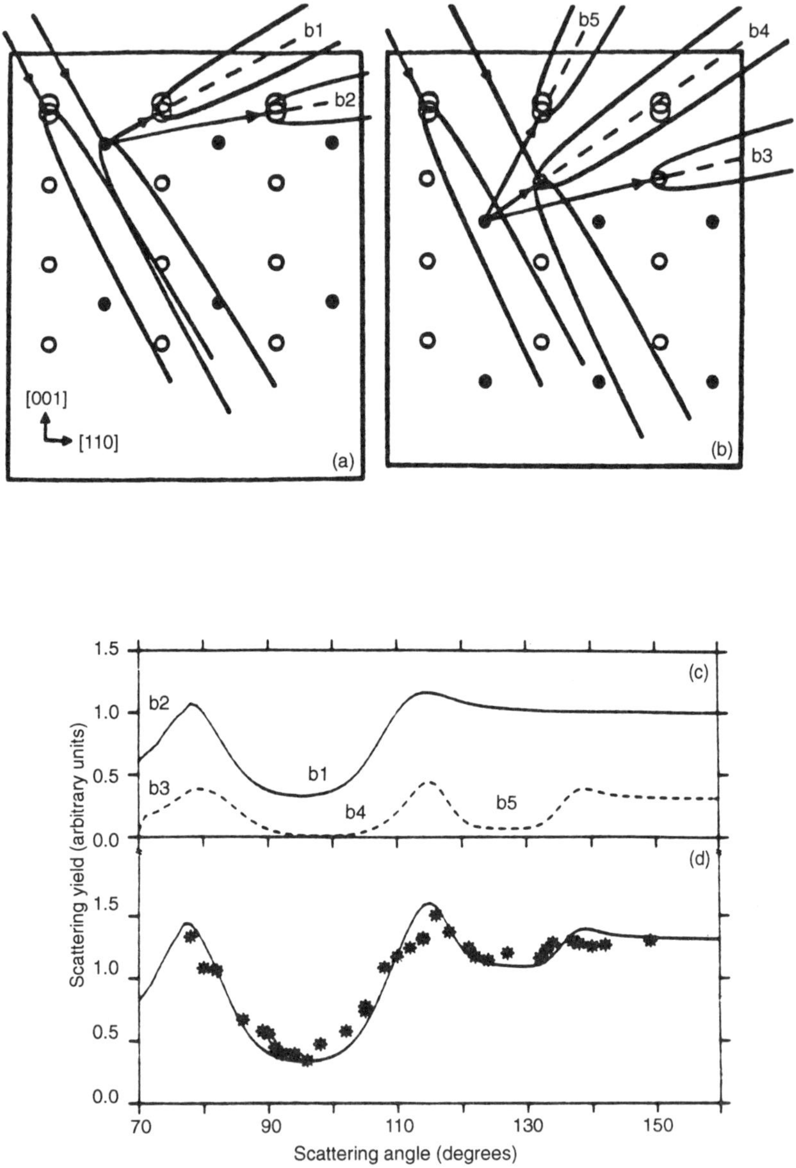

Figure 1.15 (a) and (b) The scattering geometries for the two inequivalent planes oriented along the [110] azimuth for the (100) surface of a $NiSi_2$ crystal face. The projectiles are incident normal to the surface. (c) The scattering-angle dependence of the intensity of 5 keV Li^+ ions scattered from Ni atoms in the planes represented by (a), the solid curve, and (b), the dashed curve, calculated using hitting probabilities. (d) The direct comparison of the experimental data, the stars, with the sum of the two calculated distributions in (c), the solid curve (after Huang *et al.*, 1987)

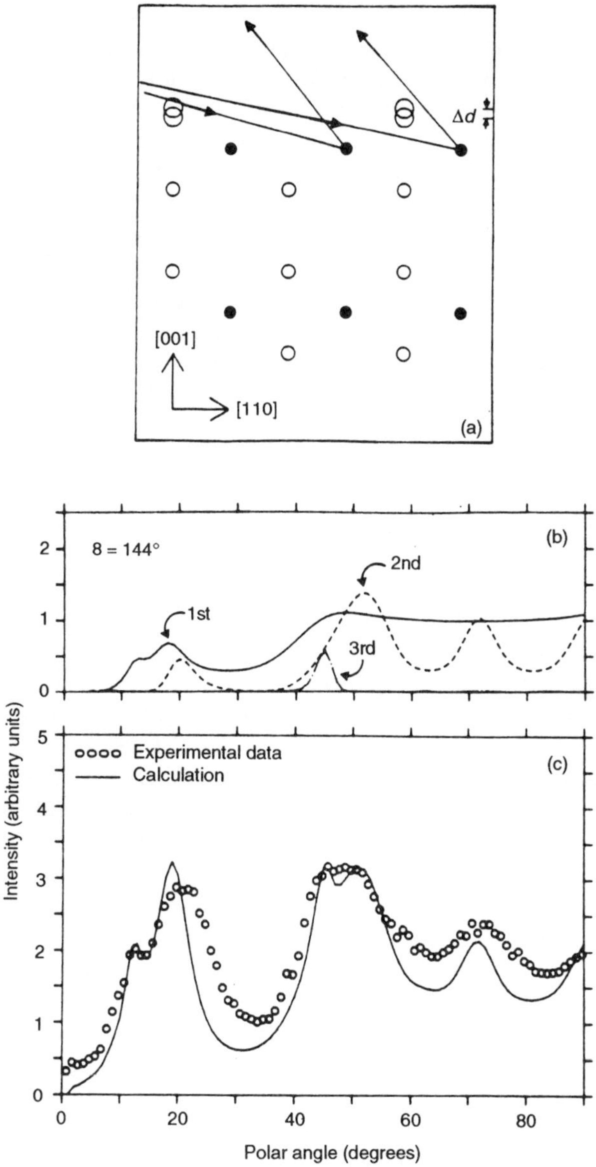

Figure 1.16 (a) The scattering geometry for an ICISS polar scan with a scattering angle of 144° along the [110] azimuth of the (100) surface of a $NiSi_2$ crystal face. (b) The polar-angle dependence of the intensity of 5 keV Li^+ ions scattered from Ni atoms in the first Ni layer, solid curve, the second Ni layer, dashed curve, and the third Ni layer, dash–dot curve, of the model in (a) calculated using hitting probabilities. (c) The direct comparison of the experimental data, the open circles, with the sum of the three calculated distributions in (b), the solid curve (after Huang *et al.*, 1987)

for the system under study by randomly choosing a series of deflections such that the total trajectory has a high probability of contributing to the scattered yield in the desired region of space. Thus, the major advantage of the Monte Carlo over the MD approaches is that in the former case, the trajectories have been preselected, and so a much smaller number of them need to be considered since they all have the *possibility* of contributing to the scattered ion spectrum. The *probability* that any one of the trajectories will contribute to the detected yield is determined from the atomic geometry and the thermal distributions of the atoms in the target crystal. An entire group of target atoms in the lattice are distorted from their equilibrium positions to yield a scattering configuration that will produce the trajectory under examination. The probability that the trajectory does contribute to the yield is determined by multiplying the probabilities for each of the scattering atoms to be found at the position required to propagate the trajectory. If atoms must be displaced a great distance from their equilibrium positions, the probability will be small, but if the required displacements are small, the probability will be large. The individual probability distributions for each target atom are usually taken to be Gaussian with second moments determined from the thermal ellipsoids of the target of interest (Barrett, 1971; Stensgaard *et al.*, 1978; Tromp and van der Veen, 1983; Frenken *et al.*, 1986), but other static or dynamic distributions can be used (van der Gon *et al.*, 1991). A major advantage to the Monte Carlo algorithms is that only one set of trajectories, which requires the largest amount of computing time to determine in this approach, needs to be generated for the system. Any reasonable structural model can be adapted to this set of trajectories to evaluate the scattering probabilities. This allows a large number of structural and non-structural parameters in a model of the surface to be simulated in parallel and with the same quality of calculation to find the optimum agreement with the experimental data, and it can handle any number of collision events. A 'typical' structural refinement for an MEIS data set collected from a GaAs(110) surface required 12 hours on a VAX 11/785 computer (Frenken *et al.*, 1986). However, this approach works best for a target geometry in which the incident beam is fixed with respect to the crystal axes and the scattering angle is allowed to vary, as is the case for standard MEIS experiments. It also requires more computing power than is currently available in most personal computers.

1.3.5 MULTI-ATOM CROSS SECTIONS

The final approach that has been developed for the calculation of scattered ion intensities is to evaluate the differential scattering cross section for the interaction of a projectile with a target that has more than one atom in it (Mashkova and Molchanov, 1985; Parilis *et al.*, 1993; Yamamura and Takeuchi, 1984; Takeuchi and Yamamura, 1984, 1992; Williams *et al.*, 1990; Daley *et al.*, 1990a,

1990b; Chaudhury and Williams, 1991; Charatan and Williams, 1992, 1993; Chaudhury *et al.*, 1993; van de Riet *et al.*, 1990a, 1990b; Kato, 1988); this is descended from the intensity calculations of Oen (1965). For this case, the derivation of the cross section presented in Section 1.2.3 above must be generalized, because such a system will no longer have cylindrical symmetry and the azimuthal angles ϕ_{in} and ϕ_{out} will not necessarily be equal to each other. In this case, the starting point is still equation (33), recognizing that ϕ_{in} and ϕ_{out} are different and that the total scattering angle in lab coordinates is θ_P (Williams *et al.*, 1990):

$$\frac{d\sigma(\theta_P)}{d\Omega_{out}} = \sum_s \frac{s}{\sin\theta_P}\left|\frac{\partial\theta_P\,\partial\phi_{out}}{\partial s\,\partial\phi_{in}}\right|^{-1} = \sum_s \frac{s}{\sin\theta_P}\begin{vmatrix}\frac{\partial\theta_P}{\partial s} & \frac{\partial\phi_{out}}{\partial\phi_{in}}\end{vmatrix}^{-1}. \tag{65}$$

If rectangular coordinates are used to define the incident impact parameter s, then

$$\frac{d\sigma(\theta_P)}{d\Omega_{out}} = \sum_s \frac{1}{\sin\theta_P}\left|\frac{\partial\theta_P\,\partial\phi_{out}}{\partial s_x\,\partial s_y}\right|^{-1} = \sum_s \frac{1}{\sin\theta_P}\begin{vmatrix}\frac{\partial\theta_P}{\partial s_x} & \frac{\partial\phi_{out}}{\partial s_y}\end{vmatrix}^{-1}, \tag{66}$$

where s_x is the component of the impact parameter in the scattering plane, s_y is the component perpendicular to the scattering plane, and the projectile is incident along the z-axis (van de Riet *et al.*, 1990a, 1990b). In equations (65) and (66), the sums are carried out over all branches of the deflection function curve that contribute to the differential cross section. If the scattering is confined to a plane, the second term in the Jacobian determinant from equation (34) vanishes by symmetry, i.e. $\partial\theta_P/\partial\phi_{in} = \partial\theta_P/\partial s_y = 0$ (Williams *et al.*, 1990; van de Riet *et al.*, 1990b). For the two-target-atom scattering case, illustrated in Figure 1.17, the Jacobian determinant in equation (66) is easier to evaluate, but once the Jacobian in equation (65) has been determined, as detailed in the appendix of Williams *et al.* (1990), then it is straightforward to apply the derivation to sequential collisions to obtain the differential cross section for the cases with more than two collisions. Kato (1988) has shown that three-dimensional Jacobians such as those in Williams *et al.* (1990) and van de Riet *et al.* (1990b) are essential to obtain reasonable agreement with experimental results, since the two-dimensional approximation to the scattering cross section (Yamamura and Takeuchi, 1984; Takeuchi and Yamamura, 1984, 1992) severely overestimates the scattering intensity at a rainbow.

The calculation of a scattering cross section from a two-atom target with a fixed internuclear separation R is very simple. As shown in Figure 1.17, the impact parameter with respect to the first atom s_1 encountered determines the entire scattering sequence. If the total scattering angle θ_P is fixed, there is only one possible value of the incidence trajectory with respect to the internuclear axis, determined by the polar orientation angle α, that can scatter the projectile into the detector, and vice versa. Given the deflection function and time integral

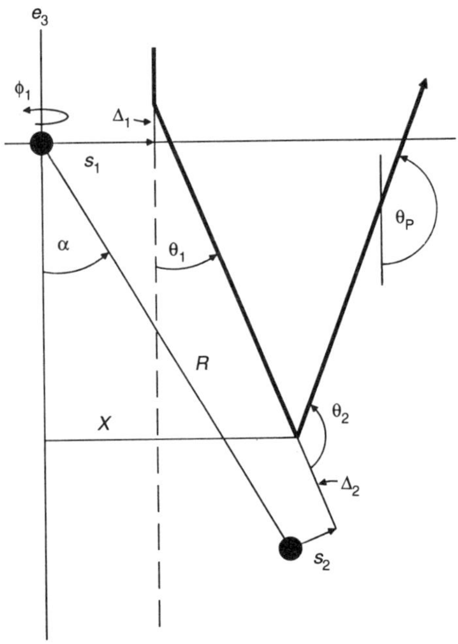

Figure 1.17 Illustration of the trajectory for a projectile that collides with two target atoms to yield a total scattering angle of θ_P. The impact parameter of the projectile with respect to the first target atom s_1 completely specifies the trajectory if either the incidence angle α or θ_P is specified.

for the first interaction, the value of s_1 determines the asymptotic trajectory of the projectile after the first collision through the values of θ_1 and Δ_1. Since the total scattering angle θ_P is fixed, the scattering from the second atom must be $\theta_2 = \theta_P - \theta_1$. Since θ_2 is determined, both s_2 (found by inverting the appropriate deflection function) and Δ_2 are also fixed. Thus, the value of α is constrained (Yamamura and Takeuchi, 1984; Williams *et al.*, 1990):

$$\alpha = \theta_1 + \arcsin\left[\frac{s_1 \cos\theta_1 - s_2 + \Delta_1 \sin\theta_1}{R}\right]. \tag{67}$$

This relation makes the calculation of ICISS distributions from a two-atom target especially simple, since the differential cross section can be calculated directly as a function of either θ_P or α, $(d\sigma/d\Omega)(\theta_P, \alpha)$. To illustrate this point, Figure 1.18 shows the relationship between s_1, θ_P and α for the scattering of 5 keV He^+ ions from a pair of Ag atoms separated by 2.89 Å. Since the effective two-atom-target deflection function has both local minima and maxima, there will be rainbows in the scattering cross section (Daley *et al.*, 1990b).

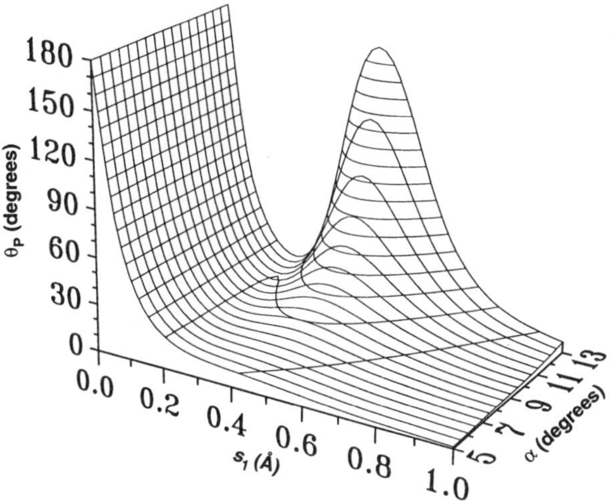

Figure 1.18 A contour plot showing the calculated classical deflection function θ_P for 5 keV He^+ ions scattering from two Ag atoms separated by 2.89 Å as a function of both the impact parameter s_1 and the incidence angle with respect to the internuclear axis α (defined in Figure 1.17). The local maximum in θ_P corresponds to the shadowing rainbow, and the local minimum to the blocking rainbow

The local minima in the $\theta_P(s_1)$ curves correspond to forward scattering and thus represent the edges of blocking cones, whereas the local maxima correspond to back-scattering and thus the edges of shadow cones. A calculation of the differential cross section as a function of α is illustrated in Figure 1.19 where $\alpha(s_1)$ is shown for a particular value of θ_P. The minimum in the $\alpha(s_1)$ curve corresponds to the critical angle for shadowing, as shown in Figure 1.19(a). The resulting two branches of the differential cross section are shown in Figure 1.19(b), with the lower-valued branch corresponding to the $\alpha(s_1)$ curve at values of s_1 smaller than the critical value for the shadowing rainbow. The divergence of the branches of the cross section at the critical angle are suppressed by determining a cutoff value of the impact parameter for calculating the classical cross section (Daley *et al.*, 1990b), as described previously in Section 1.2.3.5.

The cross section in Figure 1.19(b) is for the case of a frozen atom configuration. In order to include the effects of vibrational motion of the atoms in the crystal lattice, the frozen lattice cross section can be convoluted with a function that represents the distribution of angular orientations of the two target atoms with respect to each other. This is conveniently done by calculating the angular deviation of the internuclear axis from its equilibrium position from the thermal vibrational amplitudes of the target atoms, and using this value as the standard deviation in a Gaussian broadening of the cross section (Williams *et al.*, 1990).

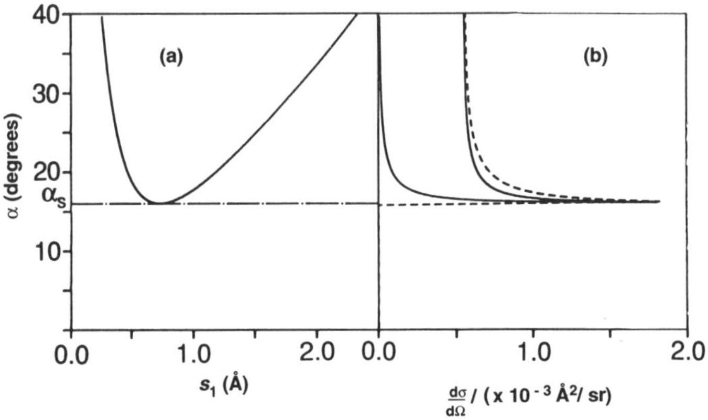

Figure 1.19 (a) The impact parameter with respect to the first atom (s_1) of a two-atom target versus the incidence angle α for 5 keV Li^+ ions scattering from two Ag atoms that are 2.89 Å apart. There are two different possible values of s_1 for each value of α except at the critical angle for scattering, α_s, at which a rainbow occurs. The differential cross section as a function of incidence angle (b) is determined by adding together the cross sections at the different values of s_1 that correspond to the same incidence angle

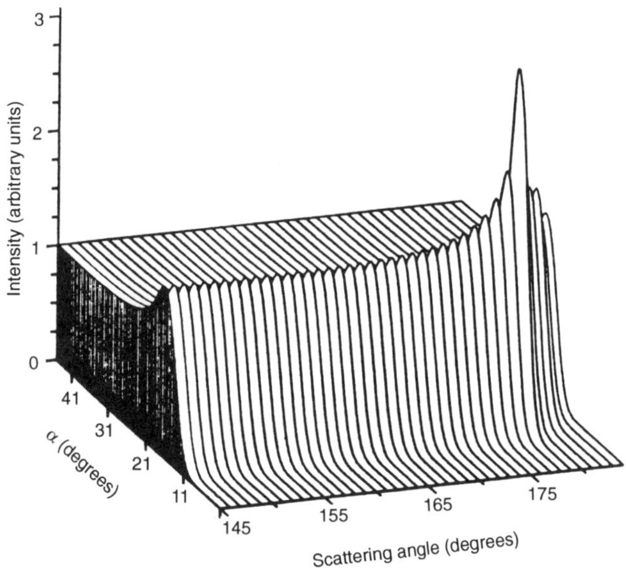

Figure 1.20 The Gaussian broadened two-atom scattering cross sections for the deflection function presented in Figure 1.18, plotted as scattered intensity versus incidence angle for a range of scattering angles. All the distributions were normalized to the same intensity at $\alpha = 90°$

A range of such thermally broadened cross sections is shown in Figure 1.20 as a function of both the polar and the scattering angles.

An important point is illustrated by the plots in Figure 1.20. The scattering intensity (i.e. the thermally broadened cross section) at the shadowing rainbow depends strongly on the scattering angle near 180°, since the projectile can interact with the shadowing atom on both the inbound and outbound trajectories after a very small impact parameter collision with the back-scattering atom (Williams *et al.*, 1990). This double-focusing effect substantially increases the scattering intensity at the critical polar angle, and leads to a very strong dependence of both the intensity and the critical polar angle on the scattering angle (Charatan and Williams, 1992). Similar correlated scattering events can also occur for scattering angles much smaller than 180° if the projectile passes between the shadowing and back-scattering atoms (Charatan and Williams, 1993). To properly simulate such scattering conditions, it is essential to treat the multiple scattering correctly, and this is easily done with the cross-section approach.

In order to simulate the scattering from an entire lattice, the broadened cross sections from all reasonable pairs of target atoms are combined using equation (64) above. This procedure has been used to simulate both ICISS polar scans (van de Riet *et al.*, 1990a, 1990b; Kato, 1988; Daley *et al.*, 1990a; King *et al.*, 1990; Katayama *et al.*, 1991; Nakanishi *et al.*, 1991) and in the special case of 180° back-scattering, azimuthal scans also (Chaudhury *et al.*, 1993). In addition, the cross-section formulation is easily adapted to calculate the angular and energy dependence of the intensity of recoils from surfaces (Chaudhury and Williams, 1991). The extension of the cross section to a three-atom target has been demonstrated (Takeuchi and Yamamura, 1992). Thus, the same basic algorithm is broadly applicable to many different geometries and events.

Calculating the scattering cross section from a multi-center target has the major advantages of being both very intuitive and extremely fast. The computer programs developed to implement the cross-section calculation were all written on PC-type computers (Williams *et al.*, 1990; Daley *et al.*, 1990b), and as the power of these computers has increased so have the capabilities of the programs. The existing routines can calculate full three-dimensional cross sections for both scattering (Williams *et al.*, 1990) and recoiling (Chaudhury and Williams, 1991) that involve either double- or triple-scattering from pairs of atoms. Several hundred cross sections per second can be calculated for arbitrary projectile and target species as functions of incident energy, scattered energy, scattering angle, and polar and azimuthal orientation of the target with respect to the incident projectile direction. Thus, it is possible to analyze large scattering or recoiling data sets with a PC in essentially real time as the data are collected. This procedure has the advantage of suggesting new experimental geometries to explore as the data are being collected, rather than after the experiments are completed.

1.4 SUMMARY AND CONCLUSIONS

For most systems of interest, accurate calculations of intensities from structural models are essential for analyzing low-energy scattering and recoiling data to determine a unique and reliable surface atomic geometry. Several computational procedures have evolved that complement the different experimental approaches and beam energies that are used in such studies. Even though the calculations are based on classical mechanics, the simulation approaches that are used are still evolving rapidly, primarily because of the limitations of current computers. The increasing power available in what are regarded as small computers and the increasingly sophisticated programming languages that are appearing will further drive developments in this area. However, in order to determine a surface structure that is generally accepted and highly accurate, it will probably be necessary to collect scattering and/or recoiling data sets in which several different parameters, such as the polar and azimuthal projectile incidence direction and the scattered or recoiled kinetic energy, are varied to provide 10^6 or more data points. This data must then be compared with the results of simulations from a large number of structural models, and the simulations must be capable of reproducing all the data.

The three major types of simulation technique all have room for improvements in algorithm development or computer implementation. The molecular dynamics approach is the most complete in terms of containing all the scattering physics and possible complex scattering trajectories. However, in order to generate a calculated intensity distribution with a large enough signal-to-noise ratio to make detailed comparisons with experimental data, approximately 10^7 trajectories must be launched. To compare many structural models to a large experimental data set will require calculations on a massively parallel computer and highly symmetrical experimental orientations for scattering and recoiling. Hitting probabilities offer two different approaches to simulate scattering and recoiling. Direct numerical evaluation of the probability integrals is only feasible for two- or three-target-atom systems. However, such calculations do yield extremely useful scattered intensity information. For the most part, they are only missing certain types of correlated scattering events. The major advantage of hitting probabilities is that they are naturally adapted to Monte Carlo calculations, in which several hundred thousand possible trajectories are preselected and their probability of contributing to the scattered yield from a particular structural model are evaluated. This places quantitative comparison of different structural models on a firm foundation. In order to minimize the number of required trajectories, the Monte Carlo approach works best if the projectile incidence angle is held fixed and the scattering angle is allowed to vary, as in MEIS experiments. Multi-center cross sections offer a very promising means for calculating both scattering and recoiling intensities, especially in the

ICISS geometry in which the scattering angle is held fixed and the sample is rotated with respect to the incident projectile direction. The cross-section formulation is readily adapted to handle certain types of correlated multiple-scattering events in which a particular target is hit twice, as is a common occurrence for certain experimental geometries, especially the 180° scattering case. Cross-section calculations have the further advantages of being both very intuitive and computationally much faster than evaluating hitting probability integrals, since one only needs to evaluate a derivative at each orientation of interest. For quantitative interpretation of scattering and recoiling data in terms of surface atomic structure, two or more of the simulation techniques may have to be combined to take advantage of their respective strengths. Cross-section calculations may be used for on-line data analysis and to select the most promising atomic geometries, while MD or extended Monte Carlo calculations may be used for optimizing the parameters of the final structural model.

1.5 ACKNOWLEDGEMENTS

My work at UCLA in the field of ion scattering has been supported since 1982 by the US National Science Foundation. The cross-section calculations were begun in 1987 during my sabbatical leave at the Frontier Materials Research Program of Rikagaku Kenkusho (RIKEN), the Laboratory for Physics and Chemistry Research in Wako-shi, Japan. I gratefully aknowledge Dr T. M. Buck and Dr L. C. Feldman for introducing me to the fields of low- and high-energy ion scattering. I am also grateful to several of my former and present students for their enthusiastic assistance: Prof. J. A. Yarmoff, Prof. R. Blumenthal, Dr D. M. Cyr, Dr J. H. Huang, Dr R. S. Daley, Dr R. M. Charatan, Dr S. Chaudhury, and most recently Mr S. J. Anz, Ms A.-M. Russell and Mr K. Nafisi. I reserve a special note of thanks for my close collaborator Dr M. Aono, as well as for my colleagues Prof. M. Kato and Dr M. Katayama. I thank Prof. J. W. Rabalais for inviting me to contribute this review and for helpful comments on the manuscript, and most recently Prof. E. Parilis for interesting conversations about the theory of ion scattering.

1.6 REFERENCES

Aono, M., C. Oshima, S. Zaima, S. Otani, and Y. Ishizawa (1981) *Jap. J. Appl. Phys.*, **20**, 1829.
Aono, M., M. Katayama, E. Nomura, T. Chasse, D. Choi, and M. Kato (1989) *Nucl. Instr. Methods Phys. Res.*, **B 37/38**, 264.
Barrett, J. H. (1971) *Phys. Rev.*, **B 3**, 1527.
Bernheim, M. and G. Slodzian (1976) *Nucl. Instr. Methods*, **132**, 615.
Boers, A. L. (1977) *Surf. Sci.*, **63**, 475.

Brongersma, H. H. and P. M. Mul (1973) *Surf. Sci.*, **35**, 393.
Buck, T. M, G. H. Wheatley, and L. K. Verheij (1979) *Surf. Sci.*, **90**, 635.
Buck, T. M., I. Stensgaard, G. H. Wheatley, and L. Marchut (1980) *Nucl. Instrum. Methods*, **170**, 519.
Chang, C.-C., N. Winograd, and B. J. Garrison (1988) *Surf. Sci.*, **202**, 309.
Chang, C. S., T. L. Porter, and I. S. T. Tsong (1989) *Vacuum*, **39**, 1195.
Charatan, R. M. and R. S. Williams (1992) *Surf. Sci. Letts.*, **264**, L207.
Charatan, R. M. and R. S. Williams (1993) *Surf Sci.*, **289**, 163.
Chaudhury, S. and R. S. Williams (1991) *Surf. Sci.*, **225**, 127.
Chaudhury, S., R. S. Williams, M. Katayama, and M. Aono (1993) *Surf. Sci.*, **294**, 93.
Child, M. S. (1974) *Molecular Collision Theory*. Academic Press, New York.
Chu, W.-K., J. W. Mayer, M.-A. Nicolet, T. M. Buck, G. Amsel, and F. Eisen (1973) *Thin Solid Films*, **17**, 1.
Chu, W.-K., J. W. Mayer, and M.-A. Nicolet (1978) *Backscattering Spectrometry*. Academic Press, New York.
Daley, R. S. and R. S. Williams (1990) *Nucl. Instr. Methods Phys. Res.*, **B 45**, 380.
Daley, R. S., J. H. Huang, and R. S. Williams (1989) *Surf. Sci.*, **215**, 281.
Daley, R. S., R. M. Charatan, and R. S. Williams (1990a) *Surf. Sci.*, **240**, 136.
Daley, R. S., D. Farrelly, and R. S. Williams (1990b) *Surf. Sci.*, **234**, 355.
Durr, H., R. Schneider, and Th. Fauster (1990) *Vacuum*, **41**, 376.
Eckstein, W. (1987) *Nuclear Instrum. Methods in Phys. Res.*, **B 27**, 78.
Fauster, Th. (1988) *Vacuum*, **28**, 129.
Feldman, L. C. and J. W. Mayer (1986) *Fundamentals of Surface and Thin Film Analysis*. Elsevier, Amsterdam.
Firsov, O. (1958) *Zh. Eksp. Teor. Fiz.*, **33**, 696; *Sov. Phys.—JETP*, **6**, 534.
Frenken, J. W. M., R. M. Tromp, and J. F. van der Veen (1986) *Nucl. Instrum. Meth. Phys. Res.*, **B 17**, 334.
Garrison, B. J. (1979) *Surf. Sci.*, **87**, 683.
Gemmel, D. S. (1974) *Rev. Mod. Phys.*, **46**, 129.
Goldstein, H. (1980) *Classical Mechanics*. Addison-Wesley, Reading, Mass.
Harrison, D. E., Jr., P. W. Kelly, B. J. Garrison, and N. Winograd (1978) *Surf. Sci.*, **76**, 311.
Heiland, W., E. Taglauer, and M. T. Robinson (1976) *Nucl. Instrum. Methods*, **132**, 655.
Helbig, H. F., M. W. Lindner, G. A. Morris, and S. A. Steward (1982) *Surf. Sci.*, **114**, 251.
Horn, T. C. M., P. Haochang, P. J. van den Hoek, and A. W. Kleyn (1988) *Surf. Sci.*, **201**, 573.
Hou, M. and M. T. Robinson (1976) *Nucl. Instrum. Methods*, **132**, 641.
Huang, J. H. and R. S. Williams (1988a) *Phys. Rev.*, **B 38**, 4022.
Huang, J. H. and R. S. Williams (1988b) *Surf. Sci.*, **204**, 445.
Huang, J. H., R. S. Daley, D. K. Shuh, and R. S. Williams (1987) *Surf. Sci.*, **186**, 115.
Hulpke, E. and K. Mann (1983) *Surf. Sci.*, **133**, 171.
Jackson, D. P., W. Heiland, and E. Taglauer (1981) *Phys. Rev.*, **B 24**, 4198.
Karpuzov, D. S. and V. E. Yurasova (1971) *Phys. Status Solidi (b)*, **47**, 41.
Karpuzov, D. S., V. A. Eltekov, and V. E. Yurasova (1967) *Soviet Phys.—Solid State*, **8**, 1726.
Katayama, M., R. S. Williams, M. Kato, E. Nomura, and M. Aono (1991) *Phys. Rev. Lett.*, **66**, 2762.
Kato, M. (1988) *A Short Note on the Self-Blocking Problem*, 5 Nov. (unpublished).
King, B. V., M. Katayama, M. Aono, R. S. Daley, and R. S. Williams (1990) *Vacuum*, **41**, 938.
Kivilis, V. M., E. S. Parilis, and N. Yu. Turaev (1967) *Soviet Phys. Dokl.*, **12**, 328.

Latta, B. M. and P. J. Scanlon (1974) *Phys. Rev.*, **A 10**, 1638.
Lehman, C. (1977) *Interaction of Radiation with Solids and Elementary Defect Production.* North-Holland, New York.
Marion, J. B. (1970) *Classical Dynamics of Particles and Systems*, 2nd edn. Academic Press, New York.
Mashkova, E. S. and V. A. Molchanov (1985) *Medium-Energy Ion Reflection from Solids.* North-Holland, New York.
Mendenhall, M. H. and R. A. Weller (1990) *Nucl. Instrum. Meth. Phys. Res.*, **B 58**, 11.
Molière, G. (1947) *Z. Naturforsch.*, **A 2**, 133.
Nakanishi, S., N. Fukuoka, K. Kawamoto, K. Umezawa, Y. Teraoka, and K. Nakahigashi (1991) *Surf. Sci.*, **247**, L215.
Newton, R. G. (1966) *Scattering Theory of Waves and Particles.* McGraw-Hill, New York.
Niehus, H. (1988) *Nucl. Instr. Methods Phys. Res.*, **B 33**, 876.
Niehus, H. and R. Spitzl (1991) *Surf. and Int. Anal.*, **17**, 287.
Nizhnaya, S. L., E. S. Parilis, and N. Yu. Turaev (1979) *Radiat. Eff.*, **40**, 43.
Oen, O. S. (1965) *Phys. Letts.*, **19**, 358.
Oen, O. S. (1982) *Nucl. Instrum. Methods*, **194**, 87.
Oen, O. S. (1983) *Surf. Sci.*, **131**, L407.
Oen, O. S. and M. T. Robinson (1976) *Nucl. Instrum. Methods*, **132**, 647.
Overbury, S. H. (1981) *Surf. Sci.*, **112**, 23.
Overbury, S. H. (1987) *Nuclear Instrum. Methods in Phys. Res.*, **B 27**, 65.
Parilis, E. S., L. M. Kishinevsky, N. Yu. Turaev, B. E. Baklitzky, F. F. Umarov, V. Kh. Verleger, S. I. Nizhnaya, and I. S. Bitensky (1993) *Atomic Collisions on Solid Surfaces.* North-Holland, New York.
Poelsema, B. and A. L. Boers (1977) *Phys. Letts.*, **64A**, 304.
Poelsema, B. and A. L. Boers (1979) *Radiat. Eff.*, **41**, 229.
Poelsema, B., L. K. Verhey, and A. L. Boers (1976a) *Surf. Sci.*, **56**, 445.
Poelsema, B., L. K. Verhey, and A. L. Boers (1976b) *Surf. Sci.*, **60**, 485.
Poelsema, B., L. K. Verhey, and A. L. Boers (1976c) *Nucl. Instrum. Meth.*, **132**, 623.
Poelsema, B., L. K. Verhey, and A. L. Boers (1977a) *Surf. Sci.*, **64**, 537.
Poelsema, B., L. K. Verhey, and A. L. Boers (1977b) *Surf. Sci.*, **64**, 554.
Porter, T. L., C. S. Chang, and I. S. T. Tsong (1988) *Phys. Rev. Lett.*, **60**, 1739.
Preuss, E. (1978) *Radiat. Eff.*, **38**, 151.
Rabalais, J. W. (1988) *CRC Crit. Rev. Sol. St. and Mats. Sci.*, **14**, 319.
Rabalais, J. W. (1990) *Science*, **250**, 521.
Robinson, M. T. (1963, 1970a) Tables of Classical Scattering Integrals, Oak Ridge National Laboratory Publications ORNL-3493 and ORNL-4556 (unpublished). (Available from the US National Technical Information Service.)
Robinson, M. T. (1970b) In D. W. Palmers, M. W. Thompson, and D. P. Townsend (eds), *Atomic Collision Phenomena in Solids.* North-Holland, New York, p. 66.
Robinson, M. T. and O. S. Oen (1963) *Phys. Rev.*, **132**, 2385.
Robinson, M. T. and I. M. Torrens (1974) *Phys. Rev.*, **B 9**, 5008.
Rutherford, E. (1911) *Phil. Mag.*, **21**, 669.
Sato, K., S. Kono, T. Teruyama, K. Higashiyama, and T. Sagawa (1985) *Surf. Sci.*, **158**, 644.
Sigmund, P. (1969) *Phys. Rev.*, **184**, 184.
Smith, D. P. (1967) *J. Appl. Phys.*, **38**, 340.
Spitzl, R., N. Niehus, and G. Comsa (1990) *Rev. Sci. Inst.*, **61**, 760.
Stensgaard, I., L. C. Feldman, and P. J. Silverman (1978) *Surf. Sci.*, **77**, 513.
Taglauer, E., W. Englert, W. Heiland, and D. P. Jackson (1980) *Phys. Rev. Letts.*, **45**, 740.

Taglauer, E., M. Beckschulte, R. Margraf, and D. Mehl (1988) *Nucl. Instr. Methods Phys. Res.*, **B 35**, 404.
Takeuchi, W. and Y. Yamamura (1984) *Nucl. Instr. Meth. Phys. Res.*, **B 2**, 336.
Takeuchi, W. and Y. Yamamura (1992) *Surf. Sci.*, **277**, 351.
Torrens, I. M. (1972) *Interatomic Potentials.* Academic Press, New York.
Tromp, R. M. and J. F. van der Veen (1983) *Surf. Sci.*, **133**, 159.
Turkenburg, W. C., R. G. Smeenk, and F. W. Saris (1978) *Surf. Sci.*, **74**, 181.
Turkevich, A. (1961) *Science*, **134**, 672.
Turkevich, A. L., *et al.* (1968) in Surveyor Project Final Report, Part II. Scientific Results, NASA Tech. Rep. 32-1265, pp. 303–387.
van de Riet, E., J. M. Fluit, and A. Niehaus (1990a) *Vacuum*, **41**, 372.
van de Riet, E., J. M. Fluit, and A. Niehaus (1990b) *Surf. Sci.*, **231**, 368.
van der Gon, A. W. D., D. Frenkel, J. W. M. Frenken, R. J. Smith, and P. Stoltze (1991) *Surf. Sci.*, **256**, 385.
van der Veen, J. F. (1985) *Surf. Sci. Reports*, **5**, 199.
van der Veen, J. F., J. B. Sanders, and F. W. Saris (1978) *Surf. Sci.*, **77**, 337.
van der Veen, J. F., R. M. Tromp, R. G. Smeenk, and F. W. Saris (1980) *Nucl. Instr. Methods*, **171**, 143.
von dem Hagen, T., M. Hou, and E. Bauer (1982) *Surf. Sci.*, **117**, 134.
Williams, R. S. (1982) *J. Vac. Sci. Technol.*, **20**, 770.
Williams, R. S. and J. A. Yarmoff (1983) *Nucl. Instrum. Meth. Phys Res.*, **218**, 235.
Williams, R. S., M. Kato, R. S. Daley, and M. Aono (1990) *Surf. Sci.*, **225**, 355.
Winograd, N., D. E. Harrison, Jr., and B. J. Garrison (1978) *Surf. Sci.*, **78**, 467.
Woodruff, D. P. (1982) *Nucl. Instr. Methods*, **194**, 639.
Yamamura, Y. and W. Takeuchi (1984) *Radia. Eff.*, **82**, 73.
Yarmoff, J. A. (1985) Ph.D. Thesis, University of California Los Angeles.
Yarmoff, J. A. and R. S. Williams (1982) *J. Phys. Chem.*, **86**, 1924.
Yarmoff, J. A. and R. S. Williams (1983) *Surf. Sci.*, **127**, 461.
Yarmoff, J. A., D. M. Cyr, J. H. Huang, S. Kim, and R. S. Williams (1986) *Phys. Rev.*, **B 33**, 3856.
Yurasova, V. E., V. I. Shulga, and D. S. Karpuzov (1968) *Can. J. Phys.*, **46**, 759.
Ziegler, J. F. and R. F. Lever (1973) *Thin Solid Films*, **19**, 291.
Ziegler, J. F., J. P. Biersack, and U. Littmark (1985) *The Stopping and Range of Ions in Solids.* Pergamon Press, New York.

2 ION SCATTERING AND RECOILING FOR ELEMENTAL ANALYSIS AND STRUCTURE DETERMINATION

Patrick G. Bertrand* and J. Wayne Rabalais
Department of Chemistry, University of Houston, Houston, Texas 77204-5641, USA

CONTENTS

* On sabbatical leave from the Laboratory of Physical Chemistry and Physics of Materials, Catholic University of Louvain, 1 Place Croix du Sud, B1348 Louvain-la-Neuve, Belgium.

Low Energy Ion–Surface Interactions
Edited by J. W. Rabalais

2.1 INTRODUCTION

2.1.1 THE BASIS OF ION–SOLID INTERACTION

When a low-energy (keV) ion beam bombards a solid, a series of complex interactions occurs by which the primary ions transfer a part of their energy to the solid. As a result of this perturbation, the solid emits photons and electrons, as well as atomic and molecular ions and neutrals. Moreover, irradiation damage is produced and ultimately, if not reemitted, the primary ions are implanted.

In ion-scattering spectrometry (ISS) and direct-recoil spectrometry (DRS), the energy distributions of particles emitted into a well-defined solid angle are measured. For ISS the primary particles scattered by the surface atoms are detected, while for DRS the target atoms emitted as a result of the initial collision of the incident ion with the solid surface are detected. Information about the topmost surface composition and structure are obtained from ISS and DRS.

Most of the secondary atomic and molecular particles result from more complex interaction schemes. When the primary ions penetrate into the solid, they undergo a sequence of collisions. During each of these collisions, target atoms can be put in motion. These recoiling atoms also generate new collisions and collision cascades are produced into the solid. The cascades develop nearly isotropically so that energy and momentum can be transferred to surface atoms. If this transfer is sufficient, secondary ions and neutrals are emitted; this is the sputtering process (Sigmund, 1981). The energy deposited into the electronic system of the target can also contribute to emission of secondary particles (mainly for insulating materials) (Johnson and Sundqvist, 1992).

The energy distribution of the sputtered particles is broad and peaked at low energy (< 10 eV) so that if the detection direction is correctly chosen, their discrimination from the direct-recoil atoms is obvious. The measurement of the mass distribution of the secondary sputtered ions comprises secondary ion mass spectrometry (SIMS). Since most of the sputtered particles are neutrals, their post-ionization in vacuum (by plasmas, electrons, or lasers) before mass analysis produces secondary neutral mass spectrometry (SNMS).

In this chapter, we will focus on the description of processes resulting from only a few collisions so that SIMS and SNMS will not be considered. For the interested reader, recent reviews and books devoted to those spectrometries are already available (Benninghoven *et al.* 1987).

2.1.2 ENERGY RANGE OF INTEREST

In the energy range considered here (0.5 to 5 keV), the primary ions do not penetrate deeply into the solid and they lose energy mainly during discrete elastic collisions with the target atomic nuclei. This is in contrast to the MeV ions used in Rutherford back-scattering spectrometry (RBS) where the energy loss is mainly electronic (Chu *et al.*, 1978). For RBS, the ions are losing their energy continuously along straight-line trajectories in the solid. Only a very small fraction undergo hard collisions leading to back-scattering. The difference between RBS and ISS is due to the decrease of the scattering cross section when increasing the primary ion energy. In RBS the dimensions of the scattering centers are of the order of 10% of the covalent or atomic radii. In the keV energy range, the scattering cross sections are much larger so that one can associate to the scattering centers dimensions close to their covalent or ionic radii. This point will be justified in detail in Section 2.3.2.4 by introduction of the shadow cone concept. If the primary ion energy is reduced to a few eV, the trajectories can no longer be considered as a succession of discrete collisions but rather as the result of interaction with a continuous surface potential (Cooper *et al.*, 1992).

Moreover if the primary ion energy is decreased further, the wavelength associated with these ions can no longer be considered as negligible when compared to the characteristic interaction dimensions (distance of closest approach, interatomic spacing). Diffraction effects have then to be considered and quantum mechanics must be used to correctly describe the ion trajectories (Engel and Rieder, 1982).

In the keV energy range, it is common to separate the elastic processes, which describe the kinetic energy transfers during the collisions, from the inelastic ones, which deal with the interaction of electronic systems and which modify the potential energy of the system. The ion trajectories are usually calculated classically and the charge exchange processes leading to ion neutralization and reionization are treated separately (see Section 2.3.4.4). However, this separation is somewhat artificial and some charge exchange mechanisms can also modify the kinetic energy of the projectile (Souda and Aono, 1986).

2.2 EXPERIMENTAL METHODS

2.2.1 BASIC REQUIREMENTS

The basic requirements for low-energy ion scattering are an ion source, a sample mounted on a precision manipulator, an energy or velocity analyzer, and a detector. The sample must be housed in an ultra-high vacuum (UHV) environment in order to prepare and maintain well-defined clean surfaces. This UHV prerequisite necessitates the use of differentially pumped ion sources. Ion scattering is typically done in a UHV chamber which houses other surface analysis techniques such as low-energy electron diffraction (LEED), X-ray photoelectron spectroscopy (XPS), and Auger electron spectroscopy (AES). The combination of these techniques provides a powerful surface-analysis system. The design of an instrument for ion scattering is guided by the type of analyzer to be used. An electrostatic analyzer (ESA) for measuring the kinetic energies of ions can be incorporated into most standard vacuum chambers. Time-of-flight (TOF) techniques for measuring the velocities of ions and neutrals require a long path length for the scattered and recoiled species. This can be accomplished by using a standard chamber with long TOF drift tubes at fixed angles or by designing an instrument for continuous variation of the scattering angle and long path length. A schematic diagram of an instrument (Grizzi *et al.*, 1990) that has been designed for low-energy ion scattering and recoiling with continuous variation of the scattering angles along with other surface-analyses techniques is shown in Figure 2.1.

Standard UHV technology with bakable stainless-steel chambers sealed with copper gaskets is used for the instrumentation. Vacuums down to the 10^{-11} torr range are usually accomplished with some combination of pumps which may

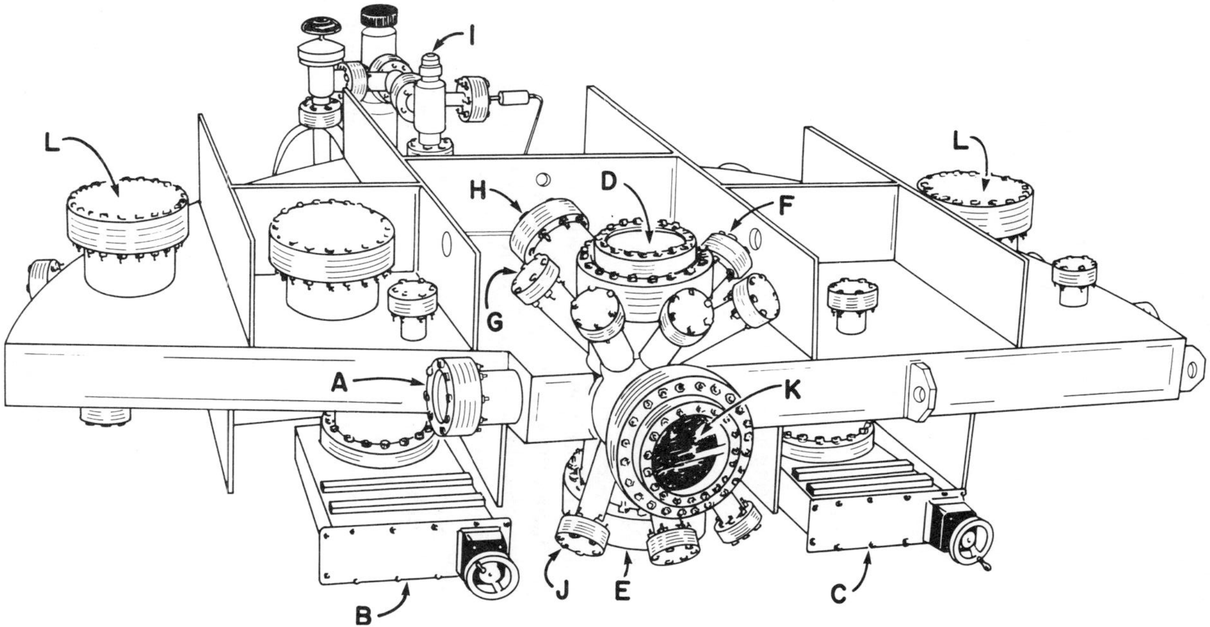

Figure 2.1 Spectrometer system designated for TOF analysis of neutrals and ions scattered and recoiled along with conventional AES, XPS, UPS and LEED analysis. A = pulsed ion beam; B = turbomolecular pump, C = ion pump; D = sample manipulator; E = detector precision rotary motion feedthrough; F = X-ray source; G = electron gun; H = 180° electrostatic hemispherical analyzer; I = sorption pumps; J = sputter ion gun; K = viewport for reverse LEED optics; L = titanium sublimation pump; M = cryopump. (Reproduced by permission of the American Institute of Physics from Grizzi *et al.*, 1990)

include turbomolecular pumps, cryopumps, ion pumps, and titanium sublimators. The sample is mounted on a precision manipulator that allows polar beam incident angle and azimuthal angle rotations, translations along three orthogonal axes, and adjustment of the sample tilt angle with respect to the incident ion beam. Stepper-motor control of the rotations is desirable. The beam incident angle and scattering angles can be aligned by a laser beam. The azimuthal angle is determined crudely from observation of the LEED pattern and then precisely by monitoring surface semichanneling effects. The sample is cleaned *in situ* by rare-gas sputtering and/or annealing to high temperature.

2.2.2 ION SOURCE AND BEAM LINE

The critical requirements for the ion source are that the ions have a small energy spread and there are no fast neutrals in the beam. The most common types of ion sources used in scattering are those that produce 1–5 keV noble-gas or alkali-metal ions with narrow energy distributions. Noble-gas ions are usually generated by electron impact sources. Such sources require differential pumping and some designs have off-axis filaments which reduce the amount of fast neutrals moving towards the sample. Alkali ions are usually generated by heating alkali aluminosilicate wafers. There are two major differences in the scattering of noble-gas and alkali-metal ions. (i) Noble-gas ions are neutralized with high probability during the scattering collisions whereas alkali-metal ions have higher survival probabilities. Therefore, for ESA analysis alkali ions provide a higher sensitivity than rare-gas ions although rare-gas ions provide the highest first-layer sensitivity due to their enhanced neutralization from subsurface layers. Although alkali ions have higher survival probabilities, there are significant variations (Overbury *et al.*, 1984) in their scattered ion yields that can affect quantitative analysis. For TOF analysis with detection of neutrals and ions, such variations in the neutralization probabilities are not important. (ii) Alkali ions are deposited on the surface during scattering, an effect which can change the work function and reconstruction among other things.

A typical beam line (Grizzi *et al.*, 1990) for these experiments is shown in Figure 2.2. A tubular Einzel lens focuses the beam as it emerges from the source regions. Mass selection is achieved by means of a Wien filter, which consists of a magnet and a pair of electrostatic deflection plates mounted such that the electric field E is perpendicular to the magnetic field B. When a beam of charged particles with velocity v and charge e pass through the filter, they are deflected in one direction by an electrostatic force $F_e = Ee$ and in an opposite direction by a magnetic force $F_m = Bev$. When these two opposing forces are equal, the charged particles with velocity v pass undeflected through the filter. The other

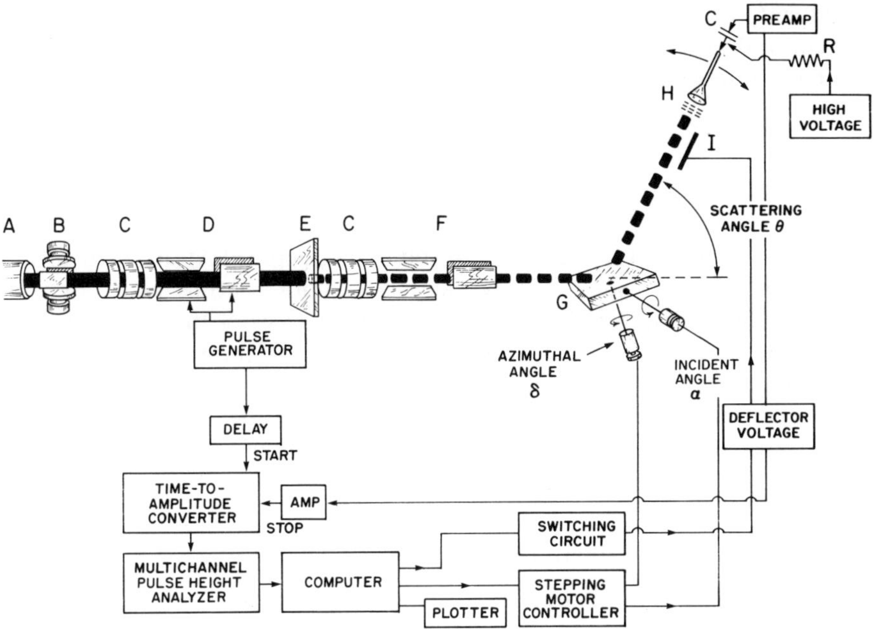

Figure 2.2 Schematic drawing of pulsed ion beam, sample, detector and associated electronics. A = ion gun, B = Wien filter, C = Einzel lens, D = pulsing plates, E = pulsing aperture, F = deflector plates, G = sample, H = electron multiplier, I = deflector. (Reproduced by permission of the American Institute of Physics from Grizzi *et al.*, 1990)

particles are dispersed on a target downstream. In practice, it is convenient to use a small fixed magnet and vary the electric field between the deflection plates.

The pulsing system consists of a pair of electrostatic plates and an aperture. The aperture facilitates collimation and differential pumping of the beam; if placed off-axis it eliminates fast neutrals from the beam. A pulsed beam is obtained by applying a voltage pulse to one of the plates in order to rapidly sweep the beam across the aperture. A second ion pulse which would normally appear on the falling edge of the voltage pulse can be removed by applying a delayed voltage pulse to the second set of orthogonal deflection plates. In practice, the duration of the ion pulse through the slit has been varied from 10 to 100 ns with ion energies of a few keV. Generation of a pulsed beam consists of producing a continuous beam of the order of 1 to 50 nA and then adjusting the deflector so that 10^3 to 10^4 ions/pulse go through the slits and reach the target. Using a repetition frequency of 5 to 50 kHz, an average ion current of 0.1 to 0.5 nA cm^{-2} is obtained. This current density is small enough that beam damage during the measurement period is negligible. TOF spectra are typically collected with a dose of $<10^{-3}$ primary ions/surface atom. For a sample area of 0.1 cm^2 with 2×10^{15} atoms per cm^2 in a monolayer, at an average ion current of 0.3 nA cm^{-2} it would take $\sim$30 h of continuous bombardment to sputter a monolayer, assuming a sputtering yield of unity.

2.2.3 ANALYZERS

Both time-of-flight (TOF) and electrostatic analyzers (ESA) are used in ion scattering. TOF provides velocity analysis of both neutrals and ions with moderate resolution. An ESA provides energy analysis of the ions with high resolution. An ESA can be used to verify the masses of the ejected particles by measuring the TOF of particles traveling through the analyzer while the analyzer is set to pass a specific kinetic energy; both the velocity and energy of the transmitted particles are known, providing a unique mass determination.

2.2.3.1 Time-of-flight analysis

The TOF analyzer is simply a long drift region. If a deflector plate, an acceleration tube lens, and a retarding grid analyzer are included in the TOF flight region, four different types of spectra can be obtained: (i) neutrals plus ions—obtained with all elements at ground potential; (ii) neutrals only—obtained with voltage on the deflector; (iii) neutrals and accelerated ions—obtained with accelerating voltage on the tube lens; (iv) neutrals plus ions with specific charges—obtained with the retarding grid at a preselected voltage. This grid consists of a biasable mesh sandwiched between two grounded meshes which can be used to transmit the scattered and recoiled neutrals plus ions with specific charges as follows. Doubly charged ions A^{2+} at kinetic energy E can be separated from singly charged ions A^+ at kinetic energy $E/2$ by placing a potential V on the grid so that $E/2e < V < E/e$. The length of the flight path is usually chosen so that it provides the required energy resolution at sufficient transmission. When the mass composition of the particles being analyzed is known, the energy of the particles (E) is connected with the flight time (t) by the relation

$$E = ML^2/2t^2, \tag{1}$$

where M is the particle mass and L is the flight length, i.e. the distance between target and detector. Differentiating this expression gives the relation between the experimentally measured flight-time distribution of particles, $\mathrm{d}N/\mathrm{d}t$, and their energy distribution $\mathrm{d}N/\mathrm{d}E$. The resulting expression is

$$\mathrm{d}N/\mathrm{d}E = -(\mathrm{d}N/\mathrm{d}t)t^3/ML^2. \tag{2}$$

The ratio of the ion pulse width to the flight time determines the energy resolution for a given length of flight path. For example, considering 3 keV Ne^+, a pulse width of 50 ns, and a flight path of 100 cm, $\Delta t/t = 8.5 \times 10^{-3}$ corresponding to an energy resolution of $\Delta E/E = 2\%$. Increasing the path length and decreasing the ion pulse width can improve this resolution considerably.

2.2.3.2 Electrostatic analysis (ESA)

In an electrostatic analyzer the energy separation is made by spatial dispersion of the charged-particle trajectories in a known electrical field. The field is produced by an analyzer whose shape and dimensions are carefully designed to provide good optical properties. Different analyzers have been used for ISS: 127° cylindrical analyzer (Smith, 1967), hemispherical analyzer with or without pre-retardation lens system (Bertrand *et al.*, 1977), cylindrical-mirror analyzer (CMA) (Sar-El, 1967), and more complex toroidal systems (Hellings *et al.*, 1985). These analyzers are either fixed in the UHV experimental chamber or mobile so that the scattering (recoiling) angle can be continuously varied from 0° to ~180°. Except for very precise energy-loss measurements (Betrand and Ghalim, 1983), energy resolution of the order of ~1% is enough for ISS surface-composition analyses. For these ESAs, the filtered energy E is proportional to the potential difference V applied across the electrodes,

$$E = eFV, \tag{3}$$

where F is a form factor depending on the analyzer shape and size. For the CMA, which is largely used in ISS owing to its large acceptance (a full annular solid angle is accepted),

$$F = 0.763 \ln (r_2/r_1), \tag{4}$$

where r_2 and r_1 are the radii of the outer and inner cylinder electrodes, respectively. With the CMA, a second-order focusing of the trajectories making an angle $\psi = 42.3°$ with respect to the spectrometer axis is obtained. For ISS, the surface is bombarded at normal incidence and the scattering angle is fixed to $180° - \psi = 137.7°$. The ion beam line must then be aligned along the spectrometer axis (see Figure 2.3). The width of the input and output slits situated in the inner cylinder fixes the solid angle and the accepted angular range $\Delta\psi$ $(\Delta\theta)$ so the energy resolution of the CMA (Rivière, 1990) is

$$R = \Delta E/E = 0.18s/r_1 + 1.39(\Delta\psi)^3. \tag{5}$$

The solid angle $\Delta\Omega$ is given by

$$\Delta\Omega = 2\pi[\cos (\psi + \Delta\psi/2) - \cos (\psi - \Delta\psi/2)]. \tag{6}$$

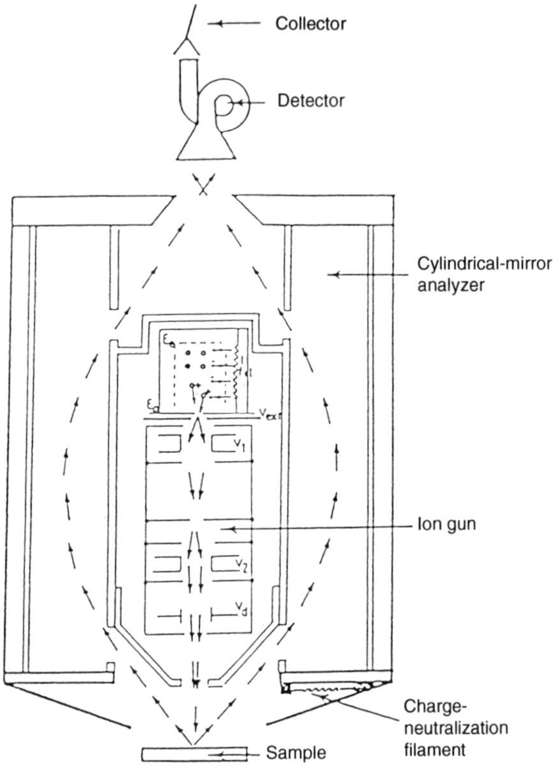

Figure 2.3 Cylindrical-mirror electrostatic-energy analyzer (CMA) with coaxial ion gun for ion scattering analysis. (Bertrand, 1986)

2.2.4 SPECTROMETER TRANSMISSION FUNCTION

For quantitative ISS analyses, it is important to understand the influence of the experimental parameters on the intensity of the measured signal. For that purpose the role of each component of the spectrometer has to be detailed. A spectrometer consists of four different parts: analyzer, detector, detection electronics, and computer/multichannel analyzer (MCA). When bombarding the solid with primary ions, only I_θ, the current reemitted into the solid angle $\Delta\Omega$ making an angle θ with the ion-beam-line direction, is accepted for analysis. The role of the analyzer is to determine the energy distribution $n_\theta(E)$ of the input current I_θ where

$$I_\theta = \int_{\Delta\Omega} \int_0^\infty n_\theta(E)\,\mathrm{d}E\,\mathrm{d}\Omega. \tag{7}$$

In fact the output current I_A of the analyzer tuned at a filtered energy E_i is given by:

$$I_A(E_i) = \int_{\Delta\Omega} \int_0^\infty n_\theta(E) T(E_i - E)\, \mathrm{d}E\, \mathrm{d}\Omega, \tag{8}$$

where $T(E_i - E)$ is the transmission function of the analyzer itself.

The analyzer output current $I_A(E_i)$ is the convolution of the initial distribution and the transmission function. This current is converted by the detector into a series of pulses with an efficiency γ which is a function of the ion energy E. These pulses, if above a fixed threshold, are amplified and shaped by the amplification/discriminator electronics. They are then counted during a specified dwell time Δt fixed by the timer and finally stored in the ith memory channel of the MCA working in the multiscaling mode. After this cycle, the analyzer energy is incremented by ΔE_{MCA} and the MCA switches to the $(i + 1)$th memory channel so that

$$E_{i+1} = E_i\, \Delta E_{\mathrm{MCA}}. \tag{9}$$

The previous procedure is repeated so as to produce a histogram as schematized in Figure 2.4. The number of counts in channel i equals

$$N_i = \Delta t \int_{\Delta\Omega} \int_0^\infty n_\theta(E) T(E_i - E) \gamma(E)\, \mathrm{d}E\, \mathrm{d}\Omega. \tag{10}$$

If the energy distribution to be measured (scattering/recoiling peak) is narrow with respect to the transmission function and is centered at an energy E_1 characteristic of the phenomenon (see Section 2.3.1), the following approximation can be applied:

$$n_\theta(E) \cong I_\theta \delta(E - E_1), \tag{11}$$

where δ is the Dirac function. If the accepted solid angle is small, equation (10) can be simplified as

$$N_i = \Delta t I_\theta T(E_i - E_1) \gamma(E_i) \Delta\Omega. \tag{12}$$

When the area A of the peak is measured in the histogram, we have

$$A = \sum_i N_i \cong I_\theta \Delta t / \Delta E_{\mathrm{MCA}}\, \Delta\Omega \int T(E_i - E_1) \gamma(E_i)\, \mathrm{d}E_i. \tag{13}$$

The summation can be replaced by an integration if ΔE_{MCA}, the energy increment of the MCA, is chosen small enough. The integration is made in the energy region of interest. If, in that region, the efficiency of the detector is constant and assuming a Gaussian transmission function,

$$T(E_i - E) = t \exp\{-[(E_i - E)/\sigma]^2\}, \tag{14}$$

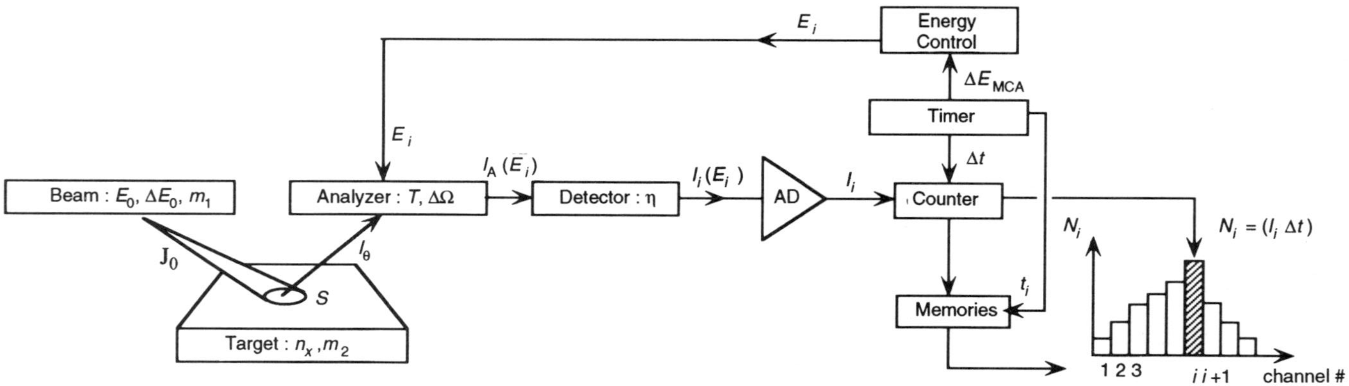

Figure 2.4 Principle of numerical acquisition of scattered ion signal

where t is supposed to be close to one and the variance σ is related to the analyzer resolution R by the relation

$$\sigma = k\Delta E_i = kE_i/R, \tag{15}$$

with

$$k = \tfrac{1}{2}(\pi/\ln 2)^{1/2}$$

and

$$\int T(E_i - E_1)\gamma(E_i)\,\mathrm{d}E_i \cong \gamma(E_1)\int_0^\infty \exp\{-[(E_i - E)R/(kE_i)]^2\}\,\mathrm{d}E_i$$

$$\cong k\gamma(E_1)E_1/R.$$

Equation (14) becomes:

$$A \cong I_\theta(\Delta t/\Delta E_{\mathrm{MCA}})\Delta\Omega k\gamma(E_1)E_1/R. \tag{16}$$

It is observed that En(E) is in fact measured. For ion energies lower than a few keV, the detector efficiency is not constant but decreases when we lower the ion energy (Burrows *et al.*, 1967).

2.3 SCATTERING AND DIRECT RECOILING SPECTROMETRIES

2.3.1 KINEMATICS OF THE SINGLE BINARY COLLISION

2.3.1.1 Binary elastic collision model (BEC)

The experimental conditions for ISS and DRS are chosen such as to detect mainly either primary particles scattered after only one single collision with the surface atoms (ISS) or secondary recoiling atoms produced from a single collision. A very simple model of *binary elastic collisions (BEC)* can be applied to interpret experimental spectra. Although this view is an oversimplification of the interaction because it neglects the influence of the surrounding atoms, it is very useful and will be justified later.

In the BEC model, the solid is assumed to be composed of individual target atoms which are considered to be free and at rest. This means that the binding energy and vibrational motion are neglected. Momentum and total kinetic energy are conserved during the collision. Solving these conservation equations produces a direct relationship between the initial energy (E_0) of the primary ions and their kinetic energy after scattering (E_1) and the kinetic energy

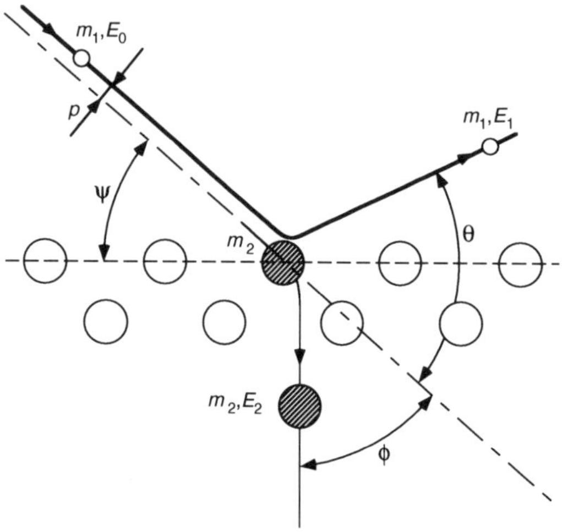

Figure 2.5 Schematic of a binary elastic collision

transferred to the target atoms (E_2). Figure 2.5 represents the scheme of the collision with the definition of the scattering angle θ and the recoiling angle ϕ.

$$E_1 = E_0 \cdot K(\theta, A), \tag{17}$$

$$E_2 = E_0 \cdot T(A) \cdot \cos^2(\phi). \tag{18}$$

Here $A = m_2/m_1$ is the ratio between the target and projectile mass, K is the kinematic factor, and T is the maximum energy transfer factor.

$$K = \left(\frac{\cos(\theta) \pm \sqrt[2]{A^2 - \sin(\theta)^2}}{(1 + A)} \right)^2, \tag{19}$$

$$T = \frac{4A}{(1 + A)^2}. \tag{20}$$

The relation (19) is represented in Figure 2.6 where K is plotted as a function of the scattering angle θ for various values of the mass ratio A. It is seen that if the projectile is heavier than the target ($A < 1$), both signs have to be considered in equation (19) and single-collision scattering is only possible for angles smaller than $\theta_{\text{lim}} = \arcsin A$. The two values for E_1 correspond to two different collisions leading to the same scattering angle θ: E_{1+} results from a glancing collision with a large impact parameter p_+ (the impact parameter p is defined as the distance between the target atom and the direction of the incoming trajectory), whereas E_{1-} corresponds to a direct collision with a small impact parameter p_-. For the limiting cases of θ equal to 0° (no scattering)

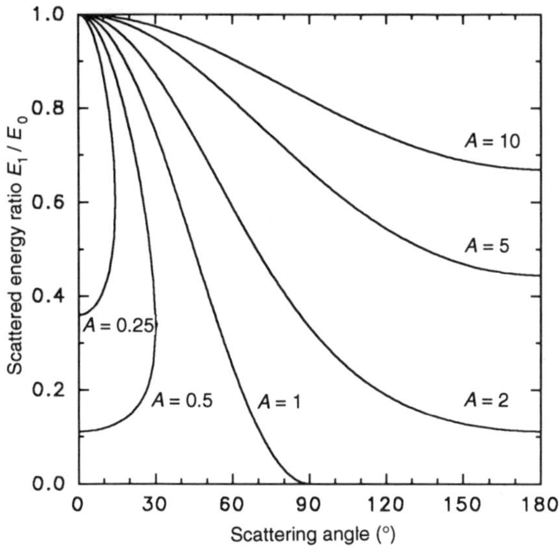

Figure 2.6 Energy ratio $K = E_1/E_0$ for the scattered particles as a function of the scattering angle θ for different values of the mass ratio $A = m_2/m_1$

and 180°, the two situations are:

No interaction, $E_{1+} = E_0$, $E_2 = 0$, $\phi = 90°$, $p_+ = \infty$;
Head-on collision, $E_{1-} = E_0 \cdot [(1 - A)/(1 + A)]^2$,
$E_2 = E_0 \cdot T$, $\phi = 0°$, $p_- = 0$.

For $A > 1$, only a single value is found for E_1. For $\theta = 90°$, the expression reduces to

$$E_1/E_0 = (m_2 - m_1)/(m_2 + m_1)$$

and for $\theta = 180°$ it reduces to

$$E_1/E_0 = [(m_2 - m_1)/(m_2 + m_1)]^2.$$

Figure 2.7 shows the energy E_2 of the dircct recoils as a function of the recoiling angle for different values of the mass ratio A. It is seen that the recoiling angle is limited to $\phi = 90°$. This situation corresponds to $E_2 = 0$.

The relations (17) and (18) are the basis for *elemental surface analysis* by ISS and DRS. Indeed, if the primary ion mass m_1 and energy E_0 are fixed, the measurement of the energy distributions of the particles emitted at a defined solid angle with respect to the primary beam direction will produce spectra in which peaked distributions appear at the different energies E_1 and E_2.

The mass m_2 of the target can be evaluated from E_1 and E_2 using equations (17) and (18). This allows one to convert the energy scale of the spectra into a

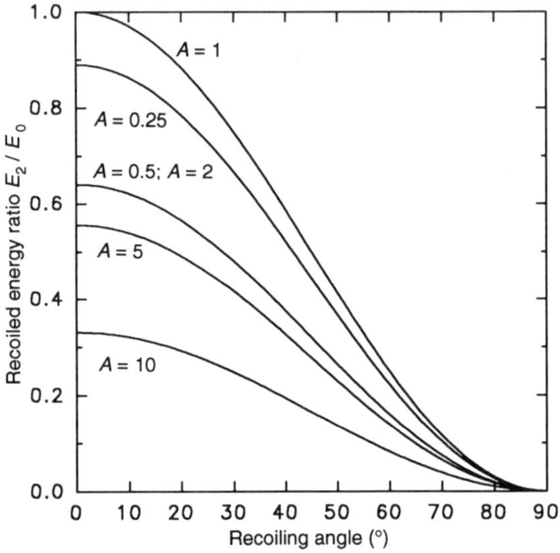

Figure 2.7 Energy ratio E_2/E_0 for the recoiled particles as a function of the recoiling angle ϕ for different values of the mass ratio $A = m_2/m_1$

mass scale. Figure 2.8 shows the scattered energy ratio E_1/E_0 as a function of the mass ratio A for different scattering angles θ. Equation (17) can also be transformed into the form

$$m_2 = m_1 \cdot (1 + K - 2K \cos \theta)/(1 - K)), \tag{17'}$$

where $K = E_1/E_0$.

Figure 2.9 shows the recoiled energy ratio E_2/E_0 as a function of the mass ratio A for different recoiling angles ϕ. From equation (18), it is seen that the same recoiling energy E_2 is obtained for two different values of m_2 if they satisfy the relation

$$m_2 \cdot m_2' = m_1^2.$$

This condition is satisfied for recoiling of H and O atoms with ^{4}He ions. This degeneracy is of course removed when a time-of-flight spectrometer is used. Indeed, since direct-recoil particles are primarily neutral atoms (Rabalais, 1988), TOF spectrometers are advantageous, producing time distributions instead of energy distributions (see Section 2.2.3.1). Relation (18) can be written very simply in the time scale as

$$t_2 = F(m_1 + m_2)$$

and

$$F = t_0/(2m_1 \cos \phi). \tag{21}$$

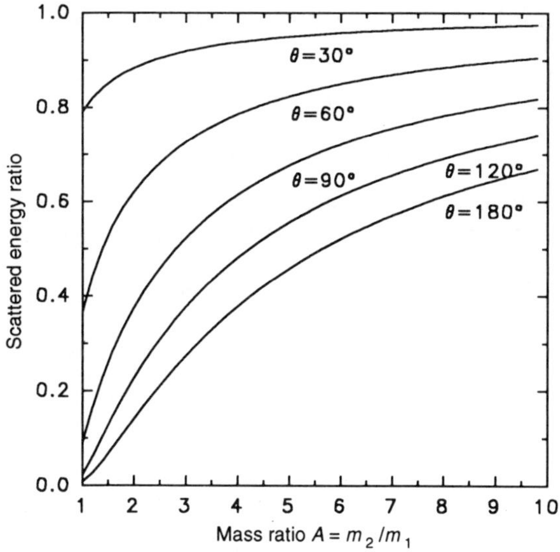

Figure 2.8 Energy ratio $K = E_1/E_0$ for the scattered particles as a function of the mass ratio $A = m_2/m_1$ for different scattering angles θ

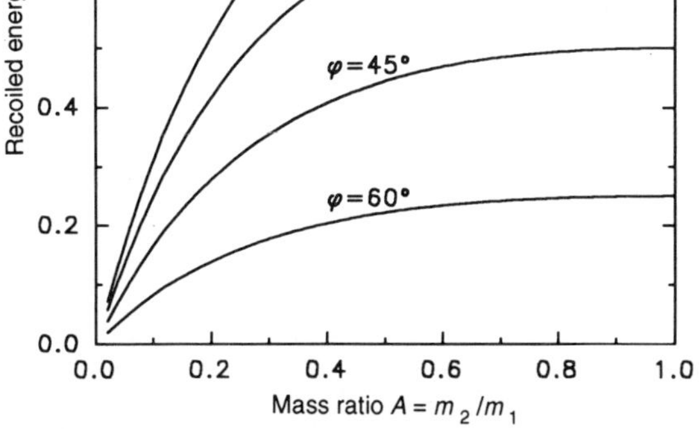

Figure 2.9 Energy ratio E_2/E_0 for the recoiled particles as a function of the mass ratio $A = m_2/m_1$ for different recoiling angles ϕ

Here t_2 and t_0 are the times required for the masses m_2 and m_1 to travel over a path length L:

$$t_2 = L/(2E_2/m_2)^{1/2},$$
$$t_0 = L/(2E_0/m_1)^{1/2}. \tag{22}$$

Both recoiling and scattering features are in principle present in the spectra if the detection angle is lower than 90°. For example, Figure 2.10(a) shows the

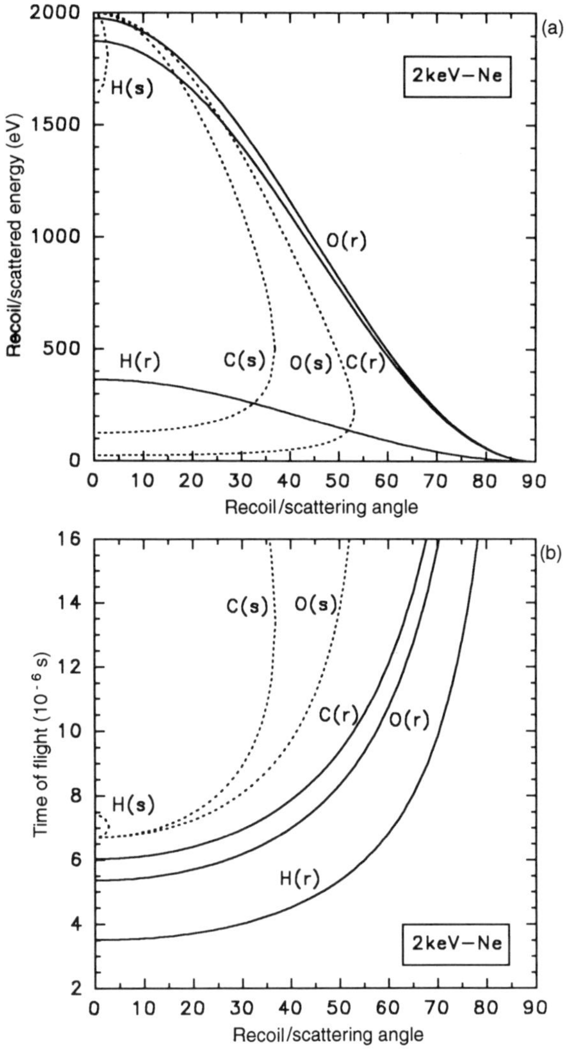

Figure 2.10 Recoiling and scattering (a) energies and (b) time of flight as functions of the detection angle for H, C and O with 2 keV Ne^+ ions. The solid lines refer to recoiling and the broken lines refer to scattering

recoiling and scattering energies for H, C and O as a function of the detection (recoiling/scattering) angle for Ne^+ (2 keV) primary ions. Figure 2.10(b) shows the same results but for flight times assuming a flight length of 0.93 m. It is seen that the detection angle has to be correctly chosen to prevent peak interferences. Also, the relative intensity of the different contributions will depend on the exact experimental conditions (see Section 2.3.4).

The BEC model is valid only for the interpretation of single-binary-collision-scattering and direct-recoiling events. In practice, the single-binary-scattering contribution is the only one present if all the trajectories more complex than a single collision with the surface atoms can be filtered before detection. This is nearly achieved for the scattering of the noble-gas ions. Indeed, owing to their high electron affinity ($1s_{He} = 24.6$ eV) as compared to the work function of the materials, they will have a very high probability of being neutralized during the interaction. This probability depends on the exact nature of the trajectories. The neutralization probability is higher for multiple-collision and penetrating trajectories than for single-collision scattering (Eckstein, 1981). An example is shown in Figure 2.11 for 90° scattering of $^4He^+$(1 keV) from a clean copper surface. Only a single scattering peak is observed and its position corresponds within a few eV to its kinematic factor (Bertrand, 1986). The neutralization filtering of complex trajectories is performed at the expense of the sensitivity; this introduces complications for quantitative analysis (see Section 2.3.4).

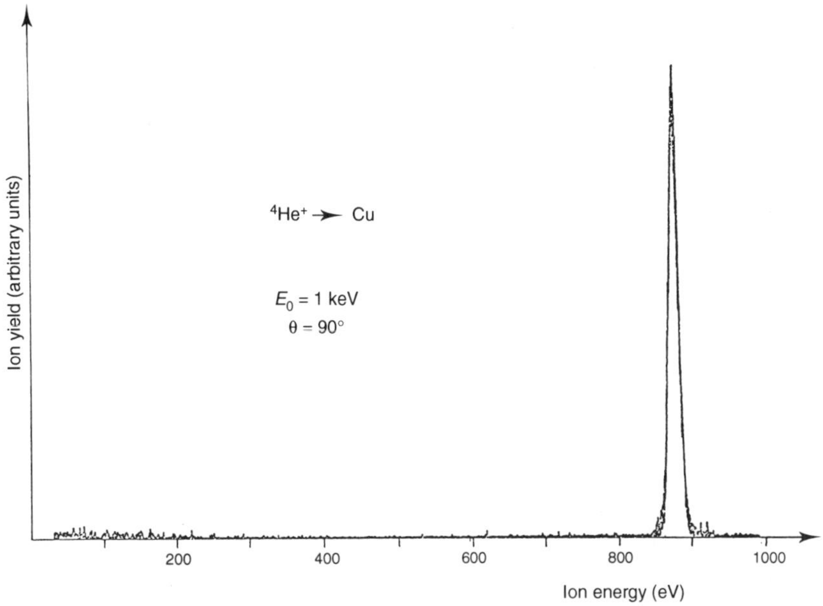

Figure 2.11 Ion scattering spectrum obtained with 1 keV He^+ scattered at 90° from a clean copper surface. (Bertrand, 1986)

More complex spectra are obtained if the total emitted flux is detected independently of the charge state of the particles as in a TOF spectrometer or if alkali ions are used as primary projectiles. Indeed, neutralization of alkali ions is much less efficient and can be independent of the trajectories (Boers, 1984). In these spectra, single-scattering and single-recoiling features can still be identified. However, other contributions resulting from multiple collisions on the surface and subsurface regions are also present with intensities depending on the exact experimental conditions.

The validity of the BEC model can be justified by the following arguments. In the energy range considered, the target atom can be considered as a free atom because the kinetic energy E_2 transferred to it (>100 eV) is much larger than its binding energy (<10 eV) in the lattice. Moreover, it can be considered at rest because the interaction time ($\sim 10^{-15}$ s) is smaller than the characteristic period for lattice vibrations ($\sim 10^{-13}$ s). The inelastic energy losses are usually small (Aono and Souda, 1987) and they can be neglected in most cases of elemental analyses. The limit of validity of a single-collison regime will be treated later.

When a recoil atom is ejected from the solid surface, it is influenced by the surface potential U. This leads to an additional energy loss and a refraction (Sigmund, 1981; Rabalais, 1988; Eckstein, 1987). The value of the surface potential U is approximated by the substrate heat of sublimation or the binding energy of the recoiled atom to the surface. The recoil energy E_2 is reduced by the surface potential and becomes

$$E_2' = E_2 - U. \tag{23}$$

The refraction modifies the polar emission angle β to the value β' where

$$\tan^2 \beta' = E_2 \sin^2 \beta/(E_2 \cos^2 \beta - U). \tag{24}$$

2.3.1.2 Mass resolution in elemental ISS/DRS analyses

According to the previous kinematic considerations, all elements should be detectable by ISS and DRS. In practice we have to take into account the achievable mass resolution, $m_2/\Delta m_2$, resulting from the finite energy resolution of the spectrometer, $E/\Delta E$, and its angular aperture $\Delta\theta$ or $\Delta\phi$. By assuming the validity of the BEC model, the mass resolution can be calculated for ISS and DRS by differentiating equations (17) and (18) respectively. This produces the following relations, for scattering

$$\Delta m_2/m_2 = \left(\frac{(1-A^2)}{A(A\cos(\theta) - \sqrt{A^2 - \sin^2(\theta)})}\right)\left(\frac{\sqrt{A^2 - \sin^2(\theta)}}{2} \cdot (\Delta E_1/E_1) + \sin(\theta)\cdot\Delta\theta\right), \tag{25}$$

and for recoiling

$$\Delta m_2/m_2 = [(1 + A)/(1 - A)] \cdot [\Delta E_2/E_2 + 2 \tan(\phi) \cdot \Delta\phi], \tag{26}$$

or, if time spectra are considered,

$$\Delta m_2 = \Delta t_2/F + (m_1 + m_2) \tan(\phi) \cdot \Delta\phi. \tag{27}$$

Figure 2.12 shows the mass resolution in ISS calculated with (25) as a function

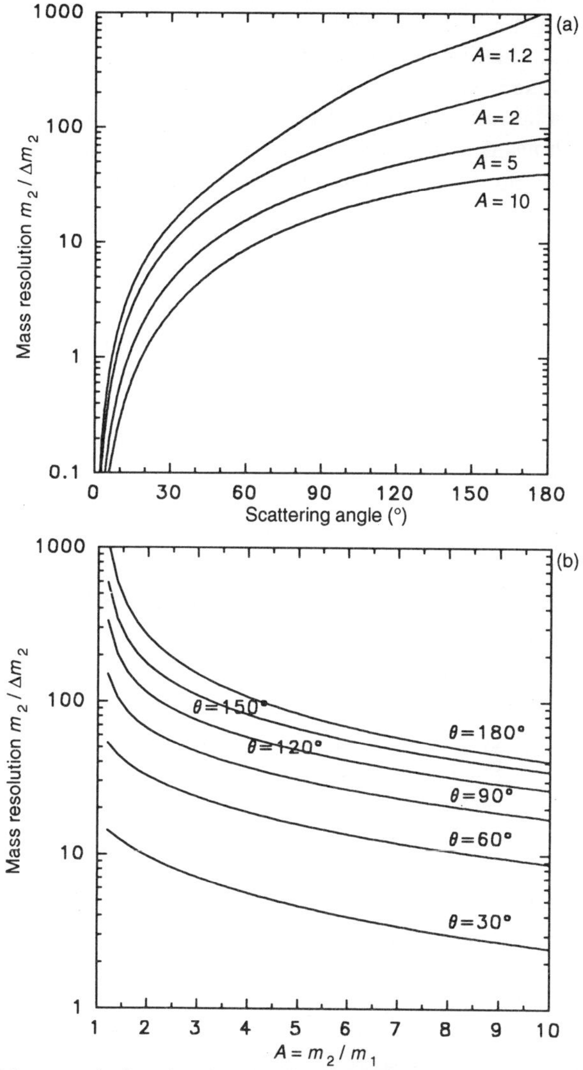

Figure 2.12 Mass resolution $\Delta m_2/m_2$ in ISS; (a) as a function of the scattering angle θ and (b) as a function of the mass ratio A. An energy resolution $\Delta E/E$ of 100 and an analyzer angular acceptance of 0.5° are assumed

of (a) the scattering angle θ and (b) the mass ratio A for an energy resolution $E_1/\Delta E_1 = 100$ and an angular aperture of 0.5°. It is seen that the mass resolution increases with the scattering angle θ and decreases with the mass ratio A. In order to increase the mass resolution in ISS, a scattering angle $\theta > 90°$ is often chosen. $^4He^+$ scattering is then adequate to detect the light elements. However, in order to have enough mass resolution at high masses, it is necessary to increase the mass of the projectile (^{20}Ne, ^{40}Ar). The elements lighter than the projectile ($A < 1$) are then no longer detected. For example, Figure 2.13 shows the improvement in mass resolution obtained by switching from $^4He^+$ to $^{20}Ne^+$ to resolve ^{59}Co and ^{52}Cr in ISS spectra measured on a $Co_{76}Cr_{24}$ magnetic thin film (Bertrand and De Puydt, unpublished).

Figure 2.14 shows Δm_2 for DRS calculated from equation (27) as a function of the recoiling angle ϕ for oxygen recoiled by different primary ions (He, Ne, Ar) with a time resolution $\Delta t_2 = 50$ ns and an angular aperture of $\Delta\phi = 0.5°$. It is seen that Δm_2 increases with the recoiling angle and with the ion mass.

DRS is particularly well adapted to detect light elements like H, C and O. With ISS, however, the detection of hydrogen is not possible because the limiting scattering angle is too small to preserve a regime of single collisions (for $^3He \rightarrow H$, $A = 1/3$ and $\theta_{lim} = 19.5°$). Figure 2.15 shows examples of TOF ISS–DRS spectra obtained at 28° with 4 keV Ar^+ on an unclean and on a clean Si{100} surface (Wang *et al.*). The different recoiling and scattering contributions are indicated.

2.3.2 DYNAMICS OF THE SINGLE BINARY COLLISION

2.3.2.1 Scattering integral

A single-elastic-collision event can be treated by straightforward classical scattering theory. The dynamics is treated in the center-of-mass coordinate system so that the two-body interaction is reduced to the equivalent one-body interaction in a central potential (e.g. Goldstein, 1950). In this frame, the collision is symmetrical with respect to the center of mass (barycenter). By solving the system of differential equations expressing the conservation of energy and angular momentum, one obtains

$$\Theta = \pi - 2p \int_{r_0}^{\infty} [r^2 F(r)]^{-1}\, dr \tag{28}$$

with $F(r) = [1 - p^2/r^2 - V(r)/E_r]^{1/2}$.

Equation (28) is the basic relation between the scattering angle in the CM system Θ and the impact parameter p for a given reduced energy E_r and

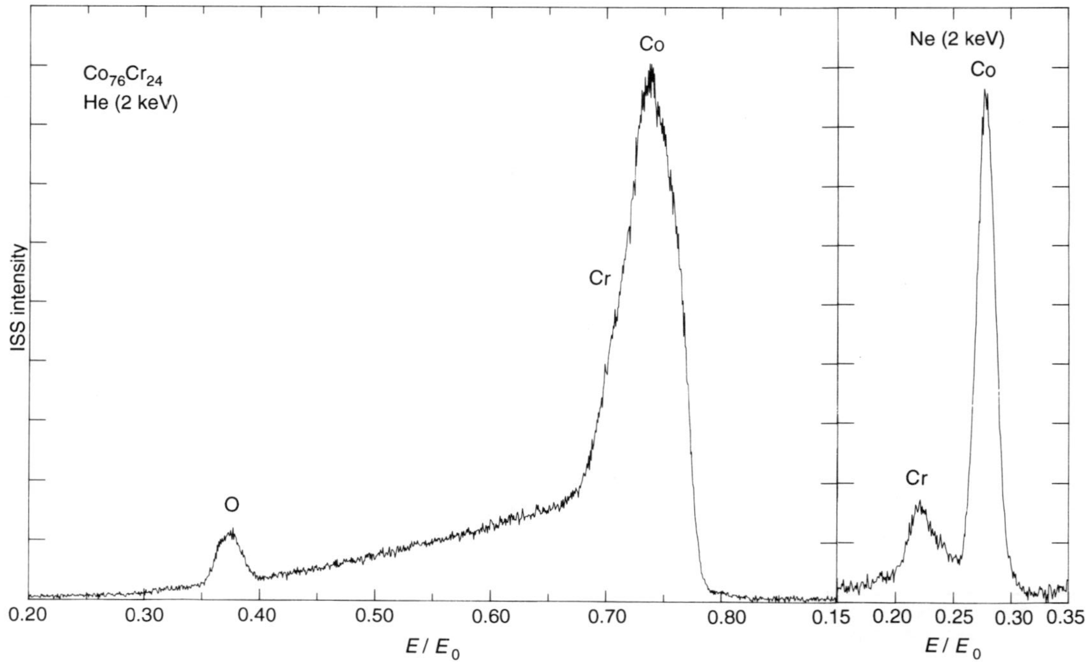

Figure 2.13 Ion scattering spectra obtained with (a) 2 keV He^+ and (b) 2 keV Ne^+ scattered at 138° from a $Co_{76}Cr_{24}$ thin-film alloy

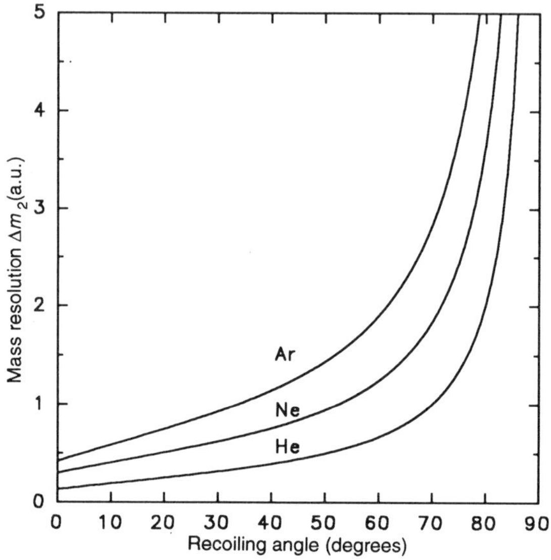

Figure 2.14 Absolute mass resolution Δm_2 in DRS as a function of the recoiling angle ϕ for oxygen recoiled by different primary ions. An absolute time resolution Δt of 50 ns and an analyzer angular acceptance of 0.5° are assumed

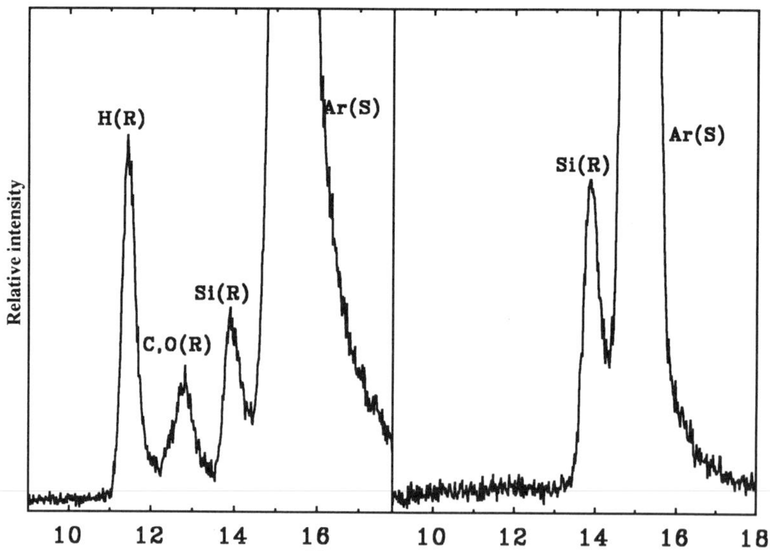

Figure 2.15 Time-of-flight spectra of an unclean (left) and clean (right) Si{100} surface obtained with 4 keV Ar^+ ions incident at 15° and detection at 28°, showing the scattered Ar and recoiled H, C, O and Si features. (Wang *et al.*)

interaction potential $V(r)$ where

$$E_r = [A/(1 + A)]E_0.$$

The variable r is the interatomic separation of the colliding atoms and r_0 is the distance of closest approach or the apsis of the collision. r_0 is defined as the positive root of the function F,

$$F(r_0) = 0.$$

Equation (28) shows that the scattering angle increases as the impact parameter decreases. For $p = 0$, a head-on collision with $\Theta = 180°$ $(A > 1)$ is produced.

The transforms of the scattering angle Θ in the CM system to the scattering angle θ and the recoiling angle ϕ in the laboratory system are given by

$$\theta = \arctan(A \sin \Theta/(1 + A \cos \Theta)$$

and

$$\phi = (\pi - \Theta)/2. \tag{29}$$

These transforms show that the recoiling angle varies from 90° to 0° when the scattering angle varies from 0° to 180° $(A > 1)$ and that corresponds to an impact parameter p varying from ∞ to 0.

2.3.2.2 Interaction potential

In order to calculate the classical trajectory, an accurate interaction potential is needed. At long distances, the potential is weakly attractive due to the van der Waals forces, but it is strongly repulsive at short distances due to the Coulomb repulsion between the nuclei. In the energy range considered ($E > 100$ eV), the ion approaches sufficiently close to the target atom that the repulsive portion of the potential dominates and the attractive part of the potential can be neglected. However, the energy is still low enough ($E < 10$ keV) that the screening effect produced by the electron clouds, which partially mask the Coulomb potential, has to be taken into account. This is done by modulating the Coulomb potential by a screening function $\Phi(r/a)$, where a represents the screening length:

$$V(r) = (1/4\pi\varepsilon_0)Z_1Z_2(e^2/r)\Phi(r/a). \tag{30}$$

Different screening functions have been proposed. One of the most commonly used is based on the Thomas–Fermi semiclassical description of the electronic charge density in the atom (Feldman and Mayer, 1986). A universal differential equation is obtained and Molière (1947) has proposed an analytical approximation for its solution:

$$\Phi(r/a) = \Phi(x) = 0.35 \exp(-0.3x) + 0.55 \exp(-1.2x) + 0.1 \exp(-6x). \tag{31}$$

Two different expressions have been proposed for the screening length a. One is due to Firsov and the other to Lindhard:

$$a_F = 0.4685/(Z_1^{1/2} + Z_2^{1/2})^{2/3}, \tag{32}$$

$$a_L = 0.4685/(Z_1^{2/3} + Z_2^{2/3})^{1/2}. \tag{33}$$

However, as small variations in the screening length can lead to large differences in structure analysis by ISS and DRS (see Section 1.4), the Firsov length is reduced by a constant factor C which is adapted so as to fit the experimental data (Grayeb *et al.*, 1987):

$$a = Ca_F. \tag{34}$$

O'Connor and Biersack (1986) proposed the following empirical expression for C:

$$C(Z_1, Z_2) = 0.54 + 0.045(Z_1^{1/2} + Z_2^{1/2}). \tag{35}$$

There is also an empirical universal function ZBL for $\Phi(r/a)$ proposed by Ziegler *et al.* (1985) which gives good results in the energy range considered.

2.3.2.3 Asymptotic trajectory

A representation of the asymptotic trajectories is very useful for visualizing the binary-elastic-collision events. This asymptotic approximation is also used in simulation codes (Robinson and Torrens, 1974) in which the complex trajectory of an ion interacting with a solid is treated as a succession of independent binary collisions. If a Cartesian coordinate system is placed at the initial position of the target atom and the x axis is chosen parallel to the incoming projectile direction (see Figure 2.16), the asymptote of the scattered trajectory

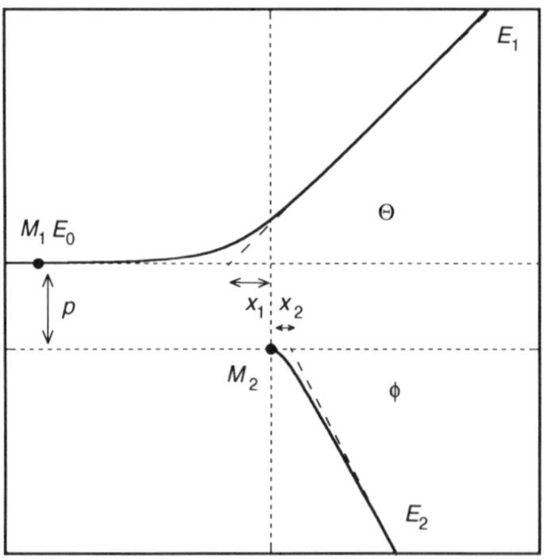

Figure 2.16 Schematic of a single collision in the laboratory coordinate system

intercepts the initial direction $y = p$ at (p, x_1). For the recoiled trajectory, the interception with $y = 0$, occurs at $(0, x_2)$:

$$x_1 = [2\tau + (A - 1)p \tan \Theta/2]/(1 + A), \tag{36}$$

$$x_2 = p \tan \Theta/2 - x_1.$$

Here τ is the time interval given by

$$\tau = (r_0^2 - p^2)^{1/2} - \int_{r_0}^{\infty} \frac{1}{F(r)} - \frac{1}{\sqrt{1 - (p/r)^2}} \, \mathrm{d}r. \tag{37}$$

For a collision between two hard spheres with radii R_1 and R_2, it is evident that $x_1 = [(R_1 + R_2)^2 - p^2]^{1/2}$ and $x_2 = 0$.

2.3.2.4 The shadow cone

The scattering (28) and time (37) integrals cannot be solved analytically with the screened Coulomb potential (30). Numerical solutions or analytical approximations are used (Severijns *et al.*, 1992). Figure 2.17 shows a series of asymptotic scattered trajectories corresponding to impact parameters ranging from 0 to 1.5 Å calculated with the ZBL potential for scattering of different ions from an isolated Ir atom (Bu *et al.*, 1991a). It is seen that a shadow cone is produced as an excluded volume behind the scattering center inside which no primary ions can penetrate. It is also observed that the deflected trajectories are focused at the edge of the shadow cone; this is called the *focusing effect*. The aperature of the shadow cone increases when the projectile energy is

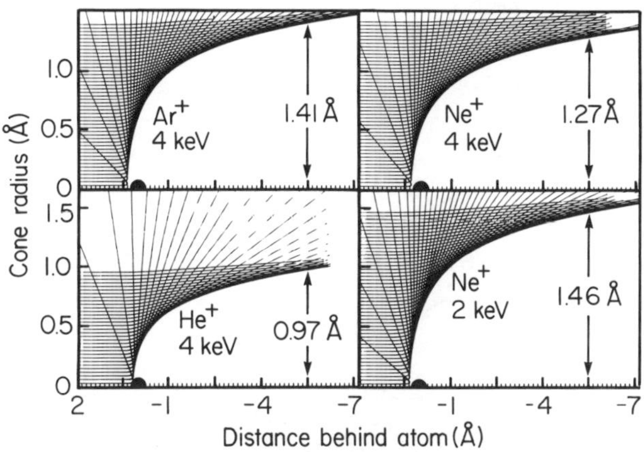

Figure 2.17 Shadow cones for 4 keV Ar^+, Ne^+ and He^+ and 2 keV He^+ computed from the envelope of the classical trajectories using the ZBL potential function (Reproduced by permission of Elsevier Science Publishers from Bu *et al.*, 1991a)

decreased. The possibility of modifying the aperture and the orientation of the shadow cones is used in ISS and DRS for surface-structure analysis; this will be developed in Section 2.4. Oen (1983) has published a universal empirical expression based on the Molière potential to calculate the radius of the shadow cone at a distance l from the target atom

$$r(l)/2(bl)^{1/2} = 1.0 - 0.12\alpha + 0.01\alpha^2 \qquad \text{for } 0 < \alpha < 4.5$$

$$= 0.924 - 0.182 \ln \alpha + 0.008\alpha \quad \text{for } 4.5 < \alpha < 100,$$

where

$$b = Z_1 Z_2 e^2/E_0, \quad \alpha = 2(bl)^{1/2}/a \tag{38}$$

and a is the screening length.

Figure 2.18 shows the radius of the shadow cone as a function of the primary energy calculated using equation (38) at a distance $l = 2$ Å for different ions (He, Ne, Ar) scattered by a Ni atom. It is seen that at low energy the radius at a distance corresponding to a typical atomic spacing is of the order of the atomic spacing. An atom situated inside the shadow cone of a previous atom cannot be hit by the primary ions; this phenomenon constitutes the *shadowing effect*. This means that if the shadow cones produced by the atoms in the first monolayer mask the atoms of the deeper layers, direct scattering and recoiling from deeper layers is not possible and ISS/DRS will probe only the first monolayer. Only contributions of the collisions with surface atoms will appear

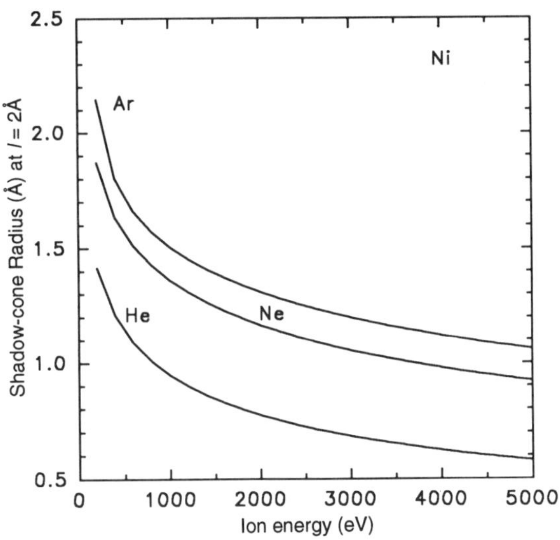

Figure 2.18 Shadow-cone radius at a distance of 0.2 nm as a function of the primary-ion energy for a Ni target bombarded by different ions (Ar^+, Ne^+ and He^+) calculated from Oen's formula. (Oen, 1983)

in the single scattering and recoiling spectral peaks. Moreover if noble-gas ions are used with only the detection of ionized particles, the contributions of the multiple collisions will be filtered by preferential neutralization, so that the spectra will only exhibit the surface scattering and recoiling peaks. It is to be noticed that this extreme surface sensitivity cannot be matched by electron surface spectrometries such as AES and XPS for which the information depth is related to the inelastic mean free path of electrons in the solid (Briggs and Seah, 1990).

2.3.3 MULTIPLE COLLISIONS

2.3.3.1 Energy

In the BEC model, only the interaction of the incident projectile with a single target atom was considered. However multiple-collision features are routinely observed in the spectra. In addition to single collisions, it is possible for the projectiles to be scattered at a fixed angle θ (or for surface atoms to be recoiled at a fixed angle ϕ) as a result of different collision sequences. This constitutes a *multiple collision*. It can be treated as a succession of n independent binary elastic collisions if, for each individual collision i, x_{1i} and x_{2i} are smaller than the interatomic spacing d. This is true as long as the primary ion energy is higher than a few tens of eV (Robinson, 1981). If the inelastic energy losses due to the electron interaction are neglected, the projectile energy after a multiple collision E_{1n} can be obtained by applying equation (17) n times, where n is the number of successive collisions.

$$\frac{E_{1n}(\theta)}{E_0} = \prod_{i=1}^{n} K(\theta_i), \tag{39}$$

where θ_i is the scattering angle associated with the ith collision. If all of the successive collisions are in the same plane, then

$$\theta = \sum_{i=1}^{n} \theta_i. \tag{40}$$

2.3.3.2 Chain model

For complex out-of-plane multiple collisions, three-dimensional computer simulations are required to calculate the scattered and recoiled energies (Hou, 1989; Boers, 1977). However, when the azimuthal direction of the incident beam is aligned along a principal-surface crystallographic direction, the simple model interaction with a chain of equally spaced atoms is sufficient to understand the main features of the scattering and recoiling in the plane perpendicular to the

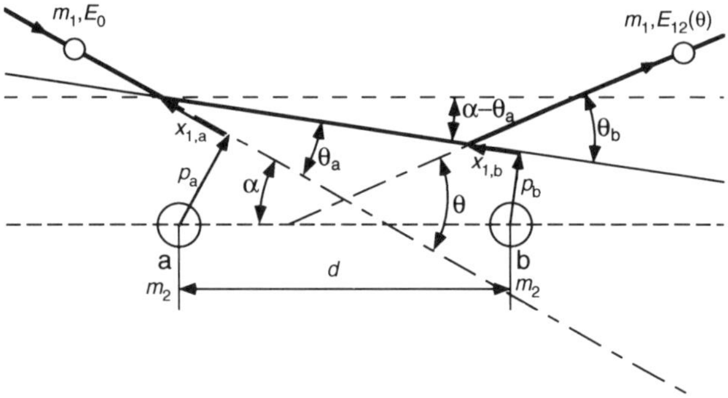

Figure 2.19 Schematic of a double collision producing a total scattering angle θ

surface (Aono, 1984). In this *chain model*, only two neighboring atoms (a and b in Figure 2.19) have to be considered. The primary ions can now interact successively with the two atoms a and b. However, for a given incident angle α with respect to the surface, the total trajectory is fully determined once the first impact parameter p_a is fixed. Indeed, due to the crystal surface periodicity, from geometrical considerations, it is seen that the impact parameter on atom b is given by

$$p_b = p_a \cos \theta_a + x_{1a} \sin \theta_a - d \sin (\alpha - \theta_a), \tag{41}$$

where d is the interatomic spacing. The total scattering angle θ is then the sum of the two scattering angles θ_a and θ_b on atoms a and b, respectively,

$$\theta = \theta_a(p_a, E_0) + \theta_b(p_b, E_1), \text{ where } E_1 = K(\theta_a)E_0). \tag{42}$$

If we neglect the trajectories for which $\theta > 180°$, because the associated total scattering cross section is very low (see Section 2.3.4), only two different trajectories can produce the same total scattering angle θ because only two couples of (θ_a, θ_b) can satisfy simultaneously both relations (41) and (42). Two values are then possible for the energy after scattering. If the incident and total scattering angles are not too small, one of the two trajectories is very close to the single collision either with atom a or atom b ($\theta_a \cong \theta$ and $\theta_b \cong 0$ or $\theta_b \cong \theta$ and $\theta_a \cong 0$) and the second is a double collision with both atoms. The two values for the energy are

$$E_1(\theta) = E_0 K(\theta), \quad E_{12}(\theta) = E_0 K(\theta_a) K(\theta_b), \tag{43}$$

with $E_{12} > E_1$. These two trajectories produce the same energy value for $\theta = 180°$.

If the scattering angle is reduced, the single-collision trajectory cannot be conserved and both become double collisions. They degenerate for a lower-limit total scattering angle $\theta_l > \alpha$, behind which scattering is no longer possible. This is the *blocking effect*. Similarly, for low incident angles, the single-collision trajectory can also not be conserved and both trajectories degenerate in the same double-collision sequence at an upper-limit total scattering angle $\theta_L <$ 180° above which scattering is also no longer possible. This is the *shadowing effect*. If these two values of the energy after scattering are plotted as a function of the total scattering angle θ and with the incident angle α as a parameter, characteristic loops are observed (see Figure 2.20) (Aono, 1984). It is worth noting that E_{12} depends on the interatomic distance d through equations (41), (42) and (43) and contains information on the surface structure. Two well-defined scattering peaks at E_1 and E_{12} are only observed on single crystals when the double-collision trajectory is not preferentially neutralized. This is for example satisfied with Na^+ scattering as compared to Ar^+ scattering if only the ions are detected but is satisfied for both Na^+ and Ar^+ with TOF detection of both neutrals and ions. Figure 2.21 illustrates clearly this difference for 70° scattering from a Au(110) surface (Overbury, 1981). The primary energies were chosen so as to produce the same value for E_1 in both cases: 1 keV K^+ and 0.976 keV Ar^+. The double-collision peak is only observed for K^+ scattering.

If the primary energy E_0 is increased, the characteristic loops expand laterally and become flatter due to the decrease of the shadow-cone size. For high E_0 as in RBS, $\theta_1 \cong \alpha$ and $\theta_L \cong 180°$ for any α.

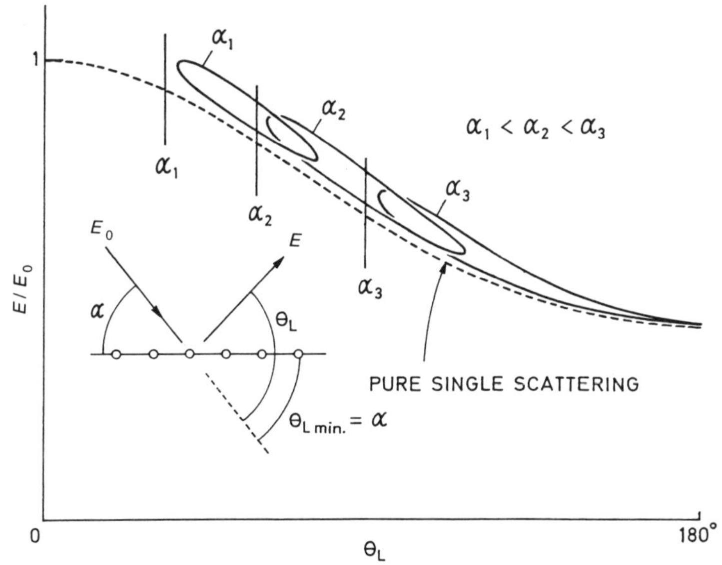

Figure 2.20 Energy of an ion scattered by an atomic chain, E, plotted against the scattering angle θ. A double-valued characteristic loop is obtained for fixed incidence angle α (solid curves). The E_1 vs θ curve for single-binary-collision scattering is also shown (broken line). (Aono, 1984)

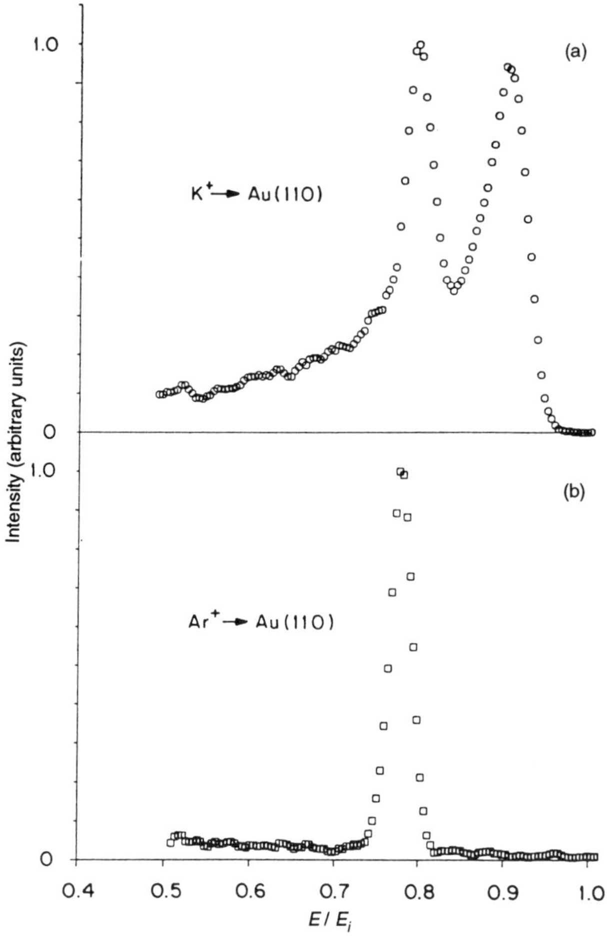

Figure 2.21 Ion scattering spectra obtained with (a) 1000 eV K^+ and (b) 976 eV Ar^+ scattered at 70° from a Au{110} surface. (Reproduced by permission of Elsevier Science Publishers from Overbury, 1981)

Pure single and double collisions are an idealization due to the possible deflections produced by the preceding and following atoms in the chain. If one considers in practice 'quasi-single' and 'quasi-double' collisions, the experimental values for E_1 and E_{12} can differ slightly from equation (43).

A single collision at θ will be possible only if the influence of the neighboring atoms is negligible. Due to the lattice periodicity, the impact parameters p' and p'' can be associated with the preceding and the following atom respectively.

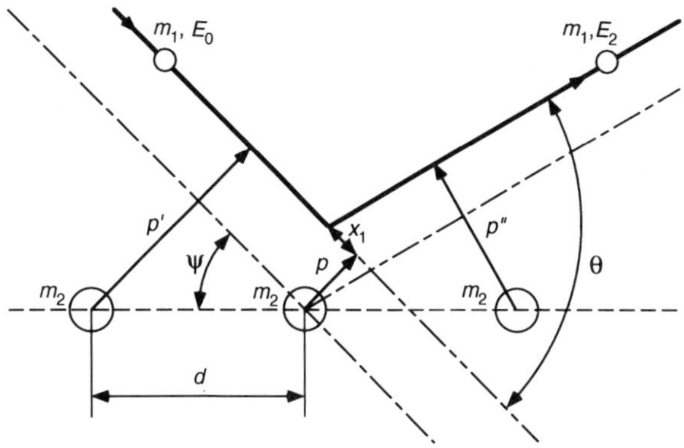

Figure 2.22 Schematic of the asymptotic trajectories for single scattering

The geometry of the scattering connects p' and p'' to p (see Figure 2.22) as

$$p' = p + d \sin \alpha,$$
$$p'' = x_1 \sin \theta + p \cos \theta + d \sin(\theta - \alpha). \tag{44}$$

The condition for single scattering requires that p' and p'' be large enough that

$$p' \to \theta' \lesssim 0.5^\circ \quad \text{and} \quad p'' \to \theta'' \lesssim 0.5^\circ. \tag{45}$$

From (44) and (45) it is seen that the contribution of multiple scattering can be reduced by increasing the primary energy E_0 (with the effect of reducing the shadow-cone aperture) and by using normal incidence and large back-scattering angles. The limiting situation is $\alpha = 90^\circ$, $\theta = 180^\circ$; then $p' = p'' = d$.

The chain model can also be applied to recoiling. Similar to scattering, the recoil atom a or b can be produced by different sequences of collisions. Indeed, if atom a is recoiled at an angle $\phi_a(p_a, E_0)$, it can afterwards be scattered by atom b at an angle $\theta_b(p_b, E_2)$ due to the impact parameter p_b, where

$$p_b = d \sin(\phi_a - \alpha) - x_2 \cos(\phi_a). \tag{46}$$

This is called surface recoiling (Bu *et al.*, 1991c), and the total recoiling angle ϕ is

$$\phi = \phi_a(p_a, E_0) + \theta_b(p_b, E_2(\phi_a)). \tag{47}$$

Atom b can also be recoiled by a primary ion previously scattered by atom a at a scattering angle θ_a, and then

$$\phi = \theta_a(p_a, E_0) + \phi_b(p_b, E_1(\theta_a)), \tag{48}$$

where p_b, given by equation (41), must be negative. This process is called deflect recoiling (Bu *et al.*, 1991c).

For a fixed detection angle, different trajectories and different total recoiling energies are possible. However, if grazing incidence is avoided, direct recoils are detected close to the energy E_2. Surface recoiling and deflect recoiling usually produce energies greater than E_2. The former is enhanced at low incident angle α and the latter at low exit angle β. There also exist, as for scattering, limiting angles due to shadowing and blocking (Chaudhury and Williams, 1991). Of course the combination of deflect and surface recoiling is possible, as is double recoiling. In this latter case, the total recoiling cross section is very low.

2.3.3.3 Shadowing and blocking effects

All of the scattering angles are not allowed in the characteristic loops (E, θ) produced by the chain model. The extent of the permitted scattering range depends on the incident angle α. Due to the influence of the neighboring atoms of the chain, surface-grazing trajectories are not possible. Indeed, for a fixed scattering angle, if the incident angle α is decreased, scattering will no longer be possible below a critical value $\alpha_c(\theta)$. For $\alpha_c(\theta)$, the position of the shadow cone produced by the preceding atom prevents the target atom being hit with the required impact parameter for scattering at θ (see Figure 2.23). For $\theta = 180°$, $p = 0$ and the shadow-cone edge crosses the position of the target atom considered. Below $\alpha_c(180°)$, no scattering is possible because each atom of the chain lies inside the shadow cone of the previous one. This is the *shadowing effect* and it is exploited for surface-structure determination in Section 2.4. Indeed, at the limit of the shadowing conditions, the following relation holds for the radius of the shadow cone:

$$r(l = d \cos \alpha_c) = d \sin \alpha_c - p(\theta). \tag{49}$$

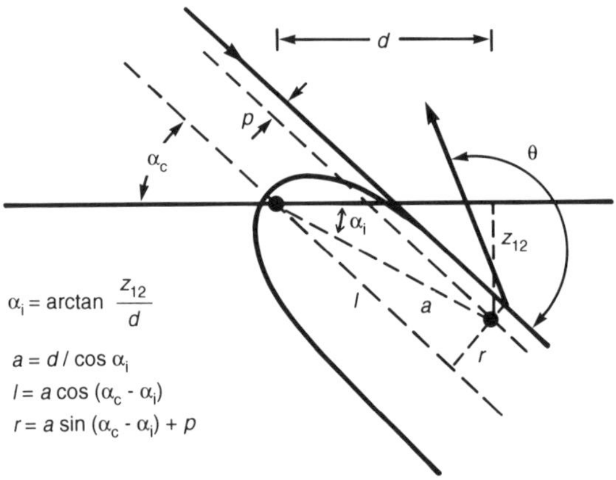

Figure 2.23 Schematic drawing illustrating the critical incident angle α_c below which no scattering at θ is possible due to the shadowing effect on the second-layer atoms

If d is known, the experimental determination of α_c allows one to calibrate the interaction potential (mainly the constant C in equation (34)) (Bu *et al.*, 1991c; Grizzi *et al.*, 1989). If the potential is known, it is possible to find the relation between α_c and d (see Section 2.4).

Surface-grazing trajectories are also not possible due to the deflections of the scattered trajectories by the neighboring atoms of the chain along the exit direction. Similarly to the shadowing produced below $\alpha_c(\theta)$ by the shadow cone of the preceding atom, there is a *blocking effect* produced below a critical exit angle β_c ($\beta = \theta - \alpha$) by the blocking cone of the next atom. Blocking cones are more difficult to calculate because they are formed by nonparallel incident trajectories having different energies.

Similar shadowing and blocking effects are also present for the recoiling atoms. In this case, the blocking effect is simpler because all of the trajectories are diverging from the same point source, i.e. the initial position of the recoil atoms. β_c gives the direction of the first undeflected trajectory which is tangent to the blocking cone. The corresponding impact parameter p_c is then such as to produce no scattering, i.e.

$$\theta(p_c) = 0^\circ, \quad p_c = d \sin \beta_c. \tag{50}$$

For two different directions, if

$$p_c(E_2) \cong p_c(E_2'),$$

where E_2 and E_2' are the recoiling energies corresponding to the angles $(\alpha + \beta_{c1})$ and $(\alpha + \beta_{c2})$, respectively, the following relation can be written:

$$d_1/d_2 = \sin \beta_{c2}/\sin \beta_{c1}. \tag{51}$$

2.3.4 INTENSITY

2.3.4.1 Scattered and recoiled flux

The flux of the primary ions scattered at an angle θ with an energy E_1 is given (Taglauer and Heiland, 1976) by

$$I(\theta) = J_0 N' \frac{d\sigma(\theta)}{d\Omega} \Delta\Omega\, T(E_1), \tag{52}$$

where J_0 (A cm^{-2}) is the incident-ion current density and N' is the effective number of scattering centers bombarded by the beam and participating in scattering without suffering any shadowing, blocking, or focusing effects. $d\sigma(\theta)/d\Omega$ (cm^2 sr^{-1}) is the differential scattering cross section. $\Delta\Omega$ (sr) is the solid angle accepted by the analyzer/detector system. It is assumed that all of the bombarded area is accepted and that $\Delta\Omega$ is small enough that $d\sigma(\theta)/d\Omega$

can be considered as constant in $\Delta\Omega$. Finally, $T(E_1)$ is the experimental factor already discussed in Section 2.2.4.

If electrostatic energy analyzers are used, only the charged fraction of the energy distribution $n(E)$ is measured and an ion survival probability P^+ (or the negative ion probability P^- if negative ions are considered) has to be introduced into equation (52) such as:

$$P^{+(-)} = I^{+(-)}(\theta)/I(\theta), \tag{53}$$

$$I(\theta) = I^+(\theta) + I^-(\theta) + I^n(\theta). \tag{54}$$

$I^n(\theta)$ represents the neutral part of the scattered intensity and is the dominant term in equation (54).

The difficulty in ISS is that N' depends on the structure and on the exact experimental conditions ($m_1, m_2, E_0, \alpha, \theta$) through the sizes and relative positions of the shadowing and blocking cones. If only the first monolayer contributes to the single-collision peak in the scattering spectrum, then

$$N' = nS, \tag{55}$$

where n (cm^{-2}) is the atomic surface density and S (cm^2) is the bombarded area. Since

$$J_0 S = I_0/\sin\alpha, \tag{56}$$

if no focusing effect is present, equation (52) can be written as

$$I^{+(-)}(\theta) = \frac{I_0}{\sin\alpha}\, nP^{+(-)} \frac{\mathrm{d}\sigma(\theta)}{\mathrm{d}\Omega}\, \Delta\Omega\, T(E_1). \tag{57}$$

2.3.4.2 Screening factor

If shadowing and blocking effects produced by other atoms are present, the surface density n has to be reduced by a screening factor η and an effective surface density n' has to be introduced in equation (57) as

$$n' = n(1 - \eta). \tag{58}$$

This screening factor can, in some cases, be very difficult to evaluate, leading to uncertainty in quantitative analyses. If more than one monolayer contributes to the signal, the same formalism can be conserved on condition that $\eta < 1$.

If more than one constituent is present in the sample, for each constituent i, $I_i^{+(-)}(\theta)$ is given by equation (57) where n_i' is found by

$$n_i' = n_i\left(1 - \sum_{i;j=1}^{n} \eta_{ij}\, n_j\right). \tag{59}$$

The summation is made over the n different elements and η_{ij} represents the screening effect produced by atoms i on atoms j.

It is to be noted that in addition to the 'elastic screening', when only the charged fraction is measured, an 'inelastic' contribution may also have to be considered in η_{ij}. Since the neutralization of noble-gas ions is trajectory-dependent, situations can occur for which, although no elastic screening is produced, the presence of one kind of atom can modify the neutralization during the scattering on the other kind of atom. This is, for example, the case for He^+ scattered on NaCl. At low incidence and along specific azimuthal directions, it has been observed that He^+ elastically scattered by Na is neutralized because it penetrates the electron charge density of Cl (Souda *et al.*, 1986; Benken *et al.*, 1989; Verbist *et al.*, 1990).

2.3.4.3 Differential scattering cross section

The differential scattering cross section is given by

$$\frac{d\sigma(\Theta)}{d\Omega} = \frac{-p}{\sin(\Theta)}\frac{dp}{d\Theta}. \qquad (60)$$

This relation comes directly from the conservation of particle flux before and after scattering. Its evaluation is made via the relation $\Theta(p)$ (see equation (29)). When expressed in the laboratory reference system, equation (60) becomes

$$\frac{d\sigma(\theta)}{d\Omega} = \frac{(1 + A\cos\Theta + A^2)^{3/2}}{A^2(A + \cos\Theta)}\frac{d\sigma(\Theta)}{d\Omega} \qquad (61)$$

and

$$\frac{d\sigma(\phi)}{d\Omega} = 4\sin\frac{\Theta}{2}\frac{d\sigma(\Theta)}{d\Omega}. \qquad (62)$$

With a screened Coulomb potential, no analytical formulation can be found and numerical derivation is performed. The general trend of the scattering cross section is to decrease with primary energy E_0 and scattering angle θ and to increase with target mass m_2. These dependences are shown in Figure 2.24 and 2.25 for He^+ scattering (Beuken, 1991). It is seen that the order of magnitude of $d\sigma(\theta)/d\Omega$ is 10^{-2} Å^2 sr^{-1}.

For a given target mass m_2, the direct recoiling cross section increases with recoiling angle ϕ (as expected since $\phi = 0°$ corresponds to $\theta = 180°$) and with incident-ion mass m_1 and decreases with primary energy E_0. For a given projectile mass m_1, it increases with target mass m_2 with an exception for H which has a higher value (Rabalais, 1988, Eckstein, 1987). These dependencies are illustrated in Figures 2.26 and 2.27 for different primary ions (3 keV) recoiling different atoms as a function of recoiling angle ϕ. The primary energy dependence is shown in Figure 2.28 for Ar^+ recoiling O and Ni at two

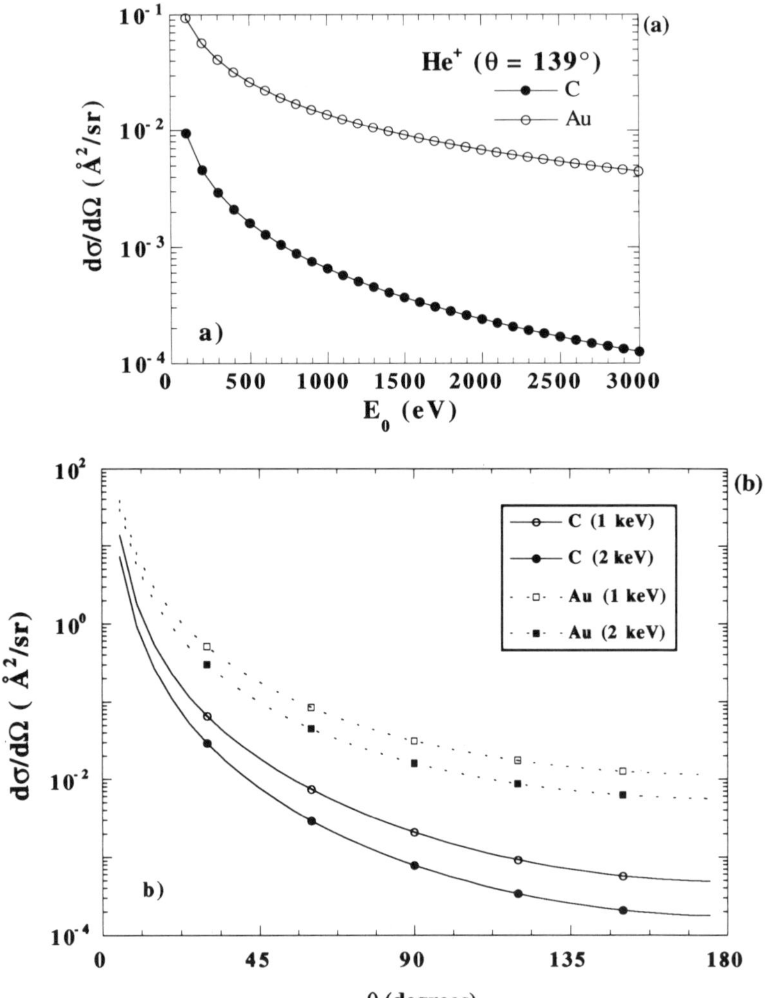

Figure 2.24 Differential scattering cross sections $d\sigma(\theta)/d\Omega$ for He^+ scattering from C and Au: (a) as a function of the primary ion energy for 139° scattering and (b) as a function of the scattering angle θ for 1 keV and 2 keV primary energies. The Molière potential has been used for the calculations. (Beuken, 1991)

different angles. At low detection angles, the cross section for recoiling is lower than for scattering.

It is worth noting that the usual opposite behavior between sensitivity and resolution is found when comparing equations (25) and (61) for scattering and (26) and (62) for recoiling. Indeed when the cross section increases, the mass resolution $m/\Delta m$ decreases.

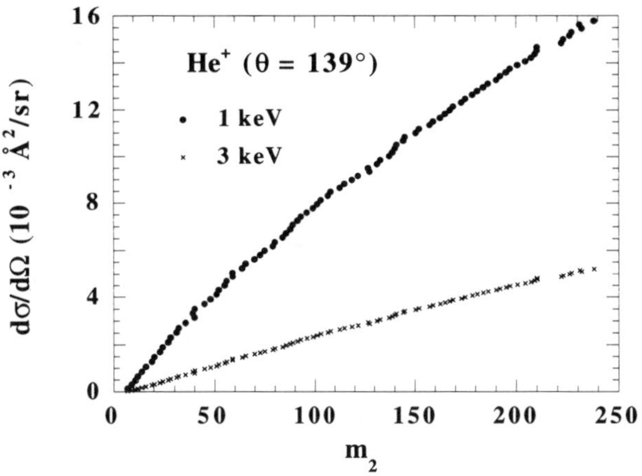

Figure 2.25 Differential scattering cross sections $d\sigma(\theta)d\Omega$ as a function of the target atom mass m_2 for 1 keV and 2 keV $^3He^+$ scattering at 139°. The Molière potential has been used for the calculations. (Beuken, 1991)

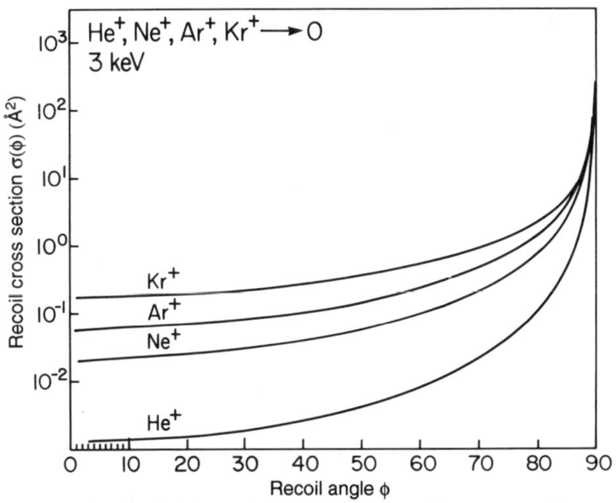

Figure 2.26 Direct-recoil differential cross sections $d\sigma(\phi)/d\Omega$ as a function of the recoil angle ϕ for oxygen atoms recoiled by 3 keV He^+, Ne^+, Ar^+ and Kr^+. (Reprinted with permission from Rabalais *et al.*, 1988. Copyright CRC Press Inc., Boca Ration, FL)

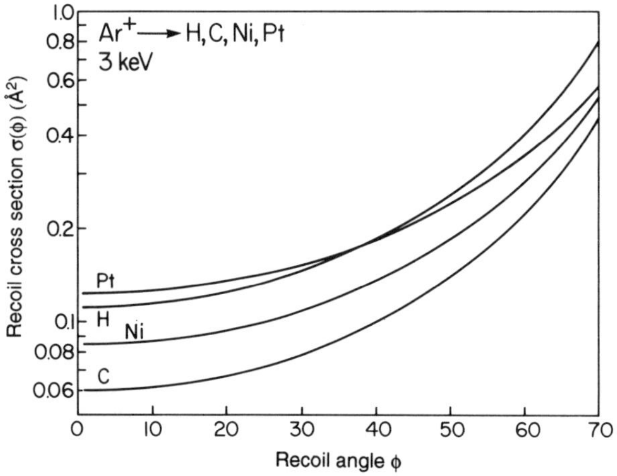

Figure 2.27 Direct-recoil differential cross section $d\sigma(\phi)/d\Omega$ as a function of the recoil angle ϕ for 3 keV Ar^+ recoiling H, C, Ni and Pt atoms. (Reprinted with permission from Rabalais *et al.*, 1988. Copyright CRC Press Inc., Boca Raton, FL)

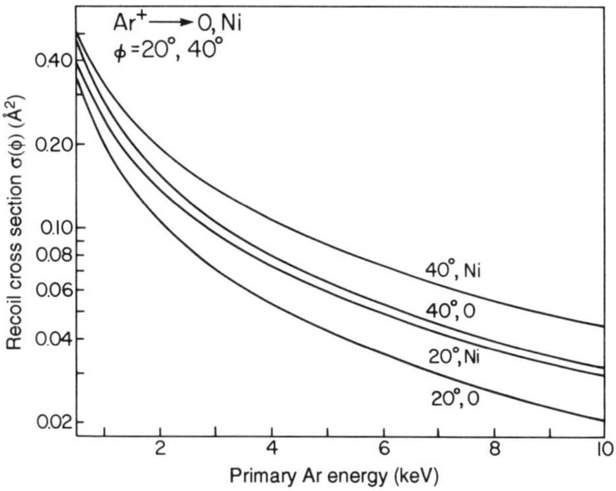

Figure 2.28 Direct-recoil differential cross sections $d\sigma(\phi)/d\Omega$ as a function of the primary energy E_0 for Ar^+ recoiling O and Ni atoms into recoil angle ϕ of 20° and 40°. (Reprinted with permission from Rabalais *et al.*, 1988. Copyright CRC Press Inc., Boca Raton, FL)

2.3.4.4 Ion-survival probability

The ion-survival probability P^+ is related to the different charge-exchange mechanisms which can occur when a low-energy ion interacts with a solid. Chapter 3 of this volume is devoted to this subject. The physics of these mechanisms is rather complex if a correct quantum formalism is applied. Although different charge-exchange mechanisms have been identified, it is not yet possible to have a universal evaluation of P^+ for the purpose of quantitative analyses. For noble-gas ions, the neutralization is sensitive to the local electron density so that the scattered intensity may depend on the chemical environment of the atoms (Woodruff, 1986). The situation is nevertheless simpler than in SIMS. Noble-gas ISS could be a useful probe for studying surface electron density.

There are three different spatial regions where different electron-exchange processes can take place. (i) In the outermost region where the ion–surface distance is large ($d > 5$ Å), resonance neutralization and ionization with the valence-band electrons of the target can take place. Their occurrence depends on the exact position of the projectile-excited levels with respect to the work function of the substrate. An upward shift and a broadening of the electronic energy levels appear when the ions approach the surface due to the image potential. It must be taken into account that energy conditions allowing resonance neutralization can reverse to resonance ionization at smaller distances (Aono and Souda, 1987; Hagstrum, 1954). (ii) In the intermediate region where the ion–surface separation is ~ 2 Å, the ion core orbitals overlap appreciably with the valence electron cloud of the target and Auger neutralization becomes dominant. The ion-survival probability P^+ depends on the exact ion trajectory and is lower if the ion penetrates regions of high valence-electron density. (iii) At very short projectile–target atom separation ($d \lesssim 0.5$ Å), core-level quasi-resonant charge exchange between the target core level and the projectile ionization level (Rusch and Erickson, 1981) and reionization of the projectile by an electron promotion mechanism to empty states in the valence band of the target are possible (Aono and Souda, 1987). Both require the formation of a pseudo-diatomic molecule during the violent part of the collision and their occurrence depends on the exact electronic structure of the collision partners. The former produces an oscillating behavior in the energy dependence of the scattered intensity (Rusch and Erickson, 1981) and the latter is associated with a constant inelastic energy loss (Aono and Souda, 1987) (see Section 2.5.2).

If only the Auger process plays a dominant role, according to the Hagstrum (1954) model, the ion-survival probability can be written

$$P^+ = \exp[-(v_c/v_p)], \tag{63}$$

where v_p is the velocity component perpendicular to the surface and v_c is a characteristic velocity for the ion–target combination obtained by integrating the Auger transition rate over the projectile trajectory. For He^+, v_c has been determined by energy and angular dependence measurements and it is in the range 10^5–10^6 ms^{-1}. Radiative neutralization is excluded since its transition rate is much smaller than those of the other processes.

An alternative way to determine P^+ is directly from experiments. This has been done with a TOF spectrometer by comparing the scattering intensity obtained with and without the charged fraction. Indeed, the charged fraction can be easily deflected or post-accelerated in the flight tube before detection (Rabalais, 1988; Boers, 1984). Reionization of the scattered neutrals in the gas phase using a stripping cell with a calibrated efficiency has also been used for this purpose (Eckstein, 1981).

2.4 STRUCTURE ANALYSES

2.4.1 INTRODUCTION

Information on surface and subsurface structures can be easily obtained with ISS and DRS. The main advantage of these techniques is that the ions probe the local atomic order and atomic positions in the 'real space' in which they are obtained. The method is much more direct than the diffraction techniques (LEED, RHEED) which are sensitive to the surface periodicity and provide an image of the reciprocal surface lattice. ISS and DRS probe local atomic structures, i.e. individual interatomic spacings, whereas diffraction techniques probe long-range order, i.e. periodicity in the range of 100–200 Å.

Different surface-structure problems have been addressed in the literature with ISS/DRS. They concern surface relaxation, surface reconstruction, either spontaneous or driven by adsorbates, adsorbate superstructures, and thin-film epitaxial growth. Recent reviews on those applications can be found in Rabalais (1990, 1991), Overbury (1987), Niehus and Spitzl (1991), Niehus *et al.* and Fauster (1988).

Several experimental procedures have been proposed, differing in the choice of primary particles and detection conditions. The most popular combinations are alkali ions with an ESA spectrometer and noble-gas ions with a TOF spectrometer. The noble-gas ion/TOF combination is known as time-of-flight scattering and recoiling spectrometry or TOF-SARS. Although used by some authors, the combination of noble-gas ions with an ESA spectrometer is not recommended for surface-structure analysis due to the added complication produced by the ion-neutralization effects. LEED is always used as a complementary technique to determine the long-range surface crystallinity and regularity of the adsorbate network.

2.4.2 EXPERIMENTAL PROCEDURE

Surface structural analyses are carried out by measuring the intensity of the different characteristic features appearing in the ISS/DRS spectra as a function of the beam orientation with respect to the sample surface. These features may be the substrate or adsorbate single-scattering peak in back-scattering (BS) or forward-scattering (FS) mode, the multiple-scattering peak (MS), the direct-recoil peak (DR) from the substrate or the adsorbate, and the surface-recoiling peak(SR).

The sample position is defined by the beam incident angle α and the azimuthal angle δ between the projection of the beam direction onto the surface and a reference crystallographic direction of the surface. The measurements are made either by scanning the incident angle α (or exit angle $\beta = \theta - \alpha$) with the azimuthal angle as a parameter or by scanning the azimuthal angle with the incident angle as parameter. The detection angle, primary energy and ion mass are chosen in accordance with kinematic and dynamic considerations in order to allow good energy separation and sensitivity for the spectral features of interest.

The interpretation of the results is straightforward and is based on the shadowing and blocking cone concepts and the focusing effect. However, for complex surface structures, the interpretation requires a trial procedure where the experimental data are compared with calculated results obtained by three-dimensional computer simulations using a model of the proposed surface structure.

The different types of measurements will now be described along with the type of information obtained and illustrated by data for clean Ni{110} and hydrogen and oxygen chemisorbed Ni{110} (Rabalais *et al.*, 1992; Roux *et al.*, 1991).

2.4.2.1 Back-scattering (BS) scans

Figure 2.29 illustrates the different methods of obtaining structural information from back-scattering intensities. They include the following.

2.4.2.1.1 BS vs incident angle α scan

At grazing incidence each surface atom lies in the shadow cone of its neighbor and BS does not occur because only large impact parameters are possible, resulting in FS. As α increases, a critical value α_c^i is reached each time the ith layer of target atoms moves out of the shadow cone, allowing the small impact parameter required for BS. The intensity I(BS) exhibits sharp rises as a function of α with well-defined maxima as schematized in Figure 2.29(a). These maxima occur each time the focused trajectories at the edge of the shadow cone pass through the neighboring atom so that equation (49) is satisfied. First-layer atomic spacings are determined by measuring α_c^1 along azimuthal directions for which specific surface-crystallographic directions are aligned with the beam. The first-second-layer spacing is obtained in a similar manner from α_c^2 measured along azimuths for which the first- and second-layer atoms lie in the same plane. Examples of such α scans are shown in Figure 2.30 for Ni{110} – (1 × 1), Ni{110} – (1 × 2)H missing row and Ni{110} – (2 × 1)O missing row surfaces along the ⟨001⟩, ⟨112⟩ and ⟨110⟩ azimuths using 4 keV Ne^+ and $\theta = 90°$. The detailed interpretations may be found in Roux *et al.* (1991).

The interatomic spacings d_i are determined from the experimental α_c^1 by the following procedure. The first step is to obtain a calibrated shadow cone. This is achieved by measuring α_c along crystal azimuths for which the atomic

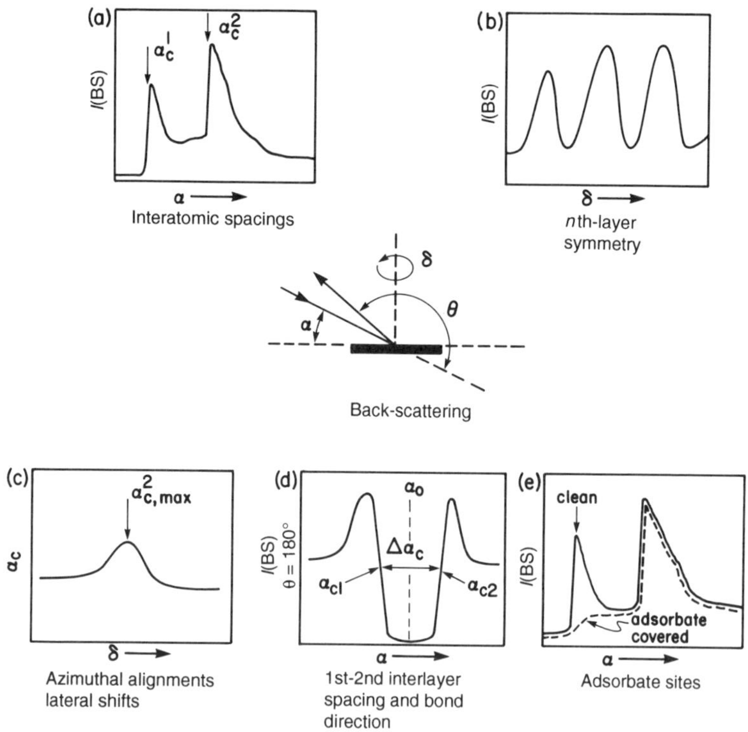

Figure 2.29 Methods of obtaining structural information from back-scattering intensities: (a) I(BS) vs α for interatomic spacings; (b) I(BS) vs δ for nth-layer periodicity; (c) α_c vis δ for azimuthal alignments and lateral shifts; (d) I(BS) at 180° vs α for first–second interlayer spacing and bond direction; (e) I(BS) vs α from a clean and adsorbate-covered surface for locating adsorbate sites. (Reproduced by permission of Elsevier Science Publishers from Rabalais *et al.*, 1992)

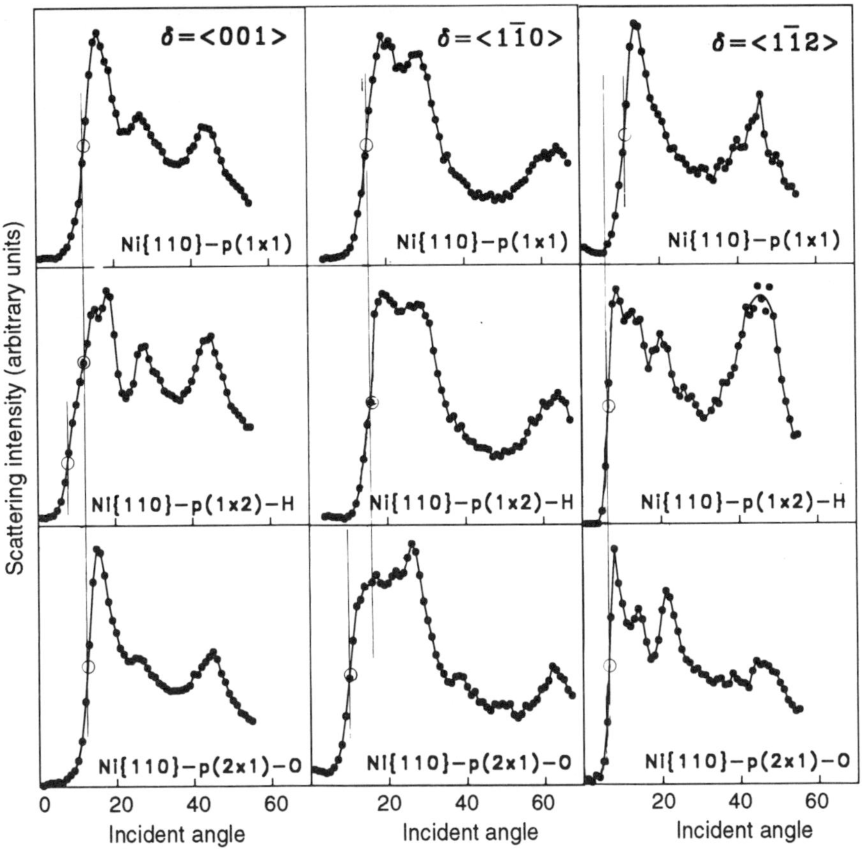

Figure 2.30 Examples of BS vs α-scans for clean Ni{110}–(1 × 1), Ni{110}–(1 × 2)–H missing row, and Ni{110}–(2 × 1)–O missing row surfaces along the ⟨001⟩, ⟨1$\underline{1}$2⟩, and ⟨1$\underline{1}$0⟩ azimuths using 4 keV Ne^+ and $\theta = 90°$. The surface structures are shown in Fig. 2.33. Note the shift of α_c towards lower values along ⟨001⟩ for H/Ni, along ⟨1$\underline{1}$0⟩ for O/Ni, and along ⟨1$\underline{1}$2⟩ for both H/Ni and O/Ni. This shift represents doubling of the lattice constant along that direction. (Reprinted by permission of Elsevier Science Publishers from Rabalais *et al.*, 1992)

spacings are known. The screening factor C in the interaction potential (equation (34)) is then adjusted so that the calculated shadow-cone radius r satisfies equation (49). With this accurate potential, equation (49) is used to calculate the calibrated $\alpha_c(d)$ curve. An unknown d can be directly read on the curve at the corresponding α_c value. An example of a calibrated $\alpha_c(d)$ curve is shown in Figure 2.31 for 4 keV Ne^+ scattered at $\theta = 90°$ from a nickel surface (Roux *et al.*, 1990). The Ni{110} unreconstructed surface with the known atomic spacings $d\langle 001\rangle = 3.53$ Å and ⟨110⟩ = 2.49 Å has been used to calibrate the

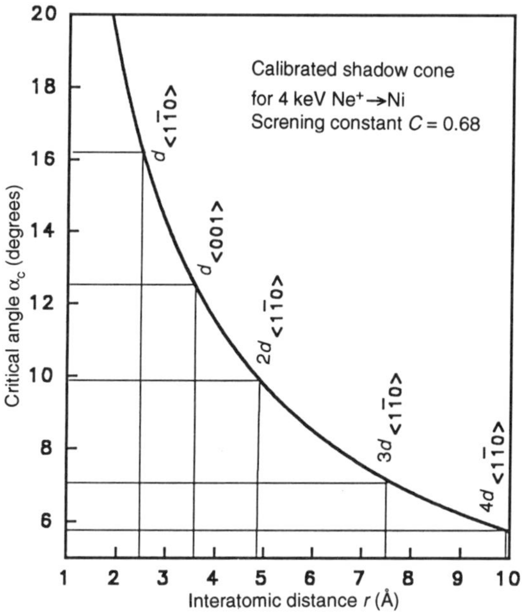

Figure 2.31 Plot of the critical incident angle α_c vs Ni–Ni interatomic distance as calculated from the shadow cone calibrated for 4 keV Ne^+ collisions with Ni. The lattice spacings d along the ⟨001⟩ and ⟨1$\underline{1}$0⟩ azimuths and multiples of the spacing along the ⟨1$\underline{1}$0⟩ azimuth are indicated. (Roux *et al.*, 1992)

potential. From the slope of the $\alpha_c(d)$ curve, it is seen that the sensitivity of the method increases for smaller distances d (larger α_c). For low α_c ($<10°$), the experimental uncertainty in determining α_c leads to poor accuracy in determining d.

2.4.2.1.2 180° BS vs incident angle α scan

A specialization of the BS vs incident angle α scans obtained by fixing the scattering angle at 180° is known as coaxial scattering. This mode is realized by focusing the pulsed primary ion beam through a central hole in a multichannel plate detector (Aono and Souda, 1985). Detection is made in the annular solid angle centered at 180° and with an aperture of a few degrees. As the ion incidence and the scattering direction are the same, the trajectories are simplified (Wang *et al.*, 1992). This straight-line probe into the crystal can sample down to several atomic layers below the surface at high incident angle (Katayama *et al.*, 1990). In this scattering geometry only head-on collisions ($p = 0$) are detected. Just above α_c an important enhancement of the $I(180°)$

signal is produced since the trajectories concentrated at the edge of the shadow cone are focused onto the shadowed atom. At α_c, the edge of the shadow cone passes through its center so that equation (45) becomes $\tan \alpha_c = r/l$; r and l are determined on the calculated shadow cone so that $d = (r^2 + l^2)^{1/2}$.

Coaxial scattering is advantageous for real-time monitoring of the early stages of film growth since the substrate and the film can be analyzed simultaneously (Aono *et al.*, 1992). This allows one to reveal any epitaxial relationship between the first layer and the layers below, which is difficult to achieve by STM (Worthington *et al.*, 1992). Also for practical considerations, only one port in the vacuum chamber is required for both the source and detector.

The I(BS) versus α scans for BS from a second-layer atom exhibit patterns similar to that of Figure 2.29(d). As α is scanned, I(BS) drops (α_{c1}) as the lower edge of the shadow cone moves through the second-layer atom, a minimum is observed when the first-layer atom shadows the second-layer atom, and I(BS) increases (α_{c2}) as the upper edge of the cone moves through the second-layer atom. α_0 corresponds to the bond direction and $\Delta\alpha_c$ is related to the bond length d by $d = r/\sin(\Delta\alpha_c)$.

2.4.2.1.3 BS vs α with adsorbates

Collecting I(BS) data as a function of α for the adsorbate-covered surface and comparing this data to that of the clean surface provides a qualitative location of the adsorption-site position (Bu *et al.*, 1989). I(BS) versus α structures that are attenuated on the adsorbate-covered surface are those for which the adsorbate atoms lie along or very near the path of the scattered primary ion. Although the primary ion cannot back-scatter from light adsorbates, the presence of the adsorbates near the scattering trajectory results in small-angle deflections of the projectile which are sufficient to produce changes as shown in Figure 2.29(e). In this figure, the first-layer scattering peak is attenuated, indicating that the adsorbate lies along this specific azimuth and interferes with the trajectories for first-layer scattering.

2.4.2.1.4 BS vs azimuthal angle δ scans

Rotating the crystal about the surface normal with fixed α provides a scan of the crystal azimuthal angles. Plots of I(BS) versus δ, as in Figure 2.29(b), reveal the surface periodicity. Using low α, scattering occurs only from the first atomic layer, and the plot reveals the periodicity of the outermost layer (Masson *et al.*, 1991).

Similar information is obtained about the second layer by using higher α values. Shifts in the first–second–layer registry (Grizzi *et al.*, 1989) can be detected by monitoring the α_c^2 values for second-layer back-scattering along directions near those azimuths for which the second-layer atoms are expected, from the bulk structure, to be directly aligned with the first-layer atoms. The α_c^2 values are maximum for those δ values where the first- and second-layer atoms are aligned, as shown in Figure 2.29(c).

For complex reconstruction, the peaks and dips in the scan are mainly produced by focusing effects so that a straightforward interpretation in terms of shadowing and blocking is difficult and the structural information can only be extracted from the data by comparison with complete calculations obtained for many different trial structures (Dorenbos *et al.*, 1992).

2.4.2.2 Forward-scattering (FS) scans

Figure 2.32 illustrates the different methods of obtaining structural information from forward-scattering intensities. They include the following.

2.4.2.2.1 FS vs azimuthal angle δ scans

When very small ($<10°$) incident angles α are used, the projectile experiences a continuous potential along low-index azimuths rather than individual atomic potentials (Bu *et al.*, 1991b; Masson and Rabalais, 1991). As a result, the outgoing trajectories are focused at the specular ejection angle β, i.e. $\alpha = \beta$. Also, surface semichanneling (Grizzi *et al.*, 1990), i.e. steering of the incident ions by troughs formed from two first- and one second-layer rows, contributes to the I(FS) enhancement along the low-index azimuths. Plots of I(FS) vs δ with $\alpha = \beta$ exhibit intensity maxima along the principal azimuths as shown in Figure 2.32(a) due to specular focusing. If a large exit angle is used such that $\beta > \alpha$, the off-specular conditions result in minima rather than maxima along the high-symmetry axes as shown in Figure 2.32(b). The surface periodicity can be read directly from these minima in the plots, thereby revealing the crystal structure. Examples of such δ scans are shown in Figure 2.33 for Ni, H/Ni and O/Ni; detailed interpretations have been published elsewhere (Roux *et al.*, 1991).

2.4.2.2.2 FS vs exit angle β scans

Varying the scattering angle θ at constant α provides a scan of the exit angle $\beta = \theta - \alpha$. At low β the structural features are determined by focusing of scattered trajectories at the edges of the blocking cones of atoms obstructing their escape along that azimuth (Shi *et al.*, 1990; Bu *et al.*, 1989). If a low α value is used so that only first-layer scattering is observed, the structures are determined by first-layer atoms blocking their first-layer neighbors. Peaks are observed, as in the BS versus α scans, for every different interatomic spacing along a given azimuth. The example of Figure 2.31(c) is typical of two different interatomic spacings along one azimuth. The critical angles β_c along with the blocking-cone dimensions allow calculation of interatomic spacings as in Section 2.3.3.3.

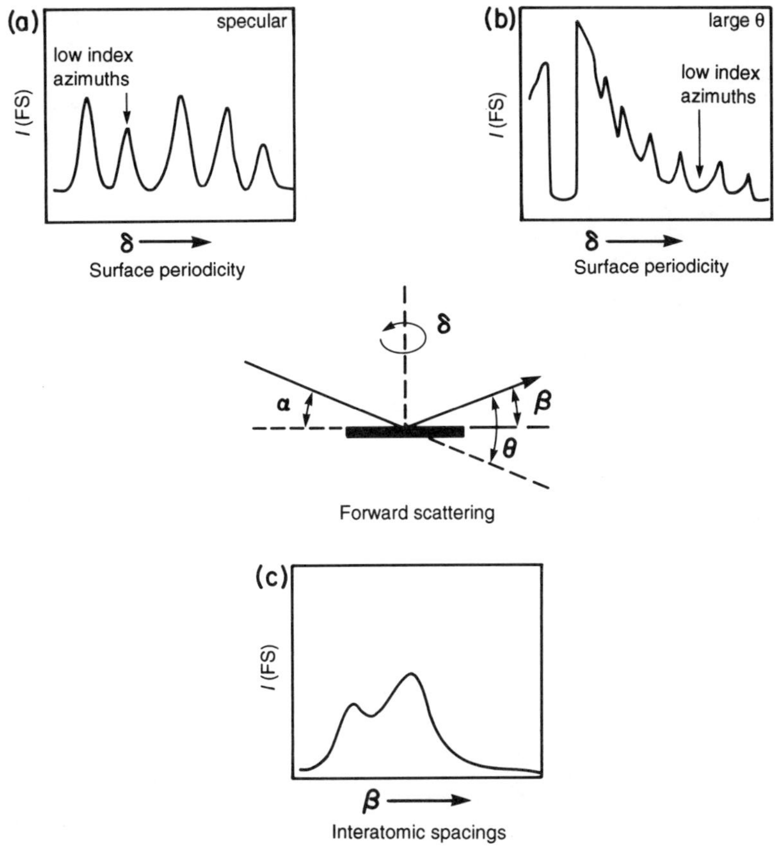

Figure 2.32 Methods of obtaining structural information from forward-scattering intensities: (a) I(FS) vs δ using specular scattering for surface periodicity; (b) I(FS) vs δ using large scattering angle θ for surface periodicity; (c) I(FS) vs β for interatomic spacings. (Reprinted by permission of Elsevier Science Publishers from Rabalais *et al.*, 1992)

2.4.2.3 Recoiling (R) scans

Figure 2.34 illustrates the different methods of obtaining structural information from recoiling intensities. They include the following.

2.4.2.3.1 R vs α scans

Plots of I(R) versus α provide two critical angles as shown in Figure 2.34(a). Here α_c corresponds to the situation where the edge of the shadow cone cast by a neighboring atom focuses the primary flux at the appropriate impact parameter for recoil into the angle ϕ. The steep drop-off at high α corresponds

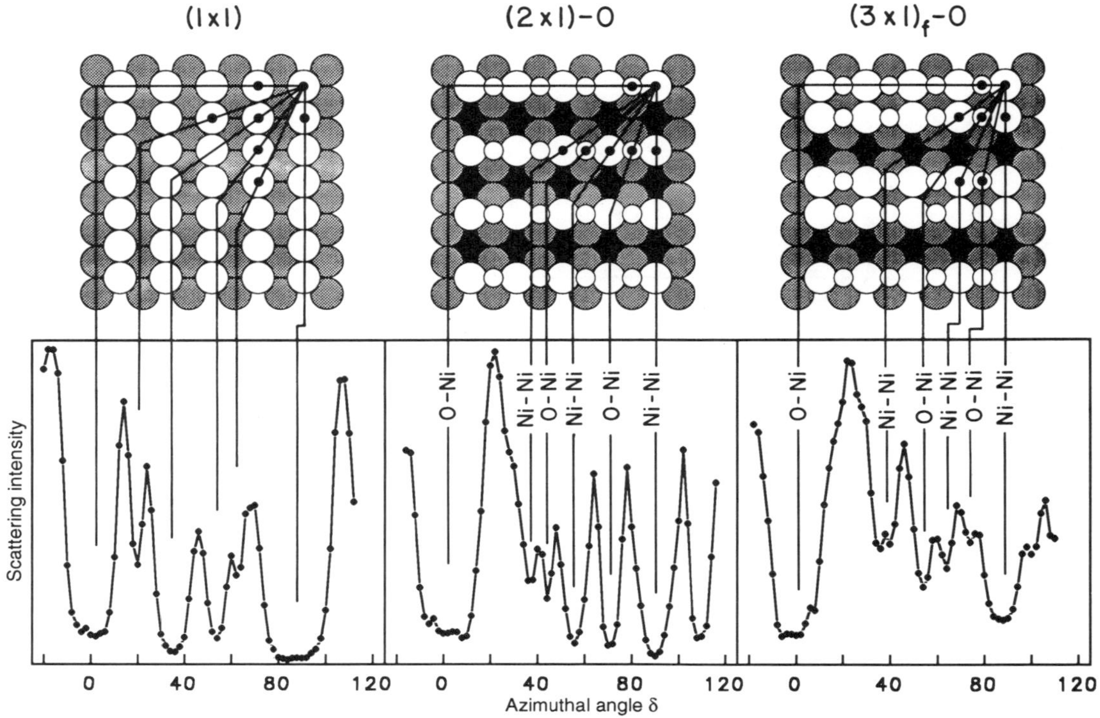

Figure 2.33 Examples of FS vs δ scans for clean Ni{110}–(1 × 1), Ni{110}–(1 × 2)–H, and Ni{110}–(2 × 1)–O missing row surfaces using 4 keV Ne^+ with $\alpha = 5^\circ$ and $\theta = 30^\circ$ (off-specular conditions). O–Ni and Ni–Ni denote the directions along which O and Ni atoms, respectively, shadow the Ni scattering center. (Reprinted by permission of Elsevier Science Publishers from Rabalais *et al.*, 1992)

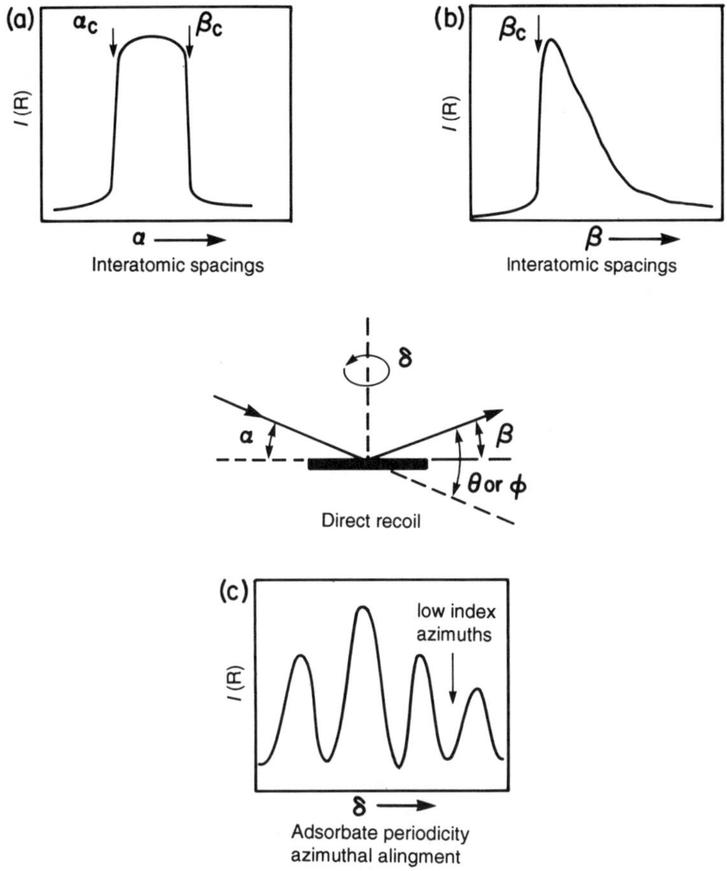

Figure 2.34 Methods of obtaining structural information from recoiling intensities: (a) I(R) vs α for adsorbate interatomic spacings; (b) I(R) vs β for adsorbate interatomic spacings; (c) I(R) vs δ for adsorbate periodicity and azimuthal alignment. (Reprinted by permission of Elsevier Science Publishers from Rabalais *et al.*, 1992)

to $\beta_c = \phi - \alpha$. Here β_c is determined by focusing of recoil trajectories by the blocking cone of a neighboring atom. This provides a direct measure of the interatomic spacings between an adsorbate atom and neighboring atoms along the azimuth (Shi *et al.*, 1989; Bu *et al.*, 1989).

2.4.2.3.2 R vs β scans

Plots of I(R) versus β, obtained by varying ϕ (Figure 2.34(b), provide a direct measure of β_c which is comparable to the previous measurement.

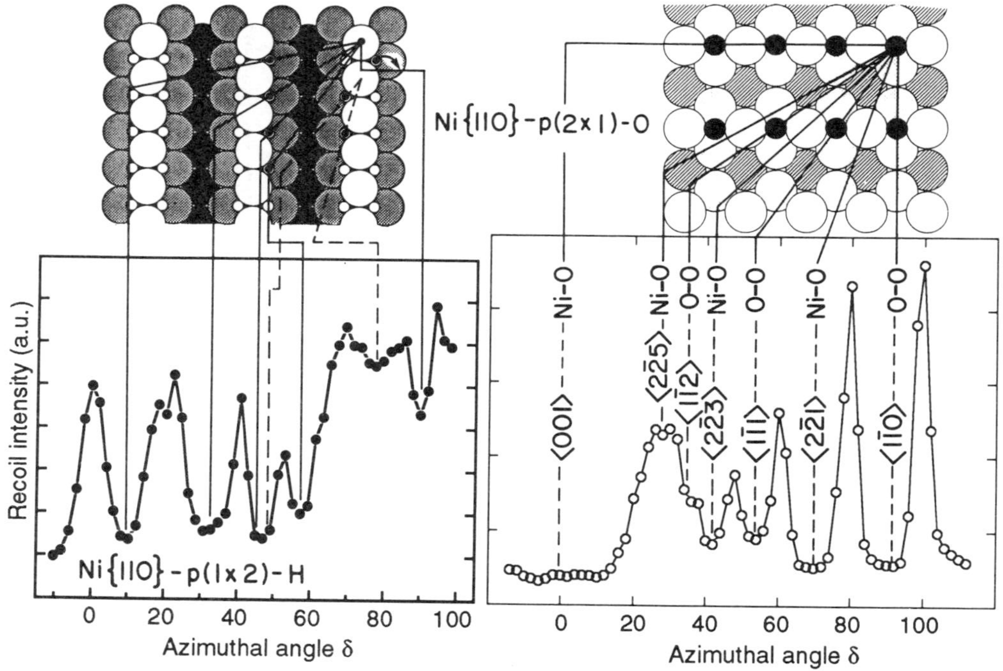

Figure 2.35 Examples of R vs δ scans for Ni{110}–(1 × 2)–H and Ni{110}–(2 × 1)–O missing row surfaces using 4 keV Ar^+ with $\alpha = 10^\circ$ and $\phi = 40^\circ$. The minima in the H/Ni scan all corresponds to Ni shadowing H; the 90° minimum corresponds to Ni shadowing the nearest H atom, even though they are not aligned along the ⟨1$\underline{1}$0⟩ azimuth. The minima in the O/Ni scan corresponds to both Ni and O shadowing O as labeled. (Reproduced by permission of Elsevier Science Publishers from Rabalais *et al.*, 1992)

2.4.2.3.3 R vs δ scans

Rotating the crystal about δ with a fixed, low value of α and monitoring $I(R)$ reveals the periodicity of the adsorbate sites on the surface (Bu *et al.*, 1989, 1991c; Shi *et al.*, 1989). Plots of I(R) vs δ exhibit maxima and minima as shown in Figure 2.34(c) resulting from shadowing and blocking of the adsorbate sites due to the different interatomic spacings as δ is scanned. This exposes the adsorbate periodicity and azimuthal alignment with respect to the substrate crystallographic axes; the adsorbate site is determined from this. Examples of such δ scans are shown in Figure 2.35; detailed interpretations have been published elsewhere (Roux *et al.*, 1991; Bu *et al.*, 1991c).

2.4.2.4 Multiple-scattering scans

The angular dependence (α and δ) of the multiple-scattering features present in the energy spectra can also be used for structure analyses (Overbury, 1987; Grayeb *et al.*, 1992). The information is not straightforward and requires a comparison of the full scattering spectra with simulation results obtained from trial models of the structure. In order to determine the most probable model of the surface that is consistent with the LEED pattern, the atom positions are modified in the simulation until agreement is obtained between the simulated spectral features and their relative intensities and the experimental scattering spectra. Using this method, an accuracy of 0.01 Å in the atomic distance determination has been reported (Grayeb *et al.*, 1992). This mode is well adapted to alkali ions and use of an ESA spectrometer because it requires a good energy resolution to identify the spectral features associated with different multiple-collision trajectories. Ideally, a structure is fully determined if the interpretation of the data from the different methods converge to the same model.

2.5 COMPOSITIONAL ANALYSES

2.5.1 INTRODUCTION

Although TOF-SARS can also be used for quantitative elemental analyses, up to now only a few studies have been published using this mode (Masson *et al.*, 1989 and references therein). TOF-SARS has been mainly restricted to surface-structure analyses (see Section 2.4). For most of the elemental-surface-composition analyses, noble-gas ions combined with an ESA spectrometer are used so that mainly single scattering from surface atoms is detected. This mode of analysis is also called LEIS (low-energy ion scattering).

However, TOF-SARS presents advantages with respect to LEIS. Detection of the particles independently of their charge state eliminates ion-neutralization effects. Also, the multichannel detection requires only primary ion doses of only $\sim 10^{11}$ ions per cm^2 or $\sim 10^{-4}$ ions per surface atom for spectral acquisition; this ensures true static conditions during analyses. The present limitations are related to a lower mass resolution than in LEIS and higher sensitivity to structural effects.

The main advantage of ISS-DRS for surface-compositional analyses is its extreme surface-sensitivity as compared to the other surface spectrometries (mainly XPS and AES) (Heiland and Taglauer, 1976; Niehus and Bauer, 1975). Indeed with a correct orientation and aperture of the shadow cone, the first monolayer can be probed selectively. However, difficulties arise when quantitative analyses are required. With LEIS, experiments give only access to the product $n'P^+$ (see Section 2.3.4) if the interaction potential is known. Assumptions have to be made on the ion-survival probability P^+ and on the screening factors η_{ij} to determine the atomic concentration n from scattered-ion intensities.

In contrast to the electron spectometries, there is no chemical shift in the peak position due to the chemical environment of the target atom because only single binary collisions are detected. The neighboring atoms may, however, influence the scattered intensity by modifying the ion neutralization and the elastic screening and specific inelastic energy losses are measured for some ion–surface combinations.

2.5.2 NEUTRALIZATION AND INELASTIC EFFECTS

Although a complete theoretical model for the neutralization is not yet available, calibration using reference samples of known composition allows some quantitative analyses. By monitoring the energy and ion-mass (^3He and ^4He) dependence of the scattered intensities, it is sometimes possible to discriminate between elastic screening and differential neutralization effects. In a multi-element sample, the problem is indeed to decide if neutralization is a surface property with a constant value for the different elements or an intrinsic atomic property which is different for each element. The answer depends on the exact ion–surface combination and the possible charge-exchange mechanisms. For metal alloys with valence-band electrons uniformly distributed, if the Auger neutralization is dominant, the ion neutralization is expected to be the same for all of the elements. For ionic compounds with localized valence electron density, it will differ for each element. Moreover the projectile reionization may be effective for some elements and not for others. By considering the energy dependence of the anion/cation intensity ratio for different ionic compounds, McCune (1979, 1980) classified the oxides into two classes depending on whether this intensity ratio decreases or remains constant with energy. This was interpreted as a structural effect due to increased screening of the anion at low energy by the surrounding cations. However, recent measurements on oxides have shown that the energy variation of the intensity ratio may be explained without this screening effect by considering the correct energy dependence of the individual ion-survival probabilities (van Leerdam and Brongersma, 1991).

In the case of gas adsorption, the scattered ion intensities given by equation (53) can be written as

$$I_s = Cn_s(1 - \eta n_a)S_s, \tag{64}$$

$$I_a = Cn_aS_a, \tag{65}$$

where the subscripts refer to the substrate (s) and adsorbate (a), respectively, and C is an experimental constant. If the ion-survival probability can be assumed to be an intrinsic property of the atom that is not modified by adsorption, $S_{a,s}$ contains only atomic properties,

$$S_{a,s} = (d\sigma(\theta)/d\Omega)_{a,s}P^+_{a,s}T(E_1)_{a,s}. \tag{66}$$

This means that there is no matrix effect. If $I_{s,0}$ represents the substrate intensity before absorption ($I_{a,0} = 0$) and $I_{a,1}$, the adsorbate intensity after completion of one monolayer adsorption ($I_{s,1} = 0$), equations (60) and (61) become

$$I_s = I_{s,0} - (I_{s,0}/I_{a,1})I_a. \tag{67}$$

The observation of this linear relationship between the substrate and adsorbate signals for different systems is evidence that matrix effects can be negligible for some ISS analyses (Brongersma *et al.*, 1981; Taglauer *et al.*, 1988). The screening and neutralization effects can be reduced by a correct choice of the experimental conditions: normal incidence, large scattering angle ($\theta > 90°$), and high enough primary ion energy ($E_0 \cong 2$–3 keV).

When considering the exact energy position of the scattering peaks, small deviations with respect to the BEC model (see Section 2.3.1.1) are observed (Bertrand and Ghalim, 1983; Heiland and Taglauer, 1981; Shoji and Hanawa, 1984). This extra inelastic energy loss is related to the interaction between projectile and target electron clouds leading to electron excitations and ionizations. At higher energy this contribution becomes dominant and governs the stopping power of the ions penetrating into the target. Two different contributions are observed. The first one applies for all ion–surface interactions and is a continuous function of the ion velocity and distance of closest approach (or impact parameter) (Firsov, 1959; Oen and Robinson, 1976; Karpuzov, 1981). The second one is only observed for specific ion–surface combinations; it has a constant value of ~20 eV for He (Souda and Aono, 1986; Bertrand *et al.*, 1988). It is related to reionization of a previously neutralized ion by an electron-promotion mechanism as described by Fano and Lichten (1965) for gas-phase collisions. The He atom must approach the target atom close enough to form a quasi-molecule so that electron transfer from He^0 to an empty state in the valence band of the target is possible. The energy required for electron promotion is obtained from the kinetic energy of the particle after scattering. It corresponds to the difference between the He ionization energy and the target work function.

At low primary energies, the usual single-scattering peak for He^+ ions escaping the neutralization is obtained along with an inelastic single-scattering peak. The inelastic peak is produced by He^+ ions that are first neutralized during the incoming approach to the surface by an Auger process, reionized during the close encounter collision ($d < 0.5$ Å), and then escape neutralization during the outgoing path. There is a threshold in the distance for which the electron promotion can take place. Reionization of He occurs for target atoms which have a large density of empty states in the valence or conduction band (transition metals) and a strong anti-bonding interaction between the He 1s level and a core level of the target atom (Tsuneyki and Tsukuda, 1986). The promotion of the He 1s orbital is minimized for target elements having a filled d band with a binding energy smaller than the He 1s orbital. Figure 2.36 shows the reionization probability of He^0 of different target elements for 1 keV He^0 scattering at 160° (Souda *et al.*, 1990a).

Since reionization can occur during a violent collision, particles penetrating into the solid and losing energy in multicollision sequences and electronic interactions can be detected in the spectra if, before being emitted, they suffer a collision violent enough to allow reionization. In this case, it will appear as a broadening and tailing on the low-energy side of the scattering peak (Pierson *et al.*, 1990). Broadening and tailing of the single-scattering peaks result from convolution of different energy-loss contributions and introduce complications in quantitative analyses (Beuken and Bertrand, 1992; van Leerdam *et al.*, 1990).

Inelastic energy losses corresponding to electron–hole pair creation and to electron excitation from core levels to the conduction band have also been observed in some ionic compounds (Souda *et al.*, 1990a, 1990b). The occurrence of inelastic events is strongly dependent on the ion–target combination. This indicates that electronic excitation is closely related to the formation of a quasi-molecular state during the collision.

2.5.3 APPLICATIONS

ISS compositional analyses have been applied in different fields of surface science mainly in situations where the knowledge of the uppermost surface composition (first monolayer) is crucial. In this section we will not review all the applications but rather suggest the broad range of possible uses by pointing out some recent results. This concerns gas adsorption (Brongersman *et al.*, 1981), surface segregation in metallic alloys (Weigand *et al.*, 1992), compounds

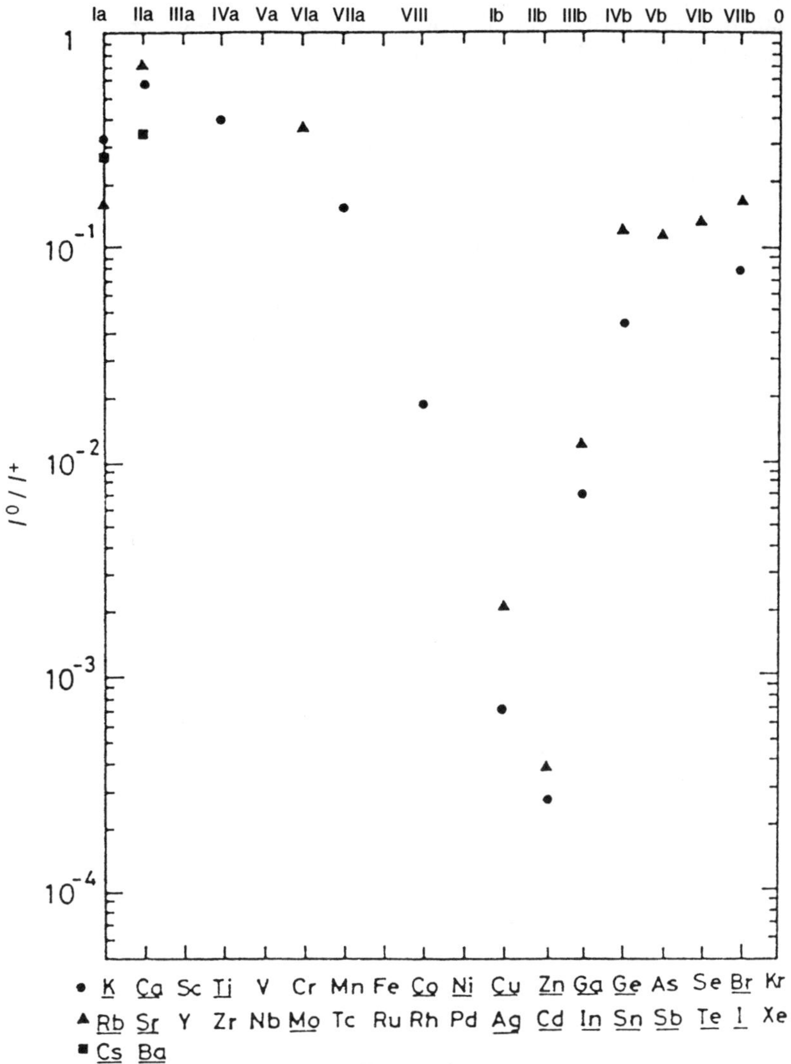

Figure 2.36 Ionization probability of He° for 29 target elements estimated from the ratio of the normalized He^+ scattering peak intensity obtained with He° incidence to that obtained with He^+ incidence, I^o/I^+, using 1 keV primary energy with $\alpha = 80°$ and $\theta = 160°$. (Reproduced by permission of Elsevier Science Publishers from Souda *et al.*, 1990a)

and polymer blends (Gardella, 1988), surface composition of real supported catalysts (Beuken and Bertrand, 1985; Brongersma and van Leerdam, 1991; Taglauer, 1991), surface modifications due to preferential sputtering by ion beams (Heiland and Taglauer, 1982), diffusion (Ackermans *et al.*, 1990), thin film growth (De Puydt *et al.*, 1991), and adhesion between thin metal films and polymers (De Puydt *et al.*, 1988). All kinds of samples can be analyzed, the only restriction being that they must be compatible with ultra-high vacuum. For insulators, compensation of the charge building up on the surface during ion bombardment is required. This is easily accomplished by flooding the surface with low-energy electrons. Concerning organic materials such as polymers, special care has to be taken in order to avoid irreversible sample damage from ion irradiation (Calgano *et al.*, 1992; Marletta, 1990). In practice, the total ion dose for spectral acquisition must be strictly controlled and must remain below $\sim 10^{13}$–10^{14} ions per cm^2 in order to preserve static conditions (De Puydt *et al.*, 1990; Bertrand and De Puydt, 1993).

The surface sensitivity of ISS combined with the sputtering and desorption-induced action of the probing beam allows one to obtain monolayer depth profiles. This is very important in catalysis where only ISS allows one to study the surface dispersion of monolayer active phases and sintering effects of thermal treatments (Bertrand *et al.*, 1983; Josek *et al.*, 1992). For deeper depth profiles, the same beam can be used for analysis and sputter etching by mixing different gases in the ion source (for example, He for analysis and Ar or Xe for sputtering).

2.6 CONCLUSIONS AND TRENDS

ISS–DRS has been shown to be an extremely powerful technique for obtaining information on surface composition and structure. The interpretation of the experimental data is straightforward and based on simple classical concepts. The structure analyses exploit the shadowing, blocking and focusing effects. The technique is complementary to the electron-diffraction techniques (LEED and RHEED) and STM. For elemental analyses ISS/DRS is complementary to AES, XPS and SIMS. It has the main advantages of selectively probing the first monolayer of the target and detection of hydrogen (DRS). This selectivity is, however, at the expense of quantification. This is mainly due to the incomplete understanding of the neutralization processes when only the charged fraction is detected. Use of the TOF technique for elemental analyses should remove this limitation but the mass resolution has then to be improved. This could be achieved by improving the ion optics in order to decrease the ion pulse width as already obtained in TOF static SIMS. Moreover by rastering a narrow focused ion beam on the surface, two-dimensional imaging of the first monolayer should be possible by ISS–DRS. On the other hand, measurement of the charged fraction of scattered and recoiled particles appears to be a promising method of observing electron density at a surface (Hsu *et al.*, 1992).

2.7 ACKNOWLEDGEMENT

P. G. Bertrand acknowledges a grant from NATO and special support from the Facultés Universitaires Saint Louis, Brussels. This material is partly based on work supported by the National Science Foundation under grant CHE-8814337.

2.8 REFERENCES

Ackermans, P. A. J., G. C. R. Krutzen, and H. H. Brongersma (1990) *Nucl. Instrum. Meth. Phys. Res.*, **B 45**, 384.
Aono, M. (1984) *Nucl. Instrum. Meth. Phys. Res.*, **B 2**, 374.
Aono, M., and R. Souda (1985) *Jap. J. Appl. Phys.*, **24**, 1249.
Aono, M., and R. Souda (1987) *Nucl. Instrum. Meth. Phys. Res.*, **B 27**, 55.
Aono, M., M. Katayama, and E. Momura (1992) *Nucl. Instru. Meth. Phys. Res.*, **B 64**, 29.
Benninghoven, A., F. G. Rudenauer, and H. W. Werner (1987) In *Secondary Ion Mass Spectrometry: Basic Concepts, Instrumental Aspects, Applications and Trends*, P. J. Elving and J. D. Winefordner (eds), Wiley, New York.
Bertrand, P. (1986) *Analusis*, **14**, 431.
Bertrand, P., and Y. De Puydt, unpublished.
Bertrand, P., and M. Ghalim (1983) *Physica Scripta*, **T 6**, 168.
Bertrand, P., F. Delannay, and J. M. Streydio (1977) *J. Phys. E: Sci. Instr.*, **10**, 403.
Bertrand, P., J. M. Beuken, and M. Delvaux (1983) *Nucl. Instrum. Meth. Phys. Res*, **218**, 249.
Bertrand, P., E. Pierson, and J. M. Beuken (1988) *Nucl. Instrum. Meth. Phys. Res.*, **B 33**, 396.
Bertrand, P., and Y. De Puydt (1993) *Nucl. Instrum. Meth. Phys. Res.*, **B78**, 181.
Beuken, J. M. (1991) Thesis, Université Catholique de Louvain.
Bejken, J. M., and P. Bertrand (1985) *Surf. Sci.*, **162**, 329.
Beuken, J. M., and P. Bertrand (1992) *Nucl. Instrum. Meth. Phys. Res.*, **B 67**, 340.
Beuken, J. M., E. Pierson, and P. Bertrand (1989) *Surf. Sci.*, **223**, 201.
Boers, A. L. (1977) *Surf. Sci.*, **63**, 475.
Boers, A. L. (1984) *Nucl. Instrum. Meth. Phys. Res.*, **B 2**, 353.
Briggs, D., and M. P. Seah (1990) *Practical Surface Analysis, Auger and X-Ray Photoelectron Spectroscopy*. Wiley, Chichester.
Brongersma, H. H., and G. C. van Leerdam (1991) In *Fundamental Aspects of Heterogeneous Catalysis Studied by Particle Beams*, H. H. Brongersma and R. A. van Santen (eds.), NATO ASI B265, Plenum, New York, p. 283.
Brongersma, H. H., G. C. J. van der Ligt, and G. Rouweler (1981) *Philips J. Res.*, **36**, 1.
Bu, H., O. Grizzi, M. Shi, and J. W. Rabalais (1989) *Phys. Rev.*, **B 40**, 10147.
Bu, H., M. Shi, and J. W. Rabalais (1991a) *Nucl. Instrum. Meth. Phys. Res.*, **B 61**, 337.
Bu, H., M. Shi, and J. W. Rabalais (1991b) *Surf. Sci.*, **244**, 96.
Bu, H., M. Shi, K. Boyd, and J. W. Rabalais (1991c) *J. Chem. Phys.*, **95**, 2882.
Burrous, C. N., A. J. Lieber, and V. T. Zaviantseff (1967) *Rev. Sci. Instr.*, **38**, 1477.

Calgano, L., G. Compagnini, and G. Foti (1992) *Nucl. Instrum. Meth. Phys. Res.*, **B 65**, 413.

Chaudhury, S., and R. S. Williams (1991) *Surf. Sci.*, **255**, 127.

Chu, W. K., J. W. Mayer, and M. A. Nicolet (1978) *Backscattering Spectrometry*. Academic Press, New York.

Cooper, B. H., C. A. Di Rubio, G. A. Kimmel, and R. L. McEachern (1992) *Nucl. Instrum. Meth. Phys. Res.*, **B 64**, 49.

De Puydt, Y., P. Bertrand, and P. Lutgen (1988) *Surf. Interf. Anal.*, **12**, 486.

De Puydt, Y., D. Leonard, and P. Bertrand (1990) *ACS Symposium Series*, **440**, 210.

De Puydt, Y., P. Phuku, and P. Bertrand (1991) *Interfaces in New Materials*, P. Grange and B. Delmon (eds), Elsevier Appl. Sci., London, p. 149.

Dorenbos, G., M. Breeman, and D. O. Boerma (1992) *Nucl. Instrum. Meth. Phys. Res.*, **B 64**, 88.

Eckstein, W. (1981) In *Inelastic Particle–Surface Collisions*, E. Taglauer and W. Heiland (eds), Springer, Berlin, p. 57.

Eckstein, W. (1987) *Nucl. Instrum. Meth. Phys. Res.*, **B 27**, 78.

Engel, T., and K. H. Rieder (1982) *Springer Tracts in Modern Physics, Vol. 91*, Springer, Berlin, p. 55.

Fano, V., and W. Lichten (1965) *Phys. Rev. Lett.*, **14**, 627.

Fauster, Th., (1988) *Vacuum*, **38**, 129.

Feldman, L. C., and J. W. Mayer (1986) *Fundamentals of Surface and Thin Film Analysis*. Elsevier, Amsterdam.

Firsov, O. B. (1959) *Sov. Phys. JETP*, **9**, 1517.

Gardella, J. A. Jr. (1988) *Appl. Surf. Sci.*, **31**, 72.

Goldstein, H. (1950) *Classical Mechanism*. Addison-Wesley, Cambridge, Mass.

Grayeb, R., M. Purushotham, M. Hou, and E. Bauer (1987) *Phys. Rev.*, **B 136**, 7364.

Grayeb, R., M. Hou, P. Mahavadi, and E. Bauer (1992) *Surf. Sci.*, **274**, 44.

Grizzi, O., M. Shi, H. Bu, and J. W. Rabalais (1989) *Phys. Rev.*, **B 40**, 10127.

Grizzi, O., M. Shi, H. Bu, and J. W. Rabalais (1990) *Rev. Sci. Instrum.*, **61**, 740.

Hagstrum, H. D. (1954) *Phys. Rev.*, **96**, 336.

Heiland, W., and E. Taglauer (1976) *Nucl. Instrum. Meth. Phys. Res.*, **132**, 535.

Heiland, W., and E. Taglauer (1981) In *Inelastic Particle–Surface Collisions*, E. Taglauer and W. Heiland (eds) Springer, Berlin.

Heiland, W., and E. Taglauer (1982) *Appl. Surf. Sci.*, **13**, 80.

Hellings, G. J. A., H. Ottenvanger, S. W. Boelens, C. L. C. M. Knibbeler, and H. H. Brongersma (1985) *Surf. Sci.*, **162**, 913.

Hou, M. (1989) *Vacuum*, **39**, 309.

Hsu, C. C., H. Bu, A. Bousetta, J. W. Rabalais, and P. Nordlander (1992) *Phys. Rev. Lett.*, **69**, 1992.

Johnson, R. E., and B. U. R. Sundqvist (1992) *Physics Today*, March, 28.

Josek, K., Ch. Linsmeier, H. Knozinger, and E. Taglauer (1992) *Nucl. Instrum. Meth. Phys. Res.*, **B 64**, 596.

Karpuzov, D. S. (1981) *Appl. Phys.*, **24**, 121.

Katayama, M., E. Momura, H. Soejima, S. Hayashi, and M. Aono (1990) *Nucl. Instrum. Meth. Phys. Res.*, **B 45**, 408.

Marletta, (1990) *Nucl. Instrum. Meth. Phys. Res.*, **B 46**, 295.

Masson, F., and J. W. Rabalais (1991) *Surf. Sci.*, **253**, 245.

Masson, F., C. S. Sass, O. Grizzi, and J. W. Rabalais (1989) *Surf. Sci.*, **221**, 299.

Masson, F., H. Bu, H. Shi, and J. W. Rabalais (1991) *Surf. Sci.*, **249**, 313.

McCune, R. C. (1979) *Anal. Chem.*, **51**, 1249.

McCune, R. C. (1980) *Appl. Surf. Sci.*, **5**, 275.

Molière, G. (1947) *Z. Naturforsch.*, **2a**, 133.

Niehus, H., and E. Bauer (1975) *Surf. Sci.*, **47**, 222.

Niehus, H., Heiland, W., and E. Taglauer (1993) *Surf. Sci. Rep.*, **17**, 213.

Niehus, H., and R. Spitzl (1991) *Surf. Interf. Anal.*, **17**, 287.

Oen, O. S. (1983) *Surf. Sci.*, **131**, L407.

Oen, O. S., and M. T. Robinson (1976) *Nucl. Instrum. Meth.*, **132**, 646.

Overbury, S. H. (1981) *Surf. Sci.*, **112**, 23.

Overbury, S. H. (1987) *Nucl. Instrum. Meth. Phys. Res.*, **B 27**, 65.

Overbury, S. H., B. M. Dekoven, and P. C. Stair (1984) *Nucl. Meth. Phys. Res.*, **B 2**, 384.

Pierson, E., J. M. Beuken, and P. Bertrand (1990) *Nucl. Instrum. Meth. Phys. Res.*, **B 45**, 374.

Rabalais, J. W. (1988) *CRC Crit. Rev. in Solid State and Mat. Sci.*, **14**, 319.

Rabalais, J. W. (1990) *Science*, **250**, 521.

Rabalais, J. W. (1991) *J. Vac. Sci. Technol.* **A 9**, 1293.

Rabalais, J. W., H. Bu, and C. Roux (1992) *Nucl. Instrum. Meth. Phys. Res.*, **B 64**, 559.

Riviere, J. C. (1990) In *Practical Surface Analysis, Auger and X-Ray Photoelectron Spectroscopy*, D. Briggs, and M. P. Shea (eds), Wiley, Chichester, p. 75.

Robinson, M. T. (1981) In *Sputtering by Particle Bombardment I*, Topics in Applied Physics, Vol. 47, R. Behrisch (ed) Springer, Berlin.

Robinson, M. T., and I. M. Torrens (1974) *Phys. Rev.*, **B 9**, 5008.

Roux, C. D., H. Bu, and J. W. Rabalais (1991) *Surf. Sci.*, **259**, 253.

Roux, C. D., H. Bu, and J. W. Rabalais (1992) *Chem. Phys. Lett.*, **200**, 60.

Rusch, T. W., and R. I. Erickson (1981) In *Inelastic Particle–Surface Collisions*, E. Taglauer and W. Heiland (eds) Springer, Berlin, p. 73.

Sar-El, M. Z. (1967) *Rev. Sci. Instr.*, **38**, 1.

Severijns, C. A., G. Verbist, H. H. Brongersma, and J. T. Devreese (1992) *Nucl. Instrum. Meth. Phys. Res.*, **B 64**, 730.

Shi, M., O. Grizzi, H. Bu, and J. W. Rabalais (1989) *Phys. Rev.*, **B 40**, 10163.

Shi, M., H. Bu, and J. W. Rabalais (1990) *Phys. Rev.*, **B 42**, 2852.

Shoji, F., and T. Hanawa (1984) *Nucl. Instrum. Meth. Phys. Res.*, **B 2**, 401.

Sigmund, P. (1981) In *Sputtering by Particle Bombardment I*, Topics in Applied Physics, Vol. 47, R. Behrish (ed) Springer, Berlin.

Smith, D. P. (1967) *J. Appl. Phys.*, **38**, 340.

Souda, R., and M. Aono (1986) *Nucl. Instrum. Meth. Phys. Res.*, **B 15**, 114.

Souda, R., M. Aono, C. Oshima, S. Otani, and Y. Ishizawa (1986) *Nucl. Instrum. Meth. Phys. Res.*, **B 15**, 138.

Souda, R., T. Aizawa, C. Oshima, and Y. Ishizawa (1990a) *Nucl. Instrum. Meth. Phys. Res.*, **B 45**, 364.

Souda, R., T. Aizawa, C. Oshima, and Y. Ishizawa (1990b) *Nucl. Instrum. Meth. Phys. Res.*, **B 45**, 369.

Taglauer, E. (1991) In *Fundamental Aspects of Heterogeneous Catalysis Studied by Particle Beams*, H. H. Brongersma and R. A. van Santen (eds) NATO ASI B265, Plenum, New York, p. 301.

Taglauer, E., and W. Heiland (1976) Appl. Phys., **9**, 261.

Taglauer, E., M. Beckschulte, R. Margraf, and D. Mehl (1988) *Nucl. Instrum. Meth. Phys. Res.*, **B 35**, 404.

Tsuneyki, S., and M. Tsukuda (1986) *Phys. Rev.*, **B 34**, 5758.

van Leerdam, G. C., and H. H. Brongersma (1991) *Surf. Sci.*, **254**, 153.

van Leerdam, G. C., K.-M. H. Lenssen, and H. H. Brongersma (1990) *Nucl. Instrum. Meth. Phys. Res.*, **B 45**, 390.

Verbist, G., J. T. Devreese, and H. H. Brongersma (1990) *Surf. Sci.*, **233**, 323.

Wang, Y., M. Shi, and J. W. Rabalais (1992) *Nucl. Instrum. Meth. Phys. Res.*, **B 62**, 505.
Wang, Y., M. Shi, and J. W. Rabalais (1993) *Phys. Rev.*, **B48**, 1678 and 1689.
Weigland, P., P. Novacek, G. van Husen, T. Neidhart, L. Z. Mezey, W. Hofer, and P. Varga (1992) *Nucl. Instrum. Meth. Phys. Res.*, **B 64**, 93.
Woodruff, D. P. (1986) *Nucl. Instrum. Meth. Phys. Res.*, **B 15**, 114.
Worthington, M. S., J. L. Stevens, C. S. Chang, and I. S. T. Tsong (1992) *Nucl. Instrum. Meth. Phys. Res.*, **B 64**, 566.
Ziegler, J. P., J. P. Biersack, and U. Littmark (1985) *The Stopping and Range of Ions in Solids, Vol. I.* Pergamon Press, New York.

3 THEORETICAL DESCRIPTION OF CHARGE TRANSFER IN ATOM–SURFACE COLLISIONS

Hongxiao Shao and David C. Langreth
Department of Physics, Rutgers University, Piscataway, New Jersey 08855-1849, USA

and

Peter Nordlander
Department of Physics and Rice Quantum Institute, Rice University, Houston, Texas 77251-1892, USA

CONTENTS

Low Energy Ion–Surface Interactions
Edited by J. W. Rabalais

3.1 INTRODUCTION

In recent years, there has been considerable theoretical and experimental interest in the charge-transfer processes that can occur when an atom or molecule collides with a solid surface (Kimmel *et al.*, 1991; Hammond *et al.*, 1992; Rau *et al.*, 1990; P. D. Johnson *et al.*, 1990; A. L. Johnson *et al.*, 1988; Hsu *et al.*, 1992; Joyce *et al.*, 1992; Los and Geerlings, 1990). Charge-transfer processes are of crucial importance for the understanding of a number of important physical processes such as sticking and desorption. From a theoretical point of view, the charge-transfer process represents an interesting physical problem in which non-adiabatic, non-equilibrium and electron-correlation effects can be strong. The relative simplicity with which charge-transfer processes can be studied experimentally makes this field a very fertile area for the study of these effects.

The conventional theory for charge transfer is based on the solution of the time-dependent Anderson model (Blandin *et al.*, 1976). The atom or molecule is assumed to follow a classical trajectory outside the surface. Since the atom–surface coupling depends on the instantaneous position of the atom, the problem of calculating the final population of the atomic levels is equivalent to integrating the Schrödinger equation for the electrons of the system in time. In developing the correct Hamiltonian for a given system, accurate knowledge of the energy shifts and broadening of the atomic levels as a function of atom–surface separation is required. Several methods for such calculations have been developed and such calculations can now be performed to high accuracy for simple systems. In this review, we will focus on the dynamic aspects of charge transfer, i.e. the solution to the time-dependent Anderson model. For details on the calculation of energy shifts and broadening of atomic resonances outside surfaces we refer the reader to some recent reviews (Nordlander, 1990; Los and Geerlings, 1990; Palmer and Rous, 1992).

To date, most of the theoretical investigations of charge-transfer reactions in atom–surface scattering have been based on the single-level Anderson model (Blandin *et al.*, 1976; Tully, 1977; Nørskov and Lundqvist, 1979; Brako and Newns, 1981). For this model, the final population of the atomic states can be obtained in analytical form. Although theoretically appealing, this model is often too crude for the proper description of realistic systems. Even for the simplest systems, there will always be a degeneracy due to the spin of the electron. In addition, recent calculations of the energy shift and broadening of atomic levels near surfaces have shown that in general several different atomic levels will shift below the Fermi energy in certain regions in the vicinity of a typical metal surface. When several atomic levels are downshifted below the Fermi energy, tunneling may occur into any of these levels. In typical situations, the intra-atomic correlation would prevent more than one of these states being occupied. The resulting charge state will thus be determined by a combination of dynamic and correlation effects. A realistic description of the interaction of atoms with surfaces therefore requires the use of the (multi-level) Anderson model including electron-correlation effects. In the following, we will refer to this model as the Anderson model except where otherwise stated. The Anderson Hamiltonian is quite complex and an exact analytic solution to the Anderson model is not possible even in equilibrium.

Several attempts to use the Anderson model for the description of charge transfer in atom–surface scattering have been made. Bloss and Hone (1978) used a projection operator technique to study the effects of intra-atomic correlation on quasi-resonant charge exchange. Grimley *et al.* (1983) studied charge-transfer by calculating the time-evolution operator within the Hartree–Fock approximation. Yoshimori *et al.* (1984) used the equation of motion method within the Hartree–Fock approximation (HF) to investigate the effects of intra-atomic correlation. Sebastian (1985) went beyond the HF approximation and adapted a time-dependent version of the coupled-cluster method for the description of charge transfer. Using a simplified version of the degeneracy expansion introduced by Gunnarson and Schonhammer (1983), Brako and Newns (1985) investigated the effects of intra-atomic correlation in charge-transfer. Kasai and Okjii (1987) studied charge transfer using an equation-of-motion method. Sulston *et al.* (1988) developed a general many-body theory for charge transfer based on the degeneracy expansion. They showed that by making a so-called local-time approximation, an approximate solution can be obtained. Marston *et al.* (1993) have further developed this method and recently applied it to charge exchange in atom–surface scattering.

The previous treatments of charge transfer using the Anderson model have resulted in rather complicated equations. The final population of an atomic state in an atom–surface scattering experiment is calculated from the solution to coupled integro-differential equations. The solution to such equations requires the use of fast computers. Recently we have developed a general method for solving the time-dependent Anderson model using slave-bosons (Langreth and Nordlander, 1991). This method is both simple and general. The Hamiltonian is solved using a non-equilibrium Green's function technique. The self-energy can be expanded to infinite order using diagrammatic perturbation methods. No approximation except the truncation of the perturbation series for the self-energy are required. In contrast to methods based on the equations of motion, which involve the decoupling of higher-order correlation functions, this is a well-controlled approximation. In particular, it can be shown that in certain physical limits, such as low velocity and finite temperature, it is possible to derive relatively simple rate equations for the calculation of the population of different atomic states. Such rate equations are physically transparent and allow a very direct interpretation of charge-transfer experiments.

In this paper, we will review the basic theoretical formalism of the non-equilibrium Green's function method. In Section 3.2, we present some theoretical details of the method and apply it to the cases where intra-atomic correlation is weak, finite and strong. We also discuss how intra-atomic correlation effects can be systematically treated in quasi-resonant charge-exchange problems. In Section 3.3, we discuss how intra-atomic correlation can qualitatively change the charge transfer in some typical atom–surface scattering situations. In Section 3.4, we present some future research directions, and Section 3.5 gives the conclusions of the paper.

3.2 THEORETICAL DESCRIPTION OF CHARGE-TRANSFER PROCESSES

The dynamic interaction between an atom and a metallic substrate can be described using the Anderson model. This is a general model Hamiltonian describing the interaction between impurity levels and the conduction band of a solid. In the application to surface scattering, the impurity levels refer to the energy levels of the scattered projectile. Since the atoms will move with respect to the substrate, the Hamiltonian will be time-dependent. The time-dependent Anderson Hamiltonian has the form:

$$H(t) = \sum_{\sigma} \varepsilon_\sigma(t) n_\sigma(t) + \tfrac{1}{2} \sum_{\sigma_1 \neq \sigma_2} U_{\sigma_1\sigma_2} n_{\sigma_1}(t) n_{\sigma_2}(t) + \sum_k \varepsilon_k n_k(t) + \sum_{\sigma k} [V_{\sigma k}(t) c_k^\dagger(t) c_\sigma(t) + \text{H.C.}]. \tag{1}$$

where the abbreviation H.C. denotes the Hermitian conjugate.

In this equation, σ labels both the spin and spatial quantum numbers of the atomic orbitals, while k labels the spin and spatial quantum numbers of the solid. The first term, $\varepsilon_\sigma(t)$, is the energy of the atomic or molecular levels, and the ε_k are the energies of the continuum eigenstates of the metal. The intra-atomic correlation term, $U_{\sigma_1\sigma_2}$, describes the Coulomb and exchange interactions associated with multiple occupancies of the impurity states. The hopping matrix element, $V_{\sigma k}(t)$, describes the interaction between the atomic states and the conduction band of the substrate. Within the present level of approximations, these hopping matrix elements will influence the population of the atomic states through the quantity

$$\Gamma_\sigma(t) = 2\pi \sum_k |V_{\sigma k}(t)|^2 \delta(\varepsilon_\sigma(t) - \varepsilon_k). \tag{2}$$

In the wide band limit, this quantity can be identified as the full width of the atomic resonance.

The solution to the Hamiltonian equation (1) is complicated by the presence of the intra-atomic correlation term U. While several methods of solution have been developed in equilibrium (no time-dependence), applications to the non-equilibrium situation have been relatively scarce.

In the following, we will present a general method for the solution to this Hamiltonian. In the first subsection, we briefly describe the basic ideas and definitions in this method. In the second subsection, we apply the formalism to the single level Anderson model. In Section 3.2.3, we discuss the effects of strong electron correlation on charge transfer. In Section 3.2.4, the case of intermediate correlation is investigated. In Section 3.2.5, we apply the method to quasi-resonant charge exchange.

3.2.1 THEORETICAL DESCRIPTION OF NON-EQUILIBRIUM PROCESSES

The theoretical description of non-equilibrium processes can conveniently be based on the Green's function technique developed by Baym and Kadanoff. In the first subsection, we briefly describe the basic ideas and definitions in this method. In the second subsection, we discuss how the semi-classical approximation (SCA) can be applied and used to derive simple rate equations for the charge-transfer processes.

3.2.1.1 Non-equilibrium Green's function formalism

The Green's function technique, introduced by Kadanoff and Baym (1962), has been reviewed and applied to a number of different systems (Langreth, 1976). The method develops exact equations for the double-time Green's function, $G_\sigma(t, t')$, which in our case is equal to

$$iG_\sigma(t, t') = \langle Tc_\sigma(t)c_\sigma^\dagger(t')\rangle, \tag{3}$$

or more accurately, equations for its analytic pieces $G_\sigma^>(t, t')$ and $G_\sigma^<(t, t')$ given by

$$iG_\sigma(t, t') = G_\sigma^>(t, t')\theta(t - t') - G_\sigma^<(t, t')\theta(t' - t). \tag{4}$$

Here the time-ordering operator T and the step functions θ operate along a contour c in the complex plane. It will not matter in the presentation given here whether c is taken to be the Kadanoff–Baym contour, the Keldysh (1965) contour, or a more general choice (Langreth, 1976).

The basic starting equations follow directly from the Dyson equations in time space for G:

$$\left[i\frac{\partial}{\partial t} - \varepsilon_\sigma(t)\right]G_\sigma(t, t') = \delta(t - t') + \int_c d\bar{t}\Sigma_\sigma(t, \bar{t})G_\sigma(\bar{t}, t'), \tag{5}$$

$$\left[-i\frac{\partial}{\partial t'} - \varepsilon_\sigma(t')\right]G_\sigma(t, t') = \delta(t - t') + \int_c d\bar{t}G_\sigma(t, \bar{t})\Sigma_\sigma(\bar{t}, t'), \tag{6}$$

plus an approximation for the self-energy Σ, and hence for its analytic pieces $\Sigma^>$ and $\Sigma^<$ defined analogously to $G^>$ and $G^<$ (i.e. symbolically replace G with Σ in (3)). By using a set of identities (Langreth and Wilkins, 1972; Langreth, 1976; Langreth and Nordlander, 1991) the equations (5) and (6) are readily continued to corresponding equations for the physical real-time correlation functions $G^<(t, t')$ and $G^>(t, t')$, whose equal-time values respectively represent the number of electrons and holes in the impurity level as a function of time. These real-time equations for $G^<$ are

$$\left[i\frac{\partial}{\partial t} - \varepsilon_\sigma(t)\right]G_\sigma^<(t, t') = \int_{-\infty}^{\infty} d\bar{t}[\Sigma_\sigma^R(t, \bar{t})G_\sigma^<(\bar{t}, t') + \Sigma_\sigma^<(t, \bar{t})G_\sigma^A(\bar{t}, t')], \tag{7}$$

$$\left[-i\frac{\partial}{\partial t'} - \varepsilon_\sigma(t')\right]G_\sigma^<(t, t') = \int_{-\infty}^{\infty} d\bar{t}[G_\sigma^R(t, \bar{t})\Sigma_\sigma^<(\bar{t}, t') + G_\sigma^<(t, \bar{t})\Sigma_\sigma^A(\bar{t}, t')], \tag{8}$$

where G_σ^R and G_σ^A are the advanced and the retarded Green's functions,

$$\begin{aligned} iG_\sigma^R(t, t') &= [G_\sigma^>(t, t') + G_\sigma^<(t, t')]\theta(t - t'), \\ -iG_\sigma^A(t, t') &= [G_\sigma^>(t, t') + G_\sigma^<(t, t')]\theta(t' - t), \end{aligned} \tag{9}$$

with analogous equations for Σ_σ^R and Σ_σ^A. In general, there will be two equations for each 'greater than' or 'less than' quantity, as in (5) and (6), and hence (7) and (8); the right side of one can be symbolically obtained from the right side of the other by switching the order of G and Σ and switching the advanced and the retarded labels. To save space, we shall henceforth write only one such equation, and refer to the other as its *conjugate* equation.

The advanced and the retarded Green's functions can be obtained by solving a conventional real-time Dyson equation:

$$\left[\mathrm{i}\frac{\partial}{\partial t} - \varepsilon_\sigma(t)\right] G_\sigma^{\mathrm{A,R}}(t, t') = \delta(t - t') + \int_{-\infty}^{\infty} \mathrm{d}\bar{t}\, \Sigma_\sigma^{\mathrm{A,R}}(t, \bar{t}) G_\sigma^{\mathrm{A,R}}(\bar{t}, t'), \tag{10}$$

along with the boundary condition that $G_\sigma^{\mathrm{A}}(t, t') = 0$ for $t > t'$, and $G_\sigma^{\mathrm{R}}(t, t') = 0$ for $t < t'$.

The population of the impurity level can be obtained directly from the Green's function, using $n_\sigma(t) = G_\sigma^{<}(t, t)$.

3.2.1.2 Semi-classical approximation

In the following discussion, we will be deriving master equations for the description of the charge transfer. The time derivative of the instantaneous population can be obtained directly from the Dyson equations using the relation

$$\frac{\mathrm{d}}{\mathrm{d}t} n_a(t) = \frac{\partial}{\partial t} G^{<}(t, t')\bigg|_{t=t'} + \frac{\partial}{\partial t'} G^{<}(t, t')\bigg|_{t=t'}. \tag{11}$$

It proves to be convenient to factor out the oscillatory part of the Green's functions. We thus write

$$\tilde{G}(t, t') = G(t, t') \exp\left[\mathrm{i}\int_{t'}^{t} \mathrm{d}\tau\, \varepsilon_a(\tau)\right] \tag{12}$$

for the respective Green's function and

$$\tilde{\Sigma}_\sigma^{\gtrless}(t, t') = \Sigma_\sigma^{\gtrless}(t - t') \exp\left[\mathrm{i}\int_{t'}^{t} \mathrm{d}\tau\, \varepsilon_a(\tau)\right] \tag{13}$$

for the self-energies.

Upon these substitutions, the Dyson equations take the form:

$$\mathrm{i}\frac{\partial}{\partial t} \tilde{G}_\sigma^{<}(t, t') = \int_{-\infty}^{t} \mathrm{d}\bar{t}\, \tilde{\Sigma}_\sigma^{\mathrm{R}}(t, \bar{t}) \tilde{G}_\sigma^{<}(\bar{t}, t') + \int_{-\infty}^{t'} \mathrm{d}\bar{t}\, \tilde{\Sigma}_\sigma^{<}(t, \bar{t}) \tilde{G}_\sigma^{\mathrm{A}}(\bar{t}, t') \tag{14}$$

and

$$\mathrm{i}\frac{\partial}{\partial t} \tilde{G}_\sigma^{\mathrm{A,R}}(t, t') = \delta(t - t') + \int_{-\infty}^{\infty} \mathrm{d}\bar{t}\, \tilde{\Sigma}_\sigma^{\mathrm{A,R}}(t, \bar{t}) \tilde{G}_\sigma^{\mathrm{A,R}}(\bar{t}, t'). \tag{15}$$

Using equations (11) and (14) the following rate equation is obtained for the population n_σ:

$$\frac{\mathrm{d}}{\mathrm{d}t} n_\sigma(t) = -2\,\mathrm{Im}\left[\int_{-\infty}^{t} \mathrm{d}\bar{t}\, \tilde{\Sigma}_\sigma^{\mathrm{R}}(t, \bar{t}) \tilde{G}_\sigma^{<}(t, \bar{t})\right] + 2\,\mathrm{Im}\left[\int_{-\infty}^{t} \mathrm{d}\bar{t}\, \tilde{\Sigma}_\sigma^{<}(t, \bar{t}) \tilde{G}_\sigma^{\mathrm{A}}(\bar{t}, t)\right]. \tag{16}$$

As will be demonstrated below, the self-energies $\tilde{\Sigma}_\sigma(t, \bar{t})$ are strongly peaked around $t = \bar{t}$. The Green's functions, $\tilde{G}_\sigma^{<}(t, \bar{t})$, vary more slowly with time. The semi-classical approximation (SCA) amounts to the assumption that the value of the Green's functions will not change significantly in the interval where the self-energies are finite. Applying the SCA to equation (14) we obtain

$$\mathrm{i}\frac{\partial}{\partial t}\tilde{G}_\sigma^<(t,t') = \tilde{G}_\sigma^<(t,t')\int_{-\infty}^{t} \mathrm{d}\bar{t}\tilde{\Sigma}_\sigma^{\mathrm{R}}(t,\bar{t}) + \int_{-\infty}^{t'} \mathrm{d}\bar{t}\tilde{\Sigma}_\sigma^<(t,\bar{t})\tilde{G}_\sigma^{\mathrm{A}}(\bar{t},t'). \tag{17}$$

The rate equation for the population of the impurity level thus takes the form:

$$\frac{\mathrm{d}}{\mathrm{d}t}n_\sigma(t) = -2n_\sigma(t)\,\mathrm{Im}\left[\int_{-\infty}^{t} \mathrm{d}\bar{t}\tilde{\Sigma}_\sigma^{\mathrm{R}}(t,\bar{t})\right] + 2\,\mathrm{Re}\left[\int_{-\infty}^{t} \mathrm{d}\bar{t}\tilde{\Sigma}_\sigma^<(t,\bar{t})\tilde{G}_\sigma^{\mathrm{A}}(\bar{t},t)\right]. \tag{18}$$

The application of the SCA results in simple rate equations for the description of the charge transfer. These rate equations will be referred to as generalized master equations (GME). The resulting integrals in the GME can be evaluated provided simple forms for the retarded and advanced Green's functions can be obtained. The accuracy of the GME therefore depends both on the accuracy of the SCA and on the accuracy of the advanced and retarded Green's function used in equation (18). In the following, we will investigate the accuracy of such rate equations by comparing the solutions from the GME and from an exact scheme for the Dyson equations. It will be shown that a particularly simple form of the GME can be obtained if additional approximations, in the spirit of the SCA, are made when calculating the advanced and retarded Green's functions. These rate equations will be referred to as simple master equations (SME).

The validity of the SCA in charge-transfer problems has been investigated in detail previously (Langreth and Nordlander, 1991). The SCA applies if one of the following criteria is satisfied:

$$\sum_\sigma \Gamma_\sigma(t) \ll 2\pi k_{\mathrm{B}}T,$$

$$\alpha v \ll 2\pi k_{\mathrm{B}}T, \tag{19}$$

or

$$\left|\frac{\alpha v - \sum_\sigma \Gamma_\sigma(t)}{\varepsilon_a(t) - \varepsilon_{\mathrm{F}}}\right| \ll 1, \tag{20}$$

where v is the perpendicular velocity of the atom and $\alpha = \mathrm{d}\Gamma(Z)/\mathrm{d}Z$.

Obviously these criteria do not apply during the time when the atomic energy levels cross the Fermi energy. However, this situation is often relevant for experiments. For the SCA to apply in the low-temperature limit, we must therefore require that only very small charge transfer occurs during the time of crossing of an atomic energy level with the Fermi energy. The crossing time,

Δt_c, can be directly obtained from equation (20), $\Delta t_c = \sum_\sigma (\Gamma_\sigma / bv)$, where

$$b = \frac{d}{dz}\,\varepsilon_a(z)\bigg|_{z=z_c}. \tag{21}$$

The charge transfer during this time is proportional to $\Gamma_\sigma(t_c)$. The condition of small charge transfer during the crossing translates into the following requirement on the velocity:

$$v > \Gamma^2(Z_c)\left[\frac{d}{dz}\,\varepsilon_a(Z_c)\right]^{-1}. \tag{22}$$

The SCA provides significantly simpler rate equations for the occupations of the impurity levels. In the following we will investigate the validity of the SCA by direct comparison with exact numerical solutions to the Dyson equations for the impurity Green's functions in different scattering situations.

3.2.2 CASE OF $U = 0$

The single-level Anderson model has been extensively used for the description of charge-transfer processes in the atom–surface scattering. Due to its simplicity, it can be solved exactly. This was first done by Blandin *et al.* (1976). For $U = 0$, the Anderson Hamiltonian takes the form:

$$H = \varepsilon_a(t)n_a(t) + \sum_k \varepsilon_k n_k(t) + \sum_k \left[V_{ak}(t)c_k^+(t)c_a(t) + \text{H.C.}\right] \tag{23}$$

In this expression, a refers to the atomic level and k represents the continuum states of the conduction band of the metal. V_{ak} is the coupling between the atom and the metal.

3.2.2.1 General equations

For the spinless Anderson Hamiltonian, an exact expression can be obtained for the self-energies:

$$\Sigma_a(t, t') = \sum_{k,k'} V_{ak}(t)g_{kk'}^{(0)}(t, t')V_{ak}(t'), \tag{24}$$

where $g^{(0)}$ is the *unperturbed* Green's function for the substrate electrons. For simplicity it is assumed that the hopping matrix element is separable, i.e. $V_{ak}(t) = V_k u(t)$ and that the band can be described by a constant density of states. With these assumptions, the density of states $\rho(\varepsilon)$ times the square of the potential V_k will be a constant function of energy ε corresponding to the quantum number k, i.e. $V_k = V$. With these simplifications we can write for the 'greater than' or 'less than' components of the self-energy

$$\Sigma_a^{\gtrless}(t, t') = u(t)u(t')\rho V^2 \int_{-D}^{D} d\varepsilon f^{\gtrless}(\varepsilon) e^{-i\varepsilon(t-t')}, \tag{25}$$

where $f^{<}(\varepsilon)$ is the Fermi function, $f^{>}(\varepsilon) = 1 - f^{<}(\varepsilon)$, and D is the bandwidth of the conduction band of the surface. In this subsection, we will assume an infinite bandwidth, except where a finite bandwidth is needed for convergence.

The integration over energy can be performed straightforwardly and the self-energies take the form

$$\Sigma_a^{\gtrless}(t, t') = \sqrt{\Gamma_a(t)\Gamma_a(t')} f^{\gtrless}(t - t'), \tag{26}$$

where

$$\Gamma_a(t) = 2\pi u^2(t) V^2 \rho, \tag{27}$$

and where the functions $f^{\gtrless}(\tau)$ are defined as:

$$f^{\gtrless}(\tau) = \int_{-D}^{D} \frac{d\varepsilon}{2\pi} f^{\gtrless}(\varepsilon) e^{-i\varepsilon\tau}. \tag{28}$$

In the wide-band limit, these expressions can be readily evaluated to yield

$$f^{<}(\tau) = [f^{>}(\tau)]^* = \tfrac{1}{2}\delta(\tau) + \frac{i e^{-i\varepsilon_F \tau}}{2\beta \sinh(\pi\tau/\beta)}, \tag{29}$$

where ε_F is the Fermi energy. For finite bandwidth, the integration can be done numerically.

It is convenient to introduce the functions

$$\tilde{f}_a^{\gtrless}(t, t') = f^{\gtrless}(t - t') \exp\left[i \int_{t'}^{t} d\tau\, \varepsilon_a(\tau) \right]. \tag{30}$$

The retarded and advanced self-energies can readily be obtained from equation (26):

$$\Sigma_a^{R}(t, t') = [\Sigma_a^{A}(t', t)]^* = -i\Gamma_a(t)\delta(t - t')\theta(t - t') = -\tfrac{1}{2} i\Gamma_a(t)\delta(t - t'). \tag{31}$$

The Dyson equation for the $U = 0$ case takes the form

$$i \frac{\partial}{\partial t} \tilde{G}_a^{<}(t, t') = -i \frac{\Gamma_a(t)}{2} \tilde{G}_a^{<}(t, t') + \int_{-\infty}^{\infty} d\bar{t} \sqrt{\Gamma_a(t)\Gamma_a(\bar{t})} \tilde{f}_a^{<}(t, \bar{t}) \tilde{G}_a^{A}(\bar{t}, t'). \tag{32}$$

The Dyson equation for the advanced Green's function is given by:

$$i \frac{\partial}{\partial t} \tilde{G}_a^{A}(t, t') = \delta(t - t') + i \frac{\Gamma_a(t)}{2} \tilde{G}_a^{A}(t, t'). \tag{33}$$

The occupation of level $|a\rangle$ can be calculated from the expression $n_a(t) = G_a^{<}(t, t)$. We discuss the solution to these equations in the next section.

3.2.2.2 Exact solution

The Dyson equation for the advanced Green's function, equation (33), can be solved directly, yielding

$$\tilde{G}_a^{\mathrm{A}}(t, t') = \mathrm{i}\theta(t' - t) \exp \int_{t'}^{t} \mathrm{d}\tau \, \frac{\Gamma_a(\tau)}{2}. \tag{34}$$

This expression can now be substituted into equation (32), resulting in the following form for the Dyson equation:

$$\begin{aligned} \mathrm{i}\frac{\partial}{\partial t} \tilde{G}_a^{<}(t, t') &= -\mathrm{i}\, \frac{\Gamma_a(t)}{2}\, \tilde{G}_a^{<}(t, t') \\ &\quad + \mathrm{i} \int_{-\infty}^{t} \mathrm{d}\bar{t} \sqrt{\Gamma_a(t)\Gamma_a(\bar{t})}\, \tilde{f}_a^{<}(t, \bar{t}) \exp\left[- \int_{\bar{t}}^{t} \mathrm{d}\tau \, \frac{\Gamma_a(\tau)}{2} \right]. \end{aligned} \tag{35}$$

The occupation of level $|a\rangle$, $n_a(t) = G^{<}(t, t)$, can be calculated from this expression or from the rate equation (16) which in this case is given by:

$$\frac{\mathrm{d}}{\mathrm{d}t} n_a(t) = -\Gamma_a(t) n_a(t) + 2 \,\mathrm{Re} \int_{-\infty}^{t} \mathrm{d}\bar{t} \sqrt{\Gamma_a(t)\Gamma_a(\bar{t})}\, \tilde{f}_a^{<}(t, \bar{t}) \exp\left[- \int_{\bar{t}}^{t} \mathrm{d}\tau \, \frac{\Gamma_a(\tau)}{2} \right]. \tag{36}$$

A formal solution can therefore be derived. We discuss the exact solution to these equations for finite bandwidth in the next section.

3.2.2.3 Master equation for charge transfer

As demonstrated in Section 3.2.2.1, the retarded self-energy for the $U = 0$ Anderson model in the wide-band limit is proportional to a delta function, equation (31), i.e. the SCA is exact. It is possible, however, to perform further approximations resulting in even simpler rate equations for the charge transfer. These approximations are within the spirit of the SCA.

The function $\tilde{f}_a^{<}(t, \bar{t})$ in equation (36) is strongly localized around equal time. The width varies relatively slowly and the integral can be approximated

$$\int_{-\infty}^{t} \mathrm{d}\bar{t} \sqrt{\Gamma_a(t)\Gamma_a(\bar{t})}\, \tilde{f}_a^{<}(t, \bar{t}) \exp\left[- \int_{\bar{t}}^{t} \mathrm{d}\tau \, \frac{\Gamma_a(\tau)}{2} \right] = \Gamma_a(t) \int_{-\infty}^{t} \mathrm{d}\bar{t} \tilde{f}_a^{<}(t, \bar{t}). \tag{37}$$

The rate equation thus takes the form

$$\frac{\mathrm{d}}{\mathrm{d}t} n_a(t) = -\Gamma_a(t) n_a(t) + 2\Gamma_a(t) \int_{-\infty}^{t} \mathrm{d}\bar{t} \tilde{f}_a^{<}(t, \bar{t}). \tag{38}$$

For a rapidly varying atomic level, the integral of $\tilde{f}$ can be approximated (Langreth and Nordlander, 1991)

$$\int_{-\infty}^{t} \mathrm{d}\bar{t} \tilde{f}_a^{\gtrless}(t, \bar{t}) \approx \tfrac{1}{2} f^{\gtrless}(\varepsilon_a(t)). \tag{39}$$

The rate equation thus takes the form

$$\frac{\mathrm{d}}{\mathrm{d}t} n_a(t) = -\Gamma_a(t)n_a(t) + f^<(\varepsilon_a(t))\Gamma_a(t). \tag{40}$$

This equation will be referred to as the SME for the $U = 0$ case. The equation can easily be derived by intuitive golden-rule considerations and is trivial to solve if the energy shifts and widths of the atomic levels are known.

In order to test the accuracy of the SME, equation (40), we have to specify the motion of the atom and its energy levels outside the surface. In the following, we will assume linear motion of the atom. The atom is assumed to start at a distance $Z_1 = 20$ a.u., and to move with constant speed v to the turning point at $Z_2 = 5$ a.u. The atom then reverses direction and moves toward the vacuum, again with constant speed v. At $Z_3 = 20$ a.u. the trajectory is completed. The energy levels of the atom are assumed to shift linearly with distance so that

$$\varepsilon_a(t) = \varepsilon_\mathrm{F} + b[Z(t) - Z_\mathrm{c}]. \tag{41}$$

The atomic level thus crosses the Fermi energy at a distance $Z = Z_\mathrm{c}$. The width of the energy level is assumed to vary exponentially, i.e.

$$\Gamma_a(t) = \Delta_a \exp[-\alpha Z(t)], \tag{42}$$

where α and Δ_a are constants that can be calculated. These assumptions qualitatively describe how the atomic affinity levels would shift for an incoming (outgoing) atom outside a perfect metal surface or how the neutral energy levels of a nearby atom would shift outside an alkali-promoted metal surface (Johnson

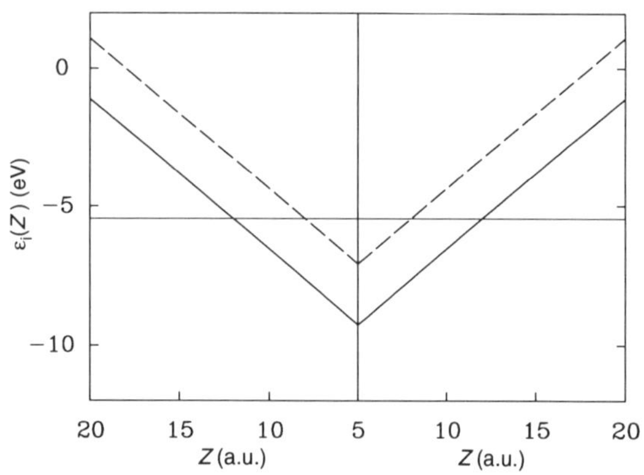

Figure 3.1 Schematic plot of the variation of the energy levels $\varepsilon_1(Z)$ (solid line) and $\varepsilon_2(Z)$ (dashed line) with distance Z. The left part of the figure represents the incoming trajectory and the right part of the figure represents the outgoing trajectory. The turning point at $Z = 5$ is indicated by a vertical line. The Fermi energy of the metal (-0.2 a.u.) is indicated by the horizontal solid line

et al., 1990). In the following, we will consider two atomic levels $|1\rangle$ and $|2\rangle$ with parameters $b_1 = b_2 = 0.02$ a.u., $Z_c^1 = 12$ a.u., $Z_c^2 = 8$ a.u., $\alpha_1 = 1.0$ a.u. and $\alpha_2 = 0.5$ a.u. These parameters are characteristic of the level-shifts and broadening of hydrogen-like atoms outside metal surfaces. In Figure 3.1, we schematically plot the energy shifts of these states as a function of the atomic position outside the surface.

In Figure 3.2, we compare the instantaneous populations $n_1(Z)$ and $n_2(Z)$

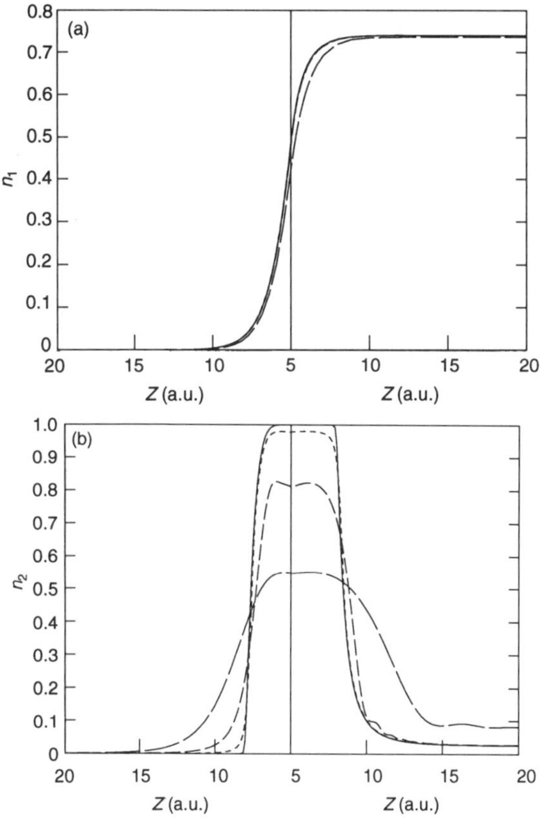

Figure 3.2 Investigation of the validity of SME for different velocities for the $U = 0$ system. The motion of the atom is linear and the atomic trajectory is assumed to start towards the surface at $Z = 20$ a.u., to turn toward vacuum at $Z = 5$ a.u., and then to propagate back to $Z = 20$ a.u. Part (a) shows the occupation $n_1(Z)$ as a function of distance from the surface. In part (b), the population $n_2(Z)$ is shown. The parameters of the system are $Z_1^c = 12$ Å, $Z_2^c = 8$ a.u., $b_1 = b_2 = 0.02$, $\alpha_1 = 1.0$ and $\alpha_2 = 0.5$. The temperature is $T = 300$ K and the Fermi energy is 0.2 a.u. The width of the impurity level has been scaled for each velocity so that similar charge-transfer occurs within each trajectory. The solid curves are the results obtained using the SME, equation (40). The dotted curves are for $v = 0.0010$, $\Delta = 0.1$. The dashed curves are for $v = 0.0100$, $\Delta = 1.0$, and the long-dashed lines are for $v = 0.1$, $\Delta = 10.0$

calculated using the exact equation (36) and the SME, equation (40), along the atomic trajectory for different velocities. To facilitate the comparison, the tunneling-matrix elements have been scaled with velocity so that similar charge transfer will occur for each trajectory. It can clearly be seen that the dynamics of level $|1\rangle$ is well described by the SME for all velocities considered. The population of level $|2\rangle$ is well described by the SME for the lower velocities but disagrees slightly for the larger velocities. This disagreement is due to the fact that the crossing of level $|2\rangle$ occurs relatively close to the surface, where the width is large. The criteria, equations (19), (20), (22), are therefore more difficult to satisfy.

3.2.3 CASE OF LARGE U

We now turn to the case of multiple levels interacting with a continuum. Because of spin degeneracy, even the simplest physically realistic examples fall into this category. The controlling difference between this case and the case considered in the previous subsection is that the intra-atomic Coulomb interaction between the electrons of the atomic species outside the surface is normally large and cannot be neglected. Thus, instead of the exactly soluble resonant-level model, one must deal with the time-dependent Anderson model.

In the large U limit, the effect of the correlation term is to restrict the occupation of the impurity levels to a maximum of one. In this context, large U means that $\varepsilon_\sigma + U \gg \varepsilon_F + kT$ and $\varepsilon_\sigma + U \gg \varepsilon_F + \Gamma$. These conditions are typically satisfied for excited neutral, ionization and affinity levels of atoms interacting with metals.

3.2.3.1 General equations

The large U limit can be effected with the use of projection operators. This procedure has the disadvantage that a conventional Green's function approach is complicated since these operators do not satisfy standard commutation relations. An alternative method for the equilibrium case (no time-dependence) was suggested by Coleman (1984) and has since been widely used. Here we adapt it to the time-dependent problem. In this method, a boson operator is introduced in the term describing the hopping between the adsorbate level and the continuum. The slave-boson Hamiltonian takes the form:

$$H(t) = \sum_\sigma \varepsilon_\sigma(t) n_\sigma(t) + \sum_k \varepsilon_k n_k(t) + \sum_{\sigma k} [V_{\sigma k}(t) c_k^\dagger(t) c_\sigma(t) b^\dagger(t) + \text{H.C.}]. \tag{43}$$

When an electron tunnels into the metal, it creates a boson, and vice versa. The sum of the impurity occupation numbers and the slave-boson occupation number, $Q_B(t) = \sum_\sigma n_\sigma(t) + n_B(t)$, is conserved, where we have let $n_B(t) = b^\dagger(t) b(t)$.

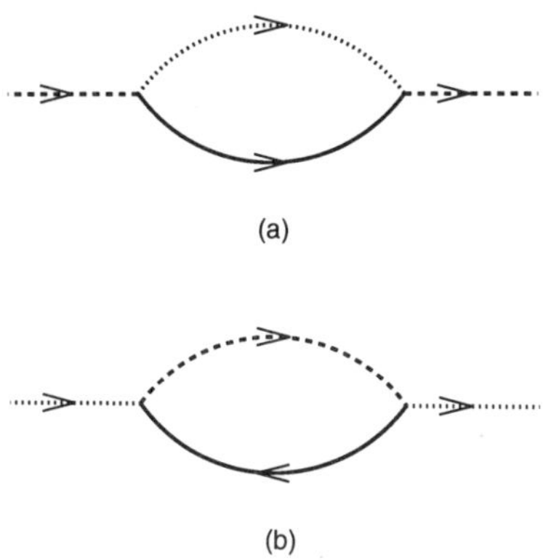

Figure 3.3 Lowest-order self-energy diagrams (a) the atomic-level propagator and (b) the slave-boson propagator. The solid line is the full propagator of the atomic level, the light dashed line is the conduction-electron propagator, and the heavy dashed line is the fully interacting propagator for the slave-boson

Therefore the diagonalization of the Hamiltonian can be performed in the subspace of the Hilbert space where $Q_{\mathrm{B}} = 1$. The intra-atomic correlation term can thus be dropped. Since the boson operator is a conventional field operator, a standard perturbative expansion can be performed and a self-energy and a set of Dyson equations can be defined. The Hamiltonian (43) is very rich and includes among other phenomena the Kondo problem and the mixed valence problem as special cases. A simple exact solution as in the single-level case can therefore not be hoped for.

The self-energies for the impurity propagators and the slave-boson propagator have a straightforward perturbation expansion in powers of the hopping-matrix element, as illustrated diagrammatically in Figure 3.3. To lowest order in the hopping-matrix element,

$$\Sigma_\sigma^{\gtrless}(t, t') = \sqrt{\Gamma_\sigma(t)\Gamma_\sigma(t')}\, f^{\gtrless}(t - t')B^{\gtrless}(t, t'), \tag{44}$$

$$\Pi_\sigma^{\gtrless}(t, t') = \sqrt{\Gamma_\sigma(t)\Gamma_\sigma(t')}\, f^{\gtrless}(t' - t)G_\sigma^{\lessgtr}(t, t'), \tag{45}$$

where $B^{\gtrless}$ are the analytic pieces of the slave-boson propagator:

$$\mathrm{i}B(t, t') = \langle Tb(t)b^{\dagger}(t')\rangle = B^{>}(t, t')\theta(t - t') + B^{<}(t, t')\theta(t - t'). \tag{46}$$

Equation (44) is thus analogous to equation (26) for the single-level case. We note that B is in principle the full self-consistent slave-boson propagator. Coleman (1984) has shown that for the case of N degenerate levels in thermal equilibrium, when equations (44), (45) and (46) are inserted into the appropriate Dyson equations, one obtains a result accurate to order $(1/N)^2$. Of course, the results are also exact to lowest order in V. The retarded and advanced Green's functions for the slave-boson are defined in the following manner:

$$B^{\mathrm{R}}(t,t') = -\mathrm{i}[B^{>}(t,t') - B^{<}(t,t')]\theta(t-t'),$$
$$B^{\mathrm{A}}(t,t') = \mathrm{i}[B^{>}(t,t') - B^{<}(t,t')]\theta(t'-t). \tag{47}$$

The sign differences between these and (9) arise from the difference in boson and fermion commutation relations.

In applying the slave-boson method, one must project onto the subspace in which $Q_{\mathrm{B}} = 1$. This is normally accomplished in the time-independent case by introducing a chemical potential associated with Q_{B} and then letting it approach $-\infty$ to achieve the low density limit in this variable. Here the same effect is accomplished by identifying the Q_{B} dependence of the various quantities, and then omitting terms that are proportional to Q_{B} to higher than the lowest order. We do this by noting that $G^{<}$ and $B^{<}$ are proportional to $(Q_{\mathrm{B}})^1$, while G^{A}, G^{R}, B^{A} and B^{R} are proportional to $(Q_{\mathrm{B}})^0$. Since the Dyson equations for $G^{<}$ will involve either $G^{<}$ or $B^{<}$ in every term, the self-energies that multiply these quantities must have all terms proportional to $(Q_{\mathrm{B}})^1$ or higher projected out. We emphasize that the omission of these terms is *not* an additional approximation, but is rather a projection procedure that is *required* by slave-boson formulation of the problem. As a result we obtain from the diagrams of Figure 3.2, the advanced and retarded self-energies

$$\Sigma_{\sigma}^{\mathrm{A,R}}(t,t') = \sqrt{\Gamma_{\sigma}(t)\Gamma_{\sigma}(t')}\, f^{>}(t-t')B^{\mathrm{A,R}}(t,t') \tag{48}$$

for the electron propagator, and

$$\Pi^{\mathrm{A,R}}(t,t') = \sum_{\sigma} \sqrt{\Gamma_{\sigma}(t)\Gamma_{\sigma}(t')}\, f^{<}(t'-t)G_{\sigma}^{\mathrm{A,R}}(t,t') \tag{49}$$

for the boson propagator. The Dyson equation for the Green's function $G^{<}$ has the form:

$$\mathrm{i}\frac{\partial}{\partial t}\tilde{G}_{\sigma}^{<}(t,t') = \int_{-\infty}^{\infty} \mathrm{d}\bar{t}\sqrt{\Gamma_{\sigma}(t)\Gamma_{\sigma}(\bar{t})}\,\tilde{f}_{\sigma}^{>}(t,\bar{t})B^{\mathrm{R}}(t,\bar{t})\tilde{G}_{\sigma}^{<}(\bar{t},t')$$
$$+ \int_{-\infty}^{\infty} \mathrm{d}\bar{t}\sqrt{\Gamma_{\sigma}(t)\Gamma_{\sigma}(\bar{t})}\,\tilde{f}_{\sigma}^{<}(t,\bar{t})B^{<}(t,\bar{t})\tilde{G}_{\sigma}^{\mathrm{A}}(\bar{t},t'). \tag{50}$$

Similarly, the Dyson equation for the boson propagator is

$$\mathrm{i}\frac{\partial}{\partial t}B^{<}(t,t') = \sum_{\sigma} \int_{-\infty}^{\infty} \mathrm{d}\bar{t}\sqrt{\Gamma_{\sigma}(t)\Gamma_{\sigma}(\bar{t})}\,\tilde{f}^{<}(\bar{t},t)\tilde{G}_{\sigma}^{\mathrm{R}}(t,\bar{t})B^{<}(\bar{t},t')$$
$$+ \sum_{\sigma} \int_{-\infty}^{\infty} \mathrm{d}\bar{t}\sqrt{\Gamma_{\sigma}(t)\Gamma_{\sigma}(\bar{t})}\,\tilde{f}^{>}(\bar{t},t)\tilde{G}_{\sigma}^{<}(t,\bar{t})B^{\mathrm{A}}(\bar{t},t'). \tag{51}$$

The Dyson equations for the retarded Green functions have the form:

$$i\frac{\partial}{\partial t}\tilde{G}^{R}_{\sigma}(t,t') = \delta(t-t') + \int_{-\infty}^{\infty} d\bar{t}\sqrt{\Gamma_{\sigma}(t)\Gamma_{\sigma}(\bar{t})}\tilde{f}^{>}_{\sigma}(t,\bar{t})B^{R}(t,\bar{t})\tilde{G}^{R}_{\sigma}(\bar{t},t'), \quad (52)$$

$$i\frac{\partial}{\partial t}B^{R}(t,t') = \delta(t-t') + \sum_{\sigma}\int_{-\infty}^{\infty} d\bar{t}\sqrt{\Gamma_{\sigma}(t)\Gamma_{\sigma}(\bar{t})}\tilde{f}^{<}_{\sigma}(\bar{t},t)\tilde{G}^{R}_{\sigma}(t,\bar{t})B^{R}(\bar{t},t'). \quad (53)$$

From the solution of these equations one can calculate $n_\sigma(t) = G^{<}_{\sigma}(t,t)$ as a function of t. The boson Hamiltonian ensures that Q_B is conserved, while the projection procedure ensures that it is unity. The unphysical states with $\sum_\sigma n_\sigma > 1$ can never become occupied if they were unoccupied at $t = -\infty$, so that $n_\sigma(t)$ is the time-dependent occupation number of the level $|\sigma\rangle$ just as in the single-level case. One can ensure that this is true simply by taking $\varepsilon_\sigma(-\infty) \to \infty$.

We close this section by reminding the reader of several facts about the interpretation of the various correlation functions above. The first is that $c^{\dagger}_{\sigma}c_{\sigma}$ is the correct electron-number operator in the $Q_B = 1$ subspace, so that $G^{<}(t,t)$ as given by the above equations is the true time-dependent occupation number. However, the operator c_σ is not the destruction operator for the true atomic state in this scheme. The operator does not commute with Q_B and will therefore overlap the $Q_B \neq 1$ subspace. This means that one cannot Fourier-transform the functions $G^{<}(t,t')$ to obtain a spectral decomposition of the real occupied atomic states, nor is a similar Fourier transform of $G^{R}(t,t')$ simply related to density of states. They are, rather, the functions related to what Coleman (1984) calls the 'pseudo' spectral density, corresponding to the destruction operator c_σ for a 'pseudo' atomic state. The true or 'real' spectral densities correspond to correlation functions where c_σ is replaced by $c_\sigma b^{\dagger}$, thus defining an effective destruction operator that does not have matrix elements with the unphysical subspace, $Q_B \neq 1$ subspace. Although it is trivial to verify that the two forms for the number of operator $(c_\sigma b^{\dagger})^{\dagger}(c_\sigma b^{\dagger})$ and $c^{\dagger}_{\sigma}c_{\sigma}$ are identical in this subspace (with corrections of order Q_B^2 which would be eliminated by the projection procedure), the spectral weight functions now involve Fourier transforms of four-operator functions. In the limit of a large number (N) of degenerate (or nearly degenerate) atomic orbitals σ, approximate forms, in similar fashion as discussed by Coleman (1984), for these functions can be derived. Specifically $G^{<}$ may be written to the accuracy of $(1/N)^2$:

$$G^{<}_{\text{real}}(t,t') \approx G^{<}_{\text{pseudo}}(t,t')b(t',t) \quad (54)$$

where $b(t,t') = B^{>}(t,t') - B^{<}(t',t)$ and $G^{<}_{\text{pseudo}}(t,t') = G^{<}(t,t')$ is defined in equation (56). The Fourier transformation of $G^{<}_{\text{real}}(t,t')$ with respect to $t - t'$ gives the so-called 'occupied spectral weight function' which is defined as the product of the spectral weight function and the Fermi function with the frequency measured from the Fermi energy. For the charge-transfer problem considered here, however, these 'real' spectral weight functions are not needed.

3.2.3.2 Exact solution

The physical quantities one tries to extract out of this theory are the instantaneous populations of the atomic states during the scattering. The retarded Green's functions are necessary input to the solution to the Dyson equations for the 'less than' Green's functions. The solution to the Dyson equations for the retarded Green's functions will be discussed first. The solution for the 'less than' Green's functions will then be presented.

3.2.3.2.1 Retarded Green's function

Before any attempt can be made to solve equations (52) and (53), it is convenient to simplify these two equations by introducing g_σ and b in the following form:

$$\tilde{G}^{R}_{\sigma}(t, t') = \frac{\theta(t - t')}{\mathrm{i}}\, g_\sigma(t, t') \tag{55}$$

and

$$B^{R}(t, t') = \frac{\theta(t - t')}{\mathrm{i}}\, b(t, t'), \tag{56}$$

where $g_\sigma(t, t') = \tilde{G}^{>}_{\sigma}(t, t') + \tilde{G}^{<}_{\sigma}(t, t')$ and $b(t, t') = B^{>}(t, t') - B^{<}(t, t')$.

With these definitions, equation (52) can be written as

$$\frac{\partial}{\partial t}\, g_\sigma(t, t') = -\int_{t'}^{t} \mathrm{d}\bar{t}\sqrt{\Gamma_\sigma(t)\Gamma_\sigma(\bar{t})}\,\tilde{f}^{>}_{\sigma}(t, \bar{t})b(t, \bar{t})g_\sigma(\bar{t}, t'), \tag{57}$$

and equation (53) can be stated accordingly as

$$\frac{\partial}{\partial t}\, b(t, t') = -\sum_{\sigma}\int_{t'}^{t} \mathrm{d}\bar{t}\sqrt{\Gamma_\sigma(t)\Gamma_\sigma(\bar{t})}\,\tilde{f}^{<}_{\sigma}(\bar{t}, t)g_\sigma(t, \bar{t})b(\bar{t}, t'). \tag{58}$$

The δ-function in equations (52) and (53) is canceled out by the derivative of the step function. Equations (57) and (58) contain less information than (52) and (53) and must be supplemented by the boundary conditions

$$\begin{aligned} g_\sigma(t, t) &= 1, \\ b(t, t) &= 1, \end{aligned} \tag{59}$$

which follow from the equal-time commutation relations for the corresponding operators. Thus $g_\sigma(t, t')$ and $b(t, t')$ are completely determined. The complex conjugate of the retarded Green's function gives the advanced Green's function.

From equations (57) and (58), one sees that the derivative of $g_\sigma(t, t')$ depends on $b(t, \bar{t})$ and $g_\sigma(\bar{t}, t')$ while the derivative of $b(t, t')$ depends on $g_\sigma(t, \bar{t})$ and $b(\bar{t}, t')$, with $t' < \bar{t} < t$. In other words, if the Green's functions were discretized and represented by a square matrix with t and t' as its rows and columns, the following fact becomes apparent: the derivatives of g_σ and b at one particular point in the matrix are completely determined by the values of the g_σ and b at the points closer to the diagonal $t = t'$. From the equal-time commutation relations, the values of the g_σ and b along the diagonal are known to be unity. Therefore, one could discretize equations (57) and (58), and solve for g_σ and b row-by-row going away from the diagonal until the lower right corner of the matrix is reached.

Equations (57) and (58) are now ready to be discretized. Using a second-order implicit scheme, one has

$$[1 + \tfrac{1}{4}\delta^2\Gamma_\sigma(m)f_\sigma^>(0)]g_\sigma(m, n) + \tfrac{1}{4}\delta^2\sqrt{\Gamma_\sigma(m)\Gamma_\sigma(n)}\tilde{f}_\sigma^>(m, n)b(m, n)$$
$$= g_\sigma(m-1, n) - G_\sigma(m, n) \tag{60}$$

and

$$\sum_\sigma \tfrac{1}{4}\delta^2\sqrt{\Gamma_\sigma(m)\Gamma_\sigma(n)}\tilde{f}_\sigma^<(n, m)g_\sigma(m, n) + \left[1 + \sum_\sigma \tfrac{1}{4}\delta^2\Gamma_\sigma(m)f_\sigma^<(0)\right]b(m, n)$$
$$= b(m-1, n) - \sum_\sigma B_\sigma(m, n), \tag{61}$$

where $G_\sigma(m, n)$ is defined as:

$$G_\sigma(m, n) = \tfrac{1}{4}\delta^2\Gamma_\sigma(m-1)f_\sigma^>(0)g_\sigma(m-1, n)$$
$$+ \tfrac{1}{4}\delta^2\sqrt{\Gamma_\sigma(m-1)\Gamma_\sigma(n)}\tilde{f}_\sigma^>(m-1, n)b(m-1, n)$$
$$+ \tfrac{1}{2}\delta^2 \sum_{i=m-1}^{m} \sqrt{\Gamma_\sigma(i)} \sum_{j=n+1}^{i-1} \sqrt{\Gamma_\sigma(j)}\tilde{f}_\sigma^>(i, j)b(i, j)g_\sigma(j, n), \tag{62}$$

and $B_\sigma(m, n)$ as:

$$B_\sigma(m, n) = \tfrac{1}{4}\delta^2\Gamma_\sigma(m-1)f_\sigma^<(0)b(m-1, n)$$
$$+ \tfrac{1}{4}\delta^2\sqrt{\Gamma_\sigma(m-1)\Gamma_\sigma(n)}\tilde{f}_\sigma^<(n, m-1)g_\sigma(m-1, n)$$
$$+ \tfrac{1}{2}\delta^2 \sum_{i=m-1}^{m} \sqrt{\Gamma(i)} \sum_{j=n+1}^{i-1} \sqrt{\Gamma(j)}\tilde{f}_\sigma^<(j, i)g_\sigma(i, j)b(j, n), \tag{63}$$

where δ is the step length, $t = m\delta$ and $t' = n\delta$.

From expressions (62) and (63), $G_\sigma(m, n)$ and $B_\sigma(m, n)$ depend on the values of $g_\sigma(i, j)$ and $b(i, j)$ with $i - j < m - n$. Therefore by solving equations (60) and (61) simultaneously one can readily obtain the values of g_σ and b at point (m, n) from g_σ and b at grid points closer to the diagonal than (m, n).

Thus, the algorithm for solving the retarded Green's function can be stated:

(a) Set the values of g_σ and b along the diagonal to be unity.
(b) Solve $g_\sigma(m, m+n)$ and $b(m, m+n)$ for all possible m with fixed n.

(c) Repeat step (b) with n being increased by 1 at a time until all matrix elements are solved.

The order between step (b) and step (c) is not unique. As long as the operations are arranged in such a way that the values of the Green's functions in equations (62) and (63) are previously obtained, the algorithm will work. The version given here is the one which would most easily be parallelized. The algorithm runs of the order M^3 in time and takes M^2 memory space where M is the number of grid points in one direction.

In the slow velocity limit, the time during which the effective charge transfer occurs becomes very large. To maintain a satisfactory convergence, more mesh points would be needed. The naive conclusion would be that the number of points would be inversely proportional to the velocity, resulting in huge matrices and large CPU time. Fortunately one can make use of the fact that $f^{\gtrless}(t - t')$ is localized along the $t - t' = 0$ direction. The localization of $f^{\gtrless}$ enables one to work on a band matrix with the bandwidth determined by $f^{\gtrless}$. Based on this fact, and the observation that expressions (62) and (63) depend only on the values of the Green's functions within the band of the matrices, a modified algorithm is developed which uses only square matrices with the side of the order of the localization length of $f^{\gtrless}$. The Green's functions are evaluated within these square matrices and updated each time step as the calculation proceeds in the $t - t' = 0$ direction. For a square matrix of size $N \times N$, with N corresponding to the width of the band matrix, and if M steps in time are needed for the integration between $t = -\infty$ and $t = \infty$, the modified algorithm will consume disk space as N^2 and CPU time as $N^2 \times M$.

In the equilibrium case, Γ_σ is independent of time and the energy levels for the impurity states are stationary, therefore g_σ and b are functions of $t - t'$ only. Equations (57), (58), (60) and (61) can therefore be simplified by setting $t' = 0$ and $n = 0$. The algorithm for solving the retarded Green's functions for the equilibrium problems would take M^2 in time and M in memory.

For the equilibrium case, Langreth and Nordlander (1991) proved that the Dyson equations for the retarded Green's functions are exactly the same as the ones derived by Coleman (1984) in his original paper. Using the equilibrium version of the algorithm and following the prescriptions by Coleman, we have managed to reproduce all the results presented in Coleman's paper.

3.2.3.2.2 'Less than' Green's function

To obtain the instantaneous populations of the atomic states involved in atom–surface scattering, the 'less than' Green's functions of the atomic states have to be solved and the value of the 'less than' Green's functions at equal time then give the population of the particular atomic state. In solving for the 'less than' Green's function, the retarded Green's functions are assumed to be known quantities, based on the discussion in the last subsection.

The two Dyson equations (50) and (51) give the time-dependence of the populations of the impurity levels. With the introduction of g_σ and b defined in equations (55) and (56), equation (50) can be written as

$$\frac{\partial}{\partial t}\tilde{G}_\sigma^<(t,t') = \int_{-\infty}^{t'} \mathrm{d}\bar{t}\sqrt{\Gamma_\sigma(t)\Gamma_\sigma(\bar{t})}\tilde{f}_\sigma^<(t,\bar{t})B^<(t,\bar{t})g_\sigma^*(t',\bar{t}) - \int_{-\infty}^{t} \mathrm{d}\bar{t}\sqrt{\Gamma_\sigma(t)\Gamma_\sigma(\bar{t})}\tilde{f}_\sigma^>(t,\bar{t})b(t,\bar{t})\tilde{G}_\sigma^<(\bar{t},t'), \tag{64}$$

and equation (51) can be expressed as

$$\frac{\partial}{\partial t}B^<(t,t') = \sum_\sigma \int_{-\infty}^{t'} \mathrm{d}\bar{t}\sqrt{\Gamma_\sigma(t)\Gamma_\sigma(\bar{t})}\tilde{f}_\sigma^>(\bar{t},t)\tilde{G}_\sigma^<(t,\bar{t})b^*(t',\bar{t}) - \sum_\sigma \int_{-\infty}^{t} \mathrm{d}\bar{t}\sqrt{\Gamma_\sigma(t)\Gamma_\sigma(\bar{t})}\tilde{f}_\sigma^<(\bar{t},t)g_\sigma(t,\bar{t})B^<(\bar{t},t'), \tag{65}$$

where g^* and b^* are the complex conjugates of g and b respectively. These terms arise from the advanced self-energies for the boson and the atomic states.

Notice that the integrals in equation (64) include $B^<(t,\bar{t})$ with $\bar{t} < t'$ and $\tilde{G}_\sigma^<(\bar{t},t')$ with $\bar{t} < t$. The same situation arises with the integrals in equation (65), where only $\tilde{G}_\sigma^<(t,\bar{t})$ with $\bar{t} < t'$ and $B^<(\bar{t},t')$ with $\bar{t} < t$ contribute. In other words, the derivatives at (t,t') are fully determined by the values of the Green's functions at (t_1,t_1') with $t_1 + t_1' \le t + t'$. Therefore, $G_\sigma^<(t,t')$ and $B^<(t,t')$ can be obtained if $G_\sigma^<(t_1,t_1')$ and $B^<(t_1,t_1')$ with $t_1 + t_1' \le t + t'$ are known.

The discretized versions of equations (64) and (65) are given by the following:

$$[1 + \tfrac{1}{4}\delta^2\Gamma_\sigma(m)f_\sigma^>(0)]\tilde{G}_\sigma^<(m,n) - \tfrac{1}{4}\delta^2\sqrt{\Gamma_\sigma(m)\Gamma_\sigma(n)}\tilde{f}_\sigma^<(m,n)B^<(m,n) = \tilde{G}_\sigma^<(m-1,n) + G_\sigma^<(m,n) \tag{66}$$

and

$$-\sum_\sigma \tfrac{1}{4}\delta^2\sqrt{\Gamma_\sigma(m)\Gamma_\sigma(n)}\tilde{f}_\sigma^>(n,m)\tilde{G}_\sigma^<(m,n) + \left[1 + \sum_\sigma \tfrac{1}{4}\delta^2\Gamma_\sigma(m)f_\sigma^<(0)\right]B^<(m,n) = B^<(m-1,n) + \sum_\sigma B_\sigma^<(m,n), \tag{67}$$

where the term $G_\sigma^<(m, n)$ is defined as

$$G_\sigma^<(m, n) = \tfrac{1}{4}\delta^2\sqrt{\Gamma_\sigma(m-1)\Gamma_\sigma(n)}\tilde{f}_\sigma^<(m-1, n)B^<(m-1, n)$$
$$-\tfrac{1}{4}\delta^2\Gamma_\sigma(m-1)f_\sigma^>(0)\tilde{G}_\sigma^<(m-1, n)$$
$$+\tfrac{1}{2}\delta^2\sum_{i=m-1}^{m}\sqrt{\Gamma_\sigma(i)}\sum_{j=1}^{n-1}\sqrt{\Gamma_\sigma(j)}\tilde{f}_\sigma^<(i, j)B^<(i, j)g_\sigma^*(n, j)$$
$$-\tfrac{1}{2}\delta^2\sum_{i=m-1}^{m}\sqrt{\Gamma_\sigma(i)}\sum_{j=1}^{i-1}\sqrt{\Gamma_\sigma(j)}\tilde{f}_\sigma^>(i, j)b(i, j)\tilde{G}_\sigma^<(j, n). \quad (68)$$

The term $B_\sigma^<(m, n)$ is given as

$$B_\sigma(m, n) = \tfrac{1}{4}\delta^2\sqrt{\Gamma_\sigma(m-1)\Gamma_\sigma(n)}\tilde{f}_\sigma^>(n, m-1)\tilde{G}_\sigma^<(m-1, n)$$
$$-\tfrac{1}{4}\delta^2\Gamma_\sigma(m-1)f_\sigma^<(0)B^<(m-1, n)$$
$$+\tfrac{1}{2}\delta^2\sum_{i=m-1}^{m}\sqrt{\Gamma_\sigma(i)}\sum_{j=1}^{n-1}\sqrt{\Gamma_\sigma(j)}\tilde{f}_\sigma^>(j, i)\tilde{G}_\sigma^<(i, j)b^*(n, j)$$
$$-\tfrac{1}{2}\delta^2\sum_{i=m-1}^{m}\sqrt{\Gamma_\sigma(i)}\sum_{j-1}^{i-1}\sqrt{\Gamma_\sigma(j)}\tilde{f}_\sigma^<(j, i)g_\sigma(i, j)B^<(j, n), \quad (69)$$

where δ is the grid spacing.

To explicitly ensure the conservation of the charge Q_B, special consideration is needed for the diagonal elements. Using equations (64) and (65) and their conjugate equations, the total derivative $d/dt = \partial/\partial t + \partial/\partial t'$, of the $G_\sigma^<(t, t)$ and $B^<(t, t)$ take the form:

$$\frac{d}{dt}\tilde{G}_\sigma^<(t, t) = 2\,\mathrm{Re}\int_{-\infty}^{t} d\bar{t}\sqrt{\Gamma_\sigma(t)\Gamma_\sigma(\bar{t})}\tilde{f}_\sigma^<(\bar{t}, t)B^<(\bar{t}, t)g_\sigma(t, \bar{t})$$
$$-2\,\mathrm{Re}\int_{-\infty}^{t} d\bar{t}\sqrt{\Gamma_\sigma(t)\Gamma_\sigma(\bar{t})}\tilde{f}_\sigma^>(t, \bar{t})b(t, \bar{t})\tilde{G}_\sigma^<(\bar{t}, t) \quad (70)$$

and

$$\frac{d}{dt}B^<(t, t) = 2\,\mathrm{Re}\sum_\sigma\int_{-\infty}^{t} d\bar{t}\sqrt{\Gamma_\sigma(t)\Gamma_\sigma(\bar{t})}\tilde{f}_\sigma^>(t, \bar{t})\tilde{G}_\sigma^<(\bar{t}, t)b(t, \bar{t})$$
$$-2\,\mathrm{Re}\sum_\sigma\int_{-\infty}^{t} d\bar{t}\sqrt{\Gamma_\sigma(t)\Gamma_\sigma(\bar{t})}\tilde{f}_\sigma^<(\bar{t}, t)g_\sigma(t, \bar{t})B^<(\bar{t}, t). \quad (71)$$

The discretized version of equations (70) and (71) have the form:

$$[1 + \tfrac{1}{2}\delta^2\Gamma_\sigma(m)f_\sigma^>(0)]\tilde{G}_\sigma^<(m, m) - \tfrac{1}{2}\delta^2\Gamma_\sigma(m)f_\sigma^<(0)B^<(m, m)$$
$$= \tilde{G}_\sigma^<(m-1, m-1) + G_\sigma^<(m, m) \quad (72)$$

and

$$-\sum_{\sigma} \tfrac{1}{2}\delta^2\Gamma_\sigma(m) f_\sigma^>(0)\tilde{G}_\sigma^<(m, m) + \left[1 + \sum_{\sigma} \tfrac{1}{2}\delta^2\Gamma_\sigma(m) f_\sigma^<(0)\right] B^<(m, m)$$

$$= B^<(m-1, m-1) - \sum_{\sigma} G_\sigma^<(m, m), \tag{73}$$

where $G_\sigma^<(m, m)$ can be expressed as

$$G_\sigma^<(m, m) = \tfrac{1}{2}\delta^2\Gamma_\sigma(m-1)[f_\sigma^<(0)B^<(m-1, m-1) - f_\sigma^>(0)\tilde{G}_\sigma^<(m-1, m-1)]$$

$$+ \delta^2 \operatorname{Re} \sum_{i=m-1}^{m} \sqrt{\Gamma_\sigma(i)} \sum_{j=1}^{i-1} \sqrt{\Gamma_\sigma(j)} \tilde{f}_\sigma^<(j, i) B^<(j, i) g_\sigma(i, j)$$

$$- \delta^2 \operatorname{Re} \sum_{i=m-1}^{m} \sqrt{\Gamma_\sigma(i)} \sum_{j=1}^{i-1} \sqrt{\Gamma_\sigma(j)} \tilde{f}_\sigma^>(i, j) b(i, j) \tilde{G}_\sigma^<(j, i). \tag{74}$$

In this equation, δ is again the grid space, with $t = m\delta$ and $t' = n\delta$. The derivation has been carried out for $t \geq t'$. For $t < t'$ the derivation can be done in an analogous fashion.

The algorithm for the calculation of the 'less than' Green's functions can be stated as follows:

(a) Initialize $G_\sigma^<(1, 1)$ with the original population of the atomic state and $B^<(1, 1)$ with 1 minus the sum of the original populations of all the atomic states.
(b) Solve $G_\sigma^<(m-n, n)$ and $B^<(m-n, n)$ for all possible n with fixed m.
(c) Repeat step (b) with m starting and increasing by 1 at a time until (M, M) is reached.

M is the number of points along one direction of the matrix and corresponds to the total time between effective $t = \pm\infty$. The entire algorithm is based on a second-order implicit scheme. The error due to each evolution step is of the order of δ^3. Like the algorithm for the retarded Green's function, steps (b) and (c) can be rearranged in many different ways provided one makes sure that the values of the Green's functions in equations (68), (69) and (74) are previously calculated. As for the algorithm for solving the retarded Green's functions, the CPU time increases as M^3 and the memory requirement increases as M^2. In the small velocity limit, a modified algorithm similar to the one discussed earlier for the retarded Green's functions can be developed. The modified algorithm takes N^2 disk space and $N^2 \times M$ time to run, where N is the number of points along one side of the square matrices.

From equations (72) and (73) one can readily prove that $\sum_\sigma G_\sigma^<(m, m) + B^<(m, m) = \sum_\sigma G_\sigma^<(m-1, m-1) + B^<(m-1, m-1)$. This shows that our algorithm does indeed conserve the charge Q_B, which guarantees that discretized representation would not cause any unphysical states with $Q_B \neq 1$ to be occupied.

The combination of the two algorithms just discussed defines an exact scheme for obtaining the solution to the Dryson equations for Green's functions of the atomic states.

3.2.3.3 Master equation for charge transfer

Applying the SCA to equations (50) and (51) we obtain

$$i\frac{\partial}{\partial t}\tilde{G}_\sigma^<(t,t') = \tilde{G}_\sigma^<(t,t')\int_{-\infty}^{t} d\bar{t}\sqrt{\Gamma_\sigma(t)\Gamma_\sigma(\bar{t})}\tilde{f}_\sigma^>(t,\bar{t})B^R(t,\bar{t}) + B^<(t,t)\int_{-\infty}^{t'} d\bar{t}\sqrt{\Gamma_\sigma(t)\Gamma_\sigma(\bar{t})}\tilde{f}_\sigma^<(t,\bar{t})\tilde{G}_\sigma^A(\bar{t},t') \tag{75}$$

and

$$i\frac{\partial}{\partial t}B^<(t,t') = B^<(t,t')\sum_\sigma\int_{-\infty}^{t} d\bar{t}\sqrt{\Gamma_\sigma(t)\Gamma_\sigma(\bar{t})}\tilde{f}_\sigma^<(\bar{t},t)\tilde{G}_\sigma^R(t,\bar{t}) + \sum_\sigma \tilde{G}_\sigma^<(t,t)\int_{-\infty}^{t'} d\bar{t}\sqrt{\Gamma_\sigma(t)\Gamma_\sigma(\bar{t})}\tilde{f}_\sigma^>(\bar{t},t)B^A(\bar{t},t'). \tag{76}$$

In order to evaluate these expressions, an approximate expression for the advanced and retarded Green's functions must be derived. Making use of the localization of $\tilde{f}^{\gtrless}$, the following Dyson equations can be derived from equations (52) and (53):

$$i\frac{\partial}{\partial t}\tilde{G}_\sigma^R(t,t') = \delta(t-t') + \tilde{G}_\sigma^R(t,t')\int_{-\infty}^{\infty} d\bar{t}\sqrt{\Gamma_\sigma(t)\Gamma_\sigma(\bar{t})}\tilde{f}_\sigma^>(t,\bar{t})B^R(t,\bar{t}) \tag{77}$$

and

$$i\frac{\partial}{\partial t}B^R(t,t') = \delta(t-t') + B^R(t,t')\sum_\sigma\int_{-\infty}^{\infty} d\bar{t}\sqrt{\Gamma_\sigma(t)\Gamma_\sigma(\bar{t})}\tilde{f}_\sigma^<(\bar{t},t)\tilde{G}_\sigma^R(t,\bar{t}). \tag{78}$$

These equations can be integrated to obtain the following expressions for the retarded Green's functions:

$$\tilde{G}_\sigma^R(t,t') = \frac{\theta(t-t')}{i}\exp\left[-i\int_{t'}^{t} d\tau\int_{t'}^{\tau} d\bar{t}\sqrt{\Gamma_\sigma(\tau)\Gamma_\sigma(\bar{t})}\tilde{f}_\sigma^>(\tau,\bar{t})B^R(\tau,\bar{t})\right],$$

$$B^R(t,t') = \frac{\theta(t-t')}{i}\exp\left[-i\sum_\sigma\int_{t'}^{t} d\tau\int_{t'}^{\tau} d\bar{t}\sqrt{\Gamma_\sigma(\tau)\Gamma_\sigma(\bar{t})}\tilde{f}_\sigma^<(\bar{t},\tau)\tilde{G}_\sigma^R(\tau,\bar{t})\right]. \tag{79}$$

Due to the localization of $\tilde{f}^{\gtrless}$, the $\mathrm{d}\bar{t}$ integral can be extended to $-\infty$. To lowest order, one can then replace the retarded Green's functions in the exponents by:

$$\tilde{g}_{\sigma}^{0^{\mathrm{R}}}(t, t') = \frac{\theta(t-t')}{\mathrm{i}} \exp\left[- \int_{t'}^{t} \mathrm{d}\tau \int_{t'}^{\tau} \mathrm{d}\bar{t} \sqrt{\Gamma_{\sigma}(\tau)\Gamma_{\sigma}(\bar{t})} \tilde{f}_{\sigma}^{>}(\tau, \bar{t}) \right],$$

$$b^{0^{\mathrm{R}}}(t, t') = \frac{\theta(t-t')}{\mathrm{i}} \exp\left[- \sum_{\sigma} \int_{t'}^{t} \mathrm{d}\tau \int_{t'}^{\tau} \mathrm{d}\bar{t} \sqrt{\Gamma_{\sigma}(\tau)\Gamma_{\sigma}(\bar{t})} \tilde{f}_{\sigma}^{<}(\bar{t}, \tau) \right]. \quad (80)$$

We thus obtain the expressions

$$\tilde{G}_{\sigma}^{\mathrm{R}}(t, t') = \frac{\theta(t-t')}{\mathrm{i}} \exp\left[-\mathrm{i} \int_{t'}^{t} \mathrm{d}\tau \int_{t'}^{\tau} \mathrm{d}\bar{t} \sqrt{\Gamma_{\sigma}(\tau)\Gamma_{\sigma}(\bar{t})} \tilde{f}_{\sigma}^{>}(\tau, \bar{t}) b^{0^{\mathrm{R}}}(\tau, \bar{t}) \right],$$

$$B^{\mathrm{R}}(t, t') = \frac{\theta(t-t')}{\mathrm{i}} \exp\left[-\mathrm{i} \sum_{\sigma} \int_{t'}^{t} \mathrm{d}\tau \int_{t'}^{\tau} \mathrm{d}\bar{t} \sqrt{\Gamma_{\sigma}(\tau)\Gamma_{\sigma}(\bar{t})} \tilde{f}_{\sigma}^{<}(\bar{t}, \tau) \tilde{g}_{\sigma}^{0^{\mathrm{R}}}(\tau, \bar{t}) \right], \quad (81)$$

for the retarded Green's functions. These expressions represent an improvement on the expressions derived previously for infinite bandwidth (Langreth and Nordlander, 1991):

$$\tilde{G}_{\sigma}^{\mathrm{R}}(t, t') = \frac{\theta(t-t')}{\mathrm{i}} \exp\left[- \int_{t'}^{t} \mathrm{d}\tau \frac{\Gamma_{\sigma}(\tau)}{2} f^{>}(\varepsilon_{\sigma}(\tau)) \right],$$

$$B^{\mathrm{R}}(t, t') = \frac{\theta(t-t')}{\mathrm{i}} \exp\left[- \int_{t'}^{t} \mathrm{d}\tau \sum_{\sigma} \frac{\Gamma_{\sigma}(\tau)}{2} f^{<}(\varepsilon_{\sigma}(\tau)) \right]. \quad (82)$$

Equations (81) can be inserted directly into equation (75) and the Dyson equations can be solved by direct integration. Notice that equation (85) is valid here as well. Therefore the populations of the impurity levels can be directly calculated from equation (75).

A rate equation based on equation (16) can be obtained directly. The equation takes the form:

$$\frac{\mathrm{d}}{\mathrm{d}t} n_{\sigma}(t) = n_{\sigma}(t) 2 \,\mathrm{Im}\left[\left[\int_{-\infty}^{t} \mathrm{d}\bar{t} \sqrt{\Gamma_{\sigma}(t)\Gamma_{\sigma}(\bar{t})} \tilde{f}_{\sigma}^{>}(t, \bar{t}) B^{\mathrm{R}}(t, \bar{t}) \right] \right.$$

$$\left. + \, n_{\mathrm{B}}(t) 2 \,\mathrm{Im}\left[\int_{-\infty}^{t} \mathrm{d}\bar{t} \sqrt{\Gamma_{\sigma}(t)\Gamma_{\sigma}(\bar{t})} \tilde{f}_{\sigma}^{<}(t, \bar{t}) \tilde{G}_{\sigma}^{\mathrm{A}}(\bar{t}, t) \right] \right], \quad (83)$$

where equation (81) is used for the advanced and retarded Green's functions. This equation will be referred to as the GME and it can be evaluated straightforwardly.

As in the $U = 0$ case, further simplifications in the spirit of the SCA are possible. Assuming that the rapid oscillation with respect to $\bar{t}$ is contained in the functions $\tilde{f}^{\gtrless}$, we obtain

$$\frac{\mathrm{d}}{\mathrm{d}t} n_\sigma(t) = -\Gamma_\sigma(t)[f^{>}(\varepsilon_\sigma(t))n_\sigma(t) - f^{<}(\varepsilon_\sigma(t))n_{\mathrm{B}}(t)]. \tag{84}$$

One could derive a similar sort of equation for $n_{\mathrm{B}}(t)$, but this is unnecessary since our approximations preserve the exact conservation of Q_{B}, so that $n_{\mathrm{B}}(t)$ is determined simply by

$$n_{\mathrm{B}}(t) = 1 - \sum_\sigma n_\sigma(t). \tag{85}$$

The simple master equation (84) coupled with (85) provides an extremely simple solution for the populations. Comparison of this equation with the corresponding equation (40) shows that the effects of intra-atomic correlation is to replace the Pauli exclusion factor $(1 - n_\sigma)$ in the $U = 0$ with the much stronger Coulomb exclusion factor $(1 - \sum_\sigma n_\sigma)$ in the $U = \infty$ case.

For a two-level problem, the equations governing the populations of the impurity states take the form:

$$\frac{\mathrm{d}n_1(t)}{\mathrm{d}t} = -\Gamma_1(t)[1 - f(\varepsilon_1(t))]n_1(t) + \Gamma_1(t)f(\varepsilon_1(t))[1 - n_1(t) - n_2(t)],$$

$$\frac{\mathrm{d}n_2(t)}{\mathrm{d}t} = -\Gamma_2(t)[1 - f(\varepsilon_2(t))]n_2(t) + \Gamma_2(t)f(\varepsilon_2(t))[1 - n_1(t) - n_2(t)]. \tag{86}$$

In the large-U limit, we can immediately see that the requirements on the SCA will be much more stringent than in the $U = 0$ situation. First, it must be required that the atomic levels cross the Fermi energy rapidly. In addition, since the charge transfer will depend on the instantaneous occupancy of all atomic energy levels in the problem, we must require that the master equation accurately represent the instantaneous population of the respective levels for all t.

In order to test the SCA for the infinite-U model, we again consider the simple two-level system introduced in the previous subsection. The distance-dependence of the shift and broadening of the two levels $|1\rangle$ and $|2\rangle$, is again assumed to be linear and exponential, respectively, following equations (41) and (42). We assume that the atom moves with constant speed v following the straight-in—straight-out trajectory described earlier (see Figure 3.1).

In Figure 3.4, the populations $n_1(Z)$ and $n_2(Z)$ calculated using the SME (84), the GME (83) and the exact method described in the previous subsection are shown as a function of distance for different velocities. It can be clearly seen that the SME only provides an accurate description of the charge transfer for the smallest velocities, $v < 0.001$ a.u. However, these velocities are rather typical in hyperthermal atom–surface scattering so it appears that one must conclude that SME provides a reasonable alternative for the description of charge-transfer processes in many cases. The GME on the other hand provides a very accurate description of the charge transfer even for as large velocities as 0.1 a.u. The numerical solution to the GME (83) is straightforward and can be performed for an arbitrary metal bandwidth.

In Figure 3.5, the influence of bandwidth on the populations $n_1(Z)$ and $n_2(Z)$ is shown for velocities $v = 0.001$ a.u. and $v = 0.03$ a.u. It can be seen that for the lowest velocity, it is primarily the population of level $|1\rangle$ that is influenced. The reason for this is simply that this level can only be formed close to the surface. For a narrow conduction band, the level shifts below the band in this region so that resonant charge-transfer cannot occur. The level $|2\rangle$ crosses the Fermi energy relatively close to the surface and is therefore much less sensitive to bandwidth. For the larger velocity, $v = 0.03$ a.u., the situation is markedly different. The trend of less charge transfer into the $|1\rangle$ state for decreasing bandwidth is obscured by an increasing population for the intermediate velocities. This is a dynamic effect and is caused by interference between the two levels. These types of interference effects occurs at large velocities and couplings in quasi-resonant charge-exchange phenomena which will be discussed further in Section 3.2.5.

3.2.4 CASE OF FINITE *U*

Atomic and molecular affinity levels shift downwards near metallic surfaces. This is due to the screening of the intra-atomic correlation. Outside a perfect metal surface $U = U - 1/2Z$, where Z is the distance from the surface. If the width of the atomic affinity level is sufficiently large, negative ion states may be formed by resonant tunneling. The dynamic description of charge transfer in a scattering event in which both negative ions and neutral excited states can be formed requires proper consideration of the intra-atomic correlation energy.

The two-level case of the finite-U Anderson model in equilibrium was solved by Kotliar and Ruckenstein (1986) using bosons to describe all four possible configurations. We have generalized the theory and applied it to the charge-exchange problem in atom–surface collisions. Our investigation shows that the 'pseudo' state describing a configuration with an odd number of electrons has to be chosen as a *fermion*, while a configuration with an even number of electrons should be represented by a 'pseudo' boson.

We will first present a detailed description of the methodology for a two-level case. We will show that the proper statistics of the 'pseudo' states is essential for obtaining the correct large-U limiting behavior. The theory for the N-level case will then be discussed. The master-equation approximation will be discussed at the end of this section.

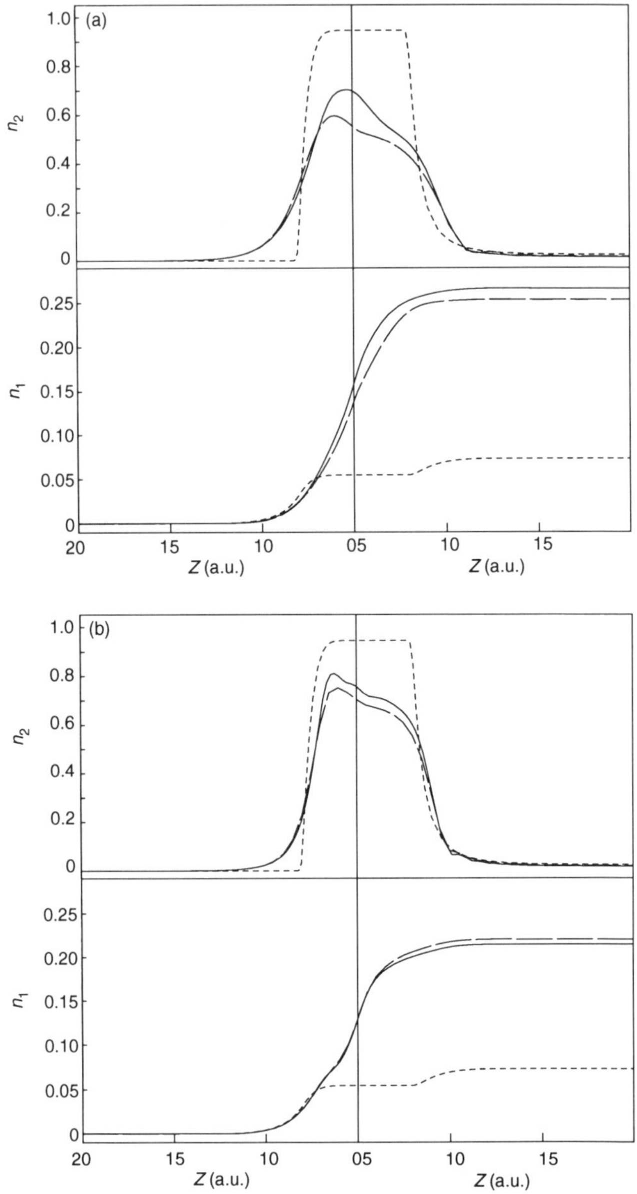

Figure 3.4 Comparison of the populations $n_i(Z)$ calculated for large U using SME, GME and the exact solution for different velocities. The parameters of the system are the same as those in Figure 3.2: $Z_1^c = 12$, $Z_2^c = 8$ a.u., $b_1 = b_2 = 0.02$, $\alpha_1 = 1.0$ and $\alpha_2 = 0.5$. The temperature is $T = 300$ K and the Fermi energy is 0.2 a.u. The widths of the impurity level have been scaled for each velocity so that similar charge-transfer occurs within each trajectory. The bandwidth of the metal is 0.368 a.u. The motion of the atom is linear and the atomic trajectory is assumed to start towards the surface at $Z = 20$ a.u., to turn towards vacuum at $Z = 5$ a.u., and then to propagate back to

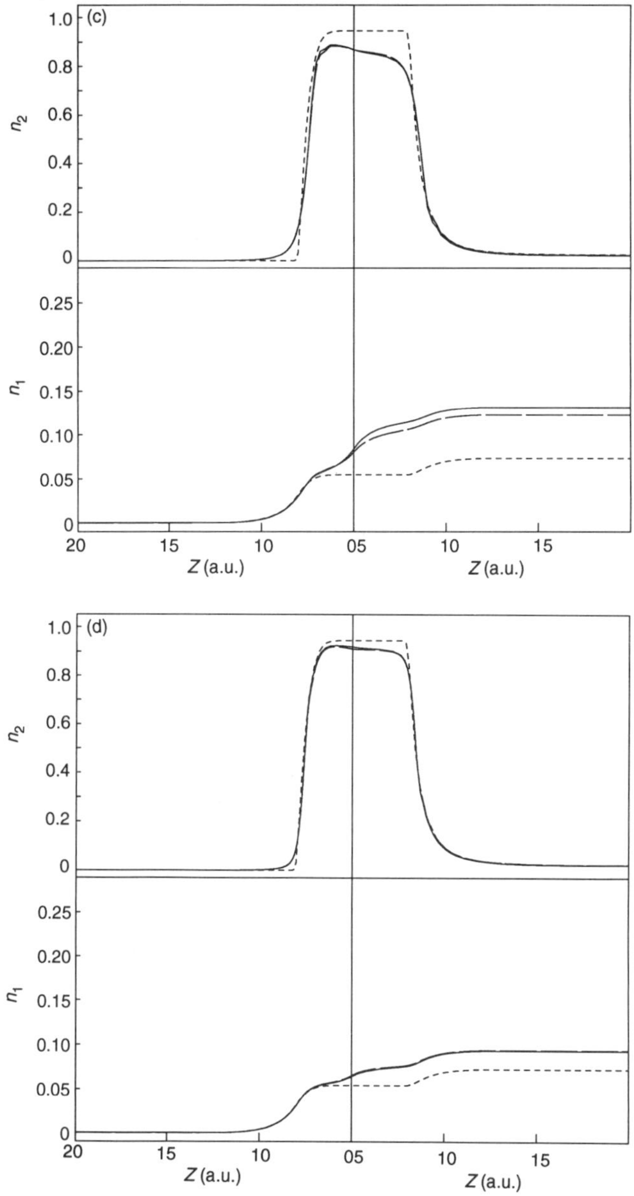

$Z = 20$ a.u. In (a), we show the results for velocity $v = 0.03$, $\Delta_{1,2} = 3$ a.u. In (b), we show the results for velocity $v = 0.01$, $\Delta_{1,2} = 1$ a.u. In (c), we show the results for velocity $v = 0.003$, $\Delta_{1,2} = 0.3$ a.u. In (d), we show the results for velocity $v = 0.001$, $\Delta_{1,2} = 0.1$ a.u. The upper part of each figure, shows the occupation $n_1(Z)$ as a function of distance from the surface. In the lower part of the figures, the population $n_2(Z)$ is shown. The dotted lines are the results using the SME, equation (84), the solid lines are the results using the GME, equation (83) and the dashed lines are the results obtained using the exact numerical solution

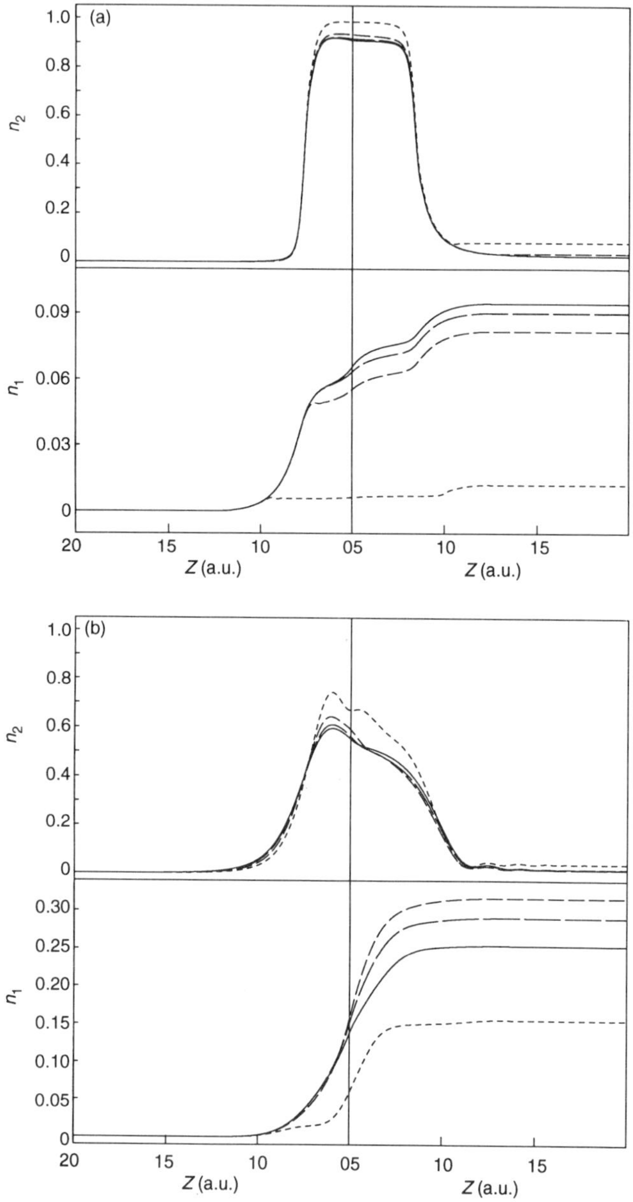

Figure 3.5 Effect of bandwidth on charge-transfer. The parameters of the system are the same as those in Figure 3.2. In the lower part of the figure, we show $n_1(Z)$ and in the upper part of the figure, $n_2(Z)$ is shown. The solid lines are the results obtained for $W = 0.368$ a.u., the long-dashed results are the results for $W = 0.184$ a.u. The dashed results are for $W = 0.92$ a.u. and the dotted lines are for $W = 0.046$ a.u. In (a), the velocity $v = 0.001$, $\Delta = 0.1$ was used. In (b), the velocity $v = 0.03$, $\Delta = 3.0$ was used

3.2.4.1 The two-level case

The Anderson Hamiltonian for a two-level system with finite U has the form:

$$H(t) = \left[\left[\sum_{\sigma} \varepsilon_\sigma(t)c_\sigma^\dagger(t)c_\sigma(t)\right] + Uc_\sigma^\dagger(t)c_\sigma(t)c_{-\sigma}^\dagger(t)c_{-\sigma}(t)\right]$$

$$+ \sum_{k\sigma} \varepsilon_{k\sigma}c_{k\sigma}^\dagger(t)c_{k\sigma}(t) + \sum_{k\sigma} [V_{k\sigma}(t)c_{k\sigma}^\dagger(t)c_\sigma(t) + \text{H.C.}] \quad (87)$$

where the first and the second terms describe the atomic states and the band electrons in the solid respectively. The subscript σ labels the two atomic states. The last term in the Hamiltonian describes the hopping between the atomic states and band electrons.

In an attempt to describe the bulk equilibrium problems with finite U, Kotliar and Ruckenstein (1986) studied a two-level system with finite correlation between them. They introduced four bosons, e, d, p_σ and $p_{-\sigma}$ representing the destruction operators for the four possible atomic configurations. The empty configuration is represented by e; p_σ and $p_{-\sigma}$ represent the singly occupied state; and d represents the doubly occupied state. These states should all be viewed as 'pseudo' configurations. Aside from the physical intuition that p_σ and $p_{-\sigma}$ should be fermions, it can be proven from the equations of motion that p_σ and $p_{-\sigma}$ have to be fermions.

There are two ways that the real atomic state created by $c_\sigma^\dagger$ can be constructed in terms of the 'pseudo' states. Specifically, the creation of the real state $|\sigma\rangle$ can be realized by (a) the creation of the 'pseudo' state $|\sigma\rangle$ and the annihilation of the 'pseudo' empty state, or (b) by the creation of the 'pseudo' double state and the destruction of the 'pseudo' state $|-\sigma\rangle$. Therefore the simplest relation between the operators for 'pseudo' states and the real ones are:

$$c_\sigma(t) = e^\dagger(t)p_\sigma(t) + p_{-\sigma}^\dagger(t)d(t),$$

$$c_\sigma^\dagger(t) = e(t)p_\sigma^\dagger(t) + p_{-\sigma}(t)d^\dagger(t). \quad (88)$$

The expressions for $c_{-\sigma}$ and $c_{-\sigma}^\dagger$ can be obtained by switching the subscript σ with $-\sigma$ in the above relations.

With these pairs of operators for the 'pseudo' states and the relations between the operators of the 'pseudo' and real states, it is straightforward to show that

$$\{c_\sigma, c_\sigma^\dagger\} = e^\dagger e + p_\sigma^\dagger p_\sigma + p_{-\sigma}^\dagger p_{-\sigma} + d^\dagger d \quad (89)$$

where $\{\cdots,\cdots\}$ is the usual anti-commutator. Therefore, a generalized charge $Q(t)$ can be defined as $Q = e^\dagger e + p_\sigma^\dagger p_\sigma + p_{-\sigma}^\dagger p_{-\sigma} + d^\dagger d$. For the proper solution, this charge should be equal to 1.

In the subspace described by e, d, p_σ and $p_{-\sigma}$, the following relation is valid:

$$c_\sigma^\dagger(t)c_\sigma(t) = p_\sigma^\dagger(t)p_\sigma(t) + d^\dagger(t)d(t). \tag{90}$$

The correlation term in equation (87) can be expressed in terms of d as $Ud^\dagger d$. Thus, the Hamiltonian in terms of the operators for the 'pseudo' states takes the following form:

$$\begin{aligned} H(t) = &\left[\sum_\sigma \varepsilon_\sigma(t)p_\sigma^\dagger(t)p_\sigma(t) + [\varepsilon_\sigma(t) + \varepsilon_{-\sigma}(t) + U]d^\dagger(t)d(t)\right] \\ &+ \sum_{k\sigma} \varepsilon_{k\sigma}c_{k\sigma}^\dagger(t)c_{k\sigma}(t) + \sum_{k\sigma} [V_{k\sigma}(t)c_{k\sigma}^\dagger(t)e^\dagger(t)p_\sigma(t) + \text{H.C.}] \\ &+ \sum_{k\sigma} [V_{k\sigma+}(t)c_{k\sigma}^\dagger(t)d(t)p_{-\sigma}^\dagger(t) + \text{H.C.}]. \end{aligned} \tag{91}$$

We note that in order to properly describe the different tunneling channels, the hopping-matrix element $V_{k\sigma}$ can be different from $V_{k\sigma+}$. This is an important generalization of the conventional Anderson Hamiltonian based on the notion that the matrix element of one tunneling channel, $V_{k\sigma}$, depends on the occupations of the remaining impurity states. An example of this is the formation of the $H^-(1s^2)$ in atom–surface scattering. The width of this state is broader and varies much more slowly (Nordlander, 1992) with atom–surface separation than the width of a singly occupied H(1s) state (Nordlander and Tully, 1988).

With the Hamiltonian equation (91), it is trivial to show that the generalized charge $Q(t)$ defined earlier is a constant of motion and thus $Q(t) = 1$. Therefore, the projection procedure presented in the infinite-U section can be applied also here when one tries to calculate the retarded and advanced self-energies for the 'pseudo' states. The Dyson equations for the Green's functions of the 'pseudo' states can be derived accordingly.

3.2.4.1.1 Self-energy

The transition from the original Hamiltonian equation (87) to (91) and the derivations associated with it are valid, independent of the statistics of p_σ. But the self-energies for the 'pseudo' states do depend on the statistics of p_σ. For convenience, we introduce a number ε_p which is defined as:

$$\varepsilon_p^< = \begin{cases} 1 & \text{if } p_\sigma \text{ is a fermion,} \\ -1 & \text{if } p_\sigma \text{ is a boson.} \end{cases}$$

where '$<$' means 'less than' self-energy. For 'greater than' self-energy, $\varepsilon_p^>$ will always be unity.

The Green's functions are defined in the same way as in Section 3.2.1. The propagator for the empty state is denoted E, P_σ refers to that of the singly

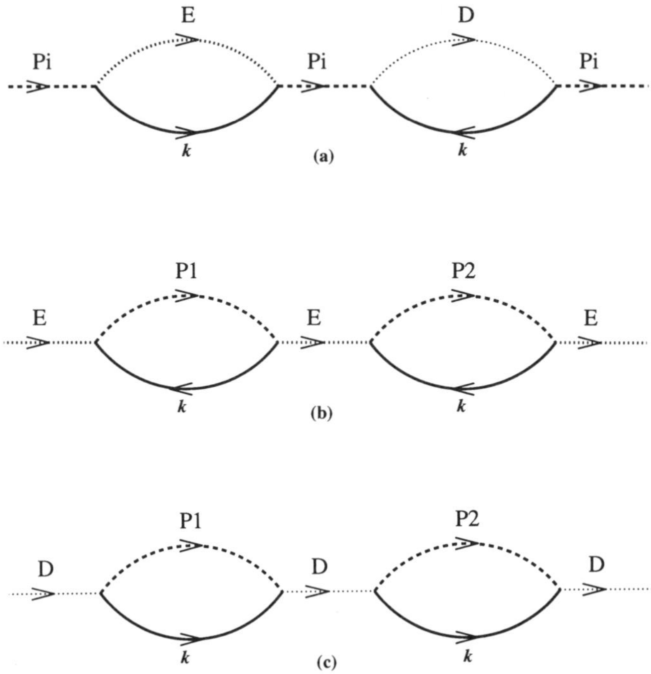

Figure 3.6 (a) Self-energy for the singly occupied atomic configuration, (b) self-energy for the empty configuration, and (c) self-energy for the doubly filled configuration. The solid line represents the Green's function of the conduction electron, dashed line that of the singly occupied configuration, the dark dotted line that of the empty configuration and the light dotted line that of the doubly filled configuration

occupied state and D represents that of the doubly filled state. The lowest-order diagrams, depicted in Figure 3.6(a), give the following expressions for the self-energies of the 'pseudo' state σ:

$$\Sigma^{\gtrless}_{p_\sigma}(t, t') = \varepsilon^{\gtrless}_p \sum_k V^*_{k\sigma}(t) V_{k\sigma}(t') E^{\gtrless}(t, t') G^{\gtrless}_{k\sigma}(t, t'),$$

$$\Sigma^{\gtrless}_{p_{\sigma+}}(t, t') = \varepsilon^{\gtrless}_p \sum_k V_{k,-\sigma+}(t) V^*_{k,-\sigma+}(t') E^{\gtrless}(t, t') G^{\gtrless}_{k,-\sigma}(t', t), \tag{92}$$

where Σ_σ and $\Sigma_{\sigma+}$ are the contributions from the two different tunneling channels in this case and $G_k(t, t')$ is the Green's function for the band electron. To the leading order, it is sufficient to approximate $G_k(t, t')$ with its equilibrium value. The self-energy for the 'pseudo' empty state illustrated in Figure 3.6(b) takes the form:

$$\Pi^{\gtrless}_e(t, t') = \varepsilon^{\gtrless}_p \sum_{k\sigma} V_{k\sigma}(t) V^*_{k\sigma}(t') G^{\gtrless}_{k\sigma}(t', t) P^{\gtrless}_\sigma(t, t'). \tag{93}$$

The self-energy for the doubly filled state is obtained from Figure 3.6(c) and takes the form:

$$\Pi_d^{\gtrless}(t, t') = \varepsilon_p^{\gtrless} \sum_{k\sigma} V_{k,\sigma+}^*(t) V_{k,\sigma+}(t') P_{-\sigma}^{\gtrless}(t, t') G_{k\sigma}^{\gtrless}(t, t'). \tag{94}$$

In terms of the level broadening Γ, defined in Section 3.2.1, the self-energies can be expressed in the following forms:

$$\Sigma_\sigma^{\gtrless}(t, t') = \varepsilon_p^{\gtrless} \sqrt{\Gamma_\sigma(t)\Gamma_\sigma(t')} f_\sigma^{\gtrless}(t - t') E^{\gtrless}(t, t'),$$

$$\Sigma_{\sigma+}^{\gtrless}(t, t') = \varepsilon_p^{\gtrless} \sqrt{\Gamma_{-\sigma^+}(t)\Gamma_{-\sigma^+}(t')} f_{-\sigma}^{\gtrless}(t' - t) D^{\gtrless}(t, t'), \tag{95}$$

$$\Pi_e^{\gtrless}(t, t') = \varepsilon_p^{\gtrless} \sum_\sigma \sqrt{\Gamma_\sigma(t)\Gamma_\sigma(t')} f_\sigma^{\gtrless}(t' - t) P_\sigma(t, t'), \tag{96}$$

$$\Pi_d^{\gtrless}(t, t') = \varepsilon_p^{\gtrless} \sum_\sigma \sqrt{\Gamma_{\sigma+}(t)\Gamma_{\sigma+}(t')} f_\sigma^{\gtrless}(t - t') P_{-\sigma}(t, t'), \tag{97}$$

where Γ_{σ^+} corresponds to $V_{k\sigma+}$.

Using the projection rules discussed in Section 3.2.3, the retarded and advanced self-energies can readily be given as the following:

$$\Sigma_\sigma^{R,A}(t, t') = \sqrt{\Gamma_\sigma(t)\Gamma_\sigma(t')} f_\sigma^{>}(t - t') E^{R,A}(t, t'),$$

$$\Sigma_{\sigma+}^{R,A}(t, t') = \sqrt{\Gamma_{-\sigma^+}(t)\Gamma_{-\sigma^+}(t')} f_{-\sigma}^{<}(t' - t) D^{R,A}(t, t'), \tag{98}$$

$$\Pi_e^{R,A}(t, t') = \sum_\sigma \sqrt{\Gamma_\sigma(t)\Gamma_\sigma(t')} f_\sigma^{<}(t' - t) P_\sigma^{R,A}(t, t'), \tag{99}$$

$$\Pi_d^{R,A}(t, t') = \sum_\sigma \sqrt{\Gamma_{\sigma^+}(t)\Gamma_{\sigma^+}(t')} f_\sigma^{>}(t - t') P_{-\sigma}^{R,A}(t, t'). \tag{100}$$

Notice that $\varepsilon_p^{\gtrless}$ does not appear in the expressions for the retarded and advanced self-energies. In the equilibrium case, almost all physics is contained in the retarded or advanced Green's functions. Therefore, the particular statistics of p_σ does not affect the final results. However, for the non-equilibrium situations, the physical information derives from the 'less than' Green's functions. Since the 'less than' Green's functions explicitly depend on the 'less than' self-energies, the statistics of p_σ will influence the response of the system. As will be shown later, to obtain the Dyson equations with the right limiting behavior for the 'less than' Green's function, p_σ must be a *fermion.*

3.2.4.1.2 General Equations

Once the self-energies are calculated, the Dyson equations for the Green's functions can readily be written down. To this end, the following notations are introduced:

$$H_\sigma(t, t') = \tilde{H}_\sigma(t, t') \exp\left[-\mathrm{i} \int_{t'}^{t} \mathrm{d}\tau \varepsilon_\sigma(\tau)\right]$$

and

$$\tilde{H}_\sigma(t, t') = \frac{\theta(t - t')}{\mathrm{i}} h_\sigma(t, t')$$

where H can be the Green's function for any of the 'pseudo' states or the Fourier transform of the Fermi function.

3.2.4.1.2.1 Retarded Green's functions The coupled Dyson equations for the retarded Green's functions are given by:

$$\begin{aligned}\frac{\partial}{\partial t} p_\sigma(t, t') = &- \int_{t'}^{t} \mathrm{d}\bar{t}\sqrt{\Gamma_\sigma(t)\Gamma_\sigma(\bar{t})}\tilde{f}^{>}_{\varepsilon_\sigma}(t, \bar{t})e(t, \bar{t})p_\sigma(\bar{t}, t') \\ &- \int_{t'}^{t} \mathrm{d}\bar{t}\sqrt{\Gamma_{-\sigma^+}(t)\Gamma_{-\sigma^+}(\bar{t})}\tilde{f}^{<}_{\varepsilon_{-\sigma}+U}(\bar{t}, t)d(t, \bar{t})p_\sigma(\bar{t}, t'),\end{aligned} \tag{101}$$

$$\frac{\partial}{\partial t} d(t, t') = -\sum_\sigma \int_{t'}^{t} \mathrm{d}\bar{t}\sqrt{\Gamma_{\sigma^+}(t)\Gamma_{\sigma^+}(\bar{t})}\tilde{f}^{>}_{\varepsilon_\sigma+U}(t, \bar{t})p_{-\sigma}(t, \bar{t})d(\bar{t}, t'), \tag{102}$$

$$\frac{\partial}{\partial t} e(t, t') = -\sum_\sigma \int_{t'}^{t} \mathrm{d}\bar{t}\sqrt{\Gamma_\sigma(t)\Gamma_\sigma(\bar{t})}\tilde{f}^{<}_{\varepsilon_\sigma}(\bar{t}, t)p_\sigma(t, \bar{t})e(\bar{t}, t'). \tag{103}$$

The equation for $p_{-\sigma}$ can be obtained by setting σ to $-\sigma$ in equation (101). With the Green's functions at equal time being unity, the coupled Dyson equations can be solved exactly using the algorithm given in Section 3.2.3.

In the large-U limit, these equations reduce to those for the infinite-U case. This can be seen directly by setting $U = \infty$. The functions $\tilde{f}^{\gtrless}_{\varepsilon+U}(t, t')$ then become zero. The remaining terms are identical to those for the infinite-U problems. The $U = 0$ limit is more difficult to derive (Shao *et al.*, forthcoming).

The relation between the Green's function of the real state and those of the 'pseudo' states, can be expressed as:

$$\begin{aligned}G^{\text{real}}_\sigma(t, t') = &E^{<}(t', t)P_\sigma(t, t') + E(t', t)P^{<}_\sigma(t, t') \\ &+ P^{<}_{-\sigma}(t', t)D(t, t') + P_{-\sigma}(t', t)D^{<}(t, t'),\end{aligned} \tag{104}$$

where $D^<$, $E^<$, $P^<_\sigma$ and $P^<_{-\sigma}$ are the 'less than' Green's function for the doubly filled state, empty state and singly occupied state, respectively; and D, E, $P_{-\sigma}$ and P_σ are the simplified retarded Green's functions for corresponding states, which are defined as $D = D^> - D^<$, $E = E^> - E^<$, $P_{-\sigma} = P^>_{-\sigma} + P^<_{-\sigma}$ and $P_\sigma = P^>_\sigma + P^<_\sigma$, accordingly.

3.2.4.1.2.2 'Less than' Green's functions The Dyson equations for the 'less than' Green's function using the simplified retarded Green's functions can be written:

$$\frac{\partial}{\partial t}\tilde{P}^{<}_{\sigma}(t,t') = -\int_{-\infty}^{t} d\bar{t}\sqrt{\Gamma_{\sigma}(t)\Gamma_{\sigma}(\bar{t})}\tilde{f}^{>}_{\varepsilon_{\sigma}}(t,\bar{t})e(t,\bar{t})\tilde{P}^{<}_{\sigma}(\bar{t},t')$$
$$+\varepsilon^{<}_{p}\int_{-\infty}^{t'} d\bar{t}\sqrt{\Gamma_{\sigma}(t)\Gamma_{\sigma}(\bar{t})}\tilde{f}^{<}_{\varepsilon_{\sigma}}(t,\bar{t})p^{*}_{\sigma}(t',\bar{t})E^{<}(t,\bar{t})$$
$$-\int_{-\infty}^{t} d\bar{t}\sqrt{\Gamma_{-\sigma^{+}}(t)\Gamma_{-\sigma^{+}}(\bar{t})}\tilde{f}^{<}_{\varepsilon_{-\sigma}+U}(\bar{t},t)d(t,\bar{t})\tilde{P}^{<}_{\sigma}(\bar{t},t')$$
$$+\varepsilon^{<}_{p}\int_{-\infty}^{t'} d\bar{t}\sqrt{\Gamma_{-\sigma^{+}}(t)\Gamma_{-\sigma^{+}}(\bar{t})}\tilde{f}^{>}_{\varepsilon_{-\sigma}+U}(\bar{t},t)p^{*}_{\sigma}(t',\bar{t})D^{<}(t,\bar{t}) \qquad (105)$$

$$\frac{\partial}{\partial t}E^{<}(t,t') = -\sum_{\sigma}\int_{-\infty}^{t} d\bar{t}\sqrt{\Gamma_{\sigma}(t)\Gamma_{\sigma}(\bar{t})}\tilde{f}^{<}_{\varepsilon_{\sigma}}(\bar{t},t)p_{\sigma}(t,\bar{t})E^{<}(\bar{t},t')$$
$$+\varepsilon^{<}_{p}\sum_{\sigma}\int_{-\infty}^{t'} d\bar{t}\sqrt{\Gamma_{\sigma}(t)\Gamma_{\sigma}(\bar{t})}\tilde{f}^{>}_{\varepsilon_{\sigma}}(\bar{t},t)e^{*}(t',\bar{t})\tilde{P}^{<}_{\sigma}(t,\bar{t}), \qquad (106)$$

$$\frac{\partial}{\partial t}D^{<}(t,t') = -\sum_{\sigma}\int_{-\infty}^{t} d\bar{t}\sqrt{\Gamma_{\sigma^{+}}(t)\Gamma_{\sigma^{+}}(\bar{t})}\tilde{f}^{>}_{\varepsilon_{\sigma}+U}(t,\bar{t})p_{-\sigma}(t,\bar{t})\tilde{D}^{<}(\bar{t},t')$$
$$+\varepsilon^{<}_{p}\sum_{\sigma}\int_{-\infty}^{t'} d\bar{t}\sqrt{\Gamma_{\sigma^{+}}(t)\Gamma_{\sigma^{+}}(\bar{t})}\tilde{f}^{<}_{\varepsilon_{\sigma}+U}(t,\bar{t})d^{*}(t',\bar{t})\tilde{P}^{<}_{-\sigma}(t,\bar{t}). \qquad (107)$$

The equation for $\tilde{P}^{<}_{-\sigma}$ can be obtained by switching σ to $-\sigma$. With the above set of equations, it is straightforward to show that the charge $Q(t)$ defined earlier is explicitly conserved.

This set of Dyson equations can be solved exactly using the algorithm presented in Section 3.2.3. The values of the 'less than' Green's functions at equal time give the populations of the 'pseudo' states. The population of a real state is then the sum of the population of the corresponding 'pseudo' state and the population of the doubly filled state.

In the limit of large U, the Dyson equations become identical to those for the large-U case if $|E\rangle$ is replaced by $|B\rangle$ and the extra factor $\varepsilon^{<}_{p}$ is unity. This immediately leads to the following conclusions: (a) $\varepsilon^{<}_{p}$ has to be unity, which means that p_{σ} has to be a *fermion*; and (b) the slave-boson not only simplifies the projection procedure, but represents a configuration where no atomic state is occupied.

The limit, $U = 0$, is more difficult to examine because the doubly filled state and the empty state still appear in the equations. In addition, the general projection rule for the 'less than' Green's function for the real state results in a complicated relation:

$$G^{<,\mathrm{real}}_{\sigma}(t,t') = E(t',t)P^{<}_{\sigma}(t,t') + P_{-\sigma}(t',t)D^{<}(t,t'). \qquad (108)$$

with E and $P_{-\sigma}$ being the simplified retarded Green's functions defined at the end of Section 3.2.4.1.2.1 (Shao *et al.*, forthcoming).

3.2.4.2 N-level Case

The extension to the N-level case is trivial. For an impurity with N levels, we introduce the following notation for the various isolated impurity configurations

$$|\{n_i\}\rangle = \prod_{n_i \neq 0} c^+_{\sigma_i}|0\rangle, \tag{109}$$

where n_i is zero or unity describing the participation of the atomic level $|\sigma_i\rangle$ in the atomic configuration. Here $\{n_i\}$ denotes the N-dimensional vector $(n_1, n_2, \ldots, n_N)$. The energy of the impurity configuration $|\{n_i\}\rangle$ is

$$\varepsilon_{\{n_i\}}(t) = \sum_i n_i \varepsilon_{\sigma_i}(t) + \tfrac{1}{2} \sum_{i \neq j} U_{\sigma_i \sigma_j}(t) n_i n_j. \tag{110}$$

We further denote by $I^+_{\{n_i\}}$ the creation (annihilation) operator of an impurity configuration $|\{n_i\}\rangle$. As in the two-level case, $I^+_{\{n_i\}}(I_{\{n_i\}})$ will be a boson if an even number of impurity levels are occupied, or a fermion if an odd number of the impurity levels are occupied. As a result of this assumption, $I^+_{\{n_i\}}(I_{\{n_i\}})$ should be seen as an operator for the 'pseudo' configuration. As in the discussion of the two-level case, it is straightforward to show that the population of a 'real' configuration is the sum of the populations of all the 'pseudo' configurations containing the corresponding 'pseudo' configuration.

Using this notation, the Hamiltonian takes the form:

$$\begin{aligned} H(t) = {} & \sum_{\{n_i\}} \varepsilon_{\{n_i\}}(t) I^+_{\{n_i\}} I_{\{n_i\}} + \sum_{ki} \varepsilon_{k\sigma_i} c^+_{k\sigma_i} c_{k\sigma_i} \\ & + \sum_{k,i} [V_{k,\sigma_i}(t) I^+_{(n_1,\ldots,n_i=0,\ldots)} I_{(n_1,\ldots,n_i=1,\ldots)} c^+_{k\sigma_i} + \text{H.C.}]. \end{aligned} \tag{111}$$

The population of impurity level $|\sigma_j\rangle$ after the scattering event can readily be obtained from the populations of the atomic configurations, $|\{n_i\}\rangle$ using

$$n_{\sigma_j}(t) = \sum_{n_j = 1} I^+_{\{n_i\}}(t) I_{\{n_i\}}(t). \tag{112}$$

It is straightforward to demonstrate that the quantity

$$Q = \sum_{\{n_i\}} I^+_{\{n_i\}} I_{\{n_i\}} \tag{113}$$

is a conserved quantity. Thus, we can look for a solution of this Hamiltonian within the subspace $Q = 1$. The projection onto this subspace will be performed as in the infinite-U situation discussed in Section 3.2.3.

To incorporate in the Hamiltonian, equation (111), the dependence of the tunneling-matrix element, $V_{\sigma k}$, on the populations of the remaining impurity states, as discussed in the previous subsection, one can define a hopping-matrix element V_{k,σ_i} according to the specific atomic configurations that are changed. The hopping term in the Hamiltonian is thus written as

$$\sum_{\{n_i\}} \sum_{k\sigma} [V_{\{n_i^+\}k\sigma}(t) I^+_{\{n_i^+\}} I_{\{n_i\}} c_{k\sigma} + \text{H.C.}], \tag{114}$$

where we have introduced the notation $\{n_i^+\}$ for an impurity configuration differing from $\{n_i\}$ by the addition of one electron. In the following, we will also use the notation $\{n_i^-\}$ for an atomic configuration derived from $\{n_i\}$ by the removal of one electron. The band index k will be taken to also include the quantum number σ.

The propagators for the impurity configurations are defined in the usual way:

$$G_{\{n_i\}}(t, t') = -i\langle T I_{\{n_i\}}(t) I^+_{\{n_i\}}(t')\rangle, \tag{115}$$

where the time-ordering is defined along the complex contour c. In continuing this equation onto the real axis, fermion or boson rules must be imposed (Langreth and Nordlander, 1991).

The self-energies for the propagators of the atomic configurations $|\{n_i\}\rangle$ can be evaluated to lowest order in the hopping-matrix elements. These lowest-order diagrams are shown in Figure 3.7 and the self-energy takes the form:

$$\begin{aligned}\Sigma_{\{n_i\}}(t, t') = \varepsilon^{\gtrless}_{\{n_i\}} \sum_{\{n_i^-\}} i \sum_k V_{\{n_i^-\}k}(t) V_{\{n_i^-\}k}(t') G_{\{n_i^-\}}(t, t') G_k(t, t') \\ + \varepsilon^{\gtrless}_{\{n_i\}} \sum_{\{n_i^+\}} i \sum_k V_{\{n_i^+\}k}(t) V_{\{n_i^+\}k}(t') G_{\{n_i^+\}}(t, t') G_k(t', t).\end{aligned} \tag{116}$$

The sum over $\{n_i^-\}$ and $\{n_i^+\}$ refers to a summation over all impurity configurations that can be obtained by the removal from or addition to $\{n_i\}$ of one electron. $\varepsilon^{>}_{\{n_i\}}$ is unity and $\varepsilon^{<}_{\{n_i\}}$ is defined as

$$\varepsilon^{<}_{\{n_i\}} = \begin{cases} 1 & \text{if } |\{n_i\}\rangle \text{ is a fermionic configuration,} \\ -1 & \text{if } |\{n_i\}\rangle \text{ is a bosonic configuration.} \end{cases}$$

The continuation of these equations to the real axis is straightforward and the following result is obtained:

$$\begin{aligned}\Sigma^{\gtrless}_{\{n_i\}}(t, t') = \varepsilon^{\gtrless}_{\{n_i\}} \sum_{\{n_i^-\}} \sum_k V_{\{n_i^-\}k}(t) V_{\{n_i^-\}k}(t') G^{\gtrless}_{\{n_i^-\}}(t, t') G^{\gtrless}_k(t, t') \\ + \varepsilon^{\gtrless}_{\{n_i\}} \sum_{\{n_i^+\}} \sum_k V_{\{n_i^+\}k}(t) V_{\{n_i^+\}k}(t') G^{\gtrless}_{\{n_i^+\}}(t, t') G^{\lessgtr}_k(t', t).\end{aligned} \tag{117}$$

In determining the expressions for the retarded self-energies, the terms propor-

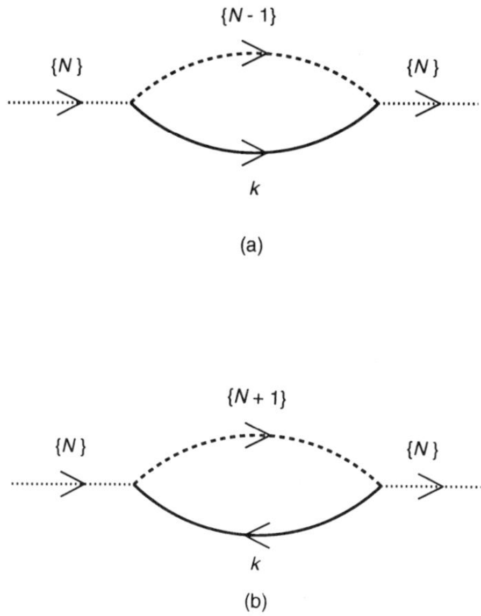

Figure 3.7 Lowest-order self-energies for the propagators $G_{\{n_i\}}(t, t')$. The notation $\{N\}$ refers to the atomic configuration $\{n_i\}$. The notation $\{N-1\}$ refers to an atomic configuration where one electron has been removed from $\{n_i\}$. The notation $\{N+1\}$ refers to atomic configurations obtained by adding an electron to the atom

tional to Q^2 must be projected out. We thus obtain

$$\Sigma^{A,R}_{\{n_i\}}(t, t') = \sum_{\{n_i^-\}} \sum_k V_{\{n_i^-\}k}(t) V_{\{n_i^-\}k}(t') G^{A,R}_{\{n_i^-\}}(t, t') G^{>}_k(t, t')$$
$$+ \sum_{\{n_i^+\}} \sum_k V_{\{n_i^+\}k}(t) V_{\{n_i^+\}k}(t') G^{A,R}_{\{n_i^+\}}(t, t') G^{<}_k(t', t). \tag{118}$$

In a similar fashion to that of the previous two subsections, the widths of the atomic configurations can be introduced in the summations over the conduction-band states:

$$\sum_k V_{\{n_i^-\}k}(t) V_{\{n_i^-\}k}(t') G^{\gtrless}_k(t, t') = \sqrt{\Gamma_{\{n_i^-\}}(t) \Gamma_{\{n_i^-\}}(t')}\, f^{\gtrless}(t - t'),$$
$$\sum_k V_{\{n_i^+\}k}(t) V_{\{n_i^+\}k}(t') G^{\gtrless}_k(t, t') = \sqrt{\Gamma_{\{n_i^+\}}(t) \Gamma_{\{n_i^+\}}(t')}\, f^{\gtrless}(t - t'). \tag{119}$$

We also introduce the notation

$$\tilde{G}_{\{n_i\}}(t, t') = G_{\{n_i\}}(t, t') \exp\left[i \int_{t'}^{t} d\tau\, \varepsilon_{\{n_i\}}(\tau) \right], \tag{120}$$

for the Green's functions and

$$\tilde{f}^{\gtrless}_{\{n_i^-\}}(t,t') = f^{\gtrless}(t-t')\exp\left[i\int_{t'}^{t} d\tau\,(\varepsilon_{\{n_i\}}(\tau)-\varepsilon_{\{n_i^-\}}(\tau))\right],$$

$$\tilde{f}^{\gtrless}_{\{n_i^+\}}(t,t') = f^{\gtrless}(t-t')\exp\left[i\int_{t'}^{t} d\tau\,(\varepsilon_{\{n_i\}}(\tau)-\varepsilon_{\{n_i^+\}}(\tau))\right], \tag{121}$$

for the Fourier transforms $f^{\gtrless}(t-t')$.

The Dyson equations for the propagators have the form

$$\begin{aligned}
i\frac{\partial}{\partial t}\tilde{G}^{<}_{\{n_i\}}(t,t') = {} & \sum_{\{n_i^-\}}\int_{-\infty}^{\infty} d\bar{t}\,[\sqrt{\Gamma_{\{n_i^-\}}(t)\Gamma_{\{n_i^-\}}(\bar{t})}\,\tilde{f}^{>}_{\{n_i^-\}}(t,\bar{t})\tilde{G}^{<}_{\{n_i\}}(\bar{t},t')\tilde{G}^{R}_{\{n_i^-\}}(t,\bar{t}) \\
& + \varepsilon^{<}_{\{n_i\}}\sqrt{\Gamma_{\{n_i^-\}}(t)\Gamma_{\{n_i^-\}}(\bar{t})}\,\tilde{f}^{<}_{\{n_i^-\}}(t,\bar{t})\tilde{G}^{<}_{\{n_i^-\}}(t,\bar{t})\tilde{G}^{A}_{\{n_i\}}(\bar{t},t')] \\
& + \sum_{\{n_i^+\}}\int_{-\infty}^{\infty} d\bar{t}\,[\sqrt{\Gamma_{\{n_i^+\}}(t)\Gamma_{\{n_i^+\}}(\bar{t})}\,\tilde{f}^{<}_{\{n_i^+\}}(\bar{t},t)\tilde{G}^{<}_{\{n_i\}}(\bar{t},t')\tilde{G}^{R}_{\{n_i^+\}}(t,\bar{t}) \\
& + \varepsilon^{<}_{\{n_i\}}\sqrt{\Gamma_{\{n_i^+\}}(t)\Gamma_{\{n_i^+\}}(\bar{t})}\,\tilde{f}^{>}_{\{n_i^+\}}(\bar{t},t)\tilde{G}^{<}_{\{n_i^+\}}(t,\bar{t})\tilde{G}^{A}_{\{n_i\}}(\bar{t},t')].
\end{aligned} \tag{122}$$

The Dyson equations for the advanced and retarded Green's function have the form:

$$\begin{aligned}
i\frac{\partial}{\partial t}\tilde{G}^{A,R}_{\{n_i\}}(t,t') = {} & \delta(t-t') \\
& + \sum_{\{n_i^-\}}\int_{-\infty}^{\infty} d\bar{t}\sqrt{\Gamma_{\{n_i^-\}}(t)\Gamma_{\{n_i^-\}}(\bar{t})}\,\tilde{f}^{>}_{\{n_i^-\}}(t,\bar{t})\tilde{G}^{A,R}_{\{n_i^-\}}(t,\bar{t})\tilde{G}^{A,R}_{\{n_i\}}(\bar{t},t') \\
& + \sum_{\{n_i^+\}}\int_{-\infty}^{\infty} d\bar{t}\sqrt{\Gamma_{\{n_i^+\}}(t)\Gamma_{\{n_i^+\}}(\bar{t})}\,\tilde{f}^{<}_{\{n_i^+\}}(\bar{t},t)\tilde{G}^{A,R}_{\{n_i^+\}}(t,\bar{t})\tilde{G}^{A,R}_{\{n_i\}}(\bar{t},t')
\end{aligned} \tag{123}$$

3.2.4.3 Master equation for charge transfer

Performing the SCA on the Dyson equations for the propagators we obtain for the population $n_{\{n_i\}}$ of the impurity configurations:

$$\begin{aligned}
\frac{d}{dt}n_{\{n_i\}} = {} & \sum_{\{n_i^-\}}\Gamma_{\{n_i^-\}}[-(1-f(\varepsilon_{\{n_i\}}-\varepsilon_{\{n_i^-\}})n_{\{n_i\}} + f(\varepsilon_{\{n_i\}}-\varepsilon_{\{n_i^-\}})n_{\{n_i^-\}}] \\
& + \sum_{\{n_i^+\}}\Gamma_{\{n_i^+\}}[-f(\varepsilon_{\{n_i^+\}}-\varepsilon_{\{n_i\}})n_{\{n_i\}} + (1-f(\varepsilon_{\{n_i^+\}}-\varepsilon_{\{n_i\}}))n_{\{n_i^+\}}].
\end{aligned} \tag{124}$$

This SME describes the instantaneous populations of the different atomic configurations. To illustrate the features of this equation, let us now consider a simple two-level problem. For this case, there are four different atomic configurations. The empty state will be denoted $|E\rangle$, the singly occupied atomic states are denoted $|P_1\rangle = |P_\sigma\rangle$ and $|P_2\rangle = |P_{-\sigma}\rangle$ and the doubly occupied state is denoted $|D\rangle$. As shown before, the energies of these states are $\varepsilon_E = 0$, $\varepsilon_{P_1} = \varepsilon_1$, $\varepsilon_{P_2} = \varepsilon_2$ and $\varepsilon_D = \varepsilon_1 + \varepsilon_2 + U_{12}$. We use the notation Γ_{i+} for the tunneling rates into level $|i\rangle$ when the other level is already populated. The SME describing the evolution of the population of these states takes the form:

$$\begin{aligned}
\frac{\mathrm{d}}{\mathrm{d}t} n_E &= -\sum_{i=1}^{2} \Gamma_i f(\varepsilon_i) n_E + \sum_{i=1}^{2} \Gamma_i (1 - f(\varepsilon_i)) n_{P_i}, \\
\frac{\mathrm{d}}{\mathrm{d}t} n_{P_1} &= -\Gamma_1 [1 - f(\varepsilon_1)] n_{P_1} + \Gamma_1 f(\varepsilon_1) n_E \\
&\quad - \Gamma_{2+} f(\varepsilon_2 + U_{12}) n_{P_1} + \Gamma_{2+} [1 - f(\varepsilon_2 + U_{12})] n_D, \\
\frac{\mathrm{d}}{\mathrm{d}t} n_{P_2} &= -\Gamma_2 [1 - f(\varepsilon_2)] n_{P_2} + \Gamma_2 f(\varepsilon_2) n_E \\
&\quad - \Gamma_{1+} f(\varepsilon_1 + U_{12}) n_{P_2} + \Gamma_{1+} [1 - f(\varepsilon_1 + U_{12})] n_D, \\
\frac{\mathrm{d}}{\mathrm{d}t} n_D &= -\sum_{i=1}^{2} \Gamma_{i+} [1 - f(\varepsilon_i + U_{12})] n_D \\
&\quad + \Gamma_{1+} f(\varepsilon_1 + U_{12}) n_{P_2} + \Gamma_{2+} f(\varepsilon_2 + U_{12}) n_{P_1}.
\end{aligned} \tag{125}$$

The populations of the levels $|1\rangle$ and $|2\rangle$ can be directly obtained from these expressions using equation (112),

$$\begin{aligned}
n_1(t) &= n_{P_1}(t) + n_D(t), \\
n_2(t) &= n_{P_2}(t) + n_D(t).
\end{aligned} \tag{126}$$

Using these relations, the following master equations can be derived for the populations of the one-particle states $|1\rangle$ and $|2\rangle$,

$$\begin{aligned}
\frac{\mathrm{d}}{\mathrm{d}t} n_1 &= (1 - n_2)[-\Gamma_1 n_1 + \Gamma_1 f(\varepsilon_1)] + n_2[-\Gamma_{1+} n_1 + \Gamma_{1+} f(\varepsilon_1 + U_{12})], \\
\frac{\mathrm{d}}{\mathrm{d}t} n_2 &= (1 - n_1)[-\Gamma_2 n_2 + \Gamma_2 f(\varepsilon_2)] + n_1[-\Gamma_{2+} n_2 + \Gamma_{2+} f(\varepsilon_2 + U_{12})].
\end{aligned} \tag{127}$$

These master equations could easily have been derived using intuitive arguments. The equations reduce to the proper rate equations in the limit of $U = 0$ and $U = \infty$. In Section 2.3, these equations will be applied to some typical atom–surface scattering situations.

3.2.5 QUASI-RESONANT CHARGE EXCHANGE

In He^+ scattering against certain surfaces, pronounced oscillations of the neutralization rate as a function of inverse velocity have been observed. It has been proposed that these are Stueckelberg oscillations, i.e. oscillations due to quasi-resonant charge exchange between the He^+ state and deep-filled d-states of the surface (Tully, 1977).

The dynamics of such a two-level problem can be described by the following simple Hamiltonian:

$$H(t) = \varepsilon_0 n_0 + \varepsilon_d n_d + V_{d0}(t) c_d^+ c_0 + V_{d0}^*(t) c_0^+ c_d \tag{128}$$

where c_d^+ and c_0^+ are the creation operators for the surface d-state and the atomic ground state respectively. This Hamiltonian can be solved trivially and results in a quantum oscillator. In the limit where the energy difference between the two levels is much smaller than the coupling-matrix element, the solution is analytic and takes the form:

$$n_0(t) = n_0(-\infty)\cos^2\left[\int_{-\infty}^{t} d\tau\, V_{d0}(\tau)\right] + n_d(-\infty)\sin^2\left[\int_{-\infty}^{t} d\tau\, V_{d0}(\tau)\right],$$

$$n_d(t) = n_d(-\infty)\cos^2\left[\int_{-\infty}^{t} d\tau\, V_{d0}(\tau)\right] + n_0(-\infty)\sin^2\left[\int_{-\infty}^{t} d\tau\, V_{d0}(\tau)\right]. \tag{129}$$

In an atomic collision, the coupling-matrix element depends on the relative separation of the atoms. The velocities under study will be relatively similar, so one can assume that the atomic trajectories during the collision are relatively similar. For an idealized trajectory, the time integral can therefore be approximated by

$$\int_{-\infty}^{\infty} d\tau\, V_{d0}(\tau) = \frac{2}{v}\int_{-\infty}^{Z_t} dZ\, V_{d0}(Z) = \frac{\alpha}{v}, \tag{130}$$

where Z_t represents the turning point of the trajectory. The post-collision population of the atomic ground state will thus oscillate as a function of inverse velocity.

Recently, He^+ scattering experiments have been performed on various clean Pb surfaces (Oelschig, forthcoming). It has been shown that the nature of these oscillations depends on the facet of the surface. These effects cannot be given a simple explanation in terms of resonant charge exchange.

In typical situations, the formation of an excited He atom, (He(1s2s) or He(1s2p)) near the surface would be a process competing with the resonant charge exchange. The formation of such a state would restrict the resonant charge exchange due to the strong intra-atomic correlation in the helium atom and results in interference effects in the charge exchange. It is possible that the experimentally observed facet-dependence could be caused by such effects. The tunneling rates between the metal conduction band and the excited He states are different for different facets of the surface.

The effects of formation of the excited He state on the resonant-charge-exchange problem have been studied previously using projection operator techniques (Bloss and Hone, 1978) and using optical potentials (Janev *et al.*, 1987). Due to the approximations involved in these calculations, the results are of limited applicability. The slave-boson approach, presented in Section 3.2.3, offers a general and accurate method of solution to this complicated charge-exchange phenomenon. In the following subsection, we will describe this application.

3.2.5.1 Model Hamiltonian

Quasi-resonant charge exchange in atomic collisions can occur if the two atoms come so close that the overlap between two near-resonant levels becomes large enough that an electron can transfer several times during the collision time. This phenomenon can occur at keV energies in the scattering of inert-gas ions against certain surfaces such as lead and gallium (Rusch and Erickson, 1976). The ionization level of the inert-gas atom lies within a few electron volts of the localized d-states of the surface. The neutral excited states of the inert-gas atom overlap with the conduction band of the metal. The charge exchange between the atom and the metal is believed to occur mainly through these two channels. Electrons can tunnel between the conduction band and the excited neutral atomic states as well as between the atomic ground state and the localized d-state of the surface. Since the inert-gas atom is a perfect example of a system with large intra-atomic Coulomb correlation, it is expected that the time-dependent Anderson Hamiltonian similar to equation (43) should well describe these types of problems.

The time-dependent slave-boson Hamiltonian for this problem has the following form:

$$H(t) = \sum_k \varepsilon_k n_k(t) + \varepsilon_\mathrm{d} n_\mathrm{d}(t) + \varepsilon_0(t) n_0(t) + \sum_\sigma \varepsilon_\sigma(t) n_\sigma(t)$$

$$+ \left[V_{\mathrm{d}0}(t) C_0^\dagger(t) c_\mathrm{d}(t) b(t) + \sum_{k\sigma} V_{k\sigma}(t) C_\sigma^\dagger(t) c_{k\sigma}(t) b(t) + \mathrm{H.C.} \right], \quad (131)$$

where n_k is the number operator for the conduction electrons, n_d for the localized surface d-state, n_0 for atomic ground state and n_σ for the atomic excited state. The index k is taken to include the spin, and labels the conduction electrons. The subscripts d, σ and 0 refer to the localized d-states of the surface, the excited atomic states, and ground state of the atom respectively. The hopping-matrix element $V_{\mathrm{d}0}$ represents the overlap between the atomic ground state and the surface d-state and $V_{k\sigma}$ represents the hopping between the excited atomic states and the surface conduction band.

As outlined in Section 3.2.3, one can define a generalized charge,

$$Q_B(t) \equiv n_0(t) + \sum_\sigma n_\sigma(t) + n_b(t), \tag{132}$$

where n_b is the number operator for the slave-boson. It can easily be shown that Q_B is a constant of motion. Thus, the projection techniques previously presented in Section 3.2.3 can be applied for the calculation of the retarded or the advanced self-energies.

The Auger neutralization processes are generally believed (Tully, 1977) to be of secondary importance when resonant tunneling can occur. Therefore this deexcitation channel is not explicitly included in the Hamiltonian. However, it can be shown that the Auger neutralization process is implicitly contained in this Hamiltonian through higher-order scattering.

The proposed Hamiltonian differs from the conventional Anderson Hamiltonian equation (1) due to the presence of the atomic ground state and the localized d-states of the surface. The populations of the atomic states are correlated so that whenever an electron tunnels into the ion, the boson associated with the atomic states will be annihilated and vice versa. In the narrow band limit, as is the case for the surface d-states, the tunneling between the atomic ground state and the surface d-states reduces to quasi-resonant charge exchange resulting in Stückelberg oscillations.

3.2.5.2 General equations

The lowest-order self-energies from the Hamiltonian equation (131) can be obtained straightforwardly. The self-energy for the atomic ground state can be expressed:

$$\Sigma_0^{\gtrless}(t, t') = V_{d0}(t)V_{d0}^*(t')G_d^{\gtrless}(t, t')B^{\lessgtr}(t, t') \tag{133}$$

where $V_{d0}^*(t)$ is the complex conjugate of $V_{d0}(t)$, $B^{\lessgtr}(t, t')$ are the analytic pieces of the Green's function for the boson and $G_d^{\gtrless}(t, t')$ are those for the surface d-state. The self-energy for the boson due to the charge-exchange channel V_{d0} can be written in the following form:

$$\Pi_0^{\gtrless}(t, t') = V_{d0}^*(t)V_{d0}(t')G_0^{\gtrless}(t, t')G_d^{\gtrless}(t', t), \tag{134}$$

where $G_0^{\gtrless}(t, t')$ are the analytic pieces of the Green's function of the atomic ground state.

Using the projection procedures described in Section 3.2.3, the retarded and advanced self-energies can be written:

$$\Sigma_0^{R,A}(t, t') = V_{d0}(t)V_{d0}^*(t')G_d^{>}(t, t')B^{R,A}(t, t'), \tag{135}$$

and

$$\Pi_0^{R,A}(t, t') = V_{d0}^*(t)V_{d0}(t')G_d^{<}(t', t)G_0^{R,A}(t, t'). \tag{136}$$

To lowest order of $V_{\mathrm{d}0}$, it is sufficient to take just the zeroth-order term for G_{d} in the expressions for the self-energy above. Therefore, the self-energies can be written as

$$\Sigma_0^{\gtrless}(t,t') = V_{\mathrm{d}0}(t)V_{\mathrm{d}0}^*(t')n_{\mathrm{d}}^{\gtrless}B^{\gtrless}(t,t')\exp\left[-\mathrm{i}\varepsilon_{\mathrm{d}}(t-t')\right] \tag{137}$$

and

$$\Pi_0^{\gtrless}(t,t') = V_{\mathrm{d}0}^*(t)V_{\mathrm{d}0}(t')G_0^{\gtrless}(t,t')n_{\mathrm{d}}^{\lessgtr}\exp\left[\mathrm{i}\varepsilon_{\mathrm{d}}(t-t')\right], \tag{138}$$

where $n_d^< = n_{\mathrm{d}}$ is the occupation number of the surface d state and $n_{\mathrm{d}}^> = 1 - n_{\mathrm{d}}$. The retarded self-energies can be expressed accordingly as

$$\Sigma_0^{\mathrm{R,A}}(t,t') = V_{\mathrm{d}0}(t)V_{\mathrm{d}0}^*(t')(1-n_{\mathrm{d}})B^{\mathrm{R,A}}(t,t')\exp\left[-\mathrm{i}\varepsilon_{\mathrm{d}}(t-t')\right] \tag{139}$$

and

$$\Pi_0^{\mathrm{R,A}}(t,t') = V_{\mathrm{d}0}^*(t)V_{\mathrm{d}0}(t')n_{\mathrm{d}}G_0^{\mathrm{R,A}}(t,t')\exp\left[\mathrm{i}\varepsilon_{\mathrm{d}}(t-t')\right]. \tag{140}$$

The self-energies for the excited atomic states, Σ_σ, and for the slave-boson due to tunneling channel $V_{k\sigma}$, Π_σ, are the same as those derived in Section 3.2.3. The self-energy for the slave-boson is the sum of the contributions from the different scattering channels, namely:

$$\Pi = \Pi_0 + \sum_\sigma \Pi_\sigma. \tag{141}$$

The Dyson equation for the retarded Green's function of the atomic ground state can be written

$$\left[\mathrm{i}\frac{\partial}{\partial t} - \varepsilon_0(t)\right]G_0^{\mathrm{R}}(t,t') = \delta(t-t') + \int \mathrm{d}\bar{t}\,V_{\mathrm{d}0}(t)V_{\mathrm{d}0}^*(\bar{t})(1-n_{\mathrm{d}})B^{\mathrm{R}}(t,\bar{t})G_0^{\mathrm{R}}(\bar{t},t')\mathrm{e}^{-\mathrm{i}\varepsilon_{\mathrm{d}}(t-\bar{t})}. \tag{142}$$

The equation for the 'less than' Green's function takes the form:

$$\left[\mathrm{i}\frac{\partial}{\partial t} - \varepsilon_0(t)\right]G_0^<(t,t') = \int_{-\infty}^{\infty} \mathrm{d}\bar{t}\,V_{\mathrm{d}0}(t)V_{\mathrm{d}0}^*(\bar{t})(1-n_{\mathrm{d}})B^{\mathrm{R}}(t,\bar{t})G_0^<(\bar{t},t')\mathrm{e}^{-\mathrm{i}\varepsilon_{\mathrm{d}}(t-\bar{t})} + \int_{-\infty}^{\infty} \mathrm{d}\bar{t}\,V_{\mathrm{d}0}(t)V_{\mathrm{d}0}^*(\bar{t})n_{\mathrm{d}}B^<(t,\bar{t})G_0^{\mathrm{A}}(\bar{t},t')\mathrm{e}^{-\mathrm{i}\varepsilon_{\mathrm{d}}(t-\bar{t})}. \tag{143}$$

The Dyson equation for the slave-boson has a similar form as in Section 3.2.3 except for the presence of the contribution from the tunneling channel $V_{\mathrm{d}0}$. The equation has the form:

$$\mathrm{i}\frac{\partial}{\partial t}B^{\mathrm{R}}(t,t') = \delta(t-t') + \int_{-\infty}^{\infty} \mathrm{d}\bar{t}\,V_{\mathrm{d}0}^*(t)V_{\mathrm{d}0}(\bar{t})n_{\mathrm{d}}G_0^{\mathrm{R}}(t,\bar{t})B^{\mathrm{R}}(\bar{t},t')\mathrm{e}^{\mathrm{i}\varepsilon_{\mathrm{d}}(t-\bar{t})} + \sum_\sigma \int_{-\infty}^{\infty} \mathrm{d}\bar{t}\sqrt{\Gamma_\sigma(t)\Gamma_\sigma(\bar{t})}\tilde{f}_\sigma^<(\bar{t},t)\tilde{G}_\sigma^{\mathrm{R}}(t,\bar{t})B^{\mathrm{R}}(\bar{t},t'). \tag{144}$$

The equation for the 'less than' Green's function for the slave-boson has the form

$$\begin{aligned}
i\frac{\partial}{\partial t}B^{<}(t,t') = &\int_{-\infty}^{\infty} d\bar{t}\, V_{d0}^{*}(t)V_{d0}(\bar{t})n_{d}G_{0}^{R}(t,\bar{t})B^{<}(\bar{t},t')e^{i\varepsilon_{d}(t-\bar{t})} \\
&+\int_{-\infty}^{\infty} d\bar{t}\, V_{d0}^{*}(t)V_{d0}(\bar{t})(1-n_{d})G_{0}^{<}(t,\bar{t})B^{A}(\bar{t},t')e^{i\varepsilon_{d}(t-\bar{t})} \\
&+\sum_{\sigma}\int_{-\infty}^{\infty} d\bar{t}\sqrt{\Gamma_{\sigma}(t)\Gamma_{\sigma}(\bar{t})}\,\tilde{f}_{\sigma}^{<}(\bar{t},t)\tilde{G}_{\sigma}^{R}(t,\bar{t})B^{<}(\bar{t},t') \\
&+\sum_{\sigma}\int_{-\infty}^{\infty} d\bar{t}\sqrt{\Gamma_{\sigma}(t)\Gamma_{\sigma}(\bar{t})}\,\tilde{f}_{\sigma}^{>}(\bar{t},t)\tilde{G}_{\sigma}^{<}(t,\bar{t})B^{A}(\bar{t},t'). \qquad (145)
\end{aligned}$$

The Dyson equations of the excited atomic states are the same as equations (52) and (50). Using equation (11), it is trivial to show that the charge $Q_{B}(t)$ is conserved.

In most problems, the localized d-state lies far below the Fermi energy. Therefore, this state will be completely filled in equilibrium. Thus, a further assumption can be made to simplify the equations above, i.e. $n_{d}=1$. We have to emphasize that although this assumption is physical, the numerical solution of the equations can be obtained for any occupancy of the d-levels. This assumption leads to the decoupling of the atomic ground state from the slave-boson and excited atomic states. The retarded Green's function of the atomic ground state can be solved directly and it is the same as that of the free electron.

In order to simplify the equations, the following notation is introduced:

$$\tilde{G}_{0}(t,t') \equiv G_{0}(t,t')\exp\left[\int_{t'}^{t} d\tau\,\varepsilon_{0}(\tau)\right] \qquad (146)$$

and

$$\tilde{V}_{d0}(t) \equiv V_{d0}(t)\exp\left[-\int_{t_0}^{t} d\tau[\varepsilon_{d}-\varepsilon_{0}(\tau)]\right] \qquad (147)$$

where t_0 is a fixed time. The Dyson equations for the retarded Green's functions of the slave-boson can be written as:

$$\begin{aligned}
\frac{\partial}{\partial t}b(t,t') = &-\int_{t'}^{t} d\bar{t}\,\tilde{V}_{d0}(t)\tilde{V}_{d0}^{*}(\bar{t})b(\bar{t},t') \\
&-\sum_{\sigma}\int_{t'}^{t} d\bar{t}\sqrt{\Gamma_{\sigma}(t)\Gamma_{\sigma}(\bar{t})}\,\tilde{f}_{\sigma}^{<}(\bar{t},t)g_{\sigma}(t,\bar{t})b(\bar{t},t'), \qquad (148)
\end{aligned}$$

where b is defined in equation (56) and the values of b and g_σ at equal time are known to be unity.

The Dyson equation for the 'less than' Green's function of the slave-boson can be derived as

$$\begin{aligned}\frac{\partial}{\partial t}B^{<}(t,t') = &-\int_{-\infty}^{t} \mathrm{d}\bar{t}\,\tilde{V}_{\mathrm{d}0}(t)\tilde{V}_{\mathrm{d}0}(\bar{t})B^{<}(\bar{t},t') \\ &+\sum_{\sigma}\int_{-\infty}^{t'} \mathrm{d}\bar{t}\sqrt{\Gamma_\sigma(t)\Gamma_\sigma(\bar{t})}\,\tilde{f}_{\sigma}^{>}(\bar{t},t)\tilde{G}_{\sigma}^{<}(t,\bar{t})b^{*}(t',\bar{t}) \\ &-\sum_{\sigma}\int_{-\infty}^{t'} \mathrm{d}\bar{t}\sqrt{\Gamma_\sigma(t)\Gamma_\sigma(\bar{t})}\,\tilde{f}_{\sigma}^{<}(\bar{t},t)g_\sigma(t,\bar{t})B^{<}(\bar{t},t'),\end{aligned} \tag{149}$$

where g_σ is defined in equation (55).

The Dyson equations for the retarded and the 'less than' Green's functions of the excited atomic states are the same as the ones discussed in Section 3.2.3, and therefore given by the equations (57) and (64).

The coupled retarded Green's functions for the slave-boson and the excited atomic states can be solved using the algorithms presented in Section 3.2.3. The solution to the coupled 'less than' Green's functions can then be obtained using the appropriate algorithm as demonstrated in Section 3.2.3. The population of the atomic ground state can be calculated using the conservation of the charge, Q_{B}, i.e. $n_0(t) = 1 - n_b(t) - \sum_\sigma n_\sigma(t)$.

The approach presented here represents an extension of the slave-boson method discussed in Section 3.2.3. It is a very convenient method for taking into account the effects of intra-atomic Coulomb correlation. In the following subsection, the numerical solutions to these equations will be discussed.

3.2.5.3 Results and discussion

In order to make the presentation simple, we will again consider an idealized two-level system consisting of a ground state located at an energy ε_0 and an excited state at an energy ε_1. The initial discussion will assume stationary levels of degeneracy 1, although the present method works as well for any degeneracy or level shifting.

The effect of intra-atomic correlation on quasi-resonant charge-transfer has previously been investigated by Bloss and Hone (1978). These authors make use of the fact that the interaction between the atomic ground state and the d-band of the solid is much more localized in time than the interaction between the excited states and the conduction electrons. This is due to the fact that the wave functions of the localized d-states in the metal are relatively compact compared to the wave functions of the conduction electrons at the Fermi energy. During the moments of impact on the surface, charge exchange involving the excited state will be very small. The integration of the Hamiltonian can therefore be performed in three different time regions: T_1: $-\infty < t < -t_0$, T_2: $-t_o < t < t_0$ and T_3: $t_o < t < \infty$. The $t = -t_0$ represents the moment the atom in scattering comes close enough to the surface for charge exchange through the d-state to occur. At $t = t_0$ during the outgoing trajectory, the atom is leaving this region. In region T_1, the Hamiltonian has the form:

$$H(t) = \varepsilon_1(t)n_1(t) \sum_k \varepsilon_k n_k(t) + \sum_k [V_{1k}(t)c_1^+(t)c_{1k}(t) + \text{H.C.}]. \tag{150}$$

Since only single occupancy is allowed on the atom, the excited state already populated must be projected out. In region T_2, the Hamiltonian therefore has the form:

$$H(t) = \varepsilon_0(t)n_0(t) + \varepsilon_d n_d(t) + [V_{d0}(t)[1 - P_1]c_0^+(t)c_d(t) + \text{H.C.}], \tag{151}$$

where P_1 is the time-independent projection operator onto state $|1\rangle$. In region T_3, to avoid double occupancy, the state in $|0\rangle$ must now be projected out. The Hamiltonian takes the form:

$$H(t) = \varepsilon_1(t)n_1(t) + \sum_k \varepsilon_k n_k(t) + + \sum_k [V_{1k}(t)[1 - P_0]c_1^+(t)c_{1k}(t) + \text{H.C.}], \tag{152}$$

where P_0 is the time-independent projection operator onto the atomic state $|0\rangle$.

The system is assumed to start with $n_0(-\infty) = 0$ and without any correlation between any of the atomic states and the metal states. The integration of the system during time interval T_1 is trivial and results in a fractional population $n_1(-t_0)$ due to the interaction between the excited atomic state and the conduction band. The integration during time interval T_2 is also trivial. The population of the excited state remains unchanged, $n_1(t_0) = n_1(-t_0)$ although, due to the projection operator, P_1, present in this time interval, the phase of this state change will change by a factor proportional to $\cos\alpha$. The population of the ground state becomes $n_0(t_0) = [1 - n_1(-t_0)]\sin^2\alpha$, where α was defined in equation (130). During time interval T_3, the integration is also straightforward. The population of the ground state remains unchanged, $n_0(\infty) = n_0(t_0)$. Because of the phase change in the excited state during time interval T_2, the final population of the excited state will contain factors proportional to both $\cos^2\alpha$ and $\cos\alpha$. The final neutralization fraction, $N = n_0(\infty) + n_1(\infty)$ is very sensitive to the details of the interaction and, depending on the situation, the dominant oscillation frequency may be either α or 2α. It is noteworthy that since the interaction between the excited state and the surface has been neglected in T_2, the final neutralization rate is independent of the functional form of $V_{d0}(Z)$ and is determined only by the quantity α, equation (130). The Bloss–Hone approach is therefore limited to a delta-function interaction.

In order to illustrate these effects, calculations of the neutralization rate will be performed for an idealized model system with parameters appropriate for

He/Pb. The localized d-states of the surface are treated as a single atomic core state with its energy being ε_d. The conduction band is modeled as discussed in Section 3.2.3. The hopping-matrix elements between the atomic states and the surface are parametrized as:

$$V_{d0}(Z) = V_{d0}^0 \exp[-0.5\alpha_0 Z],$$
$$\Gamma_1(Z) = \Gamma_1^0 \exp[-\alpha_1 Z]. \tag{153}$$

In Figure 3.8, the instantaneous populations of the ground state and the excited state calculated using the present exact approach during a typical scattering trajectory are plotted for on-resonance ($\varepsilon_0 = \varepsilon_1$) and off-resonance ($\varepsilon_0 \neq \varepsilon_1$) conditions. At large distances from the surface, only the excited state is involved in the neutralization. As the surface is approached, this state becomes partially filled. As the surface is approached further, the coupling between the atomic ground state and the d-state becomes significant and transitions between these two levels are induced. For the on-resonance case, this occurs already at

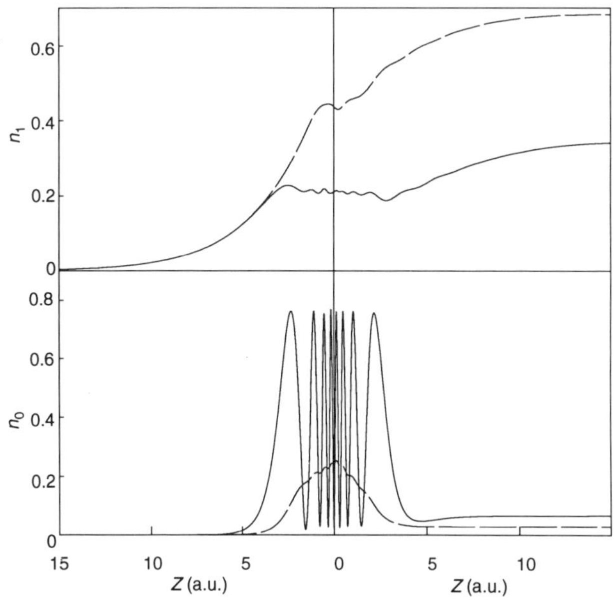

Figure 3.8 Instantaneous population of the atomic levels as a function of position outside the surface during a typical scattering event. The upper part of the figure shows $n_1(Z)$ and the lower part shows $n_0(Z)$. The kinetic energy of the atom is 1 keV. The turning point of the atom is at $Z = 0$. The bandwidth of the metal is 0.368 a.u. The energy levels are: $\varepsilon_F = -0.2$ a.u., $\varepsilon_d = -1.0$ a.u., $\varepsilon_1 - 1$ eV. The coupling parameters are $V_{d0}^0 = 31.2$ eV, $\alpha_0 = 1.82$ a.u., $\Gamma_1^0 = 1.0$ eV and $\alpha_1 = 0.368$ a.u. The solid curve is for the resonance case, $\varepsilon_0 - \varepsilon_d = 0.0$, and the long-dashed line is for $\varepsilon_0 - \varepsilon_d = -7.8$ eV

a distance of 5 a.u. from the surface. At this point, the population of the excited state is 0.2. It can clearly be seen that charge transfer is dominated by the population oscillations between the $|0\rangle$ and the $|d\rangle$ states. The amplitude of the oscillations is around 0.8, reflecting the fact that the total population of the two atomic states is restricted to 1. For the off-resonance case, significant interaction between the lower states does not take place until the relatively small atom–surface separation of 2.5 a.u. This is due to the fact that efficient coupling between the $|d\rangle$ and $|0\rangle$ levels requires a hopping-matrix element larger than the energy difference between the levels. Since the interactions between the lower levels is suppressed, the neutralization through the excited state is slightly enhanced. Only very weak oscillations are observed in this situation.

In Figure 3.9, the positive-ion fraction, $P = 1 - n_0(\infty) - n_1(\infty)$, is plotted as a function of inverse velocity for some different interaction parameters between the atomic ground state and the surface. It can be seen that the results markedly depend on the interaction parameter. Only in the near delta-function limit is the Bloss–Hone result recovered. For realistic couplings such as $\alpha_0 = 1$–5 a.u., the frequency of oscillations is twice that of Bloss and Hone.

In Figure 3.10, we show the effects of changing the resonance conditions between the atomic ground state and the d-state of the surface. It can clearly

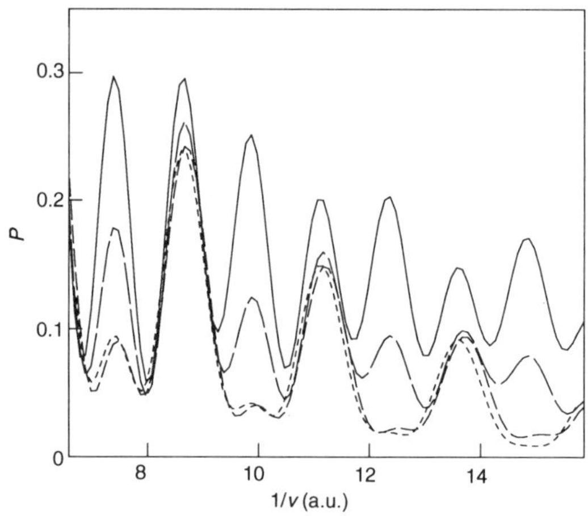

Figure 3.9 The calculated positive-ion yield P as a function of inverse velocity for various couplings between the atomic ground state and the surface. The kinetic energy ranges from 400 eV to 2.2 keV. The turning point of the atomic trajectory is at $Z = 0$ a.u. The energy levels are: $\varepsilon_F = -0.2$ a.u., $\varepsilon_d = \varepsilon_0 = -1.0$ a.u., $\varepsilon_1 - \varepsilon_F = -1$ eV. The parameters of the excited state coupling to the surface are: $\Gamma_1 = 1$ eV, $\alpha_1 = 0.368$ a.u. The parameters describing the coupling between the ground state and the surface have been chosen to give the same α in equation (130). The solid curve is for $V^0_{0d} = 2.294$ a.u. and $\alpha_0 = 3.64$ a.u. The long-dashed line is for $V^0_{0d} = 4.588$ a.u. and $\alpha_0 = 7.28$ a.u. The dashed line is for $V^0_{0d} = 36.70$ a.u. and $\alpha_0 = 58.24$ a.u. The dotted line is for the results of Bloss and Hone (1978)

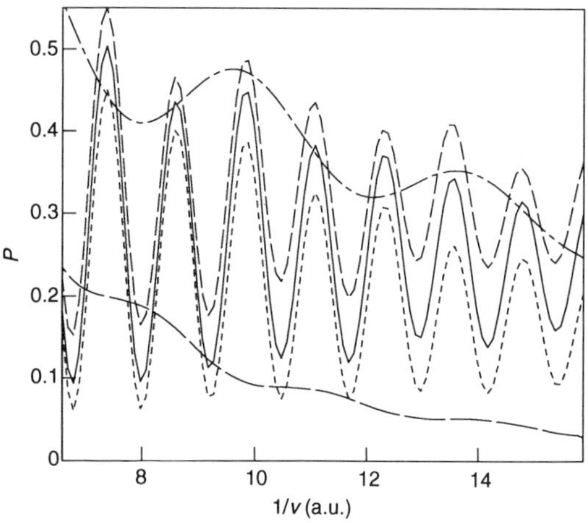

Figure 3.10 Calculated positive-ion yield as a function of inverse velocity for some different resonance conditions. The kinetic energy ranges from 400 eV to 2.2 keV. The turning point of the atomic trajectory is at $Z = 0$ a.u. The energy levels are: $\varepsilon_F = -0.2$ a.u., $\varepsilon_1 - \varepsilon_F = -1$ eV. The coupling parameters are: $\Gamma_1 = 1$ eV, $\alpha_1 = 0.368$ a.u., $V^0_{0d} = 31.2$ eV, $\alpha_0 = 1.82$ a.u. The long-dashed curve is for $\varepsilon_0 - \varepsilon_d = -7.8$ eV. The dotted curve is for $\varepsilon_0 - \varepsilon_d = -0.45$. The solid curve is for $\varepsilon_0 - \varepsilon_d = 0$. The dashed curve is for $\varepsilon_0 - \varepsilon_d = 0.45$. The dashed-dotted line is for $\varepsilon_0 - \varepsilon_d = 7.8$

be seen that, in particular, the amplitude of the oscillations depends on the relative position of the energy levels. The frequency of the oscillations is also influenced, but to a lesser degree since the hopping-matrix element in a typical atomic collision will be relatively large.

The proposed method for the study of quasi-resonant charge exchange is very general. In this section, we have illustrated its usefulness by considering a simple model system. An extensive discussion with more realistic applications will be published elsewhere (Shao *et al.*, forthcoming).

3.3 APPLICATIONS

In this section, we will apply the charge-transfer formalism to some typical atom–surface scattering situations. In the first subsection, we will discuss some general consequences of the intra-atomic correlation. We will demonstrate that the inclusion of the intra-atomic correlation can qualitatively change the charge-transfer results. In Section 3.3.2, the formalism will be applied to charge-transfer occurring in a recent experiment on direct recoil of oxygen atoms from a nickel surface. In Section 3.3.3, we will discuss charge transfer in a recent experiment concerning stimulated desorption of hydrogen atoms from a copper surface.

3.3.1 ION–SURFACE SCATTERING

We will illustrate how the effects of the intra-atomic correlation can qualitatively change the charge transfer in typical atom–surface scattering situations. The discussion will be on the level of the SME. As has been demonstrated previously, this approximation is expected to be reliable for typical velocities in experiments using hyperthermal atomic beams, i.e. $v = 0.001$ a.u.

For simplicity, we will consider a model system containing only two levels, $|1\rangle$ and $|2\rangle$, with energies $\varepsilon_i(Z)$. Although such an idealized system cannot be found in nature, it serves well for the qualitative demonstration of some important effects of the intra-atomic correlation.

Outside a clean metal surface, the shifts will be assumed image-like, $\varepsilon_i(Z) = \varepsilon_i^0 + 1/4Z$ (Nordlander and Tully, 1990). We will also study charge transfer outside alkali-covered surfaces. For such surfaces, the energy levels of an atom shift downwards (Nordlander and Lang, 1991). For these situations, the level shifts will be assumed linear, i.e. $\varepsilon_i(Z) = \varepsilon_F + b_i(Z - Z_c^i)$. The widths will always be assumed exponential, $\Gamma_i(Z) = \Gamma_i^0 \exp[-\alpha_i Z]$. The intra-atomic correlation varies with distance as $U(Z) = U^0 - 1/2Z$. The atom will start at infinity, move towards the surface with a velocity $v_\perp$ towards a turning point near the surface, after which it will move towards infinity with a velocity of $v_\perp$. The parameters chosen for the energy shifts and broadening of the levels are such that essentially no charge transfer will occur at distances Z larger than 20 a.u. from the surface. For convenience, we will take the distance Z as the independent variable instead of time.

In the absence of intra-atomic correlation effects, the populations of the two levels can be calculated from equation (40). The charge state is thus obtained by integrating the following simple equations:

$$v_\perp \frac{dn_1}{dZ} = -\Gamma_1[1 - f(\varepsilon_1)]n_1 + \Gamma_1 f(\varepsilon_1)[1 - n_1],$$

$$v_\perp \frac{dn_2}{dZ} = -\Gamma_2[1 - f(\varepsilon_2)]n_2 + \Gamma_2 f(\varepsilon_2)[1 - n_2]. \tag{154}$$

3.3.1.1 Case of large U

For large correlation, the two equations determining the populations of the impurity states take the form:

$$v_\perp \frac{dn_1}{dZ} = -\Gamma_1[1 - f(\varepsilon_1)]n_1 + \Gamma_1 f(\varepsilon_1)[1 - n_1 - n_2],$$

$$v_\perp \frac{dn_2}{dZ} = -\Gamma_2[1 - f(\varepsilon_2)]n_2 + \Gamma_2 f(\varepsilon_2)[1 - n_1 - n_2]. \tag{155}$$

We will now solve these equations for three typical cases where the intra-atomic correlation will play an important role (Langreth and Nordlander, 1991).

3.3.1.1.1 Spin-dependent scattering

The inclusion of spin effects influences the charge transfer even for the degenerate case. The inclusion of spin can lead to other important effects that could be probed using spin-polarized atom–surface scattering. In Figure 3.11, the scattering of a spin-polarized atom from a surface is modeled. The two spin levels are assumed to be degenerate and the surface is assumed to be nonmagnetic. The spin levels are denoted $|1\rangle$ and $|2\rangle$. The initial states of the atom are taken as $n_1 = 1$ and $n_2 = 0$. As the surface is approached, the level $|1\rangle$ is

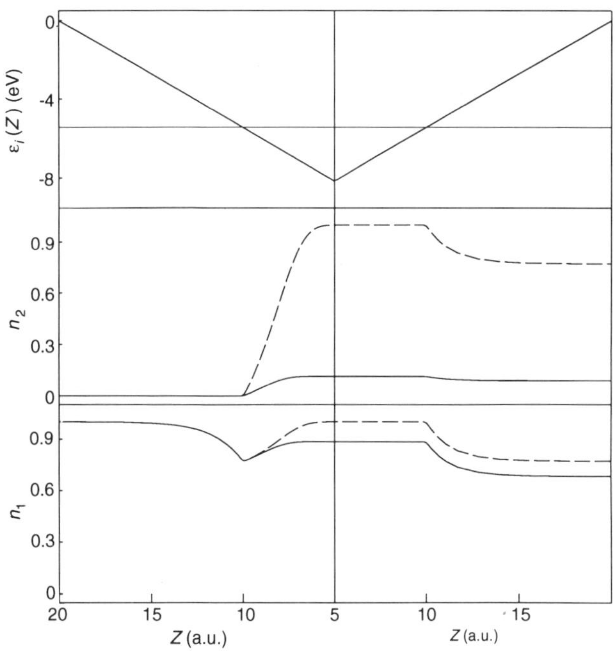

Figure 3.11 Comparison of the solutions of the master equations for infinite U and for $U = 0$ for a doubly generate atomic level. The turning point of the atomic trajectory is at $Z = 5$ a.u. The upper part of the figure shows the energy shift of the degenerate levels as a function of distance from the surface. The Fermi energy is indicated by the horizontal solid line. In the two lower parts of the figure, the instantaneous populations of the levels are shown. The solid lines refer to a calculation including the effects of intra-atomic correlation, equation (155) and the dashed lines refer to a calculation neglecting intra-atomic correlation using equation (154). The parameters of the system are: $v = 0.001$, $T = 300$ K, $Z_c^1 = Z_c^2 = 10$ a.u., $b_1 = b_2 = 0.02$, $\Delta_1 = \Delta_2 = 0.2$ a.u., and $\alpha_1 = \alpha_2 = 0.7$ a.u.

depopulated through resonant tunneling and the atom is increasingly ionized. At position $Z = 10$ a.u., where the atomic levels cross the Fermi energy, electrons can populate the atomic levels. At this distance, there exists a net spin polarization of the atom. The presence of the intra-atomic correlation restricts the total occupancy of the atomic levels to one and thus prevents the level with the opposite spin from being fully populated. As the atom moves toward the turning point, the net spin polarization is preserved since the tunneling rates into the different spin levels are the same. Thus, even at the turning point close to the surface, an ensemble of atoms will have a net spin polarization. This spin polarization is naturally maintained during the outgoing trajectory as well.

For the $U = 0$ case, both of the spin levels are completely occupied at the turning point due to the very efficient tunneling at such small distances from the surface. All spin polarization is therefore lost, which results in a complete equilibration with the magnetization of the surface.

3.3.1.1.2 Multiple levels

In the situation where several levels can be neutralized, the resulting charge state of a desorbing atom will also be determined by dynamical effects. As a rough model of proton–surface scattering against an alkali-covered metal surface, the state $|1\rangle$ will be taken to represent one of the H($n = 2$) states. This state is long-lived and crosses the Fermi energy at a relatively large distance from the surface. The $|2\rangle$ state will be taken to represent one of the short-lived H($n = 3$) states. This state crosses the Fermi energy relatively close to the surface and is very broad.

In Figure 3.12, we compare the calculated populations taking into account the intra-atomic correlation, i.e. equation (155) with a calculation using a one-electron approximation, i.e. equation (154). The initial state of the hydrogen atom is assumed to be an ion, i.e. $n_1 = 0$ and $n_2 = 0$. As the atom approaches the surface it reaches the point where the H($n = 2$) state shifts below the Fermi energy. After this point, the ion can be neutralized by resonant tunneling of an electron into this state. The tunneling rate is low, however, so the probability for neutralization of the ion in this channel is very small. At the position where the H($n = 3$) level crosses the Fermi energy, the atom is very efficiently neutralized by resonant tunneling into this state. The atom will stay in the H($n = 3$) state through the collision with the surface and during the initial part of its outgoing trajectory. The intra-atomic correlation prevents the H($n = 2$) state from being formed during this part of the trajectory. When the H($n = 3$) state again shifts above the Fermi energy, the H($n = 3$) state will re-ionize. At this point, the H($n = 2$) state can in principle be formed by resonant tunneling. However, since the tunneling rate is low at this position, the reneutralization is inefficient and most likely the atom will emerge from the surface as an ion.

These results are again in qualitative disagreement with a theory neglecting the intra-atomic correlation, in which case the two levels populate inde-

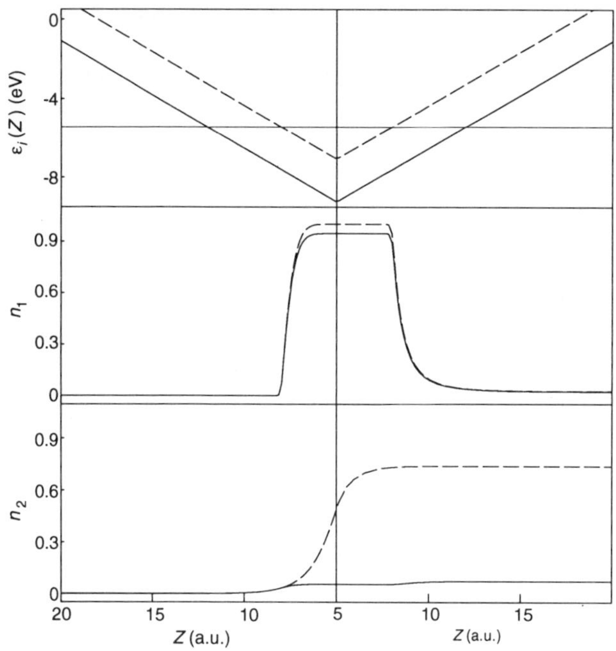

Figure 3.12 Comparison of the solutions of the master equations for infinite U and for $U = 0$, for the same levels as described in Figure 3.1. The format of this figure is the same as Figure 3.11. The parameters of the system are: $v = 0.001$, $T = 300$ K, $Z_c^1 = 12.0$ a.u., $b_1 = b_2 = 0.02$, $\Delta_1 = \Delta_2 = 0.1$, $\alpha_1 = 1.0$, $Z_c^2 = 8.0$ a.u., and $\alpha_2 = 0.5$. The solid lines refer to a calculation including the effects of intra-atomic correlation equation (155), and the dashed lines refer to a calculation neglecting intra-atomic correlation, i.e. equation (154)

pendently. The H($n = 2$) state will be almost completely filled due to the large tunneling rate into this state close to the surface. During the outgoing trajectory, as the atom passes the crossing distance where the H($n = 2$) state shifts above the Fermi energy, the state can be depopulated. However, since the tunneling rate is small at this relatively large distance from the surface, the probability for the atom to emerge with an electron in the H($n = 2$) state is very high. The H($n = 3$) state is populated and depopulated independently in the near-surface region and does not interfere with the dynamics of the H($n = 2$) state. The neglect of intra-atomic correlation thus leads to a wrong prediction of almost complete neutralization of an ensemble of atoms in this scattering experiment.

3.3.1.1.3 Almost-degenerate levels with different widths

A characteristic feature of the interaction of a hydrogen atom with metal surfaces is the surface-induced hybridization of atomic levels which leads to different geometrical orientations of the atomic orbitals. This effect can result in levels that have almost the same energy while possessing very different widths (Nordlander and Tully, 1988). Since the energies of these levels are similar, one might expect correlation effects to strongly influence the charge-transfer process. In Figure 3.13, a scattering experiment involving an atom with two almost-degenerate levels with different widths is modeled. The level $|1\rangle$ is taken to represent the long-lived state and the level $|2\rangle$ represents the state oriented towards the surface. The initial state of the atom is assumed to be ionic, i.e. $n_1 = 0$ and $n_2 = 0$. As the ion passes the crossing distance, the short-lived atomic level gets filled quickly. The occupation of the long-lived atomic level is therefore effectively blocked by the U term. When the atom passes the crossing distance during its outgoing trajectory, it starts to re-ionize. Since the tunneling rate into the long-lived state is now zero, this state can no longer be formed and the probability for the atom to emerge neutralized is vanishingly small.

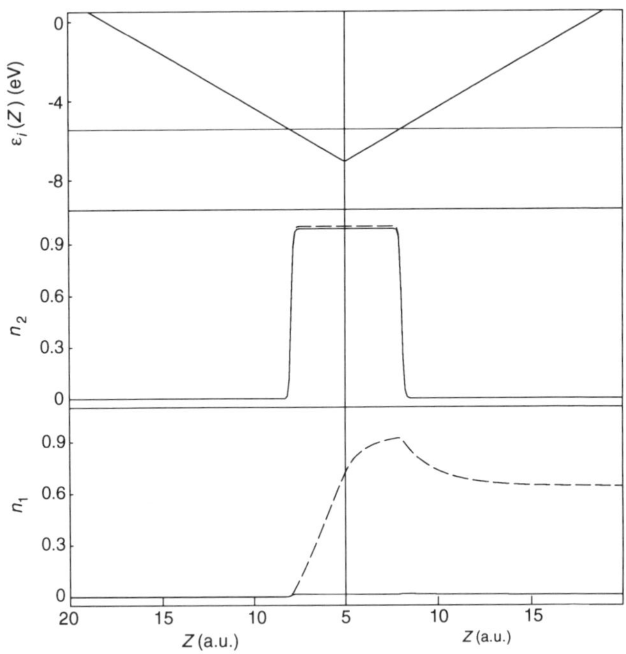

Figure 3.13 Comparison of the results obtained from the master equations for infinite U and $U = 0$, for two degenerate resonances with very different widths, displayed as in the two previous figures. The parameters of the system are: $v = 0.001$, $T = 300$ K, $Z_c^1 = Z_c^2 = 8.0$ a.u., $b_1 = b_2 = 0.02$, $\Delta_1 = 0.01$, $\Delta_2 = 1.0$ and $\alpha_1 = \alpha_2 = 1.0$. The work function of the metal is 0.2 a.u. The solid lines refer to a calculation including the effects of intra-atomic correlation, equation (155) and the dashed lines refer to a calculation neglecting intra-atomic correlation using equation (154)

The neglect of the intra-atomic correlation in this situation again would yield a qualitatively different result for an ensemble of atoms scattered against a surface. As the levels approach the surface, they will be filled independently of each other. At the turning point, both levels will be occupied. Once the atoms pass the crossing point, the atoms in the short-lived state will immediately re-ionize. Those in the long-lived state will survive, however. This result is again qualitatively different from what was obtained taking the intra-atomic correlation effects into account.

3.3.1.2 Case of finite U

For finite U, the equations describing the populations of the different atomic configurations have the form:

$$
\begin{aligned}
v_\perp \frac{\mathrm{d}}{\mathrm{d}Z} n_E &= -\sum_{i=1}^{2} \Gamma_i f(\varepsilon_i) n_E + \sum_{i=1}^{2} \Gamma_i [1 - f(\varepsilon_i)] n_{P_i}, \\
v_\perp \frac{\mathrm{d}}{\mathrm{d}Z} n_{P_1} &= -\Gamma_1 [1 - f(\varepsilon_1)] n_{P_1} + \Gamma_1 f(\varepsilon_1) n_E \\
&\quad -\Gamma_{2+} f(\varepsilon_2 + U_{12}) n_{P_1} + \Gamma_{2+} [1 - f(\varepsilon_2 + U_{12})] n_D, \\
v_\perp \frac{\mathrm{d}}{\mathrm{d}Z} n_{P_2} &= -\Gamma_2 [1 - f(\varepsilon_2)] n_{P_2} + \Gamma_2 f(\varepsilon_2) n_E \\
&\quad -\Gamma_{1+} f(\varepsilon_1 + U_{12}) n_{P_2} + \Gamma_{1+} [1 - f(\varepsilon_1 + U_{12})] n_D, \\
v_\perp \frac{\mathrm{d}}{\mathrm{d}Z} n_D &= -\sum_{i=1}^{2} \Gamma_{i+} [1 - f(\varepsilon_i + U_{12})] n_D \\
&\quad +\Gamma_{1+} f(\varepsilon_1 + U_{12}) n_{P_2} + \Gamma_{2+} f(\varepsilon_2 + U_{12})] n_{P_1}.
\end{aligned} \tag{156}
$$

Even in the two-level case, these equations can give rise to a rather complicated charge-transfer behavior. As can be seen from the equations, the system will behave as if there were four different levels, ε_1, ε_2, $\varepsilon_1 + U_{12}$ and $\varepsilon_2 + U_{12}$. The two latter energies do not correspond to real states. These show up due to the fact that the negative-ion state can be formed and decay through two different channels. To simplify our terminology in the discussions below, we will refer to these energies as negative levels. For this reason there will be four different crossings of the Fermi energy. As each of the levels crosses the Fermi energy, the charge-transfer process will be influenced.

3.3.1.2.1 Degenerate levels

In this subsection, a particularly simple system will be discussed. The levels $|1\rangle$ and $|2\rangle$ will be assumed degenerate. The intra-atomic correlation term

$U = 4$ eV. This choice of parameters is appropriate for the modeling of positive alkali ions (Li, Na) scattered against a metal surface (Nordlander and Tully, 1990; Nordlander, 1992). For this system, the state E represents the positive ion, the state D represents the negative ion and the states P_1 and P_2 represents the two different spin configurations of the ionization levels. In Figure 3.14, the instantaneous populations of these levels are plotted as functions of position outside the surface. It can be seen that the charge transfer is a relatively complicated function of distance. In the upper part of the figure, the energy shift of the ionization levels, $\varepsilon_i(Z) = \varepsilon_i^0 + 1/4Z$, and the affinity level, $\varepsilon_i - 1/4Z$, are plotted as functions of distance. The ionization levels cross the Fermi energy at a distance $Z = 2.7$ a.u. The affinity level crosses the Fermi energy at a distance of $Z = 4.5$ a.u. In the middle part of the figure, the populations of the two

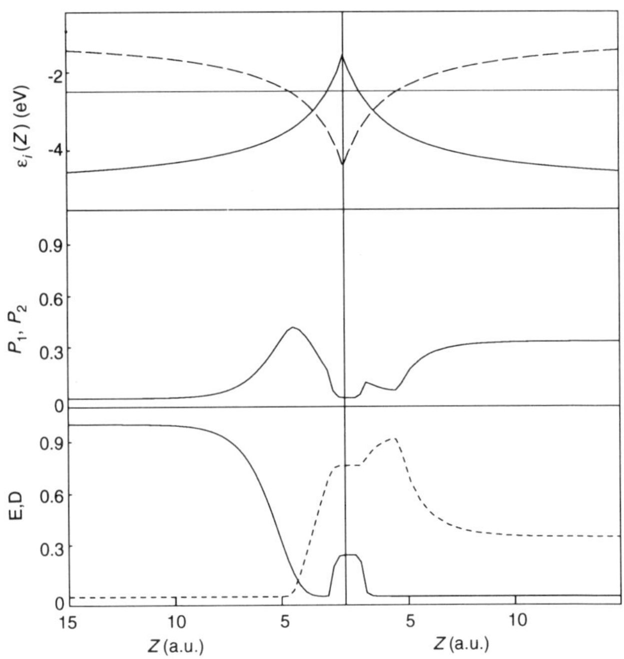

Figure 3.14 Populations of the different atomic configuration for finite U, calculated using equation (156). In the upper part of the figure, the distance dependences of the ionization levels (solid line) and the affinity levels (dashed line) are shown. The parameters for the levels are: $\varepsilon_i^0 = 5$ eV, $U^0 = 4$ eV, $\Gamma_i^0 = 10$ eV, $\alpha_i = 1.0$ a.u., $\Gamma_i^{0+} = 1$ eV and $\alpha_i^+ = 0.7$ a.u. The Fermi energy (2.5 eV) is indicated by the horizontal line. In the middle figure, the population of the degenerate singly populated states is shown. In the lower figure, the position-ion fraction (solid line) and negative-ion fraction (dotted line) are shown. The turning point of the trajectory is at $Z = 2$ a.u. and schematically indicated by the vertical line. The kinetic energy of the atom (Li) is 2 eV

different spin configurations of the ionization levels are plotted. In the lower part of the figure, the positive-ion fraction E and the negative-ion fraction D are plotted. The atom starts out as a positive ion. As the surface is approached, the populations of the ionization levels grow and, as a consequence, the positive-ion fraction decreases. At the point where the affinity level crosses the Fermi energy, the negative-ion fraction increases rapidly. The negative ions are formed by tunneling an electron into the singly occupied ionization states, the populations of which therefore decrease. At the position where the ionization levels shift above the Fermi energy, these states can depopulate and form the positive-ion state. At the same time, the negative-ion states will continue to form. In the area close to the surface, only the positive-ion and the negative-ion states will be stable. The singly occupied states will decay into one of these states by either losing an electron or forming a negative ion. The branching ratio between these two processes depends on the relative magnitude of the tunneling rates. When the energy levels of the singly occupied states shift below the Fermi energy during the outgoing trajectory, these states are again formed by neutralization of the positive ions. The figure clearly shows the increase in P_i and the decrease in E at this point. Due to the formation of the singly occupied states, there is also an increase in negative-ion formation. This continues until the affinity level again shifts above the Fermi energy. After this crossing, the negative-ion state decays into one of the singly populated states. The final charge state contains both negative ions and singly occupied states. For the work function and velocity used in this model, the positive-ion fraction leaving the surface is very small.

In Figure 3.15, we plot the final state distribution for the same system as discussed in Figure 3.14 as a function of the work function of the surface. Changing the work function means that the crossing distances are altered. Since the widths of the different levels have different spatial dependences, the branching ratios for the formation of the different states vary in a nontrivial manner. A rough understanding of the features in Figure 3.15 can be obtained from some simple arguments. For large work functions, no neutralization of the ion can occur resulting in a positive-ion fraction of 1. When the work function becomes lower than the ionization level of the atom, the probability for neutralization increases. At a work function around 3 eV, the positive-ion yield disappears. The onset for negative ions also occurs here, so all four different configurations of the atom can be observed at this work function. When the work function is decreased further, the negative-ion fraction increases at the expense of the neutral atoms.

In order to further illustrate the non-adiabatic effects that can occur when several levels are involved in the charge transfer, we have calculated the final-state populations for the system shown in Figure 3.14 as a function of the kinetic energy of the atom. The result is shown in Figure 3.16. Detailed understanding of the data plotted in the figure requires an extensive analysis.

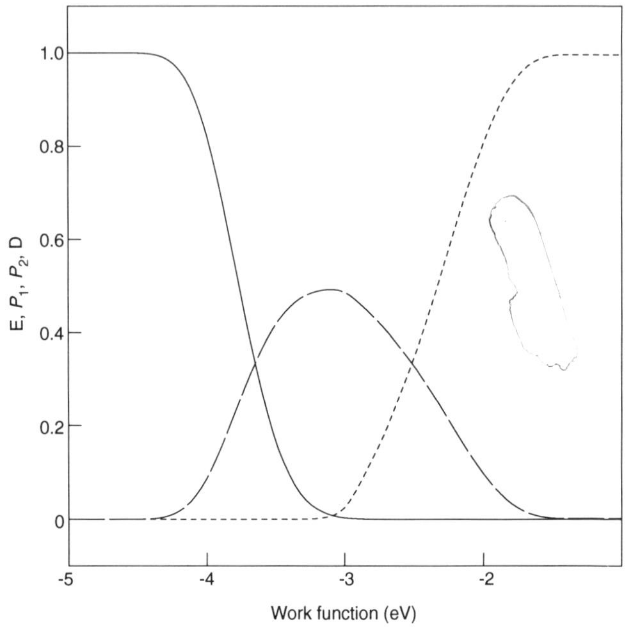

Figure 3.15 Positive-ion yield (solid line), yield of neutral atoms (long-dashed line) and negative-ion yield (dotted line) calculated using equation (156) as a function of work function. The energy-level parameters and atomic trajectory are as in Figure 3.14

A general understanding can be obtained using some relatively simple arguments. For the lowest velocities, the atoms all emerge as neutrals. Although the negative-ion state will form in the immediate vicinity of the surface, as shown in Figure 3.14, this state will be efficiently neutralized when the affinity level shifts above the Fermi energy. As the velocity increases, this neutralization of the negative-ion state becomes smaller and more negative ions survive. Consequently, the population of neutrals will decrease. At a kinetic energy around 10 eV, the negative-ion fraction decreases and the amount of neutrals starts to increase. This is simply due to the fact that at these higher velocities, the atom does not spend enough time near the surface for the negative-ion state to be formed. At even higher velocities, the population of neutrals decreases and the positive-ion fraction increases. This is due to the reduced charge transfer into the neutral levels at large velocities. For very large velocities, no charge transfer occurs at all and all atoms remain in the positive-ion state.

3.3.1.2.2 Non-degenerate levels

In order to further illustrate the complexity of charge transfer for finite U, we have performed calculations for an idealized two-level system with no degener-

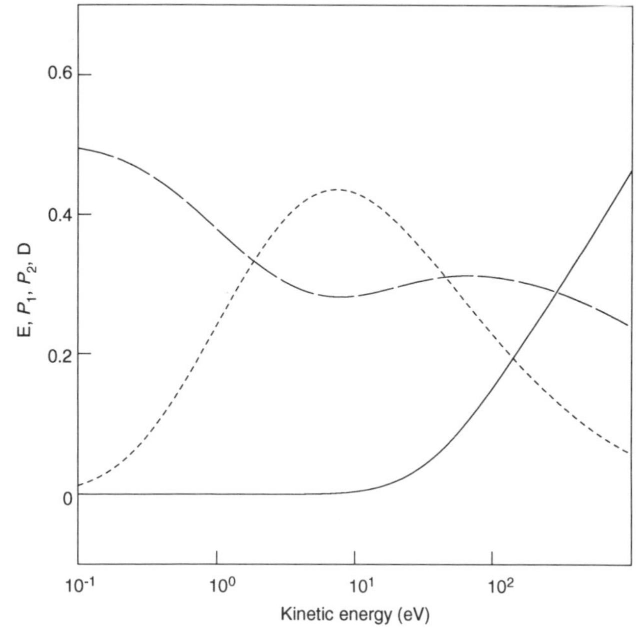

Figure 3.16 Positive-ion yield (solid line), yield of neutral atoms (long-dashed line) and negative-ion yield (dotted line) calculated using equation (156) as a function of kinetic energy of the atom. The energy-level parameters and atomic trajectory are as in Figure 3.14

acies. In Figure 3.17, the instantaneous populations of the different atomic configurations are plotted as a function of position outside the surface. As the surface is approached, the two levels P_1 and P_2 are formed and the positive-ion fraction is reduced. At a position $Z = 5$ a.u., one of the negative ion levels, $\varepsilon_1 + U$, shifts below the Fermi energy. At this point, the negative-ion fraction starts to grow. The next crossing is when the level $|2\rangle$ shifts above the Fermi energy. When this happens, the P_2 state begins to depopulate with a corresponding increase in the positive-ion fraction and a decrease in the negative-ion population. The next crossing is when the second negative-ion state shifts below the Fermi energy. At this point, there is a drastic increase in the negative-ion fraction at the expense of the population P_1. The last crossing before the turning point is when the level $|1\rangle$ shifts above the Fermi surface. This further decreases the population of P_1. At the turning point, only positive and negative ions remain. When the P_1 state shifts below the Fermi energy during the outgoing trajectory, the remaining positive ions are neutralized. When the negative-ion state shifts above the Fermi energy, the electron in $|1\rangle$ tunnels to the surface, resulting in the rapid formation of the P_1 configuration. The final state configuration of the scattered atom contains neutrals and negative ions.

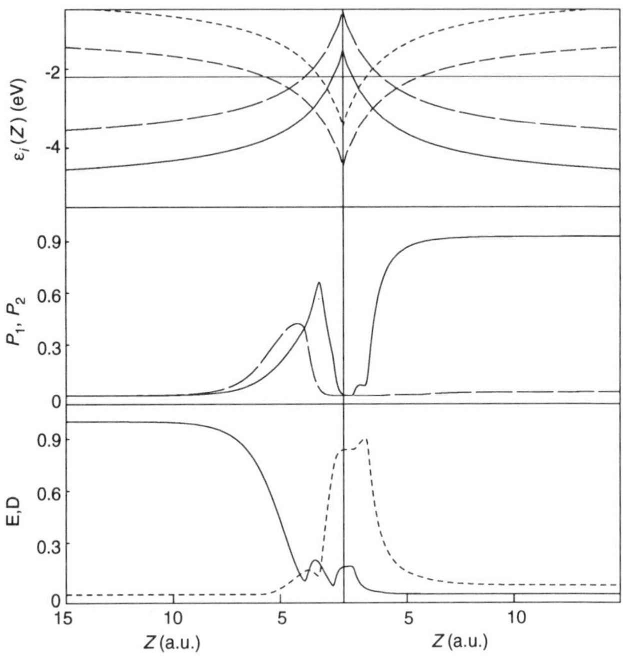

Figure 3.17 Populations of the different atomic configurations for finite U, calculated using equation (156). In the upper part of the figure, the distance dependence of the ionization levels P_1 (solid line) and P_2 (long-dashed line), and the negative-ion levels $\varepsilon_1 + U$ (dashed line) and $\varepsilon_2 + U$ (dotted line) are shown. The parameters for the levels are: $\varepsilon_1^0 = 5$ eV, $\varepsilon_2^0 = 4$ eV, $\Gamma_1^0 = 5$ eV, $\Gamma_2^0 = 10$ eV, $\alpha_i = 1.0$ a.u., $U = 4$ eV, $\Gamma_1^{0+} = 1$ eV, $\Gamma_2^{0+} = 2$ eV and $\alpha_i^+ = 0.7$ a.u. The Fermi energy (2.2 eV) is indicated by the horizontal line. In the middle figure, the population of the state $|1\rangle$ (solid line) and the state $|2\rangle$ (long-dashed line) are shown. In the lower figure, the positive-ion fraction (solid line) and negative-ion fraction (dotted line) are shown. The turning point of the trajectory is at $Z = 2$ a.u. and is indicated by the vertical line. The kinetic energy of the atom (Li) is 2 eV

In Figure 3.18, we plot the final state distribution for the system discussed in Figure 3.17 as a function of the work function of the surface. The figure is similar to Figure 3.15 with the exception of the appearance of the neutral excited state $|2\rangle$ at a work function around 3.5 eV. For work functions up to 1.5 eV, all configurations of our simple model atom can be observed. The relative magnitude of the populations depends directly on the velocity of the incoming or outgoing atom or molecule and the characteristics of the relevant level shifts and broadenings.

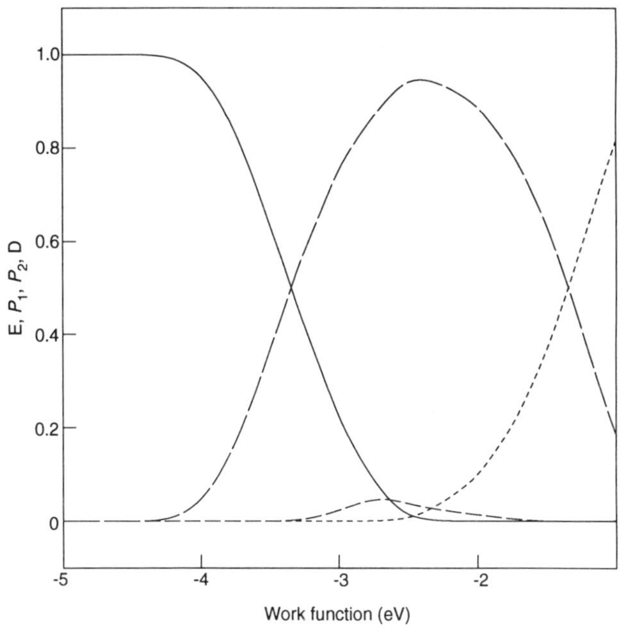

Figure 3.18 Positive-ion yield (solid line), yield of neutral atoms in P_1 (long-dashed line) and P_2 (dashed line) and negative-ion yield (dotted line) calculated using equation (156) as a function of kinetic energy of the atom. The energy-level parameters and atomic trajectory are as in Figure 3.17

3.3.2 DIRECT-RECOIL SPECTROSCOPY

An interesting application illustrating the effects of the intra-atomic correlation on charge transfer can be found in a recent experiment in which chemisorbed oxygen atoms were recoiled from a Ni(100)-c(2 × 2)O surface (Hsu *et al.*, 1992). By measuring the angular dependence of the recoiled negative-ion fraction, microscopic information concerning the spatial variation of the local tunneling rates within the surface unit cell can be obtained. In this presentation, a very brief discussion about some details of the experiment will be presented. A more extensive analysis will be published elsewhere (Hsu *et al.*, 1993).

The energy shifts and broadenings of the negative-ion states of oxygen as a function of distance outside a jellium surface have recently been calculated (Nordlander, 1992). The $O^-(^2P)$ state is of p-character and is therefore six-fold degenerate. These six states shift downwards with decreasing atom– surface separation, and split into two short-lived $m = 0$ states oriented towards the surface and four long-lived degenerate $m = \pm 1$ states oriented along the surface. Although the initial state of the oxygen chemisorbed on the surface is almost certainly a negative ion, the recoil with a positively charged argon is likely to neutralize the oxygen. However, since the negative-ion states are shifted below the Fermi energy, these states will rapidly be formed during the initial part of the trajectory. The intra-atomic correlation effects strongly influence the charge-transfer process in this region. In Figure 3.19, we show energy level shifts and instantaneous populations of the different negative-ion states during a steep and a grazing-exit trajectory calculated using equation (155). Due to the presence of the ordered oxygen overlayer, the energy levels are corrugated (Hsu *et al.*, 1992). To illustrate the effects of the intra-atomic correlation, we also include in Figure 3.19 the populations calculated using a one-electron theory,

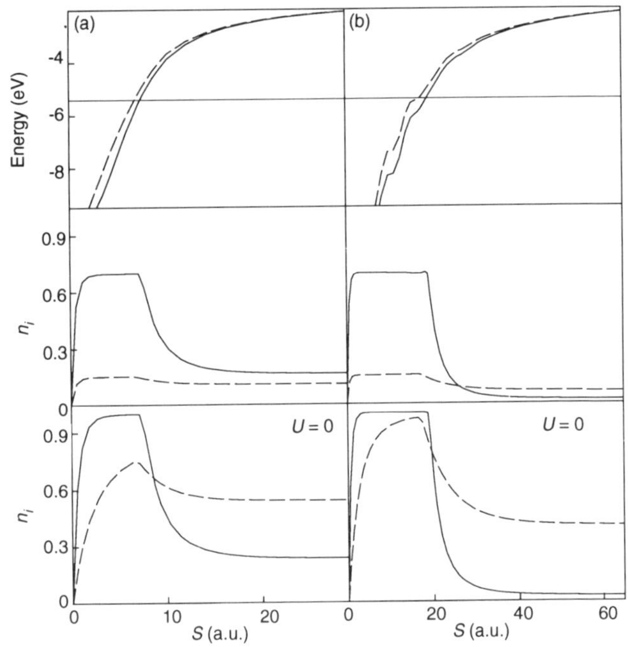

Figure 3.19 Energy-level shift and instantaneous population of the two different O^--states as a function of distance traveled for two different trajectories. The solid lines refer to the doubly degenerate $O^-(m = 0)$ states and the dashed lines refer to the fourfold degenerate O^- $(m = \pm 1)$ states. In (a), the trajectory is in the $\langle 210 \rangle$ direction with an exit angle of 29°. In (b), the trajectory is in the $\langle 210 \rangle$ direction with an exit angle of 12°. In the upper part of the figure, the energy shifts are shown. The Fermi energy is indicated by the horizontal line. The middle part of the figure shows the populations calculated using equation (155) and the lower figures show a calculation neglecting intra-atomic correlation, equation (154). The kinetic energy of the oxygen is 1800 eV. The energy shifts and widths for the oxygen levels are taken from a previous calculation (Hsu *et al.*, 1992)

equation (154). It can be clearly seen from the figure that the negative-ion states are completely formed within 1 a.u. This complete ionization is independent of the trajectory. The intra-atomic correlation prevents the formation of the $O^-(m = \pm 1)$ states resulting in negative ions predominantly in the $m = 0$ states. As the atom leaves the surface region and the level shifts above the Fermi energy, the ions begin to neutralize. The survival rate of the $O^-(m = \pm 1)$ states is very large and therefore relatively independent of the trajectory. On the other hand, the $O^-(m = 0)$ states are relatively short-lived. As can be seen from the figure, the survival rate of the $m = 0$ states strongly depends on the trajectory. The experimental observation of a strong trajectory-dependence of the recoiled negative-ion fraction is therefore due to the fact that the majority of recoiled ions are in the relatively short-lived $O^-(m = 0)$ states. The intra-atomic correlation thus played a crucial role in preventing the $m = \pm 1$ states from being formed during the initial part of the trajectory.

3.3.3 STIMULATED DESORPTION

A very direct demonstration of the effects of intra-atomic correlation on the charge-transfer process is provided by a recent stimulated-desorption experiment of hydrogen from cesium-covered copper (Tsuei *et al.*, 1993). In this experiment, the angular dependence of the Lyman-α radiation emitted from desorbing $H(n = 2)$ states was studied. It was found that most of the desorbing hydrogen will be in the two $H(m = 0)$ states. The experimental data shows that the ratio of the population of the $m = 0$ states and the $m = \pm 1$ states is 2.5.

The interaction between a hydrogen atom and a metal surface results in the hybridization of the hydrogen orbitals. The H(2s) and H($2p_z$) orbitals hybridize and form a short-lived state, ψ_2^+ oriented towards the surface and a long-lived state, ψ_2^- oriented towards the vacuum direction. Outside a smooth metal surface, the H($2p_{x,y}$) states remain degenerate and will be denoted, ψ_2^0. The widths of the ψ_2^0 states are intermediate between the ψ_2^+ and ψ_2^- states.

For a hydrogen atom outside a metallic surface, the hydrogen levels shift upwards with decreasing atom–surface separation. This upshift is due to the image potential and has been discussed elsewhere (Nordlander and Tully, 1990). Once formed by the excitation process, these states will therefore ionize rapidly. A less probable but competing process will be the formation of the negative-ion state. This state is downshifted close to the surface and decays into H(1s) by resonant tunneling at distances larger than 7 a.u. from the surface (Nordlander, 1992). Since neither the negative-ion state nor the H(1s) can emit Lyman-α radiation, the discussion can be restricted to the $H(n = 2)$ states formed by resonant neutralization of the desorbing protons.

In Figure 3.20, the energy level ε_2^+ and the instantaneous populations of the different hydrogen levels $n_2^i(t)$ calculated using equation (155) are plotted

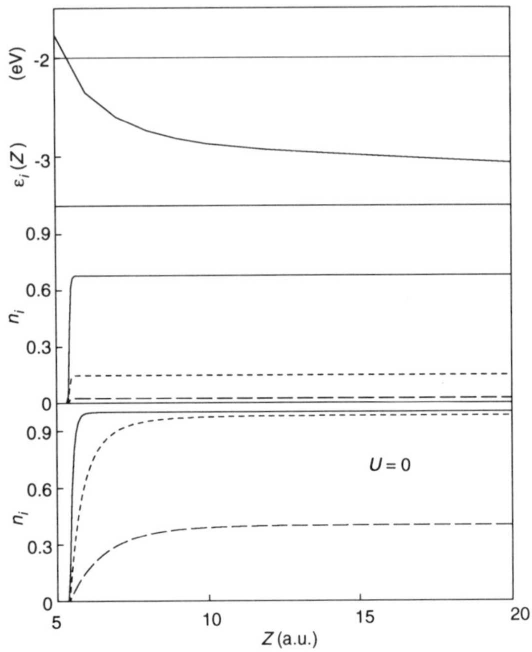

Figure 3.20 Instantaneous population of the different H ($n = 2$) states as a function of distance from the surface during desorption. The solid line refers to the doubly degenerate $|\psi_2^+\rangle$ states, the dotted line refers to the fourfold $|\psi_2^0\rangle$states and the dashed line refers to the doubly degenerate $|\psi_2^-\rangle$ state. The desorption is assumed to occur normal to the surface with a kinetic energy of 1 eV. In the upper part of the figure, the energy shifts are shown. The Fermi energy is indicated by the horizontal line. The middle part of the figure shows the populations calculated using equation (155) and the lower figure shows a calculation neglecting intra-atomic correlation, equation (154). The energy shifts and widths for the oxygen levels are taken from a previous calculation (Nordlander and Tully, 1988)

during the trajectory. For comparison, we also show the results calculated using a one-electron description, equation (154). When the hydrogen levels cross the Fermi energy at a distance of 5.5 a.u. from the surface, the $n = 2$ levels can be populated by resonant tunneling. Since the tunneling rate into the ψ_2^+ state is the largest, the population of this state increases faster than the population of the more long-lived ψ_2^0 and ψ_2^- states. At a distance of 5.7 a.u., the total population of the $n = 2$ levels is 1 and the intra-atomic correlation prevents further tunneling of electrons into the hydrogen atom. The final state occupations are $n_2^+(\infty) = 0.69$, $n_2^-(\infty) = 0.02$ and $n_2^0(\infty) = 0.14$. The ψ_2^+ state consists of equal contributions of the H(2s) and H(2p$_x$) orbitals. Lyman-α radiation is emitted from the transition H(2p $\rightarrow$ 1s). The anisotropy ratio is therefore

$R = 2.54$, in excellent agreement with the experimental results. In the calculations of the final populations of the $H(n = 2)$ states, it was crucial to include the effects of the intra-atomic correlation. If this effect had been neglected, all the $H(n = 2)$ states would have been fully populated, resulting in an anisotropy ratio of $R = 0.65$.

The results of the calculation do not depend sensitively on velocities nor on the substrate for which the widths were calculated. The crucial parameter is the anisotropy between the various $H(n = 2)$ states. Since the $H(n = 2)$ states are formed very fast, the anisotropy ratio is simply given by

$$r(Z) = \frac{\Gamma_2^+(Z) + \Gamma_2^-(Z)}{2\Gamma_2^0(Z)}, \tag{157}$$

at the position where the levels cross the Fermi energy. These anisotropies are determined by the spatial orientation of the hybridized atomic orbitals. While the absolute magnitudes of the widths of these states are sensitive to the details of the surface potential, the ratios of the widths of the different states are almost the same for all metallic substrates. Since $r(Z)$ varies only weakly with distance, i.e. $r(5) = 2.44$, $r(6) = 2.57$ and $r(7) = 2.79$, the results do not depend sensitively on the crossing distance.

3.4 FUTURE WORK

The Dyson equations describing the charge-transfer process are rather complicated coupled integro-differential equations. Although the SCA and the GME discussed in the previous sections provide an accurate solution to the coupled Dyson equations in typical cases, it is desirable to develop a more general and simple approximate method of solution.

The solution of the coupled Dyson equations is somewhat facilitated by the fact that the Dyson equations for the analytic pieces of the propagators decouple from the corresponding equations for the retarded and advanced quantities. Thus the retarded and advanced quantities can be calculated first and then inserted into the equation for the analytical pieces. The quality of the GME might improve with better approximations for the advanced and retarded Green's functions. An iterative solution to the system of equations could in principle be obtained by iterating the Dyson equations. The zeroth-order Green's functions can be used in the integrands and by integrating the time derivative, a higher-order Green's function can be obtained. These functions can in turn be inserted into the integrands and the procedure could be repeated until sufficient accuracy is obtained. Since the exact solution of the problem has been obtained, the accuracy of the approximations can readily be investigated.

Outside real surfaces, atomic level shifts and broadening will be corrugated, i.e. will depend on both the lateral and perpendicular motions of the atom with respect to the surface. To properly model the charge exchange in atom–surface scattering experiments, averages of the charge transfer over all possible trajectories consistent with the scattering and detection geometries must be performed. Since such an averaging process would be very computer-intensive, a more direct relation between the lateral corrugation and the charge state of an atom after a surface scattering event is needed.

The qualitative effects of a lateral corrugation on charge-transfer processes have been investigated using the single-level Anderson Hamiltonian (Nordlander and Tully, 1989). For a harmonically varying lateral corrugation and a random distribution of impact parameters of the scattered atoms on the surface, the problem can be solved analytically (Nordlander and Tully, 1989). The effects of the corrugation are to further localize the functions $\tilde{f}_{\sigma}^{\gtrless}$ around equal times. This additional localization makes the semi-classical approximation more accurate. Further investigations of the influence of the effects of a lateral corrugation on charge transfer will be made. In particular, we will investigate whether a similar localization of the self-energies occurs in the correlated problem. If this is the case, the SCA provides a very simple solution to the time-dependent problem, and the occupations of the atomic levels can be accurately calculated from rate equations of the type of equation (16).

An interesting application of the non-equilibrium Green's function technique is the investigation of inelastic effects. As the incoming atom scatters against the surface, electron–hole pairs are excited. This results in the loss of kinetic energy of the atom. When an atomic level crosses the Fermi energy, these effects are particularly pronounced and might lead to the sticking of the atom (Gunnarsson and Schonhammer, 1980).

Calculations of the energy dissipation due to the excitation of electron–hole pairs have been performed elsewhere (Makoshi, 1983) using a Green's function technique similar to that presented here. The evaluation of the adiabatic and the non-adiabatic contribution to the instantaneous energy of the atomic state can be done using the Green's function of the atomic state (Shao *et al.*, forthcoming). The consequences of the energy transfer on the surface conduction-band states remain to be investigated.

3.5 CONCLUSIONS

We have shown that the non-equilibrium Green's function technique provides a very convenient method for describing charge-transfer processes in atom–surface scattering. The method allows the systematic treatment of intra-atomic correlation effects. An investigation of charge transfer in typical atom–surface collisions has shown that the effects of intra-atomic correlation on charge transfer can be very strong. We have demonstrated that in situations where several levels can be populated by resonant tunneling, the conventional one-electron description is inadequate. We have further shown that for finite temperatures and low velocities, it is possible to derive simple rate equations for the charge transfer. These rate equations include intra-atomic correlation effects and provide a convenient method for the description of charge transfer.

3.6 ACKNOWLEDGEMENTS

The authors wish to acknowledge valuable discussions with and comments by Dr N. Halas. This work is supported by the National Science Foundation under grants No. DMR-91-17479 and DMR-91-03466. Acknowledgement is made to the donors of the Petroleum Research Fund, administered by the American Chemical Society for the partial support of this research.

3.7 REFERENCES

Blandin, A., A Nourtier, and D. Hone (1976) *J. Physique*, **37**, 369.
Bloss, W. and D. Hone (1978) *Surf. Sci.*, **72**, 277.
Brako, R. and D. M. Newns (1981) *Surf. Sci.*, **108**, 253.
Brako, R. and D. M. Newns (1985) *Solid St. Commun.*, **55**, 633.
Coleman, P. (1984) *Phys. Rev.*, **B 29**, 3035.
Grimley, T. B., V. C. J. Bhasu, and K. L. Sebastian (1983) *Surf. Sci.*, **124**, 305.
Gunnarsson, O. and K. Schonhammer (1980) *Phys. Rev.*, **B 22**, 1629.
Gunnarsson, O. and K. Schonhammer (1983) *Phys. Rev.*, **B 28**, 4315.
Hammond, M. S., F. B. Dunning, G. K. Walters, and G. A. Prinz (1992) *Phys. Rev.*, **B 45**, 3674.
Hsu, C. C., H. Bu, A. Bousetta, J. W. Rabalais, and P. Nordlander (1992) *Phys. Rev. Lett.*, **69**, 188.
Hsu, C. C., A. Bousetta, J. W. Rabalais, and P. Nordlander (1993) *Phys. Rev.*, **B 47**, 2369.
Janev, R. K., P. S. Krstic, and M. J. Rakovic (1987) *Phys. Rev.*, **B 35**, 7704.
Johnson, A. L., S. A. Joyce, and T. E. Madey (1988) *Phys. Rev. Lett.*, **61**, 2578.
Johnson, P. D., A. J. Viescas, P. Nordlander, and J. C. Tully (1990) *Phys. Rev. Lett.*, **64**, 942.
Joyce, S. A. *et al.* (1992) *Phys. Rev.*, **B 45**, 14264.
Kadanoff, L. P. and G. Baym (1962) *Quantum Statistical Mechanics.* Benjamin: New York.
Kasai, H. and A. Okiji (1987) *Surf. Sci.*, **183**, 147.
Keldysh, L. V. *J. Exptl. Theoret. Phys. (USSR)*, **47**, 1515; English Translation: (1965) *Soviet Physics JETP*, **20**, 1018.
Kimmel, G. A., D. M. Goodstein, Z. H. Levine, and B. H. Cooper (1991) *Phys. Rev.*, **B 43**, 9403.
Kotliar, G. and A. E. Ruckenstein (1986) *Phys. Rev. Lett.*, **57**, 1362.
Langreth, D. C. (1976) Linear and nonlinear response theory with applications, in *Linear and Nonlinear Electron Transport in Solids*, J. T. Devreese and V. E. van Doren (eds), Plenum, New York.

Langreth, D. C. and P. Nordlandler (1991) *Phys. Rev.*, **B 43**, 2541.
Langreth, D. C. and J. W. Wilkins (1972) *Phys. Rev.*, **B 6**, 3189.
Los, J. and J. J. C. Geerlings (1990) *Phys. Rep.*, **190**, 133.
Makoshi, K. (1983) *J. Phys.*, **C 16**, 3617.
Marston, B., D. R. Andersson, E. R. Behringer, B. H. Cooper, C. A. DiRubio, G. A. Kimmel and C. Richardson, *Phys. Rev.*, **B 48**, 7809.
Nordlander, P. (1990) In *Fundamental Beam–Surface Interactions*, J. Schou, P. Kruit, and D. E. Newbury (eds), Scanning Microscopy International AMF O'Hare, Chicago, p. 353.
Nordlander, P. (1992) *Phys. Rev.*, **B 46**, 2584.
Nordlander, P. and N. Lang (1991) *Phys. Rev.*, **B 44**, 13681.
Nordlander, P. and J. C. Tully (1988) *Phys. Rev. Lett.*, **61**, 990.
Nordlander, P. and J. C. Tully (1989) *Surf. Sci.*, **211**, 207.
Nordlander, P. and J. C. Tully (1990) *Phys. Rev.*, **B 42**, 5564.
Nørskov, J. K. and B. I. Lundqvist (1979) *Phys. Rev.*, **B 19**, 5661.
Oelschig, S., W. Hetterich, and W. Heiland. To be published.
Palmer, R. E. and P. Rous (1992) *Rev. Mod. Phys.*, **64**, 383.
Rau, C., C. Jin, and G. Xiang (1990) *Phys. Lett.*, **A 144**, 406.
Rusch, T. W. and R. L. Erickson (1976) *J. Vac. Sci. Tech.*, **13**, 374.
Sebastian, K. L. (1985) *Phys. Rev.*, **B 31**, 6976.
Shao, H., D. C. Langreth, and P. Nordlander. To be published.
Sulston, K. W., A. T. Amos, and S. G. Davison (1988) *Phys. Rev.*, **B 37**, 9121.
Tsuei, K. D., P. D. Johnson, P. Nordlander, and D. C. Langreth (1993) *Phys. Rev.*, **B 47**, 4142.
Tully, J. C. (1977) *Phys. Rev.*, **B 16**, 4324.
Yoshimori, A., H. Kawai, and K. Makoshi (1984) *Prog. Theor. Phys. Suppl.*, **80**, 203.

4 ELECTRON EMISSION FROM SLOW ION–SOLID INTERACTIONS

Raúl A. Baragiola
University of Virginia, Engineering Physics, Charlottesville, Virginia 22901, USA

CONTENTS

Low Energy Ion–Surface Interactions
Edited by J. W. Rabalais

ABBREVIATIONS

AC	Auger capture
AD	Auger de excitation
AU	Autoionization or autodetachment
KEE	Kinetic electron emission
PEE	Potential electron emission
RC	Resonance capture
RI	Resonance ionization

4.1 INTRODUCTION

Electron emission is one of the most general consequences of inelastic ion–surface collisions and often accompanies other processes described elsewhere in this volume, like backscattering, sputtering and desorption. As in other aspects of ion–surface interactions, the evolution of our understanding of this field is closely related to progress in atomistic descriptions. Electron emission induced by heavy particles is usually described by a three-step model as used in other studies of electron emission (e.g. photoemission and secondary electron emission under electron impact). The three steps are electron excitation, transport to the surface and escape through the surface barrier. In the specific case of low-energy ion bombardment considered here, many of the excitations occur at or outside the surface of the solid, and the separation into steps may not be justified.

Two main excitation mechanisms contribute to electron emission induced by heavy particles. In kinetic electron emission (KEE), electrons are excited as a consequence of the atomic motion of the projectile. In potential electron emission (PEE), electron excitation results from the conversion of internal energy ε brought by the projectile, through an Auger process. Both mechanisms can only occur above an energy threshold related to U, the minimum energy required to free an electron from the solid. In KEE, the center of mass energy E_{cm} has to exceed U. In PEE, the energy released in the Auger process, $\varepsilon - U$, has to exceed U (i.e. emission requires $\varepsilon > 2U$). These mechanisms are separable in two limiting cases. When $E_{cm} < U$, and $\varepsilon > 2U$, only PEE can occur. If, on the other hand, $\varepsilon < 2U$ and $E_{cm} > U$, electrons can only be ejected by the kinetic mechanism. PEE does not require the motion of the projectile but can be affected by it, whereas KEE can be affected by the degree of excitation of the projectile.

The electrons excited by these mechanisms can undergo a cascade of collisions in the solid until their energy is degraded into heat or stored in long-lived excited states. Those excited electrons which are directed outwards may cross the surface of the solid and escape before being thermalized, giving rise to electron emission. An important characteristic of the low-energy ion-impact case discussed here is that the vast majority of the initially excited electrons have very small kinetic energies, insufficient to excite additional electrons above the vacuum level.

The main observable in electron emission is the average electron yield γ (number of electrons ejected per primary particle) from the solid target, either as an integral quantity or in the form of distributions of electrons according to their energies, angles of emission, and spin state. Emission events are statistically distributed: that is, the yield fluctuates around the mean value γ and is described by probabilities of emission of n electrons per projectile, $P_n(\Sigma P_n \equiv 1; \Sigma n P_n \equiv \gamma)$.

Experiments involve measuring the dependence of electron emission on different factors. These include the projectile and target properties and the bombardment geometry. Important properties of the projectile are its excitation and ionization state, its mass, and whether it comes as a single atom, a molecule, or a cluster of atoms. Relevant target properties are the elemental composition and atomic and electronic structure in the surface region, the magnitude of surface electric fields, the temperature, and, for magnetic materials, the degree of magnetization. The relevant geometrical factors in experiments include the angle of incidence with respect to the surface and to possible preferred crystallographic planes.

The topic of ion-induced electron emission by slow ions has been surveyed recently with emphasis on observations (Brusilovsky, 1990; Hofer, 1990; Hasselkamp *et al.*, 1992) and theory (Sigmund and Tougaard, 1981; Schou, 1982; Rösler *et al.*, 1991; Baragiola, 1991, 1993b), and discussed at a specialized conference (Baragiola, 1993a). The case of Auger emission from the decay of inner-shell excitations has been treated in Baragiola (1982), Thomas (1984), Mischler and Bénazeth (1986) and Valeri (1993). A fairly complete review of PEE has been published in Baragiola (1982), and different aspects have been treated in later reviews (Hagstrum, 1988; Varga and Winter, 1992; Niehaus, (1993), especially the case of incident multiply charged ions (Varga, 1987, 1988, 1989; Winter, 1991; Andrä, 1989; Andrä *et al.*, 1992). In this chapter, I will review recent developments in both PEE and KEE from solids, especially those not covered in the work just mentioned. The review will be restricted to research on clean surfaces and on surfaces with characterized or controlled amounts of adsorbate species.

4.2 THEORETICAL CONCEPTS

4.2.1 ELECTRON STATES IN SOLIDS

We start by recalling the properties of solids that are important in understanding electron emission. Figure 4.1 shows one-electron energy diagrams useful to describe the ground-state properties of the solid. We already mentioned U, the minimum energy required to extract an electron from the solid into infinity. This quantity is the same as the photoelectric threshold, i.e. it is the energy difference between the top of the occupied valence band and the vacuum level. For metals, U is equal to the work function ϕ while for non-metals, it is the sum of the band-gap energy E_g and the electron affinity χ. Electron emission can be observed to occur below this threshold, in the presence of a significant concentration of impurity or defect states in the band gap (Ushio *et al.*, 1988). Threshold energies range from ~1.4 eV for cesiated metal surfaces to more than 20 eV for condensed He.

Another important energy quantity is the surface barrier height, defined as the energy that an electron loses when going from the solid into the vacuum. This energy equals the inner potential I, used in low-energy electron diffraction (LEED) to indicate the energy that an electron gains when entering the solid. $I = E_F + \phi$ for a free-electron model of metals and $I = \chi$ for non-metals. Typical values of I in metals, derived from LEED measurements, are around 11–15 eV (Jennings and Jones, 1988). Values of χ are, in general, poorly known; they range from -0.2 to ~4 eV.

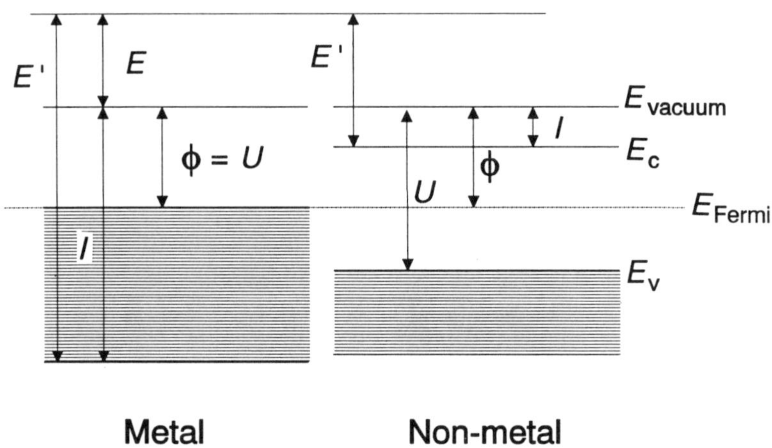

Figure 4.1 Schematic energy bands for metals and non-metals, showing energies relevant to electron emission. ϕ, work function; U, minimum binding energy; I, inner potential; E_c, bottom of conduction band; E_v, top of valence band; E (E'), electron kinetic energy outside (inside) the solid

The wave functions of electrons bound to the solid decay exponentially in vacuum with distance from the surface. The decay depends very strongly on the energy of the electron with respect to the vacuum level. Thus, an experiment which samples the solid outside the surface will preferentially sample electrons at or near the Fermi level. In the presence of an ion outside the surface, these wave functions distort considerably. It has only recently been realized that it is not a good approximation to treat the ion as a small perturbation on the metal, even for singly charged ions. The electron sees the unperturbed surface potential, plus the potential between the electron and the ion, plus the potential due to the image of the ion. Accurate description of the electron wave functions outside the surface of the solid is very important in treating electron transfer between the surface and the ion.

Theoretical concepts on the interactions of ions with valence electrons have been reviewed by Echenique *et al.* (1990). The simplest forms of electron excitations are the single electron–hole pair and the collective plasmon excitation with energy E_p. Direct excitation of plasmons is unimportant for projectiles with velocity v less than the Fermi velocity v_F, as discussed here. However, plasmons can be excited by energetic electrons (Gervasoni *et al.*, 1992), like Auger electrons from core excitations or from PEE by ions carrying sufficient potential energy. Plasmons decay into an electron–hole pair, with a lifetime of 0.1 to 1 fs. The final state of the solid after adsorbing E_p directly through an electron–hole excitation is very similar to that following the intermediate state of the plasmon creation and decay. Differences result in the spatial and angular distribution of the final electron. These differences are probably the reason for the recent finding that plasmons can contribute more to electron emission than direct electron–hole excitations of the same energy (Pijper and Kruit, 1991).

Excited electrons and holes have finite lifetimes τ due to scattering with other electrons in the solid; this is usually characterized by a width of the state: $\Gamma = \hbar/\tau$. Recent inverse-photoemission experiments (Goldmann *et al.*, 1991) have revealed that the width of s and p conduction-band states for metals seem to follow a universal curve $\Gamma(E') = 0.13(E' - E_F)$, or typically ~ 0.5 eV at the vacuum level. For d-like bands, the widths are even larger, $\Gamma(E') \sim 0.6(E' - E_F)$. The lifetime of a hole in the band depends on whether it can be filled by an Auger process or whether it cannot, like for holes located less than the gap energy below the top of the band.

Excited electrons can also be described by their scattering properties. Electrons can scatter elastically from ion cores, or quasi-elastically, losing energy to lattice vibrations. A more efficient way of energy loss is by inelastic scattering with other electrons in the solid, unless the final electron states fall in a forbidden gap. Scattering probabilities are described by elastic and inelastic inverse mean free paths. Since the majority of electrons under consideration have a low energy, an important question in the description of their motion is the role of quantum effects. This question is often neglected in theories using the classical

Boltzmann transport equation or computing classical trajectories with the Monte Carlo method.

Quantum effects appear because the associated de Broglie wavelength ($\lambda(\text{Å}) = h/mv = 12.25\ E^{-1/2}(\text{eV})$) for low-energy cascade electrons is of the order of or larger than the nearest-neighbor distance between atoms in molecules and condensed matter. Coherent scattering and interference of scattered waves will occur which, in the case of an ordered array of target atoms, will result in diffraction patterns. We will return to this problem when we discuss electron transport.

4.2.2 ELECTRONIC EXCITATION MECHANISMS

We will discuss three main mechanisms by which a heavy projectile can excite electrons above the vacuum level: (a) Auger processes involving the potential energy of the projectile, (b) direct kinetic excitations, and (c) kinetic excitations followed by Auger decay.

4.2.2.1 Potential Auger processes

The basic ideas in PEE by the Auger mechanism were developed by Hagstrum in the mid-1950s when he treated the processes as electronic transitions between the unperturbed solid and a projectile carrying a hole (vacancy) with high potential energy (positive ion or excited atom or negative ion). Basically, an Auger transition occurs before the projectile enters the solid, filling the incoming hole and ejecting an electron into vacuum. Since the process occurs outside the surface, the final electron energy is not affected much by scattering processes with the bulk of the solid. Different forms of processes can occur, depending on the origin of the two electrons participating in the Auger transition. If the two electrons are in the incoming projectile, the intra-atomic Auger process is termed autoionization or autodetachment (AU). It produces sharp lines in the electron energy spectra, with energy given by ε', which is the difference in energy between the initial and final atomic states, and the prime indicates that the atomic energies are shifted by the interaction with the surface. Since the AU electron is ejected from a moving projectile, its energy will be shifted by an amount δE, given by the Doppler effect,

$$\delta E = 0.0467\ (E_c E_a/M)^{1/2} \cos \varphi + 0.0005\ E_a/M, \qquad (1)$$

where E_c is the electron energy in the frame of the parent atom, E_a is the kinetic energy of the atom, M its mass in a.m.u., and φ the angle between the direction of emission and that of the atom motion. The last term is usually negligible at low energies. Autoionization processes in the gas phase have been reviewed by Niehaus (1990).

If one active electron comes from the projectile and the other from the solid, the process is termed Auger de-excitation (AD). Since the electron in the solid can occupy any energy in the valence band, AD produces excited electrons over a range of energies E_v wide, where E_v is the width of this band (plus a small broadening due to the lifetimes of the initial and final hole states). AD spectra reflect the surface density of states modified by the field of the ion, weighted by a transition probability that depends on the energy and symmetry of the surface electronic state. The maximum energy of ejected electrons from this process is $\varepsilon' - U$. In AD, the emitted electron may come from the target, after the intra-atomic decay of the excited state, or be emitted from the projectile, after an electron from the target tunnels to the ground state of the projectile. In the latter case, the ejected electron energy will be Doppler-shifted.

The projectile ion can also de-excite directly by a two-electron interatomic Auger decay, or Auger capture (AC), called Auger neutralization (AN) for incident singly charged ions. Since AC involves two active electrons from the solid, it gives a range of excited electron energies about $2E_v$ wide, and a maximum energy $\varepsilon' - 2U$. AC can also occur into a negative-ion state, when an atom approaches a surface. If the energy released in this attachment exceeds U, PEE can occur. This process has been called exoelectron emission when it occurs during the adsorption of electronegative gases on metals (Kasemo *et al.*, 1979).

Figure 4.2 shows potential-energy diagrams of an incident ion near a surface, with transitions corresponding to the two-elecrtron processes mentioned above,

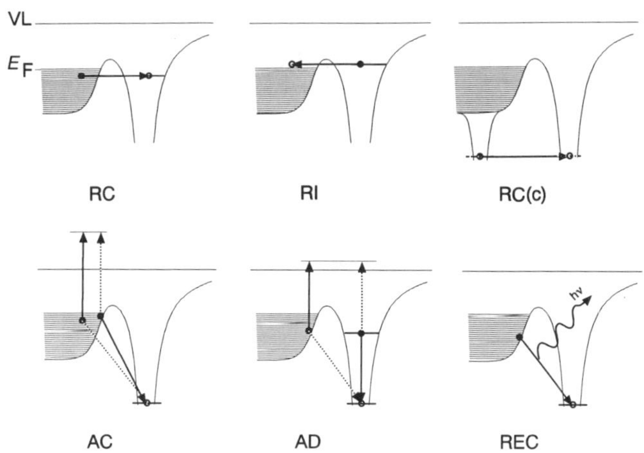

Figure 4.2 Variety of one-electron and two-electron transitions for ions and excited atoms near surfaces shown schematically as arrows connecting states in one-electron potential energy diagrams. RC, resonance capture; RI, resonance ionization; RC(c), resonant capture from core level; AC, Auger capture; AD, Auger de-excitation; REC, radiative electron capture; ●, electron; ○, hole. In the case of AC and AD, the dotted lines indicate the exchange process

and for the faster one-electron resonant transfers: resonant capture (RC) or resonance ionization (RI). RC can occur when the hole in the incident ion lies opposite to the occupied valence band of the solid; it is called resonant neutralization when it occurs for singly charged projectiles. The probability of this process depends on the overlap of the atomic wave function with the tail of the wave function of the electron in the solid. For metals, this means that the atomic states which will be preferentially populated are those lying close to the Fermi level. RC to excited states, when it can occur, is very important in determining the excited state populations of projectiles approaching surfaces and of projectiles or target species ejected from the solid as ions as a result of a violent collision.

An incident ion may de-excite by AC or undergo RC to an excited state followed by de-excitation through the AD process. The competition between AC and RC + AD is governed by the dependence of the probability for RC on the distance to the surface. The first path is dominant for singly charged ions on high-work-function surfaces, whereas the second path dominates for low-work-function surfaces. It is important to notice that resonance processes depend on the *local* work function of the surface, rather than on the macroscopic value that is measured by standard techniques (Hagstrum *et al.*, 1988). Hagstrum (1988) developed a taxonomy of seven classes of PEE processes with singly charged ions and metastable atoms, based on the magnitude of the local work function, the width of the band gap and the ionization potential of the solid. Alternatively, a simpler scheme can be obtained based on the value of the work function, and the threshold energy, as shown in Table 4.1.

RC + AD also dominates in the case of multiply charged projectiles, which can induce RC to excited states very far from the surface. The saddle point in the potential between an ion and a metal surface reaches the Fermi level at the critical distance which is given approximately (Burgdöfer *et al.*, 1991) by:

$$z_c(\text{Å}) = 20.3\ \phi^{-1}(q + 0.25)^{1/2} \tag{2}$$

Table 4.1 Potential-electron emission modes for ions and metastables incident on various surfaces

Class	Local ϕ	U	Examples	Ions	Metastables
A	High	Low	Clean metal or metal with adsorbed non-metallic monolayer. Narrow-gap semiconductor	AN	RI + AN
B	High	High	Surface with adsorbed or condensed multilayer or large-gap semiconductor	AN	AD
C	Low	Low	Metal with adsorbed alkali or with co-adsorbed alkali and other atom or molecule	RN + AD	AD

with ϕ in eV; this amounts to about 20 Å for $q = 20$. At z_c the barrier disappears and electrons from the solid rapidly spill over into the ion, populating the Rydberg levels lying near the Fermi level. This is a dynamic process; the resonant filling of the outer shells increases the screening of the projectile nuclear charge lowering the binding energy of the states and causing successive RC to occur to states of progressively lower n (Winter, 1991). A similar expression to equation (2) results (Folkerts, 1992) from fitting the classical field-emission model calculations of Bardsley and Penetrante (1991). It gives lower values than previous estimates by Apell (1987) (z_c(Å) $\approx 3q$) and Snowdon (1988) (z_c(Å) $\approx 2q + 7$).

RI can occur for negative ions which come as projectiles or are formed by RC near the surface. At short distances from the surface, the image interaction pulls the affinity level down, and ionization is no longer possible. On the other hand, excited atomic levels are promoted during approach. At sufficiently small distances, this can induce RI if the excited atomic state lies opposite to the unfilled levels in the solid. Other neutralization processes include resonance neutralization with a shallow core level in the solid (Schippers *et al.*, 1991) or the improbable radiative electron capture of a valence electron. The sequence of these steps is shown in Figure 4.3 with the aid of potential-energy curves. We show a He^+ undergoing RC to He* on approach to a metal surface (M). This produces a hole in the valence band which disappears rapidly from the intraction region due to diffusion, interband Auger decay and screening (effects of incomplete relaxation may be observed at finite velocities) (Zeijlmans van Emmichoven *et al.*, 1988; Zeijlmans van Emmichoven and Niehaus, 1990). The surface in front of the ion then cools very fast, which is indicated in the figure by a decay of the excitation to the Fermi level (wavy line). The ion–surface system then evolves along the He* + M until it undergoes RI at a crossing with a curve representing He^+ and an excited electron in the metal. Finally, AC can be explained using potential-energy curves corresponding to $(HeM)^+$ and $(HeM)^{2+}$. Transitions between these curves involve the emission of an electron with kinetic energy equal to the energy difference between the curves. Since the electronic transitions occur very fast, we can assume that the distance to the surface does not change in the process, that is, the transitions are 'vertical'. This is equivalent to the Born–Oppenheimer approximation.

Theoretical models for resonance transitions are discussed elsewhere in this volume (Chapter 3). The theory of AC was developed by Hagstrum but has been greatly advanced recently with inclusion of the perturbation of the solid produced by the incident ion. The description of ion–solid interactions at slow energies needs to consider the kinetic energy gained by the ions before hitting the surface, due to the image force (Burgdörfer and Meyer, 1993; Winter *et al.*, 1993; Aumayr *et al.*, 1993). This extra energy can be substantial for highly charged projectiles.

The Auger process is caused by the Coulomb repulsion between electrons,

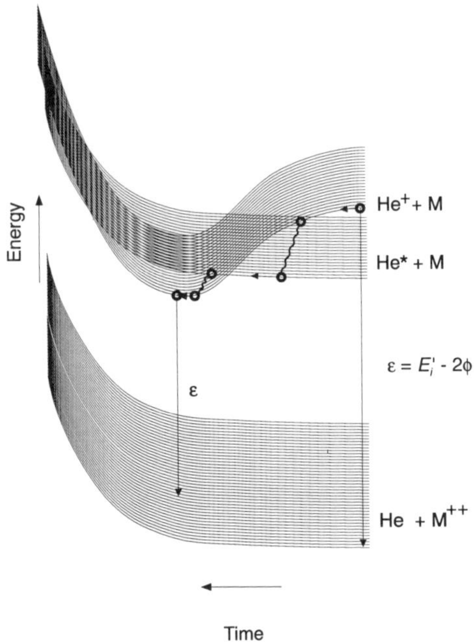

Figure 4.3 Potential-energy curves for a He^+ ion close to a surface, showing the manifolds corresponding to different excited states of a metal surface (M), and the evolution of the incoming hole. The wavy line represents the fast decay of the surface excitation to the ground state, by coupling to the bulk. The transition marked ε is the electron emission step

e^2/r_{mn}, with a transition rate given by Fermi's Golden Rule (Burhop and Assad, 1972):

$$A_{\rm if} = \hbar^{-1}|\langle\Phi_{\rm f}|\Sigma_{mn}(e^2/r_{mn})|\Phi_{\rm i}\rangle| \tag{3}$$

where $\Phi_{\rm f}$ and $\Phi_{\rm i}$ are the wave functions of the final and initial states. If we make the usual approximation that only two electrons participate in the transition (the others being spectators), and if we write the wave functions as products of one-electron orbitals, the matrix element can be written as:

$$M_{\rm if} = \iint \xi_{\rm f}^*(r_1)\xi_{\rm i}(r_1)(e^2/|r_1 - r_2|)\psi_{\rm f}^*(r_2)\psi_{\rm i}(r_2)\,{\rm d}r_1\,{\rm d}r_2 \tag{4}$$

in the absence of exchange, where the relevant wave functions are shown in Figure 4.4. The exchange contribution has been discussed by Hagstrum (1961) in some detail. This equation can be viewed as the Coulomb interaction integral between two charge clouds, $F(r_1) = \xi_{\rm f}^*(r_1)\xi_{\rm i}(r_1)$ and $G(r_2) = \psi_{\rm f}^*(r_2)\psi_{\rm i}(r_2)$. Here $F(r_1)$ will be a function localized near the position of the projectile, due to the localized character of the atomic wave function $\xi_{\rm f}$; furthermore, the

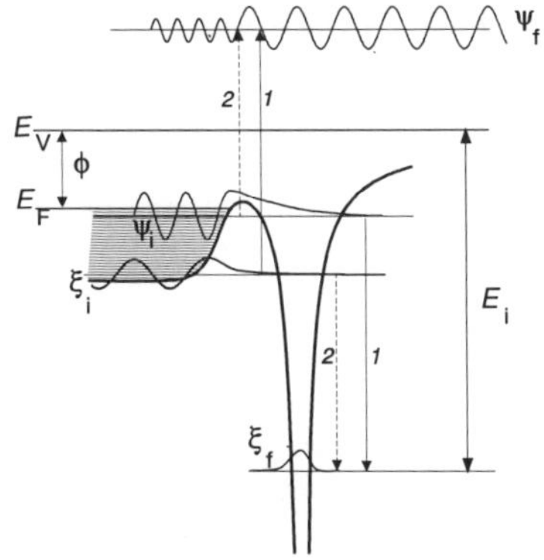

Figure 4.4 One-electron potential-energy curves for an ion next to a surface, indicating the relevant wave functions for Auger neutralization (see text)

Coulomb interaction will be screened inside the solid by valence electrons. Therefore, AC samples essentially the (perturbed) electronic density outside the solid surface. Similar considerations apply to the AD process; in this case, two wave functions are localized in the projectile, the compact final state $\xi_f(r_1)$, and the more extended excited state $\psi_i(r_2)$.

The transition rate will be larger when the 'tail' of the wave function outside the surface is larger and therefore, the overlap with the atomic wave functions is stronger. An important consideration is that at a distance from the surface where AC is likely, the perturbation of the solid by the projectile is by no means small. The ion field pulls electrons and causes a much slower decay of the electron density outside the solid than would occur for an unperturbed surface. The importance of the tail of the electron wave functions manifests in the following consequences:

(a) M_{if} decreases as the symmetry character of the wave functions ξ_i and ψ_i goes from *s* to *p* to *d* at the same energy.
(b) M_{if} increases as the electron energy in the band, *E*, increases.
(c) M_{if} is larger for initial states with wave vector **k** normal to the surface.
(d) Since the Coulomb interaction favors small momentum transfers, then, from (c), M_{if} will be larger for final states ψ_f with wave vector **k** normal to the surface. This produces the observed peaking in the angular distribution of emitted electrons along the surface normal (Klein, 1965).

The transition rate $R_t(z)$ as a function of the distance *z* of the ion to the surface image plane (Serena *et al.*, 1986) is obtained by integrating the matrix element over all initial and final states,

$$R_t(s) = \hbar^{-1} \iiint M_{if}^2 N_c(E_k)\, N_v(E_i)\, N_v(\mathrm{E_f})\, \mathrm{d}E_i\, \mathrm{d}E_f\, \mathrm{d}E_k, \qquad (5)$$

where N_v (N_c) is the density of states in the valence (conduction) band. Transition rates are often expressed as an inverse lifetime, $1/\tau$, or as the width of the excited state $\Gamma = \hbar/\tau$.

There have been several recent developments of the theory of AC. Penn and Apell (1990) included spin effects to explain the dependence of spin polarization on the energy of the electrons emitted by He* impact on Ni surfaces. Alducín *et al.* (1992) used a self-energy approach with a jellium model to calculate the Auger capture rates for He^+ as a function of the distance to the metal surface, for different electron-density parameters r_s (Ashcroft and Mermin, 1986, chap. 1). They obtained an exponential asymptotic behavior of the transition probability R similar to that proposed by Hagstrum, $R_t(z) = A \exp(-az)$. Fondén and Zwartkruis (1992) calculated $R_t(z)$ for He^+ in Na, Mg and Al using jellium wave functions in the local density approximation, without need of further approximations in the integration over states, as done in previous work (Hentschke *et al.*, 1986; Janev and Nedeljkovic, 1985). They found that although $R_t(z)$ decays approximately as an exponential, the decay constant is not the same as that for the electron density, as assumed by Hagstrum (1954) but ~1.9 times larger.

Further calculations by Lorente and Monreal (1993a) for He^+ on Cu, also used jellium wave functions ($r_s = 2.67$) but in the frame of the so-called effective medium approximation for the unperturbed metal wave functions (Nørskov and Lang, 1980; Stott and Zaremba, 1980). The results should also be valid when the ion is inside the solid. They are a factor of two smaller at large distances than those calculated by Fondén and Zwartkruis. The transition rates are found to scale with the local electron density squared, $R_t(z) \sim n^2(z)$. Figure 4.5 gives more recent, corrected results calculated with and without consideration of the effect of the perturbation induced by the He^+ on the metal wave functions (Lorente and Monreal, 1993b). The consequence of the more realistic treatment is a much lower exponential decay of the transition rate with z and the collapse of the barrier when the ion is at 0.8 Å from the image plane. Similar results were obtained by Salmi (1992) for a step barrier model.

From $R_t(z)$, one can obtain the probability that the ion survives capture after reaching the distance z, $P_o(z)$, and the probability that it will undergo AC at $z + dz$,

$$P_t(z)\,dz = R_t(z)P_o(z)\,dz/v_\perp = dP_o(z)/dz \qquad (6)$$

where $v_\perp$ is the velocity of the ion perpendicular to the surface. The results for 4 eV He^+ ions in Figure 4.6 show that AC is significant at distances in the range 1 Å $< z <$ 3 Å, close to the values derived by Hagstrum.

The energy distribution of the emitted electrons was calculated taking into account the shift of the He(1s) level on approach to the surface. Inclusion of the broadening caused by the finite lifetime of the 1s hole, produces a high-energy tail with significant intensity up to 18 eV, in rough agreement with

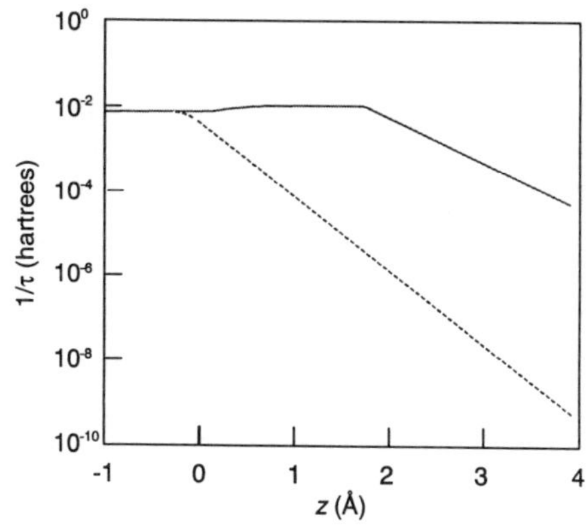

Figure 4.5 Calculated inverse lifetime or transition rate for Auger neutralization of He^+ on a Cu surface represented by an electron gas. One atomic unit equals $4.13 \times 10^{-16}\ s^{-1}$. The dashed line represent the values obtained neglecting the change in the electron distributions produced by the field of the ion. (Reproduced with permission from Lorente and Monreal, private communication)

experiment (Hagstrum, 1966). There is a discrepancy in the low-energy part of the spectrum, with the experiment giving many more low-energy electrons than theory. This could be due to processes involving more than two valence electrons. In fact, Jensen *et al.* (1992) have argued recently that in the intra-atomic Auger decay of $2p$ core vacancies in Al metal, the two-hole final states predicted by the conventional Auger model are in the minority.

In the discussion above, the ion velocity only enters in determining the distances to the surface over which de-excitation is significant. Another factor

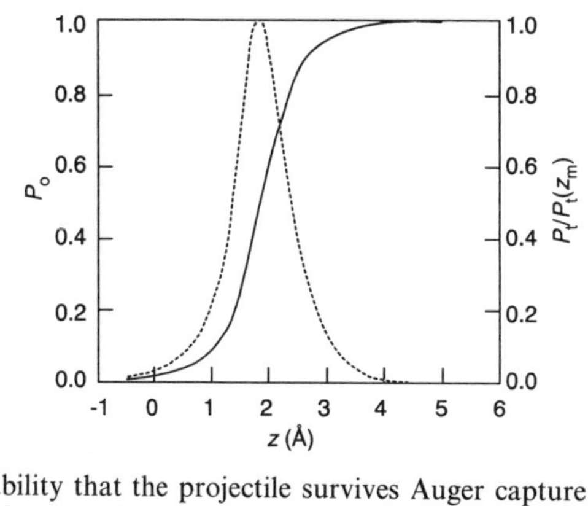

Figure 4.6 Probability that the projectile survives Auger capture on approach to the surface (——), or that it makes such a transition (– – – –) (Reproduced with permission from Lorente and Monreal, private communication)

is given by the time-dependent response of the valence electrons of the solid. This has been discussed semi-empirically by Zeijlmans and Niehaus. More elaborate, but less transparent, many-body theories have recently been developed by Makoshi and co-workers for incident ions (Kaji *et al.*, 1990; Kawai *et al.*, 1991) and excited atoms (Makoshi, 1991). They allow the treatment of the broadening caused by the finite velocity of the projectiles on the spectrum of excited electrons and Fermi edge singularity effects. The theory of AC can be extended to materials other than metals, as shown in the study by Ishii (1987) on AC on NaCl surfaces. In this binary target, it does not suffice to consider just the distance of the ion to the surface; rather, one has to consider different capture rates depending on the trajectory of the incoming ion. This type of approach could be used in the case when metal surfaces are covered with adsorbed layers. These layers hinder the tunneling of the metal electrons and present, in general, strongly bound valence levels for which AC is unlikely. This causes the observed drop of the PEE yield with adsorption of gases like O or CO (Hofer and Varga, 1984).

A semi-empirical method to unveil the contribution of different transitions to the observed electron energy distributions has been described by Niehaus and co-workers (Neihaus, 1993; Zeijlmans van Emmichoven *et al.*, 1988; Eeken *et al.*, 1992a, 1992b). Although it has adjustable parameters, it is useful since it provides a good heuristic approach to understanding the physics involved, which is not possible by other existing methods.

Niehaus's approach becomes very difficult when dealing with the more complex problem of potential excitation by incident highly charged ions. In the so-called 'ladder' model by Arifov *et al.* (1973) PEE occurs by a stepwise sequence of resonance captures to highly excited Rydberg (high-n) states followed by autoionization of the projectile. This model explains the observation that most electrons emitted by highly charged ions have low energies, since autoionization favors transitions with small energy jumps. The model also explains why the yields scale in an approximately linear way with the potential energy carried by the projectile, and why the yields increase with decreasing projectile speed normal to the surface. A consequence of Arifov's model is that a slow ion will be completely de-excited to its neutral ground state before reaching the surface, in the limit of infinite approach time (zero velocity). In fact, the approach time is limited because the ion accelerates by the image interaction with the surface. This image interaction changes dynamically as the projectile captures electrons on the way to the surface (Wille, 1991).

There is still controversy over the sequence of mechanisms leading to PEE by highly charged ions, but there seems to be consensus that the process starts near the critical distance z_c (equation (2)) where Rydberg states of the projectile are populated by mutiple electron transitions from the surface (Wille, 1992a, 1992b). The autoionization process is possible with at least two high-n electrons; it ejects one electron, increasing the charge of the ion and transferring another electron to a lower n level. However, since autoionization rates are initially slower than resonance capture rates, there is a preferential electron capture by the projectiles to a highly excited configuration, before significant autoionization occurs. For instance, Vaeck and Hansen (1991, 1992) have calculated that the autoionization rate increases from $\sim 3 \times 10^{13}\ \mathrm{s}^{-1}$ for $N^{4+}(1s7l^2)$ to $2 \times 10^{14}\ \mathrm{s}^{-1}$ for $N^{+}(1s7l^5)$ and $\sim 3 \times 10^{15}\ \mathrm{s}^{-1}$ for neutral $N(1s7l^6)$. The highly excited projectiles formed by resonance processes are termed 'hollow' because of the population inversion between the inner and outer shells.

Once the ion is almost completely neutralized to this excited configuration, a sequence of steps start to occur. Autoionization will eject low-energy electrons, increasing the charge of the projectile and populating inner shells with progressively lower n. This ionization is counteracted by further resonant capture. These successive or simultaneous steps of electron capture and autoionization will occur until the ion gets very close to the surface where Auger processes involving target electrons, like AD and AC, become important. When the ion reaches the surface it will have an electronic configuration that will depend on the speed of approach. Even at the maximum approach time, limited by the image force acceleration, slow projectiles may still reach the surface in a highly excited configuration which will then decay very fast by Auger processes due to the overlap of the multiple holes with the high electron densities in the solid. At the lowest velocities, the charge of the projectile may become negative before impact because the highly excited electrons do not completely screen the nuclear charge from the electrons at the surface of the solid (Burgdörfer and Meyer, 1993). As the projectile energy is increased, there will be an incomplete de-excitation and inner-shell holes will be able to reach and even penetrate the surface. There electrons in the more excited shells of the hollow atom will be 'peeled off' by the sudden increase in screening (Burgdörfer *et al.*, 1991).

4.2.2.2 Direct kinetic excitations

There are several ways in which slow heavy projectiles can excite electrons at the expense of their kinetic energy. The two basic kinetic-excitation mechanisms are direct binary collisions with target electrons and electron promotion due to electron–electron interactions. Light projectiles like H, He, or Li excite mainly by direct binary interactions while the heavier projectiles can excite more efficiently by the second mechanism.

The binary approximation is valid if the electron can be considered as free; that is, if its interaction with the projectile is much stronger than that with its parent atom. In metals, a free-electron model of the valence band can be used to obtain binary cross sections differential in final electron energy and momentum (Schou, 1980; Rösler, 1991b, 1993). These cross sections tend to be forward-peaked (Calera-Rubio *et al.*, 1991). The model of binary collisions with free electrons was first tested by the author and co-workers for Al bombarded with ions at velocities $v < v_F$ (Baragiola *et al.*, 1979). In this model, KEE is limited by momentum conservation due to the strong mismatch between the masses of the electron, m, and of the heavy projectile, $M \gg m$. The maximum energy of an excited electron occurs when an electron at the Fermi surface, with momentum anti-parallel to the projectile, scatters into a final state with momentum parallel to the projectile. It is given by:

$$E'_{\max} = 2mv(v + v_F). \tag{7}$$

The excited electron is directed inwards but can be observed outside due to the strong elastic scattering with the target atomic cores. This binary, free- electron model then predicts a threshold velocity v_{th} below which no emission can occur, where $E'_{\max} = U$:

$$v_{th} = 1/2\, v_F[(1 + \phi/E_F)^{1/2} - 1]. \tag{8}$$

The experimental yield curves extrapolate to zero for velocities close to this value for light-ion impact, like H, He, or Li, if one subtracts the contribution of PEE (Figure 4.7). Lower thresholds can be obtained if one takes into account phonon exchange with the lattice (e.g. umklapp processes) or if the collision occurs in the proximity of ion cores in the solid (Falcone and Sroubek, 1991), where the valence electrons have higher velocities.

Heavy ions are able to excite electrons below the velocity of equation (8), due to the compression of the electron shells during the collision. At low

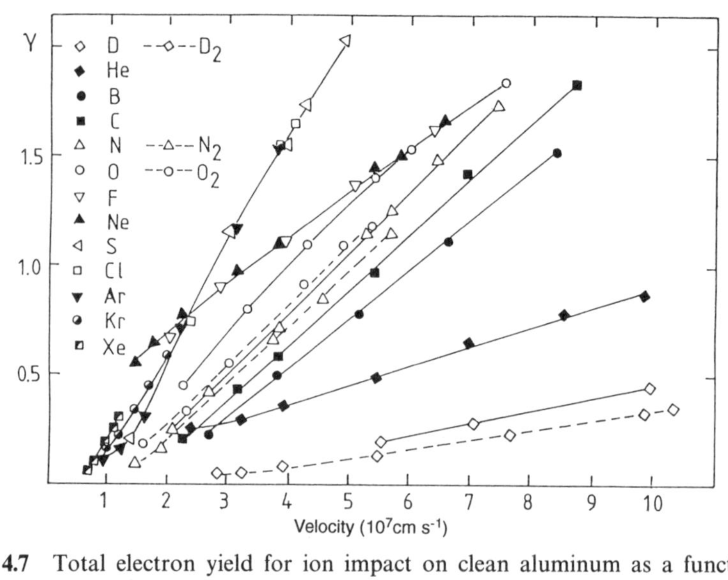

Figure 4.7 Total electron yield for ion impact on clean aluminum as a function of projectile velocity (from Alonso *et al.*, 1980). The calculated threshold velocity for emission due to binary ion–electron collisions is 1.75×10^7 cm s^{-1}

velocities, a good description of the excitation is in terms of electronic transitions in the quasi-molecule formed transiently during the collision (Krakov and Parilis, 1989). The electron states can be described in a fairly good approximation by one-electron molecular orbitals (MO). As a result of the formation of the molecule, there is the competing effect of the increased nuclear charge which tends to bind the electrons more, the increase in kinetic energy of the electrons due to localization which tends to bind the electrons less, and the limitation of the number of electrons in each orbital, imposed by the Pauli principle. As a consequence, some MO become more strongly bound when the atoms approach, and some become promoted to higher energies (Figure 4.8). At low velocities, these MO evolve adiabatically and no excitation occurs except near the avoided crossing (pseudo-crossings) of the adiabatic MO energy curves.

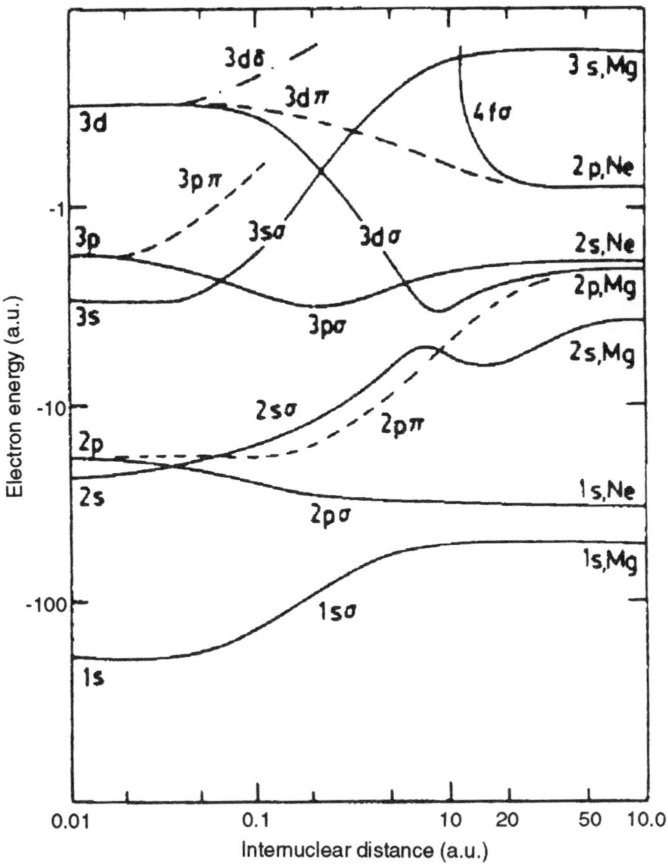

Figure 4.8 Diabatic molecular orbital diagram for the Ne–Mg collision. The energy levels at zero internuclear distance correspond to those of the united atom, Ti. (Reproduced by permission of the American Physical Society from Zampieri *et al.*, 1984)

Transition occur due to coupling between MO produced by the radial or rotational components of the center-of-mass motion. To determine which MO energy curves cross and which transitions are allowed, one uses 'diabatic' MO, where states are allowed to cross. Much can be learned from identifying the crossings with simple correlation diagrams which connect levels in the separated atoms of nuclear charge Z_1 and Z_2 to levels in the united atom of charge $Z_1 + Z_2$. Diabatic correlataion diagrams can be drawn and analyzed using prescriptions given by Barat and Lichten (1972), with the corrections by Eichler *et al.* (1976).

The apparent breakdown of adiabaticity in low-energy collisions can be described by Massey's adiabatic criterion (Massey *et al.*, 1971, chap. 17). Transitions become possible when the energy uncertainty due to the finite interaction time a/v becomes comparable to or larger than the energy separating the states at the pseudo-crossing, ΔE:

$$\hbar v/\Delta a \gtrsim \Delta E, \tag{9}$$

where v is the relative velocity and Δa is the extent of the region where the coupling between the states is significant. Transitions are possible at low velocities due to the small values of ΔE at the pseudo-crossings of energy curves.

In most cases, the achievement of the crossing distances requires small impact parameters at low velocities, i.e. violent collisions. At these distances, the ground-state band structure of the solid is not important during excitation and one can treat the collision as one between free atoms, with the outer-shell configuration which corresponds to the solid state. An improvement to this description comes from calculations that show that the motion of the atoms in the lattice induces the formation and excitation of new electronic states above and below the valence band (Sroubek, 1993a, 1993b).

A simple atomic ionization theory is Firsov's friction model of energy transfer in slow collisions (Sigmund, 1975). In this model, excitation occurs when electrons flow between the two atoms during the collision, changing their net average momentum. This builds up excitation energy which can cause ionization if it exceeds the ionization potential. Firsov's theory was developed using statistical atomic models, and has not yet been fully justified. It predicts only roughly gas-phase ionization cross sections: discrepancies with experiment (of this and other theories) can be as large as an order of magnitude (Kunc and Scon, 1991). On the other hand, Firsov's formula has had more success in describing electronic stopping powers (Sigmund, 1975). We recognized that this would allow the description of kinetic excitations induced by slow heavy ions, by assuming that the number of excited electrons produced is proportional to the stopping cross section for inelastic energy transfers larger than U, and applied the model to explain experiments and to establish quantitatively the contribution of ionizations due to recoiling target atoms (Alonso *et al.*, 1980; Ferrón *et al.*, 1981a).

In the collision process, the excited electron may originate not only from the target atom but also from the projectile. If the electron is emitted isotropically in the projectile frame, it will have an average energy $E_{eq} = (m/M)E_o$ in the laboratory system, with a noticeable forward-peaked angular distribution if E_{eq} is more than a few eV. During the collision, a loosely bound projectile electron, like in negative-ion or in Rydberg states, will have a stronger interaction with the target atom than with its parent core. It will then scatter approximately as a free electron and ionize easily for E_{eq} larger than its binding energy. I have estimated the contribution of electron loss to KEE for proton impact on Ar films (Baragiola, 1993b) by using gas-phase cross sections; it is ~30% for a beam of equilibrium charge at 10 keV, decreasing with ion energy, and 35% for a beam of 10 keV H atoms, increasing with energy. The relative importance of projectile vs. target ionization will depend, in a first approximation, on the value of the ionization potential of the projectile compared with the energy threshold U of the solid.

4.2.2.3 Kinetic Auger excitations

When inner shells of the colliding atoms are promoted in energy, electrons can be transferred into partially or totally empty outer shells, so when the atoms recede vacancies are formed in inner shells of the projectile and/or the target atom. The probability of excitation is particularly high when the promoted orbital crosses several empty MO. The physics of inner-shell excitations in slow atomic collisions has been reviewed extensively by Wille and Hippler (1986).

The requirement that a critical impact parameter r_c be reached to promote the inner-shell orbital implies a threshold center-of-mass energy for excitation:

$$E_{th,cm} = V(r_c), \tag{10}$$

where $V(r)$ is the interatomic potential. This energy can be lowered if the collision is not binary, i.e. if a third atom participates and absorbs momentum. If the promotion occurs by radial coupling between states, the probability of excitation is well approximated by a formula proposed by Cacak: $P(r) = P_m$ for $r < r_c$, and $P(r) = 0$ otherwise. This gives the excitation cross section:

$$\sigma = P_m r_c^2 [1 - V(r_c)/E]^2 \tag{11}$$

where $P_m \approx 1$ if the promoted level crosses many outer levels with vacancies.

After the inner-shell holes are formed, they can be filled in by different processes. The main decay mode for holes in the shallow inner shells that can be excited at low velocities, is the Auger mechanism. In this process, the repulsion between electrons makes an electron from an outer shell jump to fill in the inner-shell hole, and the excess energy is given to another electron, which is ejected with an energy characteristic of the intervening shells.

Another decay mode for the inner-shell vacancies is radiative filling by an electron from an outer shell, yielding a soft X-ray; it has a low probability (fluorescence yield), typically a few percent or less for transition energies of a few hundred eV or less. The excited atom can also transfer its vacancy to another atom in a subsequent close impact in the collision cascade that develops during energetic ion bombardment. Typical lifetimes of inner-shell holes $\tau = 1/R_t$ are in the range 1 to 100 fs (Chen *et al.*, 1990; Al Shama *et al.*, 1992). Auger transition rates R_t increase with the overlap between valence and inner-shell electrons. In solids, the core hole can be screened by electrons coming from neighboring atoms, resulting in higher R_t. From similar arguments one can expect larger R_t in metals than in insulators and a high enhancement of R_t for moving atoms, due to the increased overlaps between shells, during collisions.

Auger electrons can be identified as structures on the high-energy tail of the energy distribution of ejected electrons (Figure 4.9) (Grizzi *et al.*, 1990). If the core-excited atom decays inside the solid, the Auger energies will represent approximately a convolution of the density of states of the valence band at the

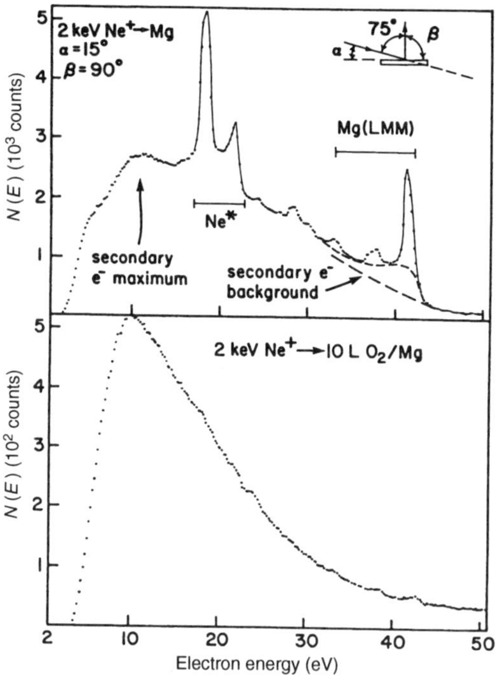

Figure 4.9 Electron energy spectrum from impact of 2 keV Ne^+ on clean and oxidized Mg surfaces 1990. (Reproduced by permission of the American Physical Society from Grizzi *et al.*, 1990)

location of the inner-shell hole. This causes a broad, 'bulk'-type spectrum similar to that obtained under electron bombardment.

The delay between the moment of excitation and that of electron emission, caused by the finite inner-shell hole lifetime, is an important characteristic of electron emission by inner-shell excitations. During the hole lifetime, the excited atom can move away from the excitation site and even be ejected (sputtered) outside the solid. In this case, the Auger energies would correspond to that of a free atom, modified somewhat by the interaction with the surface. The Auger spectra corresponding to these excited atoms or ions decaying outside the solid is composed of narrow (~ 1 eV or less) 'atomic' lines. The increase in the charge state of the sputtered particle, due to the Auger process, can lead to multiply-charged ion emission, in what has been called the *kinetic* secondary ion emission mechanism (Joyes, 1968, 1969a, 1969b; Veksler, 1983; Weller, 1989; Larson and Lauderback, 1991; Milne *et al.*, 1991; Schauer and Williams, 1992). The Auger decay of the inner-shell excitation which may accompany violent collisions can produce high mean charges of back-scattered *projectiles* (Baragiola, 1977; Hasselkamp *et al.*, 1992). Auger emission from *target* atoms can produce high ion yields in secondary ion mass spectrometry.

In ion–surface collisions, the inner-shell excitations can be produced in collisions between the projectile and a target atom or in a collision between an energetic recoil and another target atom. What type of collision dominates depends on the particular case, as we shall see in Section 4.4.3. The treatment of the excitations in the cascade and the motion of excited atoms inside the solid can be studied using computer simulations (Grizzi and Baragiola, 1984, 1987; Kitov and Parilis, 1984; Bénazeth *et al.*, 1985, 1987; Hou *et al.*, 1986, 1987, 1990; Shapiro and Fine, 1989; Herteux *et al.*, 1990; Konoplev *et al.*, 1992). They have served to clarify misconceptions in the literature, which assumed that the sputtering of excited atoms should depend on the total sputtering yield. There are two factors which are unique to the emission of core-excited atoms. First, it samples sputtering initiated by the hardest atomic collisions, those with a distance of closest approach less than the value needed to produce excitation. Thus, many collisions which can produce sputtering do not necessarily produce excitation. Second, the finite lifetime of the inner-shell hole implies that the emission of excited atoms samples the initial part of the collision cascade and preferentially the trajectories which spent the smallest times inside the solid. Therefore no *direct* correlation should be expected between Auger and sputtering yields.

Interesting effects may occur when bombarding with molecular projectiles. Ferleger and Matveev (1991) have made calculations of Auger spectra induced by bombardment with diatomic molecules (AB). In their model, one of the atoms of the molecule, A, produces the core excitation. A molecular effect would occur if the vacancy decays in the presence of the other atom in the projectile, B. Shifts and broadening on the spectra with respect to bombardment with single atoms should appear, depending on the values of the projectile velocity, the lifetime of the inner-shell hole and the screening in the solid. An exciting suggestion by Shapiro and Tombrello (1992) is that Auger emission can occur for incident cluster ions at subthreshold velocities (i.e. below that needed to reach r_c in single-atom impact). This is possible because of collisions between moving atoms in the collision 'spike' generated when the cluster suddenly stops in a heavy substrate. The calculations were made using molecular-dynamics simulations for clusters of Al with up to 10^8 atoms colliding with gold surfaces.

An additional question is what is the electronic state of the excited sputtered atom when it leaves the solid. While inside, its inner-shell hole attracts a screening charge which accompanies it until it crosses the surface. This screening may be dynamic, formed by valence electrons scattering off the ion core, or it can be a bound, core exciton state. When the ion crosses the surface the final outer-shell configuration will be determined by a competition of resonance neutralization and ionization, depending on the work function of the surface and the availability of electronic states. The Auger spectrum resulting when the excited atom finally decays in vacuum will be directly determined by this population of the outer shells.

4.2.2.4 Excitations by atomic recoils

The relative contribution of recoils increases with increasing ratio of the mass of the projectile to that of target atoms. This recoil effect has important implications in KEE at low energies. It dominates the threshold behavior of KEE if the mass of the projectile is much larger than that of the target (e.g. for Xe on Al) (Alonso *et al.*, 1980), when the recoils can be faster than the projectile. Inclusion of recoil effects in the calculation of electron emission yields requires an evaluation of the *atomic* cascade.

The theory of electronic excitations in the recoil cascade has been discussed by Caro and Victoria (1989), and by Falcone and Sroubek (1988, 1989a, 1989b), who obtain the energy distribution of electrons excited in metals. The theory uses an electron gas model and does not consider excitations in violent atomic collisions. The relative contribution of recoil atoms to KEE has been studied by Ferrón *et al.* (1981a) using Monte Carlo simulations and describing the excitation events using a modified Firsov stopping power for energy transfers larger than U.

4.2.3 ELECTRON TRANSPORT

Electron transport is important when the excited electrons are produced inside the solid, as in KEE. A proper description of electron transport requires knowledge of the probability that each electron loses energy and eventually generates new excited electrons. This problem has been studied recently by using an electron gas model to describe inelastic electron–electron scattering. In some cases, simulations have included classical electron trajectories using Monte Carlo methods between elastic scattering events with single atoms (Shimizu and Ding, 1992). The justification for doing this has not been given. The problem is that the majority of the electrons in the cascade have wavelengths of the order of or larger than the mean atomic spacing, a. The electrons will not, therefore, interact in a binary fashion with a single atom at a time, and the concept of cross section cannot be applied as is done in classical transport theory or simulations.

A different question is whether the trajectory of an electron can be defined between elastic scattering events. A classical description is valid if the uncertainty in the momentum $\hbar/a$, due to localization within a region of size a, is much less than the electron momentum p. This implies the often fulfilled condition $E' \gg 0.5$ eV far from the Brillouin zone boundaries, where E' is the electron energy inside the solid.

An important consideration in understanding electron transport is that the probability of a small inelastic energy loss is very small for low-energy electrons (Ritchie and Ashley, 1965). This is because all the states of the valence band are occupied so that only the top of the band can absorb small energy transfers. A large average-energy loss means that an inelastic interaction will most likely attenuate secondary electrons (drop their energy below the vacuum level). This in turn implies that electrons that escape suffer very few inelastic collisions. In non-metals low energy losses are more strongly suppressed; only losses larger than the band gap are allowed, neglecting excitons and other gap states.

At low projectile velocities, most of the excited electrons will not have enough energy to produce additional electron emission by cascade multiplication. The probability that an excited electron produces at least one more electron above the vacuum level is zero if the energy above the vacuum level is less than twice the threshold energy: $E < 2U$. Based on pair production in gases, an approximate multiplication factor for electrons will be given by $\sim 1 + (E - U)/U$. Thus the transition from the single-excitation to the cascade-multiplication regime will depend on the energy distribution of the excited electrons.

The diffusion of the deposited energy away from the collision site causes the number of excited electrons to vary with the distance to the surface, even if the scattering and energy loss of the projectile could be neglected (Sigmund and Tougaard, 1981; Sigmund, 1993). This is because at depth $z > R$ (the electron range) there are contributions from the region extending from $z - R$ to $z + R$, whereas at the surface there are only contributions from the region extending from $z = 0$ to $z = R$. Furthermore, for the slow projectiles considered here, energy degradation in the electron escape region is important. It causes a decrease in the primary ionization density with depth which may be compensated for by the increase due to larger path lengths produced by angular scattering and ionization by recoils. This latter contribution may increase or decrease with energy, depending on the projectile–target combination (Schou, 1980).

Since the excited electrons have low energies, we can assume that inelastic events will more likely remove the electrons from states above the vacuum level than degrade them in energy and still allow them to retain sufficient energy to escape. The part of the flux of electrons that reaches the surface with final electron energy $E' > U$ will be that which has not been attenuated in inelastic events. The probability of attenuation is related to the inverse of the inelastic mean free path λ_{in}. If the probability for elastic scattering is small, such as at high electron energies or for low-Z targets, or for electrons excited very close to the surface, the electrons are attenuated in straight paths to the surface, according to the attenuation length (Chung and Everhart, 1974):

$$L = \lambda_{in}/\cos\theta', \tag{12}$$

where θ' is the angle of the internal path with respect to the surface normal. The assumptions underlying equation (12) should hold particularly well near the kinetic emission threshold, where the electrons are excited with very small energy, and only close to the surface. In the case where elastic scattering dominates, the diffusion process will be described by an attenuation length (Sternglass, 1957):

$$L = (\lambda_{in}\lambda_{tr}/3)^{1/2}, \tag{13}$$

where λ_{tr} is the transport mean free path for elastic scattering. Monte Carlo simulations have shown that the actual attenuation as a function of depth can be more complicated than described above (Tougaard and Sigmund, 1982; Ferrón *et al.*, 1982b; Cailler and Ganachaud, 1985; Ferrón and Goldberg, 1992).

Attenuation lengths in metals are 5–15 Å when averaged over the energy spectrum of the emitted electrons (Seiler, 1983; Donath *et al.*, 1991; Zhang *et al.*, 1991). The values for insulators are usually one order of magnitude larger, but can reach thousands of Å for wide band gap insulators, such as the noble-gas solids (Gullikson, 1988; Cartier and Pfluger, 1988).

4.2.4 ELECTRON ESCAPE THROUGH THE SURFACE BARRIER

The surface electrostatic potential is corrugated on the atomic scale, i.e. small compared with the scale of the wavelength of electrons of a few eV, suggesting that it is a good approximation to consider the surface as planar. In fact, photoemission experiments in single crystals show that the momentum of the electron parallel to the surface, $\hbar k_{\parallel}$ is usually conserved during emission

(Fenerbacher and Willis, 1976). The effect of the planar barrier of height I is to refract the electrons, changing their energy from E' inside to $E = E' - I$ outside, and changing the momentum perpendicular to the surface, $\hbar k_{\perp}$. Thus, the angles of incidence and emergence with respect to the surface normal are related by: $k \sin \theta = k' \sin \theta'$. Refraction causes total internal reflection for angles larger than

$$\theta'_{max} = \arccos [(I/E')^{1/2}]. \tag{14}$$

The probability of transmission through the barrier, T, can then be calculated by assuming that all electrons in the cone with $\theta' < \theta'_{max}$ can escape. Therefore T depends on the shape of the angular distribution of the electrons reaching the surface. When elastic scattering can be neglected electrons can be considered to travel in a straight line to the surface, attenuating in number according to $\exp(-z/L)$, with L given by equation (12). Then,

$$T(E) = 1/2\, E/(E + I). \tag{15}$$

On the other hand, when elastic scattering is important, the transmission through the barrier can be calculated assuming that electrons arrive at the surface isotropically in the half-space inside the solid. In this case,

$$T(E) = 1/2\, [1 - (I/(E + I))^{1/2}]. \tag{16}$$

The escape probabilities (15) and (16) are classical. Quantum-mechanically, not all electrons with the escape condition can escape. Reflection can occur even if $E' > I$, when the uncertainty in the energy of the electron introduced by the finite interaction time τ with the surface is of the order of or larger than E'. This uncertainty is $\delta E' = \hbar/\tau$, where $\tau = b/v = b(m/2E')^{1/2}$ and b is the width of the barrier. This means that reflection can occur for $\hbar(m/2E')^{1/2}/b \gtrsim E'$. Thus quantum barrier effects will be important for $E \lesssim 0.5$ eV for a barrier width $b = 2$ Å. The quantum-mechanical reduction in the transmission above the barrier is very strong in thermionic emission, where electron energies are very small: it reduces the thermionic 'A-factor', and hence the emission current, by 25–50% (Ramsey, 1961). One should then also expect important reductions in KEE near threshold. The reflection properties of surface barriers can be treated by analyzing the short-range Fourier component of the Coulomb interaction, normal to the surface (Fano, 1987). Recent work on the transmission of electrons through barriers has focused on the effect of adsorbed layers, or on 'dirty' surfaces (Caron *et al.*, 1990).

As mentioned above, the barrier I for insulators is the electron affinity χ, which can be as low as a fraction of an eV or even negative, such as in solid Ar or in semiconductors with special surface coatings (Bell, 1973). This reduced barrier is one of the main reasons for the larger electron emission yields of insulators as compared with metals.

An interesting case is that of PEE, where the three-step model does not hold in a strict sense. The wave function of the intervening electrons extends over the barrier which is modified by the presence of the projectile. This complexity has not yet been treated in the literature. This problem should be circumvented by considering a final-state wave function which already has the correct asymptotic solution outside the solid, avoiding the artificial introduction of a surface barrier.

4.2.5 FINAL-STATE EFFECTS

So far, the description of the final state has been an electron moving away from the solid in the presence of the surface; that is, in the field of its image. In some cases, this is not a good representation because of the presence of other fields which appear when the electron leaves the solid, due to presence of other charged particles in vacuum in the vicinity of the electron. These 'final-state interactions' appear because of an incomplete description of the final-state wave functions.

It has been observed in the gas phase and for fast ions traversing thin foils that the energy and angular distributions of electrons are distorted by the field of the ionized projectile. When the velocity of the electron is very similar to the velocity of the ion in magnitude and direction, a large increase is observed in the laboratory cross section, producing a cusp-shaped peak in the doubly differential distributions. This is due to the focusing of the electron trajectories by the long-range Coulomb field of the ion. Since these electrons accompany the projectile, they have been termed 'convoy' electrons (Groeneveld *et al.*, 1984). The production of convoy electrons by ions traversing foils appear to occur only above a threshold at $v \approx 8 \times 10^7$ cm s^{-1} (Suarez *et al.*, 1992). Focusing of electron trajectories also occurs when the projectile back-scatters as an ion, as is usually the case at grazing incidence (de Ferrariis and Baragiola, 1986; Winter *et al.*, 1989) and can possibly be observed in the energy range of interest here, for very-low electron energies.

A prediction that follows from the above discussion is that in PEE with slow highly charged ions, the trajectories of the electrons will be even more severely affected by the strong Coulomb field of the positive ion and its negative image. The electric fields in this case are different from the counterpart in gas-phase collisions, where partial neutralization of the highly charged ions produces a positively charged target atom.

In many instances, projectiles can produce the emission of multiple electrons within a very narrow time interval. The electron–electron interactions or correlations will alter the energy and angular distribution of the emitted electrons, as already seen in double ionization in the gas phase (Skogvall and Schiwietz, 1988). The effect will be present in measurements of emission statistics in solids even at relatively small values of γ and in recent coincidence experiments of electron pairs ejected by primary electrons (Kirschner *et al.*, 1992). The effect should of course be more pronounced in situations which render large average electron yields (Leonas, 1991; Baragiola, 1993a; Aumayr and Winter, 1993). In these cases, the ejected electrons will move in the field of the other electrons and their images; their trajectories will be deflected, compared to those of electrons emitted alone. The ejected charge cloud will evolve in time according to the internal fields and the applied external extraction field. Electrons in the front of the charge cloud will receive extra energy while those at the trailing side will be slowed down, even pushed back to the solid (depending on the extraction field), as observed in pulsed-laser-induced photoemission (Ammosov, 1991; Petite *et al.*, 1992). The amount of energy that can be gained can be of the order of $U = \gamma e^2/r_o$, where r_o is the initial radius of the electron cloud ($U \approx 20$ eV for $r_o = 15$ Å and $\gamma = 20$). Then, this space charge should act to decrease the yields, flatten the angular distribution, broaden the energy distribution of the electrons and affect the I–V characteristics of the target-collector diode. The effects should be sought after particularly in cases where γ is very high. Electron yields in the hundreds have been seen for PEE of very highly charged ions (McDonald *et al.*, 1992, Aumayr *et al.*, 1993). In KEE from solid Ar, we have observed γ of several thousands (Grosjean and Baragiola, 1993); similar values are expected in other materials with negative electron affinity.

4.2.6 YIELDS

To calculate differential and total yields one has to develop quantitative descriptions of the steps shown above and assemble them to derive observables. No complete theory exists at this time although some special cases have been derived where simplifications are possible. An example of the latter is the theory of Kaneko (1992) for KEE by light ions on metals represented by an electron gas model, and without considering the scattering and slowing down of the projectiles. Rösler has developed a more detailed theory using the Boltzmann transport equation, and including the contribution from the Auger decay of inner-shell excitations (Rösler, 1991a; Rösler and Brauer, 1992). By comparison with Monte Carlo simulations, Devooght *et al.* (1992) have shown that the neglect of surface effects in the Boltzmann equation gives an overestimate of 25%, which they propose to correct by an approximation method.

A general finding for particle impact at high velocities is that the number of electron–hole pairs generated by the projectile is proportional to its electronic energy loss (Baragiola, 1993b). We have found (Baragiola *et al.*, 1979) a fairly universal proportionality of electron yields with the electronic stopping power, dE/dx, rather than with the stopping cross section $dE/N\,dx$, a quantity usually preferred due to its insensitivity to target density:

$$\gamma \approx B(dE/dx)_e/\cos\alpha. \tag{17}$$

B falls in the narrow band: $B = 0.10 \pm 0.03$ Å/eV for different clean-metal targets under 5–50 keV H^+ bombardment (Baragiola *et al.*, 1979) and about 20% lower for heavier ions (He through Ne) (Alonso *et al.*, 1980).

At the low impact velocities considered here, the proportionality between inelastic energy loss and number of ionizations breaks down (Lindhard *et al.*, 1968; Miller and Boring, 1974) since the fraction of the projectile energy available for electronic excitations decreases due to the generation of energetic target recoils and atomic vibrations in elastic collisions with target nuclei. A small fraction of the energy transferred to the recoils will still be delivered to the electronic system in additional ionization collisions (Knipp and Ling, 1951; Phipps *et al.*, 1964).

Kawata *et al.* (1991a, 1992) have recently made simulations similar to those of Ferrón *et al.* (1981a), but they assumed that the number of excited electrons is proportional to the total electronic stopping power. This assumption is not valid for slow projectiles (Thum and Hofer, 1984; Ferguson and Hofer, 1989, 1993), especially for the low-energy recoils. Thus, the agreement of their results with experiments may be fortuitous, especially since they were obtained by applying a fit. Nevertheless, some general conclusions should remain valid, e.g. that recoil effects are more important for heavier projectiles and that they dominate KEE at low velocities if the projectile is heavier than the target atoms. The calculated emission statistics reflect the fluctuation of the elastic collision cascade but not the fluctuations in the electron excitation processes. These types of simulations have also been used by Ohya and co-workers to study the dependence of the yields on the angle of incidence of the projectile and the roughness of the target (Kawata *et al.*, 1991a, 1992; Ohya *et al.*, 1989).

Sigmund (1993) has discussed critically some key theoretical points related to the interpretation of measurements of the distribution of probabilities P_n that a projectile ejects n electrons from the solid. This distribution originates from fluctuations in the steps that lead to electron emission. Although the statistics of collision sequences for ions and electrons follow Poisson statistics, the number of excited electrons does *not*. Experiments on generation of electron–hole pairs in semiconductors show that the fluctuations are not those corresponding to a Poisson distribution. Rather, the variance in ν, the number of electron–hole pairs, is $s^2 = F\nu$, where F is the Fano factor (Knoll, 1989). F is usually less than 1, which means that the distribution of the number of internal electrons is narrower than Poisson. On the other hand, it is usually found that the electron emission statistic is broader than Poisson. It is also found that the probability of emitting no electrons is larger than the value

$P_0 = e^{-\gamma}$ that would result from a Poisson distribution. This is very important for the use of ion electron emission in detectors used in mass spectroscopy, LEIS, etc., like channeltrons, channelplates, magnetic electron multipliers, electron converters, etc. Even if all secondary electrons and their progeny are collected, the detector counting efficiency cannot be larger than $1 - P_0$.

Ohya and Mori (1991) and Kawata *et al.* (1991b) have attempted to explain theoretically the observed emission statistics, using a semi-empirical model for the average number of electrons excited in the solid. Statistical fluctuations come from fluctuations in the escape of the excited electrons and fluctuation in the energy deposited by the ion. This latter value is obtained from a Monte Carlo simulation and includes contributions due to trajectory effects on the energy deposited by the projectile and recoils, but not fluctuations in the electronic energy loss (straggling). In later work (Ohya and Mori, 1992a, 1992b), the authors assume that the number of internal secondaries are Poisson-distributed and obtain emission statistics in fair agreement with experiment (Ohya and Mori, 1992c; Ohya *et al.*, 1992), which are broader than the Poisson distribution (Lakits *et al.*, 1989d). The enhanced P_n at large n results, according to the simulation, from back-scattered ions and energetic recoils. The agreement with experiment must be taken cautiously, however, due to the neglect of fluctuations in the primary excitation event. Electronic straggling is particularly important at keV energies and it dominates the yield near the KEE threshold, as pointed out by Alonso *et al.* (1986). An improvement upon the interesting results of Ohya and Mori can be achieved by using an impact-parameter-dependent inelastic energy loss, as done by Ferrón *et al.* (1981a). A complete calculation should also consider quantum fluctuations in the energy loss for a given impact parameter.

Once the excitation function is known, the yields can be calculated by solving the electron-transport and escape problems. As mentioned above, calculation of the electron cascade can be done using transport theory or Monte Carlo simulations starting with electron-scattering data from gas-phase collisions or free-electron models, as appropriate to the type of solid. Here the main difficulty is the lack of an adequate description of elastic scattering at the lower electron energies where quantum effects are important. Once the flow of the electrons at the surface is obtained one can calculate the transport through the barrier. In this case, the main difficulty is the estimation of the inner potential and quantum reflection effects.

To calculate the spatial distribution of excited electrons one has to follow the trajectories of projectile and target atoms (e.g. by using Monte Carlo or molecular dynamics simulations) (Ferrón *et al.*, 1981a). For each collision one calculates the inelastic energy transfer by using, for instance, Firsov's formula or improved versions (Kishinevskii, 1963; Baklitskii and Parilis, 1986). If this calculated energy transfer is larger than U, one counts an excited electron from which one builds a distribution of ionization events, $n(z)$, as a function of distance z to the surface. At low enough energies, cascade multiplication can be neglected, and one can equate $n(z)$ with the depth distribution of first-generation electrons. If one neglects inelastic electron collisions that do not attenuate the electron flux, one can calculate the energy distribution of ejected electrons using the barrier transmission coefficients from equations (15) and (16):

$$d\gamma/dE = T(E') \int n(z, E') \exp(-z/L(E'))\, dz, \tag{18}$$

where $E = E' - I$, from which one can deduce the total yield by integration.

Another simple case is that of KEE near threshold, when the projectile is lighter than the target atoms. Then, one can disregard the recoils and need to consider only inelastic collisions of the projectile with surface atoms. One can neglect electron transport in first approximation and assume that all electrons in the escape cone (equation (11)) will go into vacuum. Then the energy distribution of emitted electrons will be given by:

$$d\gamma/dE = n_S\, T(E)\, d\sigma/dE, \tag{19}$$

from which the yields can be obtained by integration. Here σ is the ionization cross section and n_S is the surface density of target atoms. The proportionality of electron-emission yields and ionization cross sections near threshold was first demonstrated by Amme (1969) for gas-covered surfaces.

An important relationship is the variation of the electron-emission yields on incident angle α with respect to the surface normal. It is found to depend on projectile energy due to different factors. At low energies, where PEE prevails, the yields depend only weakly on the perpendicular velocity. KEE yields change with α due to the change in energy deposited near the surface. At high incident energy, the energy deposited per unit path length near the surface is constant for thick layers. Therefore, the energy dependence per unit distance normal to the surface goes like $\sec\alpha$. Deviations from this behavior occur due to the anisotropy in the distribution of excited electrons (Ferrón *et al.*, 1981b). At keV energies, other factors are important in determining $\gamma(\alpha)$: nonlinear trajectories and reflection of the projectile, slowing down of the projectile, and excitation due to recoils. The last two effects are more pronounced the heavier and the slower (Oda *et al.*, 1992) the projectile. Ohya (1992) has made Monte Carlo simulations for proton and electron impact and fitted the yields to a $(\sec\alpha)^f$ dependence. For protons, the exponent f increases from 0 or small negative values at 0.1 keV, crosses unity at energies of a few keV, reaches a maximum at around 10 keV, and then decays to unity for large projectile energies. The simulation does not consider anisotropies in the ionization event.

4.3 MEASUREMENT TECHNIQUES

The general techniques for measuring electron emission has been reviewed by Hasselkamp (1992). Here we will give additional information and emphasize aspects needed to evaluate experiments and identify potential problems that may affect the reliability of the data.

4.3.1 SURFACE PREPARATION

Since electron emission is very sensitive to the state of the surface, reliable data can only be obtained under careful and specified surface conditions, under ultra-high vacuum (UHV). These must necessarily start with a clean surface. Many researchers have cleaned surfaces by sputtering with inert-gas ions. Since sputtering causes damage and some recoil implantation of the absorbed impurities, it is better to combine the method with heating to high temperatures, if possible for the type of material under study. For polycrystalline samples, it is generally found that the results obtained before heat treatment are indistinguishable from those obtained from films deposited *in situ* in UHV. High-temperature annealing of polycrystalline samples can result in preferential alignment of a major axis of each crystalline along the surface normal, while still keeping a random azimuth. If this occurs, electron yields at normal incidence would decrease due to channeling of the incoming projectiles (see Section 4.4.2.1.2).

Experiments at low energies can be extremely sensitive to minute amounts of alkali contaminants, which produce a low local work function. Published electron-emission yields smaller than $\sim 10^{-3}$ should be interpreted with caution unless the authors have analyzed the composition of the surface *in situ* using some of the standard UHV spectroscopies.

4.3.2 SURFACE-CHARGING PROBLEMS IN INSULATORS

In the study of insulators, problems may arise due to charging of the surface, especially if the insulator is thicker than the range of the ions (Cazaux, 1993). The problem can be circumvented by using pulsed beams or by discharging the sample periodically by inducing a high electrical conductivity by irradiation with light of energy larger than the band gap (Wittmaack and Homma, 1991). Some authors have heated the surface of alkali halides to induce a high conductivity, but this method changes the sample during measurements. A good discussion of charging problems and secondary electron emission from insulators is given by the work of Liehr *et al.* (1986). The effects can also happen in metals if their surface layer is converted into an insulator during bombardment by, for instance, oxygen ions (Wittmaack, 1991). Field emission due to the strong fields created by charging under electron bombardment is discussed by Fitting and Hecht (1988).

4.3.3 TOTAL YIELDS

Measurements of total electron yields are made directly by the quotient method, where the yields are calculated from the ratio of the electron current collected in a positive electrode surrounding the sample, and the ion current is obtained as the sum of the currents on the target and the collector electrode. We have discussed possible sources of errors (Baragiola *et al.*, 1979) which include spurious electrons produced by the projectiles hitting apertures, emission of negative ions from the target (back-scattered and sputtered ions), tertiary electrons ejected from electrodes by particles and photons emitted from the target, as well as stray electrons from ion pumps and nearby ion gages.

Reliable measurements require that the ion beam be mass-analyzed to eliminate unwanted impurities, including multiply charged ions with larger energy. The presence of multiply charged ion impurities in the ion beam introduces particularly large errors in measurements of energy thresholds for an emission mechanism (Baragiola *et al.*, 1991). In experiments of PEE it is essential to determine the excitation state of the ion since it is the main factor influencing the electron yields and energy distributions. A problem arises when using atomic ions (e.g. N^+) obtained from molecular gases, due to contamination with doubly charged diatomic molecular ions (e.g. N_2^{2+}), since they cannot be separated by electrostatic or magnetic analyzers. The concentration of the molecular impurities can be ascertained by the Aston bands (Hasted, 1964) produced when increasing the beam line pressure before the analyzer magnet.

Another problem often ignored is caused by the presence of neutrals in the ion beam, produced by charge exchange in the residual gas, or in collisions with collimating apertures. The magnitude of the contribution of these neutrals to the measurements can be estimated by deflecting the ion beam away using an electric or magnetic field and observing changes in the signals.

4.3.4 EMISSION STATISTICS

The statistical distribution of the number of ejected electrons is normally measured by accelerating the electrons by a voltage V to a detector which produces an output pulse proportional to the electron energy, eV. If a group of n electrons is ejected simultaneously, the detector collects n times the energy of a single electron, or neV. Analysis of the pulse-height distribution then gives the probability of multiple electron emission.

Several variations of these detectors have been discussed in the recent literature (Hofer, 1990; Lakits *et al.*, 1989d; Moshammer and Matthäus, 1989; Leonas, 1991; Aumayr *et al.*, 1991) together with the theory of their operation. Figure 4.10 shows an experimental arrangement developed by Lakits *et al.* (1990b). In this set-up, the ejected electrons are deflected by the grid and accelerated into a Si solid-state detector held at 30 kV. A possible problem with this arrangement is that the retarding field may push the electrons back into the sample. Lakits *et al.* show trajectories of electrons emitted with an energy of 10 eV; they found that only 2.5% of electrons are returned to the target.

A possible problem with measurements of the probability of emission of multiple electrons is that the trajectories are affected by the space charge. As mentioned above, this space charge can act to suppress the emission of electrons of lower energy and produce a distortion in the results which is larger the larger the number of electrons ejected simultaneously.

The measurement of electron-emission statistics have been proposed by Winter and colleagues as a method of obtaining precise total electron yields, from the expression $\gamma = \Sigma n P_n$.

4.3.5 ENERGY DISTRIBUTIONS

An apparatus and procedures used by Hagstrum to measure electron energy distributions integrated over angle have been described in detail by Chaban *et al.* (1989). Recent reviews (Ibach and Mills, 1982; chap. 2; Avery, 1987; Roy and Tremlay, 1990) have discussed instrumentation for high-resolution electron

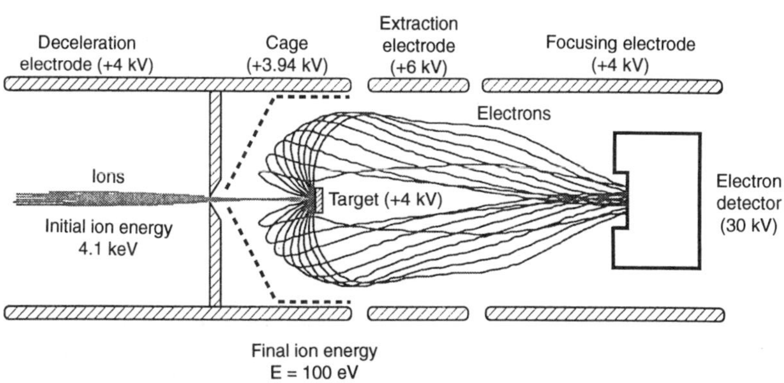

Figure 4.10 Apparatus at the Technical University in Vienna, used to study statistics of electron emission. (Reproduced by permission of Institut für Allgemeine Physik from Lakits *et al.*, 1990)

spectroscopy, while Seah and Smith (1990) have described procedures to determine the absolute efficiency of electron spectrometers. Measurement of the energy distribution of ejected electrons requires the annulment of the earth's magnetic field and of AC magnetic fields to a few milligauss or less. This can be achieved by avoiding the use of magnetic materials (including most stainless steels) and by using Helmholtz coils (with DC and AC currents) or μ-metal shields. Even if magnetic fields are canceled, there are usually problems at low electron energies due to uncontrolled surface potentials in parts surrounding the target. These fields can be minimized by avoiding the use of insulators or oxidized metals near the target, that may be hit by the ion beam. The preferred method to measure the energy distributions of very-low-energy electrons (less than a few eV) is by time-of-flight velocity spectrometry, using pulsed projectile beams (Rudd *et al.*, 1992).

A simple calibration of the energy scale of the energy analyzer can be obtained by measuring the energy of elastically back-scattered electrons, using an electron source of known energy (see Figure 4.11). The acceleration voltage V_a in the

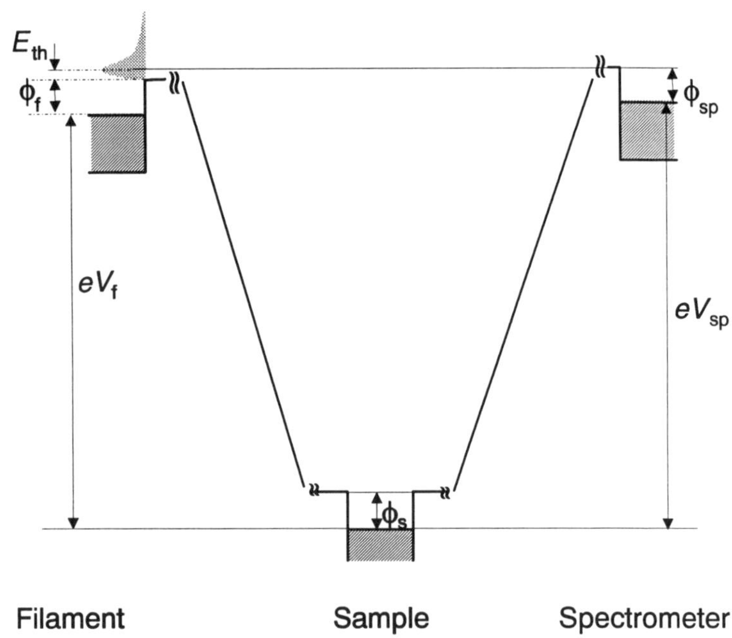

Figure 4.11 Energy diagram appropriate for calibration of the electron energies by using reflection of electrons on the sample, which is assumed to be at ground potential. V_f is electron acceleration potential, E_{th} is mean excess thermal energy of the electrons, V_{sp} is retarding potential in the analyzer which will just allow collection of the electrons with the most probable energy, ϕ_f, ϕ_s, and ϕ_{sp} are the work functions of the emitting filament, the sample and the spectrometer, respectively

electron source can be measured very accurately with a precision voltmeter, if one takes into account the voltage drop across the emitting filament to estimate the potential at the hottest region. The most probable (peak) energy of the electrons relative to the Fermi level of the spectrometer is $eV_a + \phi_f + E_{th} - \phi_s$, where ϕ_f and ϕ_s are the work functions of the filament and the spectrometer, respectively, and E_{th} is the excess thermal energy of the emitted electrons (Baragiola, 1981). For a tungsten filament at $T \approx 2700$ K, $E_{th} + \phi_f = 4.8$ eV, to within 0.1 eV.

To measure energies with respect to the vacuum level of the target one also needs to measure its work function ϕ_t. This can be accomplished by accelerating secondary electrons away from the target with a known voltage V' with respect to the spectrometer (Alov and Dadayan, 1987). The minimum electron energy measured by the spectrometer (corresponding to electrons with zero energy at the vacuum level of the sample) is given by $E_m = eV' + \phi_t + \phi_s$. Since quantum effects in the electron-escape probability at the sample surface may cause the initial slope of the energy distribution to be zero at $E = 0$, one has to be careful in taking properly into account the energy broadening caused by the analyzer while extrapolating the onset at E_m.

4.4 RECENT EXPERIMENTAL RESULTS

4.4.1 POTENTIAL EMISSION

Experimental results in the field of PEE were reviewed by the author 12 years ago Baragiola (1982). Since then, many new developments have occurred. PEE by He^+ and Ne^+ on Ni(100) and Cu(100) was studied by Hagstrum *et al.* (1988) as a function of alkali adsorption. The authors discussed the role of the local work function in determining the interplay of AD and RI + AC in the de-excitation of metastable atoms formed by RC. When the work function is small, a sharp peak appears at 16.8 eV with respect to the vacuum level, or 19–19.8 eV with respect to the Fermi level. Here, we will call this the H-peak. Following previous work using incident He metastables (Woratschek, *et al.*, 1985), Hagstrum *et al.* assigned the H-peak to AD with the emission of the K 4s electron; a similar assignment was given by Nishigaki *et al.* (1988, 1989, 1990, 1991) for metastable He impact on other alkalated surfaces. A different interpretation was offered by Hemmen and Conrad (1991, 1992), who studied PEE by thermal metastable He atoms incident on metal surfaces, as a function of K adsorption. They argue that K 4s electrons are located in a broad peak at ~1.4 eV below the Fermi level. They observed that the position of the H-peak is the same for K as for Na, Rb and Ba adsorbates, and that it also occurs with incident triplet metastables of Ne and Ar. Thus they conclude that the H-peak is due to the autodetachment of a core-excited negative ion ($He^-(1s2s^2\ {}^2S)$ in the case of He impact) formed by resonant electron capture. It is interesting to note that the use of slower metastable atoms produces a sharper higher energy cutoff in the electron spectra, compared with the case of the 10 eV He^+ used by Hagstrum *et al.* On the other hand, additional information may result from the use of higher energy He^+ ions as precursors for the metastables, since this can lead to a Doppler shift if the electron is ejected from the moving projectile. This allowed Souda *et al.* (1992) to conclude that the electron in the H-peak is ejected before de-excitation of the intermediate $He^*(^3S_1)$, in interactions of 10–500 eV He^+ with cesiated W(110) at glancing incidence.

Figure 4.12(a) shows energy spectra from Kempter's laboratory, taken with 50 eV He^+ and He^{2+} on a W(110) surface as a function of coverage with sodium (Brenten *et al.*, 1992b). The spectra have been displaced by biasing the target; the low-energy cutoff should correspond approximately to the vacuum level at the sample position, which changes as the work function is varied from a nominal value of 5.2 eV for the clean surface to ~1.4 eV. To discuss competing processes, it is convenient to use the notation $Y(h, x)$ where Y is the individual electron-transfer process (AC, AD, AU, etc.), h is the number of holes and x is the number of excited electrons in the projectile before the transition. Figure 4.12(b) indicates possible transition paths. In the case of He^+ impact, only AC(1, 0) is observed for clean surfaces:

$$He^+(1s) + M \Rightarrow He(1s^2) + M^{2+} + e^-$$

where M denotes the metal surface.

When lowering the work function by alkali adsorption, RC can occur to the excited He* states which can later produce PEE by the AD(1, 1) process, up to higher electron energies.

$$He^+(1s) + M' \Rightarrow He^*(1s2l) + M^+ \Rightarrow He(1s^2) + M^{2+} + e^- \qquad (RC + AD(1, 1)$$

The authors use the Hemmem and Conrad interpretation to assign the H-peak to double RC into a He^- state followed by autodetachment in process AU(1, 2):

$$He^+(1s) + M' \Rightarrow He^{-*}(1s2l2l') + M^{2+} \Rightarrow He(1s^2) + M^{2+} + e^- \; RC + AU(1, 2)$$

Many more transitions are observed for incident He^{2+}, due to the multiplicity of pathways for dissipation of the ~79 eV of potential energy brought by this ion. These are indicated in Figure 4.12(b) and labeled in Figure 4.12(a). For the case of the clean W surface, the spectrum has been decomposed in electrons originating in different processes, according to Niehaus's model. For low-work-function surfaces, RC to $He^{**}(2l2l')$ gives rise to sharp autoionization peaks, first reported by Hagstrum and Becker (1973) for He^+2s impact on Ni.

Recently, Brenten *et al.* (1992c) have studied the formation of the H-peak from Li-covered W(110) surfaces under the impact of thermal He* metastables and 10–1500 eV He^+, He^{2+} ions. The H-peak at 18.2 eV excitation energy appears with an intensity that is larger the lower the impact energy and the smaller the grazing angle to the surface. The H-peak shifts in agreement with the shifts in the work function of the surface, which is the behavior we showed previously for autoionization of atoms in front of surfaces (Zampieri and Baragiola, 1984). One can expect the energy of the electron from autodetaching $He^-(1s2s^2\ ^2S)$ to be shifted from the value of 19.3 eV for free atoms due to the image attraction. In a more recent study, Brenten *et al.* (1993) propose the formation of triply excited negative He^- in the configurations ($2s^22p$) and ($2s2p^2$) in collisions of slow He^{2+} with low-work-function surfaces. These states, similar to Feshbach resonances in collisions involving free electrons (Schulz, 1973), decay into electrons with energies slightly higher than those from autoionization of He**($2l\,2l'$) (see Figure 4.13). Transient negative-ion states have also been reported for other noble-gas ions (Brenten *et al.*, 1992e; Müller *et al.*, 1993b) and hydrogen ions (Müller *et al.*, 1993a) incident on partially alkalated surfaces.

PEE spectra from W(110) as a function of coverage with NaCl is shown in Figure 4.14 for incident He^+ and He^{2+} (Brenten *et al.*, 1993b). As the metal is covered by insulating layers, there is a decrease in the probability of AC of metal electrons and the spectra is consistent with AC involving two Cl3p valence electrons, causing a large drop in the average energy of the ejected electrons. Additional structures (1) and (2) may be related to AD from He* states formed by RC from W5d and Cl3p electrons, respectively. Emission for incident He^{2+} is not completely understood and may be affected by the presence of relatively long-lived, multiple final holes in the NaCl layer.

Although PEE studies have been done mostly with inert-gas ions, due to their high ionization potential, it can also occur during impact with alkali ions, provided that the work function of the target is low enough to allow capture to an excited state which can later eject electrons by AD (Brenten *et al.*, 1991a, 1991b, 1991c).

The contribution of PEE yields γ_p to total yields can be deduced in different ways. Comparison of electron emission by ground-state neutrals and ions shows that the difference in the yields remain approximately constant in the keV energy range, suggesting a constancy in γ_p with projectile velocity normal to the surface, v_o (Arifov, 1969). On the other hand, Hagstrum (1977) has shown that in many cases PEE varies with v_o due to competition of different PEE processes depending on the interaction time of the ion at the surface. A strong impact-energy dependence is found for molecular-ion impact on surfaces, where the internal energy might be dissipated in molecular breakup, depending on the ratio between the transition rates for Auger processes and dissociation (Baragiola, in press).

Figure 4.15 shows a comparison of the electron energy distribution from W as a function of K coverage under impact with 50 eV and 1 keV Ar ions (Brenten *et al.*, 1993b). In the first case, only potential emission occurs while KEE

(a)

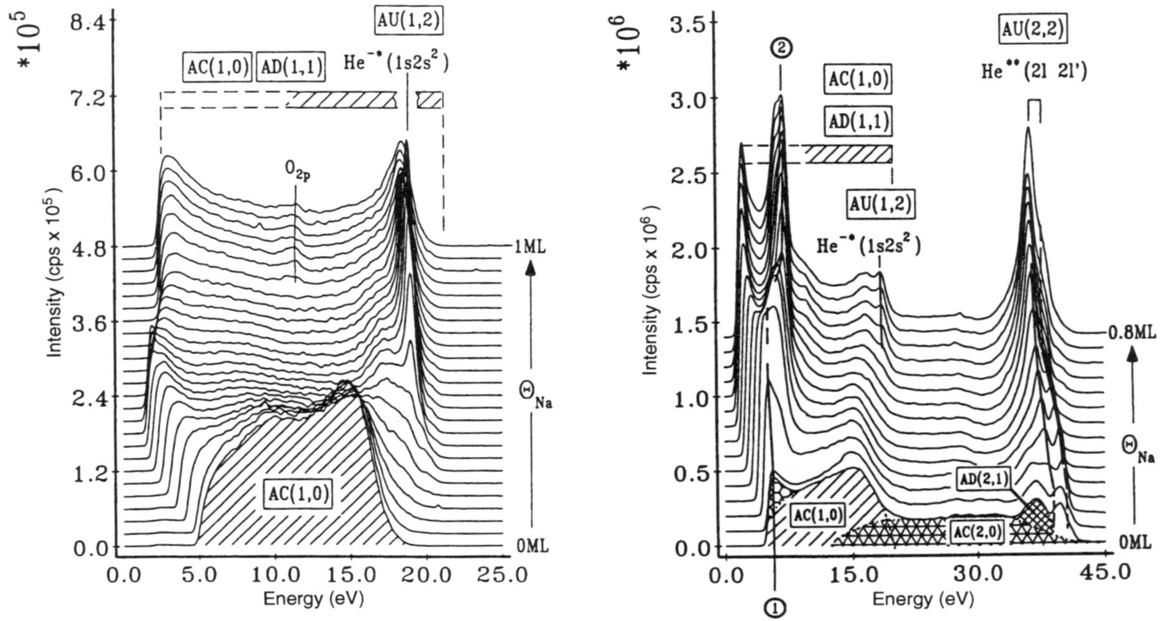
*10^5
Intensity (cps x 10^5)
AC(1,0)
AD(1,1)
AU(1,2)
He-*(1s2s^2)
O2p
1ML
ΘNa
0ML
AC(1,0)
Energy (eV)
*10^6
Intensity (cps x 10^6)
AC(1,0)
AD(1,1)
AU(1,2)
He-*(1s2s^2)
AU(2,2)
He**(21 21')
0.8ML
ΘNa
0ML
AD(2,1)
AC(1,0)
AC(2,0)
Energy (eV)

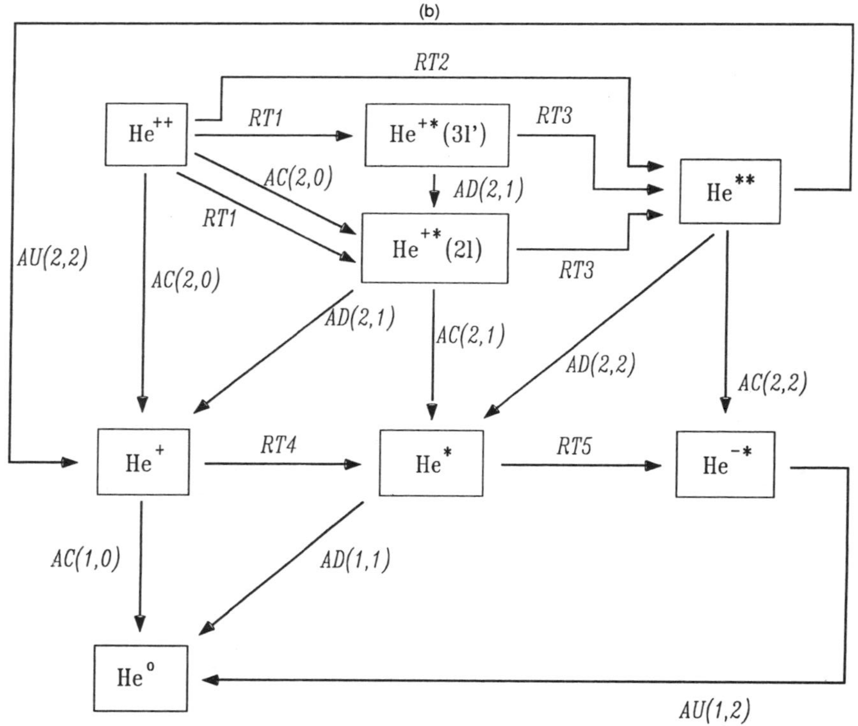

Figure 4.12 (a) Electron energy distributions from the interaction of 50 eV He^+ (left) and He^{++} (right) on a W(110) surface as a function of the coverage of Na. (b) Different pathways for the potential energy brought by the projectile. The notation $Y(h, x)$, indicates the individual electron transfer process (AC, AD, AU, etc.), where h is the number of holes and x is the number of excited electrons in the projectile before the transition. RT: resonant transition, AC: Auger capture, AD: Auger de-excitation, Au: autoionization. (Reproduced with permission from Brenten *et al.*, 1992b)

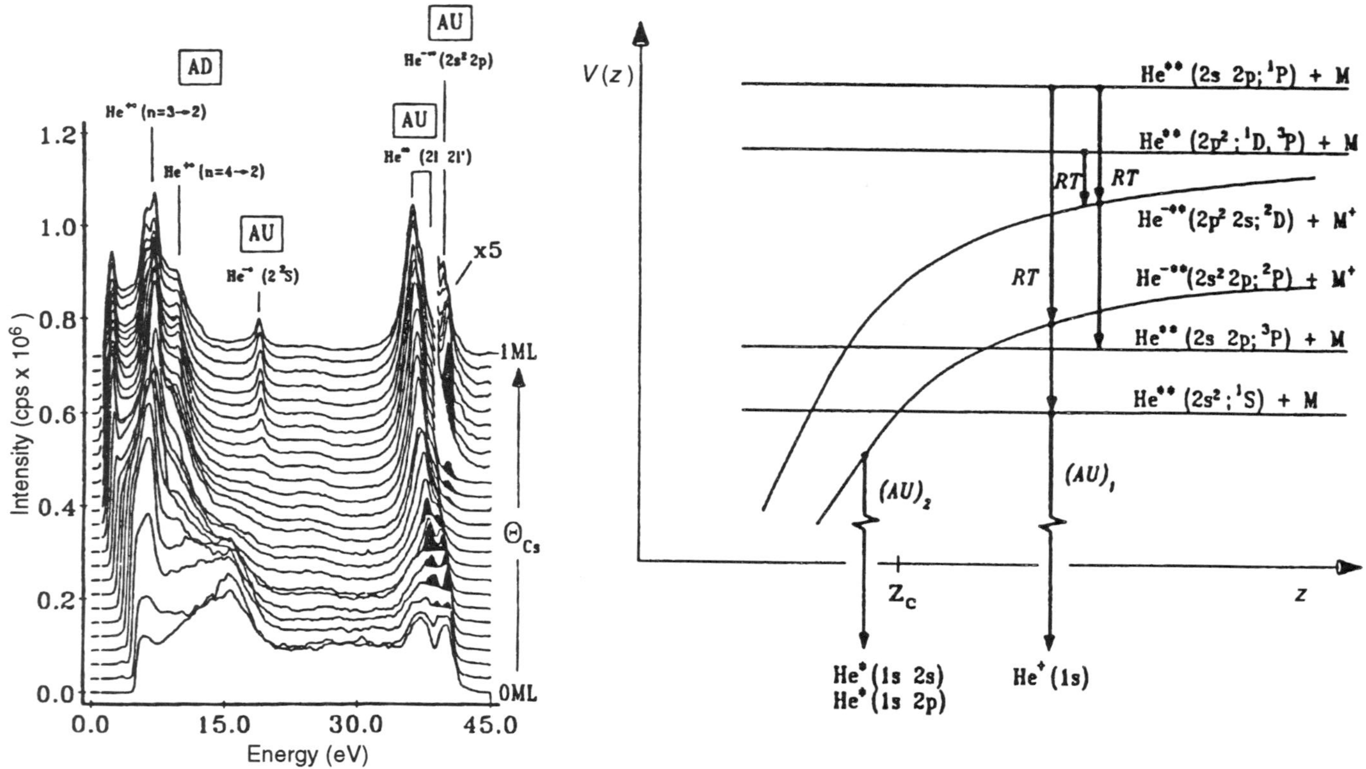

Figure 4.13 (a) Electron energy distributions from 100 eV He^{2+} impact on a partially cesiated W(110) surface as a function of Cs coverage. The angle of incidence is 5° with respect to the surface. (b) Schematic potentials for the interaction of He** and He-** with low work function surfaces as a function of z, the distance to the surface. Different decay channels are indicated by arrows. (Reproduced with permission from Brenten *et al.*, 1993a)

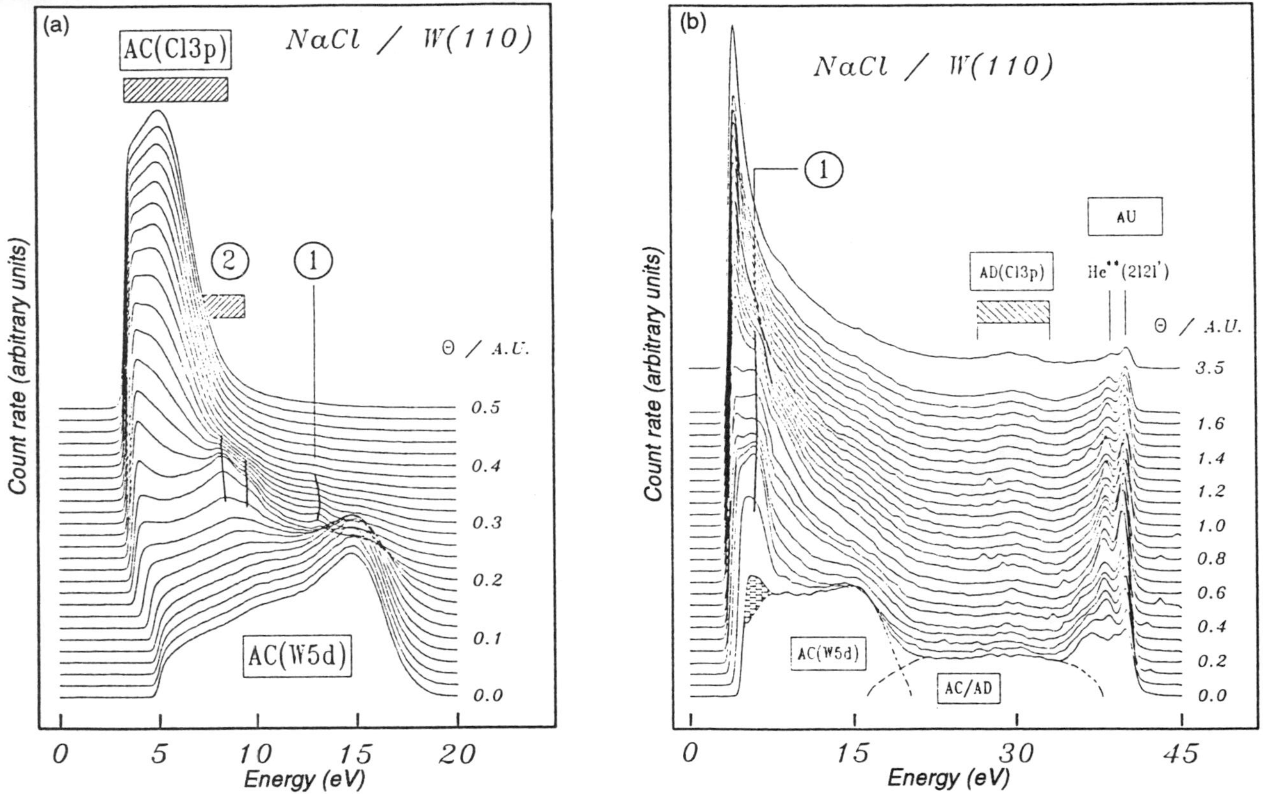

Figure 4.14 Electron energy distributions from 40 eV He^+ (left) and 100 eV He^{2+} impact (right) on a W(110) surface as a function of coverage with NaCl. The angle of incidence is 5° with respect to the surface. (Reproduced with permission from Brenten *et al.*, 1993b)

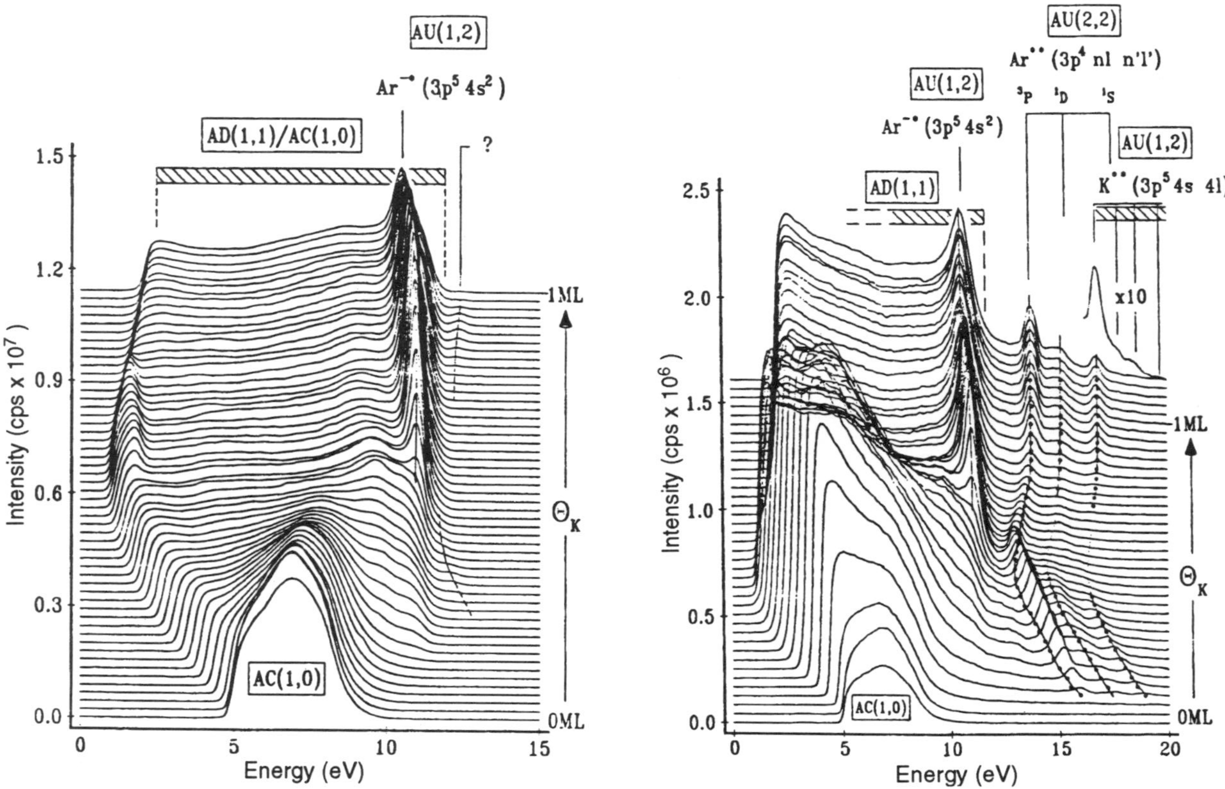

Figure 4.15 Electron energy distributions from W(110) surfaces under impact with 50 eV (left) and 1 keV Ar^+ ions (right) as a function of K coverage. The angle of incidence is 5° with respect to the surface. (Reproduced with permission from Brenten *et al.*, 1992e)

should start becoming visible at 1 keV. Additional peaks appearing at higher bombarding energies show autoionization from Ar**(3*p*4*lnl*′). The formation of these Ar** is unlikely to occur on approach, but rather from Ar^{2+} or Ar^{+*} back-scattering after a violent collision with a W atom which is accompanied by the excitation of Ar 3*p* electrons.

A related work is that of exoemission, or electron emission during adsorption. This occurs from the Auger filling of negative states of adsorbed molecules; it is discussed in recent work on oxidation of alkali films by Böttcher *et al.* (1990, 1991, 1993), of alkalated surfaces by Brenten *et al.* (1992d) and of graphite by Momose *et al.* (1992).

An interesting and complementary view of PEE is provided by recent measurements of emission statistics (Lakits *et al.*, 1989c; Töglhofer *et al.*, 1993) using singly and multiply charged ions. For Au targets and in the low kinetic energy limit, only one electron could be detected for incident He^+ or Ar^+, whereas up to three electrons can be detected per incident He^{2+} and two for incident Ar^+. This provides information that can be useful for an understanding of the importance of different pathways in the de-excitation of the incident ions, like the ones discussed above.

4.4.1.1 Metastable de-excitation spectroscopy

PEE from the impact of metastable atoms provides the basis of metastable impact electron spectroscopy, a technique to study the electronic properties of surfaces, similar to Hagstrum's (1966) ion-neutralization spectroscopy with incident ions. The technique extracts information on the surface density of states from the processing of the electron energy spectra. In the case of surfaces with high work function, the electron spectra produced by AD of metastables or by AC by ions give essentially the same information, since metastables convert to ions on approach, by resonance ionization. The use of metastable atoms offers the advantage of a small kinematic broadening in the electron spectra due to the low kinetic energies (a thermal beam can be produced with relative case) compared to the case of ions.

Most of the work on metastable-induced electron spectroscopy has been done using He* due to its high excitation energy. These metastables are prepared in either a ^1S or a ^3S state or a mixture of them. It is found that singlets convert into triplets when the metastable approaches the surface. There are several ways by which this conversion may happen. One is by two resonant tranfers, an ionization depopulating the higher-energy singlet state and a capture populating the triplet state (Lee *et al.*, 1985). Alternatively, excitation conversion may proceed by an Auger process, with an electron from the solid filling the lower-energy triplet state and ionizing the incoming singlet state (Woratschek *et al.*, 1985). This second mechanism can occur over a larger range of atom–surface distances and, unlike the resonant excitation conversion, it can also occur on low-work-function surfaces where resonance ionization is blocked up to distances where the excited state decays by AD. A third mechanism occurs through the transient formation of He^- as discussed in the work of Hemmen and Conrad mentioned above and that of Borisov *et al.* (1993).

Masuda *et al.* (1992a, 1992b) have used metastable de-excitation spectroscopy to study the Si(111) $- 7 \times 7$ surface; they found structure in the energy spectra which could be correlated with transitions involving two electrons in the dangling bond surface states. When Si surfaces are oxidized (Ishii *et al.*, 1990; Hongo *et al.*, 1991), the RI + AC channel is blocked, and the electron spectra corresponds to AD, resembling Penning ionization of oxygen in the gas phase. The case where Si(111) is doped with controlled adsorption of Li has been discussed by Nishigaki *et al.* (1988). On graphite surfaces, Masuda *et al.* (1990) found that He*, Ne* and Ar* metastables produce spectra similar to ultraviolet photoelectron spectra, rather than the self-convolution of density of states predicted for RI + AC. This indicates that the metastables de-excite directly by AD, due to a suppression of the RI channel even though the surface has empty levels resonant with the atomic metastable level. The authors argue that the reason for the suppression of RI is the small spatial overlap between the excited atomic orbital and the empty π^* bands of the solid.

Other applications of metastable-induced electron spectroscopy to characterize the electronic structure of surfaces include alkali adsorption (Maus-Friedrich *et al.*, 1991b) on W, the adsorption of CO (Maus-Friedrich *et al.*, 1991a) and oxygen (Maus-Friedrich *et al.*, 1992) on alkalated W surfaces, the electron-induced damage of thin films of alkali halides on W (Dieckhoff *et al.*, 1992), and the density of states of other materials, like ionic multilayers (Hemmen *et al.*, 1989), organic films (Harada *et al.*, 1991), and liquids (Keller *et al.*, 1986; Morgner *et al.*, 1991a, 1991b, 1991c).

4.4.1.2 Highly charged ions

Research on the interaction of multiply charged ions with surfaces has seen a revival due to the advent of efficient sources of highly charged ions. The topic has been reviewed recently (Varga and Winter, 1992; Niehaus, 1993; Varga, 1987, 1988, 1989; Winter, 1991; Andrä, 1989; Andrä *et al.*, 1992). Many of these studies have been done at energies larger than those of interest here, since it is difficult to obtain low-energy beams of highly charged ions. In a few cases, the incident velocity perpendicular to the surface is low enough that penetration is not important, and the interpretation of the results is simplified.

Delaunay *et al.* (1987b) confirmed that most electrons emitted by highly charged ions have low energies and that the yields increase with the charge of the ion and with decreasing projectile velocity. The yields are found to scale approximately linearly with the potential energy of the projectile for low charge states (up to Ar^{12+}) (Delaunay *et al.*, 1987a; Fehringer *et al.*, 1987). McDonald *et al.* (1992) have measured PEE yields for very highly charged ions (up to Th^{75+}) on gas-covered surfaces and found that they grow sublinearly with excitation energy. On the other hand, Aumayr *et al.* (1993) found a linear increase on clean Au targets for changes up to Th^{80+}.

Zehner *et al.* (1986) and Meyer *et al.*, (1987) found that in addition to the low-energy electrons, high-energy Auger electrons appear when the projectile is so highly charged that it carries inner-shell vacancies. This high-energy Auger emission was further studied by de Zwart (1987), who found Ar LMM Auger electrons when Ar^{9+} is incident on W targets at 500 eV but not at 20 keV. The author suggested that at low velocities the filling of the L-shell vacancy, originally present in the projectile, can occur before the projectile hits the surface. Later, the high-energy Auger lines were found to be Doppler-shifted, which indicated that the electrons were emitted from a source moving in the direction of incidence (de Zwart *et al.*, 1989). This was then usually interpreted to mean that emission occurred before the ion reached the surface. Under this assumption, there appeared to be a contradiction between the Arifov's ladder model and the ratios of LMM to KLL Auger intensities measured for hydrogen like C^{5+}, N^{6+}, O^{7+} and Ne^{9+} projectiles at energies below 600 eV (Folkerts and Morgenstern, 1990, 1991).

In an important development, Meyer *et al.* (1991) showed that the apparent inconsistency can be resolved by noting that the Doppler-shifted Auger lines may result from projectiles that have entered the solid but have not yet suffered much angular scattering. The study of subsurface emission of Auger electrons was recently extended by Das *et al.* (1992) down to 150 eV for N^{7+} impact on Ni(110), whereas the observation of three-electron Auger processes has been reported for 500 eV bare carbon and nitrogen ions on the same surface (Folkerts *et al.*, 1992). Further evidence for incomplete filling of the projectile L shell upon reaching the surface has been obtained by Köhrbrück *et al.* (1992) for N^{9+} and Ar^{16+} on Cu.

A consequence of the incomplete relaxation of the potential energy of the highly ionized projectile is the fall of the PEE yields with projectile velocity normal to the surface as shown, for example, in recent measurements by Kurz *et al.* (1992). These authors fit their data of yields vs. velocity to the function $\gamma(v) = \gamma_\infty + C/v^{1/2}$, where both γ_∞ and C are constants. They propose that the velocity-dependent part is due to autoionization on transit to the surface and that the velocity-independent yield, γ_∞, is due to emission when the excited hollow atom enters the surface region or the bulk of the solid.

4.4.1.3 Spin-polarized potential emission

Substantial information on the magnetic properties of solids can be obtained by analyzing the spin of the electrons, a method that has become accessible

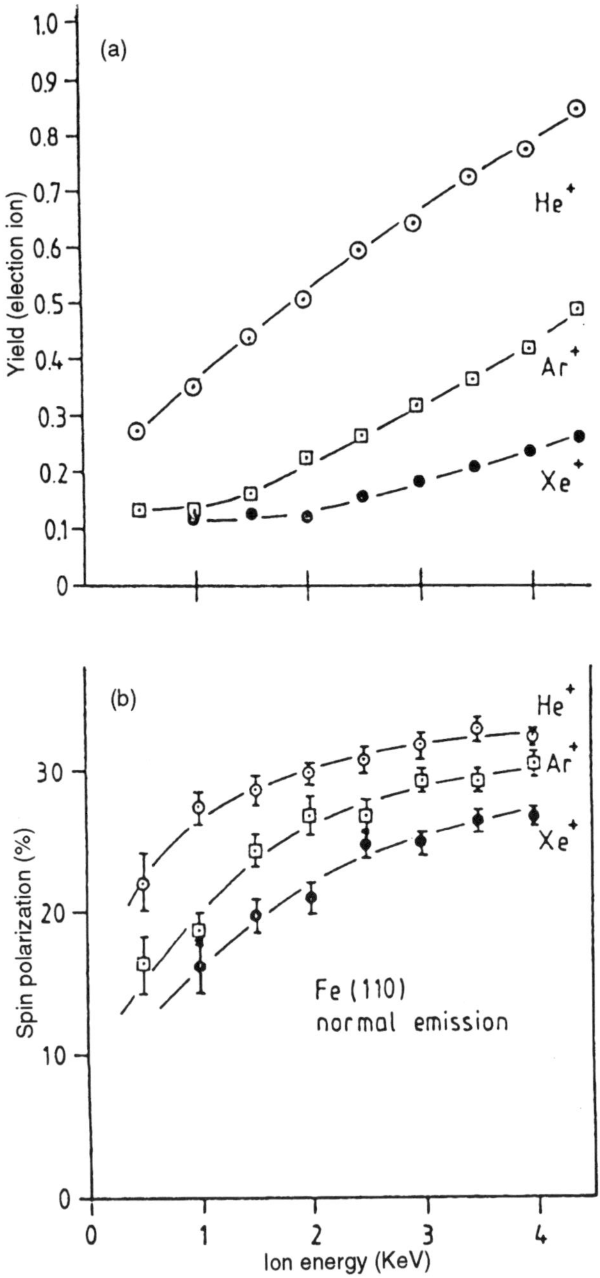

Figure 4.16 Total electron-emission yields and spin polarization of 2 eV electrons under ion impact on Fe(110). (Reproduced by permission of the American Physical Society from Kirschner *et al.*, 1987)

due to advances in detector design (Gay and Dunning, 1992). Kirschner *et al.* (1987) studied the spin polarization of electrons emitted from the surface of ferromagnetic Fe(110) excited by 0.5–4 keV He^+, Ar^+ and Xe^+ ions. The angle of incidence was 45° and the electrons were selected from a cone around the surface normal. The polarization was found to be larger for the electrons with the lowest energies and increased with the energy of the ions, as shown in Figure 4.16. This latter property may be related to a spin-dependent electron attenuation length (Koike and Kirschner, 1992) and the fact that the more energetic ions are able to excite electrons deeper into the material.

Spin-labeled PEE has been used to study the surface magnetism of Ni(110) and its dependence on temperature and gas adsorption (Onellion *et al.*, 1984). In a recent study, Hart *et al.* (1989) have measured the spin polarization of PEE induced by partially spin-polarized, triplet He*, Ne*, Ar* metastables on clean Cu(100) and Pd(110) and on these surfaces covered with K, O, or CO. The data show that the electrons involved in PEE tend to have anti-parallel spins. For He*, the data suggests a significant contribution of cascade electrons to PEE. For these systems, PEE occurs mainly by AD or RI + AC, depending on the surface conditions: both low work function and gas coverage block RI and enhance the contribution of AD. Spin polarization is smaller in the case of AC, since this process favors the situation where the two active electrons are close, which is more likely to be the case when they have opposite spins, due to the Pauli principle. AD produces higher spin polarization, which approaches unity for gas-covered surfaces. This is similar to what occurs for Penning ionization of free molecules, and provides an additional example of the usually valid approximation of treating low-energy collisions with adsorbed gas as if they occurred in the gas phase (Schol *et al.*, 1992). For clean surfaces the observed polarization was higher than expected for RI + AC, which led Dunning *et al.* (1991) to propose that even on a metal like Cu with work function higher than the ionization energy of the metastables, some projectiles may survive RI and eject electrons close to the surface by the AD mechanism. A more recent paper by the same group (Oró *et al.*, 1991) proposes that de-excitation of excited He (2^3S, 2^1S and 2^3P) occurs exclusively through the RI + AC channel, and that the observed spin polarization can be explained by assuming that the two active electrons in AC have anti-parallel spin

4.4.2 KINETIC EMISSION DATA

4.4.2.1 Total yields

4.4.2.1.1 Energy dependence and thresholds

The linear relationship between the KEE yields and electronic stopping power, observed at high energies, breaks down in the keV range, where emission is dominated by theshold effects. On clean surfaces, it is usually found that the yields depend linearly on energy in the keV range (Arifov, 1969),

$$\gamma = a(E^* - E_0). \tag{20}$$

Alonso *et al.* (1986) set up an experiment to test whether emission existed below E_0. They chose the system Xe^+ on polycrystalline Au to avoid PEE, since the recombination energy of Xe^+ is smaller than twice the work function of gold. Also, since Xe is lighter than Au, the projectiles would not generate recoils with sufficient velocity to contribute significantly to the measured electron yields. Figure 4.17(a) shows that $\gamma(E)$ has the linear dependence on energy of the type of equation (20), but also a small tail at energies lower than E_0. This tail can be enhanced by a logarithmic plot. Figure 4.16 shows that the yields below 4 keV follow the dependence

$$\gamma = b \exp(E/c) \tag{21}$$

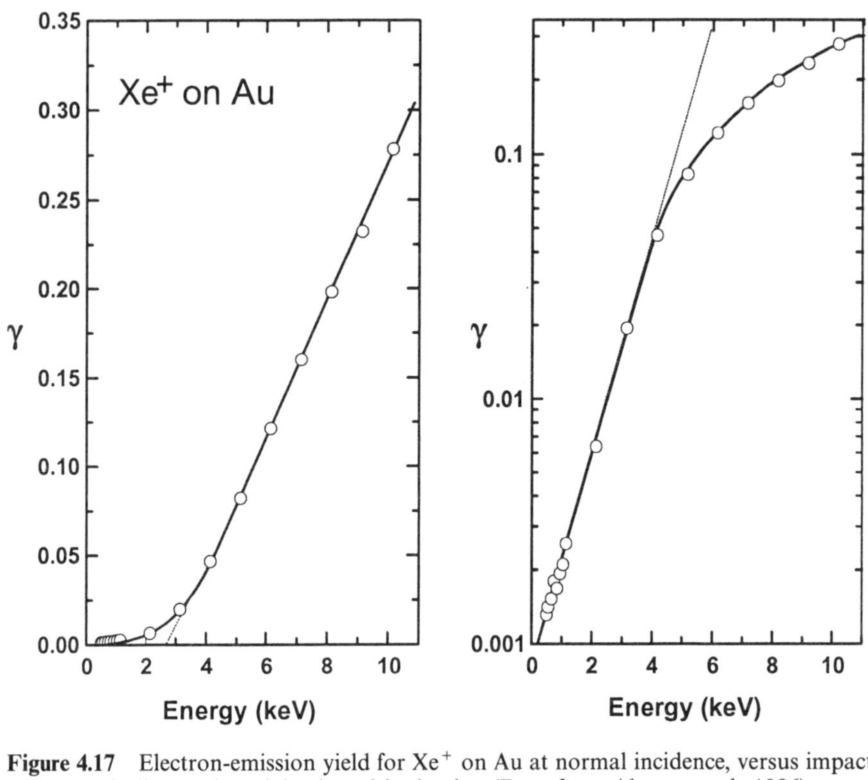

Figure 4.17 Electron-emission yield for Xe^+ on Au at normal incidence, versus impact energy. Left: linear plot; right: logarithmic plot. (Data from Alonso *et al.*, 1986)

and no KEE threshold is apparent (Alonso *et al.*, 1986). In this work, the breakdown of the proportionality of yields and stopping power was attributed to a softening of the energy distributions at low impact energies which would make the number of electrons above the Fermi level cease to be proportional to the total number of excited electrons. This work proposed to include the effect of straggling in calculations which relate yields to inelastic energy loss. The demonstration of the change in the shape of the electron energy distributions with impact energy is given in Figure 4.18. It is apparent from the figure that a new excitation mechanism appears at 3 keV, yielding electrons between 4 and 10 eV. These energies correspond to autoionization transitions in Xe (Gerber *et al.*, 1972), and so one should expect an enhanced ion fraction of back-scattered Xe correlated with this additional electron emission. Similar results were obtained by Bonanno *et al.* (1981) for Xe on W.

Lakits *et al.* (1990b) have extended the measurements on Au to lower energies and other projectiles, computing the yields from emission statistics. They assumed that electron emission at low energies is due to autoionization during the collision. The ratios of the probability of emitting two electrons to that of emitting one electron, P_2/P_1, were found to go to zero at low energies, from which the authors proposed that PEE cannot produce more than one electron (see also Aumayr and Winter, 1993). This is strictly valid only if $E_i > 3U$,

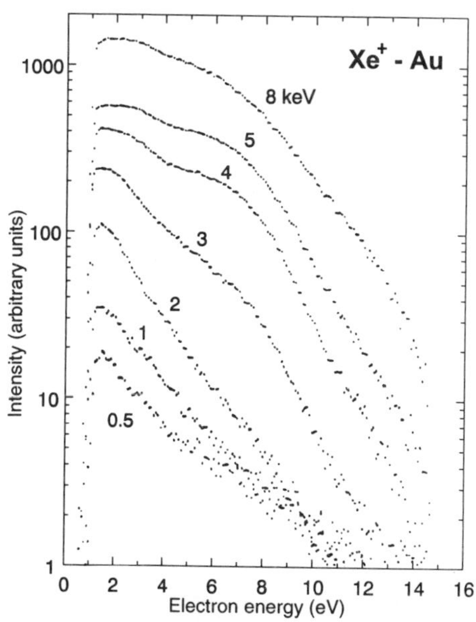

Figure 4.18 Electron energy spectra from an Au sample bombarded by Xe^+ at 45° incidence, as a function of impact energy. (Data from Alonso *et al.*, 1986)

when the excited electron cannot produce additional electron emission. The thresholds for emission of one and two electrons are different since they involve different minimum energy transfers (U and $2U$, respectively). This can be deduced from the experimental results shown in Figure 4.19 (Winter *et al.*, 1991).

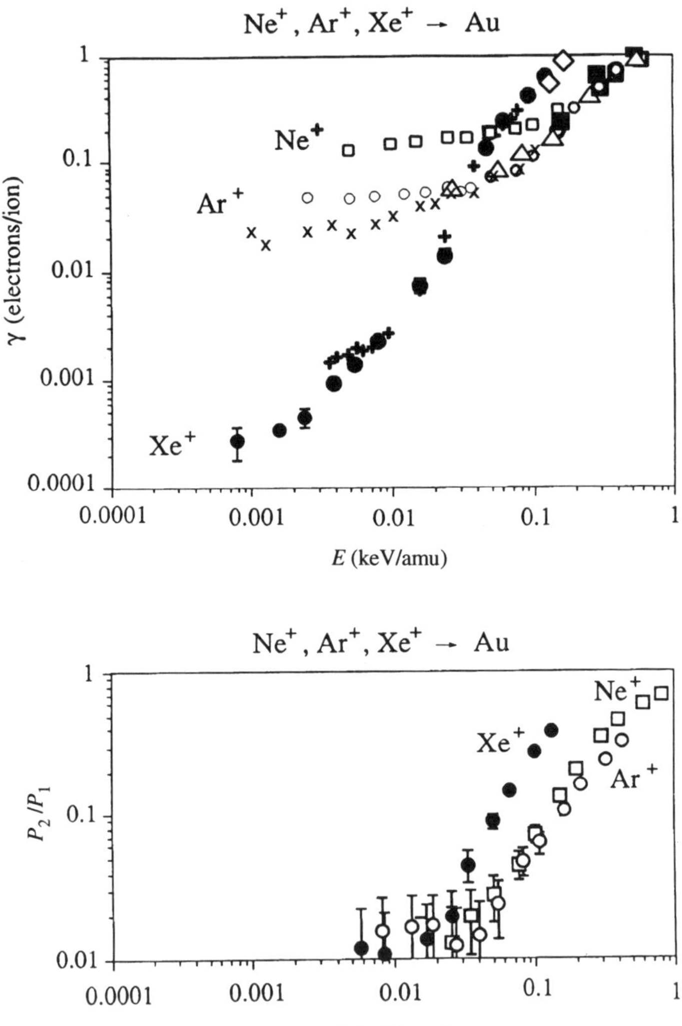

Figure 4.19 Electron yields and ratios of the probabilities for emitting two and one electron, P_2/P_1 for different ions on Au. Data from Lakits *et al.*, 1990b (○, □, ●); Töglhofer *et al.*, 1993 (×); Alonso *et al.*, 1986 (+); Alonso *et al.*, 1978, Alonso *et al.*, unpublished (◇, △, ■); Oeschner, 1978 (▲)

That KEE requires close collisions with target *atoms*, rather than nearly free electrons, has been recently demonstrated by Rabalais *et al.* (1992), who studied the impact parameter dependence of Ar^+-induced KEE from Ni(110) at 4 keV. They observed essentially no emission at grazing incidence and emission started abruptly (Figure 4.20) when the decreased shadowing of atomic planes allowed the minimum impact parameter in a collision to be below 0.3 Å, or about the sum of the maximum radial charge density of the L-shells of Ar and Ni.

Low KEE thresholds are desirable for detection of large molecules in mass spectrometry, and there is a current interest in the topic, but the studies have not been done with characterized surfaces. These experiments find extrapolated thresholds of $v_{th} \sim 10^6$ cm s^{-1} for water clusters (Beuhler, 1983), and biomolecules (Salehpour *et al.*, 1986; Geno and Macfarlane, 1989; Martens *et*

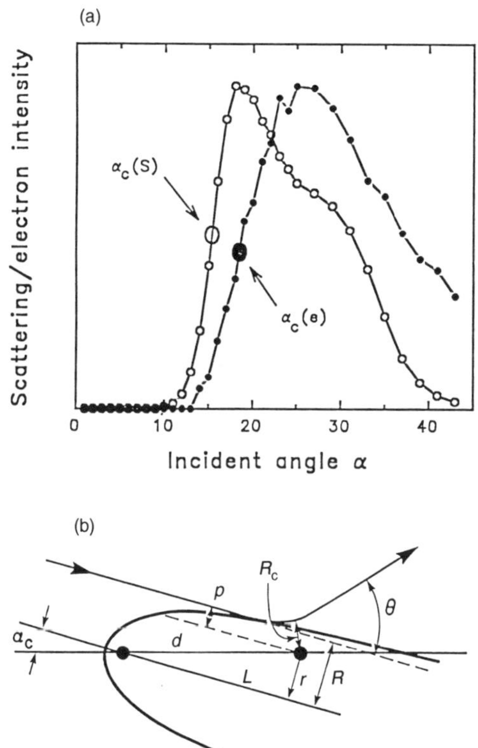

Figure 4.20 (a) Scattering intensity (○) and electron intensity (●) versus incidence angle of 4 keV Ar^+ ions along the ($1\bar{1}0$) azimuth of a Ni(110) – p(1 × 2) surface. (b) Indicates the relationship between the shadow cone, impact parameter and incidence angle. (Reproduced by permission of the American Physical Society from Rabalais *et al.*, 1992).

al., 1992). A decrease in the apparent threshold energy for electron emission may result from an intense electric field F applied to the surface. Macfarlane (private communication) has proposed this effect to improve the detection of very slow macromolecules in mass spectrometers. One can set F at a point where field emission would not be observed from the unbombarded surface but which would allow electrons excited by the projectile to tunnel out of the solid (Figure 4.21).

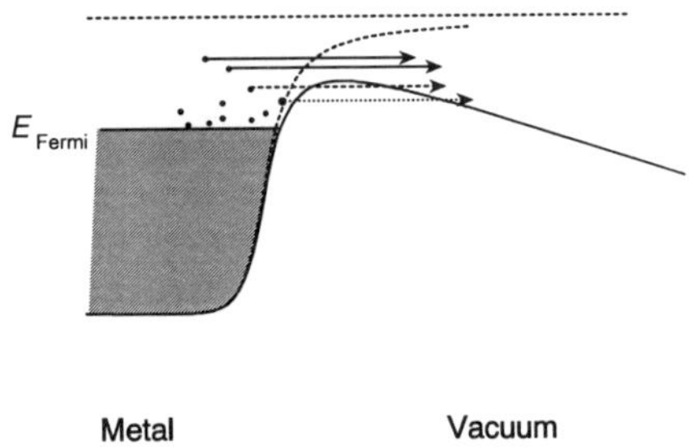

Figure 4.21 Field-assisted emission of electrons excited by an incident particle, showing preferential emission of excited electrons directly over the barrier (→), or by tunneling (⇢) through it. The solid line represents the potential seen by an electron in the presence of an electron field

4.4.2.1.2 Effects of incidence angle

PEE and KEE yields have a different dependence on incidence angle. This has been used to estimate their separate contributions to the total yields, as in the work of Kanie *et al.* (1992) for multiply charged ions. In this case, one has to take into account that neutralization of the incoming ion is incomplete; i.e. the excitation is not completely relaxed when the ion reaches the surface. The image charge attraction produces an acceleration of the ion towards the surface and limits the minimum transverse velocity and minimum effective grazing angle (Winter, 1992; Winter *et al.* 1993).

The dependence of the yields on incidence angle shows a rich structure when using single-crystal targets. Bénazeth *et al.* (1989) studied this dependence in both the total and the Auger electron yields from Si(111) bombarded with 5–80 keV noble-gas ions. They observed structure with minima at angles corresponding to low index directions. The structure was affected by the bombardment dose, due to amorphization induced by the projectiles. This effect had a temperature dependence which could be related to amorphous–crystalline phase transitions. Brusilovsky (1991) studied the electron-emission yield of Mo(100) under impact with 30 keV Ne^+ and Ar^+ ions and how it depends on angle of incidence with respect to the [100] direction and with the azimuth.

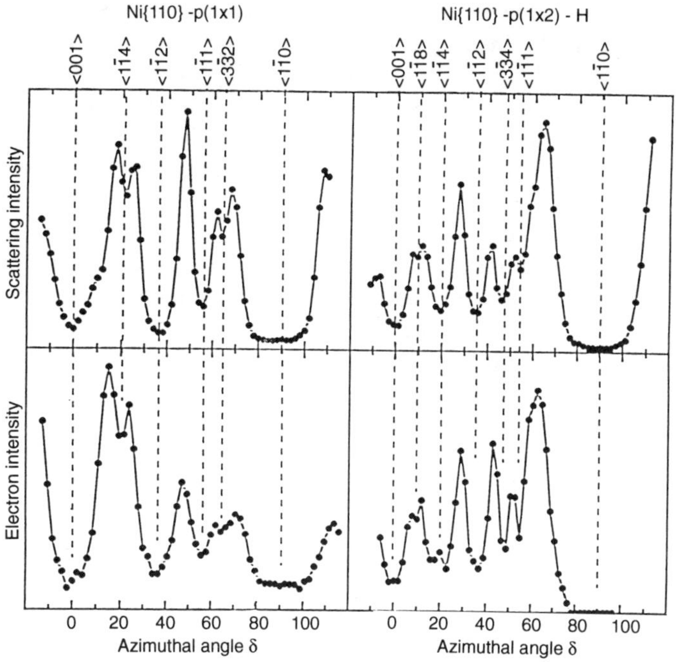

Figure 4.22 Scattering intensity and electron intensity versus azimuth for 4 keV Ar^+ incident on a Ni(110) – p(1 × 1) and a Ni(110) – p(1 × 2)-H reconstructed surface. (Reproduced by permission of the American Physical Society from Rabalais *et al.*, 1992)

Dips in the yields are observed for incidence along low index directions, which can be correlated with the transparency of the surface (or the shadowing of the second layer by the atoms in the first layer). Similar angular dependences were obtained by Susuki *et al.* (1991) on PbSe and SnTe (001) surfaces. A direct correlation between electron emission and the surface transparency as seen by back-scattered ion yields is shown by the data of Rabalais *et al.* (1992) in Figure 4.22. This behavior shows that electron yields (or target currents) can be used to align single-crystal targets in other types of study, like low-energy ion-scattering spectroscopy. Ion-induced electron emission has been used by Karolewski and Cavell (1987) for imaging surfaces by using a scanned low-energy ion beam and measuring total target currents. The imaging contrast in this case is given by the variation of the electron yield with incidence angle.

4.4.2.1.3 Effect of projectile charge on KEE

An important question in particle detection is the value of the yields for neutrals and positive and negative ions. In the low-impact-energy limit, ground-state neutrals do not produce PEE. At keV energies, near the emission threshold, KEE yields can be obtained by extracting the potential contribution from the total yields. If this is done, it is found that the KEE for the positive and neutral projectiles are essentially the same when using incident inert-gas species (Arifov, 1969). This equivalence does not necessarily hold for projectiles with lower ionization potential, where electron loss from the projectile occurs with significant probability.

Lakits *et al.* (1989a) have derived electron-emission yields for H and H^- on Au by measuring the ratios P_2/P_1 in the emission statistics and assuming that for these projectiles, the ratios correlate with the total yield γ in exactly the same way as for protons. This assumption is based on an approximate empirical correlation. In this way, they report $\sim$15% smaller yields for H than for H^+ between 1 and 15 keV. Yields for H^- are similar to those for H up to 8 keV and deviate to higher values above this energy. These differences must be viewed with caution, since the data exhibit scatter of up to 30% from the empirical relation $\gamma = f(P_2/P_1)$ (Lakits *et al.*, 1989b).

An explanation for the findings was developed in a subsequent article (Lakits *et al.*, 1990a), which included measurements on Au targets, extended to incident Ne and Ar. The authors propose that the reason for a lower yield for neutrals is the screening of the proton charge by the electron. They point to a presumably lower stopping power calculated for H^o compared to H^+, caused by screening of the proton charge in H^o. However, screening affects principally the low-energy-loss contributions to the stopping power, which are relatively unimportant in exciting electrons above the vacuum level. Furthermore, ionization cross sections at keV energies may be *larger* for H than for H^+ impact (Solov'ev *et al.*, 1962). On gas-covered surfaces, electron-emission yields are generally found to be larger for H than for H^+ (Ray *et al.*, 1979).

The data by Lakits *et al.* for incident noble-gas atoms is puzzling, since it shows that the total electron yields at energies around 1 keV are very similar for neutrals and for ions, whereas one would expect a much lower yield for neutrals, since they cannot cause potential emission. Since these findings are in conflict with all previous results for clean surfaces, their claim that *kinetic* electron emission by keV neutrals is different than that for positive ions should be taken with caution.

4.4.2.1.4 Molecular effects in KEE

An interesting question is how the electron-emission yield produced by molecular ions compares to the yields produced by its constituent atoms. It is found that for hydrogen projectiles, the ratio of yields $R_2 = \gamma(H_2^+, v)/2\gamma(H^+, v)$ at equal velocities is $R_2 > 1$ at low v and $R_2 > 1$ at v higher than about 5×10^8 cm s^{-1} (Baragiola *et al.*, 1978; Svensson and Holméen, 1982;

Hasselkamp and Scharmann, 1983; Arista *et al.*, 1986; Hasselkamp *et al.*, 1992). The reduction at low velocities occurs even after taking into account a smaller PEE yield by molecular ions (Baragiola, 1982). The origin of this molecular effect has been a matter of controversy. We have argued for an interference effect (Baragiola *et al.*, 1978), which has been demonstrated for stopping powers: molecular ions lose less energy than protons at low v and more at high v, due to an interference in the scattering of the target electrons in the molecular centers (Eckardt *et al.*, 1978). At high v, where the dynamic screening is weakened, the two protons act like a He^{2+} ion in distant collisions and give a higher yield due to the z^2 dependence of the energy losses. Additionally, electron emission may be enhanced by loss of the H_2^+ electron. An alternative view (Lakits *et al.*, 1989a; Lakits and Winter, 1990) proposes that the molecular fragments act totally independent of each other: $\gamma(H_2^+) = \gamma(H^+) + \gamma(H^o)$ (i.e. interference effects are ignored) and explains $R_2 < 1$ at low velocities by assuming that $\gamma(H) < \gamma(H^+)$ (see above). The different explanations for the molecular effect may operate at different particle velocities. For instance, an absence of correlation of the motion of the fragments may be possible due to large-angle elastic collisions immediately upon penetration of the surface.

A complete description of the molecular effect must necessarily include the (velocity-dependent) contributions of interference during excitation, electron loss, and screening effects in ionization rates.

4.4.2.2 Energy and angular distributions

Oda *et al.* (1992) have measured the energy distribution of electrons emitted from clean Ta by keV noble-gas atoms. Since the projectiles are neutral and mainly in their ground state, the distributions correspond to KEE. At 5 keV, the peak of the distributions, around 2–3 eV, occurs at progressively lower energies at the projectile is heavier, from He to Kr (i.e. as the velocity of the projectile decreases). There is also a longer high-energy tail that is more pronounced the lighter the projectile. High-energy tails have been observed in many experiments, for instance in those using Xe on Au described above. These continuum tails extend beyond the energies of the Auger peaks. The origin of these tails in the region of very high electron energies is still obscure. At high impact velocities, they have been explained as arising from binary collisions between the projectile and a target electron. At low velocities, the explanation is still a matter of debate (Clapis and Kessel, 1990); models proposed include direct coupling to the continuum and emission of Auger electrons during the collision. The measurements by the author and co-workers, using low-energy (0.5–6 keV) ions in solids, have shown that the continuum electron energy distributions extend up to a surprisingly large fraction of the center-of-mass energy (Figure 4.23) (Baragiola *et al.*, 1992). This shows that the extremely high

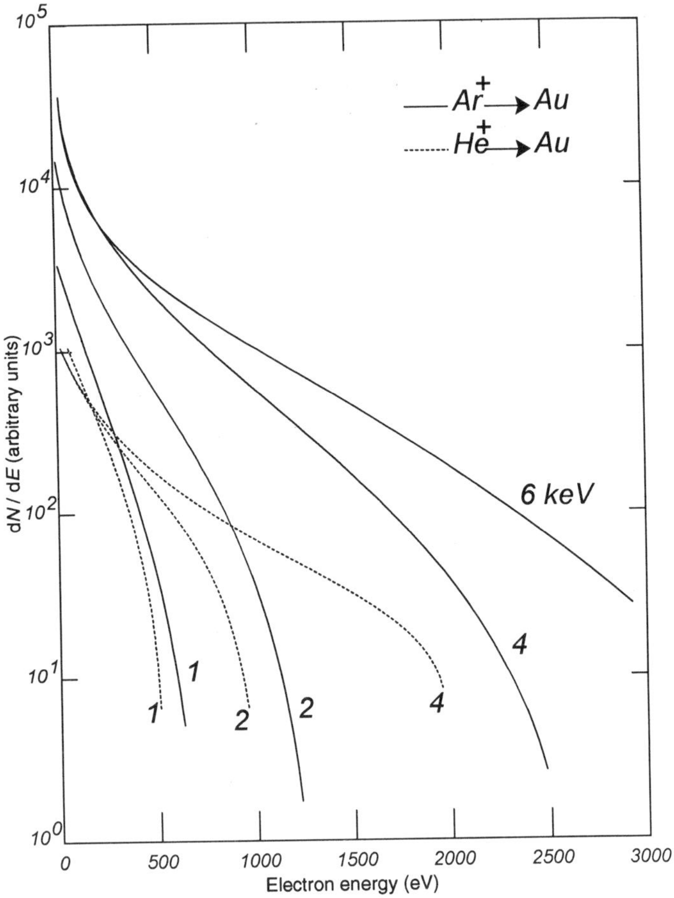

Figure 4.23 Electron energy spectra from Au under He^+ and Ar^+ impact as a function of projectile energy. Note that the tails extend to a substantial fraction of the center-of-mass energy. (From Baragiola *et al.*, 1992)

electron energies result from nearly head-on collisions between the projectile and a target atom. The high-energy tails are strongly dependent on the type of target, and generally decay slower with electron energy the higher the atomic number of the target.

The angular distribution of ejected electrons has, in general, a cosine dependence on angle with respect to the surface normal. Structure is superimposed on this dependence in the case of experiments on single crystals, due to the diffraction of the electrons (Negre *et al.*, 1985). The angular distribution of ion-induced Auger electrons can also give information about the surface roughness (Banouni *et al.*, 1985; Mischler *et al.*, 1986; Naumkin, 1991) by using the expectation of a cosine angular distribution of bulk Auger lines and isotropic angular distributions of electrons originating from decay of sputtered excited atoms in vacuum.

4.4.2.3 Emission from adsorbed layers and from insulators

It has been known for a long time that KEE from insulators is larger than from metals or semiconductors. In most of the literature this is attributed to a larger mean electron escape depth due to the reduced electron scattering for electron energies below the band gap. Another factor, often more important, is that excited electrons escape more easily from insulators due to a reduced surface barrier. The relative importance of these two factors can be discerned by analyzing the effect of the thickness of the insulating layer.

Adsorbed layers already have a very strong influence on electron yields. The lack of specification of the surface for targets exposed to air hinders the possibility of extracting information about mechanisms from electron yields. This is true for almost all the work done prior to 1950, but occasionally one finds the same shortcoming in current publications.

The adsorption of small quantities of gas (of the order of a monolayer or less) can change electron emission for many reasons. The most important seem to be (a) a different electron excitation probability for the adsorbed species as compared to the bulk, and (b) a change in the surface barrier. The change in the electron attenuation length is a second-order effect for the thinner layers. Comparative experiments on the adsorption of oxygen on clean Al and Mo surfaces showed very large effects on the yields (Ferrón *et al.*, 1982a). Since one can expect a similar electronic configuration in the oxygen in both cases, the reason for the large differences in the two materials is not likely to result from (a) above. This is further supported by the observation that the change in the yields is not proportional to the oxygen concentration. The change in the surface barrier appears to be the main factor affecting the yields, as occurs upon alkali adsorption (Brenten *et al.*, 1991b), but it is *not necessarily* related to changes in the work function. Adsorbed gas generally (but not always) causes an increase in the electron yields (Groenevelt *et al.*, 1984); the relative change in the yields is larger the lower the incident energy (Alonso *et al.*, 1979). An alternative mechanism recently proposed is the emission of electrons from the autodetachment of negative-ion species sputtered from the sample (Baker *et al.*, 1992).

Electron emission from thick insulators is an intriguing problem which is difficult to study due to surface charging and, therefore, very few studies exist.

This is unfortunate, since insulators are the most abundant type of material on earth. The topic has recently been summarized by Schou (in press). We have just reported (Grosjean and Baragiola, 1993) measurements of KEE from thin Ar films. Argon is an interesting material for electron-emission studies because it has a large band-gap, an extremely large mean electron escape depth (of the order of a micron), and a negative electron affinity (conduction band above the vacuum level). We have observed nearly complete extraction of electrons from thin films and, under some conditions, electrodeless dielectric breakdown during and after ion bombardment in the presence of a weak external electric field.

4.4.3 KINETIC AUGER EMISSION

4.4.3.1 Excitation of autoionizing states

Zampieri *et al.* (1984) studied the formation of autoionizing Ne** when 0.4–5 keV Ne^+ and Ne impact on Mg, Al and Si surfaces, and pointed out similarities and differences between collisions at surfaces and in the gas phase. In contrast with experiments in the gas phase, which show up to seventeen autoionization peaks (Andersen and Olsen, 1977), only two or three are clearly visible in the case of collisions with solid surfaces (Figure 4.24). The authors used a binary-collision model with post-collision ion–surface interaction to explain the results. The binary collision between the incoming Ne and a surface atom is similar to a gas-phase encounter in which two Ne-2p electrons are promoted mainly along the $4f\sigma$ MO; excitation can occur if the collision is violent enough that a minimum interatomic distance is reached. The excited projectile than recedes from the surface and, if ionized, is neutralized by electron transfer from the metal valence band. Only Ne** states with ionization potential less than the work function of the sample will survive resonance ionization back to the solid. This limits the number of states which can later autoionize and explains the difference from the larger number of autoionization peaks in gas-phase experiments (Zampieri and Baragiola, 1982). The existence of a band gap changes the position and number of autoionization peaks, as in experiments done on alkali halide surfaces (Fine *et al.*, 1993).

The strong autoionization peaks in the Ne–surface collisions have the configuration $Ne^{**}2p^4(^3P, {}^1D)3s^2$. Since the MO promotion produces only singlet core states, the strong triplet (lower energy) peak was explained as resulting from a two-electron rearrangement involving core–Rydberg electron exchange at a curve crossing. This process has been invoked to explain gas-phase experiments where, instead, the triplet peaks are weaker than the singlets. The proposed reasons why the rearrangements are favored near surfaces and that more electrons can participate in the transition (i.e. target

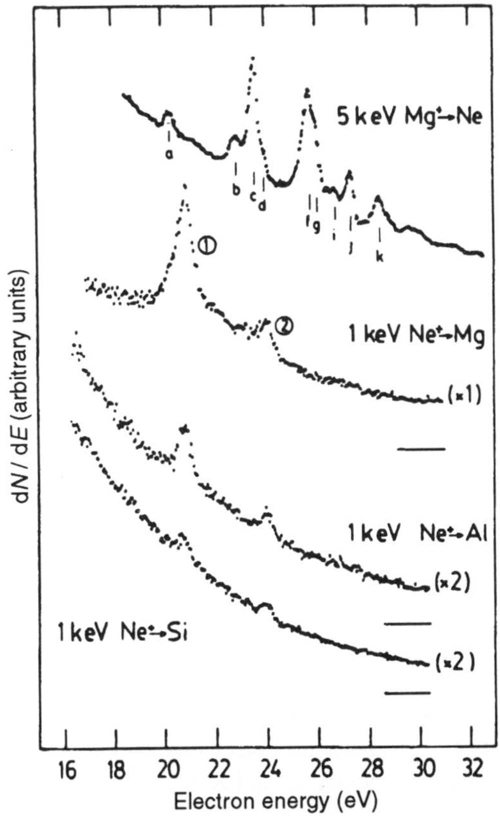

Figure 4.24 Ne autoionization spectra for collisions of Ne projectiles with gases and solids. (From Zampieri *et al.*, 1984, gas data from Andersen and Olsen, 1977)

valence electrons), more empty electronic states are available, and the transitions can occur over a wider range of internuclear distances. Alternative processes not involving the $2p$ electrons were mentioned above, when discussing singlet-to-triplet conversion of metastable He at surfaces.

Detailed kinematic studies of electrons emitted from collisions of Ne with Al surfaces were done by Pepper and Aron (1986). They found that autoionization peaks split as a result of Doppler shifts corresponding to different scattering angles, and explained it with a binary scattering model which considers only single collisions between an incident Ne and an Al surface atom (Pepper, 1986). The authors found a broad structure in the spectra (Figure 4.25) and attributed it to the decay of doubly excited Ne** in the bulk, with the excited levels broadened by the interaction with the band. The shape of the structure remains

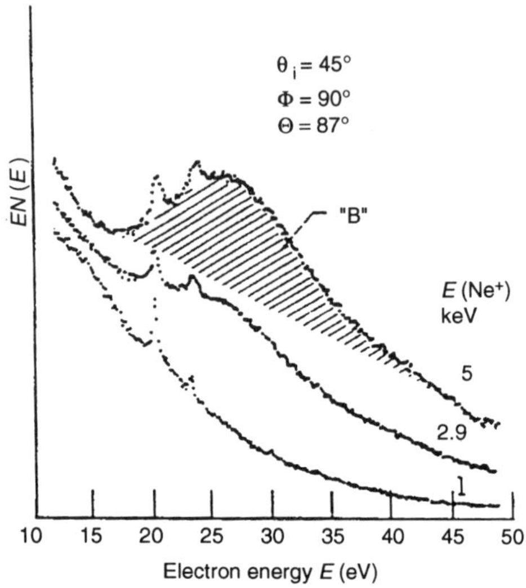

Figure 4.25 Neon Auger spectra for three different impact energies and at 45° incidence. (Reproduced with permission from Pepper and Aron, 1986)

unexplained, and it may provide important clues to the energies and lifetimes of core excitons of impurities in solids.

The emission of Ne autoionization electrons should be accompanied by two observables in the flux of back-scattered ions: a larger fraction of ions, and an inelastic energy loss due to the excitation of the two Ne-2*p* electrons. The enhanced ion fractions are expected due to the relatively long lifetime of Ne**, 1.5×10^{-14} s, which causes the decay into Ne^+ to occur relatively far from the surface where subsequent reneutralization is unlikely. Grizzi *et al.* (1990) searched for these additional observables using a time-of-flight analysis of the velocity of Ne and Ne^+ reflected from Mg surfaces. They found that the fraction of particles reflected as Ne^+, f^+, increased dramatically when the distance of closest approach in a Ne–Mg collision reached $\lesssim 0.7$ Å, the value expected for Ne-2*p* excitation along the promoted $4f\sigma$ MO. This was accompanied by an inelastic energy loss of ~30 eV in the scattered ions, attributed to the excitation of the reflected Ne**. Absorption of oxygen lowered f^+ and the inelastic energy loss and suppressed the emission of autoionization electrons, showing that Mg atoms are shadowed by the adsorbed oxygen.

Recent measurements by Buck *et al.* (1993) on Cu and Cu_3Au targets again show a correlation of energy loss, enhanced ion fractions and electron emission from the decay of $Ne^{**}(2p^4 3s^2)$. In this system, however, the energy losses were much larger, up to ~100 eV, suggesting that, for this target, the violent collisions produce excitation of the 2*s* electrons of Ne, which is then transferred to the $2p$ shell by Coster–Kronig transitions that, in this case, are possible at surfaces but not in isolated Ne atoms.

Unlike the case of Auger decay of inner-shell excitations, which often contributes to a smaller percentage of the total KEE yield, the excitation and decay of autoionizing states is a process of primary importance for Ne. The energy dependence of the total electron-emission yields on Al clearly shows the effect of enhanced emission from autoionization for Ne (and possibly also for F) projectiles (Figure 4.7, Alonso *et al.*, 1980). The increased ion fractions that result from the decay of atoms ejected from solids in autoionizing states are expected to be important also in other cases, such as secondary ion mass spectrometry and photon- and electron-stimulated desorption from surfaces. Unlike the case of Ne mentioned above, autoionization electrons may not always be identifiable since, if ejected during the collision, they can appear as broad structures in the spectra (Budzioch and Soszka, 1987; Soszka, 1988, 1989, 1990, 1991a, 1991b; Soszka *et al.*, 1989a, 1989b; Soszka and Soszka, 1992; Dorozhkin, 1993).

4.4.3.2 Auger emission from inner-shell excitations

Analysis of Doppler shifts and broadenings of the sharp atomic Auger lines (Baragiola, 1981; Baragiola *et al.*, 1982), and the observation of energy shifts which accompany work-function changes (Zampieri *et al.*, 1984) have shown that they correspond to transitions in core-excited sputtered atoms decaying in vacuum. To first order, Auger emission in vacuum is isotropic, while that from atoms in the bulk follows a cosine law with respect to the surface normal. This enhances strongly the (energy-dependent) (Valeri and Verucchi, 1991; Bonanno *et al.*, 1992b) ratio of intensity of atomic Auger lines to bulk lines for emission close to the surface (de Ferrariis *et al.*, 1986), as shown in Figure 4.26 (Baragiola *et al.*, 1991). Some of these studies have to be revised since a recent finding concerning the shape of the bulk Auger lines (Jensen *et al.*, 1992) suggests that too much secondary electron background has been subtracted from the spectra (i.e. the atomic/bulk ratios are lower than reported).

Very detailed studies of the dependence of ion-induced Auger spectra on angles of incidence and emission have been made and explained with computer simulations (Bonanno *et al.*, 1990a, 1990b; Hink and Herteux, 1992). Observations of Auger emission have been complemented by Hink *et al.* (Degel and Hink, 1986; Hink and May, 1988) with measurements of the X-ray spectra. Although this is the minority decay mode of the inner-shell vacancy, the study of X-rays can provide information that can be more easily interpreted, since it is related to the transition of just one valence electron.

Most of the studies on Auger emission have been done with clean surfaces, but in some instances, the sample has been implanted on purpose with a high

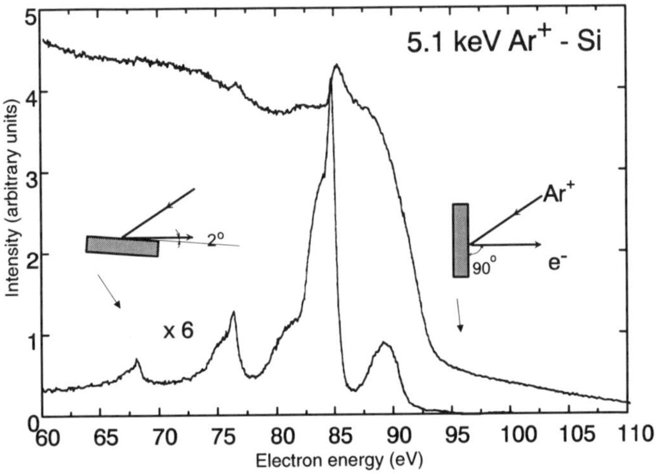

Figure 4.26 Electron energy spectra of Si bombarded by 5.1 keV Ar^+ in the region of the LVV and LMM Auger transition, for two different electron take-off angles with respect to the surface. The angle between the direction of the incident beam and the ejected electron is 146°. (Reproduced by permission of Elsevier Science Publishers from Baragiola *et al.*, 1991)

dose of projectiles and the electron emission used to understand the accumulation of the projectiles on the target (Bénazeth *et al.*, 1988, 1990, 1991; Terzic *et al.*, 1992a, 1992b). Maydell *et al.* (Milne *et al.*, 1991; Maydell, 1992; Maydell and Fabian, 1992) have explored the use of ion-induced Auger emission during high-dose bombardment for depth-profiling certain solids.

4.4.3.2.1 Symmetric versus asymmetric collisions

A topic still subject to controversy is the relative role of inner-shell excitations in asymmetric collisions between the projectile and a target atom, compared to the symmetric collisions between a fast target recoil and another target atom. Here the evidence is indirect, and comes from comparison of experiments with computer simulations which have reached a high degree of development. Our Monte Carlo analysis of Be–K emission (Grizzi and Baragiola, 1984) showed an excellent agreement with the experimental energy dependence of the Auger yield. This provided firm ground to determine the relative importance of symmetric and asymmetric processes in X-ray emission yields, more accurately than was possible with analytical theory, and to obtain information on the projectile energy dependence of the fluorescence yields and outer-shell occupation. The results show that symmetric collisions dominate shallow core excitations at the lower energies. Similar conclusions were obtained for Al–L emission induced by Ar bombardment: symmetric collisions dominate at low energies and asymmetric collisions at energies above $\sim$9 keV (Grizzi and Baragiola, 1987). On the other hand, Shapiro and Fine (1989), using a molecular-dynamics simulation method to study L-shell excitations in Al targets bombarded with Ar ions, conclude that Ar–Al collisions are largely responsible for Auger yields even at energies close to the threshold. This is in striking contrast with other recent theoretical studies (Grizzi and Baragiola, 1987; Bénazeth *et al.*, 1985), that use more realistic excitation cross sections. Additional evidence of the importance of symmetric collisions come from the observation that Auger yields in Al (Viaris de Lesegno and Hennequin, 1981) and Si (Valeri *et al.*, 1988) compounds vary with the square of the concentration of the light element which is excited. Detailed measurements by Bonanno *et al.* (1993) in Mg, Al, Si and $MgAl_{1-x}$ targets have recently confirmed that the ratio of asymmetric to symmetric contributions increases with increasing energy and also varies with incidence angle. Similar conclusions have been reached by Valeri and Verucchi (1992) using Si compounds. In some particular cases, like Ne on Al, asymmetric collisions may be important at low energies due to the sharing of vacancies during the collision (Baragiola *et al.*, 1982) (in this case, Ne-2*p* to Al-2*p*) which may occur if the relevant levels are close in energy. The topic of vacancy sharing has been reviewed by Stolterfoht (1987).

Zampieri and Baragiola (1986) studied K-3*p* Auger emission in collisions of 150–1000 eV K^+ with Be, Mg, Al and Cu surfaces. They showed that this Auger emission does not occur on clean surfaces, but requires symmetric collisions with previously implanted K. This was confirmed in subsequent work by Brenten *et al.* (1991a, 1992a) using other alkali ions colliding at nearly grazing (3°) incidence on partially alkalated W(110) surfaces. The energy dependences of the Auger yields were obtained in the symmetrical case, with surfaces covered by 0.3 monolayer of the same element as the projectile. The authors state that grazing incidence was employed to keep the interpretation of the results as simple as possible, by avoiding penetration of the ions. However, the grazing geometry introduces complications of its own, since it favors multiple collisions and it causes a critical dependence of the distribution of distances of closest approach to surface atoms on the (unspecified) flatness of the surface and the microscopic arrangement of the adsorbed atoms. An interesting aspect of the results for Cs impact is that the impact velocity threshold, $\sim 2 \times 10^6$ cm s^{-1} is the lowest observed so far for inner-shell excitations.

4.4.3.2.2 Double inner-shell excitations

A situation where asymmetric collisions predominate is in the excitation of double L-shell holes in collisions of Ar with Mg, Al and Si samples. The reason is that when two holes are excited in a $4f\sigma$ MO during symmetric collisions, they will most likely be shared after the collision—one in each collision partner—since this involves the smallest energy loss in the collision. The study of double excitations helps in determining the relevant crossing distances for *single* excitations in asymmetric collisions, since in both cases vacancies result from the promotion of the same MO and crossing with multiply vacant outer MOs. Baragiola *et al.* (1991) found a ~3.8 keV threshold for emission of Si-L^2MM Auger electrons in Ar–Si surface collisions. This threshold implies a critical excitation distance of 0.358 Å which agrees with the value of 0.344 Å calculated for single $2p$ excitation in this system (Schneider *et al.*, 1983). The authors also showed how previous measurements (Valeri and Tonini, 1989, 1990) gave artificially low threshold energies due to contamination of the ion beam with Ar^{2+}. In the interesting experiments by Bonanno *et al.* mentioned above, it was shown that 8 keV Ar bombardment produces more Mg-L^2 holes in $MgAl_{1-x}$ alloys that in pure Mg, at 8 keV. This again demonstrates the importance of excitation in the collision cascade—in this case in Al–Mg collisions—are more important than Ar–Mg collisions in producing double inner-shell excitations. Bonanno *et al.* (1992a) did an angular resolved study of the Al-L^2MM emission from Al under keV Ar^+ impact. By fitting with a simple binary-collision model, they derived critical internuclear distances for excitation which depended on projectile energy. This unexpected result is probably due to the deflection of the trajectories of the ejected excited Al by soft collisions with other surface atoms, not included in the model.

Oliva *et al.* (1991) and Xu *et al.* (1991) have recently reported detailed, angular resolved studies of Auger emission from double L excitation, for Si, Al and Mg_xAl_{1-x} alloys under bombardment with 5–15 keV noble-gas ions. The authors show that double excitations are caused by asymmetric collisions involving the projectile or, in the case of the alloy, also different target species (i.e. Mg–Al collisions). The absence of double excitation when the projectile cannot directly excite a target atom has been shown also for Cs bombardment of Mg (Milne and Fabian, 1991).

4.4.3.2.3 Analysis of atomic Auger lines

Identification of the Auger lines is difficult due to the multiplicity of possible states and uncertainties in calculated energies. The main L-shell atomic peak in the light elements Na to Si is believed to correspond to an initial neutral configuration, with the inner-shell hole and an extra outer electron. Most of the line identifications have been reviewed by Thomas, although in the last few years, several new Auger lines have been observed, but not yet identified. To understand the formation of the excited outer shells, several authors have studied the variation of the spectra induced by exposure to oxygen (Baragiola, 1981; Saiki and Tanaka, 1984; Saiki *et al.*, 1987; Baragiola *et al.*, 1991; Xu and Bonanno, 1992a; Xu *et al.*, 1992).

Analyses of the line shapes have been done for the bulk and the atomic Auger transitions. Early experiments have shown that the broad bulk Auger line observed in Mg, Al and Si is very similar to that obtained under electron bombardment, which results from LVV transitions involving two valence-band (V) electrons. A close comparison shows a discrepancy in the high-energy edge, which extends beyond the energy expected from transitions involving two valence-band electrons. Early explanations related this discrepancy to transitions from excited electronic states formed in the collision cascade created by the incident ion. Whaley and Thomas (1984) proposed, based on models for the spectra, that the discrepancy was due to an unresolved atomic line near the Fermi edge. This assumption was confirmed by de Ferrariis *et al.* (1986) for Si by doing angle resolved experiments with Ar projectiles at grazing incidence, which enhances the ratio of atomic vs. bulk intensities (see also Figure 4.25). Line-shape analysis of projectile Auger electrons shows an interesting dependence on the nature and condition of the target surface, as reported by Xu *et al.* for Ar-L Auger electrons (Xu and Bonanno, 1992b; Xu *et al.*, 1993).

Compared to the bulk feature, the atomic lines show a variety of interesting features, shifting, broadening and splitting depending on the conditions of excitation and observation. Grizzi and Baragiola (1987) have made an analysis of the different factors which contribute to the line shape. The Auger energies vary due to shifts in energy of the atomic levels resulting from the interaction with the surface and from the Doppler shift. The precise way by which these shifts affect the line shape depends on the velocity distribution of the excited sputtered particles and on the direction of observation. These distributions can be obtained by Monte Carlo simulations of the excitation events, and are strongly affected by the lifetime of the inner-shell hole. This lifetime is usually not well known; for Al, it is assumed to vary from $\sim 10^{-14}$ s inside the solid to $\sim 2 \times 10^{-13}$ s outside the solid and far from its surface.

The interaction with the surface introduces energy shifts due to shifts in the initial (usually neutral) state and to image shifts in the final ionic state, plus a small broadening. They all depend on the position of the excited particle at the time of electron emission which in turn depends on the distribution of velocities normal to the surface and the distribution of decay times. The surface interaction produces a net shift of the Auger peak to higher energies, 0.4–1 eV for 2–4 keV Ar on Al.

The Doppler shift depends on the velocity of the recoil and the direction of observation in the way shown by equation (1). If the distribution of Auger electrons is isotropic outside the solid, one can separate the Doppler from the surface shift by studying line shapes as a function of angle of emission. An implicit assumption in studies of Doppler broadening is that the angle of observation is the same as the angle at which the electron was emitted with respect to the direction of the atom. This assumption does not hold when the observed electron has back-scattered from the surface or, in the case of emission from the bulk, has undergone multiple elastic scattering events with the ion cores.

Figure 4.27 shows a comparison of experimental and calculated line shapes

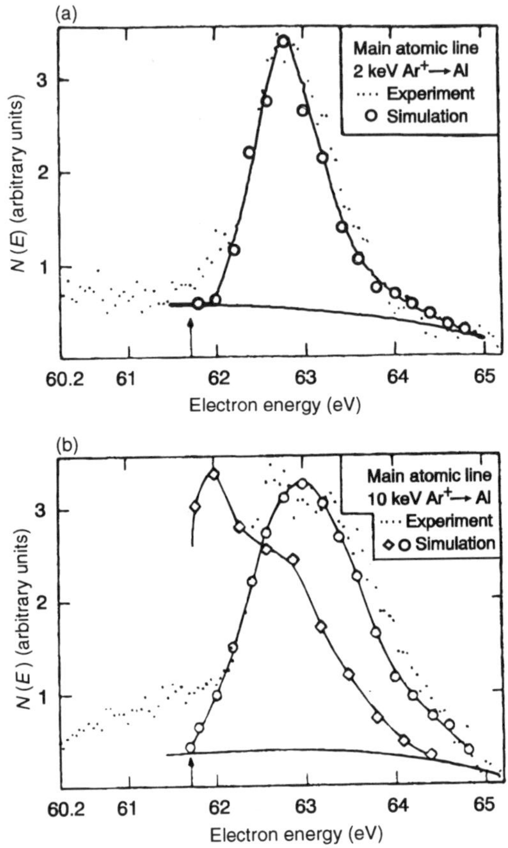

Figure 4.27 Comparison of experimental and simulated main atomic Auger line from Al bombarded with (a) 2 keV and (b) 10 keV Ar ions. The lines are broadened by the Doppler effect, recoil–surface interaction, and splitting of the initial vacancy state. In (b), the diamonds show the effect of neglecting the level shift due to the interaction with the surface. (From Grizzi and Baragiola, 1987)

for the main atomic Auger transition in Al, $Al^0(1s^22p^53s^23p^2) \Rightarrow Al^+(1s^22p^63s3p)$. A good agreement is not obtained by considering the Doppler effect alone, but requires the inclusion of the surface interaction. It is important to note that these results cannot be applied directly to other cases, since the shapes depend very much on the lifetime of the inner-shell hole, which varies strongly over the periodic table.

4.5 CONCLUSIONS

Electron emission under low-energy ion impact results from direct ionization of the projectile or target atoms with very small additional ionizations produced by energetic secondary electrons. The study of electron emission provides some of the most detailed pictures of inelastic ion–surface collisions, giving insight on mechanisms which can be applied to study other processes, such as energy loss, charge transfer, ionization damage and desorption. There is a lack of general theories and there are still many unsolved problems in the understanding of excitations by slow projectiles and by molecules or clusters, the threshold behavior of KEE, quantum effects in the motion of excited electrons, final-state interactions, the de-excitation of multiply charged ions near surfaces, and ionization and breakdown in insulators.

4.6 ACKNOWLEDGEMENTS

This work was supported by the National Science Foundation through grant DMR-9121272. I acknowledge useful discussions with D. E. Grosjean, R. E. Johnson, H. P. Winter and S. Ritzau.

4.7 REFERENCES

Alducin, M., A. Arnau, and P. M. Echenique (1992) *Nucl. Instr. Meth.*, **B 67**, 157.
Alonso, E. V. (1978) Ph.D. thesis, Institute Balserio.
Alonso, E. V., R. A. Baragiola, and J. Ferrón (1980) unpublished.
Alonso, E. V. R. A. Baragiola, J. Ferrón, and A. Oliva-Florio (1979) *Rad. Effects*, **45**, 199.
Alonso, E. V., R. A. Baragiola, J. Ferrón, M. M. Jakas, and A. Oliva-Florio (1980) *Phys. Rev.*, **B 22**, 80.
Alonso, E. V., M. A. Alurralde, and R. A. Baragiola (1986) *Surf. Sci.*, **166**, L155.
Alov, N. V. and K. A. Dadayan (1987) *Sov. Phys. Tech. Phys.*, **13**, 119.
Al Shama, F., M. Abbate, and J. C. Fuggle (1992) Appendix, In *Unoccupied Electronic States*, J. C. Fuggle and J. E. Inglesfield (eds), Springer, Berlin.
Amme, R. C. (1969) *J. Chem. Phys.*, **50**, 1891.
Ammosov, M. V. (1991) *J. Opt. Soc. Am.*, **B 8**, 2260.
Andersen, N. and J. O. Olsen (1977) *J. Phys.*, **B 10**, L719.
Andrä, H. J. (1989) *Nucl. Instr. Meth.*, **B 43**, 306.
Andrä, H. J., A. Simionovici, T. Lamy, A. Brenac, G. Lamboley, A. Pesnelle, S. Andriamonje, A. Fleury, M. Bonnefoy, M. Chasaveni, and J. J. Bonnet (1992) In *Electronic and Atomic Collisions*, W. R. MacGillivray, I. E. McCarthy, and M. C. Standage (eds), IOP Conference Proceedings, Institute of Physics, Bristol, p. 89.
Apell, P. (1987) *Nucl. Instr. Meth.*, **B 23**, 242.
Arifov, U. A. (1969) *Interaction of Atomic Particles with a Solid Surface*, Consultants Bureau, New York.
Arifov, U. A., L. M. Kishinevskii, E. S. Mukhamadiev, and E. S. Parilis (1973) *Sov. Phys. Tech. Phys.*, **18**, 118.
Arista, N. R., M. M. Jakas, G. H. Lantschner, and J. C. Eckardt (1986) *Phys. Rev.*, **A 34**, 5112.
Ashcroft, N. W. and N. D. Mermin (1976) *Solid State Physics*, Saunders, Philadelphia.
Aumayr, F., G. Lakits, and H. P. Winter (1991) *Appl. Surf. Sci.*, **47**, 139.
Aumayr, F. and H. Winter (1993) In Baragiola (1993a).

Aumayr, F., H. Kurz, D. Schneider, M. A. Briere, J. W. McDonald, C. E. Cunningham and H. P. Winter (1993), *Phys. Rev. Lett.*, **71**, 1943.
Avery, N. R. (1987) In *Vibrational Spectroscopy of Molecules on Surfaces*, J. T. Yates and T. E. Madey (eds), Plenum, New York, p. 223.
Baker, D. H., L. D. Doverspike, and R. L. Champion (1992) *Phys. Rev.*, **A 46**, 296.
Baklitskii, B. E. and E. S. Parilis (1986) *Sov. Phys. Tech. Phys.*, 31.
Banouni, M., J. Mischler, M. Negre, and N. Bénazeth (1985) *Surf. Sci.*, **163**, L720.
Baragiola, R. A. (1977) In *Proc. 7th Int. Conf. Atom. Coll. Solids*, Moscow State University, p. 151.
Baragiola, R. A. (1981) In *Inelastic Particle–Surface Collisions*, E. Taglauer and W. Heiland (eds), Springer, Heidelberg, p. 38.
Baragiola, R. A. (1982) *Rad. Effects*, **61**, 47.
Baragiola, R. A. (1991) In *Interaction of Charged Particles with Solids and Surfaces*, A. Gras-Martí *et al.* (eds), Plenum Press, New York, p. 443.
Baragiola, R. A. (ed.) (1993a) *Ionization of Solids by Heavy Particles, Proceedings of the NATO Advanced Research Workshop, Giardini-Naxos, Italy, June 1992.* Plenum Press, New York.
Baragiola, R. A. (1993b) *Nucl. Instr. Meth.*, **B 78**, 223.
Baragiola, R. A. *Nucl. Instr. Meth.*, (in press).
Baragiola, R., E. Alonso, O. Auciello, J. Ferrón, G. Lantschner, and A. Oliva-Florio (1978) *Phys. Lett.*, **67A**, 211.
Baragiola, R. A., E. V. Alonso, and A. Oliva-Florio (1979) *Phys. Rev.*, **B 19**, 121.
Baragiola, R. A., E. V. Alonso, and H. Raiti (1982) *Phys. Rev.*, **A 25**, 1969.
Baragiola, R. A., L. Nair, and T. E. Madey (1991) *Nucl. Instr. Meth.*, **B 58**, 322.
Baragiola, R. A., E. V. Alonso, A. Oliva, A. Bonanno, and F. Xu (1992) *Phys. Rev.*, **A 45**, 5286.
Barat, M. and W. Lichten (1972) *Phys. Rev.*, **A 6**, 211.
Bardsley, J. N. and B. M. Penetrante (1991) *Comments At. Mol. Phys.*, **27**, 43.
Bell, R. L. (1973) *Negative Electron Affinity Devices*, Clarendon Press, Oxford.
Bénazeth, C., N. Bénazeth, and M. Hou (1985) *Surf. Sci.*, **151**, L137.
Bénazeth, C., M. Hou, C. Mayoral, and N. Bénazeth (1987) *Nucl. Instr. Meth.*, **B 18**, 555.
Bénazeth, C., P. Hecquet, C. Mayoral, and N. Bénazeth (1989) *Rad. Eff. Def. Sol.*, **108**, 227.
Bénazeth, C., P. Cafarelli, and N. Bénazeth (1991) *Nucl. Instr. Meth.*, **B 62**, 47.
Bénazeth, N., J. Mischler, and N. Nègre (1988) *Surf. Sci.*, **205**, 419.
Bénazeth, N., N. Nègre, and C. Bénazeth (1990) *Rad. Eff. Def. Solids*, **114**, 291.
Beuhler, R. J. (1983) *J. Appl. Phys.*, **54**, 4118.
Bonanno, A., A. Amoddeo, and A. Oliva (1981) *J. Phys.*, **D14**, 1707.
Bonanno, A., F. Xu, M. Comarca, R. Siciliano, and A. Oliva (1990a) *Nucl. Instr. Meth.*, **B 48**, 371.
Bonanno, A., F. Xu, M. Comarca, R. Siciliano, and A. Oliva (1990b) *Phys. Rev.*, **B 41**, 12590.
Bonanno, A., M. Camarca, and F. Xu (1992a) *Surf. Sci. Lett.*, **264**, L213.
Bonanno, A., N. Mandarino, A. Oliva, and F. Xu (1992b) *Nucl. Instr. Meth.*, **B 71**, 161.
Bonanno, A., P. Zoccali, N. Mandarino, A. Oliva, and F. Xu (1993) In Baragiola (1993a).
Borisov, A. G., D. Teillet-Billy and J. P. Gauyacq (1993) *Surf. Sci.*, **284**, 337.
Böttcher, A., R. Imbeck, A. Morgante, and G. Ertl (1990) *Phys. Rev. Lett.*, **65**, 2035.
Böttcher, A., R. Grobecker, R. Imbeck, A. Morgante, and G. Ertl (1991) *J. Chem. Phys.*, **95**, 3756.
Böttcher, A., R. Grobecker, T. Greber, A. Morgante, and G. Ertl (1993) *Surf. Sci.*, **280**, 170.

Brenten, H., H. Müller, and V. Kempter (1991a) *Z. Phys.*, **D 21**, 11.
Brenten, H., H. Müller, K. H. Knorr, D. Kruse, H. Schall, and V. Kempter (1991b) *Surf. Sci.*, **243**, 309.
Brenten, H., H. Müller, D. Kruse, and V. Kempter (1991c) *Nucl. Instr. Meth.*, **B 58**, 328.
Brenten, H., H. Müller, and V. Kempter (1992a) *Surf. Sci.*, **217**, 103.
Brenten, H., H. Müller, and V. Kempter (1992b) *Surf. Sci.*, **274**, 309.
Brenten, H., H. Müller, and V. Kempter (1992c) *Z. Phys.*, **D 22**, 563.
Brenten, H., H. Müller, W. Maus-Friedrich, S. Dieckhoff, and V. Kempter (1992d) *Surf. Sci.*, **262**, 151.
Brenten, H., H. Müller, A. Niehaus, and V. Kempter (1992e) *Surf. Sci.*, **278**, 183.
Brenten, H., H. Müller, and V. Kempter (1993a) *Phys. Rev. Lett.*, **70**, 25.
Brenten, H., H. Müller, and V. Kempter (1993b) In Baragiola (1993a).
Brusilovsky, B. A. (1990) *Appl. Phys.*, **A 50**, 111.
Brusilovsky, B. A. (1991) *Vacuum*, **42**, 1183.
Buck, T. M., W. E. Wallace, R. A. Baragiola, G. H. Wheatley, J. B. Rothman, R. J. Gorte, and J. G. Tittensor (1993) *Phys. Rev.*, **B 48**, 774.
Budzioch, J. and W. Soszka (1987) *Mater. Sci. Eng.*, **90**, 51.
Burgdörfer, J. and F. Meyer (1993) *Phys. Rev.*, **A 47**, R20.
Burgdörfer, J., P. Lerner, and F. W. Meyer (1991) *Phys. Rev.*, **A 44**, 5674.
Burhop, E. H. S. and W. N. Assad (1972) *Adv. Atom. Mol. Phys.*, **8**, 163.
Cailler, M. and J. P. Ganachaud (1985) *Surf. Sci.*, **154**, 548.
Calera-Rubio, J., A. Gras-Martí, N. R. Arista, M. M. Jakas, Z. Sroubek, and G. Falcone (1991) *Rad. Eff. Def. Sol.*, **117**, 173.
Caro, A. and M. Victoria (1989) *Phys. Rev.*, **A 40**, 2287.
Caron, L. G., V. Cobut, G. Vachon, and S. Robillard (1990) *Phys. Rev.*, **B 41**, 2693.
Cartier, E. and P. Pfluger (1988) *Phys. Scripta*, **T 23**, 235.
Cazaux, J. (1993) In Baragiola (1993a).
Chaban, E. E., H. D. Hagstrum, and P. Petrie (1989) *Rev. Sci. Instr.*, **60**, 3647.
Chen, M. H., F. P. Larkins, and B. Crasemann (1990) *Atomic Data and Nuclear Data Tables*, **45**, 1.
Chung, M. S. and T. E. Everhart (1974) *J. Appl. Phys.*, **45**, 707.
Clapis, P. and Q. Kessel (1990) *Phys. Rev.*, **A 41**, 4766.
Das, J., L. Folkerts, and R. Morgenstern (1992) *Phys. Rev.*, **A 45**, 4669.
de Ferrariis, L. and R. A. Baragiola (1986) *Phys. Rev.*, **A 33**, 4449.
de Ferrariis, L., O. Grizzi, G. E. Zampieri, E. V. Alonso, and R. A. Baragiola (1986) *Surf. Sci.*, **167**, L175.
Degel, W. and W. Hink (1986) *Nucl. Instr. Meth.*, **B 13**, 619.
Delaunay, M., M. Fehringer, R. Geller, D. Hitz, P. Varga, and H. P. Winter (1987a) *Phys. Rev.*, **B 35**, 4232.
Delaunay, M., M. Fehringer, R. Geller, P. Varga, and H. P. Winter (1987b) *Europhys. Lett.*, **4**, 377.
Devooght, J., A. Dubus, and J. C. Dehaes (1992) *Nucl. Instr. Meth.*, **B 67**, 650.
de Zwart, S. T. (1987) *Nucl. Instr. Meth.*, **B 23**, 239.
de Zwart, S. T., A. G. Drentje, A. L. Boers, and R. Morgenstern (1989) *Surf. Sci.*, **217**, 298.
Dieckhoff, S., W. Maus-Friedrich, and V. Kempter (1992) *Nucl. Instr. Meth.*, **B 65**, 488.
Donath, M., D. Scholl, H. C. Siegmann, and E. Kay (1991) *Appl. Phys.*, **A 52**, 208.
Dorozhkin, A. A. (1993) *Vacuum*, **44**, 953.
Dunning, F. B., P. Nordlander, and G. K. Walters (1991) *Phys. Rev.*, **B 44**, 3246.
Echenique, P. M., F. Flores, and R. H. Ritchie (1990) *Solid St. Phys.*, **43**, 230.
Eckardt, J. C., G. Lantschner, N. R. Arista, and R. A. Baragiola (1978) *J. Phys.*, **C 21**, L851.
Eeken, P., J. M. Fluit, A. Niehaus, and I. Urazgil'din (1992a) *Surf. Sci.*, **273**, 160.

Eeken, P., A. Kotte, A. Niehaus, and I. Urazgil'din (1992b) *Nucl. Instr. Meth.*, **B 64**, 580.
Eichler, J., U. Wille, B. Fastrup, and K. Taulbjerg (1976) *Phys. Rev.*, **A 14**, 707.
Falcone, G. and Z. Sroubek (1988) *Phys. Rev.*, **B 38**, 4989.
Falcone, G. and Z. Sroubek (1989a) *Phys. Rev.*, **B 39**, 1999.
Falcone, G., and Z. Sroubek (1989b) *Rad. Eff. Def. Sol.*, **109**, 253.
Falcone, G. and Z. Sroubek (1991) *Nucl. Instr. Meth.*, **B 58**, 313.
Fano, U. (1987) *Phys. Rev.*, **A 36**, 1929.
Fehringer, M., M. Delaunay, R. Geller, P. Varga, and H. P. Winter (1987) *Nucl. Instr. Meth.*, **B 23**, 245.
Ferguson, M. M. and W. O. Hofer (1989) *Radiat. Eff. Def. Solids*, **109**, 273.
Ferguson, M. M. and W. O. Hofer (1993) In Baragiola (1993a).
Ferleger, V. K. and V. I. Matveev (1991) *Vacuum*, **42**, 875.
Ferrón, J. and E. C. Goldberg (1992) *Surf. Sci.*, **275**, 114.
Ferrón, J., E. V. Alonso, R. A. Baragiola, and A. Oliva-Florio (1981a) *J. Phys.*, **D 14**, 1707.
Ferrón, J., E. V. Alonso, R. A. Baragiola, and A. Oliva-Florio (1981b) *Phys. Rev.*, **B 24**, 4412.
Ferrón, J., E. V. Alonso, R. A. Baragiola, and A. Oliva-Florio (1982a) *Surf. Sci.*, **120**, 427.
Ferrón, J., E. C. Goldberg, L. S. De Bernardez, and R. H. Buitrago (1982b) *Surf. Sci.*, **123**, 239.
Feuerbacher, B. and R. F. Willis (1976) *J. Phys.*, **C 9**, 169.
Fine, J., M. Szymonski, and J. Kolodziej (1993) *Nucl. Instr. Meth.*, **B 78**, 129.
Fitting, H.-J. and D. Hecht (1988) *Phys. Stat. Sol. (a)*, **108**, 265.
Folkerts, L. (1992) Doctoral dissertation, Univ. Groningen.
Folkerts, L. and R. Morgenstern (1990) *Europhys. Lett.*, **13**, 377.
Folkerts, L. and R. Morgenstern (1991) *Z. Phys.*, **D 21**, 5351.
Folkerts, L., J. Das, S. W. Bergsma, and R. Morgenstern, *Phys. Lett.*, **A 163**, 73.
Fondén, T. and A. Z. Zwartkruis (1992) *Surf. Sci.*, **269/270**, 601.
Gay, T. J. and F. B. Dunning (1992) *Rev. Sci. Instr.*, **63**, 1635.
Geno, P. W. and R. D. Macfarlane (1989) *Int. J. Mass Spectrom. Ion Phys.*, **92**, 1955.
Gerber, G., R. Morgenstern, and A. Niehaus (1972) *J. Phys.*, **B 5**, 1396.
Gervasoni, J. L., N. R. Arista, R. O. Barrachina, and A. Gras-Martí (1992) *Nucl. Instr. Meth.*, **B 67**, 659.
Goldmann, A., W. Altmann, and V. Dose (1991) *Solid State Commun.*, **79**, 511.
Grizzi, O. and R. A. Baragiola (1984) *Phys. Rev.*, **A 30**, 2297.
Grizzi, O. and R. A. Baragiola (1987) *Phys. Rev.*, **A 35**, 135.
Grizzi, O., M. Shi, H. Bu, J. Rabalais, and R. A. Baragiola (1990) *Phys. Rev.*, **B 41**, 4789.
Groeneveld, K. O., W. Meckbach, and I. Sellin (eds) (1984) *Forward-Electron Ejection in Ion Collisions*, Springer, Berlin.
Grosjean, D. E. and R. A. Baragiola (1993) In Baragiola (1993a).
Gullikson, E. (1988) *Phys. Rev.*, **B 37**, 7904.
Hagstrum, H. D. (1954) *Phys. Rev.*, **96**, 336.
Hagstrum, H. D. (1961) *Phys. Rev.*, **122**, 83.
Hagstrum, H. D. (1966) *Phys. Rev.*, **150**, 495.
Hagstrum, H. D. (1977) In *Inelastic Ion–Surface Collisions*, N. H. Tolk, J. C. Tully, W. Heiland, and C. W. White (eds), Academic Press, New York, p. 1.
Hagstrum, H. D. (1988) In *Chemistry and Physics of Solid Surfaces VII*, R. Vanselow and R. F. Howe (eds), Springer, Berlin, p. 341.
Hagstrum, H. D. and G. E. Becker (1973) *Phys. Rev.*, **B 8**, 107.
Hagstrum, H. D., P. Petrie, and E. E. Chaban (1988) *Phys. Rev.*, **B 38**, 10264.

Harada, Y., H. Hayashi, S. Masuda, T. Fukuda, N. Sato, S. Kato, K. Kobayashi, H. Kuroda, and H. Ozaki (1991) *Surf. Sci.*, **242**, 95.
Hart, M. W., M. S. Hammond, F. B. Dunning, and G. K. Walters (1989) *Phys. Rev.*, **B 39**, 5488.
Hasselkamp, D. and A. Scharmann (1983) *Phys. Stat. Sol. (a)*, **79**, K197.
Hasselkamp, D. (1992), in Hasselkamp *et al.* (1992), Chapter 1.
Hasselkamp, D., H. Rothard, K.-O. Groeneveld, J. Kemmler, P. Varga, and H. Winter (1992) *Particle Induced Electron Emission II*, Springer, Berlin.
Hasted, J. B. (1964) *Physics of Atomic Collisions*, Butterworth, Washington.
Hemmen, R. and H. Conrad (1991) *Phys. Rev. Lett.*, **67**, 1314.
Hemmen, R. and H. Conrad (1992) *Appl. Phys.*, **A 55**, 411.
Hemmen, R., M. E. Kordesch, and H. Conrad (1989) *Phys. Rev.*, **B 40**, 2801.
Hentschke, R., K. J. Snowdon, P. Hertel, and W. Heiland (1986) *Surf. Sci.*, **173**, 565.
Herteux, M., W. Hink, J. P. Biersack, and E. Schreck (1990) *Nucl. Instr. Meth.*, **B 48**, 621.
Hink, W. and M. Herteux (1992) *Nucl. Instr. Meth.*, **B 66**, 333.
Hink, W. and W. May (1988) *Nucl. Instr. Meth.*, **B 33**, 290.
Hofer, W. O. (1990) *Scanning Microsc.*, **4**, 265.
Hofer, W. and P. Varga (1984) *Nucl. Instr. Meth.*, **B 2**, 391.
Hongo, S., T. Urano, and T. Kanaji (1991) *Appl. Surf. Sci.*, **48/49**, 143.
Hou, M., C. Bénazeth, N. Bénazeth, and C. Mayoral (1986) *Nucl. Instr. Meth.*, **B 13**, 645.
Hou, M., C. Bénazeth, and N. Bénazeth (1987) *Phys. Rev.*, **A 36**, 591.
Hou, M., C. Bénazeth, P. Hecquet, and N. Bénazeth (1990) *Nucl. Instr. Meth.*, **B 48**, 625.
Ibach, H. and D. L. Mills (1982) *Electron Energy Loss Spectroscopy and Surface Vibrations*, Academic Press, New York.
Ishii, A. (1987) *Surf. Sci.*, **192**, 172.
Ishii, H., S. Masuda, and Y. Harada (1990) *Surf. Sci.*, **239**, 222.
Janev, R. K. and N. N. Nedeljkovic (1985) *J. Phys.*, **B 34**, 915.
Jennings, P. J. and R. O. Jones (1988) *Adv. Phys.*, **37**, 341.
Jensen, E., R. A. Bartynski, R. F. Garrett, S. L. Hulbert, E. D. Johnson, and C.-C. Kao (1992) *Phys. Rev.*, **45**, 13636.
Joyes, P. (1968) *J. Phys.*, **29**, 774.
Joyes, P. (1969a) *J. Phys.*, **30**, 243.
Joyes, P. (1969b) *J. Phys.*, **30**, 365.
Kaji, H., K. Makoshi, and A. Yoshimori (1990) *Surf. Sci.*, **227**, 138.
Kaneko, T. (1992) *Nucl. Instr. Meth.*, **B 67**, 655.
Kanie, T., A. Ichimiya, S. Ohtani, H. Tawara, and T. Yasue (1992) *Nucl. Instr. Meth.*, **B 67**, 401.
Karolewski, M. A. and R. G. Cavell (1987) *Surf. Sci.*, **180**, L119.
Kasemo, B., E. Törnquist, J. K. Nørskov, and B. J. Lundqvist (1979) *Surf. Sci.*, **89**, 554.
Kawai, H., M. Nakayama, and K. Makoshi (1991) *Prog. Theor. Phys. Suppl.*, **106**, 321.
Kawata, J., K. Ohya, and I. Mori (1991a) *Jpn. J. Appl. Phys.*, **30**, 2585.
Kawata, J., K. Ohya, and I. Mori (1991b) *Jpn. J. Appl. Phys.*, **30**, 3510.
Kawata, J., K. Ohya, and I. Mori (1992) *Jpn. J. Appl. Phys.*, **31**, 2560.
Keller, W., H. Morgner, and W. A. Müller (1986) *Mol. Phys.*, **58**, 1039.
Kirschner, J., K. Koike, and H. P. Oepen (1987) *Phys. Rev. Lett.*, **59**, 2099.
Kirschner, J., O. M. Artamonov, and A. N. Terekhov (1992) *Phys. Rev. Lett.*, **69**, 1711.
Kishinevskii, L. M. (1963) *Bull. Acad. Sci. USSR, Phys. Ser.*, **26**, 1433.
Kitov, V. U. and E. S. Parilis (1984) *Surf. Sci.*, **138**, 203.

Klein, H. J. (1965) *Z. Physik*, **188**, 78.
Knipp, J. K. and R. C. Ling (1951) *Phys. Rev.*, **82**, 30.
Knoll, G. F. (1989) *Radiation Detection and Measurement*, Wiley, New York.
Köhrbrück, R., K. Sommer, J. P. Biersack, J. Bleck-Neuhaus, S. Schippers, P. Roncin, D. Lecler, F. Fremont, and N. Stolterfoht (1992) *Phys. Rev.*, **A 45**, 4653.
Koike, K. and J. Kirschner (1992) *J. Phys.*, **D 25**, 1139.
Konoplev, V. M., M. Vicanek, and A. Gras-Martí (1992) *Nucl. Instr. Meth.*, **B 67**, 574.
Krakov, B. G. and E. S. Parilis (1989) *Sov. Phys. Usp.*, **32**, 251.
Kunc, J. A. and W. H. Scon (1991) *J. Chem. Phys.*, **95**, 5738.
Kurz, H., K. Töglhofer, H. P. Winter, F. Aumayr, and R. Mann (1992) *Phys. Rev. Lett.*, **69**, 1140.
Lakits, G. and H. Winter (1990) *Nucl. Instr. Meth.*, **B 48**, 597.
Lakits, G., F. Aumayr, and H. Winter (1989a) *Europhys. Lett.*, **10**, 679.
Lakits, G., F. Aumayr, and H. Winter (1989b) *Phys. Lett.*, **139**, 395.
Lakits, G., F. Aumayr, and H. Winter (1989c) *Rad. Eff. Def. Solids*, **109**, 129.
Lakits, G., F. Aumayr, and H. Winter (1989d) *Rev. Sci. Instr.*, **60**, 3151.
Lakits, G., A. Arnau, and H. Winter (1990a) *Phys. Rev.*, **B 42**, 15.
Lakits, G., F. Aumayr, M. Heim, and H. P. Winter (1990b) *Phys. Rev.*, **A 42**, 5780.
Larson, S. A. and L. L. Lauderback (1991) *Phys. Rev.*, **B 43**, 36.
Lee, J., C. Hanrahan, J. Arias, F. Bozso, R. M. Martin, and H. Metiu (1985) *Phys. Rev. Lett.*, **54**, 1440.
Leonas, V. B. (1991) *Sov. Phys. Usp.*, **34**, 317.
Liehr, M., P. A. Thiry, J. J. Pireaux, and R. Caudano (1986) *Phys. Rev.*, **B 33**, 5682.
Lindhard, J., V. Nielsen, M. Scharff, and P. V. Thomsen (1968) *Mar. Fys. Medd. Dan. Vid. Selsk.*, **33**, No. 10.
Lorente, N. and R. Monreal (1993a) *Nucl. Instr. Meth.*, **B 78**, 44.
Lorente, N. and R. Monreal (1993b) private communication.
Makoshi, K. (1991) *Surf. Sci.*, **254**, 281.
Martens, J., W. Ens, K. G. Standing, and A. Verentchikov (1992) *Rapid Comm. Mass Spectr.*, **6**, 147.
Massey, H. S. W., E. H. S. Burhop, and H. B. Gilbody (1971) *Electronic and Ionic Impact Phenomena*, Clarendon Press, Oxford.
Masuda, S., H. Hayashi, and Y. Harada (1990) *Phys. Rev.*, **B 42**, 3582.
Masuda, S., H. Ishii, and Y. Harada (1992a) *Surf. Sci.*, **242**, 400.
Masuda, S., H. Ishii, and Y. Harada (1992b) *Sol. St. Comm.*, **82**, 587.
Maus-Friedrich, W., S. Dieckhoff, and V. Kempter (1991a) *Surf. Sci.*, **249**, 149.
Maus-Friedrich, W., S. Dieckhoff, M. Wehrhahn, and V. Kempter (1991b) *Surf. Sci.*, **253**, 137.
Maus-Friedrich, W., S. Dieckhoff, M. Wehrhahn, S. Pülm, and V. Kempter (1992) *Surf. Sci.*, **271**, 113.
Maydell, E. A. (1992) *Surf. Interf. Anal.*, **19**, 65.
Maydell, E. A. and D. J. Fabian (1992) *Nucl. Instr. Meth.*, **B 67**, 610.
McDonald, J. W., D. Schneider, M. W. Clark, and D. Dewitt (1992) *Phys. Rev. Lett.*, **68**, 2297.
Meyer, F. W., C. C. Havener, S. H. Overbury, K. J. Snowdon, D. M. Zehner, W. Heiland, and H. Hemme (1987) *Nucl. Instr. Meth.*, **B 23**, 234.
Meyer, F. W., S. H. Overbury, C. C. Havener, P. A. Zeijlmans van Emmichoven, and D. M. Zehner (1991) *Phys. Rev. Lett.*, **47**, 723.
Miller, M. S. and J. W. Boring (1974) *Phys. Rev.*, **A 9**, 2421.
Milner, R. H. and D. J. Fabian (1991) *Appl. Phys.*, **A 53**, 574.
Milne, R. H., E. A. Maydell, and D. J. Fabian (1991) *Appl. Phys.*, **A 52**, 197.

Mischler, J. and N. Bénazeth (1986) *Scanning Elect. Microsc.*, **2**, 351.
Mischler, J., M. Banouni, C. Bénazeth, M. Negre, and N. Bénazeth (1986) *Rad. Effect*, **97**, 1.
Momose, Y., T. Phaku, S. Okazaki, and Y. Fujii (1992) *J. Vac. Sci. Technol.*, **A10**, 353.
Morgner, H., J. Oberbrodhage, K. Richter, and K. Roth (1991a) *J. Electron Spectrosc. Relat. Phenom.*, **57**, 61.
Morgner, H., J. Oberbrodhage, K. Richter, and K. Roth (1991b) *J. Phys.: Condens. Matter*, **3**, 5639.
Morgner, H., J. Oberbrodhage, K. Richter, and K. Roth (1991c) *Molec. Phys.*, **73**, 1295.
Moshammer, R. and R. Matthäus (1989) *J. Physique*, **50**, C2–111.
Müller, H., R. Hausmann, H. Brenten, and V. Kempter (1993) *Surf. Sci.*, **284**, 129.
Müller, H., R. Hausmann, H. Brenten, and V. Kempter (1993) *Nucl. Instr. Meth.*, **B 78**, 239.
Naumkin, A. V. (1991) *Bull. Acad. Sci. USSR*, **55**, 54.
Negre, M., J. Mischler, and N. Bénazeth (1985) *Surf. Sci.*, **157**, 436.
Niehaus, A. (1990) *Phys. Repts.*, **186**, 149.
Niehaus, A. (1993) In Baragiola (1993a).
Nishigaki, S., M. Ohara, S. Yokoi, and S. Matsuda (1988) *Nucl. Instr. Meth.*, **B 33**, 311.
Nishigaki, S., N. Oishi, S. Matsuda, N. Kawanishii, and T. Sasaki (1989) *Phys. Rev.*, **B 39**, 8048.
Nishigaki, S., S. Matsuda, T. Sasaki, N. Kawanishi, Y. Ikeda, and H. Takeda (1990) *Surf. Sci.*, **231**, 271.
Nishigaki, S., T. Sasaki, S. Matsuda, N. Kawanishi, H. Takeda, and K. Yamada (1991) *Surf. Sci.*, **242**, 358.
Nørskov, J. K. and N. D. Lang (1980) *Phys. Rev.*, **B 21**, 2131.
Oda, K., T. Kanie, A. Ichimiya, S. Ohtani, K. Ohya, and H. Tawara (1992) *Surf. Sci.*, **262**, 437.
Oeschner, H. (1978) *Phys. Rev.*, **B 17**, 1052.
Ohya, K. (1992) *J. Phys. Soc. Jpn.*, **61**, 3013.
Ohya, K. and I. Mori (1991) *Jpn. J. Appl. Phys.*, **30**, 1084.
Ohya, K. and I. Mori (1992a) *J. Phys. Soc. Jpn.*, **61**, 2569.
Ohya, K. and I. Mori (1992b) *Jpn. J. Appl. Phys.*, **31**, 3686.
Ohya, K. and I. Mori (1992c) *Nucl. Instr. Meth.*, **B 67**, 628.
Ohya, K., J. Kawata, and I. Mori (1989) *Jpn. J. Appl. Phys.*, **28**, 1944.
Ohya, K., F. Aumayr, and H. P. Winter (1992) *Phys. Rev.*, **B 46**, 3101.
Oliva, A., A. Bonanno, M. Camarca, and F. Xu (1991) *Nucl. Instr. Meth.*, **B 58**, 333.
Onellion, M., W. Hart, F. B. Dunning, and G. K. Walters (1984) *Phys. Rev. Lett.*, **52**, 380.
Oró, D. M., Q. Lin, P. A. Soletsky, X. Zhang, F. B. Dunning, and G. K. Walters (1991) *Phys. Rev.*, **B 44**, 3246.
Penn, D. R. and P. Apell (1990) *Phys. Rev.*, **B 41**, 3303.
Pepper, S. V. (1986) *Surf. Sci.*, **169**, 39.
Pepper, S. V. and P. R. Aron (1986) *Surf. Sci.*, **169**, 14.
Petite, G., P. Agostini, R. Trainham, E. Mevel, and P. Martin (1992) *Phys. Rev.*, **B 45**, 12210.
Phipps, J. A., J. W. Boring, and R. A. Lowry (1964) *Phys. Rev.*, **A 36**, 135.
Pijper, F. J. and P. Kruit (1991) *Phys. Rev.*, **B 44**, 9192.
Rabalais, J. W., H. Bu, and C. D. Roux (1992) *Phys. Rev. Lett.*, **69**, 1391.
Ramey, R. L. (1961) *Physical Electronics*, Wadsworth, Belmont, Calif.
Ray, J. A., C. F. Barnett, and B. Van Zyl (1979) *J. Appl. Phys.*, **50**, 6516.
Ritchie, R. H. and J. C. Ashley (1965) *J. Phys. Chem. Solids*, **26**, 1689.
Rösler, M. (1991a) *Nucl. Instr. Meth.*, **B 58**, 209.

Rösler, M. (1991b) In Rösler *et al.* (1991).
Rösler, M. (1993) In Baragiola (1993a).
Rösler, M. and W. Brauer (1992) *Nucl. Instr. Meth.*, **B 67**, 641.
Rösler, M., W. Brauer, J. DeVooght, J.-C. Dehaes, A. Dubus, M. Cailler, and J.-P. Ganachaud (1991) *Particle Induced Electron Emission I*, Springer, Berlin.
Roy, D. and D. Tremlay (1990) *Rep. Prog. Phys.*, **53**, 1621.
Rudd, M. E., Y.-K. Kim, D. H. Madison, and T. J. Gay (1992) *Rev. Mod. Phys.*, **64**, 441.
Saiki, K. and S. Tanaka (1984) *Jpn. J. Appl. Phys.*, **23**, L153.
Saiki, K., I. Rittaporn, and S. Tanaka (1987) *Jpn. J. Appl. Phys.*, **26**, 45.
Salehpour, M., P. Hakansson, B. Sundqvist, and S. Widdiyasekera (1986) *Nucl. Instr. Meth.*, **183**, 278.
Salmi, L. A. (1992) *Phys. Rev.*, **B 46**, 4180.
Schauer, S. N. and P. Williams (1992) *Phys. Rev.*, **B 46**, 15452.
Schippers, S., S. Oelschig, W. Heiland, L. Folkerts, R. Morgenstern, P. Eeken, L. F. Urazgil'din, and A. Niehaus (1991) *Surf. Sci.*, **257**, 289.
Schneider, D., G. Nolte, U. Wille, and N. Stolterfoht (1983) *Phys. Rev.*, **A 28**, 161.
Schol, B., H. A. J. Meijer, M.-W. Ruf, and H. Hotop (1992) *Meas. Sci. Technol.*, **3**, 544.
Schou, J. (1980) *Phys. Rev.*, **B 22**, 2141.
Schou, J. (1982) *Scanning Microsc.*, **1**, 33.
Schou, J. (1993) In Baragiola (1993a).
Schulz, G. (1973) *Rev. Mod. Phys.*, **45**, 378.
Seah, M. P. and G. C. Smith (1990) *Surf. Interface Anal.*, **15**, 751.
Seiler, H. (1983) *J. Appl. Phys.*, **54**, R1.
Serena, P. A., J. M. Soler, and N. García (1986) *Phys. Rev.*, **B 34**, 6767.
Shapiro, M. H. and J. Fine (1989) *Nucl. Instr. Meth.*, **B 44**, 43.
Shapiro, M. H. and T. A. Tombrello (1992) *Nucl. Instr. Meth.*, **B 66**, 317.
Shimizu, R. and Z.-J. Ding (1992) *Rep. Prog. Phys.*, 487.
Sigmund, P. (1975) In *Radiation Damage Processes in Materials*, C. H. S. Dupuy (ed.), Noordhoff, Leyden, p. 3.
Sigmund, P. (1993) In Baragiola (1993a).
Sigmund, P. and S. Tougaard (1981) *Springer Ser. Chem. Phys.*, **17**, 2.
Skogvall, B. and G. Schiwietz (1988) *Phys. Rev. Lett.*, **60**, 2018.
Snowdon, K. J. (1988) *Nucl. Instr. Meth.*, **B 34**, 309.
Solov'ev, E. S., E. N. Il'in, V. A. Oparin, and N. V. Fedorenko (1962) *Soviet Phys. JETP*, **15**, 459.
Soszka, W. (1988) *J. Phys.*, **C 21**, 1155.
Soszka, W. (1989) *Phys. Stat. Sol. a*, **112**, 425.
Soszka, W. (1990) *Nucl. Instr. Meth.*, **48**, 630.
Soszka, W. (1991a) *J. Phys.*, **C 3**, 8289.
Soszka, W. (1991b) *Nucl. Instr. Meth.*, **B 61**, 547.
Soszka, W. and M. Soszka (1992) *Nucl. Instr. Meth.*, **B 67**, 632.
Soszka, W., S. Kwasny, J. Budzioch, and M. Soszka (1989a) *Phys. Lett.*, **138**, 531.
Soszka, W., S. Kwasny, J. Budzioch, and M. Soszka (1989b) *J. Phys. Condens. Matter*, **1**, 1353.
Souda, R., W. Hayami, T. Airawa, S. Otani, and Y. Ishizawa (1992) *Phys. Rev.*, **B 46**, 7315.
Sroubek, Z. (1993a) *Nucl. Instr. Meth.*, **B 78**, 140.
Sroubek, Z (1993b), *Phys. Rev. Lett.*, **70**, 4819.
Sternglass, E. J. (1957) *Phys. Rev.*, **108**, 1.
Stolterfoht, N. (1987) In *Progress in Atomic Spectroscopy*, *Part D*, H. J. Beyer and H. Kleinpoppen (eds), Plenum, New York, p. 415.
Stott, M. J. and E. Zaremba (1980) *Phys. Rev.*, **B 22**, 1564.

Suarez, S., V. Ponce, R. Pregliasco, P. Focke, and W. Meckbach (1992) *Nucl. Instr. Meth.*, **B 67**, 148.
Susuki, Y., T. Baba, S. Fukui, and M. Mannami (1991) *Surf. Sci.*, **254**, 97.
Svensson, B. and G. Holméen (1982) *Phys. Rev.*, **B 25**, 3056.
Terzic, I., Z. Rakocevic, and M. M. Tosic (1992a) *Surf. Sci.*, **260**, 200.
Terzic, I., M. M. Tosic, and Z. Rakocevic (1992b) *J. Vac. Sci. Technol. A*, 3359.
Thomas, E. W. (1984) *Vacuum*, **34**, 1031.
Thum, F. and W. O. Hofer (1984) *Nucl. Instr. Meth.*, **B 2**, 531.
Töglhofer, K., F. Aumayr, and H. P. Winter (1993) *Surf. Sci.*, **281**, 143.
Tougaard, S. and P. Sigmund (1982) *Phys. Rev.*, **B 25**, 4452.
Ushio, Y., T. Banno, N. Matuda, Y. Saito, S. Baba, and A. Kinbara (1988) *Thin Solid Films*, **167**, 299.
Vaeck, N. and J. E. Hansen (1991) *J. Phys.*, **B 24**, L469.
Vaeck, N. and J. E. Hansen (1992) *Surf. Sci.*, **269/270**, 596.
Valeri, S. (1993) *Surf. Sci. Repts.*, **17**, 85.
Valeri, S. and R. Tonini (1989) *Surf. Sci.*, **220**, 407.
Valeri, S. and R. Tonini (1990) In *Auger Spectroscopy and Electronic Structure*, G. Cubiotti, G. Mondio, and K. Wandelt (eds), Springer, Heidelberg, p. 152.
Valeri, S. and R. Verucchi (1991) *Nucl. Instr. Meth.*, **B 59/60**, 37.
Valeri, S. and R. Verucchi (1992) *Phys. Scripta*, **T 41**, 246.
Valeri, S., R. Tonini, and G. Ottaviani (1988) *Phys. Rev.*, **B 38**, 13282.
Varga, P. (1987) *Appl. Phys.*, **A 44**, 31.
Varga, P. (1988) In *Electronic and Atomic Collisions*, H. B. Gilbody, W. R. Newell, F. H. Read, and A. C. H. Smith (eds), Elsevier, New York, p. 793.
Varga, P. (1989) *Comments At. Mol. Phys.*, **23**, 111.
Varga, P. and H. Winter (1992) In Hasselkamp *et al.* (1992).
Veksler, V. I. (1983) *Vacuum*, **33**, 159.
Viaris de Lesegno, P. and J.-F. Hennequin (1981) *Surf. Sci.*, **103**, 257.
Weller, R. A. (1989) *Nucl. Instr. Meth.*, **B 40/41**, 286.
Whaley, R. and E. W. Thomas (1984) *J. Appl. Phys.*, **56**, 1505.
Wille, U. (1991) *Z. Phys.*, **D 21**, S353.
Wille, U. (1992a) *Nucl. Instr. Meth.*, **B 67**, 132.
Wille, U. (1992b) *Phys. Rev.*, **A 45**, 3004.
Wille, U. and R. Hippler (1986) *Phys. Rept.*, **3–4**, 129.
Winter, H. P. (1991) *Z. Phys.*, **D 21**, S129.
Winter, H. (1992) *Europhys. Letters*, **18**, 207.
Winter, H., P. Strohmeier, and J. Burgdörfer (1989) *Phys. Rev.*, **A 39**, 3895.
Winter, H., F. Aumayr, and G. Lakits (1991) *Nucl. Instr. Meth.*, **B 58**, 301.
Winter, H., C. Auth, R. Schuch and E. Beebe (1993) *Phys. Rev. Lett.*, **71**, 1939.
Wittmaack, K. (1991) *Nucl. Instr. Meth.*, **B 58**, 317.
Wittmaack, K. and Y. Homma (1991) *Appl. Phys. Lett.*, **58**, 2138.
Woratschek, B., W. Sesselmann, J. Küppers, G. Ertl, and H. Haberland (1985) *Phys. Rev. Lett.*, **55**, 611.
Xu, F. and A. Bonanno (1992a) *Surf. Sci. Lett.*, **273**, L414.
Xu, F. and A. Bonanno (1992b) *Phys. Rev.*, **B 46**, 11405.
Xu, F., M. Camarca, A. Oliva, N. Mandarino, P. Zoccali, and A. Bonanno (1991) *Surf. Sci.*, **247**, 13.
Xu, F., P. Riccardi, A. Oliva, and A. Bonanno (1993) *Nucl. Instr. Meth.*, **B 78**, 251.
Zampieri, G. E. and R. A. Baragiola (1982) *Surf. Sci.*, **114**, L15.
Zampieri, G. E. and R. A. Baragiola (1984) *Phys. Rev.*, **B 29**, 1480.

Zampieri, G. E. and R. Baragiola (1986) *Phys. Rev.*, **B 33**, 588.
Zampieri, G. E., F. Meier, and R. A. Baragiola (1984) *Phys. Rev.*, **A 29**, 116.
Zehner, D. M., S. H. Overbury, C. C. Havener, F. W. Meyer, and W. Heiland (1986) *Surf. Sci.*, **178**, 359.
Zeijlmans van Emmichoven, P. A. and N. Niehaus (1990) *Comm. At. Mol. Phys.*, **24**, 45.
Zeijlmans van Emmichoven, P. A., P. A. A. F. Wouters, and A. Niehaus (1988) *Surf. Sci.*, **195**, 115.
Zhang, X., H. Hsu, F. B. Dunning, and G. K. Walters (1991) *Phys. Rev.*, **A 52**, 9133.

5 SCATTERING AND CHARGE-TRANSFER DYNAMICS IN LOW- AND HYPERTHERMAL-ENERGY ALKALI-ION–SURFACE COLLISIONS

B. H. Cooper and E. R. Behringer
Laboratory of Atomic and Solid State Physics,
Cornell University, Ithaca, New York 14853-2501, USA

CONTENTS

Low Energy Ion–Surface Interactions
Edited by J. W. Rabalais

5.1 INTRODUCTION

In this chapter we will discuss the dynamics of 10 to 100 eV ('hyperthermal') ion collisions with metal surfaces as well as the charge-transfer behavior for the 5 to 1600 eV range (i.e. 'low energies'). Ion scattering at keV energies has been studied extensively for a number of years and several comprehensive discussions can be found in the literature (see, for example, Taglauer and Heiland, 1976; Boers, 1977; Kasi *et al.*, 1989; Niehus and Spitzl, 1991; and references therein). There have been relatively few studies of scattering which focus on the few eV to 100 eV range (for example, see Kasi *et al.*, 1989 and references therein). Consequently, many aspects of the fundamental interactions between particles and surfaces in the hyperthermal energy regime remain poorly understood. This lack of attention to the hyperthermal energy regime is partly an experimental issue: the ion optics for beam production and transport and the particle detectors in some cases (e.g. for detecting hyperthermal neutral atoms) must be designed specifically to handle such low kinetic energies. In addition, scattering models which work well at higher energies must be modified to treat the lower end of this energy range. Both of these issues will be discussed in more detail below.

Hyperthermal energy ion–surface interactions are of interest not only for fundamental reasons, but also because this energy regime is used in a number of surface processing and modification techniques, such as reactive ion etching, ion beam-assisted deposition, and thin-film growth (Kasi *et al.*, 1989). There are many processes one must consider in order to understand energetic ion interactions with metal surfaces. Here we will focus on two topics. One is the dynamics of scattering for incident energies in the 10 to 100 eV range. The other is charge transfer, or the exchange of electrons between the particle and surface, for alkalis in the few eV to keV range.

Alkali-ion scattering from metal surfaces was chosen as a model system for these studies. There are several reasons for this choice. First, the investigation of the interactions of alkali ions with metal surfaces has a long history and the theoretical framework exists for calculating scattering potentials and electronic interactions, both of which are necessary for a detailed interpretation of the scattering and charge-transfer measurements. There are, in addition, experimental issues that make the alkalis an attractive choice. Alkali-ion sources are easy to work with and produce ion fluxes such that, even after deceleration to the lowest energies, adequate beam currents on the sample are obtained. It is also possible to detect different electronic states of hyperthermal energy alkalis with known efficiencies, which is not the case with several other species. This latter capability enables one to make quantitative comparisons between experimental results and theory.

In this chapter we will summarize recent advances from our laboratory in the understanding of the fundamental mechanisms of hyperthermal energy alkali-ion scattering dynamics and charge transfer. We will emphasize experimental work which has led to quantitative tests of scattering potentials and charge-transfer models. While relevant references will be given, this is not intended to be a comprehensive review of the literature in the field.

In Section 5.2 we will discuss the dynamics of scattering at hyperthermal energies. This includes a brief description of experimental and modeling techniques, as well as of data and what they tell us about scattering dynamics. In Section 5.3 we will discuss charge-transfer phenomena, again including a brief description of experimental techniques. We will consider experiments that can be well modeled using a one-electron picture of charge transfer, and others that require a multi-state picture.

5.2 SCATTERING DYNAMICS

In a scattering experiment, changing the kinetic energies of the incident particles probes different aspects of the ion–surface interaction potential. At keV, or 'low energies', the scattering is determined primarily by the short-range repulsive ion–surface-atom pair potential. It is possible at keV energies to scatter from one or more atoms in the surface layer, or to penetrate the surface and scatter from deeper layers of the crystal. A particle trajectory is well represented as a series of sequential binary collisions between the incident ion and individual surface atoms. At meV, or 'thermal energies', the particle does not penetrate the corrugated electron density that exists outside the top layer of atoms in the crystal, i.e. the particle scatters from the corrugated potential of the entire surface.

The hyperthermal regime thus lies between two energy ranges which exhibit very different scattering behaviors. One might expect to see evidence of the transition, with decreasing energy, from pair-type scattering to collective interactions of the surface with the incident particles. We will see that, for alkali ions with energies of a few hundred eV and lower, the scattering can be described by simultaneous repulsive interactions of the incident ion with several surface atoms, as well as a long-range attractive interaction with the induced image charge in the metal. Measured energy and angular distributions of the scattered ions will be used to probe these interactions, i.e. the scattering potential. In turn, comparisons of measured distributions with ones calculated using model scattering potentials will be used to deduce detailed information about trajectory types, energy transfer and particle trapping. We will also see that the charge state of the particle will influence the scattered energy and angular distributions, primarily through the presence or absence of the long-range image attraction.

In Sections 5.2.1–5.2.3.3 we will consider scattering in the absence of charge transfer, i.e. ions-in to ions-out. In Section 5.2.3.4 we will see that particles with different charge states, formed via charge-transfer processes, exhibit different scattering dynamics.

5.2.1 EXPERIMENTAL TECHNIQUES

The scattering experiments were performed in an ultra-high vacuum (UHV) chamber equipped with an ion beamline. The chamber (McEachern *et al.*, 1988) and beamline (Adler and Cooper, 1989) are discussed in detail elsewhere. Only a brief summary of the important features are given here.

5.2.1.1 Beams

A schematic of the ion beamline is shown in Figure 5.1. The alkali ions used in these experiments were produced by thermal emission from a hot ($\approx 1100°C$) alkali-embedded tungsten source.[1] The source optics (Peale *et al.*, 1989) were designed to maximize the current passing through the beamline to the sample. In order to do this the phase space occupied by the beam is reduced in the source by extracting less than the space-charge-limited current from the emitter. Once extracted, the ions are mass selected using a Wien filter,[2] and neutral-particle rejection is accomplished with a 90° spherical electrostatic analyzer.

The ions are transported to the chamber using beam optics specifically designed for the few eV to several keV range. Extraction and transport at keV energies are straightforward. A typical beam at 1600 eV has a current of 5.0 nA, a diameter of 2 mm (FWHM) and an angular spread of $\pm 1°$.

Space-charge repulsion during transport at the low end of the range, below 400 eV, results in very low beam currents and a large angular divergence at the

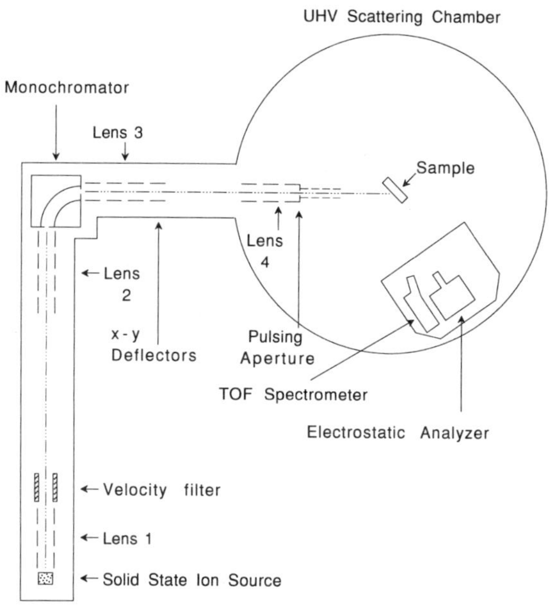

Figure 5.1 Schematic for the ion beamline and UHV scattering chamber. The alkali ions are produced in the solid-state source. The ions are extracted and transported to the scattering chamber with four electrostatic lenses and electrostatic deflectors. Apertures (not shown) are used for beam definition. The beam passes through a 90° spherical electrostatic monochromator which has a resolution of $\Delta E/E = 0.01$. The 180° electrostatic analyzer for detecting scattered ions is located on a rotatable platform. Also located on the rotating platform is the time-of-flight (TOF) spectrometer for detecting scattered alkali ions and neutral atoms. In TOF mode, the beam is pulsed using the x–y deflectors located after the monochromator to sweep the beam past the pulsing aperture located in lens 4

sample. This problem is alleviated by extracting the beam and transporting it through the beamline at 400 eV. The beam is decelerated to the desired incident energy by the final electrostatic lens in the beamline, which is located a few centimeters from the sample (see Figure 5.1). In this way low-energy ion beams can be produced with adequate currents, small angular divergences, and small beam spots at the sample. The beam current, spot size and angular divergence for a typical 15 eV beam are 0.25 nA, 2 mm (FWHM) and $\pm 2.3°$, respectively.

Details of these optics and transport modes can be found elsewhere (Adler and Cooper, 1989; DiRubio, forthcoming b). Differential pumping is used to isolate the higher pressures in the beamline from the ultra-high vacuum in the scattering chamber. The source can be changed easily for producing beams of other species.

5.2.1.2 Scattering chamber

The sample and detectors are located in a two-tiered UHV scattering chamber which has a base pressure in the high 10^{-11} torr range and typical operating pressures of (1–2) $\times$ 10^{-10} torr. The upper tier is used for sample preparation and characterization and includes a sputter gun, an Auger spectrometer, low-energy electron diffraction (LEED) optics, a Kelvin probe, gas and alkali-metal deposition capabilities, and a residual gas analyzer. The lower tier of the scattering chamber contains the scattering plane determined by the ion beam-line and the detector axes.

Figure 5.1 includes a schematic of the lower tier of the UHV scattering chamber. The scattered ions are detected using a 180° hemispherical electrostatic analyzer (ESA) with an electron multiplier, both being mounted on a rotatable platform. The ESA has an angular acceptance of approximately $\pm 1°$ and an energy resolution of $\Delta E/E = 0.02$. In principle, using the rotation of the detector and various rotations and tilts of the sample holder, scattering into any final angle could be measured. All the spectra discussed in this chapter were collected using an 'in-plane' scattering geometry, in which the axes of the incident beam and the detector lie in the same plane as the sample normal. Alignment of the optics as well as alignment procedures done before each experimental run ensure that the relative angular positions of the incident beam axis, the detector axis and the sample normal are known to within 0.5°. Linear alignments are accurate to within 0.25 mm.

5.2.1.3 Sample preparation and monitoring

All the data shown in this chapter were taken from single-crystal copper samples aligned using X-ray diffraction, then cut and mechanically polished to make 2 mm by 8.7 mm disks within 1° of the desired surface orientation. Each sample is initially cleaned by repeating cycles of argon-ion sputtering and annealing until heating the sample to elevated temperatures shows no detectable contamination, as determined by Auger. Before each experiment, and often during the experiments, as needed, shorter sputter-and-anneal cycles are performed to clean the sample. LEED is used to verify that the anneal was sufficient to restore surface order following the sputter cycle. During data collection using the alkali beams, the sample is carefully monitored both for surface damage and for contamination from the incident beam.

5.2.2 SIMULATION TECHNIQUES AND SCATTERING POTENTIALS

As discussed in the introduction to Section 5.2, because we are working at hyperthermal energies, we are concerned with the case where an ion interacts simultaneously with several surface atoms. This makes analytic trajectory calculations impractical and we must rely on computer simulations of the scattering to interpret the measured spectra. The calculations shown here were done with the trajectory calculation code SAFARI (Goodstein *et al.*, 1988), which was written specifically to treat the hyperthermal energy range. A crucial input to the scattering code is, of course, the ion–surface scattering potential.

Several recent studies have been made to probe the alkali ion–metal surface scattering potential at energies of 400 eV and lower (DiRubio *et al.*, forthcoming b; Hulpke and Mann, 1983; Tenner *et al.*, 1984, 1986a, 1986b, 1986c; van den Hoek *et al.*, 1986; Horn *et al.*, 1988; Adler and Cooper, 1991; McEachern *et al.*, 1989; Goodstein *et al.*, 1989; Cooper *et al.*, 1992). Here we will use the case of Na^+ scattering from Cu surfaces as an example. For energies of 100 eV and higher (Adler and Cooper, 1991; McEachern *et al.*, 1989; Goodstein *et al.*, 1989), and for the 10 to 100 eV range we will discuss in this chapter (see also DiRubio *et al.*, forthcoming b; Cooper *et al.*, 1992), it has been demonstrated that Na^+–Cu scattering is well described by a potential that is the sum of two terms: a repulsive term that is the sum of $(Na\text{–}Cu)^+$ pair potentials and an attractive term that represents the interaction of the incident ion with its induced image charge in the metal. The sum runs over the surface atoms nearest to the incident or scattered particle at any point in its trajectory. For all the simulations shown in this chapter, the pair potentials are calculated using the Hartree–Fock code in the quantum-chemistry package Gaussian 88 (DiRubio *et al.*, forthcoming b; Frisch *et al.*, 1988). The attractive term is of the form (DiRubio *et al.*, forthcoming b; Hulpke and Mann, 1983; Tenner *et al.*, 1986c; Adler and Cooper, 1991; Goodstein *et al.*, 1989; Cooper *et al.*, 1992):

$$V_{\text{attr}}(z) = \begin{cases} -e^2/\sqrt{16(z - z_0)^2 + e^4/V_{\min}^2}, & \text{if } z > z_0 \\ -V_{\min}, & \text{if } z < z_0, \end{cases} \tag{1}$$

where z is the perpendicular distance from the top plane of atoms. This form for $V_{\text{attr}}(z)$ reduces to the $1/4z$ classical image potential far from the surface, and saturates near the surface to the value $-V_{\min}$. The only adjustable parameters in the potential are $V_{\min}$ and z_0, where z_0 controls the distance from the surface at which $V_{\text{attr}}(z)$ saturates. In Section 5.2.3.3 we will discuss the sensitivity of the scattering to the choice of these parameters and the pair potentials used in the calculations.

The solid line in Figure 5.2 shows the total potential V, for 10 eV and lower, for Na^+ over an on-top site of the Cu(001) surface with $V_{\min} = 2.6$ eV and $z_0 = 0.8$ Å. Also shown are the repulsive term (dashed line), i.e. sum over pair potentials, and the attractive term (dotted line) from which the potential was constructed.

For a fixed scattering geometry, i.e. a particular choice of incident and final scattering angles, an energy spectrum of the scattered ions shows distinct peaks

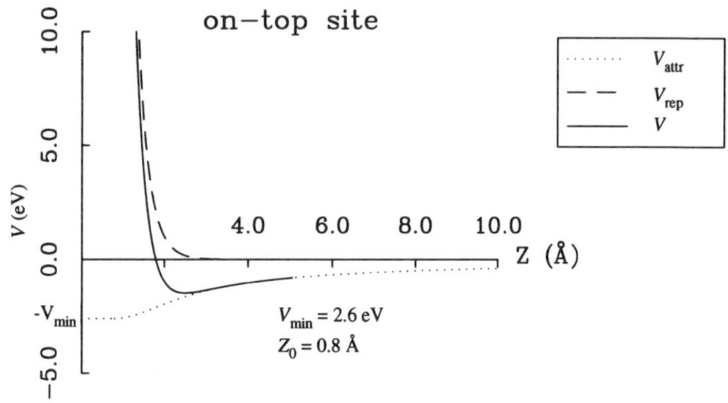

Figure 5.2 Potential energy V as a function of perpendicular distance z for a Na^+ directly above a top-layer Cu atom on the Cu(001) surface. The dashed line shows the repulsive part of the potential V_{rep}, which is the sum over nearest-neighbor surface atoms of Hartree–Fock $(Na–Cu)^+$ pair potentials (see text). The dotted line is the attractive potential $V_{attr}(z)$ with $V_{min} = 2.6$ eV and $z_0 = 0.8$ Å (see text). The plane of top-layer atoms corresponds to $z = 0$. The solid line is the total potential V, obtained by summing V_{rep} and V_{attr}

which correspond to different types of scattering trajectories. These peaks occur at characteristic energies which are determined by the recoil of the surface atoms (see below). In the simulations presented in this chapter, the energy loss is due solely to momentum transfer to the recoiling Cu atoms, with no inclusion of energy losses to electronic excitations. This is justified since the losses to electronic excitations have been generally estimated to be less than a few tenths of an eV for incident energy $E_0 \leq 100$ eV (DiRubio *et al.*, forthcoming b; Goodstein *et al.*, 1989; Brako and Newns, 1989; Inglesfield, 1983; Echenique *et al.*, 1986; Sols *et al.*, 1985; Marston *et al.*, 1993). Thermal vibrations can also be included in the simulations. Comparisons of simulations from a surface with no thermal vibrations and one including thermal vibrations indicate that, when thermal motions are included, the peaks in the energy spectra are broadened, but their positions do not change significantly (DiRubio *et al.*, forthcoming b). In the 10 to 100 eV range this broadening is due primarily to the momentum fluctuations of the surface atoms (DiRubio *et al.*, forthcoming a, b; Gerlach–Meyer *et al.*, 1979; McEachern, 1990; Burke *et al.*, 1991).

5.2.3 ENERGY AND ANGULAR DISTRIBUTIONS: MEASUREMENTS AND SIMULATIONS

Simulations have been used in a number of instances to probe scattering potentials and to gain a deeper insight into the dynamics of scattering at different energies. Here we have calculated energy and angular distributions of the scattered ions using different choices for the scattering potential; these are compared with measured distributions.

5.2.3.1 Energy spectra and trajectories

In this section we will examine in some detail the case of 10 to 100 eV Na^+ scattering from Cu(001). A schematic of this surface is shown in Figure 5.3(a). In all cases the beam is incident along the ⟨100⟩ azimuth. All the spectra shown here were collected using an in-plane scattering geometry in which the incident beam, the sample normal and the detector all lie in the same plane

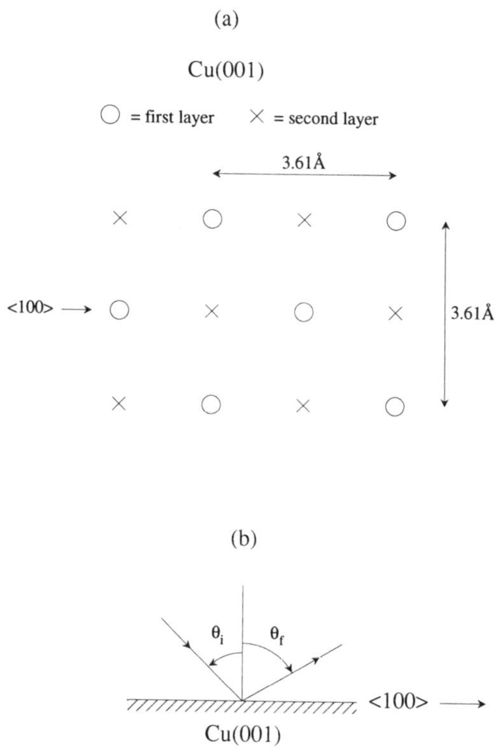

Figure 5.3 (a) Schematic top view of the first and second layers of the Cu(001) surface. The scattering was done along the ⟨100⟩ azimuth. The spacing between the first and second layer planes of atoms is 1.805 Å. (b) The scattering geometry used in these experiments. The ions were incident from the left at an angle θ_i measured from the surface normal. The detector is located at angle θ_f, also measured from the surface normal. The incident beam direction, the axes of the detector and the sample normal all lie in the same plane. This is referred to as an 'in-plane' scattering geometry

(see Figure 5.3(b)). The angle of incidence θ_i and the angle of detection θ_f are both measured with respect to the surface normal.

The solid line in Figure 5.4 shows a scattered-ion-energy spectrum for 24.5 eV Na^+ with $\theta_i = \theta_f = 45°$. The scattered intensity is plotted as a function of the reduced energy E/E_0, where E is the scattered-ion energy and E_0 is the incident energy. There are three distinct peaks in the measured spectrum. The dashed line in Figure 5.4 shows the corresponding energy spectrum calculated using SAFARI. Note that both the positions and the relative intensities of the three peaks are well reproduced by the simulation. The scattering potential used in these simulations is the same as that used in Figure 5.2 (shown for $V \leq 10$ eV).

By using simulations, we can analyze the particle trajectories which contribute to the peaks in Figure 5.4; in this case, four types of trajectories are involved. Side and top views of these trajectories are shown in Figure 5.5. There are two trajectory types that correspond to scattering along a $\langle 100 \rangle$ chain of atoms; they are the quasi-single (QS) and quasi-double (QD) trajectories, labeled 1 and 2, respectively, in Figure 5.5. In QS scattering, momentum transfer occurs to primarily one surface atom even though the ion interacts simultaneously with more than one atom at any given point in its trajectory. In QD scattering, momentum transfer occurs to primarily two neighboring atoms along the chain. There are two additional trajectory types, the double zigzag (DZZ) and the triple zigzag (TZZ), labeled 3 and 4, respectively, where primary

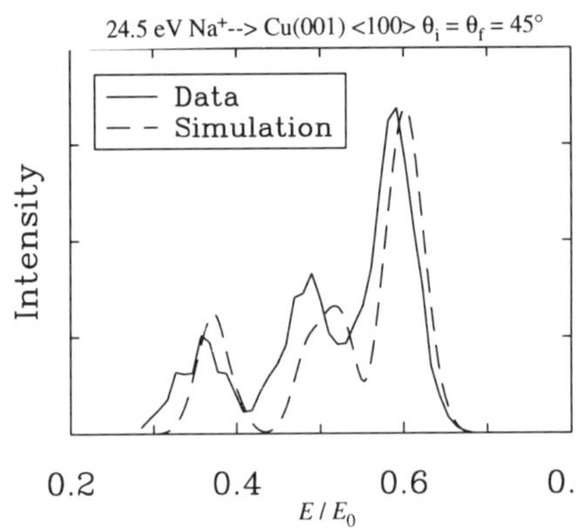

Figure 5.4 Energy spectra of 24.5 eV Na^+ scattered from Cu(001) along the $\langle 100 \rangle$ azimuth with $\theta_i = \theta_f = 45°$. The solid line is the measured spectrum. The scattering intensity is plotted as a function of the reduced energy E/E_0, where E is the energy of the scattered ions and E_0 is the incident beam energy. The sample temperature was 152 K. The dashed line is a SAFARI simulation using the potential described in Section 5.2.2 and shown in Figure 5.2 with $V_{min} = 2.6$ eV and $z_0 = 0.8$ Å, and including thermal vibrations. Note that the positions and relative intensities of the three peaks in the measured spectrum are well-reproduced by the simulation. Different types of scattering trajectories contribute to the peaks seen in this spectrum (see text and Figure 5.5)

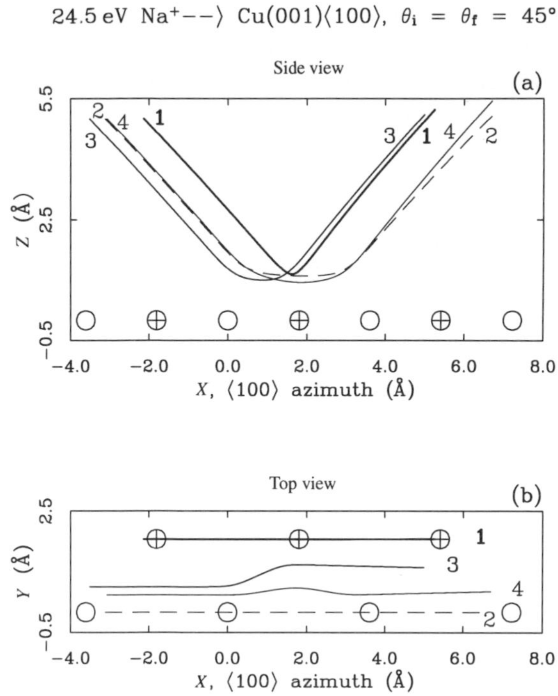

Figure 5.5 (a) Side view of trajectories for 24.5 eV Na^+ scattered from Cu(001) along the ⟨100⟩ azimuth with $\theta_i = \theta_f = 45°$. Trajectories 1 and 2 correspond to scattering along a top-layer ⟨100⟩ chain of atoms; trajectory 1 is a quasi-single (QS) collision where momentum transfer is to primarily one atom, and trajectory 2 is a quasi-double (QD) with momentum transfer to primarily two atoms. Trajectories 3 and 4 are zigzag trajectories involving atoms in two neighboring ⟨100⟩ chains of atoms; trajectory 3 is a double zigzag (DZZ) and trajectory 4 is a triple zigzag (TZZ), with momentum transfer to primarily two and three atoms, respectively. (b) A top view of the four trajectory types

momentum transfer occurs to atoms in neighboring chains, say chain A and chain B. The DZZ involves two primary collisions, one with an atom in chain A and one with an atom in chain B. The TZZ involves three primary collisions involving a sequence of collisions with atoms in chains A, B and A. In this case, by coincidence, the QD and TZZ trajectories are approximately degenerate in energy. Although energy transfer actually involves simultaneous interactions with several surface atoms and with the conduction electrons, the relative energies of these four trajectory types can be estimated from simple considerations of the collision sequences (DiRubio *et al.*, forthcoming b; McEachern *et al.*, 1989).

In general, the number of trajectory types and their energies are highly system-dependent, and vary with the incident ion species, surface structure and species, crystal azimuth, incident energy and scattering geometry. Because these trajectories involve collisions with several surface atoms, there is a region of the surface (within a unit cell) that corresponds to ion impact parameters that will give the same basic trajectory type and corresponding energy loss. The relative intensities in the final spectrum resulting from different trajectory types can be estimated using an impact-parameter plot (for example, see Adler and Cooper, 1991; McEachern *et al.*, 1989). Such a plot shows impact parameters for all trajectories that scatter into the detector. The larger the region for a given trajectory type, the larger is its relative contribution to the final spectrum. In many cases, impact-parameter plots are calculated for surfaces where no thermal motions of the atoms are included. Some caution is therefore required in interpreting such a plot since thermal vibrations may alter the impact-parameter plot significantly (Adler and Cooper, 1991).

For incident energies from 10 to 100 eV, we have performed trajectory and impact-parameter analyses for Na^+ incident along the $\langle 100 \rangle$ azimuth of Cu(001) for $\theta_i = 45°$ and $\theta_f = 45°$. In all cases, the same four trajectory types as those shown in Figure 5.5 were found to dominate the spectra. At some energies other types of trajectories appear. For example, simulations at 50 eV show trajectories with large energy losses that correspond to scattering from first- and second-layer atoms or ones involving multiple large-angle collisions with first-layer atoms.

5.2.3.2 Energy–theta plots and trapping

While it is very useful to examine individual energy spectra and trajectories, important additional information regarding the scattering dynamics is contained in the angular distributions of the scattered particles. We have measured energy spectra at a number of different final angles θ_f for 30.4 eV Na^+ incident along the $\langle 100 \rangle$ azimuth of Cu(001), with $\theta_i = 45°$. Only scattering angles of $\theta_f \geq 15°$ were used. From these data we can construct what is called an 'energy–theta' plot, shown by the open circles in Figure 5.6(a), in which the positions of the different peaks in the energy spectra are recorded as a function of the angle of detection θ_f (Boers, 1977; DiRubio *et al.*, forthcoming b; Adler and Cooper, 1991; Cooper *et al.*, 1992; Yurasova *et al.*, 1968; Taglauer and Heiland, 1972; de Wit *et al.*, 1975). In order to minimize thermal vibrations of the surface atoms, which can broaden the peaks in the energy spectra, the data shown in Figure 5.6 were collected with sample temperatures T_s of $134 \leq T_s \leq 139$ K.

As discussed above, the positions of the different peaks in the energy spectra correspond to different types of scattering trajectories. Note that the energy–

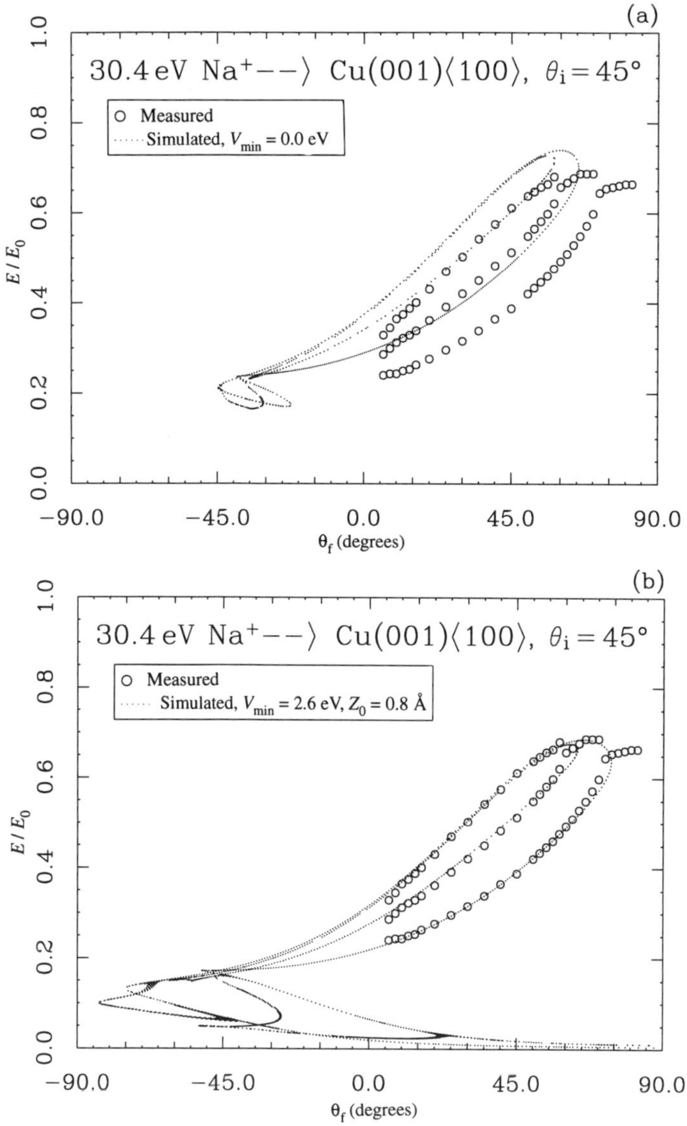

Figure 5.6 (a) A comparison of measured (open circles) and simulated (dots) energy–theta plots for 30.4 eV Na^+ incident at 45° on Cu(001) along the ⟨100⟩ azimuth. In the simulation only the repulsive component (sum of Hartree–Fock pair potentials, see text and Figure 5.2) was used to generate the plot. The simulation underestimates both the energy loss and the forward rainbow angle. (b) Same as (a) but with the attractive well included in the simulation. The parameters used to model the attractive well (see text and Figure 5.2) were $V_{min} = 2.6$ eV and $z_0 = 0.8$ Å. When the attractive well is included the simulated peak energies and forward rainbow angles are in excellent agreement with the data

theta plot in Figure 5.6(a) has three distinct branches, referred to hereafter as the *lower*, *middle* and *upper branches*, which form portions of two loops (DiRubio *et al.*, forthcoming b; Cooper *et al.*, 1992). From trajectory analysis, like that shown in Figure 5.5, we know that the larger of the two loops (i.e. lower and upper branches) corresponds to QS (lower branch) and QD (upper branch) scattering along ⟨100⟩ chains of atoms on the surface. We will refer to this as the 'chain loop'. The smaller of the two loops (i.e. the middle and upper branches) corresponds to DZZ (middle branch) and TZZ (upper branch, degenerate with the chain QD scattering) trajectory types.

The 'forward rainbow angle' is identified as the angle at which two branches of a loop merge (Boers, 1977; Yurasova *et al.*, 1968; Taglauer and Heiland, 1972; Parilis and Turaev, 1965; Horn, 1988). There are two forward rainbows in the data shown in Figure 5.6(a), one for the chain loop (at $\theta_f = 72 \pm 2°$) and one for the zigzag loop (at $\theta_f = 60 \pm 2°$). Some low-intensity scattering is apparent for $\theta_f > 72°$ (DiRubio *et al.*, forthcoming b), which is presumably due to thermal vibrations of the surface atoms or to surface defects, since no scattering is expected beyond the forward rainbow for a perfectly ordered surface with no defects (Gerlach-Meyer *et al.*, 1979; Poelsema *et al.*, 1976a, 1976b, 1977a, 1977b).

The measured energy–theta plot can be compared to ones generated from simulations. In Figure 5.6(a) the small dots show an energy–theta plot calculated for a surface with no thermal vibrations (DiRubio *et al.*, forthcoming b; Adler and Cooper, 1991; Cooper *et al.*, 1992) and using the potential described in Section 5.2.2 with $V_{min} = 0.0$ eV (i.e. no attractive well). The same data are shown in Figure 5.6(b) along with an energy–theta plot calculated using $V_{min} = 2.6$ eV and $z_0 = 0.8$ Å. These parameters were found to give the best fit to data for a range of incident energies (DiRubio *et al.*, forthcoming b; Cooper *et al.*, 1992). However, many features of the data are relatively insensitive to the specific parameters chosen (within a reasonable range, see below). From Figures 5.6(a) and 5.6(b) it is clear that the attractive well is important in determining both the energies and angles of the scattered ions.

Figures 5.7(a) and 5.7(b) show data and simulations, using $V_{min} = 2.6$ eV and $z_0 = 0.8$ Å for 14.9 and 50.0 eV, respectively. In both cases the Na^+ beam was incident along the ⟨100⟩ azimuth of Cu(001) with $\theta_i = 45°$. Note that for 14.9 eV, the two branches of the chain loop extend all the way to 90°, i.e. parallel to the surface, which corresponds to the onset of trapping for forward-scattered in-plane ions. Not surprisingly, this trapping behavior is even more pronounced at 10 eV (DiRubio *et al.*, forthcoming b). Also note that, as the incident energy decreases, one sees a corresponding increase in the fractional energy loss to the surface, or a decrease in the reduced energy E/E_0 (an effect specific to the scattering of ions). Both effects are consequences of the increasing importance of the attractive well at the lower beam energies (DiRubio *et al.*, forthcoming b; McEachern, 1990; Hulpke, 1975; Kasi *et al.*, 1988).

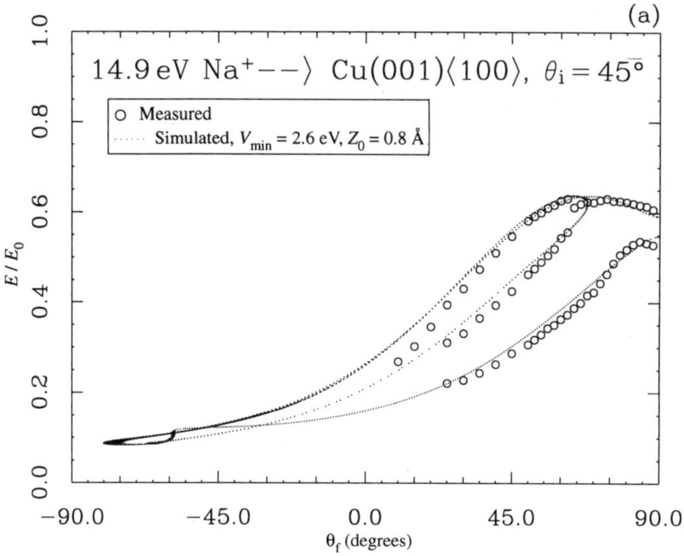

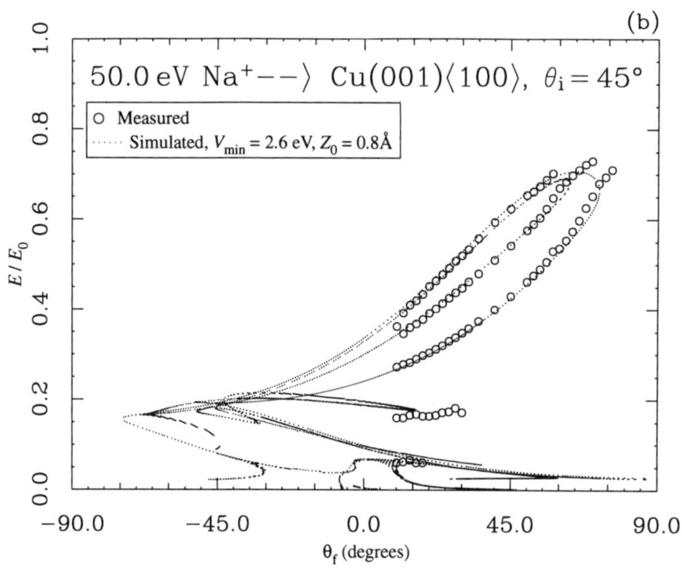

Figure 5.7 (a) A comparison of measured (open circles) and simulated (dots) energy–theta plots for 14.9 eV Na^+ incident at 45° on Cu(001) along the ⟨100⟩ azimuth. The same potential as in Figure 5.6(b) is used in the simulation. Note that the simulation reproduces the gap at $\theta_f = 90°$ in the measured energy–theta plot. The gap is due to trapping in the attractive well (see text). (b) Same as (a), only for 50.0 eV Na^+

5.2.3.3 Discussion

In order to understand some of the trends seen in the data and simulations in Figures 5.6 and 5.7, it is again instructive to examine scattering trajectories. Figure 5.8(a) and 5.8(b) show side views of simulated trajectories for 10 eV Na^+ scattering along the ⟨100⟩ azimuth of Cu(001) for $\theta_i = 45°$ (the ions are incident from the left). The impact parameters were chosen to be evenly spaced along a line directly over a top-layer ⟨100⟩ chain of atoms and spanning one unit cell. In Figure 5.8(a) no attractive well is included in the simulation,

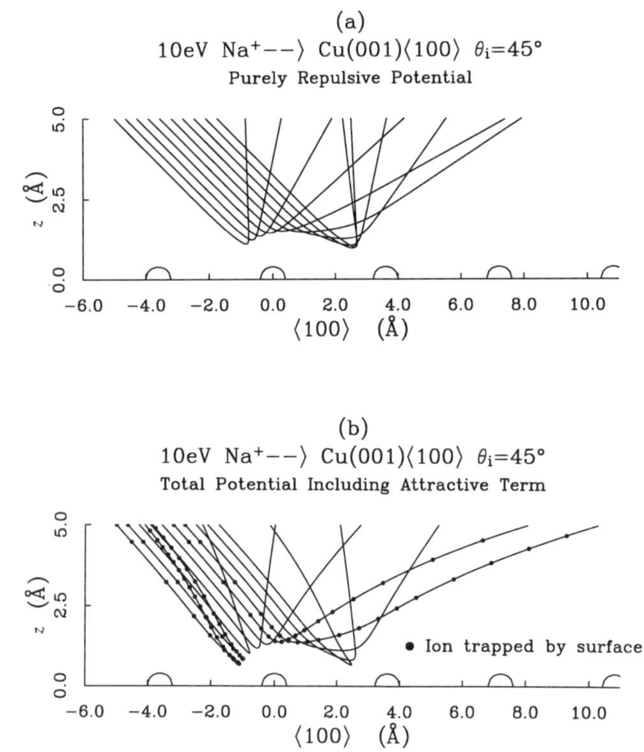

Figure 5.8 (a) A series of simulated trajectories for 10 eV Na^+ scattering from Cu(001) along the ⟨100⟩ azimuth in which the impact parameters were confined to lie along a top-layer chain of atoms parallel to the ⟨100⟩ azimuth. Only the repulsive component (sum over Hartree–Fock pair potentials, see text and Figure 5.2) of the potential was used in the simulation. The final angles of the scattered ions ranged from $-7°$ to $63°$. (b) A series of 10 eV trajectories with the same impact parameters as in (a), but using the full potential, including the attractive well (see Figure 5.2). Here $V_{min} = 3.0$ eV and $z_0 = 1.7$ Å. The attractive interaction extends the range of final scattering angles from $-90°$ to $90°$. The trajectories indicated by the dots correspond to ions that lose so much energy in the collision(s) with the surface atoms that they are unable to escape the attractive well

and in Figure 5.8(b) $V_{min} = 3.0$ eV and $z_0 = 1.7$ Å. The specific choices of V_{min} and z_0 are not important here. The dotted trajectories in Figure 5.8(b) correspond to particles which initially trap on the surface: while leaving the surface, the z-momentum of each particle decreases and then becomes negative at times longer than those shown in the figure. Whether or not a particle remains trapped on the surface depends on subsequent collisions and energy transfer to the surface atoms (see below).

It is immediately obvious from these simulations that the presence of the attractive potential strongly influences the angular distributions of the scattered ions; the z-dependent attractive force from this potential bends the ions toward the surface on both the incoming and outgoing parts of the trajectory (DiRubio *et al.*, forthcoming b; McEachern *et al.*, 1989; Cooper *et al.*, 1992; Hulpke, 1975; Kasi *et al.*, 1988). With no attractive well (see Figure 5.8(a)) the particles scatter into a finite range of angles, whereas the ions scatter into all possible in-plane angles when the full potential is used in the simulation (see Figure 5.8(b).

The attractive well also influences the scattered energies. We have already noted that, in general, the fractional energy loss to the surface increases with decreasing incident energy. This is a result of two mechanisms. First, on the incoming trajectory the ion is accelerated toward the surface, with the result that the collision(s) with the surface atom(s) occur at a higher energy than if the well were absent, giving a larger energy loss. Second, the total scattering angle of the ion in the near-surface region is larger, due to the trajectory bending, than the apparent scattering angle measured by the experiment. An increased scattering angle also increases energy transfer to the surface. Both of these effects become more pronounced as the incident energy decreases and the attractive well becomes a correspondingly larger fraction of the total potential.

It is also clear from Figures 5.6 and 5.7 that trapping in the attractive well depends sensitively on energy (DiRubio *et al.*, forthcoming b; Cooper *et al.*, 1992). Different trajectory types will have different trapping thresholds. For Na^+ incident on Cu(001) along the $\langle 100 \rangle$ azimuth with $\theta_i = 45°$, the onset of trapping for in-plane scattering near the forward rainbow angle occurs between 15 and 20 eV. Here again, simulations can be used to gain insight into the regions of the unit cell that correspond to impact parameters that lead to trapping (Horn *et al.*, 1988; McEachern, 1990).

In our analysis, the criterion we use to identify 'trapped trajectories' is relatively simple. We follow the trajectory through its first collision with the surface and ask whether the perpendicular component of the momentum of the scattered particle is sufficient to escape the attractive well; if it is not, then the particle is trapped. When the simulation is followed for times larger than those shown in Figure 5.8, the particle travels several lattice spacings across the surface and undergoes multiple collisions with the surface atoms; the atom 'bounces' along the surface (DiRubio *et al.*, forthcoming b; Snowdon *et al.*,

1988; Dodson, 1987). To accurately reproduce these multiple bounces and to calculate the probability that the particle will remain trapped on the surface, a more complete description of the scattering dynamics is required than is provided here and should include thermal vibrations, inelastic energy losses and charge transfer (DiRubio *et al.*, forthcoming b). One might also expect surface defects to play a role in determining actual trapping probabilities (DiRubio *et al.*, forthcoming b; Snowdon *et al.*, 1988).

We conclude this section with some comments about scattering potentials. We have used SAFARI to investigate the sensitivity of the scattering to changes in the potential. At the lower energies, i.e. $E_0 \leq 100$ eV, we have investigated the sensitivity of the scattering to the attractive-well parameters, V_{min} and z_0 (DiRubio *et al.*, forthcoming b). At higher energies, where $100 < E_0 < 400$ eV, we performed similar studies of the repulsive part of the potential. Our conclusions are summarized here.

The 'best-fit' attractive-well parameters $V_{\mathrm{min}} = 2.6$ eV and $z_0 = 0.8$ Å (Figures 5.4, 5.6 and 5.7) were found to give excellent agreement with the data over a wide range of incident energies ($10 < E_0 < 100$ eV) for Na^+ scattering from Cu(001). These parameters give a well depth of 1.5 eV for the on-top site (see Figure 5.2) and 1.7 eV for the hollow site. However, there are uncertainties of ± 0.5 eV in determining the scattered energies (due primarily to unknown contact potentials) and $\pm 2°$ to 3° in determining experimental rainbow angles (due primarily to the difficulty in identifying rainbow angles from thermally broadened energy and angular spectra). A surprisingly wide range of parameterizations, with well depths in the on-top site ranging from approximately 1.4 eV to 2.6 eV, have been found to give simulated scattering distributions that fall within the experimental uncertainties. An explanation for this, in terms of the influence of the attractive potential on the detailed dynamics of the scattering in both the near ($1 \leq z \leq 3$ Å) and far ($z \geq 3$ Å) surface regions is being developed (DiRubio *et al.*, forthcoming b). The attractive potential we have used here (see equation (1)) has no lateral corrugation. In reality, due to the varying electron density of the surface layer, one expects that the image-like ion–surface attraction will be laterally corrugated in the near-surface region. Such a potential can be approximated by a sum of pair potentials (Mann *et al.*, 1987). We have not used such a potential since the simpler z-dependent potential gives good agreement with the data.

We have also examined the sensitivity of the simulated energy and angular distributions to changes in the repulsive part of the potential. At higher energies, i.e. $E_0 > 100$ eV, the scattering potential is dominated by the short-range repulsive interaction between the incident ion and the surface atoms. As discussed above and elsewhere (see, for example, van den Hoek *et al.*, 1986; Goodstein *et al.*, 1989), this repulsive potential represents the interaction of the incident ion with several surface atoms simultaneously, although momentum transfer may be primarily to one or a few surface atoms.

We have found, for Na^+ and K^+ incident on Cu(110) at energies of 100 to 400 eV, that the simulated energy and angular distributions are very sensitive to the choice of pair potential used to construct the total potential. Excellent agreement with measured energy and angular distributions was obtained using Hartree–Fock $(Na–Cu)^+$ pair potentials and significantly poorer agreement was obtained with other forms of pair potentials (Adler and Cooper, 1991; McEachern *et al.*, 1989; Goodstein *et al.*, 1989). Based on a detailed trajectory analysis, this was attributed to the effect on the surface corrugation from changing the pair potential. For scattering along high-symmetry directions (e.g. $\langle 001 \rangle$ and $\langle 1\bar{1}0 \rangle$) of an open surface like Cu(110), the intensity is dominated by second-layer scattering. Ions with a relatively large range of impact parameters are focused onto the second-layer chains by lateral collisions with the tails of the potentials from top-layer atoms. This focusing is particularly sensitive to the corrugation obtained by summing pair potentials for the top-layer (and second-layer) atoms (Adler and Cooper, 1991; Goodstein *et al.*, 1989). In particular, using a pair potential that is more repulsive than the Hartree–Fock at distances corresponding to half the spacing between top-layer chains gave relatively poor agreement with the data. Physically reasonable changes in the attractive-well parameters made relatively little difference in the simulated distributions. This is not surprising since the attractive well is a small fraction of the total potential for incident energies $E_0 \geq 100$ eV. Kleyn and co-workers have found that sums of Hartree–Fock–Slater pair potentials (van den Hoek *et al.*, 1986; Horn *et al.*, 1988) give better agreement with the data than other pair potentials (Tenner *et al.*, 1984, 1986a, 1986b, 1986c) for hyperthermal-energy K^+ scattering from W(110) (Tenner *et al.*, 1984, 1986a, 1986b, 1986c; van den Hoek *et al.*, 1986) and Na^+ scattering from Ag(111) (Horn *et al.*, 1988).

Another concern about the scattering model we have used was whether the sum-of-pair potentials approximation would be adequate to describe the scattering. It has been reported for 10 to 100 eV K^+ scattering from W(110) that the ion–surface repulsion in the hollow site is underestimated using a sum of pair potentials (Tenner *et al.*, 1984; van den Hoek *et al.*, 1986). For the Na^+–Cu(001) system we have compared potentials calculated by summing $(Na–Cu)^+$ pair potentials with cluster calculations (performed using Gaussian 88) (DiRubio *et al.*, forthcoming b). Excellent agreement is found between these two potentials. Furthermore, the 'best fit' potential described above gives good agreement between a measured and a simulated energy–theta loop for 50.0 eV Na^+incident at $\theta_i = 45°$ along the $\langle 100 \rangle$ azimuth of Cu(001) (see Figure 5.7(b)). The scattering at $10° < \theta_f < 30°$ and $E/E_0 < 0.2$ are due to trajectories that probe the hollow site (DiRubio *et al.*, forthcoming b). In particular, the feature at $E/E_0 \approx 0.07$ is due to trajectories which undergo large total scattering-angle collisions with first-layer atoms and the second-layer atom in the hollow site. These features are qualitatively reproduced by the simulation.

When the energies are high enough, it is possible to get embedding of the incident ion into the crystal. Using an impact-parameter analysis similar to that discussed above for trapping, one can identify regions of the surface unit cell that correspond to impact parameters for trajectories that penetrate the surface (Horn *et al.*, 1988; McEachern, 1990). The actual fate of the particle, however, must be followed for longer times with a calculation that properly treats the surface-atom dynamics. As with trapping, embedding behaviour is highly system-dependent.

5.2.3.4 Effect of charge state on dynamics

It is clear from the energy–theta loops shown in Figures 5.6(a) and 5.6(b), and the trajectory calculations shown in Figures 5.8(a) and 5.8(b), that ionized and neutral particles will exhibit different scattering dynamics in the hyperthermal-energy range. To first order this is due to the presence or lack of the long-ranged image attraction, which depends on the charge state of the scattered particle. Thus, ion–surface charge-transfer processes that change the charge state of the particle may be important in determining scattered energy and angular distributions, and trapping thresholds. For example, if an alkali ion scattered from the Cu(001) surface neutralizes on the outgoing trajectory, one would expect it to emerge from the surface with a larger kinetic energy and at a final angle closer to the surface normal than if the neutralization had not occurred.

It has been known for many years that charge transfer between a particle and a surface can occur, e.g. for alkalis leaving a metal surface (Taylor and Langmuir, 1933). We will show in Section 5.3, for Na^+ and Li^+ scattering from clean Cu(001), that one observes both positive ions and neutral atoms in the scattered flux. Figure 5.9 shows ion (solid line) and neutral-atom (dashed line) energy spectra resulting from the scattering of 23.8 eV Li^+ incident along the $\langle 100 \rangle$ azimuth of Cu(001) with $\theta_i = \theta_f = 55°$. Note that, as expected, the ions lose more energy in the collisions than the neutral atoms. In Figure 5.9 the ion and neutral-atom peak intensities are arbitrarily normalized.

The measurements shown in Figure 5.9 were made with a time-of-flight spectrometer specifically designed for detecting hyperthermal-energy neutral and charged alkalis. This detector will be described briefly below in Section 5.3.1.1. In this spectrometer, there are experimental artefacts like contact potential differences that may influence these measurements. Although these effects have not yet been calibrated, they are expected to be relatively small. Hence, while the data in Figure 5.9 are preliminary, the differences between the ion and neutral-atom distributions are clearly visible. We turn now to a consideration of charge-transfer mechanisms for alkali scattering from metal surfaces.

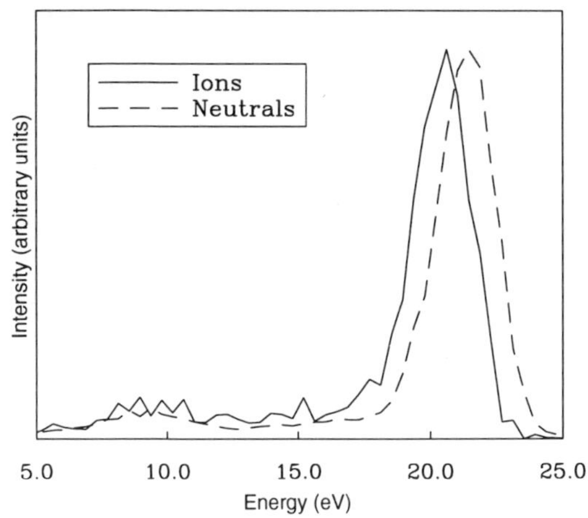

Figure 5.9 Measured ion and neutral energy spectra for 23.8 eV Li^+ scattered from Cu(001) along the $\langle 100 \rangle$ azimuth with $\theta_i = \theta_f = 55°$. The intensities are arbitrarily normalized. The peak in the ion spectrum appears at a lower energy than that in the neutral spectrum. This is due to the image potential which affects the ions but not the neutral atoms. The neutral flux is $\sim 60\%$ of the total flux

5.3 CHARGE TRANSFER

Charge transfer, or the exchange of electrons, plays an important role in many dynamical processes involving the interactions of atoms and molecules with surfaces. We have already seen in Section 5.2.3.4 that when such a charge transfer occurs, it may influence the scattering dynamics through the dependence of the potential on the charge state of the particle. Here we will restrict ourselves to a discussion of the mechanisms and dynamics of charge transfer between an energetic alkali atom and a clean or adsorbate-covered metal surface.

As discussed in the introduction, the alkalis provide an excellent opportunity to study these processes in detail. This is partly an experimental issue since, for the alkalis, it is possible to measure branching ratios to different final electronic states of the scattered particles. Furthermore, in recent years, considerable attention has been given to developing models of alkali-atom interactions with surfaces, including the charge-transfer behavior (for a recent review, see Brako and Newns, 1989), making the alkalis ideal test cases for comparison with theory. Over the past decade a large body of experimental work has amassed concerning the charge-transfer behavior of the alkali-ion surface systems and has been reviewed elsewhere (Geerlings and Los, 1990). In this chapter we will summarize some recent experiments from our laboratory which have focused on measurements of final-state branching ratios. Detailed comparisons of these data with theories of charge transfer are made.

Charge transfer in the few eV to keV range poses an interesting theoretical problem. For alkalis scattering from metal surfaces, an important charge-transfer mechanism involves the resonant transfer of electrons between the atomic level and metal states of the same energy. The relative energies of the atomic and metallic states therefore determine which transfers are energetically allowed. Hence, changes in the alkali species or surface work function may be expected to have important effects on the charge-transfer behavior. In a scattering experiment, complexities arise due to the finite velocities of the atoms and the hybridization of the atomic and metallic states. Some important effects to consider are given here. First, the energies of the atomic states vary with separation from the surface so that a given transition may be energetically allowed on some parts of the trajectory but not on others. Second, the hybridization broadens the atomic states into resonances, i.e. their lifetimes decrease with decreasing separation. Since the atomic-resonance lifetimes vary along the trajectory it becomes important to consider two time scales in the problem. One is the lifetime of the atomic resonance, which is a strongly varying function of particle–surface separation. The other is the time scale of the motion, which depends on particle velocity, and determines how much time is available for the transfer to occur. The energy range of few eV to keV represents a change in the time scale of the motion that is sufficiently large to cause significant changes in the charge-transfer probabilities.

For Li^+, Na^+ and K^+ incident on clean Cu(001) with incident energies ranging from 5 to 1600 eV, we have measured branching ratios for forming different final charge states of the scattered particles. We will briefly describe how these measurements are made and discuss the results in the context of a model that treats charge transfer to one atomic resonance, in this case the outermost valence level of the alkali. This model will be referred to as the one-electron picture of resonant charge transfer. We will also see that, through the perpendicular velocity dependence of the neutralization probability, it is possible to get information about the energies and lifetimes of the alkali resonances. In addition, we briefly discuss the dependence of the neutralization probability on the parallel velocity component of the scattered particle.

We will also present measurements made for Li^+ scattering from surfaces with different work functions, prepared by depositing alkali adsorbates on Cu(001). In this case, several final electronic states are populated in different ratios, depending on the work function and particle velocity. We will describe the experimental technique as well as a multi-state model of the electron transfer. These results will also be examined to see what they can tell us about the energies and lifetimes of the atomic resonances involved in the charge transfer, as well as the occupancies of the different resonances along a given outgoing trajectory.

5.3.1 EXPERIMENTAL TECHNIQUE

In these experiments we have measured yields of different final electronic states of the scattered alkalis. Two custom-built detectors were required for these experiments. One is for measuring velocity- and angle-resolved yields of neutral or charged (positive and negative) alkalis (see Section 5.3.1.1). The other is for detecting photons emitted from de-excitation of excited-state alkalis produced in the collisions (see Section 5.3.4.1).

5.3.1.1 Neutral detection

Quantitative detection of hyperthermal-energy neutral atoms is a difficult experimental problem. On the one hand, the velocities are sufficiently low that impact-induced secondary electron production (used in standard detectors such as electron multipliers) occurs with a low probability which varies with particle velocity in the hyperthermal-energy range. On the other hand, the velocities are high enough that secondary ionization of the neutral atoms by electron bombardment, which works well at thermal energies, is also a low-efficiency process at hyperthermal energies. Other techniques, such as those based on exciting or ionizing the particles with laser light, can be expensive to implement.

We have adapted a previously used technique for detecting low-energy alkalis (van Amersfoort *et al.*, 1985), which makes use of the fact that, regardless of their charge states, alkalis incident on high-work-function surfaces will scatter with high probability as positive ions. In our case the high-work-function surface is a Pt foil ($\Phi = 5.65$ eV) which can be cleaned by heating it to high temperatures in ultra-high vacuum. The main elements of the alkali spectrometer include a pair of biasing elements, a set of collimation apertures, the Pt foil, and a channel electron multiplier.

In our scattering experiments, when a beam is incident on the Cu sample, there are both ions (for the sake of argument, let us assume positive ions) and neutral atoms in the scattered flux. The basic operation of this spectrometer is as follows. The scattered particles (ions and neutral atoms) pass through the collimation apertures of the spectrometer, and then scatter off the clean Pt foil. At the Pt foil they are efficiently ionized. The resulting ions can then be accelerated and detected in a standard electron multiplier.

The angle-resolved neutral fraction, i.e. the neutralization probability, of the scattered atoms is then determined by the following technique. By biasing the rejection elements at the entrance of the spectrometer, only neutral atoms are transmitted to the Pt foil and counted. The total flux (ions plus neutral atoms) is counted by grounding the rejection elements. The neutral fraction is then given by the ratio of the neutral flux to the total flux. The detector sits on the rotating platform next to the electrostatic analyzer (see Figure 5.1). The angular acceptance of the spectrometer, which is determined by the collimation aperture, is 1.5° (full angle) for the data shown in this chapter. The angle-resolved neutralization probability is determined by repeating the measurement described above at a series of final angles.

As we will see, important information is contained in the velocity dependence of the neutralization probability. We use standard time-of-flight (TOF) techniques to make velocity-resolved measurements of the neutralization probability. The incident beam is pulsed using a set of electrostatic deflection plates located in the beamline (see Figure 5.1). Using standard time-to-amplitude conversion techniques, the times of arrival of the scattered particles are recorded.

Hereafter we will refer to this detector as the TOF spectrometer. Its operation relies on the fact that both ions and neutral atoms are converted to ions with the same efficiency after scattering from the Pt foil, i.e. charge transfer at the Pt foil occurs with loss of memory of the particle charge state incident on the Pt foil. This assumption is well justified for hyperthermal-energy alkalis incident on clean Pt. This detector and its operation are discussed in detail elsewhere (Kimmel and Cooper, 1993a).

5.3.2 ONE-ELECTRON TRANSFER

In this section we present measured neutralization probabilities for different alkali species scattered from clean metal surfaces and discuss these results in the context of a one-electron model of resonant charge transfer.

5.3.2.1 Li, Na, K scattering from clean Cu(001)

Using the TOF spectrometer described in Section 5.3.1.1 we have measured velocity- and angle-resolved neutralization probabilities for Li, Na and K scattered from clean Cu(001) (Kimmel and Cooper, 1993b). Figure 5.10 shows an example of how such data are obtained. The solid curve is a time-of-arrival spectrum of the total flux (ions plus neutral atoms) of Na particles resulting from scattering of 400 eV Na^+ along the $\langle 100 \rangle$ azimuth of Cu(001), with $\theta_i = \theta_f = 45°$. The dashed line shows the neutral flux only (see Section 5.3.1.1). The two spectra are arbitrarily normalized; in this case the neutral flux represents approximately 5% of the total scattered flux. The three distinct peaks in the measured spectra correspond to the four different types of scattering trajectories discussed in Section 5.2.3.1 and shown in Figure 5.5.

We have found that the differences in the neutralization probabilities for the different trajectory types are small compared to the changes in the neutralization probability when the incident energy is varied (Kimmel and Cooper, 1993b; Kimmel, 1992). Therefore, to obtain the neutralization probability from the data, we integrate the intensity in the measured spectra, say from point A

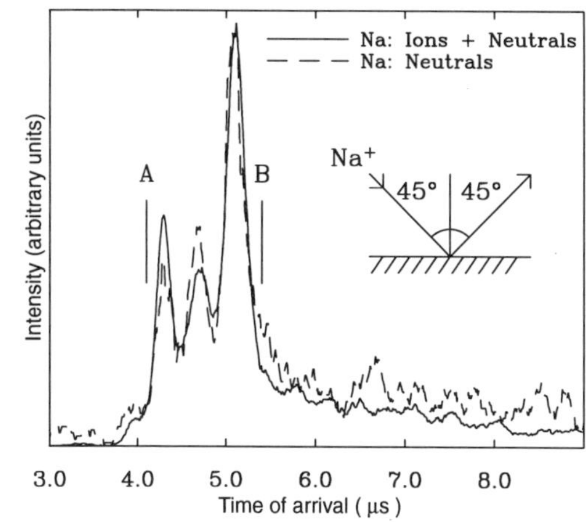

Figure 5.10 Time-of-flight spectra of scattered Na particles for 400 eV Na^+ incident along the $\langle 100 \rangle$ azimuth with $\theta_i = \theta_f = 45°$. The intensity versus time-of-arrival spectra for the neutral flux and total flux (ions plus neutral atoms) are shown. The spectra have been arbitrarily normalized. The neutral flux is $\sim 5\%$ of the total flux

to point B, as shown in Figure 5.10. The neutralization probability is the ratio of the intensity in the neutral spectrum to that in the total spectrum. The velocity associated with the resulting neutral fraction is the weighted average velocity of the spectrum in the range from A to B.

Figure 5.11 shows the measured neutralization probability for Li, plotted as a function of the perpendicular velocity of the scattered Li. The incident-ion energies ranged from 10 to 1200 eV, and three different scattering geometries were used, as indicated on the figure. Figure 5.12 shows similar data for Na and K. The incident energies ranged from 5 to 1200 eV for Na and 10 to 1600 eV for K. Note the striking differences between the neutralization behaviors of these three species. For K there is almost no neutralization over the entire range of velocities. For Li the neutralization monotonically decreases with increasing perpendicular velocity, while for Na the neutralization is *nonmonotonic*, initially decreasing with increasing velocity, then passing through a minimum and increasing again. Note also that the magnitudes of the neutralization probabilities are quite different, with Li showing the most neutralization and K the least.

The differences in neutralization behavior for Li, Na and K are related to the different lifetimes of the Li(2s), Na(3s) and K(4s) resonances, as well as their energies relative to the Fermi energy of the metal. Figure 5.13 illustrates the general behavior of the Li(2s) resonance near the Cu(001) surface. The right-hand side of the figure shows a schematic representation of the energy $\varepsilon_a(z)$ and width $\Delta(z)$ (the inverse of the lifetime) of the Li(2s) resonance as a function of separation z from the surface. On the left-hand side, the hatched region indicates the filled conduction band of the metal. The Cu(001) surface has a work function

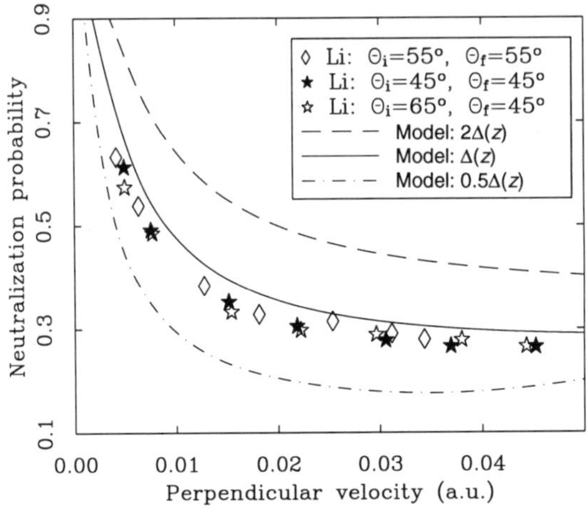

Figure 5.11 Comparisons of measured and calculated neutralization probabilities versus perpendicular velocity for scattered Li. Beams ranging in energy from 10 to 1200 eV were scattered from Cu(001) along the ⟨100⟩ azimuth for three different scattering geometries. The neutralization probabilities were calculated using equation (5). The Li(2s) resonance width $\Delta(z)$ and energy $\varepsilon_a(z)$ are from calculations by Nordlander and Tully. The model agrees with the data, and is very sensitive to the resonance widths; increasing or decreasing $\Delta(z)$ by a factor of two changes the results significantly

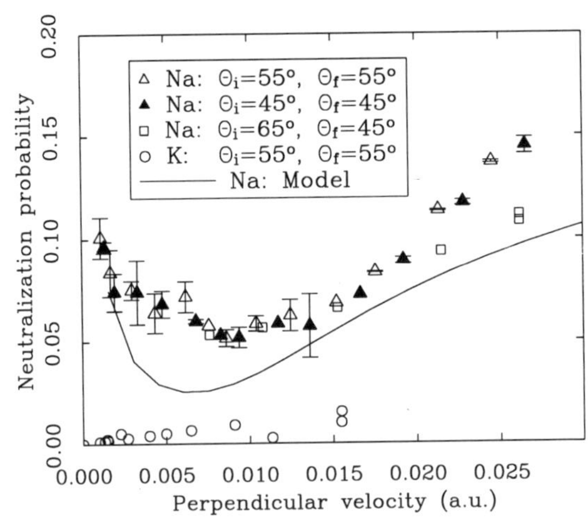

Figure 5.12 Measured neutralization probabilities versus perpendicular velocity of the scattered atoms for Na and K. For Na, the incident beam energies were 5 to 1200 eV, and for K, they were 10 to 1600 eV. Also shown is the predicted neutralization probability for Na using resonance widths calculated by Nordlander and Tully. The different neutralization behaviors for Li, Na and K reflect the high sensitivity of the charge-transfer process to the atomic resonance energies and lifetimes

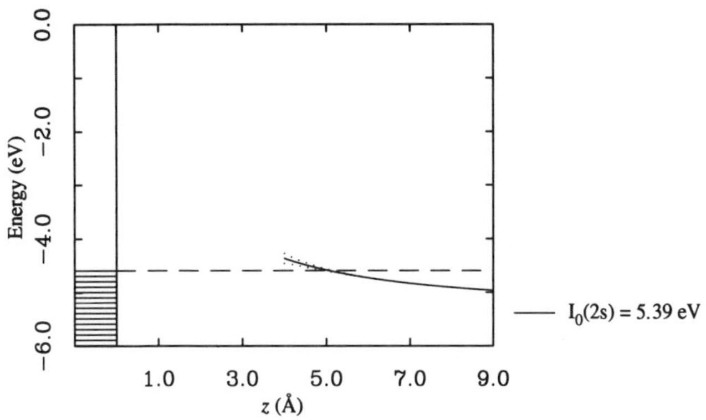

Figure 5.13 Variation in the energy $\varepsilon_a(z)$ and width $\Delta(z)$ of the Li(2s) resonance as a function of separation z from the Cu(001) surface. Here, z is referred to the jellium edge, and $\varepsilon_a(\infty)$ is 5.39 eV for lithium. At the distance where charge transfer is likely to occur, ε_a varies approximately as $1/4z$. The lifetime of the level is inversely proportional to $\Delta(z)$. The work function of Cu(001) is 4.59 eV and the hatched region on the left side of the diagram represents the filled conduction band of the metal

of 4.59 eV (Gartland, 1972; Gartland *et al.*, 1973), and far from the surface the energy of the Li(2s) level, the ionization energy, is 5.39 eV (Radzig and Smirnov, 1985) (measured from the vacuum). The upward shift in the resonance energy $\varepsilon_a(z)$ follows a $1/4z$ functional form at the distances where the charge transfer is most likely to occur (Gadzuk, 1967). The resonance width $\Delta(z)$ is due to the overlap of the atomic and metallic wave functions and thus increases approximately exponentially with decreasing ion–surface separation:

$$\Delta(z) = \Delta_0 \exp(-\alpha z). \tag{2}$$

Note that the Li(2s) resonance lies predominantly below the Fermi level far from the surface and predominantly above it close to the surface. In a static situation, with the alkali held at a distance z from the surface, the fraction of the resonance that lies below the Fermi level is precisely the charge-state occupancy of the alkali.

Because of the finite velocities of the scattered particles, the charge transfer is nonadiabatic and there are two time scales of primary interest, τ_r and τ_v. The lifetime of the resonance is $\tau_r = 1/\Delta(z)$ and $\tau_v = 1/(\alpha v_\perp)$ is the time scale for the motion of the scattered particle, which is set by the perpendicular velocity component of the scattered particle, $v_\perp$, and by the distance over which the transfer is significant, $1/\alpha$. For a given scattering trajectory, the velocity of the atom is approximately constant in the range of distances where the charge transfer is likely to occur and so τ_v is constant. However, the resonance lifetime τ_r varies roughly exponentially with distance. Far from the surface where the atom–metal coupling is weak, $\tau_r \gg \tau_v$ and charge transfer is very unlikely. Close to the surface, the coupling is very strong, $\tau_r \ll \tau_v$, and the atom loses memory of its incident charge state (Brako and Newns, 1981). The charge state of the atom is therefore determined on the outgoing trajectory, approximately at the distance where $\tau_r = \tau_v$. (By varying the scattering geometry and incident energies, we have verified for Na^+ incident on Cu(001) that the neutralization probability is independent of the incoming part of the trajectory (Kimmel and Cooper, 1993b). The distance at which $\tau_r = \tau_v$ is referred to as the 'freezing distance'. In this picture the neutralization probability is given by the fractional occupancy of the resonance at the freezing distance (Cooper *et al.*, 1992; Kimmel and Cooper, 1993b; Algra *et al.*, 1982; Rasser *et al.*, 1982).

Using this simple picture we can qualitatively understand the data in Figures 5.11 and 5.12. For Li (see Figure 5.11) at low perpendicular velocities, the charge state is determined far from the surface where ε_a lies below the Fermi level (see Figure 5.13), giving significant neutralization. As the velocity increases, the freezing distance decreases, ε_a has shifted upwards, and the neutralization decreases. For K, the ionization level is 4.34 eV, thus the K(4s) resonance lies predominantly above the Fermi level throughout the ion trajectory. Hence, as observed (see Figure 5.12), essentially no neutralization occurs. For Na, on the other hand, the neutralization is nonmonotonic (see Figure 5.12). The ionization level of Na is 5.14 eV. For low velocities, the charge state is determined far from the surface where the Na(3s) resonance is predominantly below the Fermi level. As the velocity increases and the freezing distance decreases, a decrease in the neutralization occurs as the Na(3s) resonance energy increases. As the velocity continues to increase and the freezing distance decreases further, an increased neutralization is observed. This is a result of the resonance broadening; at smaller separations more of the resonance lies below the Fermi level. Note the striking difference in behavior of the neutralization probabilities for Li and Na, whose ionization levels differ by only 0.25 eV.

5.3.2.2 One-electron model of charge transfer

To make quantitative comparisons with the data we use a charge-transfer model which is based on the Newns–Anderson Hamiltonian (Brako and Newns, 1981, 1989; Blandin *et al.*, 1976). This model assumes that a single spinless atomic orbital participates in the charge transfer. The Newns–Anderson Hamiltonian is

$$H(t) = \sum_k \varepsilon_k n_k + \varepsilon_a(t) n_a + \sum_k [V_{ak}(t) c_a^\dagger c_k + \text{H.C.}] \tag{3}$$

where ε_k and ε_a are the energies of the metal electronic levels and the atomic level, respectively, and n_k and n_a are the corresponding number operators, $n_i = c_i^\dagger c_i$, and V_{ak} is $\langle a|V|k\rangle$ with V representing the coupling of the atomic state to those in the metal. The resonance width of the atomic level (recall equation

(2)) is defined as

$$\Delta(\varepsilon, z) \equiv \pi \sum_k |V_{ak}|^2 \delta(\varepsilon - \varepsilon_k). \tag{4}$$

Within the broad-band approximation, where $\Delta(\varepsilon, z) = \Delta(z)$, the occupancy of the atomic state $n_a(t)$ is found to be (Brako and Newns, 1981, 1989):

$$n_a(t) = n_a(t_0) \exp\left(-2 \int_{t_0}^{t} \Delta(t')\,\mathrm{d}t'\right) + \frac{1}{\pi} \int \mathrm{d}\varepsilon\, f(\varepsilon, T) \times \left| \int_{t_0}^{t} \sqrt{\Delta(t')} \exp\left(-\mathrm{i}\varepsilon t' - \int_{t'}^{t} [\mathrm{i}\varepsilon_a(t'') + \Delta(t'')]\mathrm{d}t''\right) \mathrm{d}t' \right|^2, \tag{5}$$

where f is the Fermi function, and $\varepsilon_a(t)$ and $\Delta(t)$ give the time dependence of the energy and width of the atomic resonance due to the motion of the scattered atom. The energy and lifetime of the orbital are input parameters in this calculation.

The first term in equation (5) arises from the decay of the initially filled level (if $n_a(t_0) = 1$) into the continuum of metal states at a rate $\Delta(t)$. The role of this 'memory term' has been discussed elsewhere (see for example Brako and Newns, 1981). For our ion scattering experiments, $n_a(t_0) = 0$ and so the memory term is zero. (The incident particle is an ion.)

The solid line in Figure 5.11 shows results of this calculation where the lifetime and energy of the Li(2s) resonance used in the model were calculated by Nordlander and Tully (1988, 1989, 1990).[3] This model gives excellent agreement with the data. The sensitivity of the model to changes in the resonance width $\Delta(z)$ (from the Nordlander and Tully values) is also demonstrated in Figure 5.11; calculations using 2Δ and 0.5Δ are compared. The solid line in Figure 5.12 shows the same calculation for Na. Here the agreement is poorer than that for Li, but the minimum in the neutralization probability is reproduced. A more detailed discusson of these data and the modeling are given elsewhere (Kimmel and Cooper, 1993b; Kimmel, 1992).

Other calculations of resonance widths have been made recently for alkalis (for example, Geerlings *et al.*, 1987c; Teillet-Billy and Gauyacg, forthcoming; Bahrim *et al.*, forthcoming), and have been used with some success. Also, other theoretical approaches have been applied to the problem of charge transfer at surfaces (Tully, 1977; Zimny *et al.*, 1989, 1991; Zimny, 1990; Milošević, 1992; Burgdörfer *et al.*, 1987).

The model discussed above does not treat Auger neutralization and it also ignores neutralization into excited states and the formation of negative ions. These omissions will be discussed further below.

5.3.2.3 Parallel velocity

In both the qualitative picture of charge transfer presented in Section 5.3.2.1 and the more quantitative model in Section 5.3.2.2, only the perpendicular component of the velocity of the scattered atom was important in determining the neutralization probability. In previous works, the effect of the parallel component of velocity on the charge-transfer behavior has also been investigated (see, for example, Geerlings and Los, 1990 and Winter, 1991 and references therein) and found to be quite significant for some cases. Most of the previous experiments were carried out at grazing angles and with relatively high (keV) incident energies (Zimny, 1990; Winter and Zimny, 1988). Here we present data, also showing the parallel velocity effect, at lower velocities and more normal scattering geometries than discussed in the earlier works.

Figure 5.14 shows measured neutralization probabilities for Na scattered from Cu(001) with five different scattering geometries (Kimmel and Cooper, 1993b). The data are plotted as a function of the perpendicular velocity components of the scattered particles. A clear dependence on the parallel velocity is observed. Also shown in the figure are neutralization probabilities calculated using a model that includes the parallel velocity component of the scattered particle (Kimmel, 1992; van Wunnik *et al.*, 1983).

The motion of the atom with respect to the surface changes the charge-transfer probabilities because, from the rest frame of the atom, the energies of the metal electrons are shifted from their values in the rest frame of the metal,

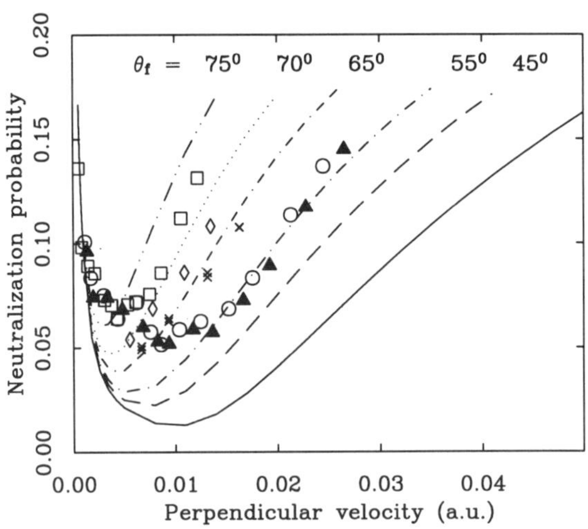

Figure 5.14 Comparison of measured Na neutralization probability versus perpendicular velocity to calculations using a charge transfer model which includes the parallel component of the velocity of the scattered particle, i.e. includes the Galilean transformation between the atom and the surface. The data were taken with five different scattering geometries: $\theta_i = \theta_f = 45°$ (solid triangles); $\theta_i = \theta_f = 55°$ (open circles); $\theta_i = 55°$, $\theta_f = 65°$ ($\times$'s); $\theta_i = 55°$, $\theta_f = 70°$ (open diamonds); and $\theta_i = 55°$, $\theta_f = 75°$ (open squares). Even at the relatively low velocities of these experiments ($v \ll v_F$, where v_F is the Fermi velocity), it is important to include the Galilean transformation in the modeling. The results of the one-electron model (solid line, see equation (5)) are shown for comparison

changing the apparent occupation of the metallic states that satisfy the resonance condition. Since the charge transfer is particularly sensitive to those electrons near the Fermi level, small changes in the apparent occupation near the Fermi level can cause significant changes in the charge-transfer probabilities. Details of the model used in Figure 5.14 are given elsewhere (Kimmel and Cooper, 1993b; Kimmel, 1992).

5.3.3 SINGLE-ELECTRON TRANSFER IN THE PRESENCE OF ADSORBATES

Alkali adsorption on metals has been known for many years to promote chemical reactions at surfaces. As a result, the study of alkali-atom adsorption on clean, single-crystal metal surfaces has an extensive literature (for an overview of alkali–metal adsorption, see Bonzel *et al.*, 1989). Alkali adsorption has several effects, such as the lowering of the work function, surface reconstruction, or relaxation of the surface. The lowering of the work function dramatically affects charge transfer between the impinging ion and the surface. A number of experiments aimed at obtaining detailed information about charge transfer in the presence of alkali adsorbates have been performed over recent years (Yu and Lang, 1983; Geerlings and Los, 1984; Geerlings *et al.*, 1987b; Kimmel *et al.*, 1989, 1991; Jiang *et al.*, 1993; German *et al.*, 1993). We will discuss these experiments in the context of the one-electron model of charge transfer presented in Section 5.3.2.2. We first discuss the influence of the work function on the ion-survival probability and then consider the effect of the inhomogeneity of the adsorbate-induced local electrostatic potential.

5.3.3.1 Basic behavior

A schematic representation of the interaction of an atom with a metal surface, within a one-electron picture, is shown in Figure 5.13. The energy of the effective ionization level corresponding to the Li(2s) level is given by:

$$\varepsilon_a(z) = -I_o + \frac{1}{4(z - z_{im})}, \tag{6}$$

where I_o is the ionization energy of the isolated lithium atom in the Li(2s) state and we have assumed that the shift of the atomic level is given by the classical image potential, with the image plane located at $z = z_{im}$. The ionization level crosses the Fermi level when the particle is a distance z_c from the surface, with z_c determined by the relation $\varepsilon_a(z_c) = -\Phi$. At z_c, an electron can transfer from the Fermi level in the metal to the empty ionization level and leave the metal in its ground state. If the particle is at a distance $z > z_c$, then a resonant electron transfer from the metal to the empty ionization level leaves an excitation in the metal. Similar transitions are possible for $z < z_c$ from a filled ionization level to empty states in the metal.

Within the limit that memory of the initial charge state of the impinging ion is erased, and assuming that the coupling has the form given in equation (2), it can be shown (Goodstein, 1990) that, under certain conditions, the ion-survival probability is given by:

$$P^+ = \exp\left(\frac{-2\Delta(z_c)}{\alpha v_\perp}\right). \tag{7}$$

Combining equations (2), (6) and (7) yields the following expression for P^+:

$$P^+ = \exp\left[\frac{-2\Delta_0}{\alpha v_\perp} \mathrm{e}^{-\alpha z_{\mathrm{im}}} \exp\left(\frac{-\alpha}{4(I_0 - \Phi)}\right)\right]. \tag{8}$$

Given the above assumptions, P^+ as expressed in equation (8) has a very strong dependence on the work function. Such a dependence has been observed by Yu and Lang (1983). Yu measured the yield of Cs^+ sputtered from different substrates where different coverages of adsorbed lithium were used to vary the work function. Other experiments (Geerlings and Los, 1984; Geerlings *et al.*, 1987b, Kimmel *et al.*, 1989, 1991) have also been done in which the work-function dependence predicted by equation (8) was not obtained. An explanation for these results lies in the fact that an adsorbate-induced work-function shift cannot, for all systems, be modeled as a uniform shift of the Fermi level. Instead, it is useful to consider what happens to the local electrostatic potential in the presence of an alkali adsorbate, which we do in the next section.

5.3.3.2 Inhomogeneities

When an alkali atom adsorbs on a metal surface, it induces charge rearrangement which changes the local electrostatic potential (e.g. see Lang, 1971 and Gonze *et al.*, 1990 and references therein). Thus, the potential varies along the surface according to the presence (or absence) of adsorbates. In a scattering experiment, the adsorbate-induced potential will cause the energy of the ionization level of the scattered atom to vary *along* the surface, and the length scale for this variation is set by the distance between the adsorbates. If this distance is comparable to z_c, then the ion-survival probability will depend upon the position of the scattering particle relative to the adsorbates. Since the incident ions sample all parts of the surface unit cell in a scattering experiment, the measured neutralization will be a weighted average over scattering trajectories which sample different local electrostatic potentials, the weights being given by the relative scattering cross sections. We note that the adsorbate-induced variation in the coupling of the atomic state to the metal (Nordlander and Lang, 1991) can also affect the ion-survival probability (Hsu *et al.*, 1992).

Many experiments for which z_c and the adsorbate spacing are comparable have been performed recently. Typically, the ion-survival probability is measured versus the work-function shift induced by low coverages of alkali atoms. The picture given above, of how the inhomogeneity of the local electrostatic potential affects charge transfer, is consistent with the data obtained in these experiments. To reconcile the results of Yu and Lang with these later experiments, it is necessary to consider the adsorbates used to produce the work-function shift and the details of adsorption in the low coverage regime for the particular adsorbate–surface system.

While the agreement between the model and the theory is encouraging (Kimmel *et al.*, 1991), there are still observed phenomena which are not included in the model. This leads us to the inclusion of multiple final states in the description of charge transfer at surfaces.

5.3.4 CHARGE TRANSFER INTO MULTIPLE STATES

As seen above, the one-electron model describes fundamental aspects of resonant charge transfer between an atomic level and a continuum of metal levels. It has been used not only to describe the filling of an empty ionization level, but also the filling of an empty affinity level (Geerlings *et al.*, 1987c, 1987d) (i.e. the production of negative ions). In the case of filling the affinity level, intra-atomic correlations are typically ignored because the production of the negative ion is considered to be a one-electron process. Indeed, qualitative agreement is obtained when describing negative-ion production with the one-electron model (i.e. equation (5)) (Geerlings *et al.*, 1987c).

However, the one-electron model discussed above does not explain a number of observations. For example, alkali ions impinging on low-work-function surfaces may scatter into excited neutral states (Hermann *et al.*, 1984, 1986; Schall *et al.*, 1989, 1990; Behringer *et al.*, 1993, forthcoming; Andersson *et al.*, 1993). Other processes such as Auger neutralization and the formation of auto-ionizing states can also occur (Brenten *et al.*, 1991a, 1991b, 1991c). Also, species other than alkalis can form negative ions; for example, oxygen ions (O^+) impinging on clean metal surfaces will sometimes scatter as O^-. In this case, the degeneracy of the negative-ion states must be accounted for. The rich set of observed phenomena described above cannot be accounted for by the one-electron model and a more complete theory must be formulated to describe ion–surface scattering.

In the following, we focus on experiments which test a theory which includes resonant transfer into multiple final states. In these experiments, alkali ions scatter from adsorbate-covered metal surfaces and we measure the charge-state fractions and relative yields of excited states in the scattered flux. We have already discussed the technique that we use to measure charge-state fractions in Section 5.3.1.1. We begin, therefore, with a description of the method used to measure the relative yields of excited states. We then describe the charge-transfer theory, and proceed to compare the experimental results to the predictions of the theory. Finally, we will briefly discuss non-resonant processes.

5.3.4.1 Excited-state detection

The technique we use to detect ions which have been scattered into excited neutral states is low-level photon counting. The excited atom can form when the ion is within a few angstroms of the surface and can subsequently decay by emitting a photon of the appropriate energy. Typical ion velocities in the scattering experiments described in this chapter are of order 10^6 cm s^{-1} and typical excited-state lifetimes are of order 10^{-8} s; therefore the excited atom is is at most 0.01 cm from the surface when it emits a photon. For an isolated lithium atom, the Li(2p) → Li(2s) transition produces a photon of energy 1.85 eV ($\lambda = 673$ nm). We have detected photons of this energy when Li^+ ions scatter from low-work-function surfaces (Behringer *et al.*, 1993, forthcoming; Andersson *et al.*, 1993).

Our detection system consists of the light-collection optics, a set of filters and a photomultiplier tube. Light produced by collisions of the incident ions with the surface passes through a sapphire window and is then focused onto the end of a fiber-optic cable by a single lens. After traversing the cable, the light passes through another lens before falling onto the photocathode of the photomultiplier. Line filters corresponding to particular atomic transitions may be inserted into the optical path to select the wavelength of the photons which fall onto the photocathode. By measuring the rate at which photons are produced and normalizing this by the incident beam current on the sample, one obtains a measure of the relative yield of ions scattered into the excited state selected by the line filter.

5.3.4.2 Theory

Since the branching ratios to excited neutral and negative ion channels are appreciable under certain experimental conditions (Geerlings *et al.*, 1987d; Schall *et al.*, 1989, 1990), we require as a minimum that excited states and intra-atomic correlations are included in a more complete theory of charge transfer. Excited states have often been neglected (with exceptions; see for example Bloss and Hone, 1978, Brako, 1984, Kawai *et al.*, 1987 and Kawai and Kawai, 1988). Recent work (Schall *et al.*, 1989, 1990) has demonstrated that they can represent a significant channel. The Hamiltonian of the one-electron model must therefore be modified to include a sum over the participating excited states. To include intra-atomic correlations, one must explicitly take spin into account and modify the Hamiltonian of Section 5.3.2.2 by adding a term describing the Coulomb repulsion between the two outer-shell electrons, i.e. the 'U-term'. A Hamiltonian which incorporates the above modifications is (noting that the sum over spin orientations, which are denoted by σ, is implicit when the spin index is repeated):

$$\begin{aligned} H(t) = {} & \sum_a \varepsilon_a c_a^{\dagger\sigma} c_{a\sigma} + \sum_k \varepsilon_k c_k^{\dagger\sigma} c_{k\sigma} \\ & + \sum_{a,k} [V_{ak} c_a^{\dagger\sigma} c_{k\sigma} + \text{H.C.}] \\ & + U(t) n_{a\uparrow} n_{a\downarrow}. \end{aligned} \tag{9}$$

Spin-conserving Hamiltonians (i.e. where no spin-flip processes are included) similar to this have been used in previous work which included correlations (Sebastian, 1985; Grimley *et al.*, 1983; Brako and Newns, 1985; Kasai and Okiji, 1987; Nakanishi *et al.*, 1988; Sulston *et al.*, 1988, 1989; Langreth and Nordlander, 1991). (We note that the work of Marston *et al.* (1993), described below, is *not* based on equation (9).) Particle-hole formation, which is important in the neutralization of the incoming ion and for losing memory of the incident charge state, is sometimes treated differently in the various approaches, leading to distinct results. Also, approximations must be made in order to keep manageable the set of equations that are developed to calculate the final occupancies of the participating states; some models neglect certain physical processes while others make convenient assumptions concerning the time dependence of various quantities. One must therefore be careful when comparing the results of different theoretical approaches. Nonetheless, a common result obtained by all approaches is that the final state occupancies are dependent on the lifetimes and energies of the participating atomic states. This was seen above in the one-electron model and it is also true for the more advanced models.

We now discuss the $1/N$ expansion (N is the number of spin orientations, with $N = 2$ for electrons), which was first applied to the ion-scattering problem by Brako and Newns (1985) and has been used with success in solving the static Anderson model. The fundamental idea on which the method is based is that the wave function of the ion–surface system is expanded into a basis containing states with increasing numbers of particle–hole pairs. This basis is truncated at a given number of particle–hole pairs and a set of coupled differential equations can be developed for the amplitudes of the various components of the wave function. This set is then solved without further approximation. The truncation keeps the number of equations manageable, and the approximation to the full wave function by the truncated wave function will be good as long as the number of particle–hole pairs formed in the ion–surface collision is small. The present expansion can be formally controlled by generalizing the usual two types of electrons to N types. In the work of Marston *et al.* (1993), it is seen that the amplitudes of those components of the true wave function containing increasing numbers of particle–hole pairs are reduced by increasing powers of $1/\sqrt{N}$ (and the corresponding probabilities, by increasing powers of $1/N$). For the rest of the chapter, we will refer to the model of Marston *et al.*, which significantly extends the work of Brako and Newns by including level crossings, excited states and negative-ion states, as the 'multiple-states model'.

Rather than provide further details of the calculation (which can be found in Marston *et al.*, 1993), we provide a qualitative overview of the model in order to obtain further physical insight. The truncated basis of the ion–surface wave function used in this work has five components, shown schematically for the case of lithium in Figure 5.15, corresponding to: (a) the ion with the metal in its ground state; (b) a ground-state neutral atom with a hole in the metal; (c) a ground-state negative ion with two holes in the metal; (d) an excited-state neutral atom with a hole in the metal; and (e) the ion with a particle–hole pair in the metal. The ion–surface interaction, which leads to the broadening and shifting of the atomic levels near the surface (Gadzuk, 1967; Gurney, 1935), couples the basis states to each other. By assuming the broad-band limit, we can determine the couplings by knowing the resonance widths of the appropri-

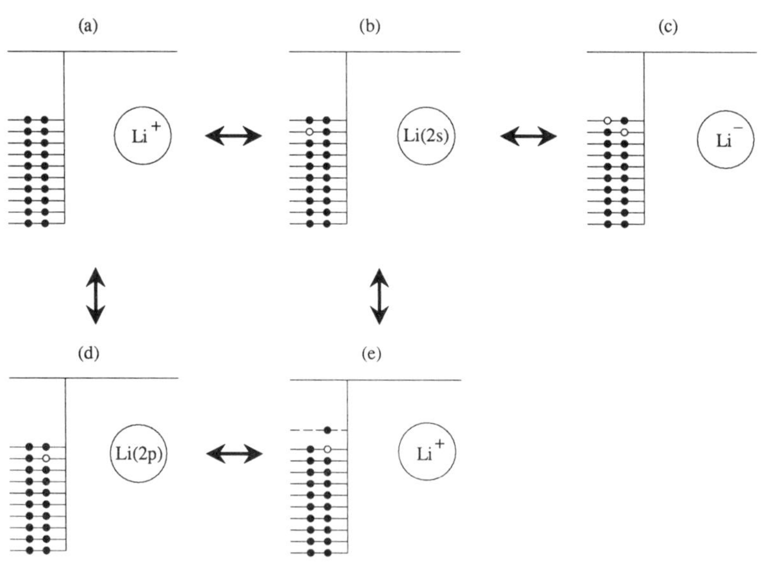

Figure 5.15 A schematic diagram of the components of the wave function of the ion–surface system. (a) The ion with the metal in its ground state. (b) The ground-state neutral atom with a hole in the metal. (c) The negative ion with two holes in the metal. (d) The excited state neutral atom with a hole in the metal. (e) The ion with a particle–hole pair in the metal. The time evolution of the amplitude of each component of the wave function depends upon the amplitude of, and coupling to, the other components, as indicated by the arrows. The choice of hole energies which are illustrated is arbitrary

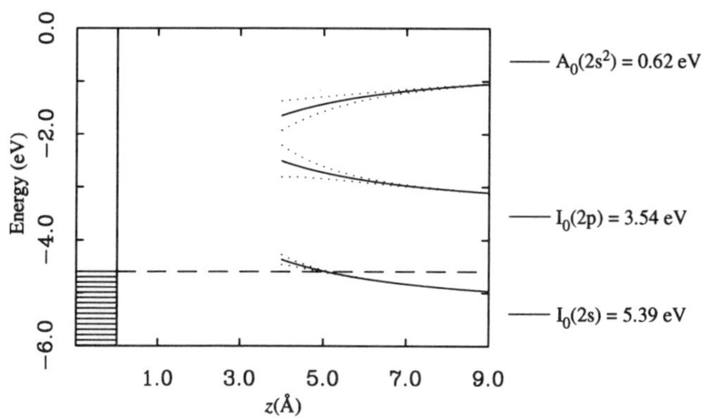

Figure 5.16 The effective affinity and ionization levels of lithium near a Cu(001) surface (for a discussion, see Kimmel *et al.*, 1991). The ion–surface interaction shifts these levels and broadens them into resonances which are indicated by the dotted lines

ate atomic states. We will initially use resonance widths calculated by Nordlander and Tully (1988, 1989, 1990; Nordlander, 1992) although we feel free to consider these as parameters which can be varied.

In Figure 5.16, we present a schematic plot of a set of atomic-state resonances for a lithium atom in the vicinity of the Cu(001) surface[4]. The level energies for the case of the isolated atom are shown at the right side of the figure. The widths of the resonances are indicated by the dotted lines. As the resonance widths increase, so do the couplings between the components of the wave function, and hence the transfer rates.

In Figure 5.17, for the case of a lithium particle scattering from a clean Cu(001) surface ($\Phi = 4.59$ eV), we present plots of the calculated position dependence (and, through the trajectory approximation, time dependence) of the probabilities with which the particle will be found in a particular charge state. For example, we denote the probability with which the atom will be found as a positive ion as P^+, recalling that P^+ is the sum of two probabilities: the probability with which the atom will be found as a positive ion with, and without, particle–hole excitations. The atom starts 2 Å from the surface (i.e. from the jellium edge) with the ion–surface system in its ground state. Subsequently, the atom moves at constant velocity ($v_\perp = 0.02$ a.u.) away from the substrate. The differences between Figures 5.17(a) and 5.17(c) are due to different assumed forms of the atomic-state lifetimes. The corresponding resonance widths used to obtain the probabilities (occupancies) are also shown. The widths shown in Figure 5.17(b), which were used to obtain the occupancies shown in Figure 5.17(a), are roughly half those shown in 5.17(d). A difference of a few percent in the final value of P^+ is obtained.

The theory for multi-electron transfer must accurately predict dependences on experimentally controlled parameters. In the following section we discuss

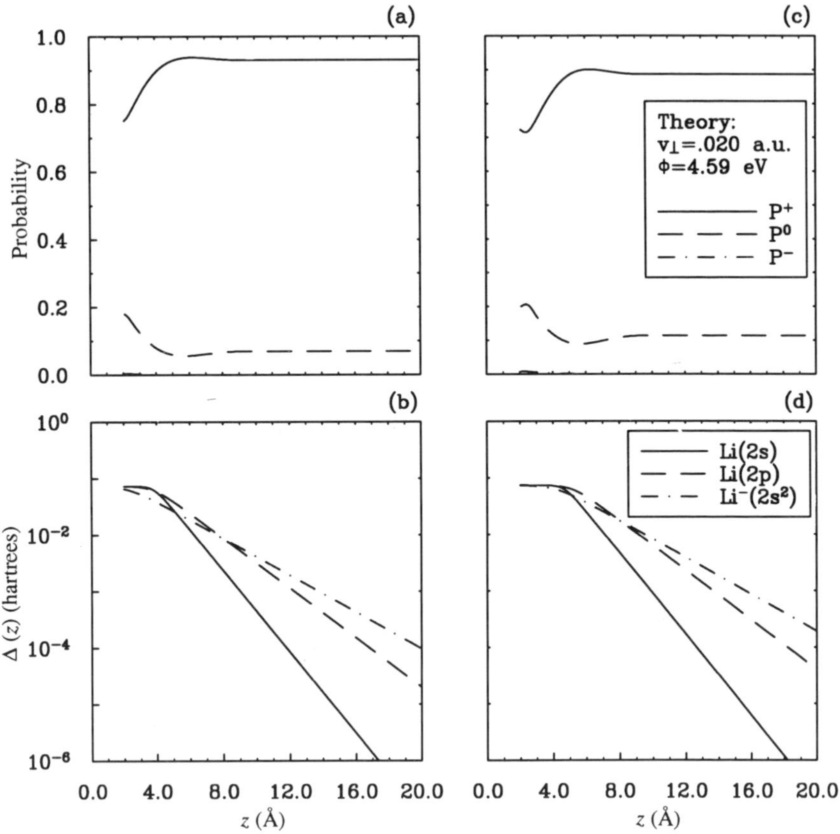

Figure 5.17 The evolution of the charge-state occupancies of lithium as the particle moves away with constant velocity from a surface characterized by $\Phi = 4.59$ eV. (a) An image shift characterized by $V_{min} = 2.6$ eV, $z_{min} = 0.78$ Å is used, along with the level widths shown in (b). (c) The same image shifts as in (a), with the level widths shown in (d). The widths in (d) are roughly twice as large as in (b) (i.e. the lifetimes are roughly half as long)

the observed work-function dependence of the charge-state fractions and the relative yields of excited states. We can predict the qualitative dependence by applying the picture given in Section 5.3.2 to the case of lithium. For the clean surface, both the affinity level and the Li(2p) ionization level are well above the Fermi level and we expect that there will be little charge transfer into these levels (see Figure 5.16). As the work function is decreased, however, the metal levels become degenerate with the affinity level and the Li(2p) level, and so we expect these atomic levels to become increasingly populated. Although these expectations are partially fulfilled, we will see that a more detailed consideration of the charge transfer dynamics is required.

5.3.4.3 Experiment and theory

A number of groups have performed experiments in which alkali ions scatter from adsorbate-covered surfaces (Geerlings and Los, 1990; Schall *et al.*, 1989; Brenten *et al.*, 1991b). Here we present recent measurements from our laboratory on the system $Li^+ \rightarrow A/Cu(001)$, with A representing potassium or cesium (Behringer *et al.*, 1993, forthcoming; Andersson 1993).

Figure 5.18 is a plot of the measured probabilities with which an impinging Li^+ ion scatters from the Cs/Cu(001) surface as a positive ion (P^+) or a negative ion (P^-) versus the work-function shift induced by the deposition of Cs. The 400 eV Li^+ ions impinge at an angle of $\theta_i = 65°$ with respect to the surface normal, and along the $\langle 100 \rangle$ azimuth. The detected particles have been scattered into a final angle of $\theta_f = 64°$ with respect to the surface normal. For the clean surface, P^+ is 0.67 ($v_\perp = 0.02$ a.u.). As the work function decreases from its clean-surface value of $\Phi = 4.59$ eV (Gartland, 1972; Gartland *et al.*, 1973) ($\Delta\Phi = 0.0$ eV) to a value of about 2.6 eV ($\Delta\Phi \approx -2.0$ eV), P^+ decreases, with a corresponding increase in the probability P^0 with which the Li^+ ion will scatter as a neutral atom (ground state or otherwise). In this range of work-function values, P^- is less than a few percent. However, as the work function is further decreased, P^- begins to increase at the expense of P^0. For work-function values less than about 2.6 eV, P^+ is less than a few percent.

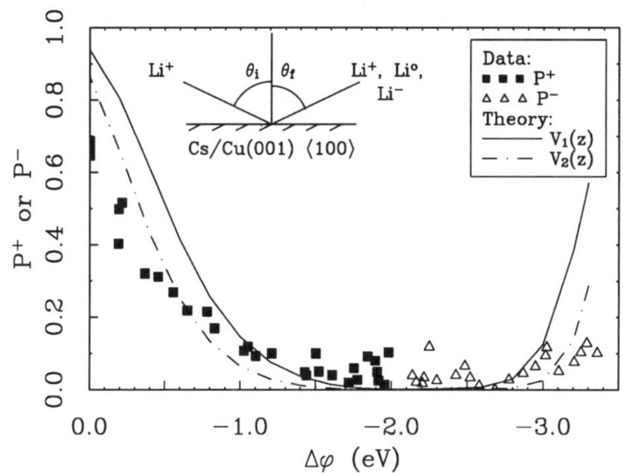

Figure 5.18 Measured and theoretical values of the positive and negative charge-state occupancies versus the work function shift induced by cesium adsorbates. Li^+ ions with a primary energy of 400 eV impinge on a Cu(001) surface along the $\langle 100 \rangle$ azimuth and with an incident angle of $\theta_i = 65°$ measured with respect to the surface normal. Particles scattered into $\theta_f = 64°$ are detected. The theoretical values were obtained assuming two different forms for the shift of the atomic levels near the surface (see note 4): $V_1(z)$ is characterized by $V_{min} = 2.6$ eV and $z_{min} = 0.78$ Å, while $V_2(z)$ is given by $V_{min} = 1.9$ eV and $z_{min} = 0.29$ Å

Also shown in Figure 5.18 are two sets of theoretical curves produced using the multiple-states model. The two sets correspond to two choices, $V_1(z)$ and $V_2(z)$ (described in the figure caption), for the energy shifts of the atomic levels in the vicinity of the surface. Both sets of theoretical curves agree qualitatively with the experimentally observed dependence on work function, but they differ quantitatively (with the experimental curve and with each other), especially in regard to their values for P^-. Near the surface, the shift $V_1(z)$ produces ionization (affinity) level energies that are higher (lower) than those produced by $V_2(z)$; i.e. the Coulomb repulsion near the surface is effectively lower for $V_1(z)$ than for $V_2(z)$, which results in the relatively enhanced values for P^- when using $V_1(z)$.

The relatively small slope of the data compared to the theoretical prediction may be due to the effect of the alkali-adsorbate-induced electrostatic potential, as discussed in Section 5.3.3.2. This effect has been discussed by Geerlings *et al.* (1987b; Geerlings and Los, 1984) and by Kimmel *et al.* (1989, 1991). Variations in the local electrostatic potential can in principle be incorporated into the multiple-states model.

Charge-state measurements for Li^+ scattering from Cs/W(110) have been reported by Geerlings *et al.* (1987a, 1987c, 1987d). These were obtained without using the time-of-flight (TOF) technique discussed in Section 5.3.1.1. The advantage of the TOF technique is that a possible source of error, namely sputtered substrate atoms, can be eliminated. Nevertheless, the data in Figure 5.18 and the data reported by Geerlings *et al.* are similar.

We note that the theoretical results of Sulston *et al.* (1989) agree in some qualitative aspects with the results we have given here. For example, the sharp decrease of P^+ with work function and the increase of P^- at very low work functions ($\Delta\Phi < -3.0$ eV) are common features of both treatments.

In addition to the overall charge-state fractions presented above, we have determined the relative yield of Li^+ ions scattered into the Li(2p) state by measuring the rate at which photons corresponding to the Li(2p) → Li(2s) transition ($\lambda = 673$ nm) are produced. A plot of the relative yield is shown in Figure 5.19. When the surface is clean, we observe no photons within the sensitivity of our detector. However, as the work function decreases, the photon yield increases, reaches a peak value, and then decreases. The distinctive feature of these data is the peak in the yield at a work-function shift of about -1.8 eV. A similar peak has been observed when 1 keV Li^+ was scattered at grazing angles from Cs/W(110) (Schall *et al.*, 1989, 1990).

Also shown in Figure 5.19 are two theoretical predictions for the Li(2p) yield which are based on the multiple-states model; they are normalized to the peak value of the measured relative yield. The only difference between the calculations is that the energy shifts of the atomic states are different (see figure caption). Both theoretical curves predict a peak in the yield at a work-function value very similar to that observed in the data. The predicted magnitude of the

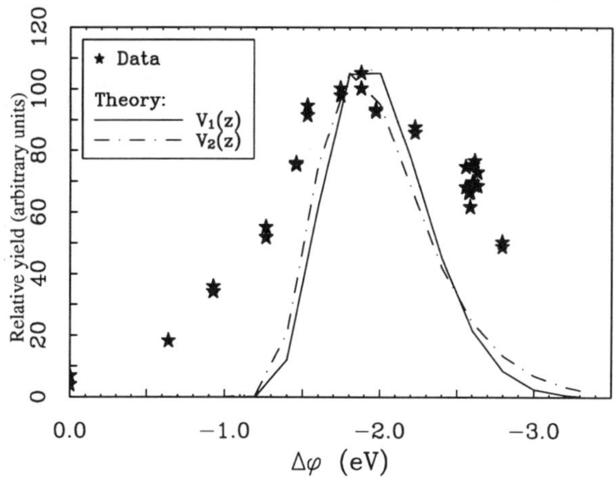

Figure 5.19 Measured and theoretical values of the relative yield of neutral atoms in the Li(2p) state versus the work function shift induced by potassium adsorbates. Li^+ ions with a primary energy of 400 eV impinge on a Cu(001) surface along the ⟨100⟩ azimuth and with an incident angle of $\theta_i = 65°$ measured with respect to the surface normal. Photons corresponding to the Li(2p) → Li(2s) transition ($\lambda = 673$ nm) are detected. The theoretical values were obtained assuming two different forms for the shift of the atomic levels near the surface (see Figure 5.18)

peak values differs by a factor of less than two, the lower value being approximately 0.014. This value is larger than the estimated experimental yield by less than a factor of four.

The existence of the predicted peak is due in part to a competition between the negative-ion state and the Li(2p) state. We explore this competition by first examining the evolution of the occupancies when the surface work function is such that the highest Li(2p) yield obtains.

In Figure 5.20, we show the position dependence of the occupancies for the case of a lithium atom moving away from a surface which has a work function of 2.79 eV. The widths shown in Figure 5.17(b) are used. Consider Figure 5.20(a), which contains a plot of the occupancies on a linear scale. We first note the dramatic difference between Figure 5.20(a) and Figure 5.17(a). By lowering the work function to 2.79 eV, the predicted majority species is now Li(2s). Most of the charge transfer appears to occur by the time the atom reaches $z = 5$ Å; however, by plotting the occupancies on a logarithmic scale (Figure 5.20(b)), we can see that the occupancies of the minority channels continue to change far from the surface. At the start of the trajectory ($z = 2$ Å), the ion–surface system is in its ground state. As the atom moves away from the surface, from $z = 2$ Å to $z = 5$ Å, the affinity-level energy increases and the ionization-level energies decrease. Meanwhile, all the level widths decrease. These changes lead to the decreasing values of P^- and $P^0(2p)$ (although the Li(2p) ionization-level energy

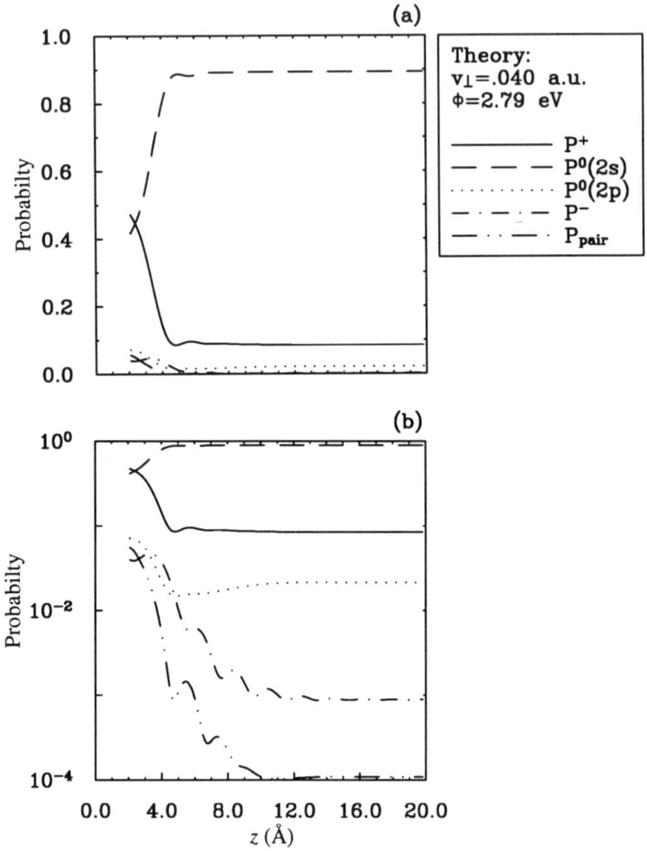

Figure 5.20 The detailed evolution of the charge-state occupancies of lithium as the particle moves away from the surface with constant velocity ($v_\perp = 0.04$ a.u.). An image shift characterized by $V_{\min} = 2.6$ eV, $z_{\min} = 0.78$ Å, and levels widths as in Figure 5.17(a) are used. (a) Linear scale. (b) Logarithmic scale. Note that the occupancies of the minority states can change considerably far from the surface

decreases, the associated resonance width decreases more quickly, bringing most of the resonance *above* the Fermi level in this region) and the steadily increasing value of P^0(2s). As the atom moves further away from the surface, P^- continues to decrease; this is because the affinity level is now well above the Fermi level and continues to empty. Meanwhile, P^0(2p) increases somewhat due to the lowering of the Li(2p) ionization level below the Fermi level, and P^0(2s) remains essentially constant because the coupling between the positive-ion ground state and the neutral ground state is relatively small, and because the occupancy of the positive-ion ground state is small. As the work function is lowered further, the negative ion state becomes more occupied and the Li(2p) state less so (i.e.

a competition occurs). A detailed discussion is given elsewhere (Behringer, forthcoming).

Evidence of this competition can be seen in Figure 5.21, which shows the results of a calculation using the model of Marston *et al.* The predicted peak in the Li(2p) yield, $P^0(2p)$, is shown along with the increase in negative ion yield, P^-, at low work-function values. The competition may be seen by considering two hypothetical systems in which only three final states are allowed. In the first system, the only allowed states are Li^+, Li(2s) and Li(2p), which are obtained with probabilities P^+, $P^0(2s)$ and $P^0(2p)$. For this system, $P^+ + P^0(2s) + P^0(2p) = 1$, or $1 - [P^+ + P^0(2s)] = P^0(2p)$. In the second hypothetical system, the only states possible are Li^+, Li(2s) and $Li^-(2s^2)$; for this system, $1 - [P^+ + P^0(2s)] = P^-$. The quantity $1 - [P^+ + P^0(2s)]$ is plotted in Figure 5.21, along with $P^0(2p)$ and P^-, as obtained from a calculation which includes all of the states. As the work function decreases from its clean surface value ($\Delta\Phi = 0.0$ eV) to about 3.39 eV ($\Delta\Phi = -1.2$ eV), almost all of the particles are predicted to be positive ions or ground-state neutral atoms. As the work function decreases further to about 2.59 eV ($\Delta\Phi = -2.0$ eV), $P^0(2p)$ increases quickly while P^- increases slowly; the coincidence of $P^0(2p)$ with

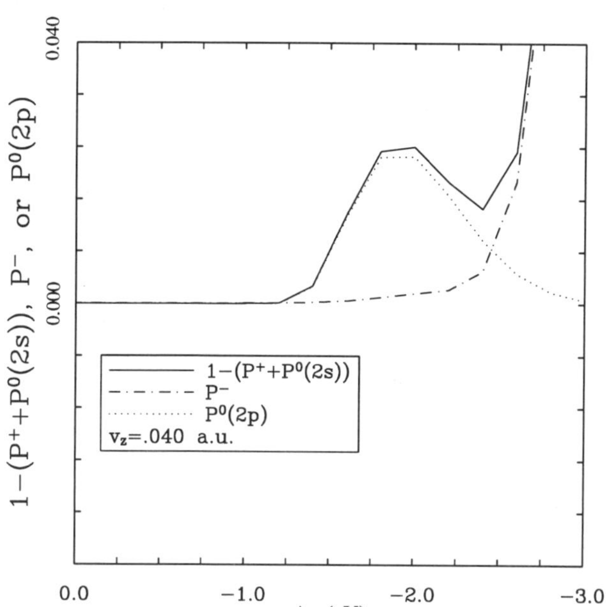

Figure 5.21 The theoretical values of the Li(2p) and $Li^-(2s^2)$ occupancies, and the quantity $1 - (P^+ + P^0(2s))$, versus the work-function shift, $\Delta\Phi$. Decreases in $P^0(2p)$ and corresponding increases in P^- and $P^0(2s)$ are seen for work-function values less than about 2.6 eV ($\Delta\Phi = -2.0$ eV), demonstrating the flow of probability between the excited state and negative-ion state

$1 - [P^+ + P^0(2s)]$ indicates the small population of the negative-ion state. As the work function decreases still more, $P^0(2p)$ decreases while P^- and $P^0(2s)$ increase correspondingly. At very low work-function values, $P^0(2p)$ is very small; this is demonstrated graphically by the coincidence of the curves for P^- and for $1 - [P^+ + P^0(2s)]$. Thus there is a net flow of probability from the excited state to the negative-ion state that occurs as the work function decreases below 2.59 eV, indicating that a competition between the two states has developed.

This indicates that it is necessary to include correlations in order to properly describe the detailed behavior of the observed yield of Li(2p). It may not be sufficient to include correlations, however; it may also be necessary to adequately describe the lifetimes and energies of the atomic states near the surface. This possibility is suggested by the work of Sulston *et al.* (1989), which included correlations but did not predict the peak in the Li(2p) yield. However, it is difficult to make a direct comparison between the work of Sulston *et al.* and that of Marston *et al.* because these differ in detail (see Section 5.3.4.2).

Finally, we add that Kempter and collaborators (Schall *et al.*, 1989, 1990) have offered a different explanation for the existence of the peak in the Li(2p) yield. This explanation also involves competition between accessible channels; however, they suggest that it is a competition between the Li(2s) state, formed by Auger de-excitation, and the Li(2p) state. They have measured electron yields which support this mechanism (Brenten *et al.*, 1991a, 1991b). These measurements indicate that Auger processes do in fact occur, along with other processes such as the formation of auto-ionizing states. We discuss these processes further in the next section.

5.3.4.4 Non-resonant mechanisms

Non-resonant mechanisms also play a role in ion neutralization during ion–surface collisions. These include, for example, the Auger (Hagstrum, 1954; Hagstrum *et al.*, 1988) and direct mechanisms, whereby the incoming ion is neutralized and the excess energy carried away by an electron or a photon, respectively. Yet another mechanism is the formation of a temporary quasi-molecule by the incident ion and a target atom (Souda *et al.*, 1990); this may lead to the formation of auto-ionizing states (Grizzi *et al.*, 1990).

Measurement of the neutralization rate due to the Auger process is very difficult since a fraction of the Auger electrons do not escape from the surface. Recently, Kempter and collaborators measured energy distributions of secondary electrons produced by collisions in a variety of alkali-ion–surface systems (Brenten *et al.*, 1991a, 1991b, 1991c, 1992). Calculations of the Auger neutralization rate have also been attempted (Zimny *et al.*, 1991; Appelbaum and Hamann, 1975; Hentschke *et al.*, 1986; Snowdon *et al.*, 1986, 1987; Fondén and Zwartkruis, 1992).

For hyperthermal ion–surface collisions, the resonant and Auger mechanisms together are thought to dominate. Therefore, as a next step in the modeling of these collisions, it seems desirable to attempt the inclusion of the Auger mechanism (Amos *et al.*, 1992). This mechanism can be incorporated into the multiple-states model discussed above and this work is in progress. In principle, the model can then be used to treat the scattering of other chemically relevant species, such as oxygen.

5.4 SUMMARY

We have presented results from recent studies in our laboratory of the collision dynamics of 10 to 100 eV ion scattering from clean metal surfaces and the charge-transfer behavior of 5 to 1600 eV alkalis scattering from clean and adsorbate-covered metal surfaces. Using a classical trajectory simulation with a model potential we have demonstrated excellent agreement between measured and simulated scattering distributions for 10 to 100 eV Na^+ scattering from Cu(001). We have shown that, for sufficiently low incident energies, the trajectories, and consequently the scattered energy and angular distributions, are very sensitive to the attractive well of the ion–surface potential. In particular, we have found that it is necessary to include the attractive potential in the simulations in order to reproduce the measured energy loss and the measured forward-rainbow angles for incident energies below 50 eV. In both the experiments and the simulations we observed evidence of particle trapping, which depends strongly on the incident energy. The sensitivity of the calculated distributions to variations in the attractive-potential parameters will be discussed in detail elsewhere. (In previous work we have already addressed the sensitivity to changes in the repulsive part of the ion–surface potential.)

Because the scattering dynamics are so sensitive to the attractive well, it is reasonable to expect that certain outcomes of the collisions, such as energy transfer and trapping, will differ depending on the charge state of the scattering particle. We have presented preliminary measurements showing a difference in the energies of positively charged and neutral Li resulting from collisions of Li^+ with Cu(001). In order to develop a quantitative understanding of these results, we have also investigated the dynamics of the charge-transfer processes that can change the charge state of the scattered particles.

We have studied the charge-transfer behavior for alkalis scattering from clean Cu(001). Velocity- and angle-resolved neutralization probabilities were measured for incident beams of Li^+, Na^+ and K^+ ranging in energy from 5 eV to 1600 eV. For K, the scattered particles are positive ions. For Li and Na, both ions and neutral atoms are observed, although these two species show very different charge-transfer behaviors. Models that describe the charge transfer in terms of a one-electron resonant transfer process give qualitative agreement with the data. Using such a model we have investigated the sensitivity of the final charge-state distributions to the energies and lifetimes of the atomic resonances participating in the charge transfer. In such models, it is often assumed that the charge-state fraction depends on the perpendicular component of the velocity of the scattered particle, which determines the time available for the transfer process to occur. Our measurements have shown, however, that the charge-transfer probability also depends on the parallel component of the scattered velocity, even at the relatively low velocities being discussed here. These results are in agreement with a model that was previously developed to treat the parallel velocity effects that have been observed in grazing-angle scattering studies.

For Li^+ incident on adsorbate-covered surfaces, we have observed positive and negative ions as well as ground-state and excited-state neutrals in the scattered flux. We have compared these results with a multi-channel charge-transfer model, which includes the interactions of several atomic states with one another and with the metal electronic states. Using the model it is possible to examine the occupancies of the different atomic resonances along the particle trajectory. Qualitative agreement with the data is obtained when only resonant transfer processes are included. Future investigations will examine the contribution of the Auger mechanism which can be included in the multi-state model, and which has been shown, for certain experimental conditions, to be a significant channel.

5.5 ACKNOWLEDGEMENTS

The work of several people has contributed in significant ways to the research presented here, as evidenced by the papers cited in this chapter. Particular thanks go to Dag Andersson, Chris DiRubio, Greg Kimmel and Brad Marston, who have contributed directly to the results presented here. We also thank Chris Myers and David Goodstein for their careful reading of the manuscript. This research has been supported by the National Science Foundation (NSF) (Grant No. NSF-DMR-9007799), the Air Force Office of Scientific Research (AFOSR) (Grant No. AFOSR-91-0137), the Cornell Materials Science Center and the Cornell Theory Center.

5.6 NOTES

1 Spectra-Mat Inc., Watsonville, CA 95076 USA.
2 Colutron Research Corp., Boulder, CO 95076, USA.
3 Nordlander calculated the energies and resonance widths for the outer s and p levels for Li and Na above a Jellium surface with $r_s = 2.6$, which is appropriate for copper. These energies and level widths differ only slightly from those calculated for $r_s = 2$.

4 As the ion approaches the surface, the atomic energy levels are shifted by the image attraction between the ion and the surface. We assume that this shift, $V_{im}(z)$, is of the form:

$$V_{im}(z) = (1/V_{min}^2 + 16(z - z_{min})^2/e^4)^{-1/2}$$

for $z > z_{min}$. Here, z is referred to the jellium edge. The levels shown in Figure 5.16 were obtained by assuming $V_{min} = 2.6$ eV and $z_{min} = 0.78$ Å.

5.7 REFERENCES

Adler, D. L., and B. H. Cooper (1988) *Rev. Sci. Instrum.*, **59**, 137.
Adler, D. L., and B. H. Cooper (1991) *Phys. Rev.*, **B 43**, 3876.
Algra, A. J., E. van Loenen, E. P. Th. M. Suurmeijer, and A. L. Boers (1982) *Rad. Eff.*, **60**, 173.
Amos, A. T., B. L. Burrows, and S. G. Davison, (1992) *Surf. Sci.*, **277**, L100.
Andersson, D. R., E. R. Behringer, B. H. Cooper, and J. B. Marston (1993) *J. Vac. Sci. Technol.*, **A 11**, 2133.
Appelbaum, J. A., and D. R. Hamann (1975) *Phys. Rev.*, **B 12**, 5590.
Bahrim, B., D. Teillet-Billy, and J. P. Gauyacq, forthcoming.
Behringer, E. R., D. R. Andersson, D. M. Goodstein, B. Kasemo, B. H. Cooper, and J. B. Marston (1993) *Nucl. Instr. Meth.*, **B 78**, 3.
Behringer, E. R., D. R. Andersson, B. H. Cooper, and J. B. Marston (forthcoming).
Blandin, A., A. Nourtier, and D. Hone (1976) *Journal de Physique*, **37**, 369.
Bloss, W., and D. Hone (1978) *Surf. Sci.*, **72**, 277.
Boers, A. L. (1977) *Surf. Sci.*, **63**, 475.
Bonzel, H. P., A. M. Bradshaw, and G. Ertl. (1989) *Physics and Chemistry of Alkali Metal Adsorption*, Elsevier, Amsterdam.
Brako, R. (1984) *Phys. Rev.*, **B 30**, 5629.
Brako, R., and D. M. Newns (1981) *Surf. Sci.*, **108**, 253.
Brako, R., and D. M. Newns (1985) *Solid State Comm.*, **55**, 633.
Brako, R., and D. M. Newns (1989) *Rep. Prog. Phys.*, **52**, 655.
Brenten, H., H. Müller, and V. Kempter (1991a) *Z. Phys.*, **D 21**, 11.
Brenten, H., H. Müller, K. H, Knorr, D. Kruse, H. Schall, and V. Kempter (1991b) *Surf. Sci.*, **243**, 309.
Brenten, H., H. Müller, D. Kruse, and V. Kempter (1991c) *Nucl. Instrum. Methods*, **B 58**, 328.
Brenten, H., H. Müller, and V. Kempter (1992) *Surf. Sci.*, **271**, 103.
Burgdörfer, J., E. Kupfer, and H. Gabriel (1987) *Phys. Rev.*, **A 35**, 4963.
Burke, K., J. H. Jensen, and W. Kohn, (1991) *Surf. Sci.*, **241**, 211.
Cooper, B. H., C. A. DiRubio, G. A. Kimmel, and R. L. McEachern (1992) *Nucl. Instrum. Methods*, **B 64**, 49.
de Wit, A. G. J., G. A. van der Schootbrugge, and J. M. Fluit (1975) *Surf. Sci.*, **47**, 258.
DiRubio, C. A., K. Burke, and B. H. Cooper (forthcoming a).
DiRubio, C. A., R. L. McEachern, J. G. McLean, and B. H. Cooper (forthcoming b).
Dodson, B. W. (1987) *Phys. Rev.*, **B 36**, 1068.
Echenique, P. M., R. M. Nieminen, J. C. Ashley, and R. H. Ritchie (1986) *Phys. Rev.*, **A33**, 897.
Fondén, T., and A. Zwartkruis (1992) *Surf. Sci.*, **269**, 601.
Frisch, M. J., *et al.* (1988) *Gaussian 88*, Gaussian, Inc., Pittsburgh, Pa.
Gadzuk, J. W. (1967) *Surf. Sci.*, **6**, 133.
Gartland, P. O. (1972) *Phys. Norv.*, **6**, 201.

Gartland, P. O., S. Berge, and B. J. Slagsvold (1973) *Phys. Norv.*, **7**, 39.
Geerlings, J. J. C., and J. Los (1984) *Phys. Lett.*, **A 102**, 204.
Geerlings, J. J. C., and J. Los (1990) *Phys. Repts.*, **190**, 133.
Geerlings, J. J. C., L. F. Tz. Kwakman, and J. Los (1987a) *Surf. Sci.*, **172**, 257.
Geerlings, J. J. C., L. F. Tz. Kwakman, and J. Los (1987b) *Surf. Sci.*, **184**, 305.
Geerlings, J. J. C., R. Rodink, J. Los, and J. P. Gauyacq (1987c) *Surf. Sci.*, **186**, 15.
Geerlings, J. J. C., R. Rodink, J. Los, and J. P. Gauyacq (1987d) *Surf. Sci.*, **181**, L177.
Gerlach-Meyer, U., E. Hulpke, and H. D. Meyer (1979) *Chemical Physics*, **36**, 327.
German, K. A. H., C. B. Weare, P. R. Varekamp. J. M. Andersen, and J. A. Yarmoff (1993) *Phys. Rev. Lett.*, **70**, 3510.
Gonze, X., P. Käckell, and M. Scheffler (1990) *Phys. Rev.*, **B 41**, 12264.
Goodstein, D. M. (1990) Ph.D. Thesis, Cornell University.
Goodstein, D. M., S. A. Langer, and B. H. Cooper (1988) *J. Vac. Sci. Technol.*, **A 6**, 703.
Goodstein, D. M., R. L. McEachern, and B. H. Cooper (1989) *Phys. Rev.*, **B 39**, 13129.
Grimley, T. B., V. C. Jyothi Bhasu, and K. L. Sebastian (1983) *Surf. Sci.*, **124**, 305.
Grizzi, O., M. Shi, H. Bu, J. W. Rabalais, and R. A. Baragiola (1990) *Phys. Rev.*, **B 41**, 4789.
Gurney, R. W. (1935) *Phys. Rev.*, **47**, 479.
Hagstrum, H. D. (1954) *Phys. Rev.*, **96**, 336.
Hagstrum, H. D., P. Petrie, and E. E. Chaban (1988) *Phys. Rev.*, **B 38**, 10264.
Hentschke, R., K. J. Snowdon, P. Hertel, and W. Heiland (1986) *Surf. Sci.*, **173**, 565.
Hermann, J., B. Welle, J. Gehring, H. Schall, and V. Kempter (1984) *Surf. Sci.*, **138**, 570.
Hermann, J., J. Gehring, and V. Kempter (1986) *Surf. Sci.*, **171**, 377.
Horn, T. C. M. (1988) Thesis, Amsterdam.
Horn, T. C. M., Pan Haochang, P. J. van den Hoek, and A. W. Kleyn (1988) *Surf. Sci.*, **201**, 573.
Hsu, C. C., H. Bu, A. Bousetta, J. W. Rabalais, and P. Nordlander (1992) *Phys. Rev. Lett.*, **69**, 188.
Hulpke, E. (1975) *Surf. Sci.*, **52**, 615.
Hulpke, E. and K. Mann (1983) *Surf. Sci.*, **133**, 171.
Inglesfield, J. E. (1983) *Surf. Sci.*, **127**, 555.
Jiang, L. Q., Y. D. Li, B. E. Koel (1993) *Phys. Rev. Lett.*, **70**, 2649.
Kasai, H., and A. Okiji (1987) *Surf. Sci.*, **183**, 147.
Kasi, S. R., M. A. Kilburn, H. Kang, and J. W. Rabalais (1988) *J. Chem. Phys.*, **88**, 5902.
Kasi, S. R., H. Kang, C. S. Sass, and J. W. Rabalais (1989) *Surf. Sci. Reports*, **10**, 1.
Kawai, R., and M. Kawai (1988) *Surf. Sci.*, **195**, 535.
Kawai, R., K. C. Liu, D. M. Newns, and K. Burnett (1987) *Surf. Sci.*, **183**, 161.
Kimmel, G. A. (1992), Ph.D. Thesis, Cornell University.
Kimmel, G. A., and B. H. Cooper (1993a) *Rev. Sci. Instrum.*, **64**, 672.
Kimmel. G. A., and B. H. Cooper (1993b) *Phys. Rev.* **B 48**, 12164.
Kimmel, G. A., D. M. Goodstein, and B. H. Cooper (1989) *J. Vac. Sci. Technol.*, **A 7**, 2186.
Kimmel, G. A., D. M. Goodstein, Z. H. Levine, and B. H. Cooper (1991) *Phys. Rev.*, **B 43**, 9403.
Lang, N. D. (1971) *Phys. Rev.*, **B 4**, 4234.
Langreth, D. C., and P. Nordlander (1991) *Phys. Rev.*, **B 43**, 2541.
Mann, K., V. Celli, and J. P. Toennies (1987) *Surf. Sci.*, **185**, 269.
Marston, J. B., D. R. Andersson, E. R. Behringer, B. H. Cooper, C. A. DiRubio, G. A. Kimmel, and C. Richardson (1993) *Phys. Rev.*, **B 48**, 7809.

McEachern, R. L. (1990) Ph.D. Thesis, Cornell University.
McEachern, R. L., D. L. Adler, D. M. Goodstein, G. A. Kimmel. B. R. Litt, D. R. Peale, and B. H. Cooper (1988) *Rev. Sci. Instrum.*, **59**, 2560.
McEachern, R. L., D. M. Goodstein, and B. H. Cooper (1989) *Phys. Rev.*, **B 39**, 10503.
Milošević, D. B. (1992) *Surf. Sci.*, **273**, 175.
Nakanishi, H., H. Kasai, and A. Okiji (1988) *Surf. Sci.*, **197**, 515.
Niehus, H., and R. Spitzl (1991) *Surf. and Int. Anal.*, **17**, 287.
Nordlander. P. (1992) *Phys. Rev.*, **B 46**, 2584.
Nordlander, P., and N. D. Lang (1991) *Phys. Rev.*, **B 44**, 13681.
Nordlander, P., and J. C. Tully, (1988) *Phys. Rev. Lett.*, **61**, 990.
Nordlander, P., and J. C. Tully (1989) *Surf. Sci.*, **211/212**, 207.
Nordlander, P. and J. C. tully (1990). *Phys. Rev.*, **B 42**, 5564.
Parilis, E. S., and N. Yu, Turaev (1965) *Sov. Phys. Dok.*, **10**, 212.
Peale, D. R., D. L. Adler, B. R. Litt, and B. H. Cooper (1989) *Rev. Sci. Instrum.*, **60**, 730.
Poelsema, B., L. K. Verhey, and A. L. Boers (1967a) *Surf. Sci.*, **56**, 445.
Poelsema, B., L. K. Verhey, and A. L. Boers (1976b) *Surf. Sci.*, **60**, 485.
Poelsema, B., L. K. Verhey, and A. L. Boers (1977a) *Surf. Sci.*, **64**, 537.
Poelsema, B., L. K. Verhey, and A. L. Boers (1977b) *Surf. Sci.*, **64**, 554.
Radzig, A. A., and B. M. Smirnov (1985) *Reference Data on Atoms, Molecules, and Ions*, Vol 31 of Springer Series in Chemical Physics, Springer, Berlin.
Rasser, B., J. N. M. van Wunnik, and J. Los (1982) *Surf. Sci.*, **118**, 697.
Schall, H., W. Huber, H. Hoermann, W. Maus-Friedrichs, and V. Kempter (1989) *Surf. Sci.*, **210**, 163.
Schall, H., H. Brenten, K. H. Knorr, and V. Kempter (1990) *Z. Phys.*, **D 16**, 161.
Sebastian, K. L. (1985) *Phys. Rev.*, **B 31**, 6976.
Snowdon, K. J., R. Hentschke, A. Närmann, and W. Heiland (1986) *Surf. Sci.*, **173**, 581.
Snowdon, K. J., R. Hentschke, A. Närmann, W. Heiland, E. Mühling, and W. Eckstein (1987) *Nucl. Instrum. Methods*, **B 23**, 309.
Snowdon, K. J., D. J. O'Connor, and R. J. Macdonald (1988) *Phys. Rev. Lett.*, **61**, 1760.
Sols, F., P. Miranzo, and F. Flores (1985) *Surf. Sci.*, **161**, 33.
Souda, R., T. Aizawa, C. Oshima, and Y. Ishizawa (1990) *Nucl. Instrum. Methods*, **B 45**, 369.
Sulston, K. W., A. T. Amos, and S. G. Davison (1988) *Phys. Rev.*, **B 37**, 9121.
Sulston, K. W., A. T. Amos, and S. G. Davison (1989) *Surf. Sci.*, **224**, 543.
Taglauer, E., and W. Heiland (1972) *Surf. Sci.*, **33**, 27.
Taglauer, E., and W. Heiland (1976) *Appl. Phys.*, **9**, 261.
Taylor, J. B., and I. Langmuir (1933) *Phys. Rev.*, **44**, 423.
Teillet-Billy, D., and J. P. Gauyacq, forthcoming.
Tenner, A. D., K. T. Gillen, T. C. M. Horn, J. Los, and A. W. Kleyn (1984) *Phys. Rev. Lett.*, **52**, 2183.
Tenner, A. D., K. T. Gillen, T. C. M. Horn, J. Los, and A. W. Kleyn (1986a) *Surf. Sci.*, **172**, 90.
Tenner, A. D., K. T. Gillen, and A. W. Kleyn (1986b) *Nucl. Instrum. Meth. Phys. Res.*, **B 17**, 108.
Tenner, A. D., R. P. Saxon, K. T. Gillen, D. E. Harrison Jr., T. C. M. Horn, and A. W. Kleyn (1986c) *Surf. Sci.*, **172**, 121.
Tully, J. C. (1977) *Phys. Rev.*, **B 16**, 4324.
van Amersfoort, P. W., J. J. C. Geerlings, L. F. Tz. Kwakman, E. H. A. Granneman, and J. Los (1985) *J. Appl. Phys.*, **58**, 2312.
van den Hoek, P. J., A. D. Tenner, A. W. Kleyn, and E. J. Baerends (1986) *Phys. Rev.*, **B 34**, 5030.

van Wunnik, J. N. M., R. Brako, K. Makoshi, and D. M. Newns (1983) *Surf. Sci.*, **126**, 618.
Winter, H. (1991) *Comm. At. Mol. Phys.*, **26**, 287.
Winter, H., and R. Zimny (1988) In *Coherence in Atomic Collision Physics*, H. J. Beyer, K. Blum, and R. Hippler (eds), Plenum Press, New York.
Yu, M. L., and N. D. Lang (1983) *Phys. Rev. Lett.*, **50**, 127.
Yurasova, V. E., V. I. Shulga, and D. S. Karpuzov (1968) *Can. J. Phys.*, **46**, 759.
Zimny, R. (1990) *Surf. Sci.*, **233**, 333.
Zimny, R., H. Nienhaus, and H. Winter (1989) *Rad. Eff. Def. Sol.*, **109**, 9.
Zimny, R., Z. L. Mišković, N. N. Nedeljković, and Lj. D. Nedeljković (1991) *Surf. Sci.*, **255**, 135.

6 THE INTERACTION OF MOLECULAR IONS WITH SURFACES

W. Heiland
FB Physik, Universität Osnabrück, D-4500 Osnabrück, Germany

CONTENTS

6.1 INTRODUCTION

The interaction of molecules with surfaces is a very broad field of research in chemistry and physics. In recent years, 'hyperthermal' beams of neutral and ionic molecules have been used to study molecule–surface interactions and have been used for surface property modifications. Hyperthermal in this context

Low Energy Ion–Surface Interactions
Edited by J. W. Rabalais

means energies from about 10 eV up to several keV. An interesting aspect of using ion beams is the possibility of varying the 'perpendicular' energy $E_{\perp}$. If we define the glancing angle of incidence ψ as being measured from the surface, the particles have velocity components $v_{\perp} = v \sin \psi$ and $v_{\parallel} = v \cos \psi$, where v is the velocity of the particle. The perpendicular energy is consequently $E_{\perp} = E_0 \sin^2 \psi$, where E_0 is the primary energy of the particle. At small angles of incidence, perpendicular energies below 10 eV can be achieved, i.e. the range of hyperthermal energies in the definition of chemical reactions can easily be achieved. Since the parallel velocity component can be quite large, the interaction times of the particles can be small compared to typical molecular vibrational or rotational times. When E_0 is in the range of 100 eV to 1 keV, the velocities of light molecules are in the range of 10^6 to 10^7 cm s^{-1}. The interaction time depends on the trajectory length, which in turn depends on the impact angle. A sensible assumption of the order of magnitude of the trajectory lengths is 10 Å. Typical interaction times are hence of the order of 10^{-14} to 10^{-15} s. Based on these estimates it is safe to consider ion-beam experiments in 100 eV to 1 keV range as not influenced by rotational effects or by solid-state phonons. The vibrations of the molecules have to be considered in each case. The most important interaction will be the electron exchange between the molecule and the surface. The electron transition times are of the order of 10^{-15} s, such that the molecule–surface interaction may be dealt with within the Born–Oppenheimer approximation.

Consequently, the interaction is divided into two aspects, the elastic scattering described by a classical trajectory and the charge exchange and vibrational events described by a potential-energy surface. The caveats and the problems of such a modelling of a very complex multistate interaction are obvious, but they are challenging for both experiment and theory.

In the following, we will discuss the basic experimental phenomena, the experimental systems, some experimental results in greater detail, and conclude with a summary of the theoretical models.

6.2 BASIC EXPERIMENTAL RESULTS

The first experiments of low-energy molecular ions scattering from surfaces were part of atomic ion scattering experiments (Panin, 1962; Smith, 1967). These experiments did not show striking effects. It was obvious that the molecules dissociated when they hit the surface—not unexpected, considering that the kinetic energies used were of the order of 1 keV and the molecular binding energies are a few eV. The situation changed drastically when in the course of experiments in the context of plasma–wall interactions in fusion reactors, D_3^+ molecular ions were scattered at grazing incidence from a clean metal surface (Figure 6.1) (Eckstein *et al.*, 1975).

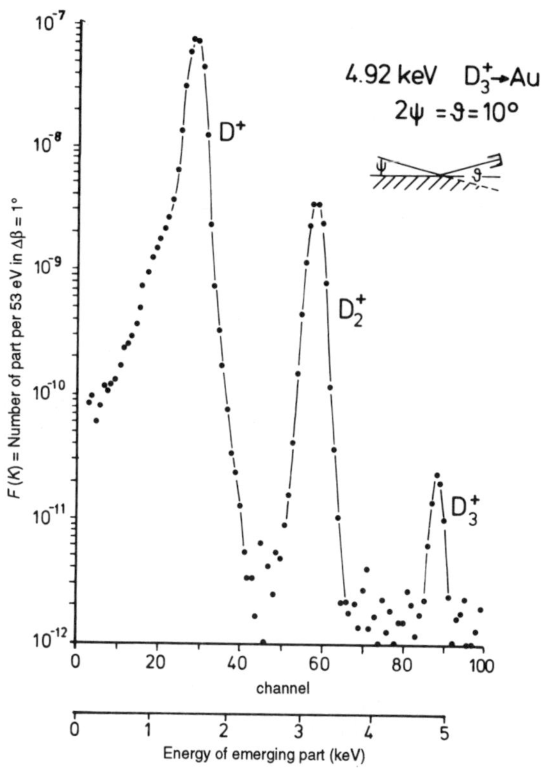

Figure 6.1 Energy spectra of D_3^+ and its dissociation products D_2^+ and D^+ incident at $E_0 = 4.92$ keV on Au, $\psi = 5°$, $\vartheta = 10°$; $\Delta\beta$ is the angle of acceptance of the detector. (Reproduced with permission from Eckstein *et al.*, 1975)

At a kinetic energy $E_0 = 5$ keV and a glancing angle $\psi = 5°$, the perpendicular energy is $E_\perp = 38$ eV. Nevertheless, surviving D_3^+ ions with nearly the full initial energy are found. The dissociation energy of D_3^+ is 4.7 eV to form $D^+ + D_2$ and 6.55 eV to form $D + D_2^+$. The peak at $\frac{2}{3}\,E_0$ is identified as D_2^+ with a dissociation energy of 2.9 eV. The third peak in Figure 6.1, positioned at $\frac{1}{3}\,E_0$, is identified as dissociation products, i.e. atomic D^+ ions. The elastic loss of the particles as estimated from the classical trajectory is very small. However, the inelastic loss, of e.g. the D_3^+ ions, is 150 eV. The fact is that D_3^+ molecules with a binding energy of about 5 eV do survive a collsion with a surface with a total kinetic energy of 5 keV and a perpendicular energy of 40 eV. Furthermore, these molecules survive a total inelastic energy loss of 150 eV.

The peaks of the D^+ and D_2^+ ions are broadened by the dissociation process as can be estimated from conservation of momentum and energy. If E_D is the binding energy released by the dissociation process, the energy distribution of the dissociated particles is given by

$$E = \tfrac{1}{2}E_0 + \tfrac{1}{2}E_D \pm \sqrt{E_0 \cdot E_D} \cdot \cos \alpha \tag{1}$$

for a diatomic molecule. The angle α defines the orientation of the molecular axis with respect to the beam direction.

Further interpretation of the experimental findings was impaired by a problem common to all experiments using an electrostatic analyser for the energy analysis of the scattered particles: it is only the ions that can be measured (see Section 6.3). The authors (Eckstein *et al.*, 1975) estimated the ion-to-neutral ratio D^+/D^0 to be of the order of 10%. The first experiment using time-of-flight (TOF) analysis corroborated this prediction (Figure 6.2) (Willerding *et al.*, 1984a, 1985), i.e. the ion peaks are of the order of 1% compared to the neutral peak. The ion peaks are separated from the neutral peak by post-acceleration. Since the flight time decreases with channel number, the lightest ions, H^+, appear at the highest channel number. The broadening of the H^+ and H_2^+ peaks according to equation (1) is obvious. The neutral peak, which is also very broad, may contain H and H_2. There will be no H_3, because H_3 is unstable in the ground state. With the diatomic primary ions separation of the dissociated and surviving neutrals is possible to a certain extent (see Section 6.4). The surviving neutrals have a narrow TOF distribution, as e.g. the H_3^+ peak in Figure 6.2, whereas the dissociated neutrals have a broad distribution, such that the neutral peak is simply the sum of a broad peak and a narrow peak. Notwithstanding the complications arising from the fact that a three-atom

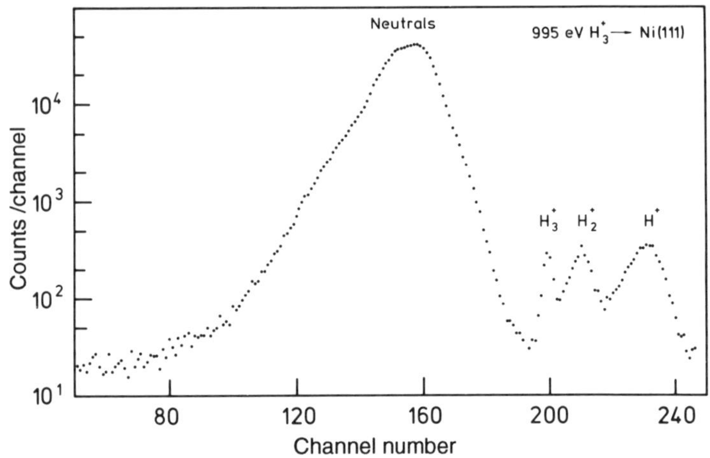

Figure 6.2 Time-of-flight spectra of H_3^+ and its ionic and neutral dissociation products H^+, H_2^+, H and H_2 incident at 995 eV on Ni(111), $\psi = 5°$, $\delta = 10°$, $\Delta\beta = 1.2°$. The time increases from right to left. The ions appear at short flight times due to post-acceleration (Reproduced with permission from Willerding *et al.*, 1985)

molecule was used, Figure 6.2 gives clear evidence that charge capture from the surface by the molecule is quantitatively the dominant process in the scattering event—more than 90% of all specularly scattered particles are neutrals.

The importance of charge capture is even more obvious in cases where negative molecular species are found. The first experiments of this kind used O_2^+ scattering rom Ag(111) (Figure 6.3) (Pan Haochang *et al.*, 1986). An electrostatic analyser was used here, too, i.e. only ionic species could be detected. The events leading to the formation of O_2^- can be listed as: (i) the neutralization of the primary positive ion, $O_2^+ + e \rightarrow O_2$, (ii) the (hard) collision with the surface, and (iii) the attachment of a second electron, $O_2 + e \rightarrow O_2^-$. The last step is a 'harpooning' reaction as known from gas-phase collisions (Herschbach,

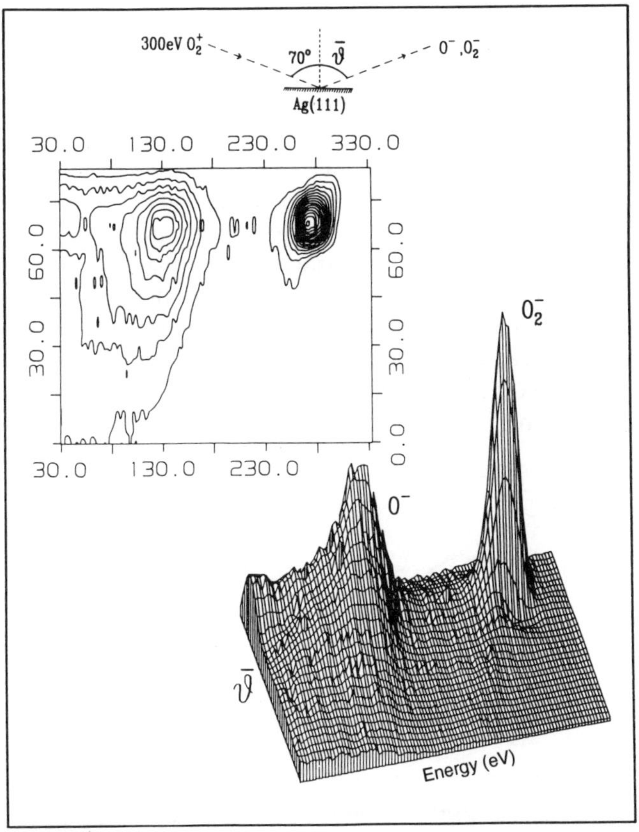

Figure 6.3 Energy spectra of the negative ions O^- and O_2^- from the scattering of O_2^+ from Ag(111) at $E_0 = 300$ eV, $\psi = 20°$ and $\bar{\delta} = 70°$ ($\delta = 40°$). Note the broadening of the O^- peak in energy and angle. The intensity scale is lincar. (Reproduced by permission of the American Physical Society from Pan Haochang *et al.*, 1986)

1966). At low velocities step (iii) is likely to precede the collision with the surface (ii). A partial coverage of the silver surface with K caused an increase of the O^- yield as may be expected from the lowering of the work function. The remarkable result is the qualitative agreement of the finding of the O_2^- with results from chemisorption experiments (Backx *et al.*, 1981; Campbell, 1986). The interpretation of the photoelectron spectroscopy data indicates an O_2^--like species as a possible chemisorbed particle on Ag(111). Later experiments using a TOF system showed that in the O_2^+ case neutralization is indeed a major event, i.e. an appreciable number of incident O_2^+ ions survive the surface interaction as O_2 neutrals (Reijnen *et al.*, 1989). The harpooning reaction also found theoretical interest. It was shown that (i) the process is indeed possible, (ii) it leads to the transfer of translational to vibrational energy, and (iii) it may enhance the dissociation (Gadzuk and Nørskov, 1984; Gadzuk and Holloway, 1985; Feibelmann, 1985).

More recently a position-sensitive detector has been used to measure the spatial distribution of the scattered particles (Figure 6.4) (Tappe *et al.*, 1991; Tappe, 1990). The comparison of the distribution obtained with H^+ and H_2^+ as the primary particles shows the expected broadening due to dissociation. From previous experiments (Willerding *et al.*, 1985), we know that there is complete dissociation of H_2^+ at a primary energy of 1500 eV. Model calculations (Imke *et al.*, 1986a, 1986b) based on a charge-capture treatment of the ion–surface interaction indicate that the dissociation probability depends on the orientation of the molecular axis with respect to the surface normal. The conservation of the total symmetry of the event (Dunn, 1962) requires that, if the dissociation proceeds via capture into the $b^3\Sigma_u^+$ state of H_2^+, the molecules oriented with their axes normal to the surface dissociate preferentially. Hence the spatial distribution is asymmetric, i.e. broadened in the direction normal

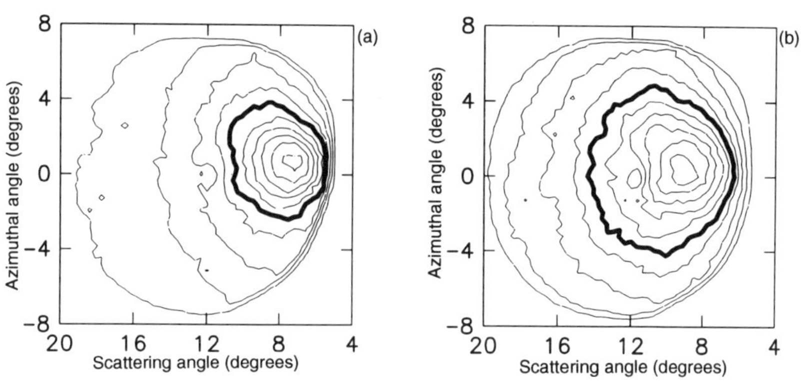

Figure 6.4 Spatial distribution of dissociated H from the interaction of H^+ and H_2^+ with a Ni(110) surface at $E_0 = 1.5$ keV per proton, $\psi = 5°$. The intensity scale is linear in steps of 0.10. The thick line marks the 50% level. The azimuthal angle is 45° of the $[1\bar{1}0]$ surface direction, i.e. 'random' (Tappe *et al.*, 1991; Tappe, 1990)

to the surface relative to the direction parallel to the surface. The binding energy released is evaluated from the widths of spatial distributions. The value of about 4 eV found for the direction normal to the surface agrees with the value expected from the $b^3\Sigma_u^+$ repulsive potential curve. Parallel to the surface, the value is only about 2.3 eV, which is interpreted as being the signature of a dissociation due to the dynamic screening of the molecules by the solid-state electrons (see Section 6.5).

In this section, we have presented a short overview of the present types of experimental results. These indicate the influence of a variety of possible electronic processes, as well as the involvement of collisional processes in the interaction of molecular ions with surfaces.

6.3 EXPERIMENTAL CONSIDERATIONS

As in all surface-science experiments, ultra-high vacuum (UHV) is a necessary requirement for the study of the interaction of molecular ions with a surface. The requirements here are even more stringent than for regular ion-scattering spectroscopy (ISS) at low energies (Smith, 1967; Heiland and Taglauer, 1985). In ISS, mainly rare-gas ions are used which do not stick to the surface, or react with adsorbed species on the target or on the walls of the vacuum vessel, thus causing changes of the rest gas composition. The interest in experiments with molecules is partly based on their reactivity, such that better differential pumping between ion source and target chamber is necessary. The pumping installed at the target must be sufficiently high to maintain a low base pressure during the scattering experiment.

In other respects the experimental arrangement is equivalent to regular ISS experiments: an ion source followed by a mass analyser, a target manipulator and a particle analysis and detection system. It is also necessary to control the target condition, cleanliness and periodicity (the latter in the case of single crystals only). A technique like ISS can be used for probing the target surface for adsorbates. LEED or surface channelling effects (Niehof and Heiland, 1990; Heiland, 1989) may be used to probe the surface structure.

Two type of systems are used for the analysis of the energy distribution of the scattered particles: electrostatic analysers (ESA) and time-of-flight (TOF) systems. Analysis of neutral particles is not possible with the use of an ESA. The reionization of the neutralized particles in the type of experiments discussed here has not been realized so far: the particles are fast (corresponding to 100 eV more) and the currents used are low, i.e. of the order below 10^{-8} A. The technical details of electrostatic analysers, ion sources and magnets for ISS have been discussed elsewhere (Heiland and Taglauer, 1985). On the other hand, TOF systems have the advantage of affording neutral-particle analysis (Rathmann *et al.*, 1985; Kilburn *et al.*, 1988). The addition of a post-acceleration stage makes

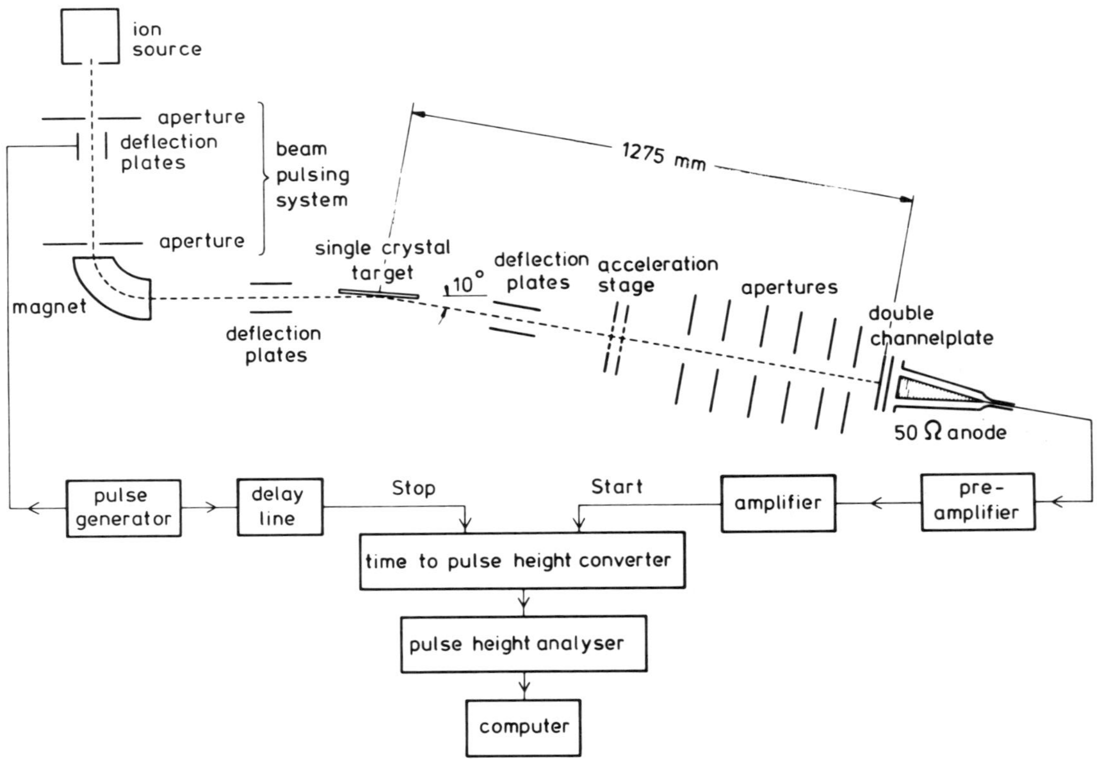

Figure 6.5 Experiment for time-of-flight analysis of molecule–surface interactions in the energy range from 100 eV to 10 keV at grazing incidence (Reproduced with permission from Willerding *et al.*, 1984b)

the separation of ions and neutrals possible (Figure 6.2). Figure 6.5 shows the schematic diagram of an experiment devoted to the analysis of molecular ions interacting with clean surfaces (Willerding *et al.*, 1984b).

Not shown in Figure 6.5 is the position-sensitive detector (PSD) mounted downstream for measurement of the angular distribution of the scattered particles (Niehof and Heiland, 1990). Using segmented anodes, such a detector can be combined with a TOF analysis. By this means, angular resolved energy analysis is possible (de Bruijn and Los, 1982; de Bruijn, 1983; Rechtien *et al.*, 1992). Hence, the relative energy of the products of a dissociation process can be measured. Angular analysis with a TOF detector is achieved simply by a movable detector (Kilburn *et al.*, 1988). It is of course necessary to provide a vacuum chamber large enough in order to reach a sufficient time resolution. The time resolution is determined by the pulse-generation system and the length of the flight path behind the target. Pulse widths of the order of a few ns ('square' pulses) and flight paths of ≤ 1 m are the approximate conditions for experiments with light molecules in the energy range from about 100 eV to a few keV (Rathmann *et al.*, 1985).

For experiments involving neutral molecules, a gas cell is installed in the primary beam path between magnet and target. Gas cell designs are amply described in the literature of atomic collision physics. These cells are essentially tubes with a diaphragm at each end. The lengths are of the order of 10 cm; diaphragms with about 1 mm diameter are used. The important parameter is the product *length* × *pressure*, by which single-collison conditions in the tube are achieved. The efficiency of the cell is given by the charge-exchange cross section (of the order of 10^{-15} cm^2) between the ion and the gas in the cell. It is sensible to use a non-reactive gas in the cell (Ar, N_2 with caution). The advantage of N_2 is its high pumping speed for all types of pumps. Surviving ions have to be deflected between cell and target. So far, three groups have used neutral swift molecules (Kleyn, 1990; Närmann *et al.*, 1992 and Schmidt, 1992; Rechtien *et al.*, 1992).

6.4 ENERGY DISTRIBUTIONS

The information about the possible dissociation mechanisms is partly contained in the energy distribution of the dissociated particles as shown by equation (1). This is a well-known phenomenon in collisonal mass spectrometry (Cooks, 1976). It is used also for the study of the structure of molecular ions by dissociation at high energies in thin foils (Vager *et al.*, 1989). At high energies (MeV), any particle is ionized rather than neutralized when passing through a solid such that the atoms of a molecule are repelled from each other. This effect causes energy distributions with widely separated peaks and ring-structured angular distributions for diatomic molecules. At low energies, charge capture is the dominant process. It has been shown (Heiland *et al.*, 1977) that a 'Coulomb-explosion' does not occur at low energies. The energy distributions measured for the atomic species dissociated from diatomic molecules are in fact caused by energy input in the centre-of-mass system, which does not have a sharp Coulomb energy but rather a distribution across part of the eV range. Such distributions can be the signature of charge-capture processes into an antibonding state, vibrational excitation, plasmon excitation, electron–hole-pair excitation, or simply a random mixture of hard collisions with surface atoms. In this section, a 'biased' selection of experimental results will be discussed in order to present a survey of the phenomena and to structure the following discussion of the models (Section 6.5).

6.4.1 HYDROGEN

Being one of the simple diatomic molecules, hydrogen is certainly of interest in the experiments discussed here. It is not accidental that some of the first studies (Smith, 1967; Heiland *et al.*, 1977) originate from a laboratory involved in 'plasma–wall interactions'. On the other hand, a basic question is one regarding the circumstances under which a molecule scatters as a structureless entity or is recognized as having a structure (Datz, 1981). The ellipticity of diatomics indeed depends on the probe. At thermal velocities, for example, it takes rather sophisticated experiments to reveal deviations from spherical symmetry.

As indicated in Figure 6.2, the ion yield in a scattering experiment at low energies is low. It tends towards $\leq 10^{-4}$, when the energy is lowered to 250 eV, which is at an angle of incidence (glancing) $\psi = 5^\circ$ is equivalent to $E_\perp = 1.9$ eV (Figure 6.6) (Schubert *et al.*, 1990; Schubert, 1989). This low ion yield is observed for both positive and negative ions. The ion yield—being so low—is extremely sensitive to contaminations and surface roughness.

When the work function of Ni(110) is lowered to 2.0 eV by K adsorption, the negative-ion yield increases by orders of magnitude (Figure 6.6). However, the energy distributions of the neutrals also change appreciably. A clear narrow peak is distinctly visible on the broad hump of neutral H atoms. The peak is even narrower than the one produced by incident H^+ at the same velocity. It is clearly elastically scattered neutralized H_2. Evidence that the broad peak is H is qualitatively obtained from the fact that it extends beyond the primary energy E_0 which is at 9.5 μs in the TOF spectrum of Figure 6.6. Model calculations give further evidence for this interpretation (Heiland *et al.*, 1977; Schubert *et al.*, 1990; Schubert, 1989).

In cases where there are suriviving ions, atoms and molecules, their energy spectra can be separated (Panin, 1962; Smith, 1967; Eckstein *et al.*, 1975). In those cases, the widths of the ionic peaks are equal to the widths of the neutral

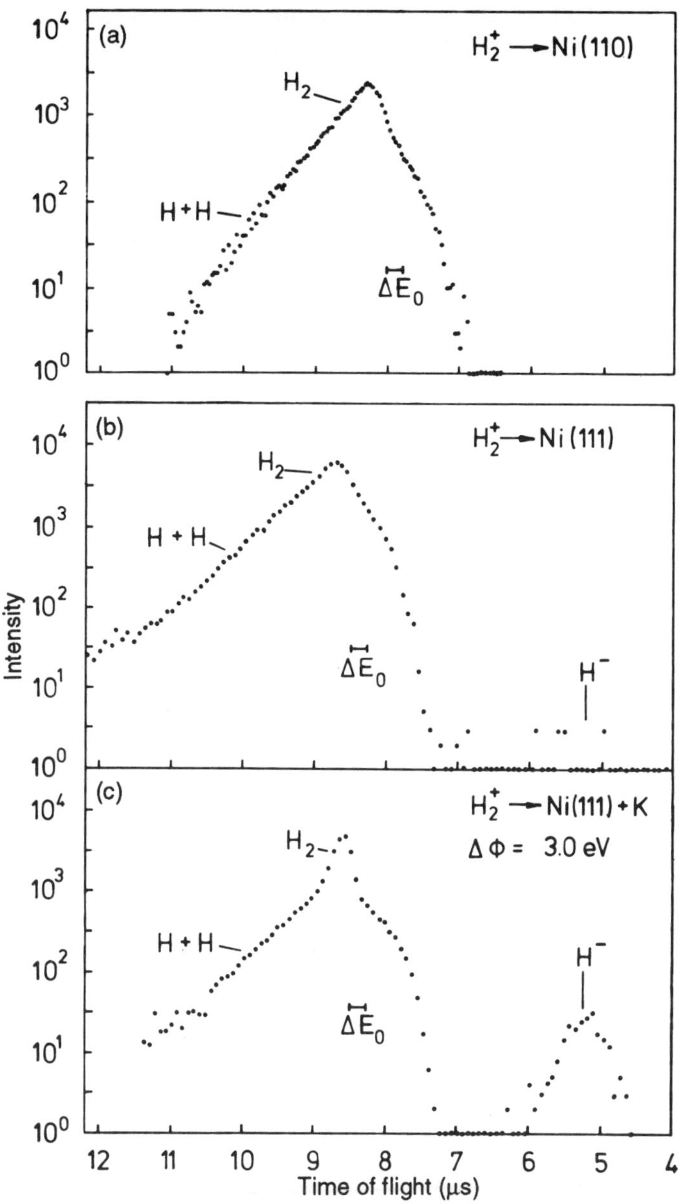

Figure 6.6 Time-of-flight spectra of H_2^+ and its ionic and neutral products H^-, H_2 and H incident at $E_0 = 250$ eV on Ni(110), Ni(111) and Ni(111) + K, $\psi = 5°$, $\delta = 10°$ and $\Delta\beta = 1.2°$. The H^- are post-accelerated (Schubert *et al.*, 1990; Schubert, 1989)

peaks—an observation giving further qualitative evidence for the interpretation of the energy or TOF spectra. It is important to note that not only does the ion yield change by changing the work function, as found in a previous ESA experiment (Pan Haochang *et al.*, 1986), but also the neutral yields change quantitatively, i.e. molecular survival yield is affected.

Another type of experiment using hydrogen was aimed at probing the influence of the electronic structure of the surface by comparing scattering from Al and Ni (Heiland *et al.*, 1987). Within the experimental error, the energy distributions are identical for the two surfaces. This finding does exclude the influence of plasmons—not unexpected at energies below 1 keV and grazing incidence (Lucas, 1979). Figure 6.7 shows the TOF spectra of the ions only. The neutral parts of the spectra are identical too. The ions ought to be most sensitive to the differences of the plasmon properties of Ni and Al.

Still another 'failure' was the search for electronically excited H_2 molecules due to their interaction with the surface. The Balmer series of H is easily measured up to main quantum numbers of the upper state 9 or even 10 (Neumann *et al.*, 1987), but with H_2 no optical emission was detected in that experiment.

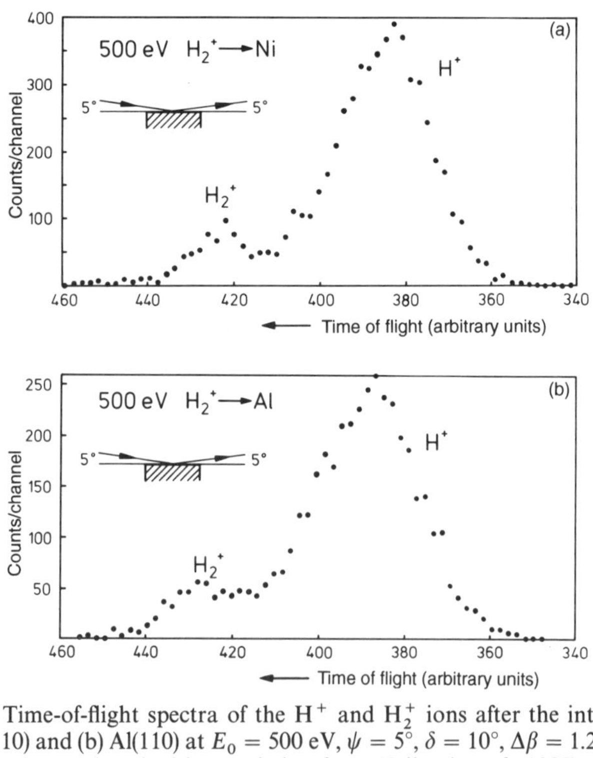

Figure 6.7 Time-of-flight spectra of the H^+ and H_2^+ ions after the interaction of H_2^+ with (a) Ni(110) and (b) Al(110) at $E_0 = 500$ eV, $\psi = 5°$, $\delta = 10°$, $\Delta\beta = 1.2°$. The intensity scale is linear (Reproduced with permission from Heiland *et al.*, 1987)

A new series of experiments was started recently by the use of neutral primary molecules (Harder *et al.*, 1992; Kleyn, 1990; Närmann *et al.*, 1992; Schmidt, 1992; van Slooten *et al.*, 1991; Schmidt *et al.*, 1992a). With H_2 impinging on Ag(111), the energy spectra of the back-scattered H_2^+ and H^+ were measured using an ESA. the primary energy range is 70 eV to 1 keV. The glancing angle of incidence is varied between 10° and 20°. Thus, the perpendicualr energy $E_\perp$ is actually varied from 2 eV to 50 eV. At all energies the ionic–molecular yield is lower for incident H_2^+ compared to incident H_2^0. In the case of incident H_2^0, the dissociated fraction $F_{\text{dis}} = (H^+/(H^+ + H_2^+))$ scales with $E \cdot \Theta^2$, where Θ is the total scattering angle.

The neutral fraction was measured in new experiments with H_2^+ and H_2^0 scattering from Ni(110) and Pd(110) (Schmidt *et al.*, 1992). Here, the neutral-molecular yield is also larger in the case of the neutral beam compared to the ionic beam. The positive- and negative-ion yields are in the 10^{-3} range for both Ni and Pd; these are much lower ion yields compared to the H_2–Ag case, estimated at about 1% (van Slooten *et al.*, 1992).

6.4.2 NITROGEN AND CARBON MONOXIDE

Compared to hydrogen, the isoelectronic molecules N_2 and CO are in some respect simpler systems. In surface scattering experiments at low energies and grazing incidence, N_2 and CO show very little dissociation (Willerding *et al.*, 1984a; Schubert *et al.*, 1990; Schubert, 1989; Heiland *et al.*, 1987). At clean well-prepared Ni surfaces, the molecular survival probability is close to 1.0. The energy spectra exhibit a clear separation of the peak of surviving molecules form the (here) low broad peak of dissociated molecules (Figure 6.8) (Schubert *et al.*, 1990; Schubert, 1989). In the case of CO^+ primary ions, a small yield of negative O ions is measured. It is quite obvious that these negative ions are post-dissociation products. The ion yields from clean Ni(111) are very low.

The dissociation probability of N_2^+ and CO^+ increases when the energy is raised (Willerding *et al.*, 1984b; Sass and Rabalais, 1987; Kang *et al.*, 1988; Kasi *et al.*, 1989; Balashova *et al.*, 1982; Balashova and Molchanov, 1982; Balashova, 1991; Heiland and Taglauer, 1982). Furthermore, trapping of the dissociation products is observed (Kang *et al.*, 1988; Kasi *et al.*, 1989; Heiland and Taglauer, 1982). This is especially interesting in cases in which the atoms or molecules do not absorb at a given surface or do not adsorb dissociatively. The effects of trapping and thin-film growth will be discussed by Marton in Chapter 9. Of course, increasing kinetic energy also leads to a transition from trapping to implantation.

At the low-energy end of molecule–surface interactions, results have been reported which show a reasonance-like feature for the dissociative adsorption of N_2^+ on Ni (Figure 6.9) (Akazawa and Murata, 1988a, 1986b). The observed

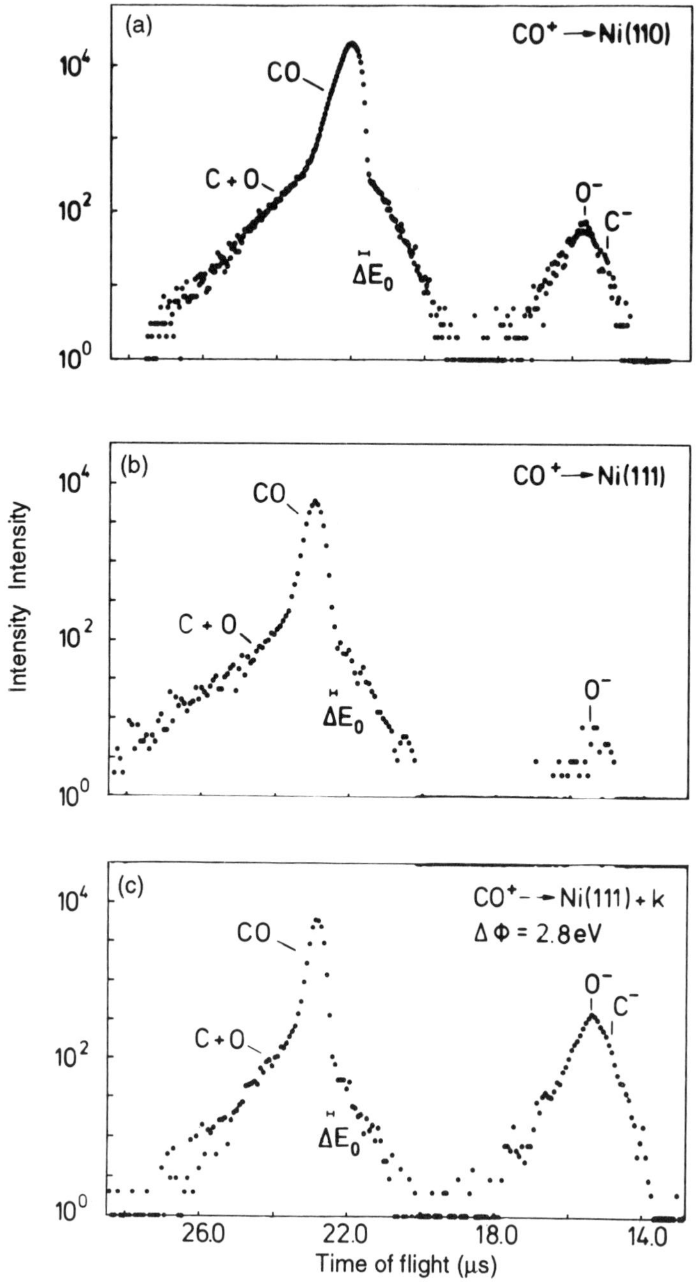
(a)
CO⁺ → Ni(110)
CO
C + O
ΔE0
O⁻
C⁻
(b)
CO⁺ → Ni(111)
(c)
CO⁺ → Ni(111) + k
ΔΦ = 2.8eV
Intensity
26.0
22.0
18.0
14.0
Time of flight (μs)

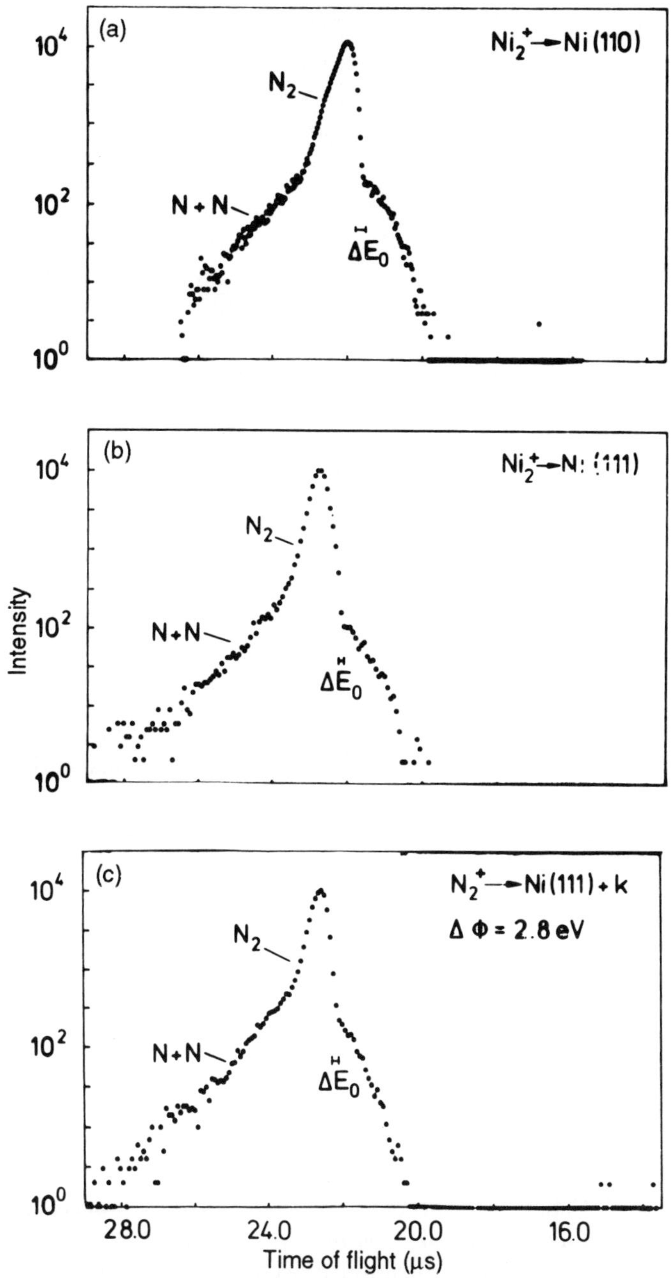

Figure 6.8 Time-of-flight spectra of CO^+ and N_2^+ scattering from Ni(110) at $E_0 =$ 480 eV, $\psi = 5°$, $\delta = 10°$, $\Delta\beta = 1.2°$ (Schubert *et al.*, 1990; Schubert, 1989)

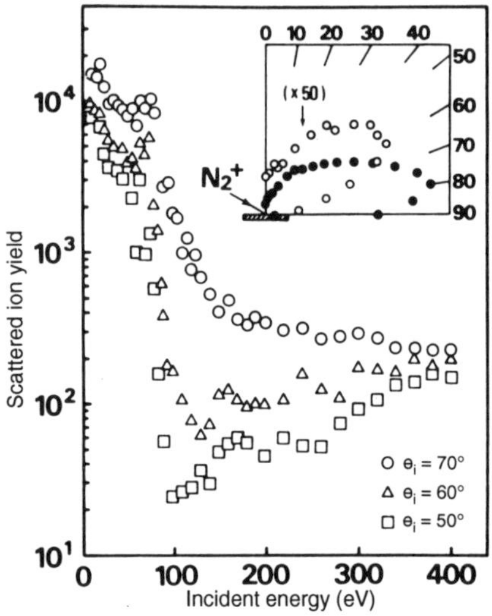

Figure 6.9 Ion yield of N_2^+ scattered at different angles of incidence Θ_i (measured from the surface normal) as a function of primary energy. For each Θ_i the detector is at the specular position. The inset shows the angular distributions of N_2^+ at $E_0 = 200$ eV (open circles) and at $E_0 = 50$ eV (filled circles) at angle of incidence $\Theta_i = 60°$. (Reproduced by permission of the American Physical Society from Akazawa and Murata, 1988a)

effects are accompanied by irregularities in the ion yields of Ar^+ and N^+ at the same low energies (below 100 eV). The angular dependence of the dissociative adsorption of N_2^+ on Ni above 100 eV leads to the conclusion that total energy scaling prevails over the normal energy scaling. This finding differs from the H_2–Ag case where $E\Theta^2$-scaling has been observed (see Section 6.4.1 and van Slooten *et al.*, 1991).

6.4.3 REACTIVE MOLECULES AND HARPOONING

The molecular species discussed in this section are distinct from those discussed previously in that they have a stable negative state, i.e. negative affinity energy As shown in Figure 6.3 for O_2 scattering from Ag, negative atomic ions and molecules are found (Pan Haochang *et al.*, 1986). These findings have been interpreted in terms of the classical 'harpooning' process (Herschbach, 1966), (see Section 6.5). Figure 6.10 shows the dependence of the molecular yield on the surface work function (Schubert *et al.*, 1990; Schubert, 1989). It is obvious that there is a dominant neutralization process, such that the role played by the O_2^- state in the dissociation process is not immediately evident. As in the case of N_2 and CO, there is a distinct neutral O_2 peak. However, in the case

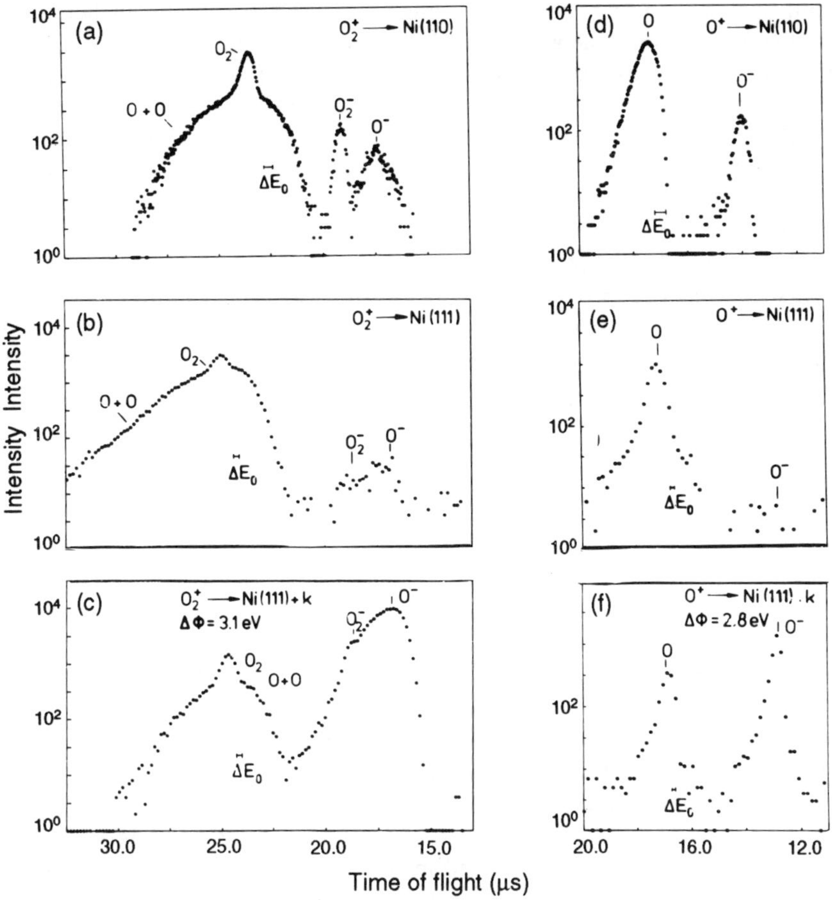

Figure 6.10 Time-of-flight spectra of O_2^+ and O^+ with post-acceleration of the negative species, incident on Ni(110), Ni(111) and Ni(111) + K. Primary energies vary between 450 eV and 490 eV; geometry is as in Figure 6.8 (Schubert *et al.*, 1990; Schubert, 1989)

of O_2^+ primary ions, there are two possible paths for surviving O_2. The following scheme gives a list of possible occurring 'reactions':

(a) $O_2^+ + e^- \to O_2$	Auger neutralization
(b) $O_2 + e^- \to O_2^-$	resonant capture (harpooning)
(c) $O_2^- - e^- \to O_2$	resonant loss
(d) $O_2^+ + e^- \to O + O$	dissociative attachment (Auger or resonant capture)
(e) $O_2^- \to O + O^-$	rovibrational excitation
(f) $O + e^- \to O^-$	resonant capture
(g) $O^- - e^- \to O$	resonant loss

Additional information has to be sought in order to obtain further insight into the mechanism. One way is the dependence on the primary energy, i.e. the search for thresholds; the second way is the study of angular dependences.

The spectra of incident O^+ are also shown in Figure 6.10. The dependence of the negative-ion yields on the work function follows the same pattern as in the case of incident O_2^+. The comparison of the O^- peak of Ni(110) with the O_2 and O_2^- peak and the O^- distribution from O_2^+ shows the significant differences of the respective peak widths. The O_2 and O_2^- are clearly due to primary projectiles, whereas the O^- from the O_2^+ shows the broadening due to dissociation. This broadening is obvious for the O distribution as well. The changes of the shape of the O spectra with work function are not yet understood. However, the changes occur at rather low intensity levels, i.e. below the 10% level.

Table 6.1 gives a list of thresholds observed for different systems. The observation of thresholds is necessary evidence for the occurrence of harpooning (Kashihira *et al.*, 1974). But on the other hand, a threshold is not necessarily the signature of harpooning. We know that in the case of N_2^+ and CO^+, no negative molecular species are found (Schubert *et al.*, 1990; Schubert, 1989), hence the thresholds in those cases are not related to harpooning. The cases of I_2, O_2^+ and CO_2^+ are comparably conclusive, i.e. the combined observation of negative molecular ion and thresholds allows one to assume that the negative molecular state is indeed involved in the surface reaction. In the case of CO_2^+, the CO_2^- species was found in an independent experiment (Schubert *et al.*, 1988). Even though the latter experiments used Ni(110) and not Pt(100), it is safe to assume that CO_2^- is formed on Pt, too.

Based on these findings, there is at least a channel of dissociation (and trapping) given via the harpooning reaction followed by rovibrational excitation. Further corroboration is found in the study of other sytems, i.e. O_2 and NO on Ag[111] and Pt(111) (Reijnen *et al.*, 1991), O_2^+ on Si(111) (Rechtien *et al.*, 1990), and further studies of C^+, O^+, O_2^+, N_2^+, CO^+ and CO_2^+ on

Table 6.1 Thresholds for ionic species observed in the interaction of swift molecules with surfaces

Incident particle	Observed particle	Surface	Threshold $E_\perp$ (eV)	Reference
I_2	I_2^-	C(111)	3.0	Danon and Amirav (1988)
O_2^+	O_2^-	Ag(111)	1.5 ± 0.5	Reijnen *et al.* (1989)
O_2^+	O^-	Ag(111)	15.4 ± 4	Reijnen *et al.* (1989)
N_2^+	N^+	Pt(100)	10	Akazawa and Murata (1990b)
CO^+	C^+	Pt(100)	17.5	Akazawa and Murata (1990b)
CO_2^+	CO^+	Pt(100)	6.25	Akazawa and Murata (1990b)

Pt(100) (Akazawa and Murata, 1990a, 1990b). The Si case is particularly interesting since, owing to the band structure of a semiconductor, resonant-loss processes are less likely than in the case of a metal (Table 6.1). In other words, the neutral O and O_2 yields in the case of metals may indeed by partly due to these processes—a conclusion supported by the observation that the negative-ion yields O_2^- and O^- from Si are ten times higher than from Ag.

The angular dependences (Reijnen *et al.*, 1991; Akazawa and Murata, 1990a) indicate that, especially at the very-low-energy end, a chemisorption interaction between the molecule and the surface occurs. For example, C^+ exhibits an angular distribution which is shifted towards the surface compared to the rare-gas ion Ne^+, as does CO_2^+ compared to Ar^+ (Akazawa and Murata, 1990a).

6.4.4 THE INFLUENCE OF THE WORK FUNCTION

As mentioned before, the molecule–surface interaction is influenced by the work function (Pan Haochang *et al.*, 1986; Schubert *et al.*, 1990; Schubert, 1989). In the case of hydrogen, the lowering of the work function causes an increase of the molecular survival yield (Figure 6.6). In a systematic study with H_2^+, N_2^+, CO^+, O_2^+ and CO_2^+, the influences of K adsorption on Ni(111) on the molecular yield and the ion yield were shown to be significantly different in the following way. Whereas in the case of H_2, the molecular yield increases when the work function decreases, in the case of N_2 and CO there is no great change, and in the case of O_2 and CO_2 a large decrease of the molecular survival yield is found (Figure 6.11). The molecular yield is calculated from $Y_M = \Sigma$(molecular peaks)/Σ(all peaks). The difference of the detection probability for these ions and neutrals is taken into account. For N_2 and CO, the molecular yield varies from about 0.95 to 0.80 at $\Delta\phi = 3$ eV. This is followed by an increase after the work-function minimum at $\phi = 1.8$ eV is reached. At the monolayer coverage of K on Ni(111), the molecular survival yield is back to the value for the clean surface. These data suggest that there is a small contribution of collisional dissociation, since at lower K coverages the surface is atomically rough. For H_2, the low K coverage causes some scatter of the data indicating a small decrease of Y_M, but at about $\Delta\phi = 3$ eV, the molecular survival yield increases up to about 0.60. This is a 50% increase from the value of 0.40 for the clean surface. This result is clear evidence for an electronic effect. We cannot expect the K-covered surface to be smoother than the original Ni(111) surface.

In the case of O_2 and CO_2, the molecular yield decreases with decreasing work function to about $\Delta\phi = 3$ eV. There is a levelling off followed by a further decrease of Y_M after the work-function minimum has been reached. It should be noted that Y_{CO_2} decreases from a value between 0.30 and 0.35 down to about

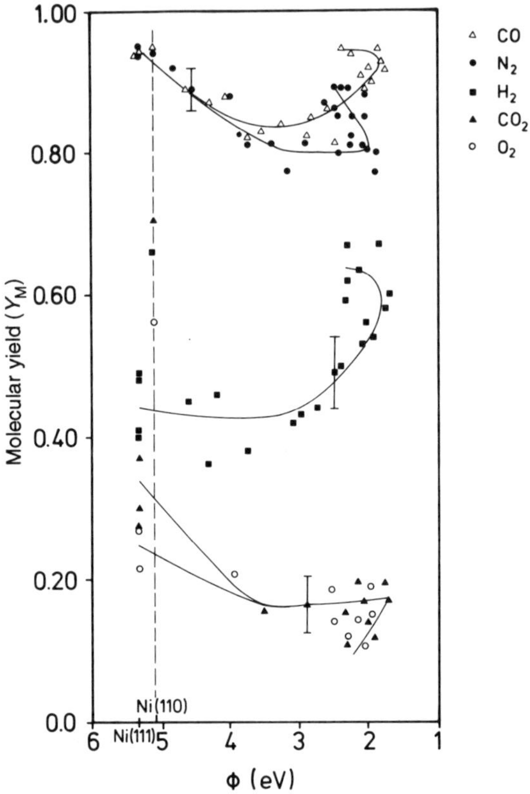

Figure 6.11 Molecular yield of N_2^+, CO^+, H_2^+, O_2^+ and CO_2^+ as a function of the work function of Ni(111) changed by K-adsorption. Data for clean Ni(110) are marked by the dashed line. Primary energies are between 450 eV and 500 eV for N_2^+, CO_2^+ and CO_2^+, and are 250 eV for H_2^+. Geometry is as in Figure 6.9 (Schubert *et al.*, 1990; Schubert, 1989)

0.15 at $\Delta\phi = 3.0$ eV. This is a change of 100%, and is followed by a further change of 50% when completing the K monolayer. Qualitatively these data support the harpooning model and the assumption that negative molecular states play an important role in dissociative chemisorption, as known from chemisorption experiments (Herschbach, 1966; Backx *et al.*, 1981; Bonzel, 1988; Freund and Messmer, 1986).

However, there are some effects that are not simple to comprehend in detail. Figure 6.10 shows the O_2 TOF spectra from Ni(110), Ni(111) and Ni(111) + K at $\Delta\phi = 3.1$ eV in comparison with O TOF spectra (Schubert *et al.*, 1990; Schubert, 1989). Besides the change of the molecular yields, we find a tremendous increase of the negative-ion yield with decreasing work function. Furthermore, there is a puzzling difference between the results from clean

Ni(111) and clean Ni(110) for both O_2 and O. The work functions for Ni(111) and Ni(110) are $\phi = 5.35$ eV and 5.04 eV, respectively. Both the molecular survival yield and the negative-ion yields change between Ni(111) and Ni(110). As plotted in Figure 6.11, the yields are $Y_{O_2} = 0.55$ from Ni(110) and $Y_{O_2} = 0.25$ from Ni(111), which contradicts the simple work-function interpretation. Similar numbers are found for H_2 and CO_2, i.e. $Y(\text{Ni}(110)) \gg Y(\text{Ni}(111))$. For CO and N_2 there are no differences between the two surfaces. In the cases of Ag(111) and Pt(111), differences in the interaction with O_2 and NO have been found which could not be related to the work functions of 4.8 eV and 5.8 eV, respectively (Reijnen *et al.*, 1991).

The oxygen-ion yields for Ni(111) + K surfaces are summarized in Figure 6.12 (Schubert *et al.*, 1990; Schubert, 1989). The primary ions are O^+, O_2^+, CO^+ and CO_2^+, all at primary energy about 450 eV to 500 eV. Whereas O,

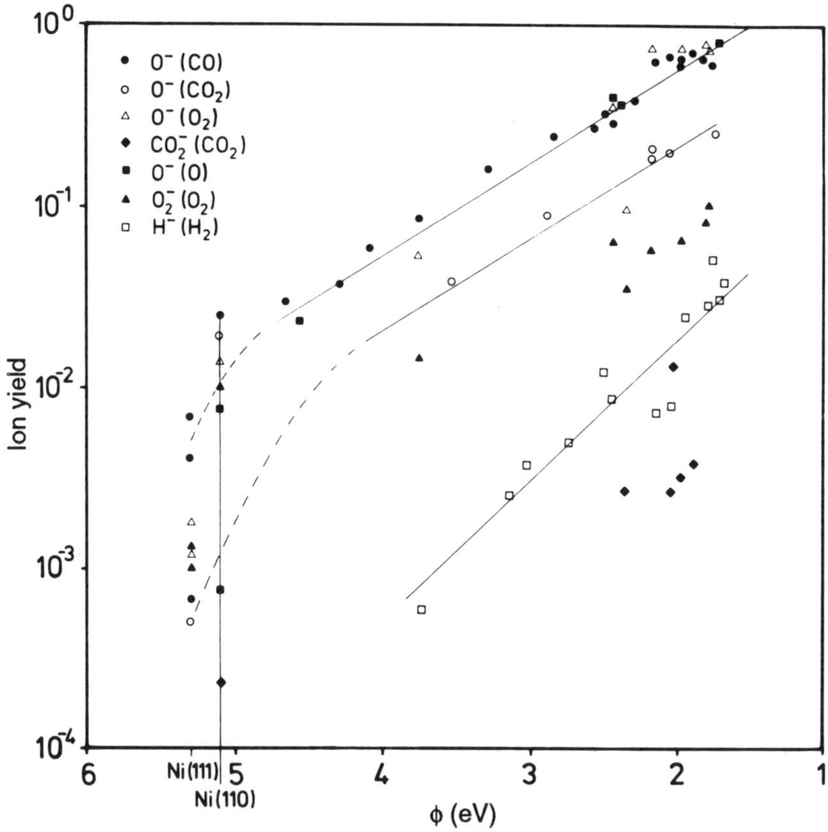

Figure 6.12 Negative-ion yields of different molecular species scattering from Ni + K as a function of the work function (Schubert *et al.*, 1990; Schubert, 1989)

O_2 and CO show the same amount of O^-, the yield from CO_2 is about a factor two lower althought it has the same slope. Even though the molecular yield for O_2 and CO_2 exhibit very similar behaviour (Figure 6.10), the ion yields are very different. This is even more evident when considering the O_2^- and CO_2^- yields (Figure 6.12) which differ by an order of magnitude. A comparison of the molecular survival yields based solely on the ion yields would be, in the case discussed here, rather misleading. On the other hand, in the case of oxygen, the situation from clean to K-covered surface completely reverses the relative importance of the neutrals and the ions (Figure 6.10).

Another example of high negative-molecular-ion yield is the scattering of C_2^+ from clean Ni(110) (Figure 6.13) (Franke *et al.*, 1992b). The dissociation rate is not very high; the C^- yield is moderate. The high negative-molecular yield may be related to the comparably large electron affinity of C_2, which is $E_A = -3.54$ eV. However, there is no simple scaling with the electron affinity. It is rather 'trends' which are observed. In the case of NO^- and O_2^-, with $E_A = 1.24$ eV and $E_A = 0.74$ eV, respectively, the yields from Pt(111) and Ag(111) are about equal (Reijnen *et al.*, 1991). When comparing C_2 with C^- and CH^-, with $E_A = 1.24$ eV and $E_A = 0.74$ eV, we find lower yields for both C^- and CH^- (Franke *et al.*, 1992b). In a qualitative sense, the situation is comparable to the work-function dependences, i.e. a large electron affinity favours large negative-ion yields as a small work function favours large negative-ion yields. In detail, there is an obvious dependence, such as on the electron structure, as shown very clearly in an experiment using atomic C and O beams (van Pinxteren *et al.*, 1989). In that case, it is the structure of the atomic state, i.e. the sublevels of the l-states, which causes the yield of O^- to be higher than that of C^-. The electron affinities are 1.46 eV (O) and 1.22 eV

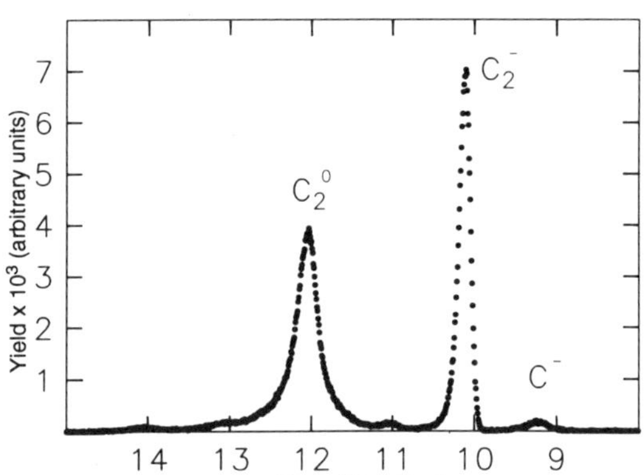

Figure 6.13 Time-of-flight spectra of C_2 and C_2^- from C_2^+ scattering from Ni(110) at a primary energy of 1045 eV. Geometery is as in Figure 6.9. The negative ions are separated from the neutrals by post-acceleration (Franke *et al.*, 1992b)

(C). A similar clear dependence of the ion yields or dissociation probabilities on the electron structure of the surface in question has yet to be established experimentally.

6.4.5 PROTON HARPOONING

The capture of a hydrogen atom from a surface by a fast molecular beam was first observed in case of the scattering of DABCO (1.4 diazobicyclo-222-octane) from H-covered Pt(111) (Kuipers *et al.*, 1991). In heterogeneous catalysis, two types of reactions are important, namely the Eley–Rideal and the Langmuir–Hinshelwood mechanisms (Gates *et al.*, 1979; Knözinger, 1991). The Eley–Rideal mechanism describes the interaction of a species chemisorbed on a surface and a species impinging from the gas phase by a direct reaction between the two species.

In the Langmuir–Hinshelwood scheme, the gas-phase particle adsorbs and equilibrates on the surface before reacting with the other species. Information as to whether one or the other mechanism prevails is difficult to deduce from studying any catalytic reaction. The importance of observing proton capture by DABCO in an experiment with a hyperthermal (9 eV) molecular beam is clear evidence for an Eley–Rideal reaction. The detailed analysis (Figure 6.14) (Kuipers *et al.*, 1991) shows the dependence of the positive-ion yield, i.e. DABCO $- 1e^-$, and the positive-ion yield for the protonated DABCO, i.e., DABCO $+ H^+$. The Pt surface is partly covered with H; only ionization but not protonation is found for a clean surface and vice versa. Since the thresholds are equal for both processes, i.e. $E_{th} = 1.55$ eV, and since $E_{th} = E_i - \phi_{Pt}$ it is concluded that a two-step process takes place: ionization followed by hydrogen capture. E_i is the ionization energy of DABCO ($E_i = 7.25$ eV) and ϕ_{Pt} is the work function of Pt at a hydrogen coverage of $\Theta = 0.20$, which is $\phi_{Pt} = 5.7$ eV. Because of the large mass of DABCO ($M = 112$ amu) the capture of a H atom does not change the translational energy of the primary molecule (Kuipers *et al.*, 1991).

In our series of experiments with CH-containing molecules, we found a case of hydrogen capture with an appreciable change of the translational energy of the primary ion (Franke *et al.*, 1992a). In the TOF spectra of CH^+ scattered from Ni(110), we observe the expected neutral peak and CH^- and a peak which is identified as CH_2^- (Figure 6.15). The intensity of this peak decreases with increasing energy. The 'perpendicular' energies vary from about 7 eV to about 30 eV in the range shown for a grazing angle of incidence of 5°. The acceleration of the captured H by the CH costs translational energy as can be estimated from the conservation of energy and momentum. The time of flight of the charged particles is estimated from the velocity of the particles, the applied post-acceleration voltage and the length of the flight paths. If the particles

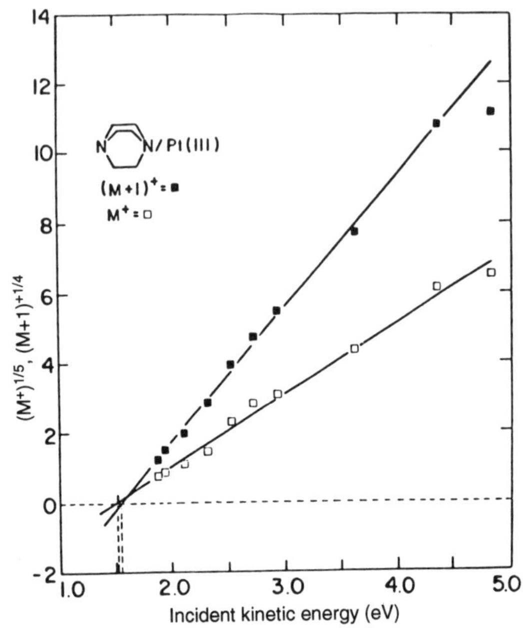

Figure 6.14 Proton harpooning and ionization (□) of DABCO with mass $M = 112$ amu from H-covered Pt(111) as a function of primary energy ($T_{surf} = 386$ K, $P_{H_2} = 5 \cdot 10^{-6}$ mbar, coverage $\Theta_H \simeq 0.20$). The lines are fits of the form $k(E_i - E_{th})^n$ yielding threshold energies of $E_{th} = 1.55$ eV for the protonation and $E_{th} = 1.52$ eV for the electron transfer. The parameters k and n are 223, 38.2 and 4.5, respectively. (Reproduced by permission of the American Physical Society from Kuipers *et al.*, 1991)

change their mass by dissociation or by proton capture, their energy and flight time are estimated from conservation of energy and momentum.

The distance from the crystal to the TOF detector is $S_a + S_b$, where S_a is the distance from the crystal to the post-acceleration stage where the voltage V_a is applied. The total time of flight is

$$t = t_a + t_b = \sqrt{\frac{m}{2E_{kin}}} \cdot S_a + \sqrt{\frac{m}{2(E_{kin} + eV_a)}} \cdot S_b,$$

where E_{kin} is the kinetic energy of the particle with mass m at the target. For particles which conserve their mass, E_{kin} is equal to E_0, the primary kinetic energy corrected for possible inelastic losses. If the particle picks up a proton, the velocity changes from v_0 to v_1. Since momentum is conserved the new velocity is

$$v_1 = \frac{m}{m+1} v_0.$$

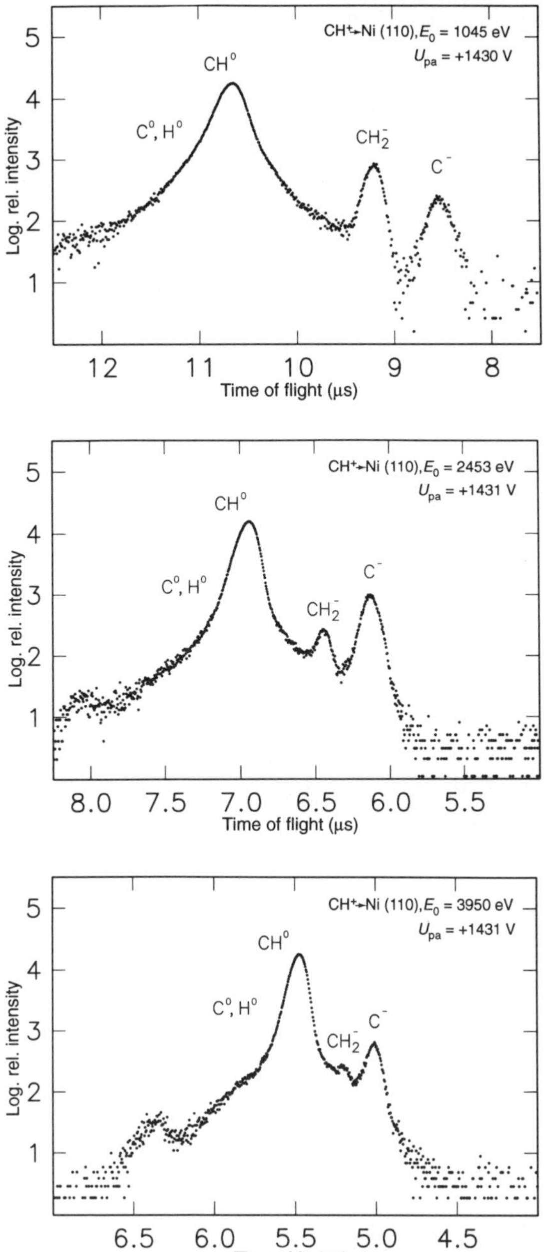

Figure 6.15 Time-of-flight spectra of neutral and negative species of CH^+ scattering from Ni(110) at three different primary energies E_0. U_{pa} is the post-acceleration voltage applied to separate the ions from the neutrals

The energy of the particles is hence

$$E_{\mathrm{kin}}^{m+1} = \left(\frac{m}{m+1}\right) E_{\mathrm{kin}}^{m} = \left(\frac{m}{m+1}\right) E_0.$$

With the values $S_A = 0.58$ m and $S_B = 0.70$ m and $m = 13$, the flight time for a particle with $m = 14$ is $t = 9.03$ μs at an acceleration voltage of 1430 V and $E_0 = 1000$ eV. Note that $E_0 = 1000$ eV is reduced by 45 eV with respect to the beam energy of 1045 eV. This reduction represents the inelastic energy loss as evaluated from the position of the CH peak.

6.4.6 MISCELLANEOUS RESULTS

In Table 6.2 we list references for the cases in which both in ion-scattering experiments and in chemisorption experiments, negative-molecular ions have been found. However, in chemisorption experiments it is rather difficult to establish the existence of negative-molecular states. The reason is that in the reported cases, these states are precursor states, i.e. they initiate dissociative chemisorption or they precede associative desorption. Lifetime estimates are of the order of 10^{-13} s (Luntz *et al.*, 1988), agreeing in order with the scattering experiments (Schubert *et al.*, 1990; Schubert, 1989). In ion scattering, the lower limit for a lifetime is relatively simple to estimate from the velocity of the particle and the trajectory length.

Another aspect of swift molecule experiments and chemisorption is the possibility of studying the effects of symmetry conservation (Tappe *et al.*, 1991; Tappe, 1990; van Pinxteren *et al.*, 1989). Figures 6.4 and 6.16 show the spatial

Table 6.2 References for negative molecular ions in scattering and chemisorption

Molecule	Surface	Scattering	Chemisorption
O_2	Ag(111)	Pan Haochang *et al.* (1986) Reijnen *et al.* (1989) Schubert *et al.* (1990) Schubert (1989) van Slooten *et al.* (1991)	Backx *et al.* (1981) Campbell (1986)
CO_2	Ni(110)	Schubert *et al.* (1988, 1990) Schubert (1989)	Bartos *et al.* (1987)
O_2	Pt(111)	Reijnen *et al.* (1991)	Luntz *et al.* (1988) Wurth *et al.* (1990)
NO	Pt(111)	Reijnen *et al.* (1991)	Kiskinova *et al.* (1984)
O_2	Ni(111)	Schubert *et al.* (1990) Schubert (1989)	Shayegan *et al.* (1984)

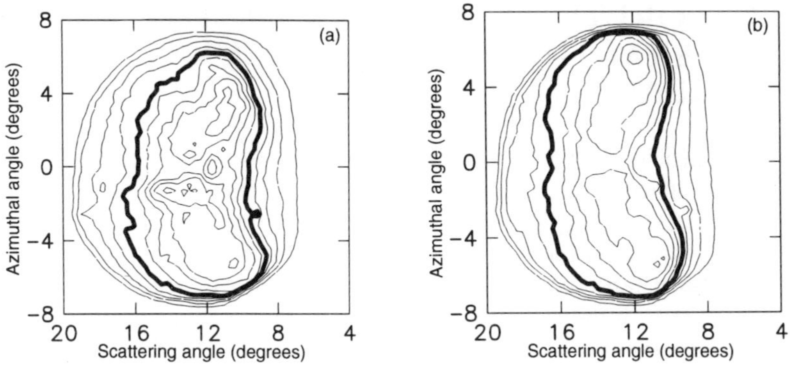

Figure 6.16 Spatial distributions of H scattered of Pd(110) for H^+ and H_2^+ incident at 1.25 keV per proton. The incident beam is aligned parallel to the $[1\bar{1}0]$ surface channels. The grazing angle of incidence is $\psi = 5°$. The scale of the contours is linear, the thick line marks the 50% level. (Schlathölter, 1993)

scattering distribution of H^+ and H_2^+ from Ni(110) and Pd(110) respectively. In Figure 6.4, the incident beam direction is parallel to a direction at an angle of 45° to the $[1\bar{1}0]$ surface channels, i.e. it is a random orientation. We know from previous experiments that the dissociation probability is close to 1.0 and that the ion yields of H^+ and H_2^+ are low. The comparison of the spatial distributions of the neutral H products as measured by the position-sensitive detector shows anisotropic broadening of the H_2^+-produced pattern. The quantitative analysis based on equation (1) gives for the direction normal to the surface $E_\perp = 4$ eV and parallel to the surface $E_\parallel = 2.3$ eV (Tappe *et al.*, 1991; Tappe, 1990). The value of 4 eV agrees quite well with a dissociation via the non-bonding $b^3\Sigma_u$ triplet state H + H. From the low value of 2.3 eV we conclude that the dissociation is partly caused by dynamical screening.

When the incident beam is oriented parallel to the [110] surface channels a half-moon-shaped intensity with a two maximum distribution is found for incident H_2^+ (Figure 6.16). This pattern is the signature of surface channelling (Sizmann and Varelas, 1976; Lindhard, 1965). The data shown are for a Pd(110) surface. We find a small additional broadening of the channelling pattern as in the random case for Ni(110) when comparing the H_2^+ pattern with the H^+ pattern. Small irregularities in the respective patterns may be an indication of a particle interaction in the molecular case. Obviously the reading of direct information from the pattern is obscured by straggling effects.

Finally it is worth mentioning scattering experiments with 'large' molecules (Danon and Amirav, 1988) including the newly found fullerenes (Busmann, 1990). These experiments are of great interest for the production of thin films.

6.5 THEORETICAL ASPECTS OF THE INTERACTIONS OF MOLECULES WITH SURFACES

As in any particle–solid interaction study, it is useful to separate the elastic and the inelastic effects (Heiland and Taglauer, 1976; Heiland, 1991). In the energy range of interest here, it is safe to assume that the particles follow a classical trajectory—at least in the case of atomic particles. This point is under discussion. For the inelastic part of the interaction, several models are in use based on experience with atomic ions, transferred from atomic-collision physics, or extended from theories successful in surface chemistry at thermal energies.

6.5.1 COLLISIONAL MODELS

The first model to describe the dissociative interaction of molecules with surfaces was developed by the Tashkent group (Bitensky and Parilis, 1984; Bitensky *et al.*, 1988, 1990). It is a classical trajectory model. The basic idea is shown in Figure 6.17. The incident diatomic molecule with an internuclear distance l hits first atom 1 with impact parameters b_{12} and b_{11}. The atoms of the molecule thereafter follow independent trajectories, namely, the asymptotes to the scattering angles Θ_{11} and Θ_{21}. Further collisions may follow, in Figure 6.17, with a second atom 2 with impact parameters b_{21}, b_{22} and scattering angles Θ_{12}, Θ_{22}. In each collision, the atoms suffer different elastic energy losses and have velocities $\vec{v}_1$ and $\vec{v}_2$ after the collision. The condition for a molecule to survive is then

$$\frac{m_1}{2}(\vec{v}_1 - \vec{v}_2)^2 \ll E_D,$$

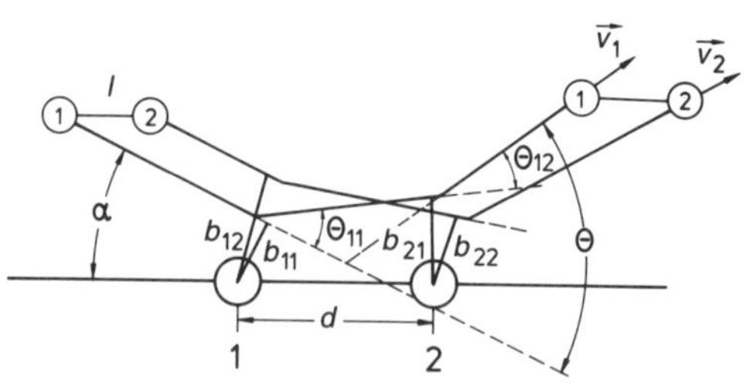

Figure 6.17 Scheme of elastic scattering of a diatomic molecule with atoms 1 and 2 and length l from two atoms of a surface chain with lattice constant d. The individual impact parameters are b_{ij} and scattering angle Θ_{ij} (Reproduced with permission from Bitensky and Paritis, 1984)

where m_1 is the mass of one of the atoms, and E_D the dissociation energy. Estimates of the molecular survival yield of 30 keV N_2^+ interacting with a Cu surface agree quite well with experimental data (Balashova *et al.*, 1982; Balashova and Molchanov, 1982; Balashova, 1991). Possible explanations for the agreement are: (i) the high kinetic energy, and (ii) the fact that a polycrystalline surface was used in the experiments. The collisions may be violent, i.e. the energy transfer in a single collision may be large enough to cause dissociation. The chance of single collisions even at grazing incidence is expected to be large at grain boundaries, steps and macroscopic defects. Nevertheless, the model contains all necessary concepts and is useful for gaining a first qualitative insight. It is not too difficult to introduce improvements of the model molecule beyond just a dissociation energy E_D, for example by taking vibration and rotation into account (Jakas and Harrison, 1985; Bitensky and Parilis, 1985; van den Hoek *et al.*, 1988; van den Hoek, 1989, van den Hoek and Kleyn, 1989). Another setback of the model is in a way more difficult to overcome; namely, that it is basically a 'string model'. The solid is represented by a string or chain of atoms. However, experiments accompanied by computer simulations have shown that the probability for string-scattering events is in general very low (Taglauer *et al.*, 1980; Jackson *et al.*, 1981). Multiple scattering events prevail, and the majority of these trajectories have three-dimensional character, i.e. many out-of-the-plane scattering events occur. These trajectories dominate even when a small solid angle detector positioned in the plane of scattering is used. So far there has been no computer code which fully combines molecular properties and solid structure.

Another problem in the elastic scattering part is the choice of interaction potential. As indicated schematically in Figure 6.18, the interaction between a molecule and a surface is not purely repulsive (Jackson and Morgan, 1974). The trajectories are bent towards the surface by an attractive potential. Furthermore, the potential may be quite different for the second half of the trajectory due, for example, to dissociation of the molecule. Or, if harpooning occurs, a strong image potential is switched on, which is switched off again when the dissociation is combined with an electron loss, i.e. $A_2 + e^- \rightarrow A_2^- \rightarrow A + A + e^-$. Again, Figure 6.18 is schematically simplified to a string model; nevertheless, even in that case quite different trajectories are possible.

In order to reduce the variety of possible trajectories, surface channelling conditions can be chosen. In that case, the interaction between the particle and the surface is such that when using an in-plane detector a rather uniform trajectory type contributes to the scattering (Figure 6.19) (Höfner *et al.*, 1992). The trajectories are calculated using the MARLOWE code (Robinson and Torrens, 1974; Hou and Robinson, 1976). Thermal vibrations, inelastic losses and crystal structure are taken into account. The azimuthal direction is chosen 45° off the $[1\bar{1}0]$ surface direction which is considered to be a 'random' direction. Figure 6.19(b) shows the depth distribution of the trajectories. There

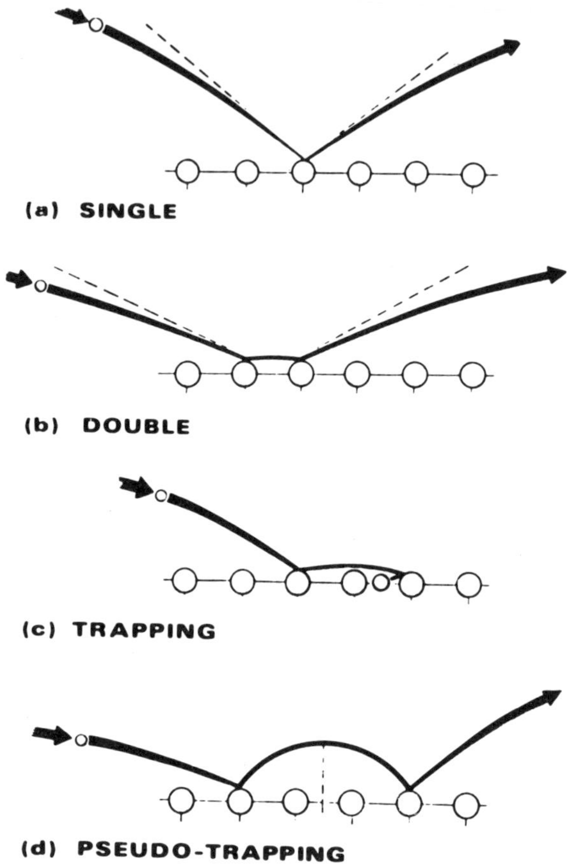

Figure 6.18 Scheme of different possible trajectories of a particle interacting with a chain of atoms. The action of an attractive potential is indicated by the curvature of the trajectories (Reproduced with permission from Jackson and Morgan, 1974)

is an average depth of 0.7 Å above the plane defined by the ion cores of the top Ni layer. The width of the depth distribution is of the order of 0.15 Å or 20%. The width is due to the statistics of the thermal motion of the surface atoms and the statistics of the impact points. Even at 0 K the surface potential is not completely planar and the statistics of the impact points leads to different trajectory lengths. Therefore, the inelastic interactions will have a distribution, thinking in terms of an overlap of wave functions, even under the idealized conditions of planar channelling. When changing the azimuthal angle such that the particles impinge parallel to the $[1\bar{1}0]$ surface channels, three trajectory types with quite different penetration depths can be distinguished (Figure 6.20). The type with the smallest penetration depth are string collisions from the top

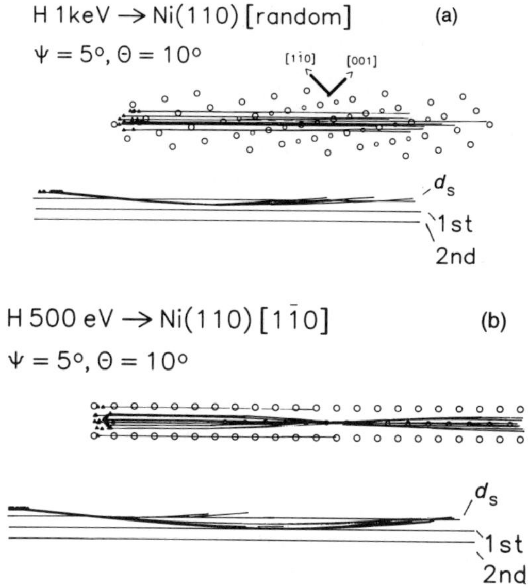

Figure 6.19 (a) Calculated trajectories of H scattering from Ni(110) at $E_0 = 1$ keV, $\psi = 5°$ and a scattering angle $\Theta = 10°$. The azimuthal direction is 'random'. The detector has an acceptance angle of 5°. (b) Same as (a) but for $E_0 = 500$ eV and orientation along the [1$\bar{1}$0] surface channels. The larger circles represent the top-layer atoms, the smaller ones the second-layer atoms (top view). $d_s = 1.3$ Å in the side view is used to define the trajectory length (Höfner *et al.*, 1992)

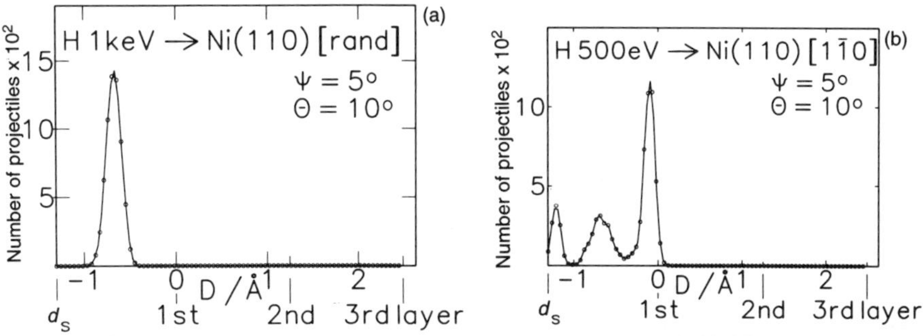

Figure 6.20 Depth distribution of H back-scattered from Ni(110) under the conditions of Figure 6.19. 'Negative' depth means that the particles scatter from above the top-layer (1st) ion core positions at 0.0 Å. The three peaks for the [1$\bar{1}$0] case correspond to the trajectory types of Figure 6.19(b), i.e. straight scattering from the top rows, zigzag scattering and scattering from the troughs (Höfner *et al.*, 1992)

rows of Ni atoms. There is a class of zigzag collisions and another class of almost straight trajectories guided in the bottom of the potentials of the surface troughs. Comparable calculations using 'real' molecule have not been published so far, but it takes little imagination to see that, especially in the case of Figure 6.20, a molecule will cause some complications. When using a wide-angle detector more of the out-of-plane scattering trajectories contribute, as shown experimentally in Figure 6.16.

For the case of high-energy (MeV) molecules which are Coulomb-exploded in thin foils, first calculations show correlation effects for the motion of the dissociation products (Sigmund, 1992a, 1992b). These calculations may serve as a guideline for the low-energy regime.

6.5.2 NEUTRALIZATION MODELS

Based on the experience that low-energy atomic ions are neutralized at surfaces with high probability, it was proposed that charge capture also plays a role in the dissociation of molecular ions (Willerding *et al.*, 1984a; Heiland *et al.*, 1977). This proposal is essentially an extension of the Hagstrum model for the neutralization of atomic ions (Hagstrum, 1954a, 1954b, 1977). Figure 6.21

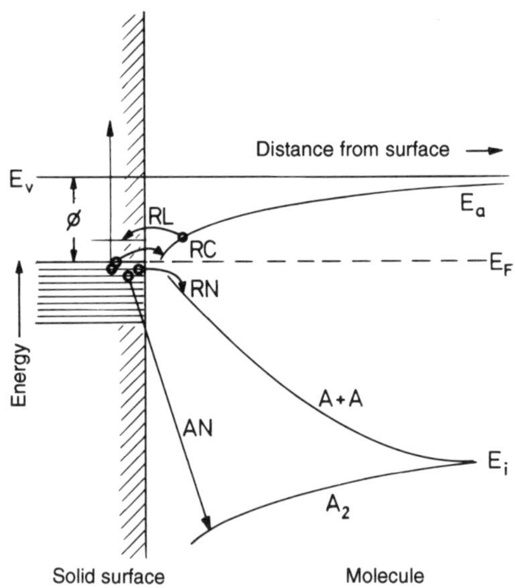

Figure 6.21 Scheme of possible electronic interactions between a molecule and a metal surface. RL is resonant loss of an electron from a state A_2^- with electron affinity E_a, RN is resonant neutralization of an ionic molecule ($A_2^+ + e^- \rightarrow A + A$), RC is resonant capture into the affinity level A_2^-, AN is Auger neutralization of an ionic molecule $A_2^+ + e^- \rightarrow A_2$, with an ionization energy E_i

presents the scheme of the model. For ionization energies of the molecules $E_i \geq 2\phi$, Auger neutralization (AN) is possible (ϕ is the work function of the solid). Non-bonding states and electronically excited states of the molecule are accessible by resonant processes. The dependence of the electronic potential curves on the intermolecular distances as shown in Figure 6.21 results as the difference between the potential curve of the upper state A_2^+ and the final state in question, A_2, $A + A$ or $A + A^-$ etc. Also the harpooning process, i.e. the formation of A_2^-, is a resonant process in this model. The charge-capture model has been treated theoretically on the basis of the Anderson–Newns Hamiltonian (Imke *et al.*, 1986a, 1986b).

The task of the theory is to calculate an energy distribution which by comparison with experimental data can be the test of the theory. Dissociation occurs in the centre-of-mass (CM) system with a kinetic-energy distribution $dN(\varepsilon)/d\varepsilon$ (Figure 6.22). For hydrogen, the potential curve in question, $b^3\Sigma_u^+$, is well known. The Franck–Condon region defined by the dashed lines together with the initial vibrational excitation defines the range of potential energy which is then transformed into kinetic energy. The energy distribution in the laboratory system is calculated via

$$\frac{dN(E)}{dE} = w(E) \int dE' v(E - E') \int_{\Delta\Omega_s} d\Omega_s \frac{dN(\varepsilon)}{d\varepsilon} \tag{2}$$

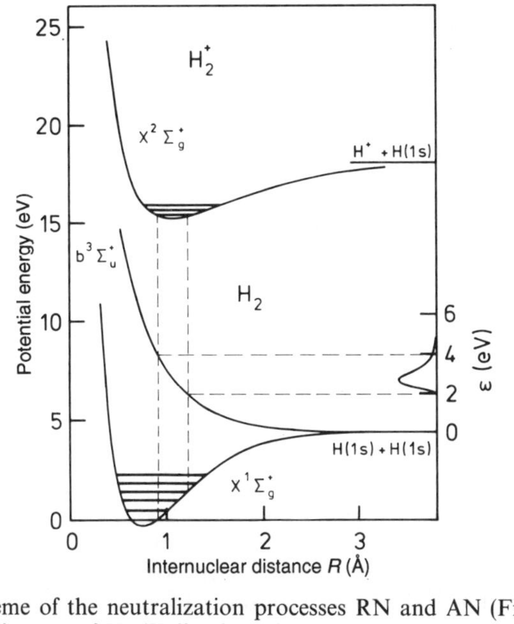

Figure 6.22 Scheme of the neutralization processes RN and AN (Figure 6.21) in the potential-energy diagram of H_2 (Heiland *et al.*, 1987)

where $w(E)$ is the detection probability, $v(E)$ is the inelastic-energy loss function and $\Delta\Omega_s$ is the acceptance angle of the detector in the centre-of-mass system. The inelastic-energy loss function is obtained from surviving H_2 molecules. This assumption means that with respect to inelastic loss, the separating H atoms behave like the surviving H_2 molecules. Figure 6.23 shows an example of the application of the model discussed here (Imke *et al.*, 1987). The broad peak of dissociated H atoms is fitted by dN/dE (equation (2)), the difference peak is identified as surviving H_2. This result is quite satisfactory, although there is a discrepancy if the experiments are extended to higher energies with respect to the molecular survival probability. The theoretical estimate is 0.5 to 0.3 in good agreement with the experimental data at low beam energies for Al (Imke *et al.*, 1987) and Ni (Figure 6.11) (Schubert *et al.*, 1990; Schubert, 1989). But the yield predicted by the theory is essentially independent of the energy—naturally, since it essentially is an adiabatic model.

Further qualitative support for the model is the angular anisotropy (Figure 6.16) related to the symmetry character of the $b^3\Sigma_u^+$ and the $X^1\Sigma_g^+$ hydrogen wave functions (Tappe *et al.*, 1991; Tappe, 1990; Imke *et al.*, 1986a, 1986b; Dunn, 1962). Results for the interaction of O_2 and H_2 with Ag(111) are also partly explainable by charge-capture-induced dissociation (van Slooten *et al.*, 1991, 1992). Recent results concerning an isotope effect in the O_2–Ag interaction also agree with a resonant-capture model for the formation of the O_2^- molecule initiating the dissociation (Flores, 1991). On the other hand, the same systems gives ample evidence for collision-induced dissociation. Resonant capture into excited states and electron promotion into excited states has been invoked for $N_2^+ \rightarrow$ Pt, Au and C (Akazawa and Murata, 1988b, 1990a, 1990b; Kasi *et al.*, 1988).

The extension of the Hagstrum model to the interaction of molecular ions has proven very useful, at least for a qualitative understanding of many experimental findings. For the interpretation of atom–surface interaction experiments, we have recently adopted theoretical approaches from particle–solid interactions (Närmann *et al.*, 1990, 1992; Schmidt, 1992; Flores, 1991). This theory is essentially dynamic; the response of the solid to the moving particle is understood in terms of a dielectric function. At low kinetic energies ($v \ll v_0$), the main events are the creation of electron–hole pairs at the Fermi edge of the solid and an effective dynamic screening of the moving particle. The moving particle is hence neutral or even a negative entity in the case of H. The bulk dielectric function becomes effective when the particle has penetrated the image plane; this is clearly the case for H_2. That is, within the time scale of the dynamic screening, any hydrogen molecule will be dissociated and its atomic parts will continue to travel as H^- along the surface. The observation of surviving H_2 and H_2^+ with yields of 0.5 and 0.01, respectively, does not contradict the model; it is a matter of the velocity and the trajectory length compared to the characteristic lifetimes of the processes occurring. For a particle with

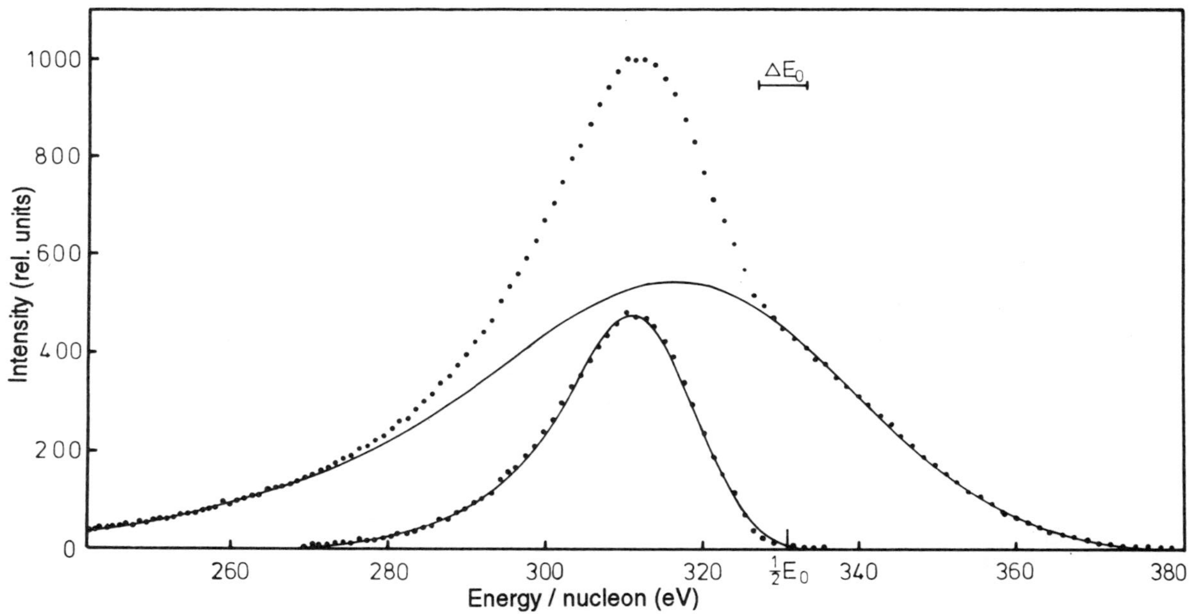

Figure 6.23 The model analysis of an experimental energy spectrum (dots, upper curve) of species neutralized into H (upper line) and H_2 (dots and lower line) at $E_0 = 600$ eV. The target is Al(110), $\psi = 6°$, $\delta = 10°$. Curve 2 can be fitted by using a Franck–Condon distribution for $dN(\varepsilon)/d\varepsilon$ (equation (2)) and the potentials shown in Figure 6.22. (Heiland *et al.*, 1987)

$v = 0.1v_0 \simeq 2 \times 10^{-7}$cm s^{-1} and a trajectory length of 20 Å, the dwell time along the surface is of the order of 10^{-15} s. Estimates for the characteristic times for resonant and Auger neutralization produce values of the order of 10^{-15} s (Närmann *et al.*, 1990, 1992; Schmidt, 1992). Hence, about equal yields of H and H_2 are to be found and ion yields of less than 1/e.

Experimentally, the ion yields are in fact of the order of 10^{-3} and the dissociation probability increases with increasing energy, i.e. increasing trajectory length. This observation alone suggests a third process, besides the capture into the triplet state and rovibrational excitation, to be at work. Also, the observation of the spatial distribution of the dissociated hydrogen (Figure 6.16) supports the model of dissociation by screening. Nevertheless, further experiments will be necessary to clarify the relative importance of electronic and collisional effects in the molecule–surface scattering process.

Evidence for electron–hole-pair excitation in the solid is obtained from the energy-loss measurements (Flores, 1991; Närmann *et al.*, 1990). Direct evidence is also found from transient conductivity increases of Ge(100) and InP(100) when bombarded with hyperthermal ($\sim$10 eV) Xe and Kr (Amirav *et al.*, 1986; Amirav and Cardillo, 1986; Weiss *et al.*, 1988).

It is worth noting that the dissociation mechanism via the dynamic screening discussed here is based on the same principal ansatz as the mechanism involving plasmon excitation (Lucas, 1979). The plasmons and the electron–hole-pair excitation are solid-state properites described by the dielectric function. Here we relate the charge capture and the dynamic screening to the dielectric function as well, naturally as part of the energy-loss behaviour (Flores, 1991; Närmann *et al.*, 1990). The equality of the spectra of H^+ and H_2^+ scattered from Al and Ni (Figure 6.7) is due to the facts that (i) plasmons are not excited at these low energies, and (ii) the particle–solid interaction concerns the s-electrons mainly. Since the main difference between the electronic structures of Al and Ni is the d-bands of Ni, whereas there is little difference in the s-electron density of states, thc result of Figure 6.7 is in qualitative agreement with the model discussed here.

6.5.3 POTENTIAL-ENERGY SURFACES

Another approach to the interpretation has its origin in classical gas dynamics, i.e. the interaction of thermal molecules with surfaces (Freund and Messmer, 1986; Tully, 1980; Harris, 1989).

In a one-dimensional representation, the molecule–surface interaction can be schematically represented by Figure 6.24. Usually a thermal atom or molecule will interact with a surface via some physisorption potential well, representing van de Waals forces acting on the polarizability of the molecule. For the interaction to proceed, activation barriers, the interaction with the phonon, and the vibrational and rotational state of the molecule become important.

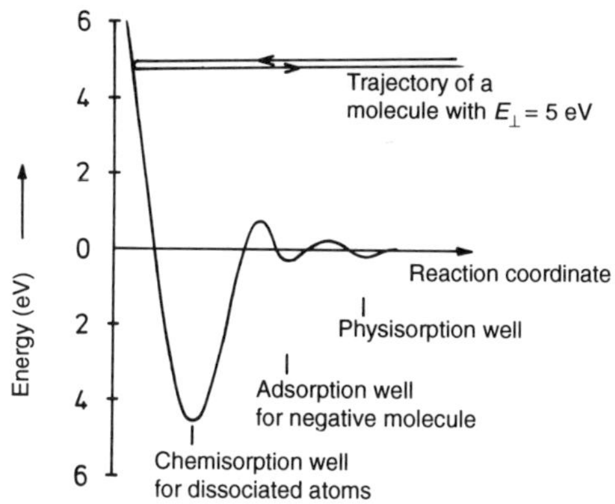

Figure 6.24 Scheme of the interaction potential between a molecule and a surface

These properties will decide whether the molecule will proceed or dissociate and its atoms be bound in a chemisorption well, giving up any remaining kinetic energy. In reality, the potential is at least a two-dimensional surface (Figure 6.25) (Freund and Messmer, 1986; Bartos *et al.*, 1987). This representation allows the discussion of different 'trajectories' on that surface, e.g. dependent on the initial state of the vibrational excitation. Figure 6.25 shows the potential-energy surface (PES) of the CO_2 interaction with Ni. The flatly adsorbed CO_2, the bent $CO_2^{\delta-}$ and the dissociation products have been identified by photoelectron spectroscopy and high-resolution electron-loss spectroscopy (Bartos *et al.*, 1987). The different bound states are well understood, i.e. good agreement between measured and calculated binding energies, their symmetry properties and dispersion is obtained (Freund and Messmer, 1986; Bartos *et al.*, 1987). The finding of CO_2^- in the scattering experiments is certainly in qualitative agreement with the PES. As indicated in Figure 6.23, a particle with appreciable kinetic energy may pass very well over all potential barriers and reflect from the inner wall of the potential; however, this 'violent' collision is not part of the scenario of regular chemisorption. Any effort put into the understanding of these collisions would be worthwhile (Kleyn, 1990).

Another aspect of the theoretical treatment related to the PES modelling concerns total-energy-level diagrams (see e.g. Kang *et al.*, 1988; Freund and Messmer, 1986). These are essential to the understanding of surface chemical reactions. In our context, they are also useful for the discussion of work-function dependences.

The PES calculations have been elaborated especially for the 'harpooning' cases (Gadzuk and Holloway, 1985). These treatments invoke quantum mechanics, wave-packet dynamics and classical mechanics approaches. An important result of the calculations is the 'seams' on the PES which separate, e.g. the A_2

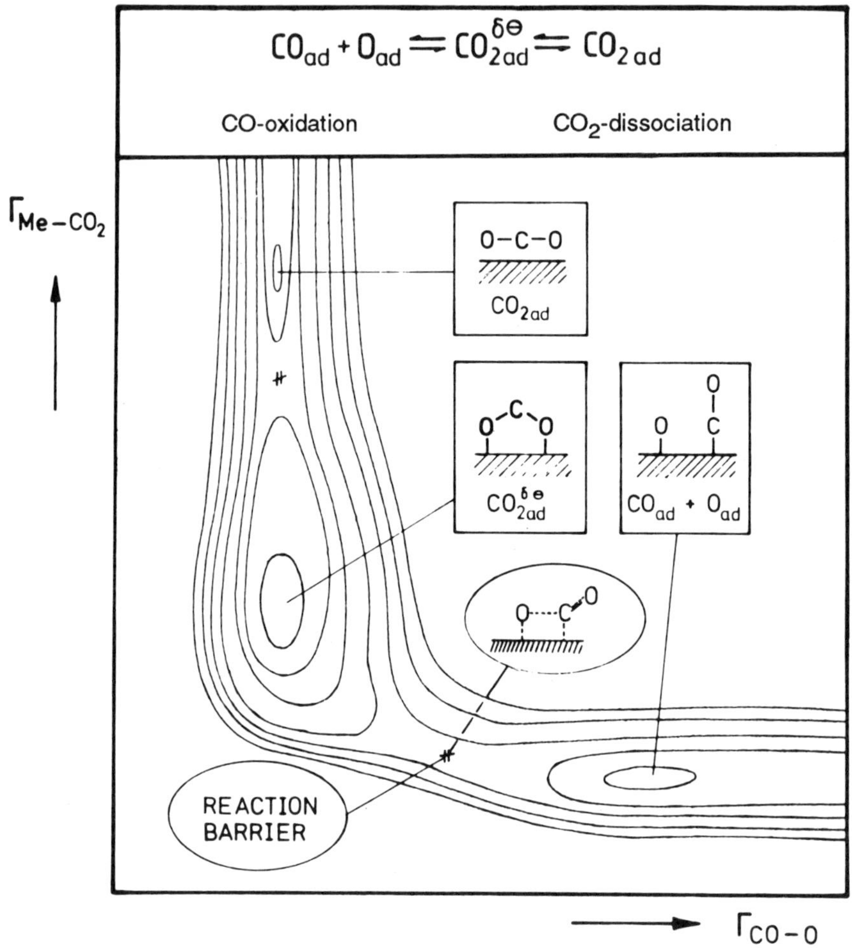

Figure 6.25 Potential-energy surface for the interaction of CO_2 with Ni(110). (Reproduced with permission from Bartos *et al.*, 1987)

and A_2^-. At the same time, these states are degenerate in a one-dimensional representation at a point where $\phi - E_A \simeq e^2/4z_c$, with work function ϕ and vertical electron affinity E_A. The charge exchange is accompanied by a change of the vibrational state of the molecule: kinetic energy T is transformed into potential energy V. It has yet to be elaborated how these models change when particles are faster and the finite electron-capture rate causes a broadening of the seam from a line to a surface.

So far these theories give qualitative guidelines for the experiments. The experiments with fast beams afford the option of focusing onto electronic interactions and separating these from rovibrational effects. Independent of the quantitative understanding of the experiments they are the basis for the tailoring of thin films.

6.6 CONCLUSION

The interaction of swift molecular ions with surfaces is an active field for research. Some of the most recent publications are listed here without further discussion (Rabalais and Marton, 1992; Schmidt *et al.*, 1992b; Kleyn, 1992; Schins *et al.*, 1992). These papers corroborate some of the essential findings discussed and point towards the field to which molecule–surface interactions are related, i.e. chemisorption and thin-film growth. With respect to chemisorption, the experiments discussed here provide results for the electronic interaction under almost ideal Born–Oppenheimer conditions, i.e. the intermolecular motion is frozen during the time interval of the interaction of the molecule with the surface. Therefore electronic processes, such as 'harpooning', can be studied rather selectively. For the thin-film field making use of low-energy ion implanting with atoms, molecules and clusters, the results discussed here provide some understanding of the underlying elementary processes.

6.7 ACKNOWLEDGEMENTS

Contributions to this paper by H. Derks, H. Franke, C. Höfner, U. Imke, A. Närmann, A. Niehof, K. Schmidt, M. Schleberger, S. Schubert, S. Speller and W. Tappe are gratefully acknowledged. I thank A. Kleyn, U. van Slooten and K. J. Snowdon for helpful discussions. The work was in part suppported by the Deutsche Forschungsgemeinschaft and the Deutsche Akadamischer Austauschdienst.

6.8 REFERENCES

Akazawa, H., and Y. Murata (1988a) *Phys. Rev. Lett.*, **61**, 1218.
Akazawa, H., and Y. Murata (1988b) *J. Chem. Phys.*, **88**, 3317.
Akazawa, H., and Y. Murata (1990a) *J. Chem. Phys.*, **92**, 551.
Akazawa, H., and Y. Murata (1990b) *J. Chem. Phys.*, **92**, 5560.
Amirav, A., and M. J. Cardillo (1986) *Phys. Rev. Lett.*, **57**, 229.
Amirav, A., W. R. Lambert, M. J. Cardillo, P. L. Trevor, P. N. Luke, and E. E. Haller (1986) *J. Appl. Phys.*, **59**, 2213.
Aono, M., and R. Souda (1987) *Nucl. Instr. Meth.*, **B 27**, 55.
Backx, C., C. P. M. de Groot, and P. Biljoen (1981) *Surf. Sci.*, **104**, 300.
Balashova, L. L. (1991) *Rad. Effects and Defects in Solids*, **115**, 289.
Balashova, L. L., and V. A. Molchanov (1982) *Surf. Sci.*, **119**, L378.

Balashova, L. L., A. I. Dodonov, and Sh. N. Garin (1982) *Vestnik Mosc. State Univ. Ser. Phys.*, **23**, 85.
Bartos, B., H. J. Freund, H. Kuhlenbeck, M. Neumann, H. Lindner, and K. Müller (1987) *Surf. Sci.*, **179**, 59.
Bitensky, I. S., and E. S. Parilis (1984) *Nucl. Instr. Meth.*, **B 2**, 384.
Bitensky, I. S., and E. S. Parilis (1985) *Surf. Sci.*, **161**, L565.
Bitensky, I. S., Ya. S. Gilenko, and E. S. Parilis (1988) *Sov. Phys. JETP*, **67**, 470.
Bitensky, I. S., E. S. Parilis, and I. A. Wojciechowski (1990) *Nucl. Instr. Meth.*, **B 47**, 246.
Bonzel, H. P. (1988) *Surf. Sci. Rep.*, **8**, 1.
Busmann, H.-G., Th. Lill, B. Reif, and I. V. Hertel (1990) *Surf. Sci.*, **272**, 146.
Campbell, C. T. (1986) *Surf. Sci.*, **173** L641.
Cooks, R. G. (1976) *Collision Spectroscopy*. Plenum Press, New York.
Danon, A., and A. Amirav (1988) *Phys. Rev. Lett.*, **61**, 2961.
Datz, S. (1981) In *Inelastic Particle–Surface Collisions*, E. Taglauer and W. Heiland (eds), Springer Series in Chemical Physics **17**, 142.
de Bruijn, D. P. (1983) Thesis, Amsterdam.
de Bruijn, D. P. and J. Los (1982) *Rev. Sci. Instr.*, **53**, 1020.
Dunn, G. H. (1962) *Phys. Rev. Lett.*, **8**, 62.
Eckstein, W., H. Verbeek, and S. Datz (1975) *Appl. Phys. Letters*, **27**, 527.
Feibelmann, P. (1985) *Surf. Sci.*, **160**, 139.
Flores, F. (1991) In *Interaction of Charged Particles with Solids and Surfaces*, A. Gras-Marti, H. M. Urbassek, N. R. Arista, and F. Flores (eds), Plenum Press, NATO ASI Series B: Physics **271**, 3.
Franke, H., K. Schmidt, and W. Heiland (1992a) *Surf. Sci.*, **212**, 189.
Franke, H., K. Schmidt, and W. Heiland (1992b) *Surf. Sci.*, **269/270**, 219.
Freund, H. J., and R. P. Messmer (1986) *Surf. Sci.*, **172**, 1.
Gadzuk, J. W., and S. Holloway (1985) *Physica Scripta*, **32**, 413.
Gadzuk, J. W., and J. K. Nørskov (1984) *J. Chem. Phys.*, **81**, 2828.
Gates, B. C., J. R. Katzer, and C. C. A. Schuit (1979) *Chemistry of Catalytic Processes*. McGraw-Hill, New York.
Hagstrum, H. D. (1954a) *Phys. Rev.*, **96**, 325.
Hagstrum, H. D. (1954b) *Phys. Rev.*, **96**, 336.
Hagstrum, H. D. (1977) In *Inelastic Ion–Surface Collisions*, N. H. Tolk, J. C. Tully, W. Heiland, and C. W. White (eds) Academic Press, New York, p. 1.
Harris, J. (1989) *Surf. Sci.*, **221**, 335.
Heiland, W. (1989) *Vacuum*, **39**, 367.
Heiland, W. (1991) In *Interaction of Charged Particles with Solids and Surfaces*, A. Gras-Marti, H. M Urbanek, N. R. Arista, and F. Flores (eds), Plenum Press NATO ASI Series B, Physics Vol. 271, p. 253.
Heiland, W., and E. Taglauer (1976) *Nucl. Instr. Meth.*, **132**, 535.
Heiland, W., and E. Taglauer (1982) *Nucl. Instr. Meth.*, **194**, 667.
Heiland, W., and E. Taglauer (1985) In *Experimental Methods in Physics*, *Vol. 22*, R. L. Park and M. Lagally (eds), Academic Press, New York, p. 299.
Heiland, W., U. Beitat, and E. Taglauer (1977) *Phys. Rev.*, **B 19**, 1677.
Heiland, W., U. Imke, S. Schubert, and K. J. Snowdon (1987) *Nucl. Instr. Meth.*, **B 27**, 167.
Herschbach, D. R. (1966) *Adv. Chem. Phys.*, **10**, 319.
Höfner, C., A. Närmann, and W. Heiland (1992) *Nucl. Instr. Meth.*, **B 72**, 227.
Hou, M., and M. T. Robinson (1976) *Nucl. Instr. Meth.*, **132**, 641.
Imke, U., K. J. Snowdon, and W. Heiland (1986a) *Phys. Rev.*, **B 34**, 41.

Imke, U., K. J. Snowdon, and W. Heiland (1986b) *Phys. Rev.*, **B 34**, 48.
Imke, U., S. Schubert, K. J. Snowdon, and W. Heiland (1987) *Surf. Sci.*, **189/190**, 960.
Jackson, D. P., and D. V. Morgan (1974) *Contemp. Phys.*, **14**, 25.
Jackson, D. P., W. Heiland, and E. Taglauer (1981) *Phys. Rev.*, **B 24**, 4198.
Jakas, M. M., and D. E. Harrison (1985) *Surf. Sci.*, **149**, 500.
Kang, H., S. R. Kasi, O. Grizzi, and J. W. Rabalais (1988) *J. Chem. Phys.*, **88**, 5894.
Kashihira, N., T. Schmidt-Beck, and S. Datz (1974) *J. Chem. Phys.*, **612**, 160.
Kasi, S. R., M. A. Kilburn, H. Kang, and J. W. Rabalais (1988) *J. Chem. Phys.*, **88**, 5902.
Kasi, S. R., H. Kang, C. S. Sass, and J. W. Rabalais (1989) *Surf. Sci. Rept.*, **10**, 1.
Kilburn, S. M., H. Kang, W. Rabalais, L. Tavernini, and P. Hochmann (1988) *J. Chem. Phys.*, **88**, 59.
Knözinger, H. (1991) In *Funamental Aspects of Heterogeneous Catalysis Studies by Particle Beams*, H. H. Brongersma and R. A. van Santen (eds), NATO ASI Series, B: Physics Vol. 265, p. 7.
Kiskinova, M., M. Pirug, and H. P. Bonzel (1984) *Surf. Sci.*, **136**, 285.
Kleyn, A. W. (1990) *Vacuum*, **41**, 248.
Kleyn, A. W. (1900) *Comments At. Mol. Phys.*, **24**, 187.
Kleyn, A. W. (1992) *J. Phys. Cond. Matt.*, **4**, 8375.
Kuipers, E. W., A. Varid, A. Danon, and A. Amirav (1991) *Phys. Rev. Letters*, **66**, 116.
Lindhard, J. (1965) *Mat.-Fys. Medd. Kong. Vidensk. Selsk,*, **34**, No. 15.
Lucas, A. A. (1979) *Phys. Rev. Lett.*, **43**, 1350.
Luntz, A. C., M. D. Williams, and D. S. Bethune (1988) *J. Chem. Phys.*, **89**, 4381.
Närmann, A., R. Monreal, P. M. Echenique, F. Flores, W. Heiland, and S. Schubert (1990) *Phys. Rev. Lett.*, **64**, 1601.
Närmann, A., H. Franke, K. Schmidt, A. Arnau, and W. Heiland (1992) *Nucl. Instr. Meth.*, **B 69**, 158.
Neumann, J., S. Schubert, U. Imke, P. Varga, and W. Heiland (1987) *Europhys. Lett.*, **3**, 859.
Niehof, A., and W. Heiland (1990) *Nucl. Instr. Meth.*, **B 48**, 306.
Pan Haochang, T. C. M. Horn, and A. W. Kleyn (1986) *Phys. Rev. Letters*, **57**, 3035.
Panin, B. V. (1962) *J. Exptl. Theoret. Phys. (USSR)*, **42**, 313.
Polanyi, M. see Herschbach (1966), p. 367.
Rabalais, J. W., and D. Marton (1992) *Nucl. Instr. Meth.*, **B 67**, 287.
Rathmann, D., N. Exeler, and B. Willerding (1985) *J. Phys. E.*, **18**, 17.
Rechtien, J. H., U. Franke, K. J. Snowdon, P. H. F. Reijnen, P. J. van den Hoek, A. W. Kleyn, and A. Namiki (1990) *Surf. Sci.*, **227**, 35.
Rechtien, J. H., G. Harder, G. Herrmann, C. Röthig, and K. J. Snowdon (1992) *Surf. Sci.*, **269/270**, 213.
Reijnen, P. H. F., P. J. van den Hoek, A. W. Kleyn, U. Imke, and K. J. Snowdon (1989) *Surf. Sci.*, **221**, 427.
Reijnen, P. H. F., A. Rankema, U. van Slooten, and A. W. Kleyn (1991) *Surf. Sci.*, **253**, 24.
Reijnen, P. H. F., U. van Slooten, and A. W. Kleyn (1991) *J. Chem. Phys.*, **94**, 695.
Robinson, M. T., and I. M. Torrens (1974) *Phys. Rev.*, **B 9**, 5008.
Sass, C. S., and J. W. Rabalais (1987) *J. Chem. Phys.*, **89**, 3870.
Schins, J. M., R. B. Vrijen, W. J. van der Zande, and J. Los (1992) *Surf. Sci.*, **280**, 145.
Schlatholter, T. (1993) Diploma Thesis, Osnabrück.
Schmidt, K. (1992) Thesis, Osnabrück.
Schmidt, K., A. Närmann, and W. Heiland (1992a) *J. Phys. Cond. Matt.*, **4**, 9869.
Schmidt, K., T. Schlatholter, A. Närmann, and W. Heiland (1992b) *Chem. Phys. Lett.*, **200**, 465.

Schubert, S. (1989) Thesis, Universität Osnabrück.
Schubert, S., U. Imke, W. Heiland, K. J. Snowdon, P. H. F. Reijnen, and A. W. Kleyn (1988) *Surf. Sci.*, **205**, L793.
Schubert, S., U. Imke, and W. Heiland (1990) *Vacuum*, **41**, 252.
Shayegan, M., J. M. Cavallo, R. E. Glover, and R. L. Park (1984) *Phys. Rev. Lett.*, **53**, 1578.
Sigmund, P. (1992a) *Nucl. Instr. Meth.*, **B 67**, 11.
Sigmund, P. (1992b) *Phys. Rev.*, **A 46**, 2596.
Sizmann, R., and C. Varelas (1976) *Nucl. Instr. Meth.*, **132**, 633.
Smith, D. P. (1967) *J. Appl. Phys.*, **18**, 340.
Taglauer, E., W. Englert, W. Heiland, and D. P. Jackson (1980) *Phys. Rev. Lett.*, **45**, 740.
Tappe, W. (1990) Diploma Thesis, Universität Osnabrück.
Tappe, W., A. Niehof, K. Schmidt, and W. Heiland (1991) *Europhys. Letters*, **15**, 405.
Tully, J. C. (1980) *Adv. Chem. Phys.*, **42**, 63.
Vager, Z., R. Naamann, and E. Kanter (1989) *Science*, **244**, 429.
van den Hoek, P. J. (1989) Thesis, Amsterdam.
van den Hoek, P. J., and A. W. Kleyn (1989) *J. Chem. Phys.*, **91**, 4318.
van den Hoek, P. J., T. C. M. Horn, and A. W. Kleyn (1988) *Surf. Sci.*, **198**, L335.
Pinxteren, H. M., C. F. A. van Os, R. M. A. Heeren, R. Rodink, J. J. C. Geerlings, and van J. Los (1989) *Europhys. Letters*, **10**, 715.
van Slooten, U., D. Andersson, A. W. Kleyn, and E. A. Gislason (1991) *Chem. Phys. Lett.*, **185**, 440.
van Slooten, U., D. R. Andersson, A. W. Kleyn, and E. A. Gislason (1992) *Surf. Sci.*, **274**, 1.
Weiss, P. S., P. L. Trevor, and M. J. Cardillo (1988) *Phys. Rev.*, **B 39**, 9928.
Willerding, B., W. Heiland, and K. J. Snowdon (1984a) *Phys. Rev. Lett.*, **53**, 2031.
Willerding, B., H. Steininger, K. J. Snowdon, and W. Heiland (1984b) *Nucl. Instr. Meth.*, **B 2**, 453.
Willerding, B., K. J. Snowdon, and W. Heiland (1985) *Zeitschr. f. Physik*, **B 59**, 435.
Wurth, W., J. Stöhr, P. Feulner, X. Pan, K. R. Bauchspiess, Y. Baba, E. Hudel, G. Rocker, and D. Menzel (1990) *Phys. Rev. Lett.*, **65**, 2426.
Yang, H., and J. L. Whitten (1991) *Surf. Sci.*, **255**, 193 and ref. therein.

7 SPUTTERING OF METALS AND INSULATORS WITH HYPERTHERMAL SINGLY AND DOUBLY CHARGED RARE-GAS IONS

P. Varga and U. Diebold*
Institut für Allgemeine Physik, Technische Universität Wien, Wiedner Hauptstraße 8–10, A-1040 Vienna, Austria

CONTENTS

* Permanent address: Department of Physics, Tulane University, New Orleans, Louisiana 70118, USA

Low Energy Ion–Surface Interactions
Edited by J. W. Rabalais

7.1 INTRODUCTION

In the field of ion–surface interactions, many aspects have been studied quite extensively. One of them is sputtering, where a great variety of target/projectile systems have been investigated. This has resulted in an impressive amount of data, including typically sputtering yields as well as yields for the production of secondary ions or excited atoms. Most of these data have been collected by bombarding metal targets with rare-gas ions, the energy of the projectiles usually ranging from several hundred eV up to several hundred keV. The interest in these processes is driven by applications like surface analysis and surface preparation; impact energies above 500 eV are of importance (Behrisch, 1981, 1983; Benninghoven *et al.*, 1987). Very little is known about ions at lower energy and/or insulating targets.

Besides sputtering and formation of secondary ions by energetic primary ions, neutralization of ionic projectiles near surfaces has also been studied. This was done either by measuring the survival probability of scattered ions or by studying the electron emission which results from electronic transitions. It has been shown that detailed information on the electronic processes which occur during ion–surface interaction can be received by using very-low-energy ions and by comparing results obtained with singly charged projectiles with results from multiply charged ions. In particular, processes which are governed by the potential energy of the projectiles can be tackled by this approach (see review by Varga and Winter, 1991).

The present chapter describes experimental results for ion-induced desorption and sputtering of metallic and insulating surfaces by impact of singly and doubly charged ions with very low energy. It refers to one of the few studies where the energy range of the ions is in the so-called hyperthermal range of 10–500 eV. The goal is to obtain insight into some of the fundamental processes of ion–surface interaction.

It should be mentioned that interest in this low-energy range is also driven by technical applications like plasma-assisted deposition techniques (see for example Cuomo *et al.*, 1989). Atoms and ions with energies above thermal energy and well below 500 eV play an important role in the quality of deposited thin films. Processes like sputtering and resputtering, deposition and formation of metastable compounds occur during deposition. A further understanding of these processes will be of help in controlling the properties of thin films.

The main emphasis of this work is the investigation of the basic aspects of ion–surface interaction. In particular, the possibility of the conversion of potential energy brought to a surface by the projectile into kinetic energy of the desorbed or sputtered secondary ions and atoms has been a controversial issue. So-called DIET (desorption induced by electronic transitions) processes, which have mainly been studied using photons and electrons, see e.g. (Tolk *et al.*, 1983), can also be induced by ion bombardment; their occurrence is examined in this chapter.

The influence of the charge state of the projectile (i.e its ionization energy) on desorption and sputtering should be visible at very low impact velocity where all processes induced by ballistic sputtering are drastically reduced. (Note that 'desorption' refers to the removal of an adsorbate by either collisional or electronic interaction whereas 'sputtering' is more often used for removal of substrate surface atoms. The terms are used interchangeably throughout this chapter.) As a possible mechanism a Coulomb-explosion model has been proposed (Bitenskii *et al.*, 1979; Bitensky and Parilis, 1989). It is known from experiments on metal surfaces that even multiply charged ions are already completely neutralized at a distance of several ångstroms from the surface (Varga and Winter, 1991). During this neutralization, an appreciable number of electrons may be emitted from the surface depending on the charge state and the potential energy of the projectile. For insulating or semiconducting surfaces, this loss of electrons leads to a positively charged region in the vicinity of the impact point of the ion because of the lack of delocalized and loosely bound valence electrons. According to Bitenskii *et al.* (1979) and Bitenskii and Parilis (1989), the repulsive forces that are built up in the surface region will be sufficient to cause emission of the ionized lattice atoms, resulting in the so-called 'Coulomb explosion'. A few experiments have been performed to examine the validity of this concept but no conclusive results are yet available.

Since experiments in the very-low-energy range are still sparse, a description of the concept and characterization of a low-energy beam is given in Section 7.2. By the same token, the precautions which have to be taken to perform reliable experiments with insulating surfaces are discussed in some detail in Section 7.4.2. The results of ion-induced desorption from CO-covered Ni and from LiF are given in Sections 7.3 and 7.4, respectively.

7.2 EXPERIMENTAL SET-UP

The challenge is to build a high-intensity low-energy ion source, well characterized and compatible with surface-sensitive experiments. The experimental set-up used in our work is described here in some detail. At present such a set-up is not commercially available but it is a crucial requirement for experiments answering the questions raised above.

All experiments on surfaces are extremely sensitive on the residual gas pressure. Therefore, ultra-high vacuum (UHV) is a *conditio sine qua non.* Three different pumping stages allow us to run plasma ion sources with a typical gas pressure of $\sim 10^{-2}$ mbar and at the same time hold UHV ($\sim 10^{-10}$ mbar) in the target region.

The primary-ion beam must be well defined in mass, charge state, energy, and beam spot size. The first three requirements are met by mass selection and suppression of the ion beam neutrals, which are produced to some extent by collisions of the ions with the residual gas particles. Since the quality of a low-energy ion beam suffers severely from space-charge effects, the transport of ions has to be performed in a high-energy region as long as possible. This can be obtained by an acceleration–deceleration set-up. Additionally, as our interest is in measuring sputtering as a function of energy, tuning the final energy must not mean changing all the supply voltages of the beam transport system. For easier operation, the target should be held near ground potential. For desorption experiments, a homogeneous ion current density over the whole target is a crucial requirement.

The complete system (Figure 7.1) consists of three stages which are pumped differentially. Stage 1 houses a plasma ion gun and is insulated by ceramic tubes from pumping stage 2 and an oil diffusion pump. A duoplasmatron ion source was found to produce doubly charged ions with satisfactory current intensity. In this part of the system, a pressure of 10^{-5} mbar is maintained. The vacuum houses an Einzel lens (I) and a Wien filter where the beam is separated according to the ion velocity. A flexible bellows allows a bend of 3° in the beam line to get rid of neutral particles in the ion beam. The neutrals are stopped at the next aperture (between the first and second pumping stages) while the ions are deflected by the horizontal plates. A second Einzel lens (II) focuses the ion beam onto the next aperture (2 mm diameter), which separates the second from the third pumping stage. Right after this aperture, a deceleration lens is mounted and the beam line ends with x–y-deflection plates.

The residual gas pressure in the target chamber is of the order of 10^{-10} mbar. The target is mounted on a manipulator and can be optionally cooled down to the temperature of liquid nitrogen or resistively heated up to 1000 °C. The total distance from the ion source to the target is 160 cm. An Auger electron spectrometer and a quadrupole mass spectrometer are mounted at the level of and perpendicular to the ion beam for analysing the target surface and measuring desorbed species, respectively.

To obtain the above-mentioned convenience in tuning the final ion energy with the target at ground potential, we have separated the power supplies for the different components of the beam transport system by isolation transformers. Figure 7.1(b) shows an energy schematic of the ion beam; we start with (nearly) zero energy at the anode of the plasma ion source, which is floated by the voltage which determines the final energy of the ions. The ions are accelerated by a maximum voltage of -3.5 kV (operated mostly at -2.85 kV). The common of this high-voltage power-supply unit is defined by the final-energy voltage and therefore follows when tuning the final energy. This 'beam voltage' represents the common of all the power supplies for the beam transport facilities including the Einzel lenses, the deceleration lens, the x–y-plates in all

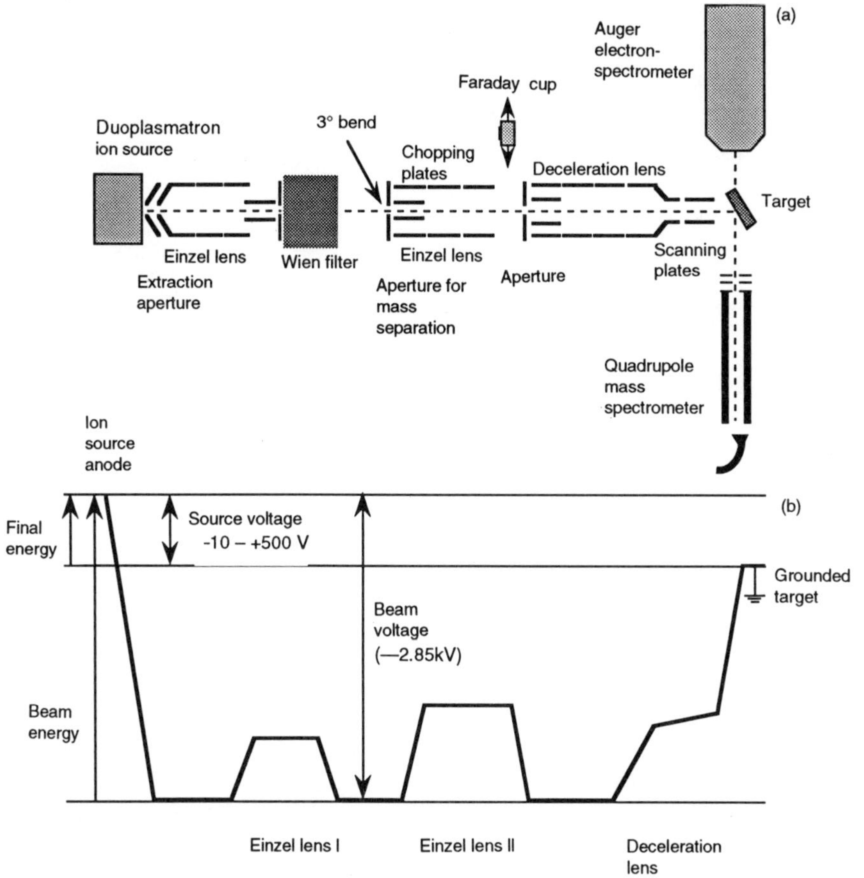

Figure 7.1 (a) Schematic of the beam transport components and of the low-energy beam, (b) Supply voltages of the beam transport components, the beam energy and final energy are given for singly charged ions

the lenses and the Wien filter. The Einzel lenses are operated in retarding mode. As the ions have a kinetic energy defined by the beam voltage only and independent of the final energy while passing through the beam transport system, a change in the final energy affects the focusing done by the deceleration lens only.

The deceleration lens is decisive for beam quality. It is a four-element cylindrical lens with the first electrode at beam voltage and the last at ground potential. The fourth electrode is of conical shape for holding stray fields off the target region. The focusing behaviour was modelled by a lens program.

For the measurements, the ion-beam current is the decisive factor. Figure 7.2 shows the Ar^+ ion-beam current as measured at the target position. The

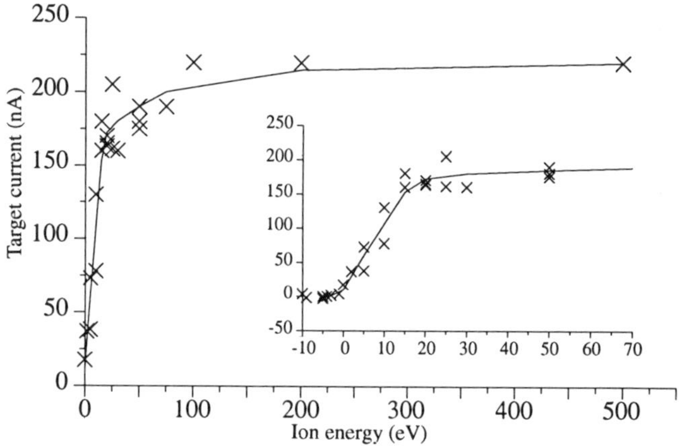

Figure 7.2 Ar^+ ion-beam current as measured at the target position. For convenience, the very-low-energy data are shown in the inset. Negative source voltages have been applied to show the zero crossing. The lines are drawn to guide the eye

relatively wide scatter of the values is due to different adjustments of the beam-transport supply voltages. Usually, this 'target current' amounts to 200 nA for Ar^+ ions. The ion beam starts to lose intensity and to spread significantly in size for final energies <20–30 eV. The widening of the beam is due to space-charge effects, typical values of Ar^+ beam diameters as measured with the Faraday cup at the target position are given in Figure 7.3.

The target current depends on the ions used, average values for rare-gas ions are listed in Table 7.1. The current density J_{max} is given for the focused ion beam (and values of the ion beam FWHM as in Figure 7.3), $J_{scanned}$ shows the values when scanning the focused ion beam over an area of $6 \times 6\ mm^2$. These values are typical of most of our measurements. The dependence of the current density on the ion species is due to the specific ionization efficiency of the ion source, being of course much lower for doubly charged ions. Generally, higher

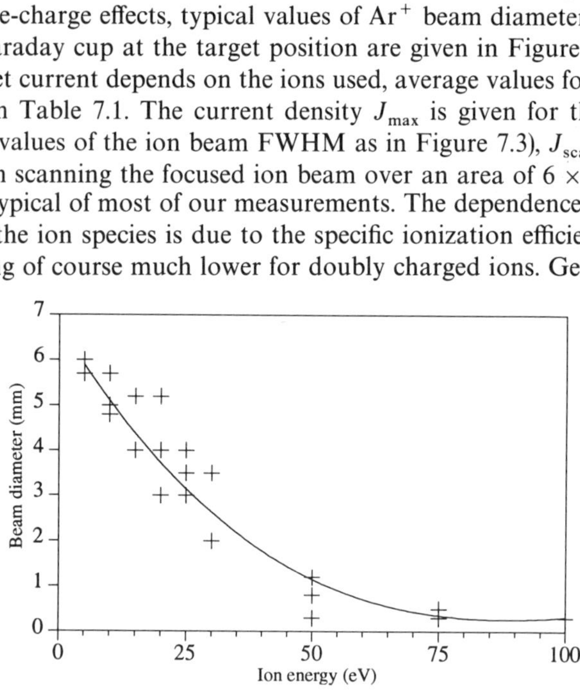

Figure 7.3 Ar^+ ion-beam diameter (FWHM) of the retarded beam at the target position

Table 7.1 Typical ion-beam current (J_{target}) measured at the target for a final energy of 100 eV

	J_{target} (nA)	J_{max}[a] ($\mu A\ cm^{-2}$)	$J_{scanned}$[b] ($\mu A\ cm^{-2}$)
$^4He^+$	700	16	1.6
$^3He^{2+}$	1	0.18	0.012
Ne^+	250	6	0.48
Ne^{2+}	10	0.8	0.014
Ar^+	200	10.6	0.4
Ar^{2+}	36	3.8	0.09
Kr^+	120	4	0.32
Kr^{2+}	12	1.2	0.02

[a] J_{max} = maximum current density for focused ion beam
[b] $J_{scanned}$ = current density with ion beam scanned over 6 × 6 mm^2

current densities can be achieved with lighter ions, since ions with a higher velocity suffer less from space-charge effects.

Due to the 3° bend in the beam line, neutrals in the ion beam can be largely suppressed. Assuming resonant neutralization of ions at the residual pressure of the ion-beam gas as the most probable process, the proportion of ions which were neutralized when passing pumping stages 2 and 3 was estimated to be less than 0.1% of the total particle current.

The beam energy and the energy spread of the duoplasmatron ion source have been determined by measuring the current in the Faraday cup when biasing its aperture with a positive voltage. The typical energy spread is 2–3 eV, depending on the ion-source parameters as shown in Figure 7.4. This

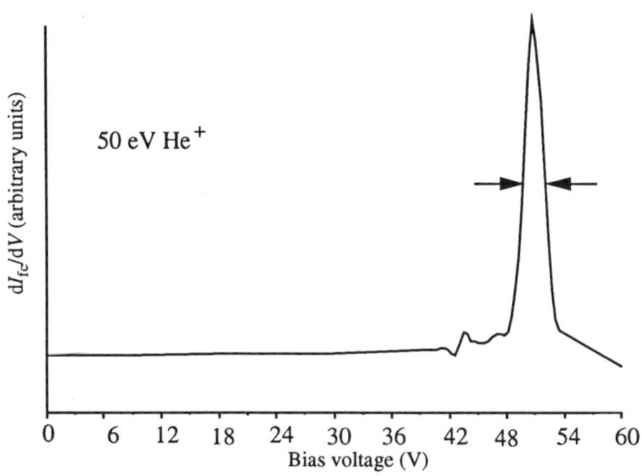

Figure 7.4 Energy spread of a 50 eV He^+ ion beam. I_{fc} is the ion current as measured in the Faraday cup when the positive bias voltage V is applied to its aperture. The change of I_{fc} with bias voltage V is shown. The ions are produced in a duoplasmatron ion source

method is quite simple and, as the retarding field is not corrected for any inhomogeneities, the 'real' energy spread is probably smaller than the energy distributions obtained.

Determination of the final energy of the ion beam is of great importance for measurements at very low kinetic energy. Since the ion yield decreases severely with the primary-ion energy in the very-low-energy regime, artefacts could be caused by slight inaccuracies in the kinetic energy of the ion beam. Such a systematic error is excluded by measuring the primary-ion energy and energy spread for every single experiment.

7.3 BOMBARDMENT OF CO-COVERED Ni(111)

As the first system for studying the influence of the potential energy of an ionic projectile on the sputtering mechanism, we chose a Ni(111) single crystal covered with a monolayer of CO. For this surface, many data on ion-induced desorption at higher impact energy are available and indicate a relatively large cross section for desorption ($\sigma \sim 10^{-14}$–10^{-16} cm^2 (Taglauer *et al.*, 1979, 1980; Winters and Sigmund, 1974). These values could be confirmed by us and have been extended to lower energies close to the sputtering threshold (Diebold and Varga, 1990a, 1990b; Diebold *et al.*, 1991). Electron- and photon-stimulated desorption have been studied quite extensively for CO adsorbed at different metal surfaces (Netzer and Madey, 1982; Campuzano *et al.*, 1981; Froitzheim and Köhler, 1987). Whereas a Coulomb explosion is expected to be hindered at a metal surface (Bitenskii *et al.*, 1979; Bitenskii and Parilis, 1989), DIET processes induced by the potential energy of an incoming low-energy ion might be possible. For the first time, ion-induced DIET processes on adsorbate-covered metal surfaces are separated using projectiles with different charge states (Diebold and Varga, 1990a, 1991).

7.3.1 EXPERIMENTAL DETAILS

The Ni(111) single crystal was spot-welded on two 0.1 mm Mo wires which could be resistively heated. The target was prepared by standard methods (Grunze *et al.*, 1988), involving several cycles of 700 eV Ar^+-ion sputtering and subsequent annealing at up to 600°C. Remaining impurities of carbon could be removed by adsorbing 0.5 langmuir O_2 with subsequent annealing. This procedure was repeated regularly after each desorption measurement in order to anneal the target from defects caused by sputtering and to remove all carbon produced by ion-induced dissociation.

For the desorption measurements, the target was loaded with 4 langmuir CO (exposure time of 112 s at a pressure of 4×10^{-8} torr) at room temperature. This resulted in a c(4 × 2) CO layer (Froitzheim and Köhler, 1987) with a saturation coverage of 0.5 monolayers as checked by thermal-desorption spectroscopy (TDS). For determination of the ion yields (Figures 7.6–7.9) the partial pressure of CO was kept at 5×10^{-7} mbar during the measurements. Adsorption of CO is known to increase the secondary-ion yield of a metal surface. By holding the ambient pressure high, 'steady-state' conditions are achieved, where ion-induced desorption is balanced by adsorption from the gas phase. This is necessary to provide a fully CO-covered surface and the same work function conditions during all the measurements. The same particle densities for singly and doubly charged ions have been used for data collection at each energy. With the exception of the ion charge state all experimental conditions have been kept the same, allowing us to exclude any other reasons for deviation in the secondary-ion yields.

7.3.2 SECONDARY-PARTICLE EMISSION FROM CO/Ni(111)

Figure 7.5(a) shows the desorption cross sections σ of neutral CO desorbed from Ni(111) by Ne^+ bombardment at different impact energies. The values are quite large, e.g. for an impact energy of 300 eV σ amounts to 3.9×10^{-15} cm^2, corresponding to 3.6 desorbed molecules per incoming ion. This comparably high desorption yield can be explained by the low binding energy of CO in bridge sites ($E_B = 0.7$ eV (Froitzheim and Köhler, 1987), combined with a high collisional cross section for the diatomic CO molecule (Taglauer *et al.*, 1980). The angular dependence of the desorption cross section at a fixed impact energy of 50 eV shows a pronounced arccos φ behavior as shown in Figure 7.5(b).

Figure 7.6 shows the *ratio* of the initial desorption yields, i.e. desorbed CO molecules per incident Ne^{2+} ion over desorbed CO molecules per incident Ne^+ ion for the fully covered surface. As the ratio is unity in this energy range, one can conclude that either the charge state of the primary ion does not influence the desorption of neutral CO or that such an influence cannot be detected because desorption by momentum transfer is much more effective, even at these low impact energies.

This picture changes if one considers the desorbed CO^+ ions. For bombardment with Ne^{2+} at impact energies below 200 eV, the CO^+ signal is larger than that induced by Ne^+. The ratio of the CO^+ secondary-ion yields induced by doubly and singly charged Ne ions, increases up to a factor of five for 50 eV impact energy as shown in Figure 7.7. The absolute amount of desorbed CO^+ ions decreases with decreasing energy for both projectiles and reaches the detection limit around 40 eV (Diebold and Varga, 1990a, 1991).

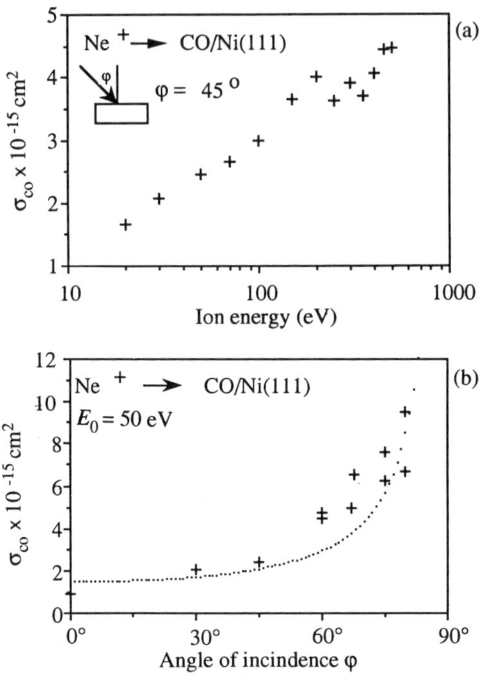

Figure 7.5 (a) Desorption cross sections σ for neutral CO induced by bombardment of CO/Ni(111) with Ne^+. Initial coverage was 0.5 monolayer and impact angle was 45° with respect to the target normal. (b) Measured desorption cross sections for neutral CO for bombardment with 50 eV Ne^+ at different impact angles φ. The dotted curve corresponds to an arccos φ distribution

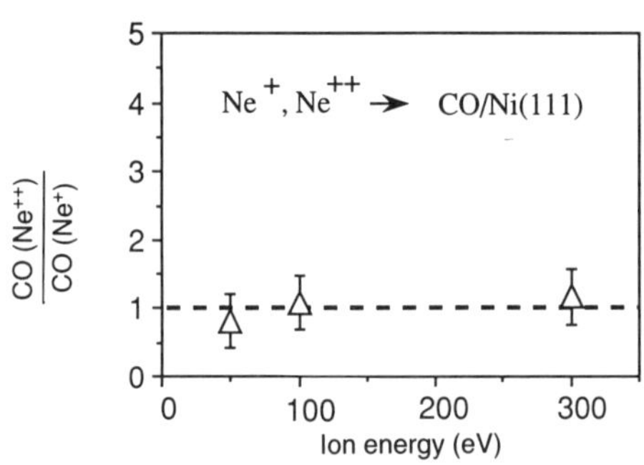

Figure 7.6 Ratio of the yields of neutral CO produced by singly and doubly charged Ne ions, respectively

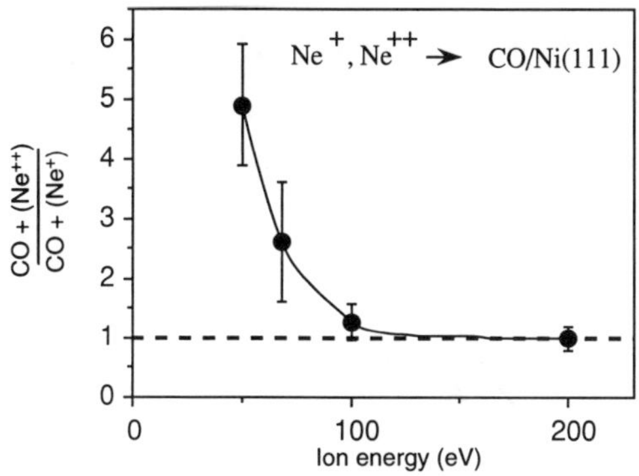

Figure 7.7 Ratio of the yields of CO^+ ions produced by singly and doubly charged Ne ions, respectively

In addition to CO^+, Ni^+, $NiCO^+$, Ni_2^+ and O^+ were also found to desorb from the CO-covered Ni surface. As compared to CO^+, these signals were larger by a factor of ~3000 (Ni^+), ~30 ($NiCO^+$ and Ni_2^+) and ~0.3 (O^+), respectively. The variation of the secondary-ion yield of Ni^+ with impact energy was determined for singly and doubly charged Ne ions (Figure 7.8). These measurements have been performed using a very low ion flux of 6×10^{10} ions $cm^{-2}\,s^{-1}$ for Ne^+ as well as for Ne^{2+}. Again, a decrease of the secondary-ion yield with decreasing impact energy of Ne^+ as well as Ne^{2+} was found, but an increase of the ratio of the secondary-ion yields in favour of Ne^{2+} was apparent towards low impact energy.

Figure 7.9 shows the Ni^+ yield with Kr^+ and Kr^{2+} primary ions under the same experimental conditions as for Ne ions. Again, bombardment with low-energy (≤ 300 eV) doubly charged primary ions results in a higher

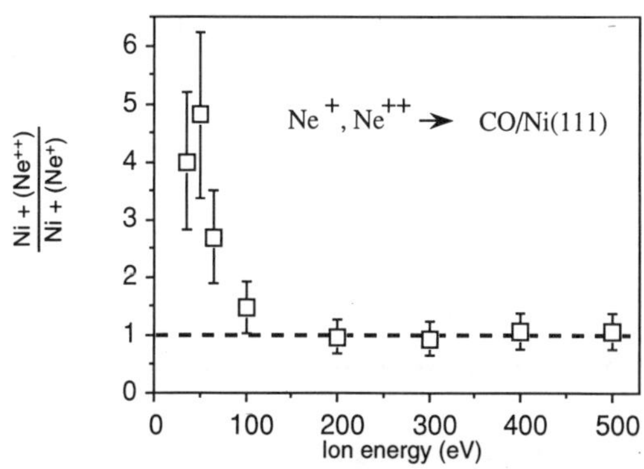

Figure 7.8 Ratio of the yields of Ni^+ ions produced by singly and doubly charged Ne ions, respectively

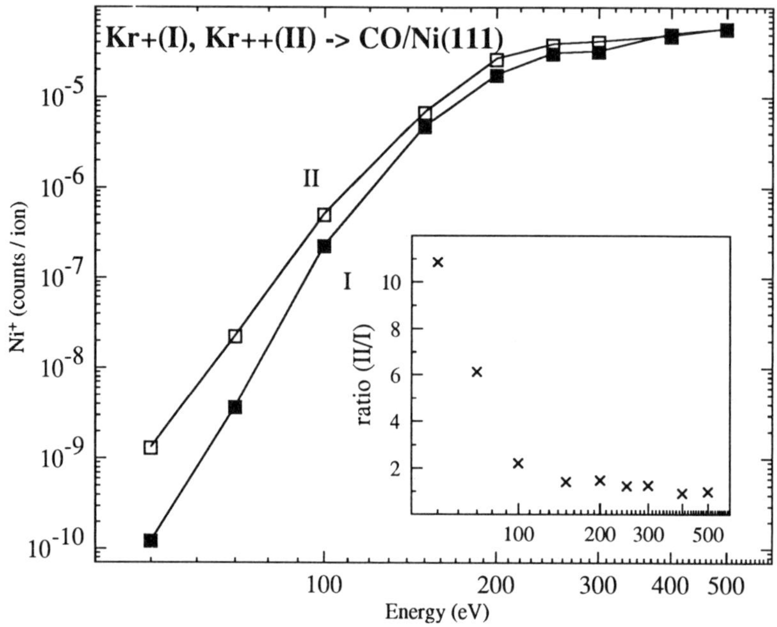

Figure 7.9 Ni^+ yield for Kr^+ and Kr^{2+} impact on CO-covered Ni(111). The inset shows the ratio of the yields

secondary-ion yield. The yield ratio (inset Figure 7.9) increases with decreasing bombardment energy up to a factor of ~ 10. The lowest impact energy used was 50 eV, which is close to the theoretical sputtering threshold of 33 eV for a Ni target and Kr projectiles (Matsunami *et al.*, 1984). Also small amounts of scattered Kr^+ and Kr^{2+} could be detected; scattering of singly and doubly charged ions has also been observed for Ne (Diebold and Varga, 1991). This indicates that even at these low energies, the ion-survival probability is not zero during scattering.

Results with Ar bombardment show a similar effect of charge-dependent secondary-ion production.

7.3.3 A MODEL FOR CHARGE-DEPENDENT ION EMISSION FROM A METAL SURFACE

Direct knock-off caused by a collision between the primary ion and the adsorbed molecule is known to be the dominating desorption process at low ion-impact energies. For direct knock-off, Winters and Sigmund (1974) gave as a first approximation for the desorption cross section

$$\sigma \approx \frac{\ln E_0}{\cos \varphi}. \tag{1}$$

To obtain equation (1), a Born–Mayer potential and an isotropically bound adsorbate molecule have been assumed. The latter is hit by the incoming ion with a primary energy E_0 and then reflected with zero energy loss from the substrate. These are very rough approximations; however, the observed energy dependence of the desorption cross sections is described quite satisfactorily (see Figure 7.5(a) where σ has been plotted against E_0 in a semilogarithmic fashion). Furthermore, good agreement is found by fitting the results for different impact angles to an arccos φ dependence (Figure 7.5(b)). Due to the almost equivalent masses of CO and Ne, a major part of the kinetic energy can be transferred during the first collision. Accordingly, the effect of cascade sputtering is of minor importance. It should be noted that model calculations using the computer code TRIDYN (Möller and Eckstein, 1984; Möller, 1989) reproduced the experimentally determined cross sections for desorption of neutral CO very well (Diebold *et al.*, 1991). A statistical evaluation of the collisional processes which give rise to the desorption event shows the dominant contribution of direct ion–adsorbate interactions and the minor contribution of collision-cascade processes at low impact energies. This is shown in Figure 7.10 for the case of Ne. The fraction

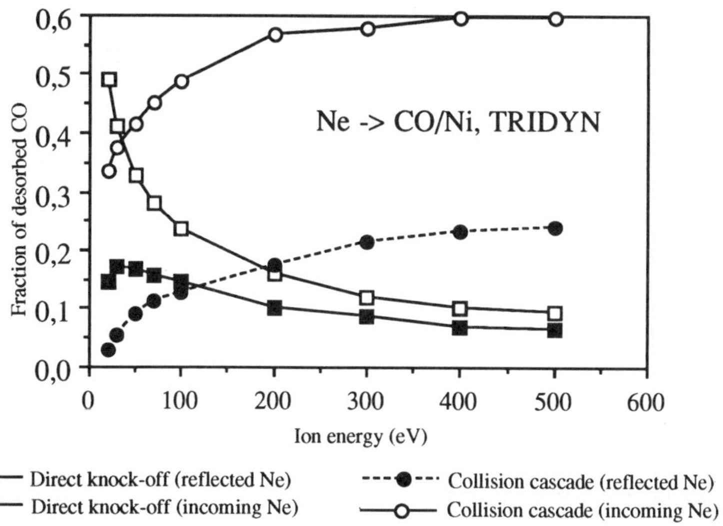

Figure 7.10 Calculation of the fraction of CO molecules emitted by direct knock-off and collision cascade processes, respectively. The primary particle was Ne. The calculations have been performed with the TRIDYN code. (Möller and Eckstein, 1984; Möller, 1989)

of emitted particles induced by the development of a collision cascade decreases with decreasing impact energy whereas the fraction of particles emitted in a binary collision with an incoming or reflected Ne ion increases.

The above mentioned Coulomb explosion, which has been proposed as a possible mechanism causing additional sputtering from insulators due to the potential energy of the impinging particle (Diebold *et al.*, 1991; Diebold and Varga, 1991) is believed to be unlikely for the CO/Ni(111) system. Because of the high mobility of the conduction electrons in the metal substrate, electron vacancies in the adsorbate created by charge-exchange processes with the projectile will be recombined before the ejection mechanism is started.

In Behrisch (1981, 1983) and Benninghoven *et al.* (1987) and in Wittmaack (1977), overviews of the possible processes leading to secondary-ion emission during particle bombardment are given. Neither of the proposed mechanisms involves a dependence on the primary-ion charge state.

From our results, two conclusions can be drawn directly. First, the charge state of the projectile does *not* influence significantly the *amount* of desorbed or sputtered particles. The potential energy of the impinging ion influences only the *charge state* of the emitted particles, at least at very low kinetic impact energy (compare Figures 7.6 and 7.7). Second, the main production process of desorbed (respectively sputtered) species is due to the *direct collisional interaction between primary ion and adsorbate* or *substrate* since the yield decreases very strongly with decreasing kinetic impact energy. This is also plausible in the light of the simulation results discussed above.

The secondary-electron yield is known to increase with the potential energy of an incoming low-energy ion (Varga and Winter, 1991). Thus, one could invoke an ionization of the sputtered particles by secondary electrons in the gas phase for the observed phenomenon. However, the timing of such a process is somewhat hard to imagine. The ion currents used in our experiment are very small, far below the value where secondary electrons produced by one primary ion might cause ionization of particles sputtered by a different ion. Only the secondary electrons which are caused by that ion which is also responsible for the secondary-particle emission could cause such an effect. Moreover, only the secondary-electron yield due to the neutralization of the primary ion is influenced by the primary-ion charge state. This neutralization event usually occurs along the incoming path, in close proximity to the surface. This means that the secondary electrons (the majority of them also at an energy far below the ionization threshold) will certainly have left *before* the sputtered particle starts to move, making ionization by secondary electrons unlikely.

As a likely explanation, we propose a secondary-ion production process where a secondary particle is sputtered and ionized simultaneously by the primary ion in a binary collision (a primary knock-off process which is very probable at low impact energy). During the collision, the primary ion forms a quasi-molecule with a target atom. If a doubly charged ion is involved in the

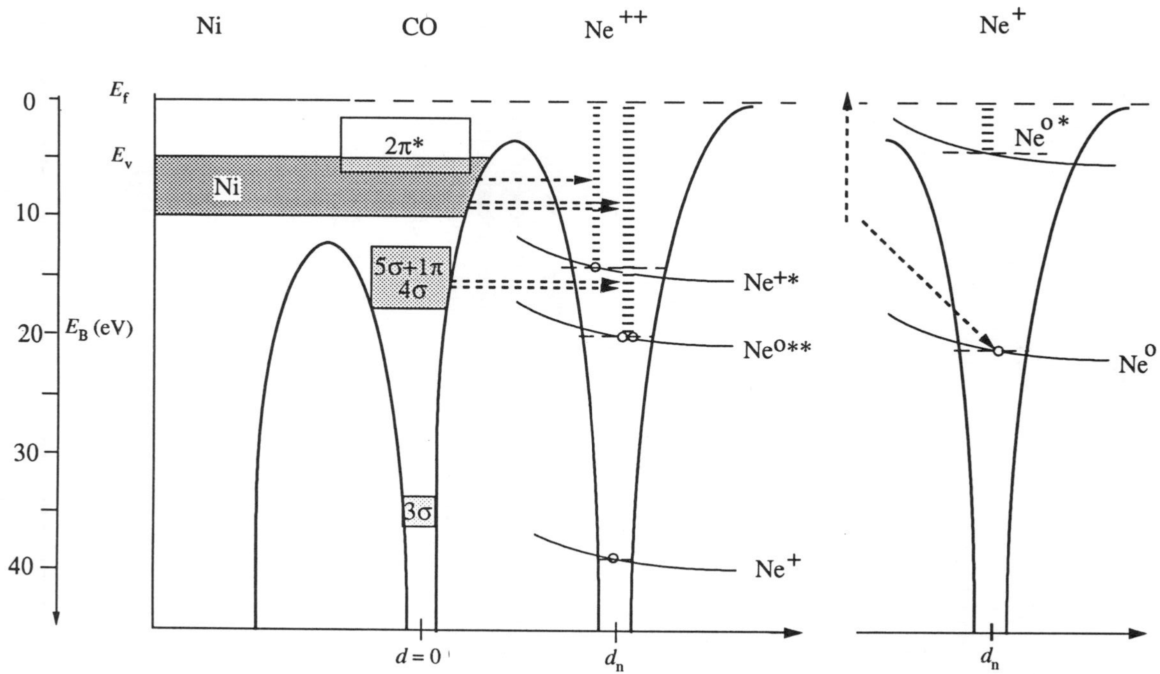

Figure 7.11 Electron energy diagram of a doubly and singly charged Ne ion in front of a CO-covered Ni surface. A shift to lower binding energy of the electronic states of the incoming ions due to the image forces is drawn schematically. Ne^{+*} and Ne^{o*} are excited states of the ion and the neutral atom, respectively. If occupied by resonance neutralization processes, Ne^{o**} are doubly excited states of the Ne atom, which will de-excite by autoionization. They will de-excite by Auger de-excitation

sputtering process, a higher electronic excitation energy will be supplied and electrons can be captured resonantly by the projectile from the target. Considering the distance of closest approach ($\sim$1.5 Å for a head-on collision with 50 eV impact energy), the velocity of the particle, and the distance where electronic transitions start to become important ($\sim$3–5 Å), the collision time is of the order of 10^{-14} s. During that time the ion and the surface atom are in their 'quasi-molecular state'. This is also the typical time scale for Auger and resonance transitions. The decay of the quasi-molecule may result either in an ionized or a highly excited state of the secondary particle (see the potential-energy diagrams in Figure 7.11).

At higher impact energies, where a sputter cascade is developed and the emitted secondary particles originate mainly from the end of the cascade, no vacancy sharing between primary ion and emitted atom is possible. This causes the vanishing of the charge effect at higher impact energies.

An alternative explanation can be given following the model of Sroubek (1989) where it is assumed that the secondary-ion production depends on the electronic excitation of the substrate. At low impact energy the electronic excitation of the substrate by Auger transitions between target atoms and projectiles may become comparable to the excitation by kinetic processes. Doubly charged primary ions have a higher potential energy and can transfer more excitation energy to the substrate and consequently an increase in secondary-ion production is expected.

How the presence of the CO on the Ni surface affects the charge-dependent ionization process is not resolved yet, since the experiment could not be performed with an uncovered target. For a clean metal surface, the yield of secondary ions decreases several order of magnitude and was therefore not measurable at these low impact energies. In a first approach, we assume that the CO coverage is only important for suppressing the neutralization of the primary particle and also the positive secondary ions by increasing the work function.

7.4 LITHIUM FLUORIDE

7.4.1 CONCEPTS FOR ION-INDUCED DIET PROCESSES ON LiF

On the basis of experimental results (Radzhabov *et al.*, 1976; Morozov *et al.*, 1979), Bitenskii *et al.* (1979) have developed the 'Coulomb-explosion' model for sputtering of insulators by highly charged ions, i.e. the ejection of positively charged microscopic domains of the surface, resulting from step-by-step neutralization of the ion which leads to the ejection of several Auger electrons from the lattice atoms.

In Eccles *et al.* (1986), it was demonstrated that bombardment of Si with ions increases the sputter yield by more than a factor of 2 as compared with bombardment by neutral atoms of the same mass and energy. In contradiction to these results, de Zwart *et al.* (1986) have shown that only the yield of ionized particles emitted from Si increases noticeably with increasing charge state of the projectile, whereas the total sputter yield remains unaffected for an ion charge state of up to 9+. In Bitenskii and Parilis (1989) it was argued that the B-doped Si which has been used as a target in the latter experiment had a too high carrier concentration which would make a Coulomb explosion unlikely.

Williams (1981) reported on ion-stimulated desorption of F^+ where an interatomic Auger process, similar to the Knotek–Feibelmann (1978) mechanism in electron-stimulated desorption was proposed to explain the relatively high energy of emitted F^+. In (Schultz *et al.*, 1983), a relationship between the potential energy of the primary ion and the F^+ yield has been demonstrated for LiF sputtering. For bombardment with singly charged He and Ne ions, a constant sputtering yield down to zero impact energy was extrapolated from measurements with ion energies between 200 and 2000 eV. The virtual absence of a sputter threshold has been explained by an interatomic Auger transition between the projectile and the F^- ion at the surface. In this process, the F atom changes from its negatively charged ground state in the LiF lattice into F^+ and becomes desorbed by the now repulsive Madelung potential.

In Thompson and Taylor (1986) a higher F^+ yield has been found if doubly charged Ar ions instead of singly charged ions were used for LiF sputtering in the projectile energy range of 1–5 keV.

To obtain further insight into electronic effects in sputtering of insulators we measured the interaction of singly and doubly charged noble-gas ions with LiF surfaces at energies between 10 and 500 eV (Varga *et al.*, 1991; Wutte *et al.*, 1992; Neidhard *et al.*, 1993a,b). We could not confirm (Varga *et al.*, 1991) the formerly extrapolated constant sputtering yield for F^+ down to zero impact energy for ions with sufficiently large potential energy (Schultz *et al.*, 1983).

If an ion approaches a surface with low kinetic energy, neutralization by electron transitions between the surface and the projectile takes place. For a metal surface, these electron transitions will influence the charge state of the projectile, cause electron emission from the valence band of the surface into the vacuum, and induce a higher probability for secondary-ion emission as discussed in Section 7.1. For insulators, these electronic transitions will also change the charge state of surface atoms.

In the case of an alkali halide, it is expected that the main part of emitted particles consists of the lattice ions, i.e. F^- and Li^+ for LiF. These ions are emitted by momentum transfer in a collision cascade and may be neutralized on the way out of the surface. Additionally, positively and negatively charged clusters and molecular particles will also be emitted.

For F^+, the situation is quite different. Because of its very high ionization energy, it can be produced only in a violent collision or by electronic transitions. Especially at low impact energy, effects of electronic transitions should be observable. The mechanism which might possibly lead to F^+ production during bombardment of LiF is shown schematically in Figure 7.12. The electron energy diagram of LiF (Piacentini and Anderegg, 1981) and of the projectile ions in different charge states in front of the surface are indicated. Auger neutralization from the F 2p band with subsequent electron emission is possible for impact of He^+, Ne^+, and all doubly charged ions because the potential energy of the respective ground states is large enough, i.e. at least twice the energy of the band gap.

Let us now consider possible transitions which will result in the production of F^+ ions. For He^+, this process can be described in the following way,

$$He^+ + F^- \text{—/AN/}\rightarrow He^0 + F^+ + e_{kin}. \tag{2}$$

with AN is the symbol for Auger neutralization and e_{kin}, the emitted electron carrying away the surplus energy. For all doubly charged rare-gas ions, resonance neutralization (RN) in excited ion states with subsequent Auger de-excitation (AD) into the ground state of the singly charged ion can also occur, e.g. for Ne^{2+}:

$$Ne^{2+} + F^- \text{—/RN/}\rightarrow Ne^{+*} + F^0 \text{—/AD/}\rightarrow Ne^+ + F^+ + e_{kin}. \tag{3}$$

Furthermore, double resonance neutralization (2RN) into doubly excited atom states Ne^{0**} is energetically possible. These states de-excite by autoionization (AI):

$$Ne^{2+} + F^- \text{—/2RN/}\rightarrow Ne^{0**} + F^+ \text{—/AI/}\rightarrow F^+ + Ne^+ + e_{kin}. \tag{4}$$

For Ar^+ and Kr^+, only one resonance transition is possible, which leads to a neutral F^0 atom, and F^+ production is impossible:

$$Ar^+ + F^- \text{—/RN/}\rightarrow Ar^0 + F^0. \tag{5}$$

Since the transition probability for resonance processes is much higher than for Auger transitions, RN will always take place if energetically allowed, i.e., if empty projectile states become energetically resonant to the surface valence electrons.

7.4.2 CHARGING EFFECTS

A severe difficulty in ion-beam experiments with insulating surfaces is the generation of electrostatic charge at the surface. Both the impact of positively charged ions and the subsequent electron emission will form a positively

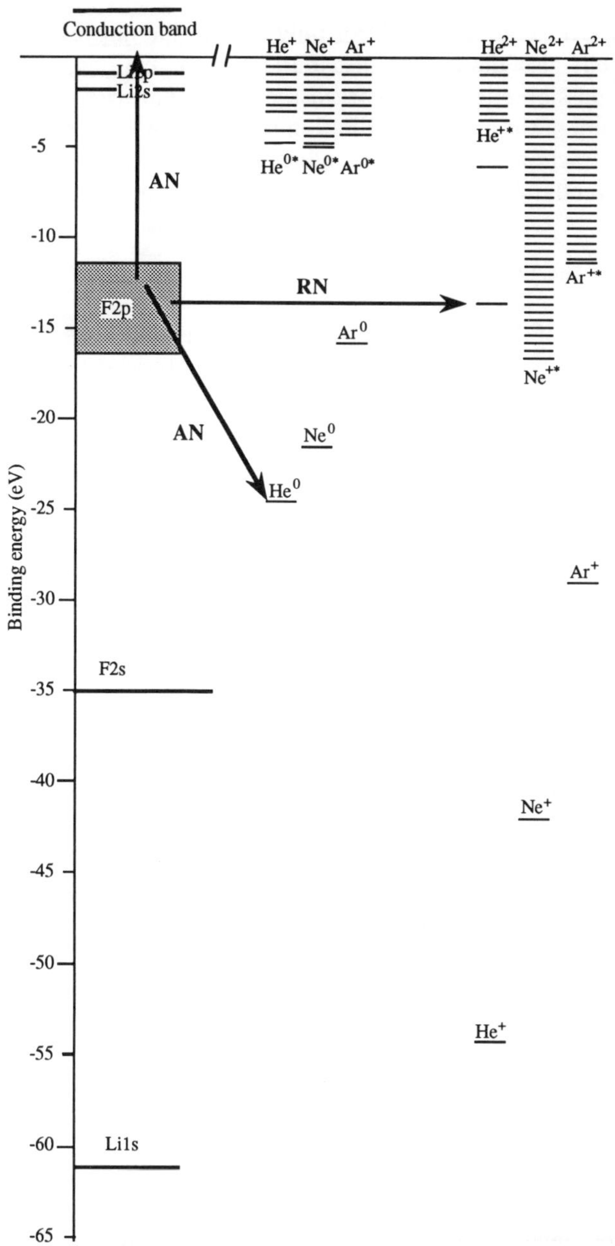

Figure 7.12 Electron-energy diagram of LiF (Piacentini and Anderegg, 1981) and of singly and doubly charged He, Ne and Ar ions in front of the surface. Parallel lines indicate allowed unoccupied levels. As an example, possible Auger neutralization and resonance neutralization processes from the valence band of the surface into the ions are shown by arrows

charged surface layer. This will not only change the impact energy and beam geometry, but also the energy distribution of the emitted, charged particles. Since the emission energy of charged secondary particles is usually very low (the maximum of the energy distribution is only a few eV), even an electrostatic charge of a fraction of one eV can influence the total ion yield.

Several standard methods can be used to overcome the charge build-up at the surface:

(i) flooding the target with carriers of appropriate polarity (electrons in our case);
(ii) deposition of the insulating target material as thin films (μm) or ultra-thin films (nm) on metal substrates, to reduce the resistance of the surface;
(iii) heating of the sample to a temperature where it becomes a good ionic conductor.

In principle, the deposition of a metallic grid to the surface could help to decrease charging effects. This method has not been tested since it involves a serious risk of surface contamination, especially in sputtering experiments. We have used the other three methods in order to study the influence of electrostatic charges on the emission of charged secondary ions. After performing extensive tests, we are reasonably confident that we are able to distinguish between artefacts induced by charging up and 'real' phenomena.

Two different LiF targets have been used, a LiF(100) single crystal and thin films of LiF. The single crystal ($5 \times 5 \times 1\ mm^3$) has been cleaved outside of the vacuum chamber and then clamped onto a sheet of Ta. On the backside of this Ta holder two 0.25 mm Mo wires have been spot-welded for heating the target. The crystal could be heated up to 700 °C by resistive heating of the Ta sheet; the temperature was measured by a thermocouple fixed on the holder. The substrate for the thin LiF films was stainless steel. The thickness of the films was 0.25 μm and the target could be heated by a coaxial resistance heater. A tungsten filament installed in the vicinity of the surface has been used for emission of electrons to bombard the surface with low impact energy. The bias voltage of the filament was -5 eV since this energy lies below the threshold for electron-induced desorption which is about 7 eV (Wurz *et al.*, 1991).

At room temperature, the LiF(100) crystal charged up to an extent where it performed as electrostatic mirror for the ion beam. By increasing the temperature, charging started to decrease. At a temperature above 350 °C, the yield of the emitted positively and negatively charged particles stabilized. All data taken from LiF(100) have been measured at a temperature of 400 °C. At this temperature, LiF becomes a good ionic conductor, and a stoichiometric surface is restored because radiation-induced lattice defects are annealed (Wurz *et al.*, 1989).

Heating experiments have also been performed with the thin LiF film targets. The emission of Li^+, F^+ and F^- was measured at different target temperatures

(20 °C and 400 °C) with or without electron flooding. The yields of these species might be influenced by surface charging as well as changes in surface composition, the latter being a product of different sample treatments. To separate the two effects the following test procedure has been developed.

As a first step, the ratio of Li^+ to F^+ yields was monitored to control the surface composition. Since both emitted particles are positively charged, this ratio should be independent of the electrostatic potential of the surface and each change should indicate different surface composition. If the result of the first step showed an unchanged surface composition for the cold and the hot surface, the ratio of Li^+ to F^- was taken to determine any charging up of the surface. This ratio should increase if the target is at positive potential (i.e charged up) because positively charged particles (Li^+) will be repelled and negatively charged particles (F^-) will be attracted by the surface.

Figure 7.13, the Li^+/F^+ ratio is plotted for the impact of He^+ ions on a thin film of LiF at 20 °C and 400 °C. No significant difference can be seen. This was only the case if the target had been carefully annealed and then brought to room temperature before taking the data. For a thoroughly sputtered cold surface, quite different results have been observed, showing Li segregation because of sputter-induced lattice defects (Wurz *et al.*, 1989).

On the other hand, the ratio of emitted Li^+ and F^- yields (Figure 7.14) is clearly higher for the cold target than for the hot one; this indicates that even the thin LiF film is slightly positively charged at room temperature under positive-ion bombardment. The difference between the ratios *increases* with *decreasing* impact energy because the energy distribution of the emitted secondary ions moves towards lower energy. Slower ions are more likely to feel any field induced by the charging of the surface and the yields may be changed substantially.

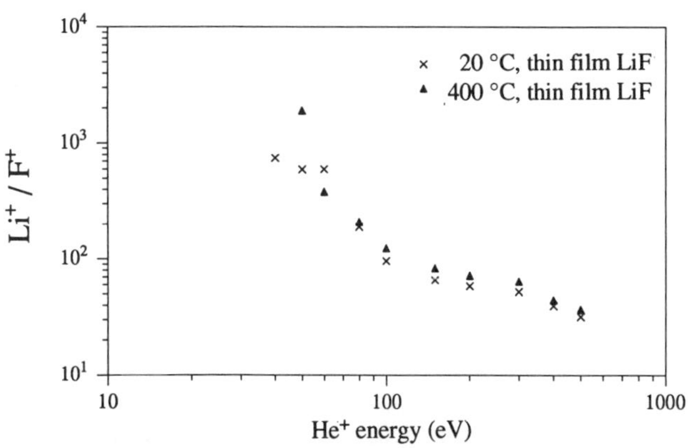

Figure 7.13 Li^+ yield divided by F^+ yield for impact of He^+ on a thin LiF film at 20 °C (×) and 400 °C (▲), respectively

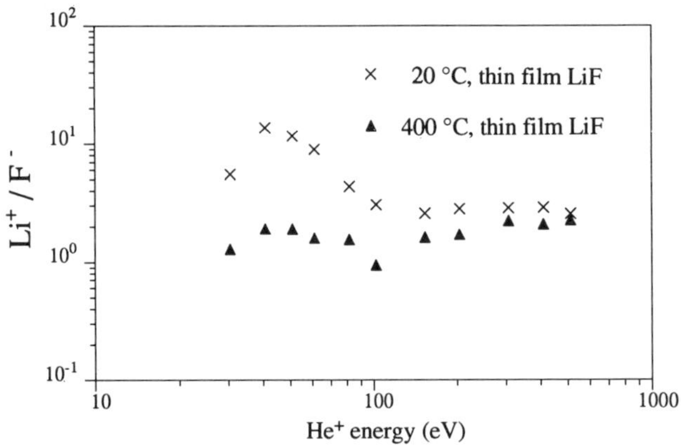

Figure 7.14 Li^+ yield divided by F^- yield for impact of He^+ on a thin LiF film at 20 °C (×) and 400 °C (▲), respectively

The last point has been confirmed further by flooding the cold target with 5 eV electrons which also left the Li^+/F^+ unchanged (i.e. no change in the surface composition), but brought the Li^+/F^- ratio back to the value for the hot surface, provided careful balance between the ion and electron currents was maintained. Unfortunately, the charge balancing with low-energy electrons was not very stable and the results were thus not reproducible in a satisfactory fashion.

The conclusion drawn from these tests was that heating LiF at 400 °C is a convenient method for getting rid of the electrostatic charging. At this temperature, LiF becomes a good ionic conductor and also a stoichiometric surface is restored. These experiments also showed that artefacts in ion-yield measurements can be introduced very easily and that only great experimental care can avoid wrong conclusions. For example, at room temperature charging cannot be suppressed completely for a film of 0.25 μm thickness. Ultra-thin films of several atomic layers are known to show no charging (Dieckhoff *et al.*, 1992) but are not well suited for sputtering experiments.

The following results have been obtained using a hot (400 °C) LiF(100) single-crystal surface.

7.4.3 SECONDARY-ION EMISSION FROM LiF

For all projectile species used, the Li^+ and F^- dominated the secondary-ion spectra, since they constitute the lattice ions of the alkali halide. Furthermore, F^+ ions have been detected as well as cluster ions like Li_2^+, LiF^+, Li_2F^+, LiF^-, F_2^- and LiF_2^-.

As already discussed in the previous section, the secondary-ion yields induced by singly and doubly charged rare-gas ions have been compared to study the influence of the projectile potential energy. All other parameters like impact angle, particle density, etc., have been kept the same for measurements at one particular impact energy. Therefore, any difference in the yield has to be attributed to the potential energy and the charge state of the projectile.

The results on Li^+ emission for impact of singly and doubly charged He, Ne and Ar, respectively, are presented in Figure 7.15(a)–(c). The ion-impact energy was kept below 500 eV. Below 100 eV, a slightly higher Li^+ yield can be observed for doubly charged ions. These data show that the emission of Li^+ particles is due to a classical sputtering process, dominated by momentum transfer. The influence of the potential energy of the projectile is comparable to that of a metal surface. The difference at low impact energy may be caused by the same processes which are mainly responsible for formation of F^+ (as discussed below).

The absolute yields are of the same order of magnitude for the different projectiles. The yield is largest for He followed by Ne and Ar. The difference becomes more pronounced for lower impact energy. All yields decrease with impact energy and reach their threshold around 10–30 eV depending on the ion mass and charge state.

In Figure 7.16, similar results are shown for the emission of F^-. Again, the absolute yields are comparable for the three ions. As for Li^+ emission, the yield induced by doubly charged ions is larger compared to the one for singly charged projectiles at impact energies below 100 eV. The emission thresholds are about 10 eV higher than for Li^+.

The whole situation changes dramatically for the yield for F^+ (Figure 7.17). This yield is largest for Ne impact which is certainly because of the perfect mass match (this is similar to the F^- emission). This fact and the decrease of all yields with decreasing impact energy indicate that sputtering is responsible for particle emission. It is remarkable that yields for Ar are lower than for He; this is not the case for F^- emission and can be taken as an indication that electronic transitions are of some influence. Much stronger evidence for such an influence is provided by the large difference between yields obtained by singly and doubly charged ions. The ratio of these yields is about 2–3 at 500 eV, but at lower energy the yield of doubly charged ions becomes higher by up to two orders of magnitude (e.g. for Ne projectiles with 60 eV).

For He projectiles (Figure 7.17(a)), the situation can be described with the help of the energy diagrams depicted in Figure 7.12. The binding energy of the fluorine 2p valence band is between 11 eV and 16 eV (Piacentini and Anderegg, 1981). Therefore He^+ can ionize F^- to F^+ by one AN, whereas He^{2+} can be neutralized to He^+ either by RN into the $2s^2S$ state, followed by AD or by two RN into a doubly excited neutral atom which de-excites by autoionization (AI). Both possibilities have been confirmed by measuring the electron emission of slow He^{2+} scattered on a W(110) surface covered with a few layers of NaCl (Dieckhoff *et al.*, 1992).

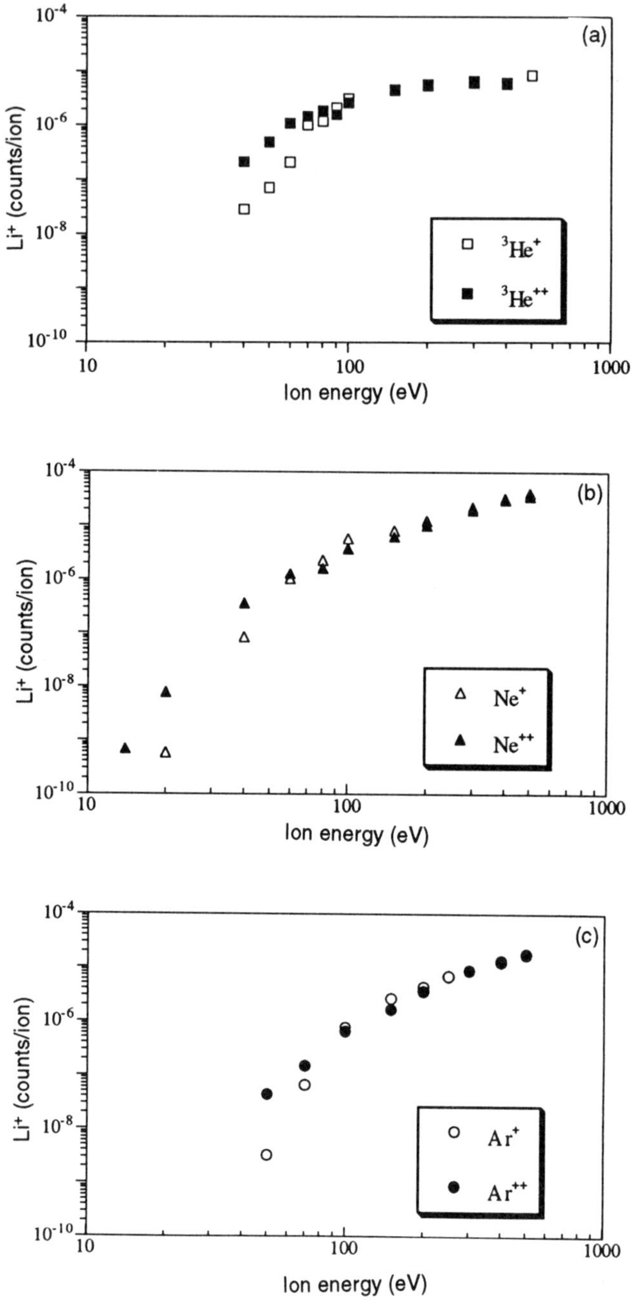

Figure 7.15 Li^+ yield for impact of singly (open symbols) and doubly (full symbols) charged He (a), Ne (b) and Ar (c) ions on LiF (100). The target temperature was 400 °C

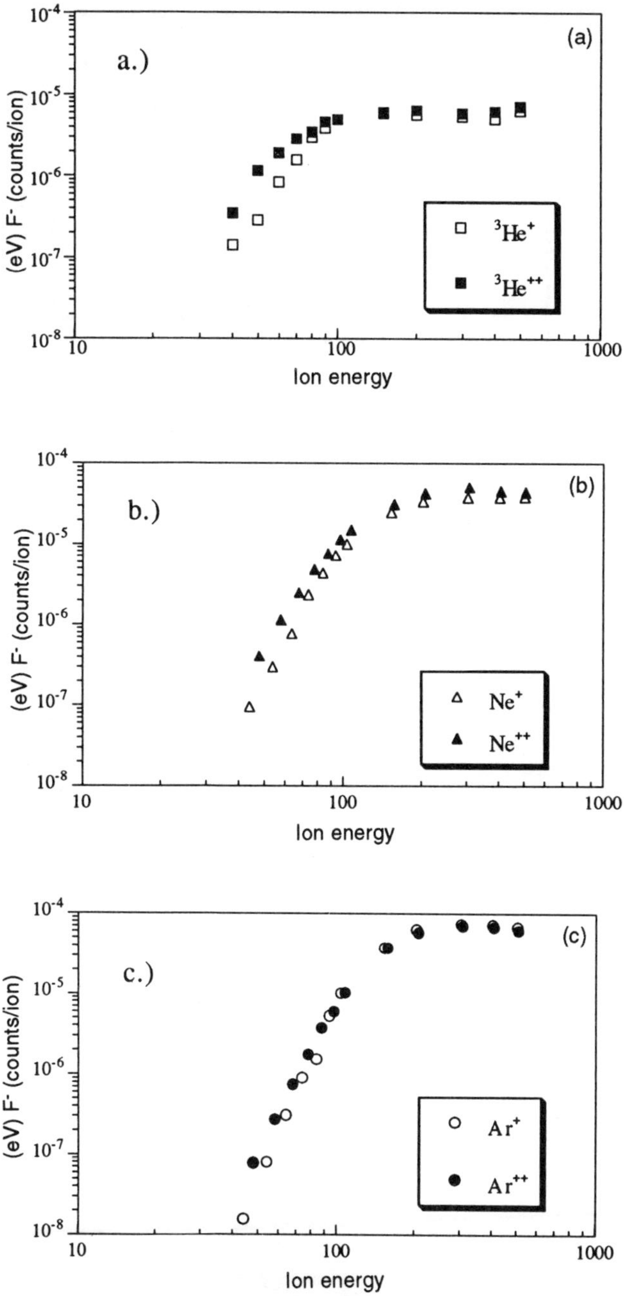

Figure 7.16 F^- yield for impact of singly (open symbols) and doubly (full symbols) charged He (a), Ne (b) and Ar (c) ions on LiF (100). The target temperature was 400 °C

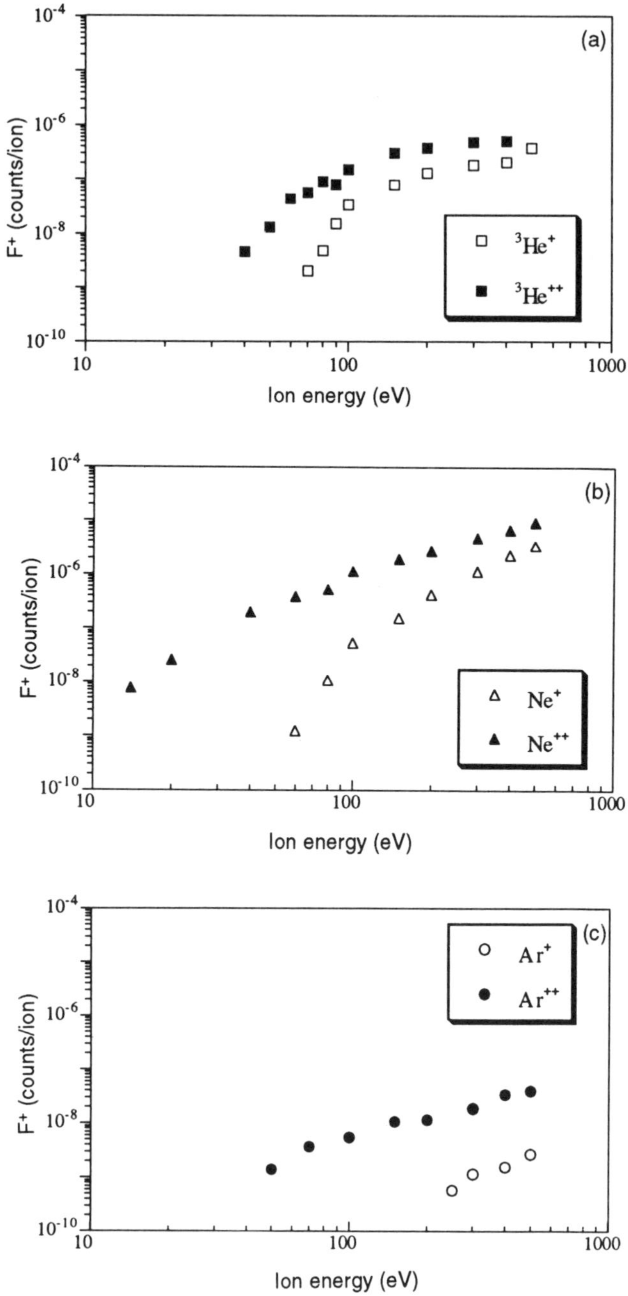

Figure 7.17 F^+ yield for impact of singly (open symbols) and doubly (full symbols) charged He (a), Ne (b) and Ar (c) ions on LiF (100). The target temperature was 400 °C

This additional neutralization step for He^{2+} can explain why the yield for He^{2+} is larger than for He^+. Nevertheless, both yields decrease monotonically with decreasing impact energy. It is evident that the potential energy of the projectile is of great importance for the production of F^+. In addition to the electron exchange between the sample and the ion, a certain momentum transfer seems to be necessary to emit the F^+ ion. If this were not the case, the yields should keep constant as the impact energy is further decreased.

Ne^+ can change F^- to F^+ by an AN with low probability since the available energy is at the limit for such a process (almost exactly twice the bandgap value of LiF). For Ne^{2+}, similarly to He^{2+}, several transitions like AN, RN followed by AD and 2RN followed by AI are easily possible. This can explain why the yield for Ne^{2+} is by about one order of magnitude higher than for Ne^+. The sputtering threshold for Ne^{2+} is below 10 eV which is the lowest value we observed in any of the experiments performed with singly and doubly charged rare-gas ions.

Ar is a very interesting case (Figure 7.17(c)) because for Ar^+ no electronic transition which can produce F^+ is possible (see Figure 7.12 and formula (5)). Therefore, only the kinetic energy of the projectile can produce F^+ by an electron-promotion process during the impact. This results in a very low yield and an energy threshold around 200 eV, the highest value observed in all our experiments. For Ar^{2+} similar processes as for Ne^{2+} are possible, and therefore the yield is much larger as compared with Ar^+.

A small fraction of the emitted ions are positively and negatively charged clusters and molecules. In Figures 7.18 and 7.19, the yields for Li_2^+ and Li_2F^+ for impact of singly and doubly charged Ne ions are shown as an example. The number of emitted molecular ions is about two orders of magnitude lower than

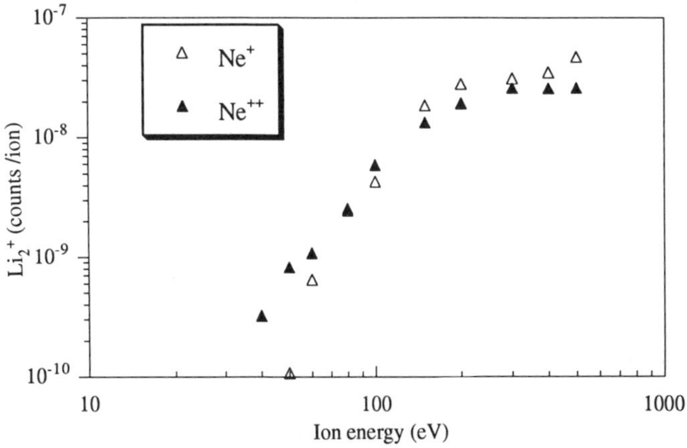

Figure 7.18 Li_2^+ yield for impact of singly (open symbols) and doubly (full symbols) charged Ne on LiF (100). The target temperature was 400 °C

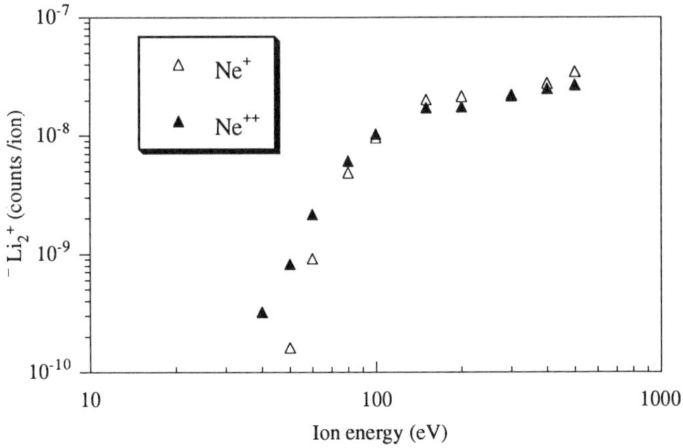

Figure 7.19 Li_2F^+ yield for impact of singly charged (open symbols) and doubly charged (full symbols) Ne on LiF (100). The target temperature was 400 °C

the yield for atomic ions. Collisional processes are mainly responsible for the emission and no influence of the potential energy can be observed.

7.4.4 THE MODEL OF DYNAMICAL BINDING ENERGY

Two conclusions can be drawn on the basis of our data. First, electronic transitions are indeed important for the creation of F^+ ions. Second, additional momentum transfer is necessary to cause emission of F^+. If electrostatic repulsion alone were sufficient, the yield for F^+ should be constant down to zero impact energy; this is in contrast to the energy thresholds observed.

We propose the following model of dynamical binding energy during sputtering which combines both electronic transitions and ballistic processes (Varga *et al.*, 1991; Wutte *et al.*, 1992; Neidhart *et al.*, 1993a,b). It is based on a scenario originally suggested by Walkup and Avouris (1986) in order to explain the results of electron sputtering from alkali halides. They performed classical trajectory calculations on the behaviour of an alkali-halide lattice if the negative halide ion (e.g. F^- in NaF) is suddenly changed into a positive one. These calculations show that the initial large repulsive energy of 11 eV between F^+ and the NaF lattice becomes slightly attractive within 15–30 fs, due to lattice distortions where the Na^+ ions are pushed back and the F^- ions are attracted. This time is too short for desorption of the F^+, which therefore becomes trapped in a stable state. The F^+ ion finally occupies a bridge position between two F^- surface atoms, about 2.5 Å above the surface and with a (reduced) binding energy of 4 eV.

The same processes may also be applied to low-energy ion sputtering, because electron transition processes (AN and double RN) which create a F^+ ion on the surface take place when the projectile ion is still several ångstroms away from the surface. Therefore, the sudden change from F^- into F^+ can be compared with formation of F^+ by electron impact, with a consequently similar reaction of the lattice. However, one has to consider the subsequent impact of the neutralized particle which can change the rearrangement of the lattice. The comparably low current densities used in the present experiments exclude formation and emission of F^+ by two different, subsequently arriving projectiles. Assuming that NaF is not, in principle, different from LiF as already indicated in Walkup and Avouris (1986), we want to apply the time scale given above to our experiment.

For metal surfaces, the transition rate for the above mentioned electron transfer from the surface valence band to an ionized particle in front of the surface depends exponentially on the surface–particle distance, because it is governed by the overlap of the respective atomic wave functions. The transition rate is independent of the ion-impact energy, but the position where the process takes place most probably moves closer to the surface with increasing particle velocity. In a rough approximation (Varga and Winter, 1991; Appell, 1987; Varga, 1987), the transitions start at a distance in front of the surface which is about three times the initial charge state in angstroms. This means about 3 Å for singly charged ions and 6 Å for doubly charged ones. This distance may be smaller for alkali halides where more localized wave functions have to be assumed because of the larger binding energy of the F 2p electrons as compared to valence electrons in metals.

Ions with an energy of 100 eV touch the surface within about 10 to 40 fs after they have been neutralized. The exact value depends on the mass, for He it is near 10 fs whereas for Ar a value of 30–40 fs can be assumed if different impact angles are also considered. Therefore, impact of He^0 actually takes place when the lattice is still reacting, whereas Ar^0 reaches the surface when the lattice rearrangement is already finished and the F^+ formed is already in a bound state. *This means that, depending on the velocity of the projectile, different binding energies of F^+ to the surface have to be considered.* For lower impact velocity, the binding energy of the F^+ ion has already increased and additionally the available kinetic energy of the projectile is also lower. As a consequence, the emission yield decreases and reaches its threshold. At medium impact velocities, the projectiles hit the surface while the F^+ is still in a repulsive state. When going to higher velocities, the kinetic processes will dominate because the electronic transitions will become more and more unlikely (not enough time for the electronic transitions—Varga and Winter, 1991; Appell, 1987; Varga, 1987). Therefore, a steep increase of the emission yield with impact energy is observed and the differences between singly and doubly charged projectiles become smaller and smaller.

The influence of the potential energy on the emission of Li^+ and F^- is much weaker and only observable for impact energies below 100 eV as already mentioned (see Figures 7.15 and 7.16). However, the rearrangement at the LiF surface after the production of a F^+ ion might also be responsible for this effect. The formation of the F^+ ion forces Li^+ to move away and the F^- ions to come closer to F^+. Because of this rearrangement not only is the binding energy of the F^+ ion changed but also so are the distances and therefore the binding energies of the surrounding Li^+ and F^- ions (Walkup and Avouris, 1986). The influence of the slightly lowered binding energy is seen only in the vicinity of the sputtering threshold. At higher impact energy where sputtering is mainly caused by the momentum transfer at the end of a sputter cascade, a slightly reduced binding energy of Li^+ and F^- ions at the impact point contributes less to the total yield.

7.5 CONCLUSIONS

We have performed measurements of secondary-ion emission near the threshold-energy region. By comparing results of singly and doubly charged ions, the role of the potential energy in the production process was examined.

The interaction of projectile ions with a CO-covered Ni(111) surface causes desorption of neutral CO molecules with a high efficiency. Momentum transfer from the primary ions by binary collisions with the adsorbates is mainly responsible for the high desorption yield at low impact energies. The influence of impact energy and angle can be explained by applying conventional sputtering theory.

For very low impact energies, the secondary-ion yield is influenced by the charge state of the primary ion. We propose a new mechanism for the production of secondary ions where momentum transfer and simultaneous ionization of the secondary particle takes place in a binary collision with the primary ion. Ionization of the secondary particle is much more effective for doubly charged than for singly charged projectiles. At low impact energy, binary collisions start to dominate sputtering and a charge dependence of the ion yield becomes observable.

On insulating surfaces, some essentially different processes can take place. We have shown that F^+ emission from a LiF surface is strongly influenced by the potential energy of the impinging projectiles for impact energies as high as 500 eV. Some momentum transfer is necessary to emit the F^+ ions which indicates that pure 'Coulomb explosion' is unlikely. We suggest that F^+ is produced by electronic transitions from the LiF surface induced by the projectile. The F^+ ion which is initially in a repulsive state is recaptured due to the relaxation of the surface lattice. This causes a rearrangement of the lattice in the vicinity of the impact point and a lowering of binding energies of surface atoms. This explains the increased sputter yield for F^+ and also the slight increase of the sputter yield for all other particles below 100 eV.

As a final conclusion, one has to say that a clear confirmation of the Coulomb-explosion model is still pending and more experiments need to be focused on the total sputter yield for projectiles in even higher charge states.

7.6 ACKNOWLEDGEMENT

This work was supported by Austrian Fonds zur Förderung der wissenschaftlichen Forschung (Projects Nr. S-43/03, and P8969).

7.7 REFERENCES

Appell, P. (1987) *Nucl. Instr. Meth.*, **B 23**, 242.

Behrisch, R. (ed.) (1981) *Sputtering by Particle Bombardment 1*, Topics Applied Physics, Vol. 47, Springer, Berlin.

Behrisch, R. (ed.) (1983) *Sputtering by Particle Bombardment 2*, Topics Applied Physics, Vol. 52, Springer, Berlin.

Benninghoven, A., F. G. Rüdenauer, and H. W. Werner (1987) Secondary Ion Mass Spectrometry, Wiley, New York.

Bitenskii, I. S., M. N. Murakhmetov, and E. S. Parilis (1979) *Sov. Phys. Tech. Phys.*, **24**, 618.

Bitenskii, I. S. and E. S. Parilis (1989) *Journal de Physique*, **50**, C2.227.

Campuzano, J. C., R. Dus, and R. G. Greenler (1981) *Surf. Sci.*, **102**, 172.

Cuomo, J. J., St. M. Rossnagel, and H. R. Kaufmann (eds) (1989) *Handbook of Ion Beam Processing Technology*. Noyes, NJ.

de Zwart, S. T., T. Fried, D. O. Boerma, R. Hoekstra, A. G. Drentje, and A. L. Boers (1986) *Surf. Sci.*, **177**, L939–L946.

Diebold, U. and P. Varga (1990a) In *Desorption Induced by Electronic Transitions, DIET IV*, Springer Series in Surface Science Vol. 19, G. Betz and P. Varga (eds), Springer, Berlin.

Diebold, U. and P. Varga (1990b) *Vacuum*, **41**, 210.

Diebold, U. and P. Varga (1991) *Surf. Sci. Lett.*, **241**, L6.

Diebold, U., W. Möller, and P. Varga (1991) *Surf. Sci.*, **248**, 147.

Dieckhoff, S., H. Müller, W. Maus-Friedrich, H. Brenten, and V. Kempter (1992) *Surf. Sci.*, **279**, 233.

Eccles, A. J., J. A. van den Berg, A. Brown, and C. Vickerman (1986) *Appl. Phys. Lett.*, **49**, 188.

Froitzheim, H. and U. Köhler (1987) *Surf. Sci.*, **188**, 70.

Grunze, M., H. Ruppender, and O. Elshazly (1988) *J. Vac. Sci. Technol.*, **A 6**, 1266.

Knotek, M. L. and P. J. Feibelmann (1978) *Phys. Rev. Lett.*, **40**, 964.

Matsunami, N., Y. Yamamura, Y. Itikawa, N. Itoh, Y. Kazumata, S. Miyagawa, K. Morita, R. Shimizu, and H. Tawara (1984) *Atomic Data and Nuclear Data Tables 31*.

Möller, W. (1989) In *Materials Modification by High-Fluence Ion Beams*, R. Kelly and M. F. Da Silva (eds), Kluwer, London.

Möller, W. and W. Eckstein (1984) *Nucl. Instrum. Meth.*, **B 2**, 814.

Morozov, S. N., D. D. Gruich, and T. U. Arifov (1979) *Iz. Akad. Nauk SSR, Ser. Fiz.*, **43**, 612.
Neidhart, T., M. Schmid, and P. Varga (1993a) In *Desorption Induced by Electronic Transitions, DIET V*, A. R. Burns, E. B. Stechel, and D. R. Jennison (eds), Springer Series in Surface Sciences, Vol. 31, p. 129.
Niedhart, T., M. Schmid, and P. Varga (1993b) In *Ionization of Solids by Heavy Particles*, R. A. Baragiola (ed), Plenum Press, New York, p. 447.
Netzer, F. P. and T. E. Madey (1982) *J. Chem. Phys.*, **76**, 710.
Piacentini, M. and J. Anderegg (1981) *Sol. State Comm.*, **38**, 191.
Radzhabov, Sh. S., R. R. Rakhimov, and D. Abdusalomov (1976) In *Izv. Akad. Nauk SSSR, Ser. Fiz.*, **40**, 2543.
Schultz, J. A., P. T. Murray, R. Kumar, Hsin-Kuei Hu, and J. W. Rabalais (1983) In *Desorption Induced by Electronic Transitions, DIET I*, Springer Series in Chemical Physics 24, N. H. Tolk, M. M. Traum, J. C. Tully, and T. E. Madey (eds), Springer, Berlin, p. 191.
Sroubek, Z. (1989) *Spectrochimica Acta*, **B 44**, 317.
Taglauer, E., W. Heiland, and U. Beitat (1979) *Surf. Sci.*, **89**, 710.
Taglauer, E., W. Heiland, and J. Oonsgard (1980) *Nucl. Instru. Meth.*, **168**, 571.
Thompson, P. M. and J. W. Taylor (1986) *Surf. Sci.*, **176**, 610.
Tolk, N. H., M. M. Traum, J. C. Tully, and T. E. Madey (eds) (1983) *Desorption Induced by Electronic Transitions, DIET I*, Springer Series in Chemical Physics 24, Springer, Berlin.
Varga, P. (1987) *Appl. Phys.*, **A 44**, 31.
Varga, P. and H. Winter (1991) In *Particle Induced Electron Emission II*, Springer Tracts in Modern Physics, Vol. 123, Springer, Berlin, p. 149.
Varga, P., U. Diebold, and D. Wutte (1991) *Nucl. Instr. Meth.*, **B 58**, 417.
Walkup, R. E. and Ph. Avouris (1986) *Phys. Rev. Lett.*, **56**, 524.
Williams, P. (1981) *Phys. Rev.*, **B 23**, 6187.
Winters, H. F. and P. Sigmund (1974) *J. Appl. Phys.*, **45**, 4760.
Wittmaack, K. (1977) In *Inelastic Ion Surface Collisions*, N. H. Tolk, J. C. Tully, W. Heiland, and C. W. White (eds), Academic Press, New York.
Wurz, P. and C. H. Becker (1989) *Surf. Sci. Lett.*, **224**, L559.
Wurz, P., J. Sarnthein, W. Husinsky, G. Betz, P. Nordlander, and Y. Wang (1991) *Phys. Rev.*, **B 43**, 6729.
Wutte, D., U. Diebold, M. Schmid, and P. Varga (1992) *Nucl. Instr. Meth.*, **B 65**, 167.

8 CHEMICAL REACTIONS DURING THIN-FILM SYNTHESIS BY LOW-ENERGY IONS: ION-BEAM OXIDATION (IBO) AND ION-BEAM NITRIDATION (IBN) OF SEMICONDUCTOR SURFACES

Nicole Herbots, Olof. C. Hellman*,
Peihua Ye, Xiaodong Wang
Department of Physics and Astronomy, Arizona State University, Tempe, Arizona 85287-1504, USA

and

Olivier Vancauwenberghe†
Department of Materials Science and Engineering, Massachusetts Institute of Technology, Cambridge, Massachusetts, USA

CONTENTS

To the memory of Jacqueline Xantippe and Danielle Herbots.

* Present Address: NTT Basic Research Laboratories, 3-9-11, Midoricho, Musashino-shi, Tokyo, 180, Japan
† Present Address: Ecole Supérieure d' Ingénieurs en Electrotechnique et Electronique (ESIEE) Département de Physique, Cité Descartes, B.P. 99, 2 Blvd. Blaise Pascal F- 93162 Noisy-Le-Grand, France

Low Energy Ion–Surface Interactions
Edited by J. W. Rabalais

8.1 INTRODUCTION, CONCEPTS AND SCIENTIFIC MOTIVATION

8.1.1 OVERVIEW

This chapter discusses low-energy (0.1–1 keV) ion-beam synthesis (IBS) of oxide and nitride thin films at the surface of semiconductors. Low-energy IBS includes a variety of techniques using low-energy ions such as ion-assisted deposition (IAD), ion-beam deposition (IBD) (Colligan *et al.*, 1976), ion-beam epitaxy (IBE) (Zalm and Beckers, 1982), as well as the two techniques of interest in this chapter: ion-beam oxidation (IBO) (Harper *et al.*, 1981; Herbots *et al.*, 1988; Todorov and Fossum, 1987) and ion-beam nitridation (IBN) (Taylor *et al.*, 1978; Vancauwenberghe *et al.*, 1991a). For the sake of simplicity, 'low-energy IBS' will refer in this chapter to low-energy IBO and IBN only. IBD and IBE are discussed by Marton in Chapter 9 of this book. This text emphasizes the commonality of mechanisms and effects between IBD, IBE, IBN and IBO. Characteristic properties that set them apart will be highlighted. The nature of low-energy IBS, the scientific motivation for investigating it, and its potential in materials synthesis and processing are described later in this section.

Section 8.2 outlines the technological motivation for applications of low-energy IBS to Si, Ge, GaAs, and Si_xGe_{1-x}. Conventional techniques for the formation of dielectric thin films on these semiconductors are briefly outlined for comparison with low-energy IBS. Various high-energy ion-beam processing techniques are also compared with low-energy IBS and the differences in thermal budget (Section 8.1.2) between high-energy and low-energy IBS is emphasized. The literature on high- and low-energy IBS for insulating films on semiconductors (i.e. oxides and nitrides of Si, Ge, Si_xGe_{1-x} and GaAs) is reviewed. Section 8.3 presents our current understanding of IBS and the framework specific to the model of IBO and IBN developed in this work. While the modeling and computation of purely collisional processes are fairly well established (Robinson, 1993; Biersack and Haggmark, 1980; Biersack and Eckstein, 1984) subsequent chemical reactions and diffusion mechanisms have not been incorporated into a model for low-energy IBS and compared with experiments. This chapter, therefore, outlines recent progress in both modeling and experimental observations.

Section 8.3 describes both analytical and quantitative computational models. The physical phenomena incorporated into these models are identified; the kinetic pathways and the identification of the rate-limiting step during growth are outlined. In Section 8.4, key variables identified in the kinetics of IBS are explored experimentally in a systematic fashion: ion energy, ion dose, angle of incidence and, in the case of nitrides, substrate temperature. The effects of these variables on thin-film stoichiometry, thickness, microstructure and properties are assessed. In order to isolate the role of ion energy, ultra-high-vacuum deposition from a monoenergetic, directional, collimated ion *beam* is used for these experiments. Thin-film growth using a combination of molecular-beam epitaxy (MBE) and IBS where the ion energy, direction of incidence, ion dose and substrate temperature are controlled is described. The technique combining MBE with low-energy IBS, either sequentially or concurrently, is referred to as combined ion and molecular-beam deposition (CIMD) (Herbots *et al.*, 1988, 1992; Herbots and Hellman, 1989). Characterization of samples grown by CIMD is carried out both *in situ* and *ex situ* by X-ray photoelectron spectroscopy (XPS), Rutherford backscattering spectrometry (RBS) combined with ion channeling, cross-sectional transmission electron microscopy (TEM) including scanning transmission electron microscopy (STEM) for laterally resolved compositional analysis and high-resolution electron microscopy (HRTEM) electrical characterization and stylus profilometry. This combination of synthetic and analytical techniques provides for the first time a comprehensive analysis of the composition, microstructure and physical properties of IBO and IBN thin films, which can be correlated with ion-beam parameters, especially ion-beam energy.

Section 8.5 compares results obtained from experimentation and simulations. On the basis of these comparisons, conclusions on the mechanisms of growth and chemical reactions by IBO and IBN are drawn, the regimes of applicability for the various models are discussed, and perspectives on applications are proposed. The latter aims at providing the reader with insights regarding the use of models and direction for future experiments and applications. Section 8.6 summarizes the results presented in the chapter.

8.1.2 LOW-ENERGY ION-BEAM SYNTHESIS AND INDUCED CHEMICAL REACTIONS

8.1.2.1 Ion-beam synthesis: the concept

Ion-beam synthesis (IBS) consists in the formation of a thin film where the growth mechanism depends on the kinetic energy of ions. Ions can be the principal source of the constituents from which the film is made. Alternatively, their only function can be to impart kinetic energy to other species that are incorporated into the growing film. In either case, the ion kinetic energy is critical in activating rate-limiting steps in the chemical reaction turning the source materials into the desired phase. The ion kinetic energy acts as a substitute for thermal activation. The elimination of thermal activation can modify the sequence of kinetic steps followed by the reaction, and hence the kinetic regime. Since the kinetic pathway of such reactions can differ from that of materials made by other means, the resulting phases can also have different and unique properties. Examples discussed here are oxides and nitrides of Group IV and III–V semiconductors, based on the results of several authors: P. D. Augustus (IBO of Si), S. Todorov (IBO of Si), J. A. Taylor and J. W. Rabalais (IBN of Si and Ge), J. R. Troxell (IBN of Si), N. Herbots, O. Vancauwenberghe and O. C. Hellman (IBO and IBN of Si, Ge, Si_xGe_{1-x} and GaAs), J. L. Alay and W. Vandervorst (IBO of GaAs, AlGaAs), G. Hubler and K. Grabowski (IBN of Si), L. DeLouise (IBN of GaAs) and W. M. Lau (IBN of GaAs).

Ion-beam oxidation (IBO) is IBS in which an element or compound is converted into an oxide. *Direct* ion-beam oxidation specifies that an ion beam provides the oxygen incorporated into the film through the ion-beam-induced reaction. Here IBO refers uniquely to the particular situation where the ion beam provides both the source material for the film and the kinetic activation. This distinction is important as the ion beam can also enhance absorption, decomposition and recoil implantation of residual gases during deposition. This also results in a low-temperature reaction, but with different kinetics and thus different properties. IBN designates direct ion-beam nitridation, where nitrogen is incorporated by direct impingement of nitrogen ions into a substrate.

8.1.2.2 Motivation for low-energy ion-beam synthesis: thermal budget and low-temperature kinetics

The IBS techniques defined here can be applied at high energies where they are known under specific names such as SIMOX (separation by implantation of oxygen) for oxides formed 'deep' (i.e. 0.1–0.2 μm below the surface) into Si substrates through the high energy employed. Mesotaxy (White *et al.*, 1987) is an application of high-energy IBS to the formation of heteroepitaxial silicides. All high-energy IBS techniques require, however, significant processing temperatures (1100 °C annealing for SIMOX, 400–600 °C for mesotaxy of silicides) to remove damage and provide adequate mass transport between reactants. This chapter emphasizes that for low-energy IBS, *high temperatures* ($T >$ 400 °C) are not needed to obtain the same reactivity, sharpen interfaces, or eliminate defects. For instance, fully reacted, uniform SiO_2 (Todorov and Fossum, 1987; Vancauwenberghe *et al.*, 1991a; Herbots *et al.*, 1992) and $Si_xGe_{1-x}O_2$ films (Vancauwenberghe *et al.*, 1991a) of well-controlled thickness (Todorov and Fossum, 1987; Vancauwenberghe *et al.*, 1991a, 1992a; Augustus *et al.*, 1988) can be obtained at room temperature. Both experimental data and theoretical predictions (Vancauwenberghe *et al.*, 1992a) demonstrate that room-temperature diffusivity plays a significant role (Vancauwenberghe *et al.*, 1991a; Herbots *et al.*, 1992), when combined with the appropriate ion energy (Vancauwenberghe *et al.*, 1992b). Obtaining such properties with low-energy IBS depends on the balance between diffusivity and recombination at room temperature and the low density of radiation damage produced by low-energy ion beams. Room-temperature diffusion and reaction are sufficient to remove point defects and incorporate them into the reacted layer rather than letting defects coalesce into extended damage. Examples of both elimination of irradiation damage and damage retention are shown in the experimental section (Hellman *et al.*, 1992b).

When low-energy IBS is compared to purely thermal treatment, the same point can be made: low-energy IBS eliminates the need for elevated temperatures. This is where the motivation for investigating low-energy IBS ultimately residues. Steps in the kinetic pathway during IBS are not all thermally activated. The resulting kinetics has different rate-limiting steps than in the purely thermal case. New kinetic paths are created. Fundamental understanding of IBS kinetics leads to new insights into a chemical reaction and into point-defect chemistry. It establishes new degrees of freedom in the design of a kinetic path, as demonstrated for nitrides in Section IV.1.b. Thermal budgets (also known as the *D.t* product or integrated temperature as a function of thermal processing time) are drastically reduced. This leads to better control of diffusion, segregation and other thermal relaxation effects which limit the manufacture of very small structures in microelectronics. This is why low-energy IBS has been mostly investigated in semiconductors. Also, since defect chemistry is relatively simple and well characterized in Si and Ge, these are good model systems for such investigations.

8.2 TECHNOLOGICAL BACKGROUND AND MOTIVATION FOR LOW-ENERGY IBS

8.2.1 OXIDATION OF SEMICONDUCTORS

The motivation for IBO and IBN of Si, Ge, Si_xGe_{1-x} alloys, and GaAs resides in the limitations of thermal oxidation and nitridation. Oxidation of a semiconductor occurs through oxygen transport to the oxide/semiconductor interface, followed by an interfacial reaction. Steps involved in transport and reaction are shown in Figure 8.1(a). After an initial stage limited by the interfacial reaction rate and linear with time, diffusion of the oxidizing species is the rate-limiting step, and the oxide growth rate is parabolic with time. The exact nature of the oxidizing species is still an object of debate. While diffusion of

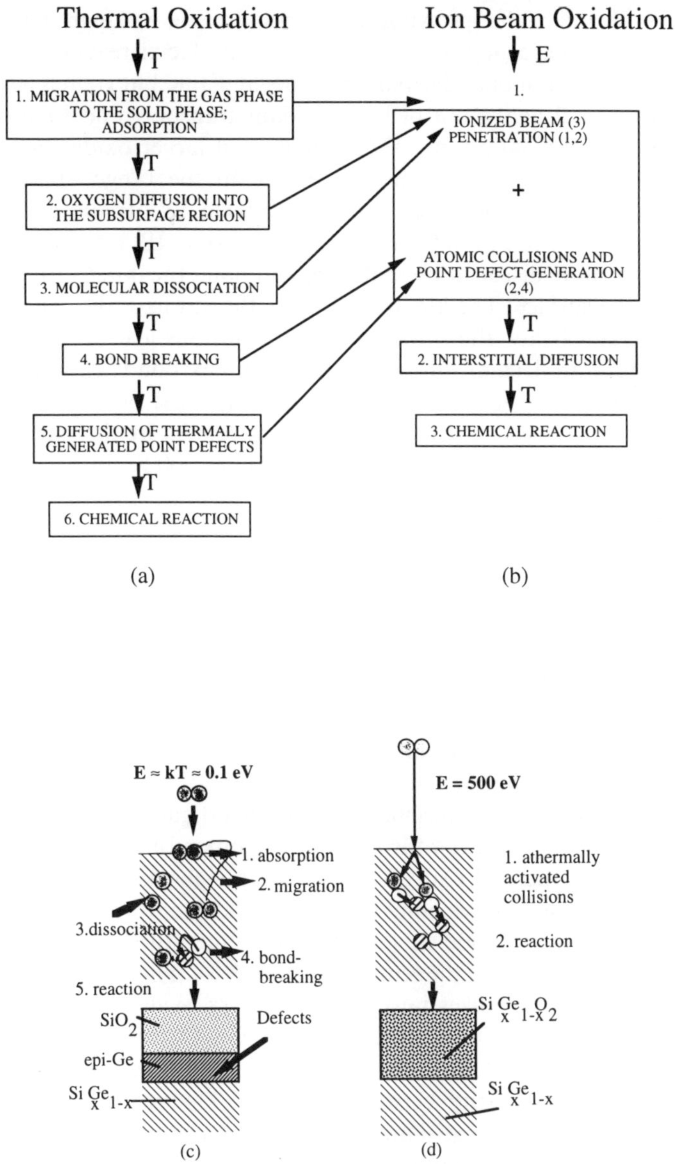

Figure 8.1 Comparison between the kinetic paths for thermal oxidation (a) and direct ion-beam oxidation (b) for semiconductors. Arrows indicate how thermally activated steps in thermal activation are substituted by other steps or activated by atomic collisions during ion-beam oxidation. Note that diffusion in the case of IBO occurs to oxidant species which have not yet been trapped by recombination. Illustration of the kinetic steps depicted in (a) and (b) in the case of Si_xGe_{1-x} for thermal oxidation (c) and direct IBO (d)

the oxidant through the growing oxide requires fairly high temperatures (>800 °C for silicon, >300 °C for GaAs), the interfacial reaction takes place at 300 K: the formation of native oxides occurs at 300 K but is thickness-limited (1–2 nm). If thermal activation of both oxidant transport and bond breaking are circumvented, oxidation at 300 K would yield larger oxide thicknesses. If IBO is used, the oxidized thickness depends on the range of the ion and diffusion, including the enhanced diffusion created by atomic collisions (see Figure 8.1(b) and Section 8.6). In semiconductors, IBO thus significantly bypasses absorption, dissociation, diffusion and bond breaking within a specific depth below the surface. The kinetic energy of the ion fixes the range at which oxygen penetrates below the surface. Atomic collisions lead to decomposition of O_2^+ into O^+, if molecular ions are used, and break bonds between substrate atoms. The generation in the semiconductor of intrinsic point defects and the deposition of oxygen lead to an oxide thickness defined by the rate of reaction and the diffusion of oxygen and of point defects at 300 K. In Section 8.3 and 8.5, a model including atomic collisions, defect generation by the ion beam, a high reaction rate and diffusion of the reactants is shown to account for the thicknesses and the interface widths observed between the substrate and the reacted film. When the reaction rate is comparable to the diffusion rate, as in IBN, the same model can account for broader interfaces and damage accumulation. Such a model can also explain the temperature dependence of these features. Understanding of IBS enables the selection of conditions where kinetics favors chemical reaction and eliminates radiation damage. New oxides that are by nature metastable can be formed without the phase separation or decomposition usually observed in thermal oxidation of compound semiconductors and alloys. Two specific cases are investigated where the volatility of an oxide can be overcome by the use of low-temperature synthesis: Si_xGe_{1-x} and GaAs. During thermal oxidation, one component, Ge or As respectively, oxidizes into a volatile binary compound rather than incorporates into a ternary oxide phase. The binary oxide sublimes at the temperature of oxidation and is metastable in the presence of the semiconductor phase. The less volatile element, Si or Ga respectively, oxidizes into a stable binary oxide. Thermal treatment leads thus to the stable binary oxide (SiO_2 or Ga_2O_3 respectively) while the volatile element is rejected from the stable oxide and piles up at the interface (Figure 8.1(c)). The similarity between oxidation kinetics of Si_xGe_{1-x} alloys and the GaAs compound is striking, and provides a motivation to investigate IBO of these semiconductors in parallel.

Thermal oxidation of $Si_{1-x}Ge_x$ alloys does not result in a $Si_{1-x}Ge_x$-oxide film (LeGoues *et al.*, 1989; Holland *et al.*, 1987), but decomposes $Si_{1-x}Ge_x$ into a SiO_2 film with segregated Ge (Figure 8.1(c)). Recently, steam oxidation in a narrow range of low temperatures (700 °C) has produced an oxide with reduced Ge pile-up at the interface (Liu *et al.*, 1991). However, the oxide/SiGe interfaces are defective, and pure Ge oxide crystallites are found in the oxide (Liu *et al.*,

1992). Phase separation cannot be climinated during thermal processing. Homogeneous, smooth $Si_{1-x}Ge_x$-oxides, stable up to at least 500 °C, with interfacial perfection comparable to that of thermal SiO_2/Si interfaces can be grown at room temperature by IBO, as shown in Section 8.5. The systematic study of $Si_{1-x}Ge_x$-oxides as a function of ion energy is an illustration of the capability of IBO to access kinetic regimes that are unreachable thermally. IBO of pure Si and pure Ge have also been conducted in parallel. IBO of $Si_{1-x}Ge_x$ at 300 K is motivated by microelectronic technologies using heteroepitaxial strained layers of $Si_{1-x}Ge_x$ (People, 1986; Wang and Woo, 1992; Hoffmann, 1991).

Similarly to $Si_{1-x}Ge_x$, the difficulty of growing a dielectric film on GaAs has hindered GaAs MOS technology. Thermal oxides of GaAs have been studied

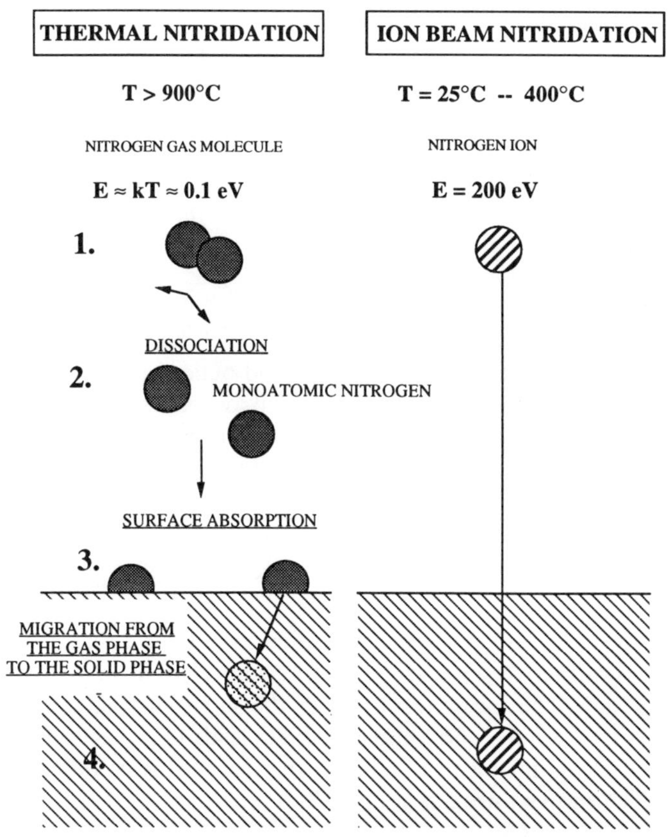

Figure 8.2 Initial steps in the kinetic path of direct IBN. Similarly to the case of IBO, steps are either substituted or activated by collisions. The last steps of the kinetic path are not illustrated because they are concurrent rather than sequential

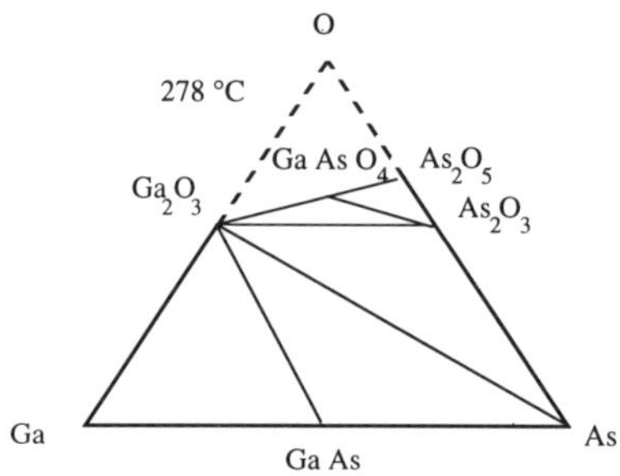

Figure 8.3 Ternary phase diagram of Ga–As–O below the melting point of As_2O_3 (278 °C). (After Le Goues *et al.*, 1989)

for more than 15 years (Croydon and Parker, 1981; Wilmsen, 1985). In the ternary phase diagram of Ga–As–O (Figure 8.3) (Thurmond *et al.*, 1980), several oxides of Ga and As exist in equilibrium. Ga_2O_3 and As_2O_3 are thermodynamically stable oxides, but As_2O_3 is more volatile than Ga_2O_3 and As. $GaAsO_4$ and As_2O_5 are oxides richer in oxygen: they are even more volatile than As_2O_3 and decompose into the lower oxides upon annealing. Also, no arsenic oxide is thermodynamically stable in the presence of GaAs. Thus, forming a good dielectric on GaAs requires the formation of a metastable phase. Oxidation techniques studied for GaAs involve thermal oxidation, plasma oxidation and anodic oxidation. Each uses a processing step above 300 °C, where As_2O_3 decomposes through the reaction:

$$2\,As_2O_3 \rightarrow 4\,As + 3\,O_2.$$

Arsenic is thus depleted in the oxide film and builds up at the interface. The oxide exhibits poor insulating and structural properties; the Fermi level is pinned at the oxide/semiconductor interface and inversion of electronic carriers is inhibited. This limits the development of high-density GaAs-based MOS technology.

8.2.2 NITRIDATION OF SEMICONDUCTORS

The motivation for IBN of semiconductors is similar to that for IBO. A comparison is attractive since in thermal nitridation, diffusion is not the only rate-limiting step. The lower reactivity of the nitrogen species coupled with limited diffusivity (Kijima and Shirasaki, 1976) leads to a reaction that is more complex and more difficult to obtain, by thermal means or otherwise. Silicon can be nitrided in dry nitrogen above 900 °C (Belyi *et al.*, 1988). The thermal reaction is

$$3Si + 2N_2 \rightarrow Si_3N_4.$$

with a heat of formation of $\Delta H = -745\ \mathrm{kJ\ mol^{-1}}$ and an entropy of formation of $\Delta S = -326\ \mathrm{J\ mol^{-1}\ K^{-1}}$. The kinetics of nitridation is very slow in comparison with oxidation, even above 1100 °C: the growth rate is less than $3 \times 10^{-12}\ \mathrm{nm\ s^{-1}}$, compared to about $0.3\ \mathrm{nm\ s^{-1}}$ for oxidation. Films of Si nitride are produced through other chemical pathways such as chemical vapor deposition (CVD):

$$x\,\mathrm{SiH_4} + (1-x)\mathrm{SiCl_4} + 4\mathrm{NH_3} \rightarrow \mathrm{Si_3N_4} + 12x\,\mathrm{H_2} + 12(1-x)\mathrm{HCl}\,|\quad 0 < x < 1$$

in which x can be controlled to change the rate of reaction or to create films of varying composition (Kijma and Shirasaki, 1976). CVD produces films of unlimited thickness, but with a stoichiometry and atomic density dependent on the deposition pressure, gas ratios and temperature. The kinetic steps include dissociation, adsorption, diffusion and reaction, and require thermal activation. 'Si nitride' rather than Si_3N_4 is used here because the stoichiometry is variable and the phase is not crystalline. Si_3N_4 refers to crystalline Si nitride. 'Stoichiometric' nitride refers to an amorphous phase of Si_3N_4.

Ge nitride has only been shown to exist in bulk form (Ge_3N_4). The heat of nitridation of Ge is relatively small, $-15.6\ \mathrm{kcal\ mol^{-1}}$ for Ge_3N_4, whereas the heat of oxidation of Ge is $-128.4\ \mathrm{kcal\ mol^{-1}}$ for GeO_2.

By contrast with thermal nitridation, IBN can be viewed as a process with few kinetic steps, some of which are athermally activated (Figure 8.2). The nitrogen ion, if it is a N_2^+ molecule, dissociates by collision; nitrogen is carried beneath the surface, bypassing diffusion through an interface boundary layer and into the substrate. Bonds in the substrate are broken, creating reaction sites. The low reactivity of nitrogen leads to a different kinetic regime than that of IBO. These differences are interesting to test experimentally.

High-temperature nitridation is not satisfactory for the same reasons pointed out for thermal oxidation. Purely thermal nitridation takes place at temperatures where Si_xGe_{1-x} alloys relax and phase-separate. In addition, low-temperature (20–420 °C) nitridation offers more degrees of freedom in choosing synthesis conditions for tailoring properties. IBS allows the formation of conventional nitrides in conditions never used before, with new properties. IBN Si nitrides are used as oxidation barriers (Troxell, 1985; Taniguchi *et al.*, 1989; Chiu *et al.*, 1982), gate insulators (Hezel and Lieske, 1982; Sodini *et al.*, 1984), optical coatings (Netterfield *et al.*, 1986), and SOI insulators (Petruzello *et al.*, 1985; Chang *et al.*, 1987; Skorupa *et al.*, 1987; Meekison *et al.*, 1986, 1991; Bourguet *et al.*, 1980) but have a reliability problem due to the stresses created by thermal expansion mismatch. IBS films have different stresses due to their thermal history since ion energy and deposition temperature can be chosen to control these stresses. The low diffusivity of many species in Si nitrides makes them attractive for passivation layers in devices and for packaging. The high density

IBS films further minimizes diffusion. The low temperatures used make IBS applicable to end-of-process steps such as packaging. The high stability of nitrides is attractive for optical coatings where the refractive index is adjusted through the stoichiometry of the film. Ion energy and dose can be chosen to grade stoichiometry. In summary, IBN of semiconductors is pursued for somewhat different motivations than IBO. Besides, the formation of new or metastable compounds that cannot be grown thermally, such as Si_xGe_{1-x} nitride, a stable compound such as Si_3N_4 formed in far-from-equilibrium conditions exhibits new strain properties that can be controlled.

The motivations for the investigation of low-energy IBS can be summed up as follows: (i) new kinetic paths (IBO, IBN) can be activated using athermal means, (ii) new or metastable compounds (such as Si_xGe_{1-x} dioxide) which cannot be grown thermally can be formed by athermal activation of rate-limiting steps, (iii) stable compounds (Si_3N_4) can form in far-from-equilibrium conditions, and exhibit new material properties, and finally (iv) insights into low-temperature kinetics derived from IBS can be used to explore new synthetic schemes.

8.3 MODELS FOR THE KINETICS OF IBS

8.3.1 A NEW KINETIC VARIABLE: ION ENERGY

IBS kinetics is most often investigated to establish the relationship between ion energy and the resulting phase. The role of ion energy can be investigated theoretically, based on the mechanism shown in Figure 8.1(b) and (d) and Figure 8.4, at room temperature, since metastable oxides constitute its most attractive application. Oxide growth rates are expected at saturation to become independent of oxygen dose, although radiation damage keeps accumulating. IBN kinetics is more complex, since reactivity and diffusivity compete. The effects of ion energy, ion dose and substrate temperature should thus all be considered.

Using ion energy as a processing variable is an attractive prospect. In ion implantation, the ion kinetic energy controls the depth and distribution of a buried film or dopant. It catalyzes chemical reactions as well. Several effects should be accounted for to gain an understanding of the relationship between the kinetics and ion energy (Figure 8.4) (Herbots *et al.*, 1988, 1991, 1992; Vancauwenberghe *et al.*, 1992a, 1992b): (i) The range should be large enough to avoid immediate desorption of oxygen from the surface; (ii) the sputtering yield of the semiconductor and the oxide formed should be low enough to obtain net oxide growth; (iii) point-defect generation should be compensated for by the reaction rate; (iv) diffusivity and reactivity of ions and intrinsic point defects should be such that all are consumed into fully reacted layer.

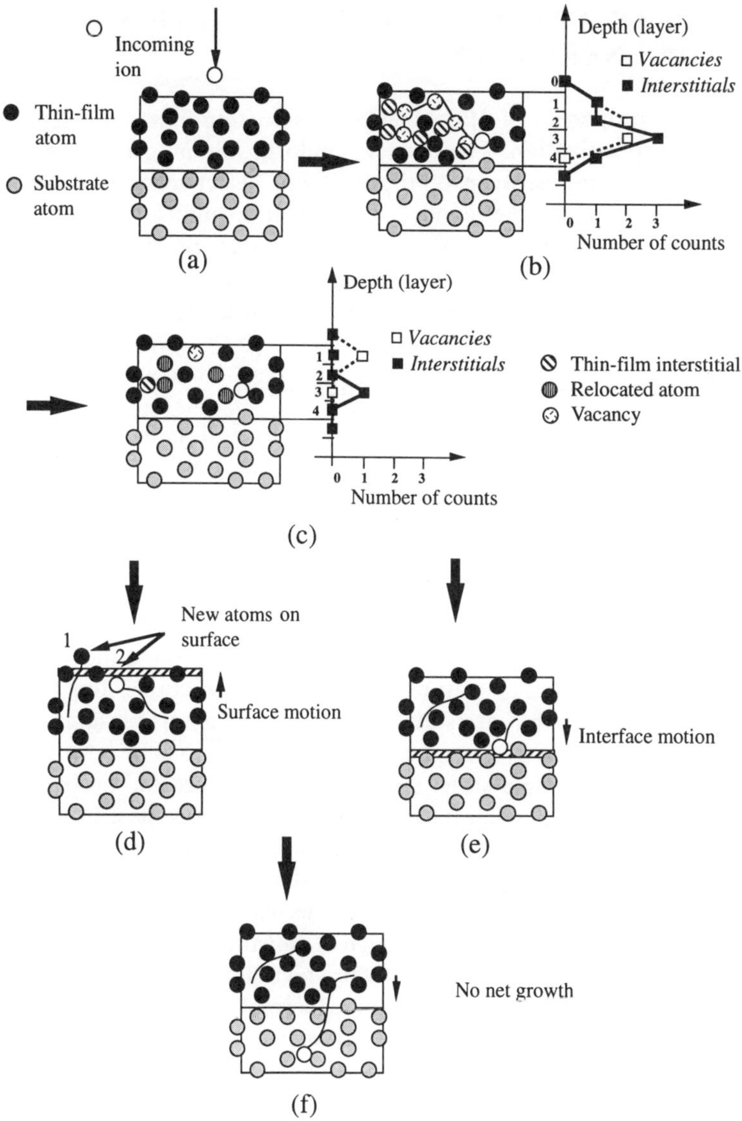

Figure 8.4 Kinetic steps during IBS in the present model. The first steps (a)–(c) are calculated by TRIMCSR, and the diagrams next to the structure illustrate the redistribution of different species. Last steps (d)–(f) during IBS illustrating two possible paths for the incoming ion. Recombination or reaction at the surface (d) can lead to surface growth as shown or lead to desorption in the final stages of IBO and IBN after thickness saturation. Recombination or reaction at the interface leads to thin-film growth as well (e). A third possibility is for a nonreactive defect to escape into the substrate, which is observed when diffusivity competes with reactivity (as in the case of IBN), or when reactivity is absent (f)

8.3.2 UNIQUE FEATURES OF THE KINETICS OF LOW-ENERGY IBS

Two essential properties of IBS kinetics lead to thin-film homogeneity and low-temperature activation. These are (i) uniform inhomogeneous nucleation and (ii) time separation between, on the one hand, defect generation and species introduction, and on the other hand, defect relaxation, diffusion and recombination. These two properties are not unique to low-energy IBS in the narrow sense used here, but are found in low-energy IBS techniques such as IBD, IBE and IAD (Chapter 9). The kinetic model developed is relevant to these also.

8.3.2.1 Uniform inhomogeneous nucleation

A problem in the growth kinetics of thin films as well as in microelectronics fabrication is small-scale inhomogeneities, or inhomogeneous nucleation. Thin films often grow as islands rather than layer-by-layer even when thermodynamic considerations favor layer-by-layer growth. Surface inhomogeneities, contamination, or crystal defects alters the growth mode. For example, surface geometry strongly affects CVD, since absorption and desorption depend on surface structure. In general, growth and deposition are governed by inhomogeneities.

Homogeneity in films deposited by IBS results from the uniformity of the ion beam and the stochastic nature of ion–solid interactions. Ion beams are generated by processes that are statistical in nature such as plasma ionization and extraction. As a result, the spatial distribution of ions in an ion beam is stochastic on a microscopic scale. Small-scale defects on the ion-beam target leading to heterogeneous nucleation in thermal processes have little effect on the ion beam or the collision cascades it generates and lead to quasi-homogeneous nucleation in IBS films. Each species is introduced through collisions acting as nucleation events, rather than thermal growth. IBS thus favors 'uniform heterogeneous nucleation', which is indistinguishable from homogeneous nucleation (Herbots *et al.*, 1988, 1992). This is less apparent in high-energy IBS because it includes thermal annealing: thermally activated heterogeneous nucleation dominates the kinetics, even though growth rate is enhanced by defect generation due to irradiation damage.

8.3.2.2 Time separation between atomic collisions and thermal effects: the foundation of a simple kinetic model

A second interesting property of IBS kinetics is the separation in time of defect generation, transport, recombination and relaxation due to atomic collisions on the one hand, and of thermally activated migration, relaxation, reaction and recombination on the other. This property enables a relatively simple mathematical treatment of an IBS kinetics comprising both ion-induced atomic collisions and thermally activated atomic transport and reaction. The IBS kinetic model presented here is based on the identification of the sequence and time scale of atomistic events leading to thin-film growth (Herbots *et al.*, 1988, 1991, 1992; Vancauwenberghe *et al.*, 1991b). The model is specific to direct deposition (including reaction) from a low-energy ion beam used with a well-defined energy ranging between the atomic displacement energy E_d up to a few keV,[1] and a normal angle of incidence. Normal incidence can be shown theoretically to provide maximum efficiency of ion incorporation (Herbots *et al.*, 1988, 1992). Recent low-energy IBO experiments by Alay and Vandervorst (1992a) have confirmed that normal incidence also maximizes the amount of chemical reaction taking place. The limit of 1 keV used here has been established as an approximate limit where sigificant thin-film growth and/or damage annihilation ceases, as shown in the experimental section. This threshold is material- and temperature-dependent. Obviously, the present model can be applied beyond that energy limit for fundamental studies of ion-irradiation effects. This model can also be used when other beams contribute to the deposition.

The time scale of specific atomistic events depends upon the atomic and electronic structure of the material, making the model specific to a class of materials. For instance, atomic packing determines free volume and, thus, the net rate of atomic displacements, while the rigidity of the lattice and the temperature of plastic deformation determine the rate of elastic relaxation. Both parameters determine point-defect generation, migration and thermal diffusivity. The time scale and model described here are specific to semiconductors below the temperature of plastic deformation, although recent experiments indicate that it could also be applicable to metals held at temperatures below which vacancies are immobile (Breeman and Boerma, 1993), as occurs in Cu *below* 300 K.

8.3.3 MODELS FOR THE KINETICS OF LOW-ENERGY IBS

The modeling of low-energy IBS can be conducted step-by-step. One motivation for doing so is that several models of increasing complexity can be constructed, starting by considering collisional effects alone, and then adding, for instance, thermal effects. The comparison between experimental data and results from these different models pinpoints which effects are dominating the kinetics. For instance, the ion-energy dependence of IBO of Ge can be quantitatively accounted for by atomic collisions alone. For other materials, thermal diffusion has to be considered to account for the results of experiments. Three kinetic steps are identified and consequently three types of kinetic models are presented: IS (for implantation and sputtering), ISR (for implantation, sputtering and relocation), and ISRD (for implantation, sputtering, relocation, reaction and diffusion). IRSD includes two additional steps with respect to ISR: reaction and diffusion. A separate model can be constructed to include diffusion alone (DRIVIC), but reactions should be accounted for simultaneously in low-energy IBS.

8.3.3.1 Step I: atomic collisions and ballistic generation of point defects

When low-energy ions interact with a substrate, the first step leading to ion incorporation is the creation of vacancy/interstitial pairs by direct collisions. The time scale is less than 10^{-13} s (Figure 8.5). Thus, such pair creation can be considered as 'instantaneous'. Monte Carlo methods combined with the binary collision approximation (BCA),[2] and a universal atomic potential can be used to calculate instantaneous defect profiles, provided the surface binding (Vancauwenberghe *et al.*, 1990) and displacement energies (Brice *et al.*, 1989) are chosen appropriately with respect to bulk values. Molecular dynamics (MD) is in principle more accurate for E close to or below E_d (Dodson, 1987, 1988, 1989;

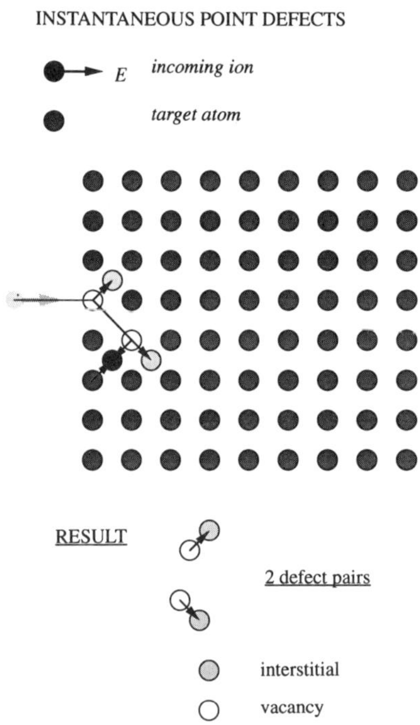

Figure 8.5 IS model: example of low-energy collision cascade *prior to any relaxation* for an ion into a Group IV substrate. Note that the reactive ion is deposited several monolayers below the surface

Garrison *et al.*, 1989). The lattice structure and packing play a key role in defect generation and migration. In the open diamond structure, interaction tends to be short-range. The free volume available facilitates the generation of point defects with fairly little distortion in the lattice position of first neighbors. Cubic diamond is thus a structure that lends itself fairly well to the BCA. On the contrary, in a close-packed metal, interactions are long-range. The smaller free volume leads to larger and more numerous atomic displacements. Calculations of ballistic defect generation for metals usually require MD, while for cubic diamond and zincblende structures one can rely fairly on a Monte Carlo BCA-based code such as TRIMCSR (Herbots *et al.*, 1988, 1992; Biersack and Haggmark, 1980; Biersack and Eckstein, 1984; Vancauwenberghe *et al.*, 1990; Vancauwenberghe, 1991).

The idea of an amorphous medium having point defects may seem contradictory because point defects are commonly defined as local regions of disorder in an ordered matrix. Although amorphous solids lack long-range order, short-range order is found in most amorphous media. For example, in SiO_2, both amorphous and crystalline forms exhibit fourfold coordinated Si with similar bond angles (Kingery *et al.*, 1984). In addition, radial distribution functions for amorphous SiO_2 closely matches that for crystalline SiO_2 for short distances. Thus point defects can exist in amorphous solids, and can be quantified in thermodynamic terms like any point defect in any medium. A calculation of point defects in solids where atomic potentials are short-range can then be fairly independent of whether a crystalline or amorphous target is being modeled.

These speculations have been experimentally verified. The creation of point defects in amorphous silicon by ion implantation has been shown by Roorda to scale with defect production as calculated by TRIM (Roorda, 1990 and references therein). Differential scanning calorimetry (DSC) was used to determine thermodynamic parameters for point defects in amorphous silicon. Roorda (1990) puts an upper limit on the entropy of such defects as $S_{def} <$ 0.15 k/defect and the configurational entropy S_{config} is estimated by Spaepen (1974) to be equal to 0.2 k/defect. Point-defecting annealing as observed in DSC was within experimental error of these values (1990).

8.3.3.1.1 The IS model (implantation/sputtering)

This model is presented in the context of IBO and IBN. Simulations are compared with IBS experiments in Section 8.5. The kinetics in the IS model consists only of sputtering and implantation, which are competing effects. Once an ion is incorporated, it bonds with substrate atoms to form a molecule. Upon reaction, the volume expands, leading to surface growth in the initial stage of IBS. The saturation of the growth rate, observed experimentally, results from the balance between the sputtering rate of the reacted layer at the surface and the reaction rate of the substrate accompanied by volume expansion at the back interface. This results in a stable thickness which depends solely on the ion–solid interaction in IS-type models. The thickness is determined by the relative magnitude of the range and straggling of the ion and of the sputtering yields of the different elements. In the IS model, ions incorporated after saturation of the growth rate are assumed to diffuse quickly to the surface and to desorb completely.

An IS model was developed by Todorov and Fossum (1987) for IBO of Si with similar assumptions for the chemical reaction, volume expansion and excess oxygen. An analytical formula was used to calculate the implantation profile and experimental values for the sputtering rate of SiO_2 and Si. In the IS model developed here for IBS of single- (Si, Ge) and multi-element ($Si_{1-x}Ge_x$, GaAs) targets, TRIMCSR is used to calculate the ion–solid interactions, i.e. the ion implantation profile and the 'sputtering profile' or depth distribution of ejected atoms. This profile usually involves only the first two to three monolayers. The total sputtering yield Y can be obtained by integration of the individual sputtering yield of each monolayer as a function of the depth z: $Y = \int_0^\infty y(z)\, dz$.

The IS model can thus predict stoichiometry modification at the surface due to variation of the individual sputtering yield with depth, $y(z)$. From the implantation and sputtering profiles calculated by TRIMCSR for elemental and compound targets, an interpolation with normalization to the ion dose is used to determine the implantation and sputtering profiles in the target when the composition varies during IBS. This was found to be acceptable when compared with the more advanced ISR model which includes a dynamic version of TRIMCSR.

8.3.3.2 Step II: defect relaxation by elastic recombination

8.3.3.2.1 The ISR model (implantation, sputtering and relocations)

After instantaneous defects are generated by atomic collisions, elastic recombination of point defects, amorphization, or extended defect-formation occur. Local electronic and atomic structures affect the relaxation of instantaneous defects. In semiconductors below the temperature of plastic deformation ($T < 540$ °C for Si), the rigidity and directionality of the bonds make point defects relax back into the lattice structure, if the instantaneous defect density in the cascade is low enough and the dose rate is below the limit for cascade overlap. Relaxation events are called 'replacements' if an ion recombines into a lattice position and 'relocations' if substrate atoms return into lattice sites (Herbots *et al.*, 1988). This type of relaxation favors epitaxial deposition and crystallization. If the point-defect density is large, then random relaxation and damage generation or amorphization occur. If the ion reacts with substrate atoms, another path can be followed when the defect density is below the damage-generation threshold. Chemical bonds are formed and distort the lattice according to the geometry of the new phase. Depending on whether this distortion can be accommodated or not, a transformation occurs into a crystalline or amorphous phase. An example of the latter case is the formation of amorphous SiO_2 or stoichiometric Si nitride when Si is subjected to IBO and IBN.

Besides recombination and reaction of ions, intrinsic point defects generated by the ion beam relax too. The elastic relaxation for a Si point defect has been computed by Landman *et al.* (1989) using MD, and has been shown to last about 3×10^{-13} s. MD calculations are thus necessary to establish the time scale of elastic relaxation and provide the assumptions necessary to construct the present model. Given the small spatial extension of the cascade, a very high dose rate would be necessary for overlap during elastic relaxation. For relaxation times lower than 10^{-12} s and a collision cascade area of 10 nm^2, the upper dose rate limit is 10^{24} ions-$cm^{-2}s^{-1}$, well above the present limits on ion current densities (Herbots *et al.*, 1988, 1992). Within these boundaries, relaxation leads to ordering if the initial matrix is crystalline and reduces the instantaneous defects population by about 75%. Depending on the location of relaxation, intrinsic point defects either relax into the crystalline matrix, or into the region disturbed by the introduction of a reactive species, depending on diffusion and reaction rates. Since the time scale of elastic relaxation is short in semiconductors, relaxation can be computed independently of thermal diffusion (10^{-9} s) (Herbots *et al.*, 1988, 1991, 1992; Vancauwenberghe *et al.*, 1991b) and different computational techniques used for each step. In materials which are more closely packed, more isotropically bonded and more 'plastic' at the temperature of interaction, a different relaxation process takes place. A large number of atoms remain displaced from their original sites, leading to extended defects which affect subsequent diffusion and annealing behavior. The microstructure of IBD Cu is an example of such an occurrence (Armour, 1988). MD calculations are then more appropriate. IBO and IBN of metals at room temperature and above thus require MD computations.

8.3.3.3 Step III: defect thermal diffusion, recombination and reaction

8.3.3.3.1 Thermal diffusivity during IBS: the DRIVIC model

The thermally activated last steps in the sequence are described for the case where net surface growth occurs, but can easily be adapted for an etching process. Two species are now present: the reactive species deposited by the ion beam and intrinsic point defects. For the sake of simplicity they are both referred to as 'defects'.

Thermal diffusion and reaction of point defects lead to annihilation of defects at recombination sinks. Annihilation via reaction or lattice recombination can contribute to thin-film growth, epitaxy, and damage generation or phase precipitation. Sinks are other point defects (Figure 8.6), or the surface (Figure 8.7). If the depth where extrinsic defects are created is comparable to the diffusion length, defects migrate to the surface. They participate in surface growth by recombination in the case of simple IBD or desorb in the case of IBS when free oxygen and nitrogen diffuse back to the surface. Thus in IBS, in contrast to IBD, precipitation with point defects below the surface favors thin-film growth. Since the ion range is energy-dependent and diffusion length is temperature-dependent, diffusion to the surface instead of incorporation in the substrate depends upon the energy–temperature combination. If a chemical reaction occurs, the growth rate is fixed by the competition between reaction and diffusion, either to the surface or towards the interface and the substrate. Simulations of this mechanism, shown in Figure 8.7, can be conducted using the model in DRIVIC (Herbots *et al.*, 1991; Vancauwenberghe, 1991) when no chemical reactivity is present, and ISRD when a chemical reaction occurs. The principle of the calculation is straightforward: a continuity equation (1) for the point-defect concentration is integrated by finite differences. The relaxed defect profiles computed by TRIMCSR are used as the generation term in (1) and (2). Thermal recombination and diffusion rates are derived from Gösele *et al.* (1983)

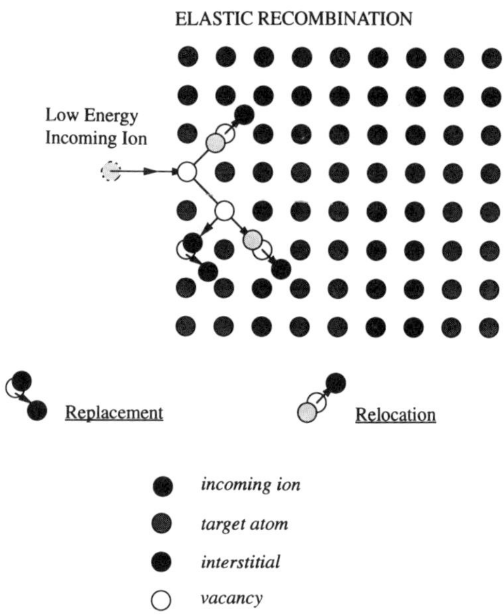

Figure 8.6 ISR model: low-energy collision cascade *after* elastic relaxation for one incoming ion into a Group IV structure

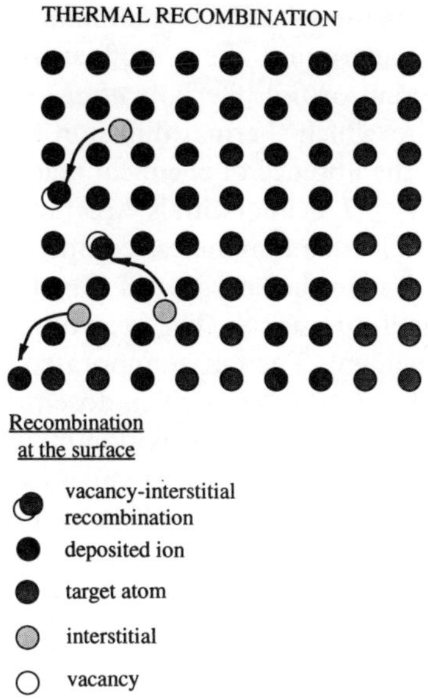

Figure 8.7 ISRD model: thermal diffusion, vacancy annihilation and surface recombination *after* collision and elastic recombination

for Si. The equations integrated in DRIVIC are (Herbots *et al.*, 1991; Vancauwenberghe, 1991): the vacancy diffusion equation

$$\frac{\partial C_v}{\partial t} = D_v \frac{\partial^2 C_v}{\partial z^2} + G_v(z)\phi_{ion} - k_{V-I}C_V C_I \tag{1}$$

and the interstitial diffusion equation

$$\frac{\partial C_I}{\partial t} = D_I \frac{\partial^2 C_I}{\partial z^2} + G_I(z)\phi_{ion} - k_{V-I}C_V C_I \tag{2}$$

where C_V and C_I are the instantaneous local point-defect concentrations rates, G_V and G_I the point-defect generation rates from TRIMCSR scaled with the dose rate, D_V and D_I the diffusion coefficients (Gösele *et al.*, 1983), and k_{V-I} the recombination coefficient. The growth rate is

$$\frac{1}{\Omega}\frac{dz_s}{dt} = G_m + J_I(z_s) - J_V(z_s). \tag{3}$$

The growth rate (3) depends on the sputtering rate $J_I(z_s) - J_V(z_s)$ the dose rate ϕ_{ion}, the molecular beam flux, G_m, and the diffusive fluxes, J_I and J_V, of point defects recombining at the film surface. $J_I(z_s) - J_V(z_s)$ obviously reduces the growth rate as energy increases, but is not the sole source of the energy dependence. The increase in range with energy leads to more subsurface recombination at the expense of surface recombination (Vancauwenberghe *et al.*, 1991b). This effect determines the cut-off energy for epitaxy whether a chemical reaction is present or not (Herbots *et al.*, 1992; Vancauwenberghe *et al.*, 1991b). The extent to which thermal diffusion controls the growth rate is shown in Figure 8.8, in the absence of chemical reaction (Herbots *et al.*, 1992). At high temperature ($T \geq T_1$), interstitials are mobile enough to reach the surface and recombine to form a film, unless competition with bulk recombination takes over, in which case the energy and thus the range is too high. T_1 is very close to room temperature for Si. The growth rate $V(E, T, \Phi)$ at $T > T_1$ is thus temperature-independent. The curve annotated T_1 gives the growth rate as a function of energy for $T \geq T_1$ for a given dose rate ϕ_{ion}. The upper energy limit where defect recombination in the bulk dominates due to increasing defect density is $E_{max}(T_1) = 200$ eV for Si. At lower temperatures (T_2), the cut-off energy $E_{max}(T_2)$ decreases due to the shortened diffusion length. The kinetics of IBS are thus controlled by a thermally activated kinetic step, the transport of relaxed defects towards the surface and the interface. This is a dramatic alteration from thermal-deposition kinetics. It is still temperature-limited, but in a lower temperature window, $T < T_1$, than the purely thermal case due to athermal-defects generation. This window can be controlled by a temperature independent parameter, energy.

8.3.3.3.2 Enhanced diffusivity during IBS

The diffusivity $D = D_0 \exp(-E_{diff}/kT)$ during IBS is larger than the equilibrium diffusivity. The activation energy E_{diff} is the sum of the defect-formation energy E_{form} and migration energy E_{mig}. Since defect creation by ions is

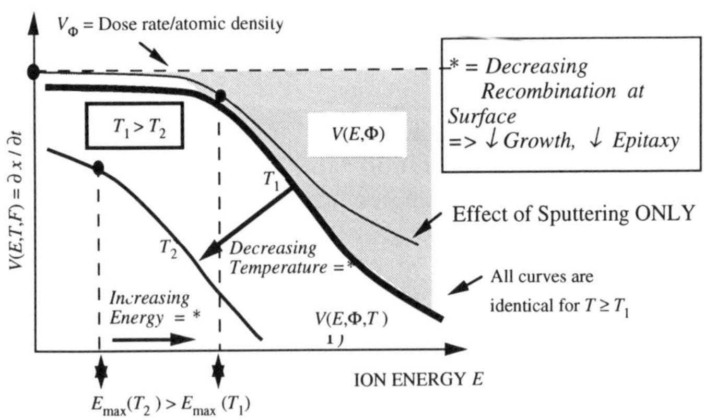

Figure 8.8 Energy–temperature dependence on the thin-film growth rate V_Φ. The regime where V_Φ becomes independent of T due to competition for recombination in the bulk is shaded

athermal, E_{form} is effectively zero, and the factor D_0 depends on the ion-beam parameters. An example of atomic motion enhanced by ions is solid-phase epitaxial growth, the crystallization of an amorphous solid epitaxially on a substrate. For Si on Si the solid-phase epitaxial activation energy is 2.7 eV over a wide range of temperatures (450–800 °C) (Priolo *et al.*, 1988). Under ion-beam irradiation, it drops to 0.25 eV (Goleci *et al.*, 1979). The enhanced diffusivity of Si under ion-beam irradiation has been theoretically investigated by both Vancauwenberghe *et al.* (1991b; Vancauwenberghe, 1991) and by Kitabatake *et al.* (1991). Vancauwenberghe *et al.* (1991b) showed that the enhancement of growth kinetics can be explained via the above equilibrium concentration of Si interstitials created; the activation energy for diffusion of interstitials is taken to be identical to that measured at high temperatures, 0.4 eV. This is evidence for ion-enhanced diffusivity being simply ion-enhanced interstitial creation. The evidence for athermally enhanced diffusivity comes from experimental observations of fast diffusion during high-energy implantation of both oxygen and nitrogen in Si. In the absence of reactivity or if reactivity is low, the only way for diffusion enhancement to cease is for interstitials to coalesce into dislocations in the substrate or to migrate to the surface. Herbots *et al.* (1986a, 1986b, 1986c) have reported the formation of interstitial dislocation loops in substrates during IBD and IBE. O. C. Hellman reported similar observations for IBN in (Hellman *et al.*, 1992b). Kitabatake *et al.* (1991) reached similar conclusions in a recent calculation: interstitial diffusion towards the surface, and annihilation of vacancies mediate growth as predicted in (Vancauwenberghe *et al.*, 1991a, 1991b; Herbots *et al.*, 1986c, 1988, 1991, 1992). If reactivity is taken into account, a third mechanism traps the defects and competes with both surface recombination and coalescence in the bulk. From Figure 8.8, the energy threshold favoring the formation of a chemically reacted layer is the cross-over between surface recombination and coalescence in the substrate. However, the reaction rate has to be much higher than the diffusion rate to trap point defects. In summary, a given ion–substrate combination leads to a specific length of ion penetration and a defect-generation rate. The substrate temperature determines the transport of the generated point defects at the surface or at defect clusters, and consequently, whether thin-film growth occurs or not. The relative magnitude of reactivity and diffusivity determines whether a chemical reaction traps most of the injected species into a reacted layer. Low-energy IBS growth is not controlled by step flow, but by transport back to the surface or into the bulk. Below the cross-over energy, each ion impact can lead to the recombination of an atom at the surface so that growth occurs by surface recombination without requiring lateral diffusion. This mechanism also allows for mobile point defects to escape into the substrate as the energy or the temperature increases. The ion energy where cross-over from surface recombination to substrate injection occurs is the optimum for low-energy IBS. IBS depends thus on the ratio between reaction rate and diffusion rate, which is temperature-dependent.

8.3.3.3.3 Reactivity: the ISRD model for the case of IBO

The continuity equations solved numerically to describe the IBO of a two-element target (elements s and g) (Vancauwenberghe *et al.*, 1992a; Vancauwenberghe, 1991) are:

unreacted s: ss
$$\frac{\partial C_{ss}}{\partial t} = -k_{so}C_{ss}C_{oo} + g_{ss}; \tag{4a}$$

unreacted g: gg
$$\frac{\partial C_{gg}}{\partial t} = -k_{go}C_{gg}C_{oo} + g_{gg}; \tag{4b}$$

oxide phase of s: so
$$\frac{\partial C_{so}}{\partial t} = +k_{so}C_{ss}C_{oo} + g_{so}; \tag{5}$$

oxide phase of g: go
$$\frac{\partial C_{go}}{\partial t} = +k_{go}C_{gg}C_{oo} + g_{go}; \tag{6}$$

free oxygen: oo
$$\frac{\partial C_{oo}}{\partial t} = \frac{\partial}{\partial z}\left(D_o(z)\,\frac{\partial C_{oo}}{\partial z}\right) + g_{oo} - 2k_{so}C_{ss}C_{oo} - 2k_{go}C_{gg}C_{oo}. \tag{7}$$

Substrate and oxygen atoms are either free, unbonded (ss, gg, oo) or bonded into an oxide phase (so, go). The equations (4)–(7) contain three terms: a reaction rate k for oxidation, a generation rate g for defects, and a diffusion rate D. It is assumed that all bonded oxygen is consumed, that molecules dissociate upon impact, and that oxidation is of first order in substrate atoms and free-oxygen concentrations.

Ion–solid interactions are calculated by the same dynamic version of TRIMCSR as in ISR. The generation term g provides the creation rate of vacancies and interstitials for each species. For example, an oxygen vacancy corresponds to an oxygen taken away from an oxide molecule. The oxide molecule is assumed to decompose while a substrate atom and an oxygen atom are freed. In the version of ISRD for IBO, only the diffusion of free oxygen is modeled. The diffusion of unbounded substrate atoms and oxide molecules is not expected to be significant. Ion–solid interactions are observed experimentally to enhance the diffusion of oxygen both for high- and low-energy IBS. Oxygen diffusivity in SiO_2 during high-energy (15–200 keV) implantation has been reported by many authors to be enhanced. In high-dose oxygen implantation into Si, the oxygen distribution in the film is close to that predicted by ion–solid collisions calculations up to the dose where a stoichiometric oxide forms at the concentration maximum. The oxygen distribution then flattens due to the inability of oxygen to react at that location in the film. Oxygen diffuses to the edges of the distribution at a rate much higher than expected from thermal oxygen diffusivity in SiO_2, due to excess oxygen interstitials. Littlewood and Kilner (1988) show that oxygen interstitials diffusing during implantation readily exchange with oxygen atoms on substitutional sites. Namavar *et al.* (1985) show that enhanced diffusion ceases when the temperature is low during implantation while oxygen bubbles form at the oxygen peak concentration. Oxygen bubbles also form at very-high-current implantation (Kreissig, 1982). IBS kinetics is thus still thermally activated. Free-oxygen diffusion can be assumed to occur through two mechanisms:

$$D_o = D_{\text{thermal}} + D_{\text{rad.enh.}} \tag{8}$$

$$= D_{o,sg} x_{sg} + D_{o,ox} x_{ox} + D_{\text{enh.}}(n_{\text{vac}} + n_{\text{istl}})/n_{\text{ion}}. \tag{9}$$

Thermal diffusivity (D_{thermal}) is constant for oxygen implanted in the substrate ($D_{o,sg}$) and the oxide layer ($D_{o,ox}$) and is determined by linear interpolation as the composition in the structure evolves under bombardment. The second component is 'radiation-enhanced' diffusion ($D_{\text{rad.enh.}}$). This is directly proportional to the number of defects (vacancies and interstitials) created by ions at a given depth, and is equal to 0 beyond the end of the range damage. Free oxygen diffuses towards the interface to react or towards the surface to desorb. The net surface flux is either positive (incorporation) or negative (desorption). The concentration of free oxygen at the surface can be 0: this models a 'fast' desorption limited only by the arrival of oxygen from the bulk by diffusion. ISRD also models the association of free oxygen atoms into O_2 molecules which then diffuse. The continuity equations are:

free oxygen: oo
$$\frac{\partial C_{oo}}{\partial t} = \frac{\partial}{\partial z}\left(D_o(z)\,\frac{\partial C_{oo}}{\partial z}\right) + g_{oo} - 2k_{o2}C_{oo} - 2k_{so}C_{ss}C_{oo} - 2k_{go}C_{gg}C_{oo}, \tag{10}$$

O_2 molecules: o2
$$\frac{\partial C_{o2}}{\partial t} = \frac{\partial}{\partial z}\left(D_{o2}(z)\,\frac{\partial C_{o2}}{\partial z}\right) + k_{o2}C_{oo}, \tag{11}$$

where k_{o2} is the rate constant for O_2 formation. Equations (10) and (11) form a system of nonlinear coupled differential equations. A finite-element method is used for discretization in space and an implicit forward Euler scheme for discretization in time. The numerical equations are then solved using a self-consistent iteration method.

8.3.3.3.4 The TRIMIBN model for the case of IBN

For high-energy nitrogen implantation into Si, effects of ion current density and temperature similar to those seen in SIMOX are observed (Petruzello *et al.*, 1985; Chang *et al.*, 1987; Skorupa *et al.*, 1987; Meckison *et al.*, 1986; Bourguet *et al.*, 1980). However, high doses result in the formation of N_2

bubbles even at moderate substrate temperatures (20–500 °C) and current densities (20 μA/cm^2) (Petruzello *et al.*, 1985; Chang *et al.*, 1987; Skorupa *et al.*, 1987; Meckison *et al.*, 1986; Bourguet *et al.*, 1980; Namavar *et al.*, 1986a). By comparing high-energy IBN with IBO one can conclude that either the diffusion rate of nitrogen is much smaller than that of oxygen, or the reaction of nitrogen to form N_2 is much faster than the reaction of oxygen to form O_2. Both of these conclusions are likely: Si nitride is a very good diffusion barrier, the N–N bond is stronger than the O–O bond, and the Si–N bond in Si_3N_4 is weaker than the Si–O bond in SiO_2. The rate of reaction of formation of an N_2 molecule should thus be higher than that of O_2.

The reaction can be modeled (Hellman, 1991; Hellman *et al.*, 1992a) as

$$[\mathrm{S_{unr}}] + x[\mathrm{N_{int}}] \Rightarrow \mathrm{SN}_x \quad (12)$$

where $[S_{unr}]$ is the total concentration of substrate atoms minus the fraction nitrided by the nitrogen atoms, and $[N_{int}]$ is the nitrogen interstitial concentration. This boundary conditions avoids the formation of a superstoichiometric compound, since it is never observed experimentally. The reaction is taken to be of order n with respect to S_{unr},

$$\mathrm{d}[\mathrm{N_{int}}]/\mathrm{d}t = k[\mathrm{N_{int}}][\mathrm{S_{unr}}]^n. \quad (13)$$

The simplest reaction rate between diffusing nitrogen interstitials and Si is a reaction rate in first order of the concentrations of both reactants. To prevent the formation of a superstoichiometric nitride, the concentration of unreacted silicon is be taken as

$$C_{\mathrm{unreacted\,Si}} = [\mathrm{Si}] - 1.33\,[\mathrm{N}]. \quad (14)$$

This model ignores the case of a reaction involving more than one Si since the redistribution of N would then be slow. One might expect a higher reaction order if more than one unreacted Si must participate; however, this is unlikely since reacted nitrogen redistributes rapidly, as observed in Si oxides by Littlewood and Kilner (1988).

The model for desorption from a surface given by Langmuir can be invoked to model the behavior of N leaving the surface: desorption is assumed to be of first order in the concentration of desorbing species on the surface. Thus the desorption flux J_d is expressed as

$$J_\mathrm{d} = -C_{\mathrm{surf}}\,k_\mathrm{d} \quad (15)$$

where k_d is the reaction-rate constant. For desorption from IBN films, this model assumes that the activation energy for desorption does not change as the concentration of the film evolves. Wang *et al.* (1986) have proposed such a model for ion beam mixing. The equations are simple enough that continuum models can be used in first approximation to calculate nitride growth (Figure 8.9).

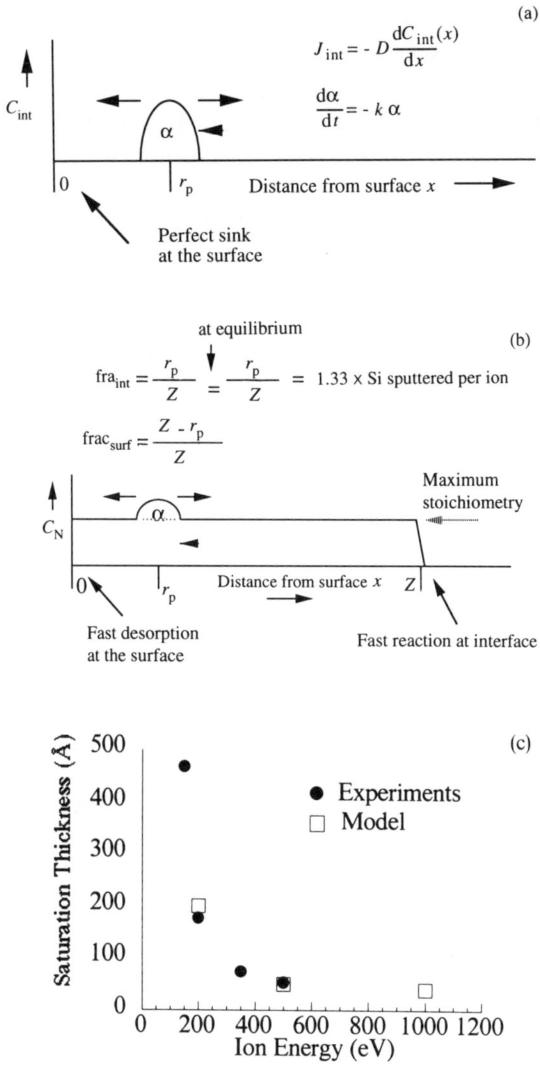

Figure 8.9 (a) IBN kinetics at low nitrogen dose. A small quantity of defects α is created at a distance r_p from the Si surface. After time t, the defects undergo two processes. They react and become trapped in the Si with a characteristic rate constant k, and they diffuse through the silicon with a characteristic diffusion coefficient D. (b) IBN kinetics at high nitrogen dose. A small quantity of defects α is created at a distance r_p from the surface of a completely reacted Si nitride film of thickness Z. The defects will either diffuse to the surface and desorb, or diffuse to the Si/Si nitride interface and react. Equilibrium is reached when the fraction of Si atoms reacting with the nitrogen diffusing to the interface is equal to the sputtering yield for Si. (c) Calculated saturation thickness vs energy obtained with the steady state (high dose) analytical model (□), compared to experimentally observed film thickness (●).

8.3.3.3.5 Diffusion and reaction at low doses during IBN

At low doses, a Gaussian distribution of N interstitials builds up, characterized by the distance from the surface r_p and a half-width at half-maximum dr_p, analogous to the projected range and range straggling of ions in solids. The surface can be assumed to be a perfect sink for interstitials, which diffuse with a diffusivity D, react with Si, and become trapped with a first-order rate constant k (Figure 8.9(a)). An analytical expression provides the concentration of unreacted interstitials:

$$C_{\text{int}}(x, t) = \left(\frac{\alpha_o e^{-kt}}{2\sqrt{\pi D(t + t_o)}}\right)\left[\exp\left(\frac{-(x - r_p)^2}{4D(t + t_o)}\right) - \exp\left(\frac{-(x + r_p)^2}{4D(t + t_o)}\right)\right] \tag{16}$$

where t_0 is the time of diffusion corresponding to the original distribution dr_p, and α_0 is the initial areal concentration of interstitials. Their relationship is given by

$$t_o = \frac{dr_p^2}{2D}. \tag{17}$$

From the first-order rate equation,

$$\frac{dC}{dt} = -kC, \tag{18}$$

the depth profile of reacted interstitials is obtained by substituting (4) in (6) and integrating:

$$C_N(x) = \int_0^\infty -k\left(\frac{\alpha_o e^{-kt}}{2\sqrt{\pi D(t + t_o)}}\right)\left[\exp\left(\frac{-(x - r_p)^2}{4D(t + t_o)}\right) - \exp\left(\frac{-(x + r_p)^2}{4D(t + t_o)}\right)\right] dt. \tag{19}$$

This integral can be evaluated numerically for a given set of D, k, r_p, dr_p and α_0.

8.3.3.3.6 Analytical model for the high-dose limiting case in IBN

This model is shown in Figure 8.9(b). A Gaussian distribution of nitrogen interstitials is assumed. Both the surface and the silicon nitride/silicon interface are considered to be perfect sinks for interstitials. When the layer is already fully reacted, the high-dose model is used to determine for a given ion range and film thickness how many interstitials diffuse to the surface and how many diffuse towards the substrate and increase the thickness of the film. When the nitridation reaction has reached completion in the near-surface where these interstitials are created, film growth is due exclusively to the diffusion of N interstitials towards unreacted Si beneath the film. The growth rate depends on the distance between the interstitials and the surface or interface.

The growth rate depends on the rate of sputtering and desorption caused by the ions: when these rates are equal, no growth occurs. Thus, this calculation used in conjunction with TRIMCSR provides the limiting thickness of the nitride layer at which steady state is reached, as a function of ion energy.

Another assumption is made when steady state is reached: since there is no change in the number of N vacancies near the surface, every N vacancy created by the collision cascade must subsequently trap one N interstitial to annihilate it. The number of N interstitials introduced in the surface at steady state is then simply the number of ions introduced by the ion beam. The number of interstitials which migrate to the interface is a trivial function of the ion range and the nitride thickness,

$$\text{frac} = \frac{t - r_p}{t}. \tag{20}$$

These N interstitials react with Si at the interface. Assuming a stoichiometric film growth, the number of Si atoms reacting per incoming N ion is:

$$\frac{[\text{Si}]}{\text{ion}} = \frac{3(t - r_p)}{4t}. \tag{21}$$

The number of Si atoms contributing to growth decreases by the number of Si ions sputtered. For a given ion energy, the sputtering yield does not change with the thickness of the nitride film. Therefore, a steady-state thickness can be defined by

$$\frac{[\text{Si}]}{\text{ion}} = \frac{3(t - r_p)}{4t}, \tag{22}$$

$$\text{Si}_{\text{sput.yield}} = \frac{3(t_s - r_p)}{4t_s} \tag{23}$$

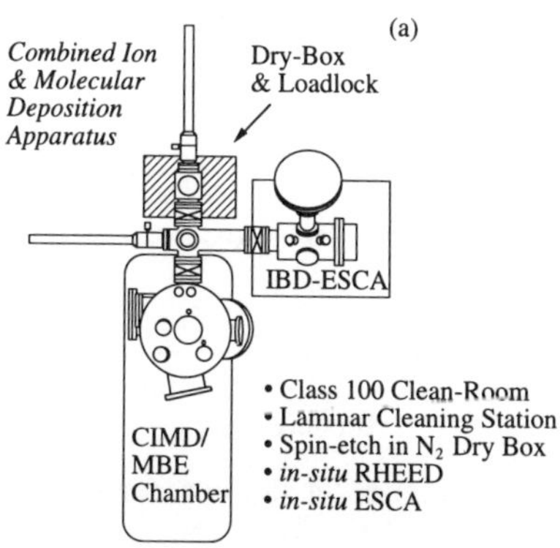

Figure 8.10 (a) CIMD apparatus.

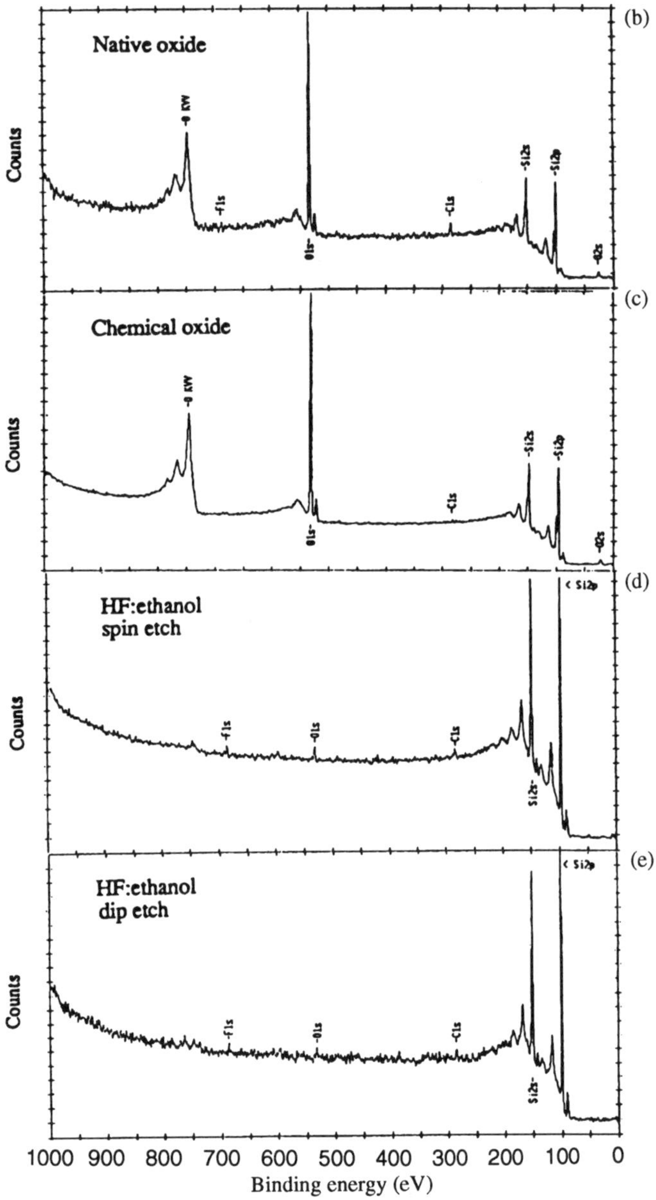

Figure 8.10 (b) to (e) XPS spectra of Si(100) surfaces at different stages of the cleaning procedures described in 8.4.1.1.

where $t = t_s$ is the steady-state thickness, which is plotted in Figure 8.10(b) as a function of ion energy.

From the simulations shown in Figure 8.9(c) it can be concluded that the balance of sputtering and nitrogen reaction produces *thicker films at lower ion energies after steady-state growth has been established.* While the ion range increases with energy, promoting thicker film growth, the increase in sputtering offsets this effect. The accuracy of these continuum models is discussed in Section 8.5 where experimental results are compared with this simulation.

The diffusivity of N at thermal equilibrium has been measured experimentally both in Si and Si_3N_4, and is very low. Even at 1200 °C, the diffusivity of N in Si_3N_4 is less than 10^{-20} cm^2 s^{-1} (Kijima and Shirasaki, 1976). In Si, N is slightly more mobile; at 750 °C, diffusivities reach 6×10^{-17} cm^2 s^{-1}. These diffusivities are greatly enhanced in the region of deposition during IBS.

8.4 EXPERIMENTAL RESULTS

Based on the models discussed above, the experimental goal of this work is to investigate film formation and properties as a function of ion energy, ion dose and substrate temperature during deposition for oxides and nitrides of Si and the formation of phases that cannot be grown thermally: oxides of GaAs, oxides and nitrides of Ge and Si_xGe_{1-x} heteroepitaxial on Si(100). The results are examined in the light of the models in Section 8.5.

8.4.1 EXPERIMENTAL PROCEDURES

Differential preliminary cleaning methods are used for Group IV and III–V surfaces. The first involves an *ex situ* chemical cleaning and the latter an *in situ* low-energy sputter cleaning without subsequent annealing. The technique for cleaning Group IV is described in brief because of its recent development.

8.4.1.1 Preliminary surface cleaning

8.4.1.1.1 The spin-etch technique

For epitaxy on Si, the initial surface has to be free from impurities. This is also critical for the study of the IBS growth kinetics. A 'clean' Si surface in air has a 2 nm surface oxide (10^{16}O/cm^2) contaminated with carbon, Si_xGe_{1-x} epitaxy requires surface contamination of less than 10^{13} at/cm^2 Fig. 8.10(b). In CIMD (Figure 8.10(a)), isotropic ^{18}O ions are used to discriminate ions introduced by IBO from atmospheric ^{16}O in oxides and nitrides. Chemical etching with hydrofluoric acid (HF) in ethanol removes *ex situ* native SiO_2 and passivates Si by hydrogen termination. H passivation prevents ^{16}O from recontaminating Si before insertion in vacuum. Fenner *et al.* (1989) show that the H-terminated surface is stable in air. *In situ* cleaning involves heating Si to 200 °C, which is lower than the temperatures used after sputter-cleaning ($T > 850$ °C) and for oxide sublimation ($T > 1000$ °C). *Ex situ* cleaning has two steps (Grunthaner *et al.*, 1990). One produces carbon-free SiO_2, and can be accomplished by various recipes (Figure 8.10(b)). The second step uses a substrate spinner in a glove box maintained under a positive pressure of N_2 (50 mm H_2O) while the chemically grown SiO_2 is etched by droplets of HF in ethanol.[3] The solution of HF in ethanol rather than in water and the positive pressure of N_2 inhibit the reoxidation of Si and provide almost complete passivation by H termination of the dangling bonds of etched Si. The substrate is then transferred into UHV under N_2. Residual contamination (O, C, F) as determined by XPS (Figure 8.10(d)) is low: $<3\%$ of a monolayer for O, $<5\%$ for C, $<1\%$ for F (a monolayer at the Si(100) surface is 6.8×10^{14} atoms/cm^2). O, C and F are physisorbed: no chemical shift is observed in the Si 2p signal at 99.6 eV. They are desorbed by annealing at 175 °C for 30 minutes. Temperature is then raised slowly (15 °C min^{-1}) up to the growth temperature (500–600 °C). Around 400–450 °C, H desorbs, the surface reconstructs as monitored by reflective high-energy electron diffraction (RHEED) in the CIMD chamber and a 100 nm-thick epitaxial film can be grown without annealing above 500 °C. Thin films of Si and Si_xGe_{1-x} can be cleaned by this procedure: no significant change in stoichiometry is observed by XPS. After MBE growth, a sharp 2×1 RHEED pattern indicates epitaxy in Si and $Si_{1-x}Ge_x$. If the surface is not clean, the 2×1 pattern is not observed: a 1×1 spotty pattern characteristic of SiC precipitates appears instead. The level of remaining contaminants at the original surface as determined by SIMS (Grunthaner *et al.*, 1989 and references therein) is comparable to procedures using high temperatures. STM observation (Grunthaner *et al.*, 1989) reveals that the smoothness of the surfaces improves when spin-etch cleaning is used. Slow heating of the wafer up to the growth temperature is particularly important. If the wafer is heated up too quickly, the contaminants do not desorb, but chemisorb, requiring then an anneal at 850 °C or above to desorb.

8.4.1.1.2 The dip-etch technique

Prior to performing IBS directly onto a substrate rather than following MBE, another low-temperature cleaning procedure can be used, shorter than the spin-etch technique but nearly as efficient since IBS sputters loosely bonded contaminants off the surface (as seen by *in situ* XPS). This shorter cleaning consists of a 2 min methanol rinse, a 15 s dip-etch in HF:ethanol (HF:H_2O: anhydrous alcohol, 1:1:10), a 1 min methanol rinse, drying with N_2, and loading in UHV. The surface cleanliness obtained is slightly lower than with the spin-etch method (Figure 8.10(e)) but satisfactory for IBS. The first rinse removes most of the carbonaceous contaminants. The HF:ethanol dip-etch eliminates the reoxidation of Si by water present in aqueous $HF:H_2O$ etch and provides H termination of Si. When water is used instead of alcohol, a Langmuir–Blodgett (LB) film of contaminants forms on the surface when the wafer is retrieved because water does not wet Si. In the dip-etch cleaning, however, the alcohol wets Si and eliminates the formation of an LB film. When dried with N_2, methanol leaves little residue on the surface, the contaminants are physisorbed, and desorb during the low-temperature anneal (175 °C, 30 min). The Si surface exhibits the characteristic 2 × 1 pattern at 450 °C. The key point for a low-temperature cleaning is thus to provide an H-passivated surface with little physisorbed contaminants. A dip-etch clean of 30 s is also used for Ge. However, residual oxygen amounts to 25% of a monolayer, much higher than for Si. $Si_{1-x}Ge_x$ layers are grown by MBE and then directly transferred to the IBO/XPS chamber without breaking vacuum. The pressure is at all times below 10^{-9} torr, avoiding contamination of the surface. The same conditions prevailed for GaAs (Section 8.4.2.4).

8.4.1.2 Experimental apparatus: CIMD with *in situ* XPS

8.4.1.2.1 Design considerations

Two types of configurations can be used to obtain monomodal energy distributions of ions relative to a target surface. The first consists of attaching a low-energy (10 eV to 3 keV) ion source to a deposition chamber and running it at the desired energy. The other is more complex and involves the use of either an ion source or an ion-implantation accelerator running at its extraction voltage and a deceleration lens to slow ions from their extraction voltage down to the desired energy. Several considerations make the second option preferable. In this work, a deceleration scheme using a simple 5 keV ion source is used. Indeed, despite its simplicity, the use of an ion source at the desired low energy is problematic because space-charge effects limit the theoretical brightness of an ion source. The effect of space charge increases with decreasing extraction energy and increasing time of flight from ion source to target. The combination of low energies and high current density can result in extreme defocusing and current losses before the ions reach the target. In addition, low energies limit magnetic energy analysis and especially mass analysis, making difficult the deposition of high-purity or monoisotopic films. Higher energy ion beams combined with a deceleration lens provide much higher deposition rates (>100 nm per hour) (Herbots *et al.*, 1986a), which is quite competitive for oxidation. The ions are decelerated only after mass analysis and transport to the target, in the immediate vicinity of the substrate by biasing the sample holder: the sample holder is a flat electrode in the same plane as the sample held at the extraction voltage minus the desired deposition voltage. The energy distribution in the decelerated ion beam can be considered monomodal, with an asymmetry in the distribution due to the fact that ions with an initial energy lower than $E_{extraction} - E_{deposition}$, where $E_{deposition}$ is the energy selected for deposition, cannot reach the target. Such ions are reflected at the end of the deceleration when their energy approaches zero with respect to the substrate holder and are reaccelerated when they turn around. This reflected beam is strongly defocused. A highly undesirable consequence of this is the sputtering of material from the chamber walls onto the film. The problem of film contamination can be drastically reduced by coating the chamber walls with material identical to the material being deposited on the substrate. This occurs naturally in a CIMD deposition chamber. Hence CIMD has the attractive feature of eliminating cross-contamination problems. In conclusion, two essential features of an IBS or CIMD system are (i) the use of a deceleration scheme and (ii) the use of all surfaces in direct line of sight of the substrate material compatible with the deposit, which is the case for the data shown here.

8.4.1.2.2 Apparatus description: CIMD with in situ XPS

A characteristic of the methodology used here is the *in situ* combination of three techniques: MBE, IBD and XPS, the latter having the alternative name 'electron spectroscopy for chemical analysis' (ESCA) (Figure 8.10(a)). Surfaces to be subjected to IBS are prepared by MBE in uninterrupted UHV. After IBS is completed, the chemical nature of the compound formed and the degree of reaction are measured *in situ* also. The compound formed can be identified without interference from native oxide formation or atmospheric contamination. Deposition is interrupted at regular time intervals to monitor by *in situ* XPS the reaction kinetics, the presence of intermediate phases, and the growth-rate saturation in IBS kinetics (Section 8.3.3).

Prior to IBS of Ge or Si_xGe_{1-x} alloy, Ge and Si_xGe_{1-x} films are grown on Si(100) by MBE. After spin-etch cleaning, a 100 nm Ge film is deposited at 500 °C on Si(100) for study of IBS of pure Ge. For IBS of Si_xGe_{1-x}, 130 nm-thick $Si_{0.89}Ge_{0.11}$ layers are deposited by MBE at 500 °C on top of a Si buffer layer (130 nm) grown at 600 °C. The samples directly transfer to the XPS chamber through a UHV transfer system (Figure 8.10(a)).

8.4.1.3 Ion source

IBS is performed in a UHV chamber pumped by an ion pump assisted by titanium sublimation. The ion source is a hot-filament electron-impact ion gun (PE 04-300). This ion source produces mainly molecular ions when N_2 or O_2 is the source gas (Cuomo and Rossnagel, 1987) and operates between 1 and 4 keV. For energies less than 1 keV, a special XPS sample holder with a floating ground is used to perform IBS at room temperature in a base pressure less than 10^{-9} torr. For IBO, $^{18}O_2^+$ instead of $^{16}O_2^+$ is used: the sensitivity of RBS being higher for heavy isotopes; incorporation of ^{16}O after exposure to atmosphere can be detected thanks to the good mass separation between ^{18}O and ^{16}O. The ion current density ranges from 0.5 to 2 μA cm^{-2} depending on the ion energy. The beam incidence is normal; the beam is rastered over an area of 2 cm^2 (Figure 8.10). The ion energy is varied between 100 eV to 1 keV to study its effects on phase formation and film properties. At fixed intervals, the sample is grounded to characterize the growing film by *in situ* XPS, until the oxygen 1s or the nitrogen 1s photoelectron signal reaches a saturation level. The same procedure is used by Alay and Vandervorst (1992a) in their work on IBO. The ion dose is estimated by combining characterization of the ion beam, the ion current during IBS, and the IBS spot area. A significant error with regard to the ion doses is expected due to the difficulty of suppressing secondary electrons. The saturation dose varies between 1×10^{16} and 3×10^{16} O/cm^2 as a function of the ion energy and the substrate. After IBS, *in situ* angular-resolved XPS is used to profile the oxide stoichiometry as a function of depth. The IBS area is mapped by SEM and small-area XPS.

8.4.1.4 Depth profiling by X-ray photoelectron spectroscopy

One interesting features of XPS is depth sensitivity. A 1.25 keV Mg Kα X-ray beam incident on Ge generates 2p photoelectrons of energy $E = 1.25$ keV − 1.225 keV and 3d photoelectrons of energy $E = 1.25$ keV − 0.03 keV. The low-energy 2p photoelectrons have a correspondingly shorter escape depth and reflect near-surface concentration. Escape depths have been measured for Si (Gossmann, 1984). The XPS signal as a function of depth being exponential, the contribution to the XPS signal from atoms deeper than any given depth can thus be calculated. To use XPS as a depth-profiling technique, XPS intensity is measured as a function of the angle away from the sample surface normal at which photoelectrons are collected. Angles further away from normal decrease the projected escape depth, making the measurement more sensitive to species near the surface. This analysis, called angular-resolved ESCA (AR–ESCA) is highly dependent on a controlled geometry during spectroscopy and requires a flat, smooth and laterally homogeneous surface. IBS films on polished Si(100) are thus ideal for AR-ESCA (Chiu *et al.*, 1982). The main interest in using *in situ* XPS during IBS is the monitoring of the chemical shift in the binding energy which depends on the chemical state of the element in the solid. The magnitude of this shift can be as great as 5 eV, and thus easily observable with a conventional electron analyzer. The XPS apparatus was a Perkin-Elmer ESCA 5400, with a spherical capacitor analyzer 10-360, equipped with a small-area Omni Focus lens and a single channel detector. The analyzer is operated in the small-area and fixed-analyzer transmission (FAT) mode, resulting in an energy resolution given by $\Delta E/E = 0.015$, where E is the pass energy for the acquisition. The width of the measured XPS signal arises from the energy spread of the X-ray source, of the photoelectron emission, and of the resolution of the analyzer. Because the typical FWHM of the measured peaks is 2 eV, peaks separated by less than 1.4 eV are deconvoluted. Because of the relative nature of the chemical shift measurements, they are self-calibrating and reproducible to within ± 0.05 eV which is critical for the observations reported here. Absolute measurements of the binding energy are reproducible only to ± 2 eV which is generally attributed to the instability of the electronics and of the electrostatic analyzer.

8.4.1.5 Other characterization methods

Ex situ characterization includes scanning electron microscopy to detect the presence of an insulating film and the dimensions of the IBS area; it is performed in a CAMSCAN Series 2 SEM operated in secondary electron imaging mode, with a 20 keV electron beam. A Gaertner L117 ellipsometer operated at an angle of 70° with a He–Ne laser light (632.8 nm) measures thicknesses. Conventional cross-section TEM is also used to measure film thickness and uniformity, interface widths, microstructure and the presence of defects in the underlying substrate. The cross-sections are thinned mechanically, polished, dimpled, and ion-beam-milled by Ar at 1 keV (Hellman, 1991).[4] A TEM JEOL 200-CX operated at 200 keV is used for conventional cross-section TEM with a spatial resolution of 0.4 nm. A high-resolution 200 keV EMOO2B Topcon(Akashi)TEM with a resolution of 0.18 nm provides for lattice imaging.[5] 2 MeV $^4He^+$ Rutherford back-scattering (RBS) combined with ion channeling is used to quantify the incorporation of $^{18}O_2^+$ (in the case of IBO), N_2^+ (in the case of IBN), the number of displaced atoms, damage and the possible presence of ^{16}O. The FWHM is 25 keV, corresponding to a depth resolution of 30 nm in Si at a fixed angle of 10° between the incident $^4He^+$ and the detector.

8.4.2 EXPERIMENTAL DATA

The growth by IBS of thin oxide and nitride films of Si and Ge and their alloy $Si_{1-x}Ge_x$ and of oxides of GaAs is described. The dependence of phase formation and film properties upon ion energy is investigated.

8.4.2.1 Ion-beam synthesis of oxides and nitrides of Si

8.4.2.1.1 IBO of elemental Si

Stoichiometry of the film In Figure 8.11, time-resolved XPS spectra show the progressive oxidation of Si under 100 eV $^{18}O_2^+$ bombardment as a function

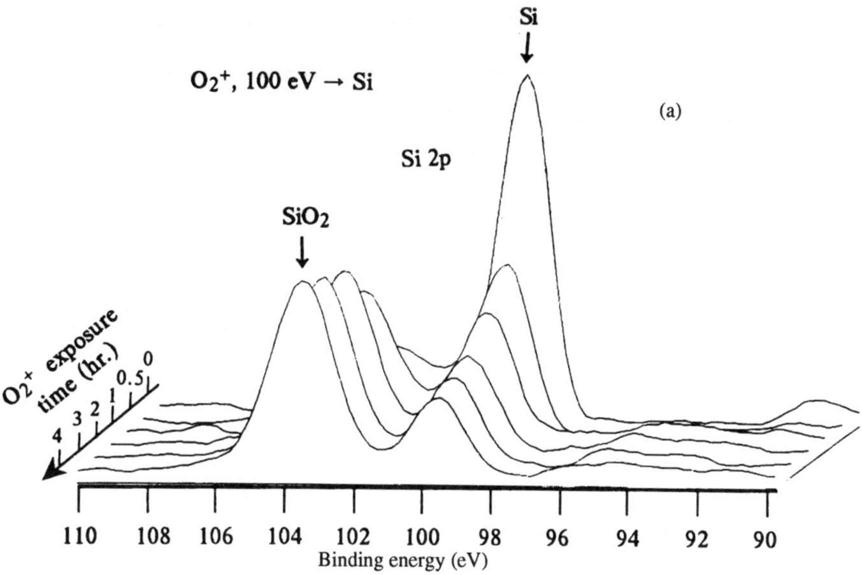

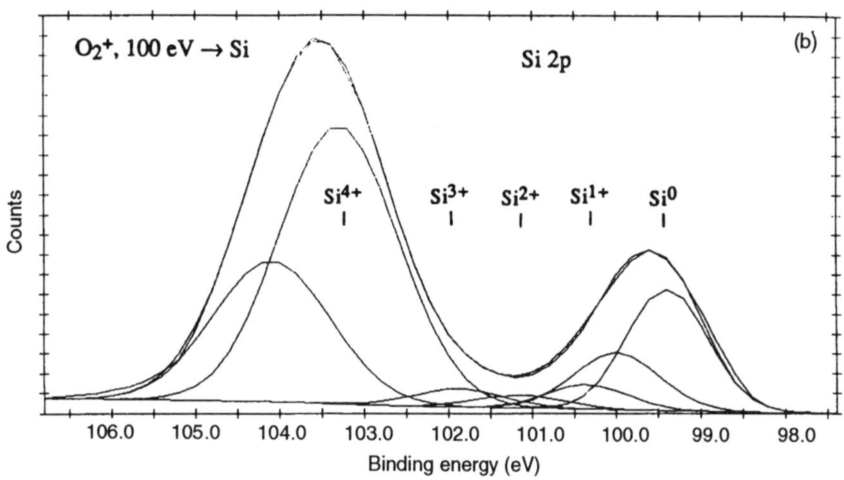

Figure 8.11 (a) Time-resolved XPS spectra of the Si 2p during IBO of Si(100) with 100 eV $^{18}O_2^+$ as a function of exposure time to the $^{18}O_2^+$ beam. (b) Curve fitting of the Si 2p signal of SiO_2 formed by IBO of Si(100) with 100 eV $^{18}O_2^+$ depicting the contribution of the suboxides tabulated in Table 8.1

of time. As the ion dose increases, the Si 2p photoelectron signal of unreacted Si (Si°, 99.6 eV) originating from the substrate decreases in intensity while a signal at higher binding energy (103.6 eV) appears and increases due to the formation of stoichiometric SiO_2 (Si^{4+}). In the initial stages of IBO, Si suboxides are formed, as seen in the XPS spectrum. But as the $^{18}O_2^+$ dose increases, the

Table 8.1 Estimation of the suboxide contribution in the Si 2p photoelectron signal after IBO of Si with 100 eV O_2^+. The XPS chemical shift and FWHM for the suboxide peaks are based on the work of Himpsel *et al.* (1988)

Peak	BE (eV)	ΔBE (eV)	FWHM (eV)	Area (%)	
Si^0 $2p_{3/2}$	99.4		1.11	14.20	21.30
Si^0 $2p_{1/2}$	100.0	0.6	1.19	7.10	
Si^{1+}	100.35	0.95	1.13	2.86	6.76
Si^{2+}	101.15	1.75	1.22	1.56	
Si^{3+}	101.88	2.48	1.24	2.34	
Si^{4+} $2p_{3/2}$	103.30	3.90	1.65	47.97	71.94
$Si^{4}+$ $2p_{1/2}$	104.10	4.70	1.62	23.97	

suboxides disappear and only stoichiometric SiO_2 grows on the Si substrate. The amount of suboxides in the IBO oxide after saturation is estimated by curve fitting of the Si 2p signal, based on the data from Himpsel *et al.* (1988), as shown in Figure 8.11 and Table 8.1. Table 8.1 lists the characteristics of each contributing peak: the suboxides amounting to less than 7% of the total intensity of the Si 2p peaks. This is a characteristic of the transition region between SiO_2 and Si observed in both thermal and native oxides of Si, due to the bridging between SiO_2 and Si(100) through a SiO layer (trydimite) along the interface.

Figure 8.11 illustrates another characteristic of IBO: the growth rate of SiO_2 saturates (Vancauwenberghe *et al.*, 1991a, 1992a, 1992b; Herbots and Hellman,

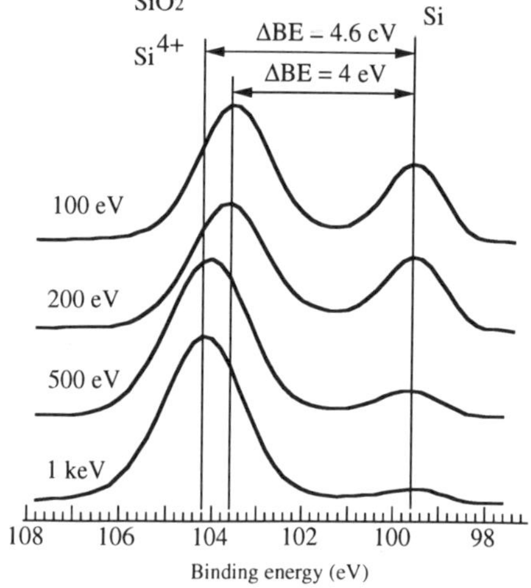

Figure 8.12 XPS spectra of the Si 2p signal as a function of $^{18}O_2^+$ energy for IBO of Si(100) after signal saturation

1989; Herbots *et al.*, 1992), as expected (Section 8.2.1) and observed in previous IBO experiments (Todorov and Fossum, 1987). The quantitative model proposed in Section 8.3.3.3 is compared with the Figure 8.11 data in Section 8.5.1.1.

Atomic structure of the interface Figure 8.12 shows the Si 2p signal at saturation as a function of ion energy (E_{ion}). With E_{ion} increasing from 100 eV to 1 keV, the intensity of the SiO_2 signal at higher binding energy (Si^{4+}) increases from 70% to 90%, while the signal of unreacted Si (Si°, 99.6 eV) originating from the substrate decreases correspondingly. This is due to an

Figure 8.13 (a) HRTEM micrograph of thermal oxide of Si(100). (b) HRTEM micrograph of an IBO oxide of Si(100) and ion beam oxidation.

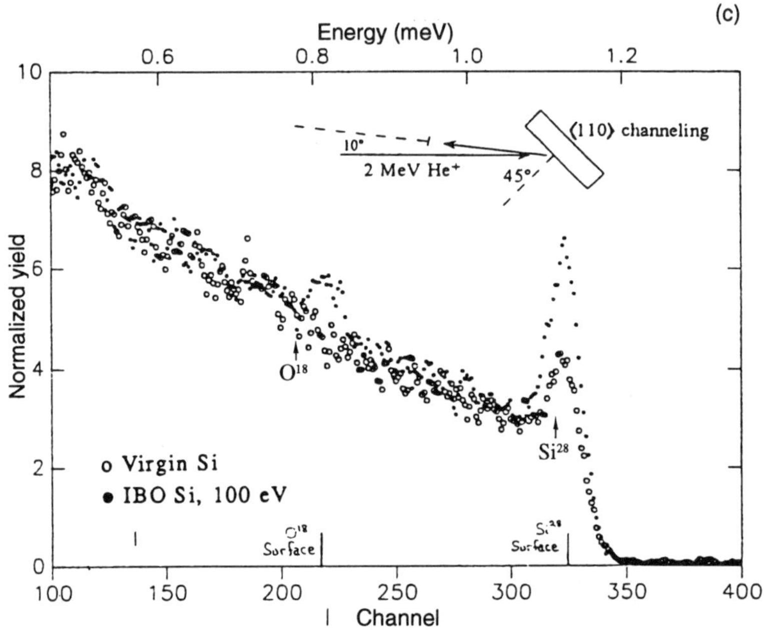

Figure 8.13 (c) 2 MeV $^4He^+$ RBS combined with ion channeling in the Si⟨110⟩ direction of SiO_2 formed by IBO of Si(100) with 100 eV $^{18}O_2^+$

increase of the SiO_2 film thickness with increasing E_{ion}, as confirmed by cross-section TEM. TEM also reveals that IBO films were laterally uniform. Lattice imaging by HRTEM demonstrates that the SiO_2 film has an 'atomically' sharp interface (Figure 8.13(a,b)) with the substrate: in other words an interface confined to the width of one atomic layer. The atomic structure of the IBO interface matches closely that of a thermally grown oxide (Figure 8.13(b)).

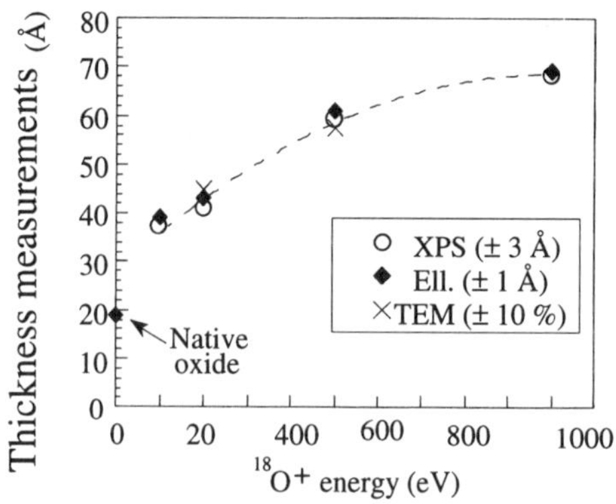

Figure 8.14 Thickness of IBO SiO_2 films as a function of IBO energy, as measured by XPS (±0.3 nm), ellipsometry (±0.1 nm), and cross-section TEM (±10%)

The chemical structure of the IBO oxide interface matches that of a thermal oxide as well, as discussed below. The SiO_2 thickness is computed from the relative intensities of the SiO_2 and the Si XPS signals (Hill *et al.*, 1976), assuming an escape depth in SiO_2 of 3 nm. Figure 8.14 shows the excellent agreement between the thicknesses derived from XPS, those measured by cross-section TEM, and those from ellipsometry assuming an index of refraction of 1.46. This index is that of stoichiometric SiO_2, which is what XPS established the IBO oxides to be.

Stoichiometry of the IBO oxide/Si(100) interface Figure 8.12 shows that the SiO_2 signal shifts to a slightly lower binding energy as thickness decreases (Table 8.2): from 104.2 eV for the 7 nm film and 103.6 eV for the 3.9 nm oxide. The latter shift is characteristic of very thin oxide films (Grunthaner and Grunthaner, 1986). According to the Grunthaners' thorough analysis of thin thermal oxides, the change in atomic density and bridging bond angle distribution in the SiO_2/Si interfacial region results in a change in the valence charge distribution in the Si–O bond and thus in an XPS chemical shift. SiO_2 interfaces formed by IBO are thus similar to those formed by thermal oxidation. Schulze *et al.* (1983) have already established this fact for high-energy IBO: a 160 nm surface oxide grown using $^{16}O^+$ ions at 30 keV and gradually etched exhibits a decrease in the chemical shift when its thickness falls below 4 nm. In the low-energy IBO oxides discussed here, no damage in the underlying Si(100) is detected by HRTEM and RBS (Figure 8.13(b,c), in contrast to Schulze *et al.* (1983). By channeling in the ⟨110⟩ direction during RBS, ^{18}O incorporation is measured with minimal background noise and no ^{16}O is seen.

Temperature dependence IBO of Si at 200 eV was performed at 300 K and 700 K. Little differences were noted between the two films. The thicknesses determined by *in situ* XPS differed by only 0.2 nm, the 700 K oxide being slightly thicker, while ellipsometry gave the same thickness for both films. This indicates a weak temperature-dependence of the IBO growth rate over the 300–700 K range, in agreement with results previously reported by Todorov *et al.* (1987).

Table 8.2 Chemical shifts measured by XPS for the oxide signals obtained after IBO of Si, Ge and $Si_{1-x}Ge_x$ as a function of IBO ion energy

E_{ion} (eV)	IBO Si	IBO Ge	IBO $Si_{1-x}Ge_x$	
	Si 2p	Ge 3d	Si 2p	Ge 3d
100	4.0	3.0	3.8	4.0
200	4.0	3.2	3.9	4.0
500	4.4	—	4.0	4.2
1000	4.6	—	4.2	4.6

Role of background pressure When Si is bombarded with energetic ions (Ar^+ and N^+ at 40 keV) in 10^{-8} to 10^{-5} mbar of O_2, a thin SiO_2 film is formed (Holmen and Jacobsson, 1988) by ion-beam-induced oxidation (IBIO). The growth rate of SiO_2 film formed by IBIO saturates quickly. The saturation thickness strongly depends on O_2 partial pressure (0 nm at 10^{-8} mbar, 10 Å at 10^{-7} mbar, 4 nm at 10^{-6} mbar, and 10 nm at 10^{-5} mbar) but not much on ion-beam current density, species, or energy (in the keV range). With the highest working pressures used here (8×10^{-8} torr $= 6 \times 10^{-8}$ mbar), a slight contribution from background $^{18}O_2$ to the IBO SiO_2 could occur. Also, the present IBO experiments were performed with much lower energies than in the case of IBIO in Schulze *et al.* (1983) and in Holmen *et al.* (1988). Low ion energy favors direct oxidation by incorporation of $^{18}O_2^+$ instead of recoil implantation of background $^{16}O_2$. The evolution with time of the oxide formation at different energies, i.e. slower initial oxidation rates at higher energies, supports the occurrence of a subsurface process (IBO) rather than a surface-driven process (IBIO). The strong dependence of the film thickness observed upon ion energy while the O_2 partial pressure remains constant further demonstrates that oxidation is governed by IBO rather than IBIO. Lastly, the SiO_2 film thicknesses reported here can be compared to those reported for other IBO experiments done in lower vacuum conditions ($<10^{-9}$ torr), either at the same low energy (200 eV O^+) (Herbots *et al.*, 1992) or at slightly higher energy (4 keV O_2^+) (Augustus *et al.*, 1988), and are found to be in extremely good quantitative agreement with them (Figure 8.12(b)).

8.4.2.1.2 IBN of elemental Si

To study the energy dependence of IBN at low dose, Si(100) substrates were nitrided at ion energies of 100, 150, 200 and 350 eV at room temperature using a dose of 1.5×10^{16} N_2^+/cm^2 or less. Another set of Si(100) substrates were nitrided at 200, 500 and 1000 eV for a larger dose of 3×10^{17} N_2^+/cm^2. To investigate the role of temperature on IBN, a third set of Si samples was nitrided at a low dose of 1.5×10^{16} N_2^+/cm^2 and fixed ion energy, 200 eV, at deposition temperatures of 20, 120, 220, 320 and 420 °C. For this latter experiment, IBN is terminated when the saturation of the chemical reaction at the surface is detected by XPS: the dose at which saturation occurs varies between 6×10^{15} and 2×10^{16} N_2^+/cm^2.

Characterization is done in the following order. Sample curvature is measured before and after deposition as the first *ex situ* characterization step to characterize the stress before and after IBN growth. Chemical bonding is monitored by *in situ* XPS during IBN. Samples are then cleaved into two pieces. One of these pieces is used for making TEM samples, the other sample first for RBS analysis, then for other analysis.

XPS measurements and IBN nitrides stoichiometry The near-surface composition of the films is measured by XPS during and after IBN. The increase in nitrogen content as a function of irradiated dose is detected in the XPS spectra during deposition. As the N 2s photoelectron signal intensity increases, the binding energy of the Si 2p photoelectron shifts by 2.35 eV from 99.7 eV to 102.05 eV as is also found for bulk Si_3N_4. The ratio of the unshifted Si 2p peak area to the shifted peak area is taken as a measure of near-surface stoichiometry.

Figure 8.15 shows the XPS Si 2p signal for 200 eV IBN of Si(100) at 20 °C as a function of deposition time. The Si 2p signal has an energy of 99.6 eV before IBN, then shifts in the IBN initial stage, indicating an immediate reaction of N with Si. Both the initial increases in the N 1s and in the shifted Si 2p signals are found to be linear with dose. The magnitude of the chemical shift increases from 1.4 eV to a value of 2.4 eV in a nonlinear fashion.

The intensities of the N 1s and shifted Si 2p signals gradually reach a steady state, indicating that as IBN proceeds, the nitridation of Si near the surface saturates. Because of the limited escape depth of Si 2p photoelectrons, reactions occurring deeper than 5 nm beneath the surface are not detected by XPS. Near-surface stoichiometry of IBN films is measured by the ratio of the shifted to the unshifted Si 2p signals. Characteristic XPS spectra after IBN at different energies are shown in Figure 8.16. Figure 8.17 shows the stoichiometry of Si nitrides as a function of deposition energy as determined from the signal ratios obtained from Figure 8.16. Films grown within energies $\geq$500 eV have a stoichiometry close to Si_3N_4, whereas those produced under 500 eV exhibit a significant excess of unnitrided silicon.

Deposition temperature is not observed to influence the final XPS spectra at all. 200 eV IBN at temperatures between 20 °C and 420 °C all exhibit a Si 2p chemical shift of 2.38 ± 0.05 eV, and a nitrided fraction of Si of 0.8. However, during IBN at higher temperatures, the magnitude of the nitrogen signal and of the chemical shift are found to increase more quickly than at lower temperatures.

Microstructure of IBN nitrides and its temperature dependence Films are examined after IBN by cross-section TEM. This thickness measurement does not depend on a corollary assumption while measurements by ellipsometry assume knowledge of the refraction index and the structure of the films. Thickness measurements by RBS rely on the knowledge of thin-film density and composition while XPS is limited to uniform films thinner than the escape depth of the photoelectrons.

A VGHB5 scanning transmission electron microscope (STEM) operating at 100 keV profiled N throughout the thickness films by focusing the electron beam to a size of 1 nm. X-rays generated by the electron beam are collected with an energy-dispersive analyzer, and reflect the composition of the area illuminated by the electron beam. The resolution of STEM is essentially the spot size of the electron beam, namely 1 nm, but is not quantitative because of X-ray absorption and fluorescence, and thickness and concentration fluctuation

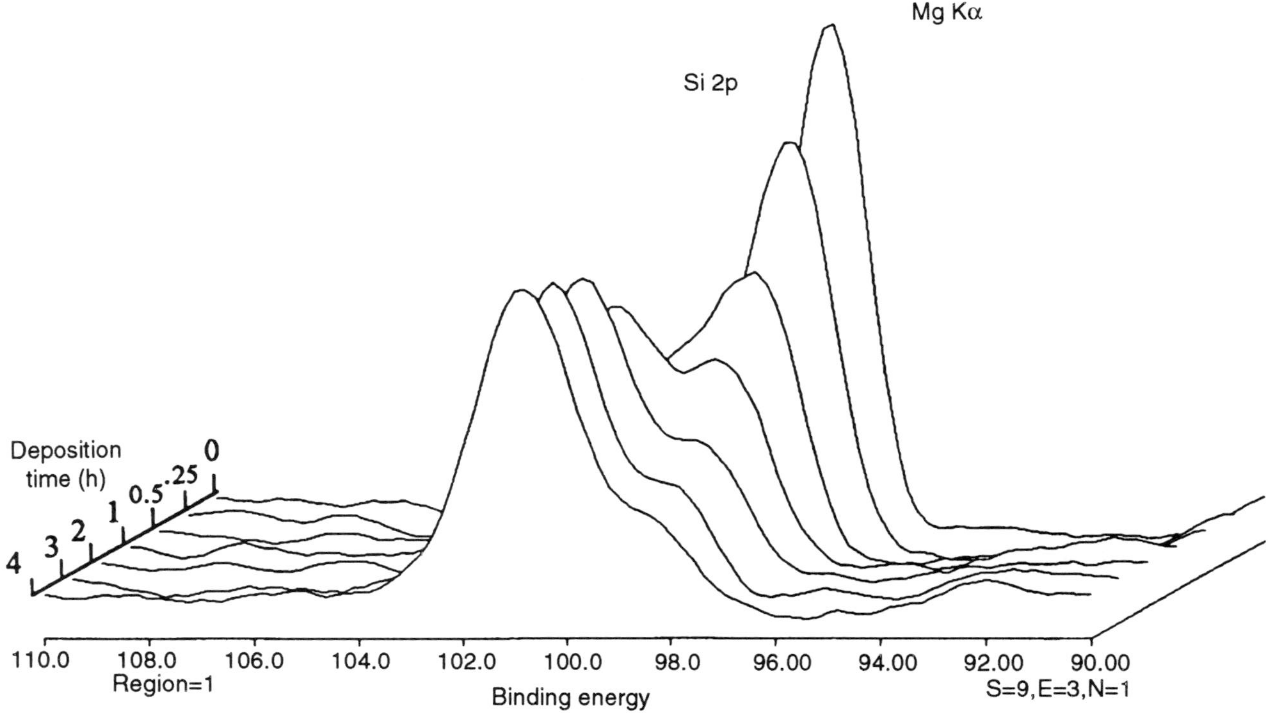

Figure 8.15 Time-resolved XPS spectra of Si 2p during IBN of Si(100) with 200 eV N_2^+ as a function of exposure time to the N_2^+ beam

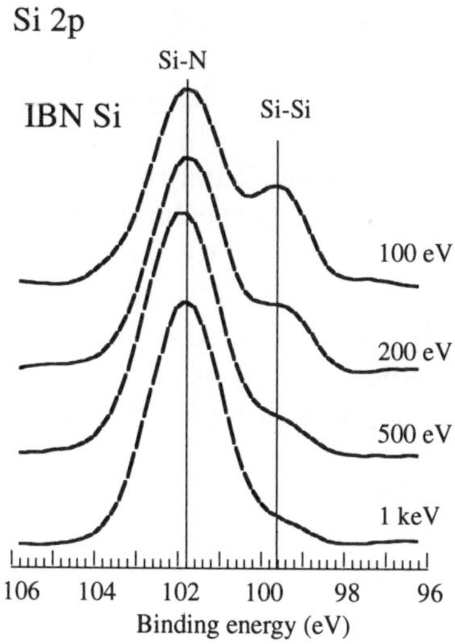

Figure 8.16 XPS spectra of the Si 2p signal as a function of N_2^+ energy for IBN of Si(100) after signal saturation

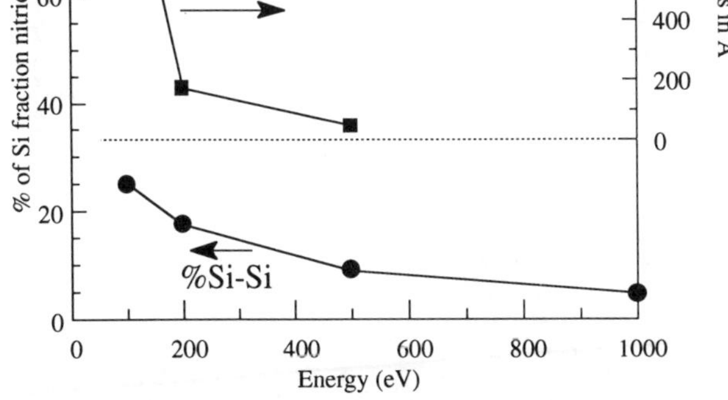

Figure 8.17 Stoichiometry of IBN Si nitride films: amount of nitrided Si (% Si–N) and unreacted Si (% Si–Si) as measured by XPS, and TEM thickness of the IBN film as function of IBN energy

in the films. Furthermore, the sample being a thin section, its whole surface becomes heavily contaminated by exposure to air.

For a dose of $3 \times 10^{17}\ N_2^+/cm^2$, nitrides grown between 200 eV and 1 keV are found to be amorphous on an otherwise undamaged substrate. Thicknesses measured by cross-section TEM are found to increase (from 4 to 20 nm) with *decreasing* ion energy (Figure 8.18) (Hellman *et al.*, 1992a, 1992c). These thicknesses are well beyond the range of the ions, in contrast to what XPS data may suggest, as in Taylor *et al.* (1978). That the amorphous-layer thickness as measured by cross-section TEM far exceeds that of the ion range shows that IBN involves more than atomic collisions.

For smaller doses, the opposite is found: the thickness of the amorphous films increases with ion energy, although all these films are, as expected, thinner (2.5–4 nm) (Figure 8.18). TEM reveals a subsurface region of strain contrast in these samples, indicative of point-defect coalescence. At higher doses, this coalescence would lead to the creation of an amorphous layer beneath the nitride film, as is observed. No dependence of thickness on film temperature is found at low doses.

For IBN at temperatures >200 °C, TEM reveals the presence of small regions of strain below the nitride film (Figure 8.19). These strained regions most likely correspond to dislocation loops which have formed with the coalescence of nitrogen and silicon interstitials. They can be imaged in TEM by aligning the electron beam with the [110] direction in the plane of the film, which is a high-diffraction condition, then tilting the sample away from the [110] direction by 2.5° while remaining in the plane of the film. This moves the beam away from the high-diffraction condition while maintaining an

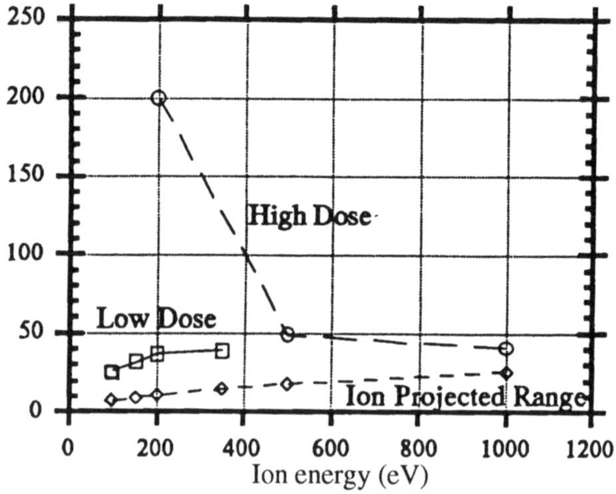

Figure 8.18 IBN film thickness measured by cross-section TEM for low-dose IBN

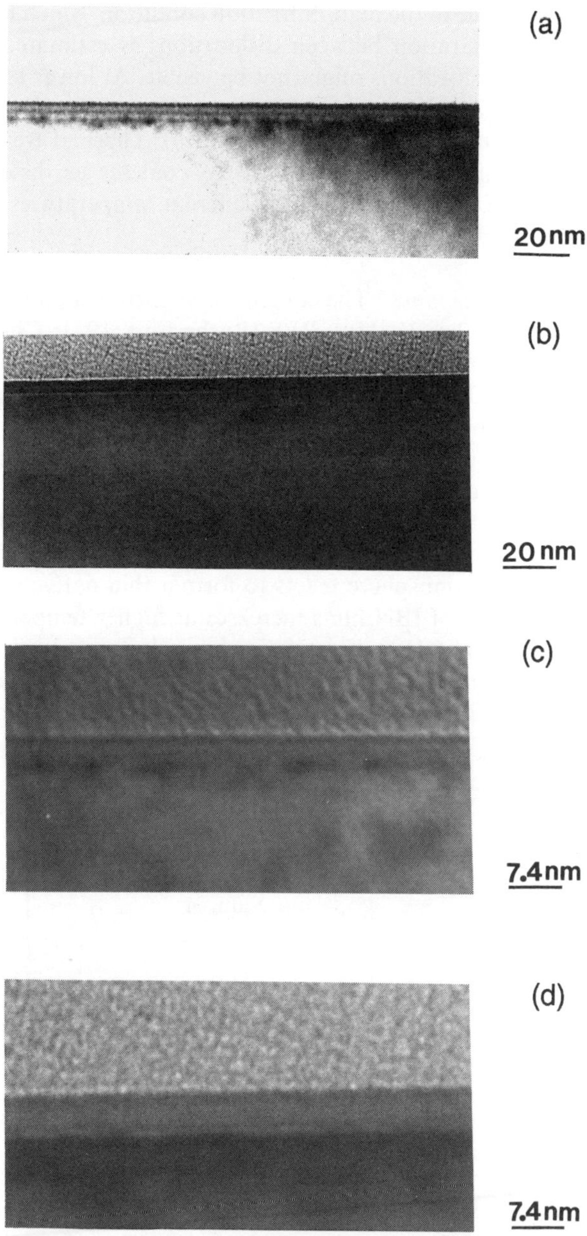

Figure 8.19 Imaging of defects by cross-section TEM after IBN of Si at (a) 420 °C, (b) 320 °C, (c) 120 °C and (d) room temperature

edge-on view of the film. Strain causes a local lattice distortion, which aligns a small part of the lattice in the high-diffraction condition, which hence becomes easily visible. The separation between dislocations is estimated to be about 20 nm, although all dislocations might not be visible. At lower temperatures of deposition, strain regions are not seen. Contrast is present at the nitride/Si(100) interface, yet no defects are detected (Figure 8.19). Thermal treatment during deposition is thus sufficient for interstitials to coalesce as dislocation loops. TEM micrographs of IBN films grown at different temperatures clearly depict this effect (Figure 8.19).

Reactivity of IBN nitride films The content of Si, nitrogen and oxygen in IBN layers is measured by 2 meV $^4He^+$ RBS (Rutherford, 1911; Chu *et al.*, 1978) and the RBS analysis code RUMP (Doolittle, 1983). Although *in situ* XPS detects no oxygen, ^{16}O is found to incorporate into the films as a native oxide after IBN films are removed from vacuum. RBS shows that ^{16}O incorporation decreases with increasing ion energy (Figure 8.20). This is consistent with the measurements of nitride stoichiometry described above. At high energies, the surface stoichiometry is close to that of Si_3N_4, so there is no free Si at the surface to react with air. At lower energies, where there is unnitrided Si at the surface, ^{16}O from the atmosphere reacts to form a thin native oxide.

The nitrogen content of IBN films increases at higher temperatures (Figure 8.21). This indicates that IBN kinetics is thermally activated, either in the diffusion of nitrogen interstitials, or in the coalescence of point defects into a nitride phase. This is in contrast with the invariance of IBN film thicknesses

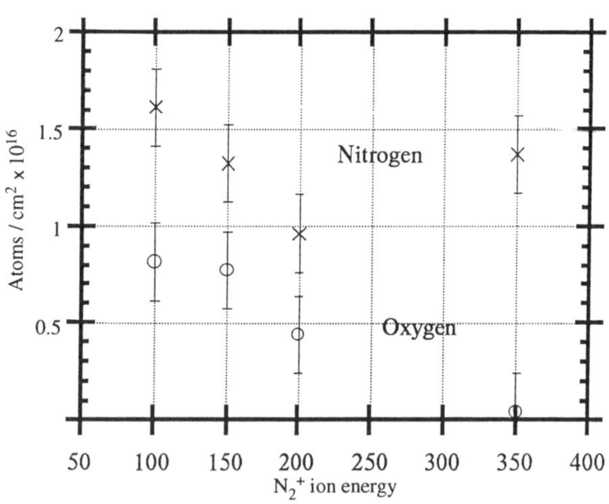

Figure 8.20 Concentrations of N and ^{16}O after IBN as a function of IBN energy as measured by 2 MeV $^4He^+$ RBS combined with ion channeling. Note that no ^{16}O is not detected by *in situ* XPS after IBN before exposure to air

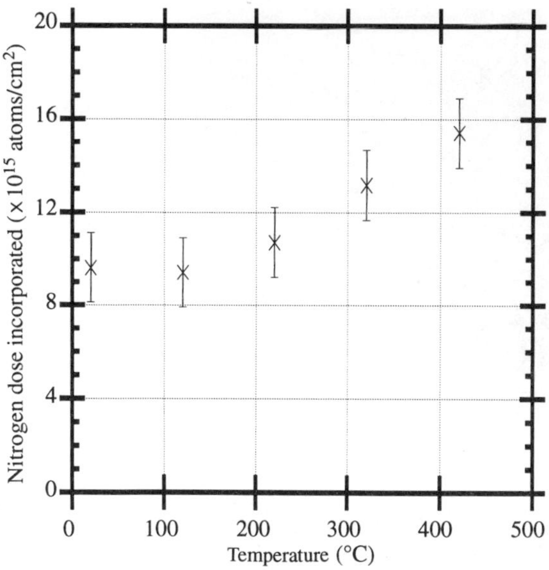

Figure 8.21 2 MeV $^4He^+$ RBS measurement of N incorporation in IBN films as a function of temperature for a fixed dose of 6×10^{15} atoms/cm^2

observed by TEM and shows that the low-temperature films are Si-rich, but reach the same thickness as high-temperature nitrides.

STEM is used to profile nitrogen as a function of depth in films nitrided at 200 eV and 500 eV at doses of 3×10^{17} N_2^+/cm^2. Relative fluctuations in nitrogen concentrations are found. In 200 eV IBN films, nitrogen is observed to be confined to the top half of the 20 nm of the amorphous film (Figure 8.22). Thus, the amorphous Si region below nitride films as observed by Thomas *et al.* (1982) for 2 keV N^+ ions is also found here. In contrast, the 500 eV IBN film shows an even concentration of nitrogen throughout the amorphous film, and a sharp decrease of nitrogen at the amorphous/crystalline boundary.

Strain dependence upon energy and temperature Two applications of nitride films are packaging and micromechanical devices, and one of their critical properties is the strain present after deposition. IBN provides a new degree of freedom in controlling strain, through the choice of ion energy, in addition to deposition temperature. As shown below, the temperature dependence of IBN is a strong function of ion energy: a 200 eV deposit with N_2^+ exhibits compressive stress at low temperatures and tensile stress at high temperatures, pointing to an unparalleled flexibility of IBN nitrides.

Stylus profilometry characterizes the curvature of Si after IBN. The radius of curvature of the wafer can be converted to an absolute amount of stress in

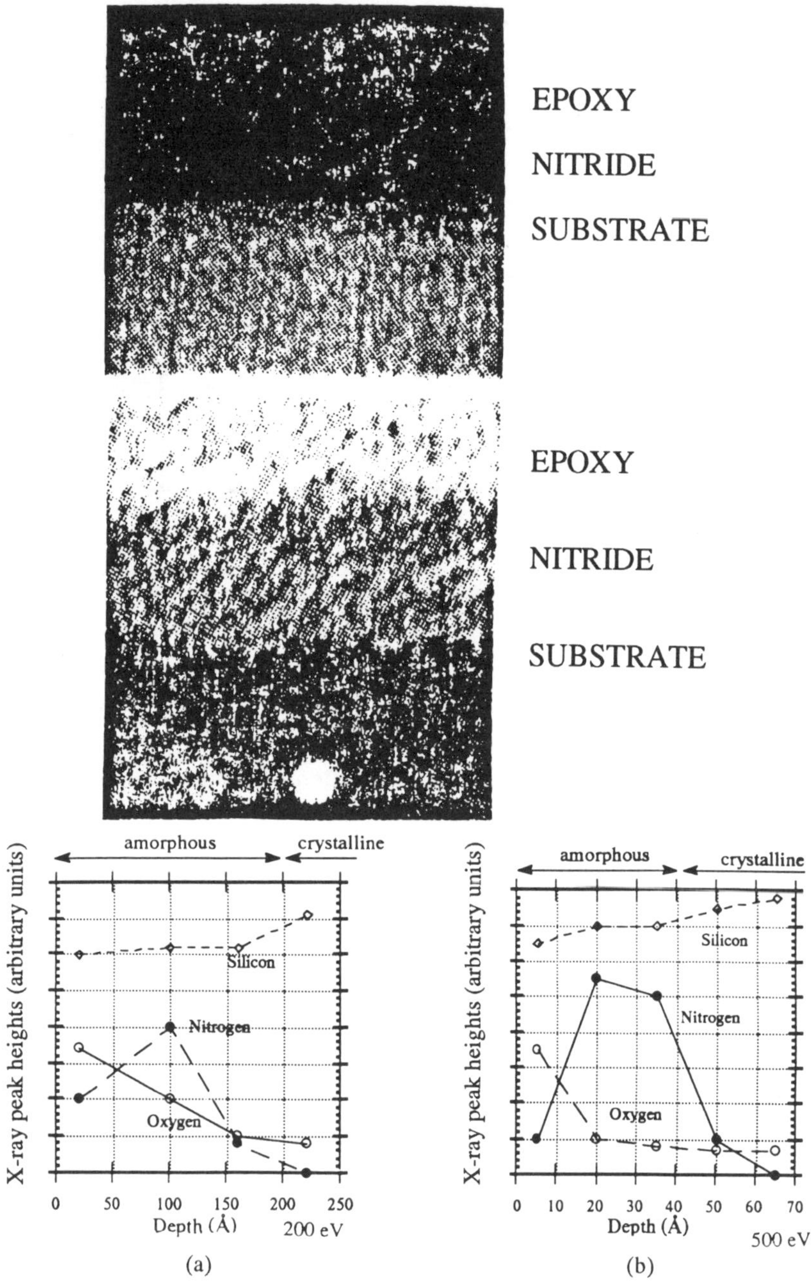

Figure 8.22 STEM micrographs for IBN 6×10^{15} atoms/cm^2 at 200 eV (a) and 500 eV (b). Relative heights of electron-stimulated X-ray signal as a function of depth are shown underneath

Si(100), using the well-known mechanical constants of Si and the thin-film approximation of a strip thermal stress calculation given by Timoshenko (1930). Stress in the film is calculated, based on the substrate curvature and the thicknesses of the substrate and the film.

Film stress is expressed as

$$\sigma = \left(\frac{E}{1-\nu}\right)\frac{t_{\mathrm{Si}}^2}{t_{\mathrm{SiN}}^2}\frac{1}{6r} \tag{24}$$

where E and ν are the Young's modulus and the Poisson ratio of Si(100), t_{Si} and t_{SiN} are the thicknesses of the Si(100) wafer and the film, and r is the measured radius of curvature. For the calculations of stress presented here, the film thickness is measured by TEM. The thickness of Si(100) stems from wafer specifications (300 μm). The radius of curvature is measured using a Tencor Alpha-Step 1000 stylus profilometer. The curvature of the substrate after IBN is the sum of the curvature of Si(100) before IBN and of the curvature caused by the film stress. Curvature is thus measured before and after IBN. The net change in curvature is attributed wholly to the film stress. The Tencor instrument is designed to measure step heights: the vertical deflection of the stylus in contact with the sample is thus measured during scanning of a horizontal distance of 5 mm. Samples are prepared by cleaving a Si(100) wafer into 1 cm × 1 cm squares and scribing hash marks at the midpoint of each edge to serve as centering guides. The samples are subjected to a dip-etch (Section 8.4.1.1) and positioned so that the stylus contacts the sample at its center as defined by the scribed has marks (Tencor). Noting that the stylus height before and after each scan is a measure of mechanical drift during a scan, it is used to improve the resolution of the measurement (Figure 8.23).

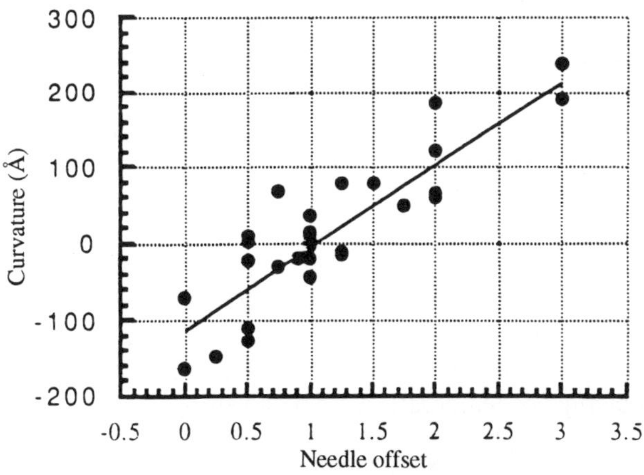

Figure 8.23 Calibration of needle offset using the stylus height

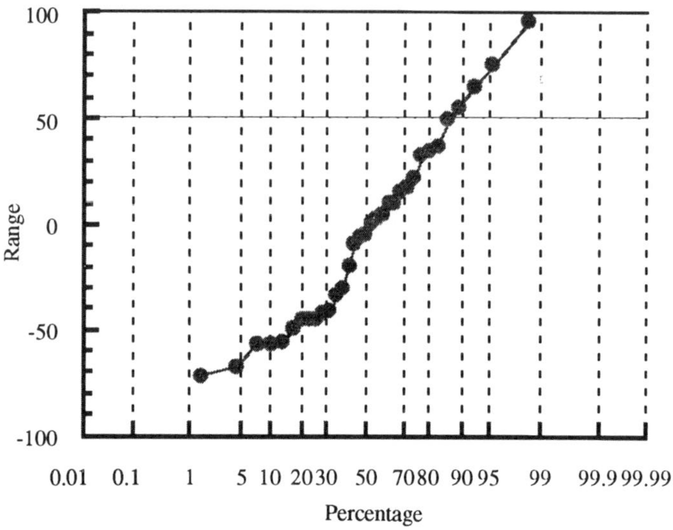

Figure 8.24 Statistical reproducibility of curvature measurements

Figure 8.24 shows a set of 33 data points taken on the same surface. The radius of curvature of the samples is then extracted: an *arc height h* is defined as the deviation of the wafer surface from a straight line extending from two points on the surface a distance of L apart where L is the distance scanned, 5 mm, and the measured arc height ranges from 0 to 300 nm $\pm$ 10 nm. The radius of curvature is

$$\rho = \left(\frac{L^2}{8\eta}\right). \tag{25}$$

Film stress is found to vary for IBN at different temperatures. Since films grown above room temperature are cooled after IBN, the stress in these films is the sum of the stresses due to IBN and to thermal expansion mismatch between silicon nitride and Si(100). The values for film stresses are calculated using the thicknesses measured by TEM. Nitrides prepared at room temperature have a stress two orders of magnitude greater than comparable films prepared by a CVD technique (Figure 8.25). Such high levels of stress in IBN films can explain the results of Grabowski *et al.* (1989), who deposited nitrides at $T < 70\,^\circ\mathrm{C}$ with 500 eV N_2^+ ions combined with Si evaporation. Upon annealing at 950 °C, 'blisters' appear in nitrides that cannot be attributed to thermal-expansion mismatch alone. Indeed, thermal stress in a thin film subjected to a change of temperature ΔT is given by

$$\frac{F}{A} = \left(\frac{\alpha_{\mathrm{Si}} - \alpha_{\mathrm{SiN}}}{2}\right)\left(\frac{E_{\mathrm{SiN}}}{1 - \nu_{\mathrm{SiN}}}\right)\Delta T \tag{26}$$

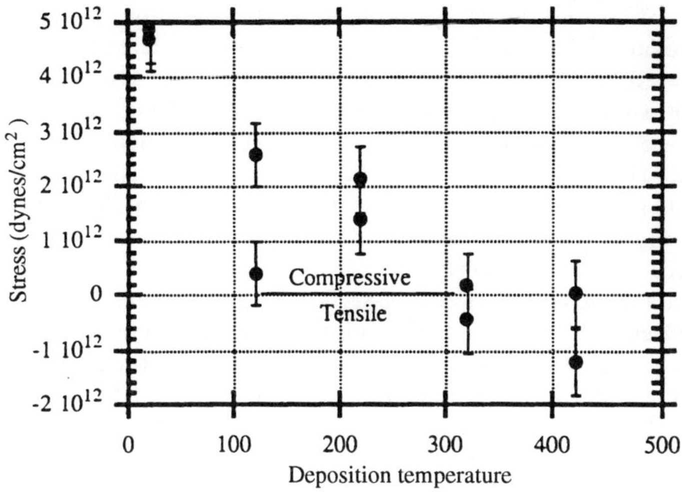

Figure 8.25 Stress in IBN films as a function of deposition temperature

where α_{Si} and α_{SiN} are the thermal-expansion coefficients of Si and Si nitride, and E and v are the Young's modulus and Poisson ratio of the film. The thermal stress does not depend on the thickness of the film, and the only mechanical property of Si that affects the stress is the thermal-expansion coefficient. To separate thermal stress effects from as-deposited stress, thermal stress is calculated and subtracted from the measured stress. However, the mechanical properties of the IBN nitride are not known. Instead, they are approximated by existing measurements on nitride films prepared by other means. These thermal stresses are small in comparison to the as-deposited stress. Taking $E/(1 - v)$ as 4×10^{11} Pa, ΔT as 200 K and α_{SiN} as 5×10^{-6} K^{-1}, the thermal stress is 2×10^8 Pa. This is two orders of magnitude below the measured film stress. The effect of thermal-expansion mismatch is thus small compared to stress due to the deposition method. The dependence of stress on temperature is thus attributed to a thermal component of IBN kinetics.

One possibility is that the appearance of the interstitial loops as seen in TEM is responsible for relaxation in the film. This implies that stress induced by IBN is not limited to the nitride film itself, but extends to Si(100) as well. The assumption that the amorphous layer and the nitride layer alone are under stress, may then not be viable.

8.4.2.1.3 Comparison between IBO and IBN of Si

In summary, oxides grown by IBS exhibit an increase in thickness with ion energy, while nitrides grown by IBS exhibit the reverse trend. This is easily explained and accounted for by modeling the differences in reactivity and diffusivity. The stoichiometry of IBO oxides at saturation is that of a fully reacted dioxide at all energies and temperatures used while the stoichiometry of nitrides is a function of both temperature and energy. Also, the nitrides do not exhibit a saturation thickness as the oxides do. This is accounted for by the differences in reactivity and diffusivity. Finally, the microstructure of IBO oxides and IBN nitrides is drastically different. While IBO oxides exhibit high-quality interfaces and no substrate damage, nitrides exhibit damage below the nitride layer, either in the form of an amorphous phase at low temperature, or coalesced dislocation loops at higher temperature. Comparison in Section 8.5 between the models developed in Section 8.3 and the present data accounts for these effects. Finally, while IBO exhibits little temperature dependence up to 400 °C, IBN nitrides exhibit significant dependence in stoichiometry, microstructure and stress upon temperature. Finally, it should be noted that the dependences of IBO and IBN as a function of ion energy look similar when explained by surface-sensitive XPS, but are found to be in fact radically different when complementary techniques such as RBS and cross- section TEM are used.

8.4.2.2 IBS of oxides and nitrides of elemental Ge

8.4.2.2.1 IBO of elemental Ge: energy dependence of phase formation

Figure 8.26 illustrates the chemical shift observed for Ge 3d photoelectrons from 29.7 eV to 33 eV as GeO_2 is formed under $^{18}O_2^+$ bombardment at 200 eV,

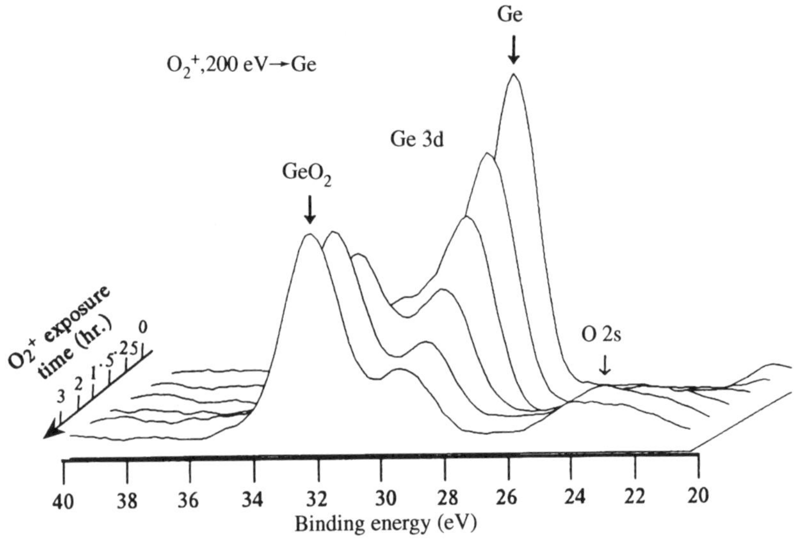

Figure 8.26 Time-resolved XPS spectra of Ge 3d during IBO with 200 eV $^{18}O_2^+$ as a function of exposure time to the $^{18}O_2^+$ beam

as a function of time. Features are similar to those observed in IBO of Si (Figure 8.12, Section 8.4.2.1). However, oxide formation depends upon ion energy for the metastable GeO_2/Ge structure, which is not the case for IBO of Si. IBO of elemental Ge leads initially to the formation of Ge suboxides. Then the growth of stoichiometric GeO_2 occurs until all the Ge being oxidized turns into GeO_2, at which point the growth rate saturates and the GeO_2 thickness stops increasing. As pointed out already, phase formation depends dramatically upon ion energy (Figure 8.27). Below 200 eV, GeO_2 forms while at 500 eV and above, the chemical shift of the Ge 3d signal does not correspond to GeO_2 (Ge^{4+}), but to the monoxide GeO (Ge^{2+}). The signal is quite broad, and the shift can only be estimated to be 2 eV. GeO_2 is a volatile oxide. At higher ion energies, it is sputtered away as it is formed (Section 8.3.3.1). The high sputtering rate of GeO_2 inhibits partially (500 eV) or totally (1 keV) the oxidation of Ge. For $E_{ion} < 200$ eV, the sputtering rate is lower and GeO_2 is formed. The relative intensity of the Ge^{4+} to Ge^0 signals increases when the ion energy increases from 100 to 200 eV. The chemical shift observed for 100 eV IBO GeO_2 is slightly lower than for 200 eV IBO (Table 8.2). This indicates, as for Si, an increase of oxide thickness with ion energy and the existence of an interfacial region between the oxide and the substrate, which is similar to the reduction in chemical shift ΔBE observed in both thin thermal SiO_2 films in (Grunthaner and Grunthaner, 1986) and in thin IBO SiO_2 (Section 8.4.2.1.1, Figure 8.12 and

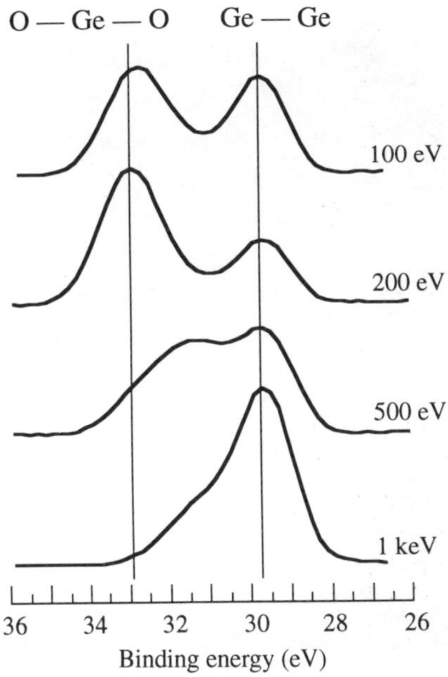

Figure 8.27 XPS spectra of the Ge 3d signal as a function of ion energy for IBO of elemental Ge after signal saturation

Table 8.2). Phase decomposition plays an important role at higher energies. The insulating properties of IBO Ge oxides, as measured by SEM, are found to be directly related to the presence of GeO_2. While Ge oxides grown with ion energies of 100 and 200 eV charge electrostatically during SEM, only faint charging occurs for 500 eV IBO oxides and none for 1 keV. IBO oxides are insulating only when their stoichiometry is that of GeO_2.

TEM shows that the epitaxial quality of Ge/Si(100) is good, although some large stacking faults are present (Figure 8.28). The surface, however, is rough and makes difficult the detection of an amorphous layer. TEM sample preparation may have affected the microstructure: amorphous GeO_2 is soluble in water which is used during the dimpling of the samples; ion-beam-milling effects are also observed.

8.4.2.2.2 IBN of elemental Ge

The behavior of elemental Ge subjected to IBN is quite different from IBO. When exposed to an N_2^+ beam at room temperature, the Ge 3d photoelectron signal in XPS *broadens* at each ion energy from 100 eV to 1 keV (Figure 8.29). The signal broadening at the lowest IBN eneryg peak can possibly be deconvoluted into an unreacted Ge^0 signal and a 'subnitride' signal. For IBN at 100 eV,

Figure 8.28 Cross-section TEM micrograph of Ge after 200 eV IBO

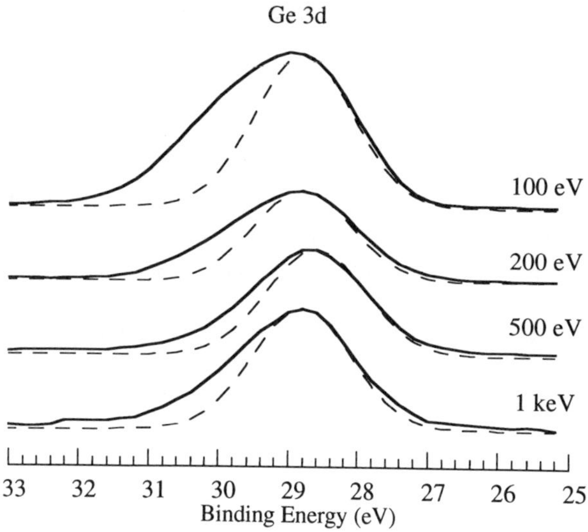

Figure 8.29 Ge 3d after XPS signal saturation as a function of IBN energy

the chemical shift is estimated to be 1.33 eV, with 39% of the total emission from Ge 3d photoelectrons being shifted. At 200 eV, the magnitude of the chemical shift is estimated to be 1.24 eV, with 26% of the Ge 3d being shifted. Even at these low energies, the error in signal deconvolution due to overlap is large. At higher energies of IBN, the magnitude of the chemical shift is even smaller, and overlap makes a deconvolution meaningless. There is no clear evidence of nitride formation at these energies: chemical reaction of elemental Ge and nitrogen cannot be driven by IBN between 100 eV and 1 keV. However, a trend towards more successful incorporation below 100 eV is observed. While IBO is successful in growing metastable oxides of Ge, IBN is not successful in growing metastable nitrides of Ge above 100 eV.

8.4.2.3 IBS on strained $Si_{1-x}Ge_x$ heteroepitaxial thin films on Si(100)

8.4.2.3.1 IBO of strained $Si_{1-x}Ge_x$ heteroepitaxial thin films on Si(100) stoichiometry of IBO $Si_{1-x}Ge_x$ oxides

During IBO of $Si_{1-x}Ge_x$ layers under $^{18}O_2^+$ bombardment, both Si and Ge within the alloy fully oxidize into a ternary dioxide. This is to be contrasted with high-energy IBO of $Si_{1-x}Ge_x$ where Ge oxidizes only after Si is completely oxidized (Castle *et al.*, 1992) and with IBO of elemental Ge (Vancauwenberghe *et al.*, 1992a, 1992b), where IBO occurs only at the lowest energies. During IBO of $Si_{1-x}Ge_x$ the Ge 3d signal shifts even when ion energies reach 1 keV. Figure 8.30 illustrates the gradual oxidation by IBO of $Si_{1-x}Ge_x$ at 200 eV. As for IBO of elemental Si and Ge, $Si_{1-x}Ge_x$ dioxide forms in three stages. Initially, substoichiometric oxides grow as seen in the smaller chemical shifts in XPS (at 0.5 h of $^{18}O_2^+$ exposure in Figure 8.30). In the second stage, rapid growth of the dioxide occurs, the suboxides disappear and contribute to less than 5% of the XPS signals, after 2 h of $^{18}O_2^+$ exposure (Figure 8.30). The IBO rate then saturates and net ^{18}O incorporation does not increase anymore (XPS spectra at 2, 3 and 4 h in Figure 8.30). Since Ge 2p photoelectrons have a lower kinetic energy than the Ge 3d photoelectrons, therefore originating from a more shallow depth than the Ge 3d signal, the Ge 2p signal from unoxidized Ge disappears almost entirely when ^{18}O incorporation saturates, indicating that the top surface is fully oxidized within the escape depth of Ge 2p in $Si_{1-x}Ge_x$ dioxide.

The chemical shift for the Si^{4+} and Ge^{4+} states observed in the $Si_{1-x}Ge_x$ dioxide differ from the Si^{4+} and Ge^{4+} states in elemental SiO_2 and GeO_2 (Figures 8.12 and 8.27). The chemical shift for Si^{4+} is consistently lower in $Si_{1-x}Ge_x$ dioxide than in SiO_2 by 0.1 to 0.4 eV, while ΔBE for Ge^{4+} is *larger* by 0.8–1.0 eV in $Si_{1-x}Ge_x$ dioxide than in GeO_2. These differences in chemical shift indicate the formation of a ternary phase, $Si_{1-x}Ge_xO_2$ rather than a mixture of binary SiO_2 and GeO_2 phases. As the ion energy increases (Figure 8.31), the relative intensity of Si^{4+} an Ge^{4+} versus Si^0 and Ge^0 increases, just as in IBO of elemental Si. This increase is stronger for Si 2p (Figure 8.31(b)) and Ge 2p (Figure 8.31(a)) than for Ge 3d (Figure 8.31(c)), which suggests that $Si_{1-x}Ge_xO_2$ thickness increases with ion energy. However, the difference in intensity between the Ge 2p and the Ge 3d signals demonstrates an increasing non-uniformity of the stoichiometry with depth at high energies. In Figure 8.31, the chemical shift in the fully oxidized states of the Si 2p and Ge 3d is slightly smaller at low and high energies. This is similar to the reduction in chemical-shift ΔBE observed in both thin thermal SiO_2 films in Grunthaner and Grunthaner (1986), in thin IBO SiO_2 (Section 8.4.2.1, Figure 8.12 and Table 8.2) and in thin IBO GeO_2 (Section 8.4.2.1, Figure 8.27 and Table 8.2). An interfacial region similar to that of SiO_2/Si(100) exists between amorphous $Si_{1-x}Ge_xO_2$ and crystalline $Si_{1-x}Ge_x$. High-resolution TEM micrographs, discussed in the next section appear to confirm this hypothesis as well.

Another important feature is to be noted. During IBO $Si_{1-x}Ge_x$, Ge *fully* oxidizes into the Ge^{4+} state at all energies used, even at 1 keV. No lower oxide state (such as Ge^{2+}) is detected. This is to be contrasted with IBO of elemental Ge (Figure 8.27). The alloying of Si with Ge thus enhances the oxidation of Ge during IBO. However, as the energy increases, the Ge fraction in $Si_{1-x}Ge_x$ is lower than the original Ge fraction of $Si_{1-x}Ge_x$ (Figure 8.32) when energies exceed 200 eV or if the $Si_{1-x}Ge_x$ heteroepitaxial layers are defective or rough. At 200 eV and below, both composition uniformity and Ge retention and oxidation are achieved.

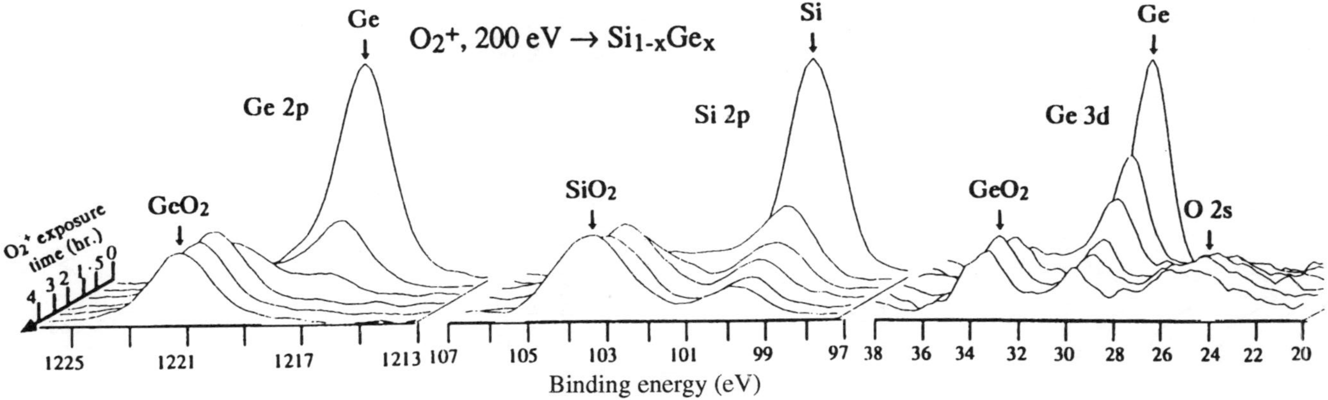

Figure 8.30 Time-resolved XPS spectra of IBO of $Si_{1-x}Ge_x$ with 200 eV $^{18}O_2^+$ as a function of exposure time to $^{18}O_2^+$

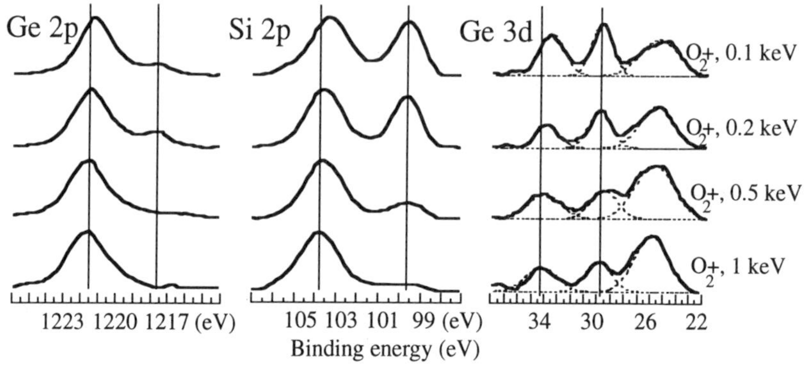

Figure 8.31 XPS spectra of the Ge 2p, Si 2p and Ge 3d XPS signals as a function of IBO energy for $Si_{1-x}Ge_x$ alloys after signal saturation

Microstructure of IBO $Si_{1-x}Ge_x$-dioxide Several observations can be made. High-resolution cross-section TEM of $Si_{1-x}Ge_xO_2/Si_{1-x}Ge_x$ interfaces (Figure 8.33(a)) demonstrates that at low energies, where $Si_{1-x}Ge_xO_2$ is most uniform, the $Si_{1-x}Ge_xO_2/Si_{1-x}Ge_x$ interface exhibits ideal sharpness, confined to one interatomic distance, and comparable to thermal SiO_2 (Figure 8.13(a)) and IBO SiO_2 (Figure 8.13(b)). Z-contrast imaging indicates a highly uniform composition through $Si_{1-x}Ge_xO_2$ and demonstrates the absence of Ge segregation and GeO_2 cystallites observed in thermally grown $Si_{1-x}Ge_x$ oxides (LeGoues *et al.*, 1989; Holland *et al.*, 1987; Liu *et al.*, 1991, 1992). These data also suggest that the difference between the Ge 3d and Ge 2p signals may be due to differences in thickness across the area analyzed by XPS. Indeed, XPS samples a macroscopic area of about 1 mm^2 while the Z-contrast samples a much smaller region (1 nm by 1 nm). Further evidence of the role of lateral non-uniformity is shown

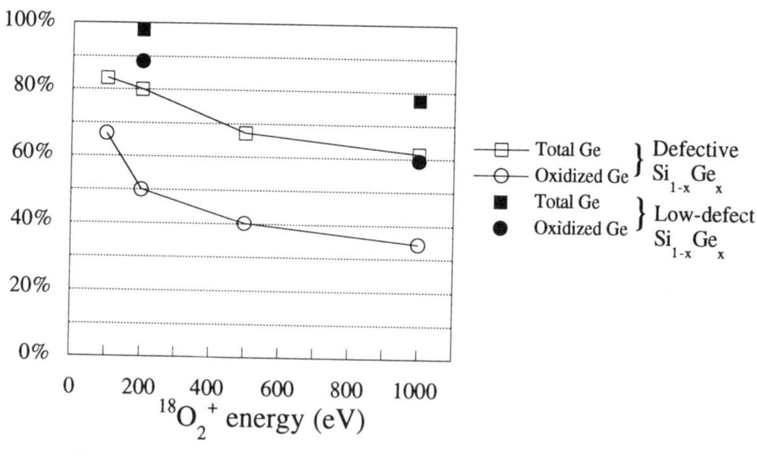

Figure 8.32 Retention of Ge after IBO of $Si_{1-x}Ge_x$ as a function of ion energy

below. At higher ion energies, HRTEM micrographs show that the interface loses its sharpness, and extends over three atomic layers rather than one (Figure 8.33(b)). The broadening of the interface with increasing ion energy observed by HRTEM is qualitatively in agreement with XPS data, but it is not clear whether a 1 nm wide interface would account completely for the non-uniformity detected by XPS. Rather, a combination of the increase of interface width with energy and of surface roughness and defects in certain regions (as shown in Figure 8.34) would account for this data.

Figure 8.33 (a) HRTEM micrograph of a smooth $Si_{1-x}Ge_xO_2$ formed by IBO with 200 eV $^{18}O_2^+$ depicting the 'atomically sharp' $Si_{1-x}Ge_x/Si_{1-x}Ge_xO_2$ interface. (b) HRTEM micrograph of a smooth $Si_{1-x}Ge_xO_2$ formed by IBO with 1 keV depicting the $Si_{1-x}Ge_x/Si_{1-x}Ge_xO_2$ interface extending over four atomic layers

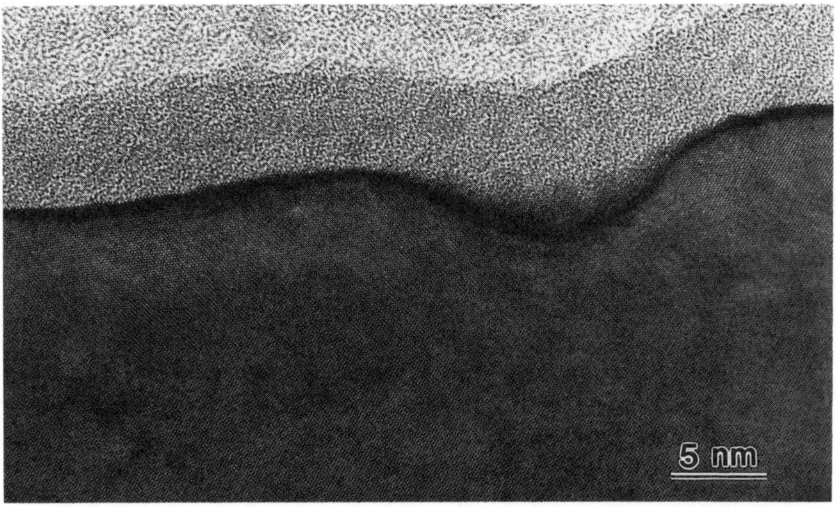

Figure 8.34 Cross-TEM of a 200 eV IBO $Si_{1-x}Ge_xO_2$ grown on a rough $Si_{1-x}Ge_x$ heteroepitaxial layer. The oxide exhibits thickness fluctuations

Conventional cross-section TEM at smaller magnification, i.e. below atomic imaging conditions, reveals the presence of stacking faults in $Si_{1-x}Ge_x$ heteroepitaxial layers that reach the top surface, and the roughness of the $Si_{1-x}Ge_x$ surface. Both lead to differences in morphology of $Si_{1-x}Ge_xO_2$. On defective and rough $Si_{1-x}Ge_x$ heteroepitaxial films, IBO $Si_{1-x}Ge_xO_2$ closely follows the morphology of the alloy layer and exhibits local thickness fluctuations (Figure 8.34). IBO $Si_{1-x}Ge_xO_2$ is thicker where a defect intersects the surface. On smooth, low-defect-density layers, 200 eV IBO $Si_{1-x}Ge_xO_2$ exhibits a sharp interface with the $Si_{1-x}Ge_x$ MBE layer and a uniform thickness (Figure 8.33(a)).

The composition of $Si_{1-x}Ge_xO_2$ is also strongly affected by the microstructure. Ge retentions by $Si_{1-x}Ge_xO_2$ after IBO is much higher in low-defect-density $Si_{1-x}Ge_x$ than in defective or rough layers (Figure 8.32). A possible explanation is that Ge is more easily sputtered away in these regions. Nevertheless, in both cases (defective and defect-free $Si_{1-x}Ge_x$ films), sputtering decreases with ion energy and the Ge/Si ratio in $Si_{1-x}Ge_xO_2$ is close to the original Ge/Si ratio in the $Si_{1-x}Ge_x$ alloy. Low ion energies (100–200 eV) thus favor both uniformity in composition and Ge retention (Figure 8.32). Figure 8.35 shows a 2 MeV $^4He^+$ RBS spectrum of $Si_{1-x}Ge_x$ oxidized by IBO. The RBS spectrum taken along the $\langle 110 \rangle$ channeling direction demonstrates the incorporation of ^{18}O into the $Si_{1-x}Ge_x$ alloy. Again, no segregation of Ge is detected, as already observed by HRTEM (Figure 8.33(a)). The substrate channeling yield is also higher under an unoxidized region of the $Si_{1-x}Ge_x$ layers than under the oxidized region. Given the presence of an additional

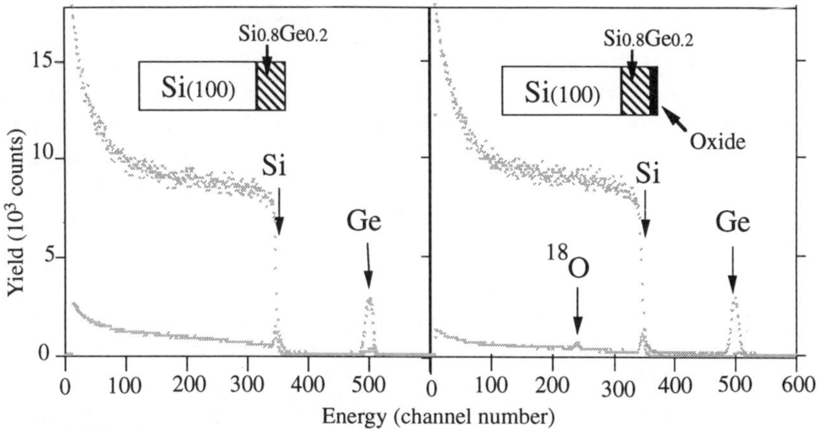

Figure 8.35 2 MeV $^4He^+$ RBS spectra in the ⟨110⟩ channeling direction for unoxidized $Si_{1-x}Ge_x$ and after 200 eV IBO

amorphous $Si_{1-x}Ge_xO_2$ on top of the $Si_{1-x}Ge_x$ heteroepitaxial strained layer after IBO, one would intuitively expect the opposite effect at first, since dechanneling in the amorphous $Si_{1-x}Ge_xO_2$ would increase the minimum yield (Chu *et al.*, 1978, Chap. 8). However, strain relief in the $Si_{1-x}Ge_x$ heteroepitaxial layers is either due to the presence of an oxide cap replacing the free surface on top of the film or to an effective thinning of the $Si_{1-x}Ge_x$ heteroepitaxial layer upon formation of the oxide film. In either case, the steering of the $^4He^+$ beam away from the substrate channel is reduced, since the tetragonal distortion in $Si_{1-x}Ge_x$ heteroepitaxial layers is decreased (Pixraux *et al.*, 1982). The decrease in dechanneling then results in a lower minimum channeling yield in the substrate under the IBO region.

Thermal stability of IBO $Si_{1-x}Ge_x$ dioxides The thermal stability of IBO $Si_{1-x}Ge_xO_2$ is assessed by subjecting $Si_{1-x}Ge_xO_2$ grown with 200 eV $^{18}O_2^+$ to an *in situ* thermal annealed monitored *in situ* XPS. 200 eV IBO $Si_{1-x}Ge_xO_2$ is annealed at 200 °C for 30 min, 300 °C for 30 min, 400 °C for 30 min and finally 500 °C for 2 h. XPS spectra taken at each anneal temperature and after sample cool-down do not detect changes in the composition and chemical state of either Si or Ge. This good thermal stability of $Si_{1-x}Ge_xO_2$ up to 500 °C indicates that $Si_{1-x}Ge_xO_2$ may be able to withstand further device processing, as required for applications in SiGe technology. It should also be noted that cross-section TEM samples were repeatedly prepared on these films over a period of 18 months. No significant changes in microstructure were observed over time.

8.4.2.3.2 IBN of $Si_{1-x}Ge_x$ heteroepitaxial thin films on Si(100)

Stoichiometry of IBN $Si_{1-x}Ge_x$ nitrides When a $Si_{1-x}Ge_x$ heteroepitaxial layer is nitrided by IBN, a chemical shift is observed for the Ge 3d signal at each ion energy investigated (Figure 8.36). This is in marked contrast with IBN of elemental Ge (Figure 8.29, Section 8.4.2.2.2) where no significant chemical shift is observed. But it is quite similar to the effect of Si alloying upon IBO of Ge. Thus the alloying of Si with Ge also enhances the reaction of Ge with nitrogen by IBN. However, in each case a fraction of the Ge 3d signal remains unshifted, indicating that either complete nitridation does not occur or that the nitride film is thinner than the escape depth of Ge 3d photoelectrons in $Si_{1-x}Ge_x$ nitride. Further, the magnitude of the chemical shift of the Ge 3d increases with decreasing ion energy, from 1.9 eV for 1 keV ions to 2.3 eV for 100 eV ions. The latter shift is equal to that observed for Si_3N_4 and indicates a much more significant nitridation than in the case of IBN of elemental Ge where the chemical shift measured at 100 eV was only about 1.33 eV and was the largest shift obtained. In IBN of $Si_{1-x}Ge_x$, the chemical shift increases

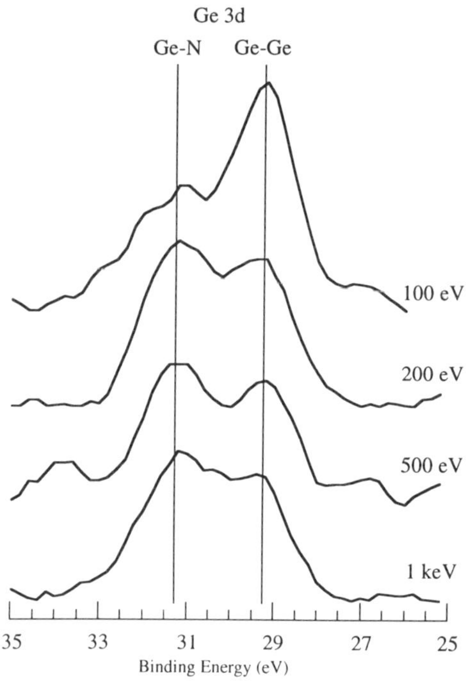

Figure 8.36 XPS spectra of the Ge 3d signal as a function of ion energy for IBN of elemental Ge after signal saturation

with decreasing energy just as for elemental Ge, but is always much larger. This is different from IBO $Si_{1-x}Ge_xO_2$ (Figure 8.31, Section 8.4.2.3.1), where the Ge 3d chemical shift increases with increasing ion energy. The chemical shift $\Delta BE = 2.35$ eV of the Si 2p peak in IBN $Si_{1-x}Ge_x$ nitrides is independent of the ion energy and is equal to that observed for elemental silicon subjected to IBN (Figures 8.16 and 8.17, Section 8.4.2.2).

From these XPS data, Si is clearly preferentially nitrided during IBN of $Si_{1-x}Ge_x$ alloys. The fraction of the shifted Si 2p signal is invariably larger than the fraction of shifted Ge 3d signal. As for pure Si, the near-surface stoichiometry is a function of ion energy. In all cases, the fraction of Si that is nitrided is somewhat lower than for IBN of pure Si at the same energy. This fraction increases from about 70% nitrided Si at 100 eV to almost 90% nitrided Si at 1 keV, while it ranged from 75% to 95% in the case of elemental Si (Figures 8.16 and 8.17). The corresponding fraction of nitrided Ge is about 50%.

The near-surface region is increasingly depleted of Ge with increasing ion energy, according to XPS data, in similar fashion to IBO of $Si_{1-x}Ge_xO_2$ (see Figure 8.32, Section 8.4.2.2). The Ge 3d to Si 2p signal ratio remains nearly constant before and after IBN at 100 eV. As the energy increases, the Ge to Si ratio drops by as much as a factor of two (Figure 8.37(a)). Although the XPS measurement is not rigorously quantitative, the Si 2p and Ge 3d photoelectrons are of similar energy, and thus absorption and fluorescence effects are not expected greatly to affect the ratio of these two signals. Thus, the change in signal ratio is attributed to the preferential removal of Ge from the surface during IBN, either by sputtering or ion-beam-induced decomposition, and this preferential removal increases with energy. A profile of the evolution of the Ge 3d peak as a function of deposition time is shown in Figure 8.38. It is important to note here the relatively large error ($\pm 15\%$) in the measurement of the fraction of nitrided Ge. Because of the initial small fraction of Ge ($x = 0.2$) in the $Si_{1-x}Ge_x$ films and the preferential removal of Ge from near the surface, the deconvolution of such weak signals carries a large error.

Chemically enhanced sputtering during IBN of $Si_{1-x}Ge_x$ The observed depletion of Ge from the surface of $Si_{1-x}Ge_x$ corresponds to a preferential sputtering rate of Ge by nitrogen twice as large as the sputtering rate of unreacted Ge. However, TRIMCSR calculations using surface binding energies (E_{SBE}) 2.8 eV for Si and 1.8 eV for Ge, based on measurements for elemental phases (Ghose *et al.*, 1983; Chereckdjian and Wilson, 1986) (Figure 8.37(a)), gives comparable sputtering yields for Si and Ge in the low-energy regime used in IBN (Ghose *et al.*, 1983; Chereckdjian and Wilson, 1986). Further, the sputtering of $Si_{1-x}Ge_x$ alloys by noble-gas ions does not lead to enrichment of the surface by either Si or Ge (Betz and Wehner, 1983). This is evidence that in binary $Si_{1-x}Ge_x$ alloys, sputtering is only physical sputtering, and that the binding

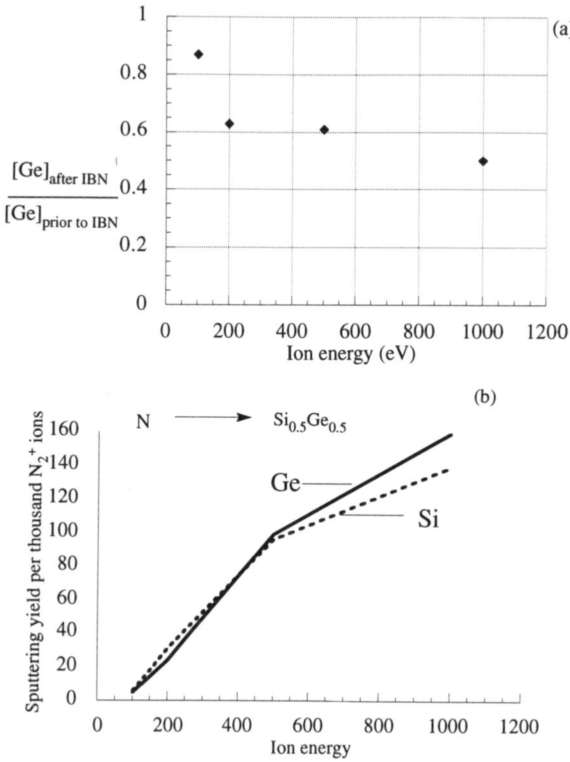

Figure 8.37 Retention of Ge after IBN of $Si_{1-x}Ge_x$ as a function of IBN energy, as measured by XPS. The fraction of Ge remaining in the film after IBN is normalized to the initial fraction present in the film (0.2)

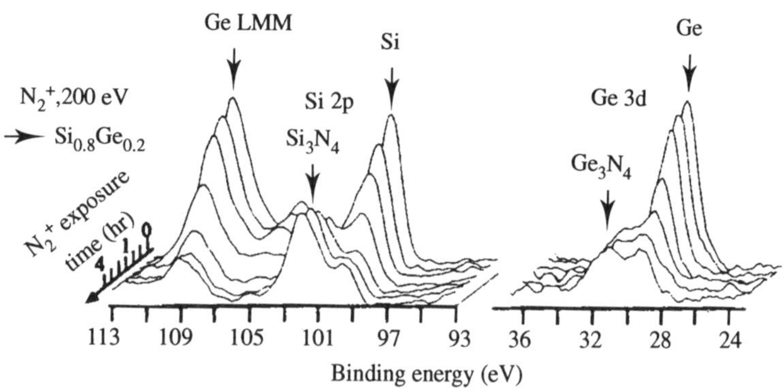

Figure 8.38 Time-resolved XPS spectra of IBN of $Si_{1-x}Ge_x$ with 200 eV N_2^+ as a function of exposure time to N_2^+

energies for elemental Si and elemental Ge can be applied to the calculation of sputtering of binary $Si_{1-x}Ge_x$ alloys.

The observation of preferential sputtering from $Si_{1-x}Ge_x$ alloys during IBN leads to the conclusion that the mechanism for sputtering is not controlled merely by physical sputtering, but that there is also a chemical effect involved in the preferential sputtering of Ge during IBN. That is, in bonding with nitrogen near the surface, the sputtering yield of Ge is enhanced over the sputtering yield of Si.

This increase in the sputtering yield of Ge can be attributed to two factors. The first is *chemical sputtering*: molecules containing Ge are formed by the nitrogen near the surface. If these molecules have a low sublimation energy, they evaporate easily, leading to a chemically included desorption of nitrided Ge from the surface. This is a two-step process: the ion-induced formation of volatile molecule followed by thermal desorption of that molecule.

The second possibility is *chemically enhanced physical sputtering*. In this case, the chemical environment of Ge at the surface is modified by the presence of nitrogen in such a way that the binding energy of Ge is lowered, or that a Ge–N complex with a low binding energy is formed. The preferential removal of Ge from the surface is then a single-step mechanism due to sputtering enhanced by the chemical conversion of the surface to a nitride.

8.4.2.4 IBS on GaAs(100)

8.4.2.4.1 IBO of GaAs(100)

Ion-beam oxidation of GaAs at room temperature is investigated as a function of O_2^+ energy between 100 eV and 8 keV, in three sets of experiments (Vancauwenberghe *et al.*, 1991c). Below 1 keV and at normal incidence, a thin oxide film is formed: it is composed of Ga_2O_3 and As_2O_3, and exhibits insulating properties. As the ion energy increases, preferential sputtering of As and decomposition of As_2O_3 increase and prevent the formation of an insulating film. No damage in the substrate is detected by RBS combined with ion channeling for IBO below 1 keV.

Experimental procedure for IBO of GaAs(100) The same Czohrkralski-grown, low-resistivity GaAs(100) 3″ substrate is used for all experiments by cleaving it into small pieces. In the first experiment, IBO of GaAs is explored for ion energies between 1 keV and 8 keV using a VG duoplasmatron ion source (VG DP 2001/C) with *in situ* SIMS analysis. In this case the ion beam does not have a normal incidence but impinges at 30° off the normal. The O_2^+ ions are extracted at 10 keV and decelerated by biasing the target. For the experiment at 500 eV, an electron-impact sputtering ion gun (Perkin-Elmer 04-303) is used to produce the IBO film, with *in situ* scanning Auger microscopy (SAM). The O_2^+ ions are produced at 1 keV and decelerated to 500 eV at the target at normal incidence. The samples are sputter-cleaned *in situ* prior to IBO with 4 keV Cs^+ in the SIMS, and O_2^+ in the SAM, for a duration between 12 s and 1 min with a beam incidence of 30° to favor sputtering over incorporation. The duration is optimized by experimentation prior to IBO and removes surface carbon and oxygen, while maintaining the GaAs surface as stoichiometric as possible. The first two IBO experiments are also conducted in a UHV environment (5×10^{-9} torr during IBO). With the ion sources used above, ions are predominantly O_2^+. The ion current density ranges between 10 and 20 μA cm^{-2}. In each experiment, the same total dose of O_2^+ is used: 10^{17} ions/cm^2. The samples are characterized by *ex situ* XPS, SEM and RBS combined with ion channeling.

The third set of experiments is conducted in the CIMD system (Figure 8.10) in the conditions described in Section 8.4.2.1–3. A dip etch of 30 s HF:ethanol solution removes carbon but oxygen contamination is still high (0.25–0.3 ml) and mostly bonded to As in As_2O_3, as was also observed for Ge (Section 8.4.1.1). This is probably due to a less efficient and less stable passivation of the GaAs surface by H-termination. Nevertheless, the dip etch provides a stoichiometric GaAs surface. *In situ* XPS is used to monitor the phase formation as a function of exposure to O_2^+. IBO is terminated when oxygen incorporation saturates. O_2^+ energy ranged between 100 eV and 1 keV in this last set of experiments to investigate the role of ion energy on GaAs-oxide formation by IBO.

Stoichiometry of IBO GaAs oxides XPS analysis of the GaAs IBO layers is more complex than that of Group IV semiconductors. Several photoelectrons as well as Auger electrons contribute to the XPS spectrum. Several oxide phases can be formed and identified. The peak position and chemical shift in the oxide phases for the 3d photoelectrons and L3M4sM4s Auger electron signals of both Ga and As are summarized in Table 8.3 (Schwartz *et al.*, 1979). The chemical shift in the Ga 3d signal between GaAs and Ga_2O_3 is fairly small ($\Delta BE = 1.0$ eV). For very thin oxides, this chemical shift is even smaller (Grunthaner *et al.*, 1979), as found in thin thermal SiO_2 (Grunthaner and

Table 8.3 Binding energies of the 3d photoelectrons and L3M45M45 Auger electrons for Ga and As in GaAs, and the chemical shifts observed by XPS for the known oxides of GaAs (Schwartz *et al.*, 1979)

	Ga 3d (eV)	As 3d (eV)	Ga LMM (eV)	As LMM (eV)
GaAs	19	41	$h\nu - 1062$	$h\nu - 1225.6$
Ga_2O_3	+1		+ 4.9	
As_2O_3		+3.5		+ 5.8
$GaAsO_4$	+1.9	+4.7		
As_2O_3		+4.7		

Grunthaner, 1986) and thin IBO oxides (Section 8.4.2.1.1, Table 8.1, Section 8.4.2.3.1, Table 8.2). This makes the contribution of oxidized and unoxidized Ga in the Ga 3d signal difficult to deconvolute. Fortunately, the shift in the Ga LMM Auger signal is quite significant ($\Delta BE = 4.9$ eV) and easy to deconvolute. There is no such difficulty for As and its oxides since the chemical shift in the 3d signal is large enough ($\Delta BE = 5.8$ eV) to resolve the different contributions easily.

The results of the first two sets of experiments are summarized in Figure 8.39. The first spectrum in Figure 8.39 is taken on virgin GaAs with a native oxide intact on the surface. The three other spectra (Figure 8.39) correspond to oxidized GaAs with three different ion energies: 500 eV at normal incidence, and 1 keV and 3 keV at 30° incidence. By comparing the binding energy in Figure 8.39 with the values in Table 8.3, the oxidized states of Ga and As are found to be Ga_2O_3 and As_2O_3, respectively, with no detectable presence of $GaAsO_4$ or As_2O_5. At 500 eV, a dramatic difference is observed between the native oxide and the IBO film: in the IBO film, Ga is completely oxidized into Ga_2O_3, with no Ga bonded to As, while most of the As (80%) is oxidized into As_2O_3. As the ion energy increases and the incidence angle moves away from

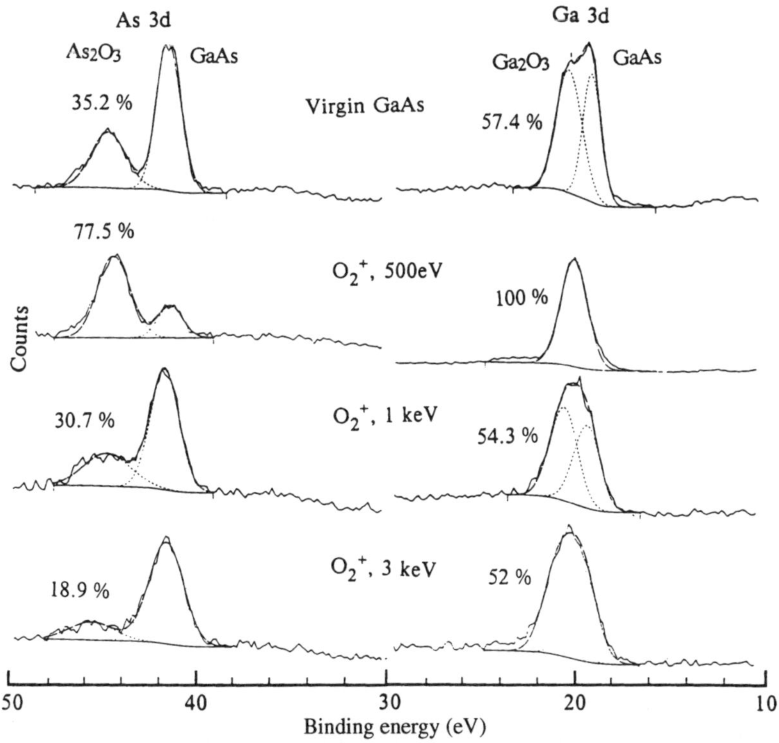

Figure 8.39 As 3d and Ga 3d XPS signal for virgin GaAs, 0.5 keV IBO at normal incidence, 1 keV IBO and 3 keV IBO with O_2^+ at 30° incidence

normal incidence, the oxidation decreases: the amount of oxidized Ga drops to 50% while the As_2O_3 content drastically decreases to 30% at 1 keV and 20% at 3 keV. These results demonstrate that decomposition of As_2O_3 increases with increasing ion energy and off-normal incidence, preventing the formation of a GaAs-oxide. IBO at low energy (< 500 eV) and at normal incidence favors the oxidation of As over its removal as well as the formation of a more stoichiometric GaAs oxide. This is also confirmed by a more accurate study of GaAs IBO performed with *in situ* XPS analysis.

IBO of GaAs in the CIMD system (Section 8.4.1.2, Figure 8.10) of GaAs is performed with $^{18}O_2^+$ ions at energies ranging from 100 eV to 1 keV (Figure 8.40). A strong dependence of phase formation and stoichiometry is measured by *in situ* XPS. When the ion energy decreases, the fraction of oxidized Ga decreases in the Ga 3d signal and in the Ga LMM Auger lines (Figure 8.41). This suggests that the oxide films are thinner at lower energies, as for IBO of Group IV semiconductors (Section 8.4.2.1, Figure 8.12, 8.4.2.2, Figure 8.27 and 8.4.2.3, Figure 8.31). On the other hand, the fraction of oxidized As decreases significantly with increasing ion energy (Figure 8.40). Similarly, the Ga : As ratio

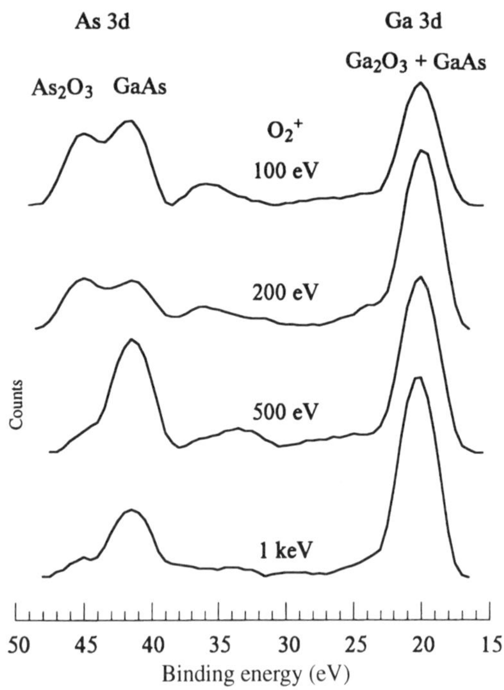

Figure 8.40 As 3d and Ga 3d XPS signal for 100 eV, 200 eV, 500 eV and 1 keV IBO with O_2^+

(Figure 8.42) increases dramatically during IBO at higher energies. Arsenic is depleted, especially in the oxide, which can be due to chemically enhanced sputtering of As through As_2O_3 formation, similar to IBN of $Si_{1-x}Ge_x$ (Section 8.4.2.3). At 100 eV, As oxidizes significantly (Figure 8.40) and the Ga : As ratio in the oxide layer (Figure 8.42) is close to 1, indicating the formation of a more stoichiometric and uniform oxide at 100 eV.

Insulating properties and substrate damage The insulating nature of the oxide films is assessed by SEM observations: IBO GaAs oxides charge electrostatically when grown at energies below 500 eV, while above 500 eV, charging is weak or not observed. Optically, oxides grown below 500 eV are not visible, while those grown with $E_{ion} > 500$ eV are easily observed.

RBS combined with ion channeling provides information about film formation and damage in the underlying substrate. The surface peak of GaAs increases with decreasing energies from 3 keV to 500 eV, indicating the formation of an amorphous oxide film at lower energies (≤ 500 eV). However, the depth resolution is not sufficient to measure its thickness with a reasonable degree of confidence. The minimum yield (χ_{min}) in the substrate is observed to decrease with decreasing energy. Below 1 keV, it is equal to the χ_{min} of GaAs(100) or 4%. This establishes that low IBO energies (<500 eV) not only minimize sputtering and oxide decomposition, but also avoid defect creation in the underlying substrate. These observations are very similar to those made for IBO of $Si_{1-x}Ge_x$ in Section 8.4.2.3. The concentration of the more volatile element (Ge in $Si_{1-x}Ge_x$ and As in GaAs) decreases with increasing ion energy (compare Figure 8.42 with Figure 8.32 in Section 8.4.2.3) and damage appears at higher energies only, as by comparing Figure 8.33(a) (200 eV IBO) with Figure 8.33(b) (1 keV IBO) in Section 8.4.2.3. The evolution of composition in reduced elements is similar to that of IBN of $Si_{1-x}Ge_x$ (see Figure 8.37 in Section 8.4.2.3).

Effect of the ion-beam angle of incidence upon stoichiometry Alay and Vandervorst have extensively investigated oxidation of Si and III–V compounds (including GaAs, AlGaAs and InGaAs) by IBO using *in situ* XPS, at energies ranging between 1 and 5 keV, and at angles of incidence ranging from normal incidence to 55° off normal (Alay and Vandervorst, 1992a, 1992b; Alay, 1993; Vandervorst, 1993). They show through this systematic study that as the angle of incidence decreases, the formation of substoichiometric oxides increases at the expense of fully reacted oxides during IBO for Si, GaAs and AlGaAs (Figure 8.43). This is in agreement with the discussion presented in Section 8.3.2.2. Alay and Vandervorst also show that in ternary compounds such as AlGaAs, Al oxidizes first, then Ga, and lastly As. Their results also confirm that lower energies favor more uniform oxidation and higher retention of volatile elements.

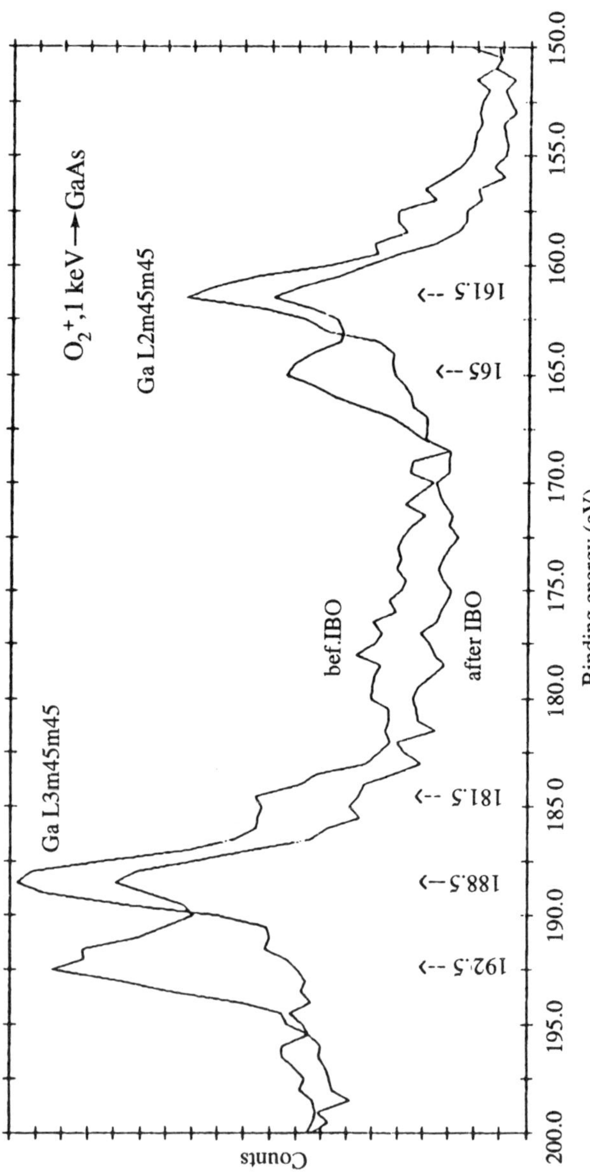
O_2^+,1 keV → GaAs
Ga L2m45m45
Ga L3m45m45
bef.IBO
after IBO
161.5 -->
165 -->
181.5 -->
188.5 -->
192.5 -->
150.0
155.0
160.0
165.0
170.0
175.0
180.0
185.0
190.0
195.0
200.0
Binding energy (eV)
Counts

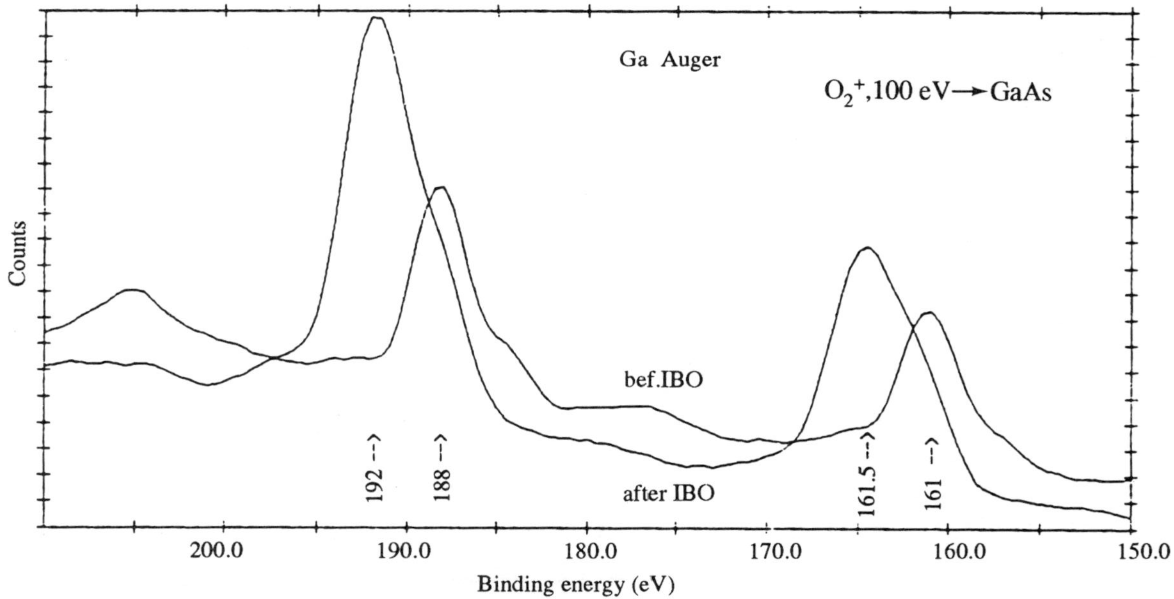

Figure 8.41 Ga $L_2M_{45}M_{45}$ and Ga $L_3M_{45}M_{45}$ Auger signals for IBO of GaAs at 100 eV and 1 keV

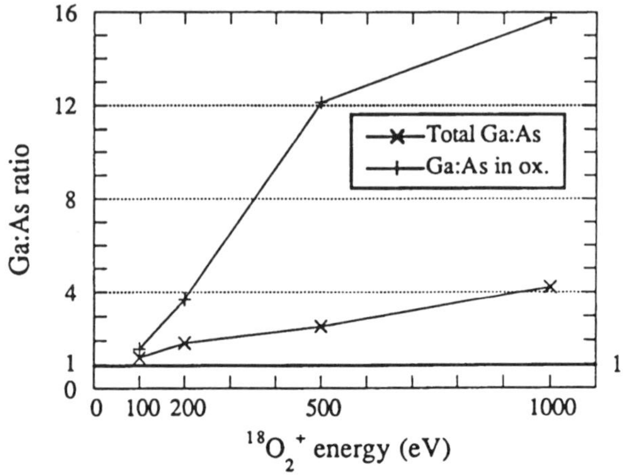

Figure 8.42 Ga:As ratio after IBO as a function of IBO energy measured by XPS from the Ga 3d and As 3d signals

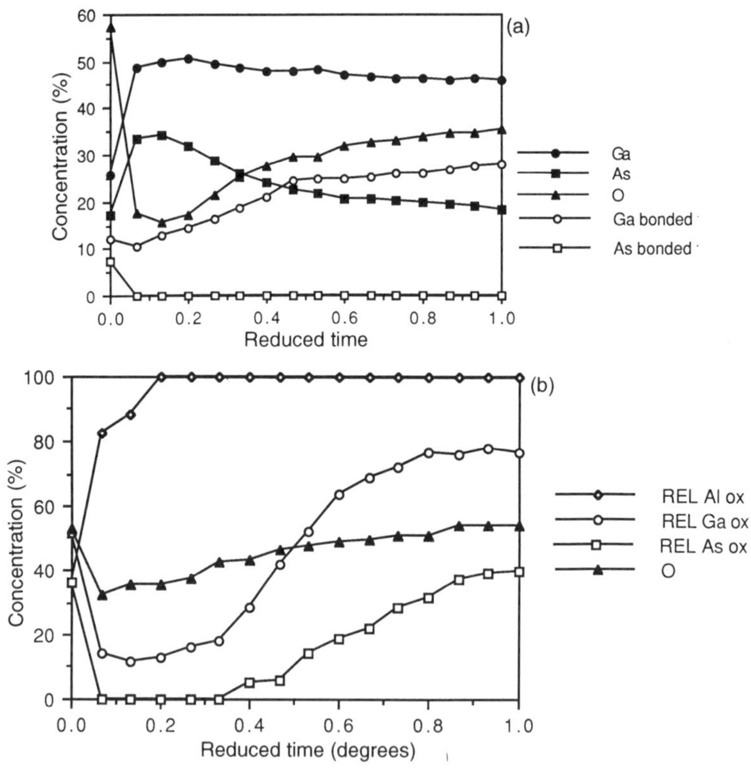

Figure 8.43 (a) and (b). XPS time-resolved profile of O_2^+ 5 keV IBO of GaAs (a) and AlGaAs (b), giving the relative magnitude of unoxidized and oxidized Ga and As 3d, and O as a function of reduced time (=time/saturation time) in (a), and the relative concentration of oxidized Al, Ga and As 3d in (b)

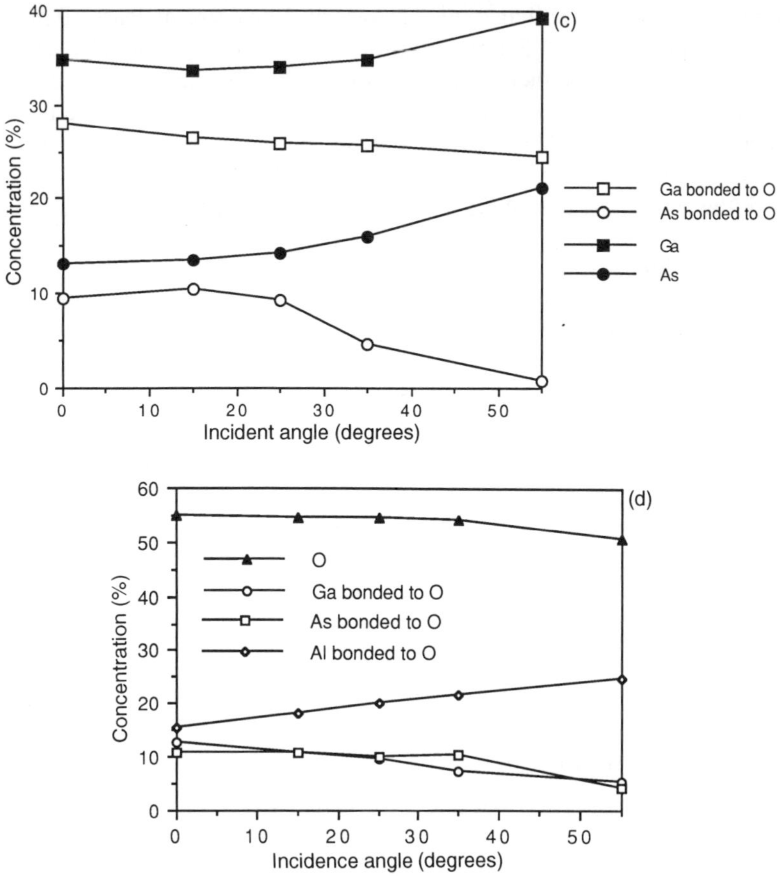

Figure 8.43 (c) and (d). XPS signal as a function of the ion beam angle of incidence with respect to the surface formal for O_2^+ 1 keV IBO after XPS signal saturation for GaAs (c) and AlGaAs (d), giving the relative magnitude of unoxidized and oxidized Ga and As 3d, and O in (c), and the relative concentration of oxidized Al, Ga and As 3d in (d). (From Alay and Vandervorst (1992a))

Reproducibility of the composition of IBO oxides of GaAs While results in Alay and Vandervorst (1992a, 1992b), Alay (1993) and Vandervorst (1993) are all in good qualitative agreement with those presented here, the absolute amount of oxidation and the relative concentration of As retained after IBO in GaAs differ, while results for IBO for Si are always in perfect quantitative agreement with those shown in Section 8.4.2.1. The origin of this discrepancy for GaAs probably results from differences in initial surface-preparation procedures, which strongly affect As stoichiometry. However, no systematic study of the effect of surface cleaning of GaAs upon IBO has yet been conducted. Alay and

Vandervorst also show that oxides grown at 1 keV are at least 10 nm thick, demonstrating that the IS (implantation and sputtering) model discussed in Section 8.3.3.1 is not sufficient to account for the thickness observed, but that an IRSD model, which includes diffusion (Section 8.3.3.3) has to be used. This in turn makes IBO of GaAs dependent on the GaAs defect chemistry, which controls diffusion. Since defect chemistry depends upon the conditions used during the crystal growth of GaAs and GaAs surface processing and polishing, fluctuations in the composition of the IBO oxides prepared by different sources can be attributed to the variability of GaAs substrates.

In summary, observations for IBO of GaAs are similar to the results obtained for IBO and IBN of $Si_{1-x}Ge_x$ with respect to the dependence of oxide stoichiometry and Ga/As ratio upon incoming ion energy. Substrate damage appears in both cases at energies of 1 keV (Figure 8.33(a) and (b), Section 8.4.2.3). It should be noted that HRTEM (Figure 8.33(a) and (b) is a more sensitive technique for the detection of damage than ion channeling: the amount of damage detected by HRTEM is of the order of 10^{12} displaced atoms per cm^3 while ion channeling can detect only about 10^{17} displaced atoms per cm^3 (Figure 8.35 in Section 8.4.2.3). Since only ion channeling is used to characterize damage in GaAs by IBO, a low level of defects may be present in the substrate as well as electronic damage. Further studies are necessary to investigate the presence of low-level damage.

8.4.2.4.2 IBN of GaAS(100)

DeLouise (1991a, 1991b, 1992) has investigated IBN of GaAs as a function of nitrogen energy using N_2^+ between 0.5 and 3 keV. The incorporation of nitrogen increases with decreasing energy, in the same fashion as in IBN of elemental Ge (Section 8.4.2.2, Figure 8.29) and for Ge in Si_xGe_{1-x}. (Section 8.4.2.3, Figures 8.36 and 8.37). However, the more volatile As is not nitrided at all in GaAs. This is contrary to the behavior of the more volatile element (Ge) in Si_xGe_{1-x} where alloying with Si enhances nitridation. In IBN of GaAs, only Ga is nitrided: DeLouise observes no shift in the As 3d (binding energy = 40.7 eV) while a ΔBE of 2 eV should occur if an arsenic nitride is formed (Gourrier *et al.*, 1983). The chemical shift for Ga 3d photoelectrons when Ga bonds to N (Ga–N binding energy = 19.5 eV) is found to be ΔBE = 0.6 eV with respect to the Ga–As binding energy for 3d electron whose binding energy BE = 18.9 eV. This is also quite different from IBO of GaAs where both As and Ga oxidize when energy decreases (Section 8.4.2.4.1, Figures 8.39, 8.40, 8.42 and 8.43). DeLouise also established that the stoichiometry of the nitride is independent of IBN energy. This has not been established for IBO of GaAs, different oxidation states are too difficult to separate (see Table 8.3): $GaAsO_4$ is virtually indistinguishable from a mixture of GaAs and As_2O_5.

8.5 COMPARISON BETWEEN IBS EXPERMENTAL RESULTS AND IBS MODELS

IBS experiments and models are compared in the following manner. Each experiment is first compared to the simplest predictive model, the IS (implantation and sputtering) model (Section 8.3.3.1). If good agreement is found, then IBS kinetics is identified with the two-step mechanism of the IS model. If not, the results are compared to ISR simulations which include additional effects such as relaxation, relocations and replacements, in addition to implantation and sputtering (Section 8.3.3.2). Finally, if there is still a discrepancy between the data and ISR, the ISRD model, which includes reactivity and diffusion, is used, if point-defect diffusivities and reactivity rates are known or can be estimated. Using this methodology, the rate-limiting steps can be identified. For instance, one can determine whether thermal activation plays a role or not in the kinetics.

8.5.1 COMPARISON BETWEEN IS SIMULATION RESULTS AND EXPERIMENTS

8.5.1.1 IS prediction of ion-energy dependence for IBO of Si(100)

8.5.1.1.1 Film thickness

The IS model accounts for the three stages observed experimentally during IBO of Si: (i) formation of substoichiometric oxides, (ii) conversion into SiO_2, and (iii) saturation of IBO growth rate (Figure 8.44). Due to volume expansion upon oxidation, significant surface growth occurs until saturation (Figure 8.45). The balance between sputtering at the surface and volume expansion due to oxidation of the substrate results in a smaller erosion rate of the surface. This effect is most significant with higher ion energies. The time to reach saturation should increase with ion energy, as experimentally observed. This can be accounted for by the increase of range straggling of the oxygen profile with ion energy. Lastly, the oxide thickness increases with ion energy, both experimentally and in the IS model. However, the calculated thicknesses are much lower than the experimental ones (Figure 8.45). This is in contrast to the model of Todorov and Fossum (1987) where a good correlation was found between calculated and experimental thicknesses for ion energies between 60 and 100 eV. However, the implantation profiles in Todorov and Fossum (1987) were based on analytical formulas that are known to overestimate the range of the ions, especially at low energy. In the case of IS, the range is more accurately simulated through the Monte Carlo method used in the TRIM program (Section 8.3.3.1).

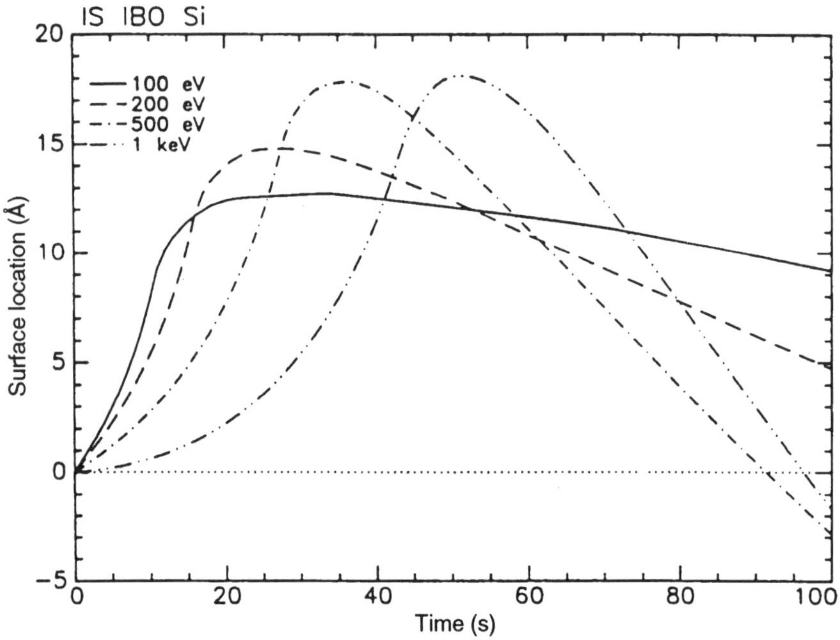

Figure 8.44 Oxide thickness (Å) as a function of time as calculated by IS for IBO of Si between 0.1 and 1 keV

It should be emphasized that the TRIM version used for the present IS model follows all recoils in the collision cascade, which is much more accurate than relying on the Kinchin–Pease approximation used in PC versions of TRIM.

8.5.1.1.2 Interface width

Another important point is that the interface width calculated with IS for 100 eV IBO is 0.8 nm, i.e. 25 monolayers thick, and increases with ion energy (Figure 8.46). The interface width calculated by IS is determined solely by the range straggling of the ions. But the interface width measured experimentally by HRTEM for 100 and 200 eV is less than 0.2 nm, which is less than one monolayer thick or 'atomically' sharp (Figure 8.13).

In summary, the rather large discrepancy between SiO_2 thicknesses observed between experiments and IS calculations indicates that the IBO of Si is not governed by the competing processes of implantation and sputtering, but by other mechanisms.

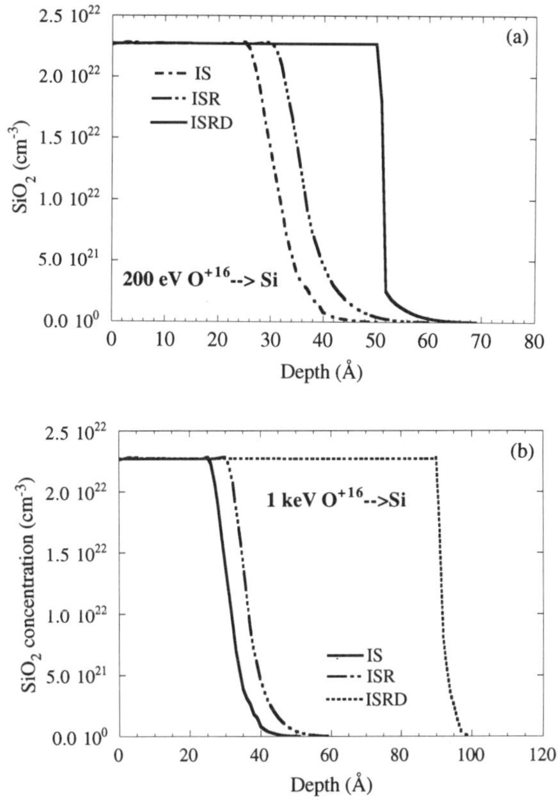

Figure 8.45 Comparison between the SiO_2 concentration profile computed by IS, ISR, ISRD for (a) 200 eV O_2^+ and (b) 1 keV O_2^+. The interface width computed for the 200 eV O_2^+ is 'atomically' sharp while the widths computed by IS and ISR are wider than the widths found experimentally

8.5.1.2 IS prediction of ion-energy dependence for IBO and IBN of elemental Ge

Conclusions drawn from the comparison between IS simulations and experimental data for IBO of elemental Ge are quite different from the conclusion drawn for IBO of Si. The main difference resides in more pronounced sputtering effects. Significant erosion of the surface is predicted by IS at all energies for both IBN and IBO of elemental Ge, even though the volume expansion upon GeO_2 formation ($\Delta V/V = 2.33$) is slightly larger than upon Si oxidation ($\Delta V/V = 2.2$). At low energy (100 eV), a very thin but continuous layer of stoichiometric GeO_2 can be formed (Figure 8.47(a)), while at higher energy (1 keV), the high sputtering rates of both Ge and GeO_2 prevent the formation

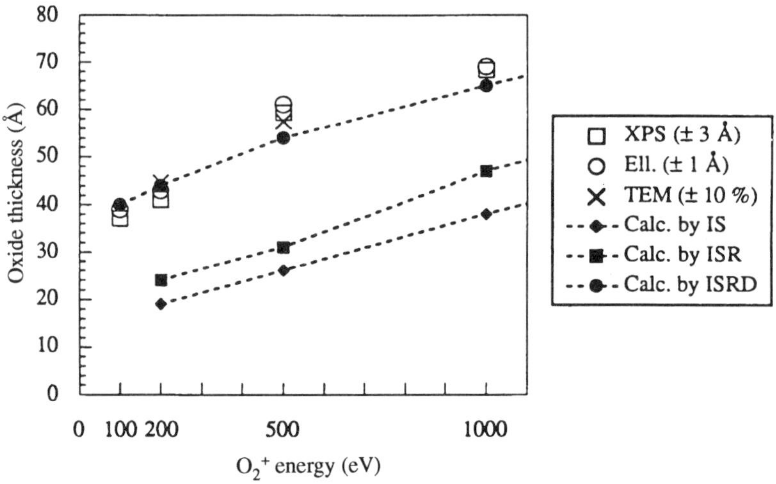

Figure 8.46 Comparison between IBO SiO_2 thickness dependence upon IBO energy as measured by XPS (□), ellipsometry (○), cross-section TEM (×) and as simulated by IS (◆), ISR (■) and ISRD (●). Only the latter agrees with the experimental data

of a continuous layer, although there is some incorporation of O (Figure 8.47(b)). This agrees qualitatively with the experimental dependence upon ion energy of IBO of Ge. The IS model, although relatively simple, thus provides insights into the ion-energy dependence of phase formation during IBO of Ge. The same conclusions can easily be reached for the case of IBN of elemental Ge.

8.5.1.3 IS predictions for the thickness dependence upon ion energy for IBO of $Si_{0.8}Ge_{0.2}$

8.5.1.3.1 Ge/Si fraction

Due to the large content of Si in the Si_xGe_{1-x} alloys used here, both experimental and IS simulation results for IBO of $Si_{0.8}Ge_{0.2}$ are closer to those for IBO of elemental Si than of Ge. However, the presence of Ge, even at a relatively low concentration, alters significantly the simulation results compared with IBO of Si(100). The predicted oxide thicknesses are smaller due to the smaller ion range in the alloy layer. Also, the surface erosion rates are larger due to the large sputtering yields. A uniform $(Si_{0.8}Ge_{0.2})O_2$ film (Figure 8.47(a)) is predicted for all energies, even at 1 keV. This is in qualitative agreement with the experimental observation that the oxidation of Ge in the $Si_{1-x}Ge_x$ layer is enhanced as compared to the case of Ge. However, the variation with depth of the Ge/Si fraction at 1 keV in $Si_{0.8}Ge_{0.2}O_2$ is only predicted at the very surface and results from the difference in sputtering rates between Si and Ge.

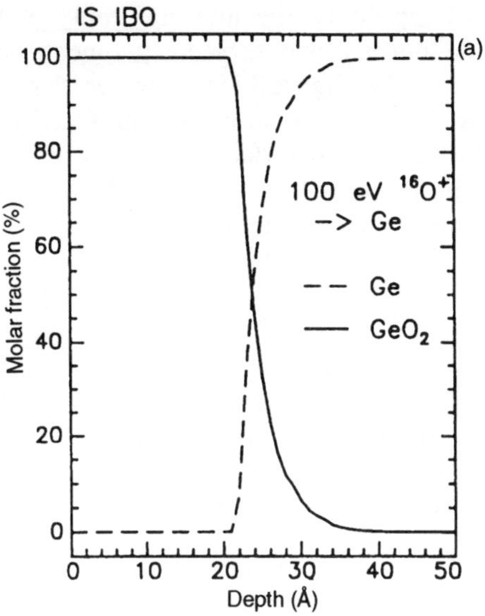

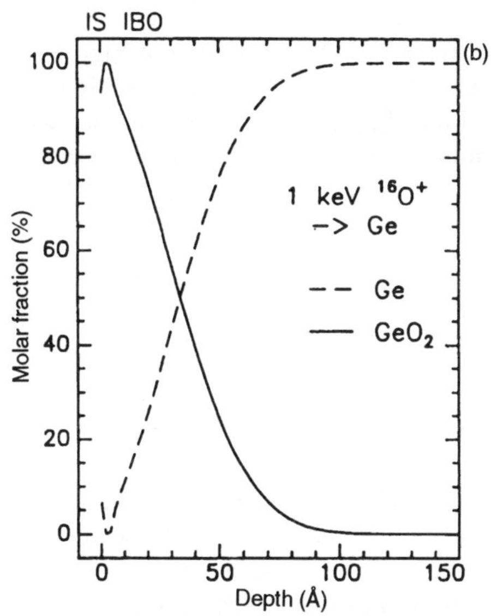

Figure 8.47 (a) Ge oxide concentration profile as calculated by IS for IBO Ge at 100 eV, depicting the growth of thin GeO_2 layer, as found experimentally. (b) Ge oxide concentration profile as calculated by IS for IBO Ge at 1 keV, showing no net growth of thin GeO_2 layer, as found experimentally

IS simulations do not reproduce the non-uniformity of the Ge/Si fraction throughout the dioxide film that is observed experimentally by XPS for 'high' IBO energies (Section 8.4.2.3.1, Figure 8.31). However, as pointed out in Section 8.4.2.3.1, it is not clear whether this non-uniformity is due to fluctuations of composition within the film, or to surface roughness of the original Si_xGe_{1-x} heteroepitaxial thin film (Figures 8.32 and 8.34). Further experimental study with well-controlled smoothness of the original Si_xGe_{1-x} heteroepitaxial thin film is necessary before this point can be addressed.

8.5.1.3.2 Thickness

While the thickness fluctuations of $Si_{0.8}Ge_{0.2}O_2$ films when roughness or surface defects are present affect the reproducibility of thickness as a function of ion energy, the width of Si_xGe_{1-x} dioxide films at 1 keV is found to be at least 7.5 nm on average, which is close to that for Si at the same energy. The thickness predicted by IS for that energy (6.5 nm) is lower than that value, but the discrepancy is less dramatic than for IBO of elemental Si.

8.5.1.3.3 Prediction on interface width

The interface widths predicted by IS simulations are much wider (2 nm at 1 keV in Figure 8.48) than those measured experimentally by HRTEM (less than 0.2 nm in Figure 8.33(a) for 200 eV and 1.2 nm in Figure 8.33(b) for 1 keV). The simulated interface widths are mainly determined by ion straggling, which is an effect of implantation, while the measured interface widths are significantly narrower and thus must be due to another effect. The shortcomings of the IS model are thus similar for IBO of Si_xGe_{1-x} as for IBO of elemental Si.

8.5.1.4 IS prediction dependence upon ion energy for IBO of GaAs

Alay and Vandervorst (1992a) have shown by XPS that the thickness of IBO GaAs oxides exceeds that expected from the range and range straggling of the O^{2+} ions, and that the simple IS model is thus inadequate. Also, the IS model cannot take into account chemical effects in sputtering, such as chemically enhanced physical sputtering, as pointed by Vancauwenberghe *et al.* (1991c). The stoichiometry of the compound IBO oxides cannot be predicted by IS, as physical sputtering cannot account for the composition observed in Vancauwenberghe *et al.* (1991c) and Alay and Vandervorst (1992a). Present work by Alay and Vandervorst (Alay, 1993) aims at applying the IRSD model developed by Vancauwenberghe *et al.* (1992a) for IBO of Si and Si_xGe_{1-x} alloys to the case of GaAs and AlGaAs.

8.5.1.5 Conclusions on the accuracy of the IS model

In summary, the IS model is a simple model which gives insights into the mechanism of phase formation resulting from the competition between oxidation and sputtering. It successfully predicts the kinetics of IBO and IBN of elemental Ge. However, it cannot account for the thicknesses observed experimentally for Si and GaAs and the interface widths and composition of IBO oxides of Si, Si_xGe_{1-x} and GaAs.

8.5.2 THE ROLE OF ION-BEAM MIXING

The ISR model takes into account replacement and relocation events. These events correspond to ion-beam mixing and lead to mass transport over several atomic distances. The light elements are expected to move deeper into the substrate while heavier elements accumulate at the surface. However, differences in displacement (E_d) and surface binding (SBE) energies may affect this conclusion. The ISR model requires a dynamic version of TRIMCSR in order to update the amount of ion incorporation and the vacancy, interstitial and replacement/relocation distributions as the substrate is exposed to increasing doses of oxygen. The same assumptions as in the IS model are made, however, regarding the oxidation process itself: each oxygen species is bonded to the available substrate atoms and the excess oxygen diffuses to the surface and desorbs. ISR is used to simulate IBO of Si and $Si_{0.8}Ge_{0.2}$ to examine the effect of the ion-beam mixing upon composition, thickness and interface width.

8.5.2.1 Dependence upon ion energy for IBO of elemental Si

8.5.2.1.1 Film thickness

The main effect of ion-beam mixing in IBO of elemental Si is a significant increase of SiO_2 thicknesses, i.e. by about 20% (Figure 8.45). Some of the implanted oxygen is recoiled deeper into the substrate by new incoming oxygen ions. The calculated thicknesses are, however, still significantly smaller than the observed thicknesses, indicating that other processes are taking place, such as oxygen diffusion.

8.5.2.1.2 Interface width

The interface width predicted by ISR is still about 1 nm thick (Figure 8.45) and thus well below the thickness observed experimentally in Figure 8.13(a). Ion-beam mixing thus does not account for the sharpness of IBO SiO_2 interfaces.

8.5.2.2 Dependence upon ion energy for IBO of $Si_{0.8}Ge_{0.2}$

8.5.2.2.1 Ge/Si fraction

As in the case of IBO of Si, oxide thicknesses increase with respect to IS simulations (by 11%). The composition of the SiGe-dioxide now varies with depth for 'high' IBO energies such as 1 keV (Figure 8.48). Ion-beam mixing leads to significant redistribution of Si and Ge in the oxide. The surface region is depleted in Ge, while deeper in the oxide there is an enrichment in Ge. Such a redistribution is not what is intuitively expected from simple mass-effect considerations. It is due to the higher sputtering rate for Ge and GeO_2 and the lower displacement energy for Ge (13 eV compared to 15 eV for Si). The depth non-uniformity increases in magnitude as the ion energy increases. ISR also predicts that the net amount of Ge in the SiGe-dioxide decreases with increasing ion energy. This correlates qualitatively with the experimental observation of Ge retention after IBO of $Si_{1-x}Ge_x$ films with a low defect density (Figure 8.32, Section 8.4.2.3). In defective heteroepitaxial $Si_{1-x}Ge_x$ films, the increased loss of Ge is attributed to defects, not accounted for in the ISR model.

8.5.2.2.2 Interface width

The interface widths, however, are still very significantly different from these observed by HRTEM (Figure 8.33): 0.4 nm at 100 eV instead of atomic sharpness (<0.2 nm) at 1 keV and 2 nm (Figure 8.33a) instead of 1 nm (Figure 8.33b). This points out that in heteroepitaxial $Si_{1-x}Ge_x$, just as in the case of Si, the interface sharpness must be determined by diffusion and reaction rather than atomic-collision effects.

8.5.2.3 Conclusions

Several experimental features of IBO can thus be accounted for only by considering the ion–solid interactions, namely ion incorporation, sputtering and ion-beam mixing effects. However, there is poor agreement between the experimental thicknesses and interface widths of IBO of SiO_2 and of $Si_{0.8}Ge_{0.2}$-dioxide and the computed ones. Oxygen diffusion appears thus to play a significant role in the IBO process. One might wonder if ion–solid interactions and ion-implantation profiles calculated by TRIMCSR are accurate. The universal potential has been proven adequate to describe the ion–solid interaction down to low energies (Tenner, 1986), while the BCA only breaks down below 50 to 100 eV. The only probable source of error in the TRIMCSR calculations is channeling effects which can be significant at very low ion energies due to increasing collective effects on the ion trajectory (Dodson, 1987). However, these channeling effects should decrease rapidly with increasing ion energy. The deviation between the computed energy dependence of oxide

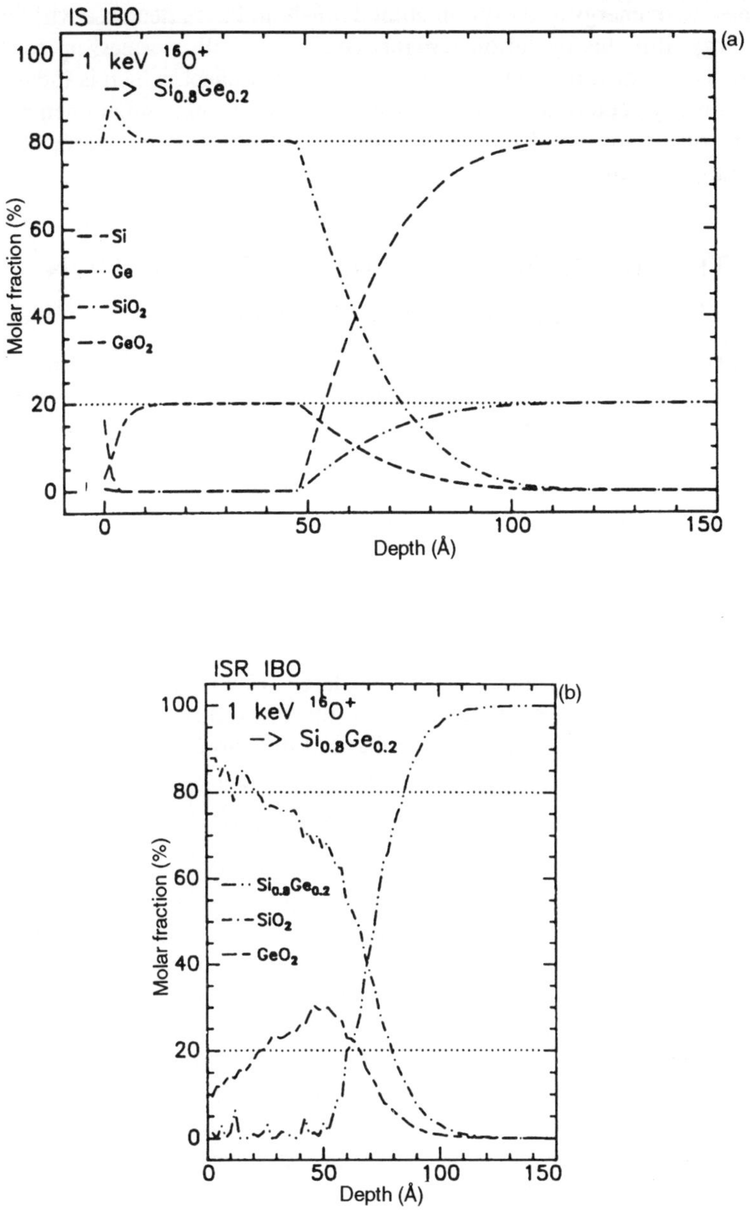

Figure 8.48 (a) Si, Ge and oxide profiles calculated by IS for 1 keV IBO of $Si_{1-x}Ge_x$ showing a uniform composition, which is not found experimentally at 1 keV, but only at 200 eV and below. (b) Si, Ge and oxide profiles calculated by ISR for 1 keV IBO $Si_{1-x}Ge_x$ showing nonuniform Ge distribution as found experimentally. The interface width is at least 2 nm, though, which is larger than found experimentally in Figure 8.33(b)

thickness with energy and experimental data should then decrease with increasing energy. But this deviation remains constant at all energies, up to 45 keV (Figure 8.46). This observation shows that another effect, which is independent of ion energy, determines the thickness. An additional phenomenon, independent of the ion–solid interaction, such as diffusion or reaction, has to be taken into account.

8.5.3 THE ROLE OF DIFFUSION AND REACTION

8.5.3.1 ISRD prediction for IBO of elemental Si

The ISRD model is used primarily to account for the experimental observations made during IBO of Si(100), and specifically to obtain quantitative predictions for the thickness of IBO SiO_2 as a function of ion energy. In these simulations, a large reaction constant k_{so} for oxidation is assumed ($k_{so} = 10^4/\Omega_{so}$ $cm^3\ s^{-1}$, where Ω_{so} is the molecular density of SiO_2). This choice is based on present knowledge of oxidation, and experimental evidence, such as the formation of buried oxides by high-energy oxygen implantation (SIMOX) (Kilner, 1991 and references theirin) performed at liquid nitrogen temperature (77 K) (Namavar *et al.*, 1986b). The effects of oxygen diffusion, both thermal and enhanced by the ion bombardment, and of the O desorption at the surface are investigated in the simulations. Three cases are considered (Figure 8.49): (a) only thermal diffusion and no enhanced diffusion, (b) no thermal diffusion and only enhanced diffusion, and (c) both thermal and enhanced diffusions.

8.5.3.1.1 ISRD with only thermal diffusion and no enhanced diffusion

If free oxygen only diffuses thermally in the oxide ($D_{o,ox} > 10^{-15}$ $cm^2\ s^{-1}$), the oxide grows almost linearly with time at a rate proportional to the diffusion coefficient but no saturation regime of the IBO growth is observed (Figure 8.49). Even if the oxygen desorption flux is increased in the simulations, saturation is never observed. If, on the other hand, oxygen thermal diffusion is small ($D_{o,ox} < 10^{-15}$ $cm^2\ s^{-1}$), the oxide thickness does not increase significantly with time and remains equal to the thickness given by the ISR model. However, the concentration of free oxygen increases in the subsurface region to very high levels (up to 5×10^{23} O/cm^3), which is physically unrealistic. Again, assuming a high desorption rate does not change this conclusion either, since the free oxygen cannot diffuse. However, it is important to note that the SiO_2/Si interface is much sharper when computed by ISRD than by IS or ISR (Figure 8.45). In IS and ISR, the interface width is determined mostly by the range straggling. In ISRD, the oxygen atoms diffuse and react with the substrate atoms where they are available. The interface width is then fixed by the chemical-reaction rate and is much sharper due to the strong driving force for

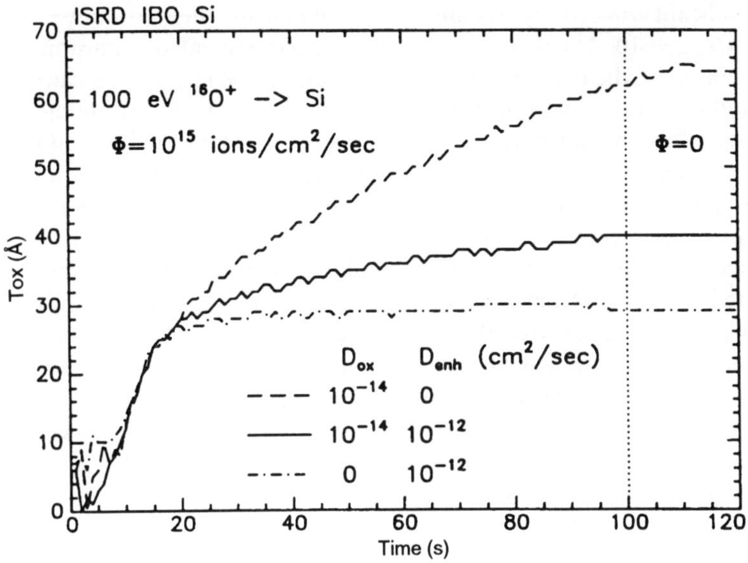

Figure 8.49 IBO SiO_2 thickness as a function of time as calculated by ISRD assuming different thermal and enhanced diffusivities for oxygen. Only the case assuming finite values for both diffusivities correctly predicts the kinetics observed experimentally

oxidation. In other words, the SiO_2/Si interface acts as an efficient recombination sink for the excess oxygen, similar to the surface where desorption occurs. This explains also why the oxide film keeps growing when oxygen is allowed to desorb significantly. This is actually the very mechanism used in the analytical model for high-dose nitridation developed by O. C. Hellman (1991), and described in Section 8.3.3.3. The growth rate is then only dependent on the sputtering rate and diffusion coefficient in the nitride.

8.5.3.1.2 No thermal diffusion and only enhanced diffusion

When only enhanced oxygen diffusion through fast-moving points defects is considered, oxygen is not transported beyond the end of range of the ion. The oxide thicknesses are then similar to those predicted by ISR and thus significantly smaller than experimentally observed (Figure 8.49).

8.5.3.1.3 Both thermal and enhanced diffusion

By considering only one type of O diffusion, either thermal or point-defect enhanced, one cannot account simultaneously for both the IBO saturation regime and the oxide thicknesses. This indicates that both types of diffusion are significant. Indeed, by assuming both significant thermal diffusions in the oxide ($D_{o,ox} = 10^{-14}\ cm^2\ s^{-1}$) and point-defect-mediated diffusion ($D_{enh} = 10^{-12}\ cm^2\ s^{-1}$), the oxide thicknesses are found to be larger than calculated by ISR, and do saturate with time (Figure 8.49), due to the following mechanism. Thermal diffusion transports oxygen over some distance, making SiO_2 thicker than predicted by ion–solid interactions (ISR). However, the creation of defects close to the surface favors the migration of free oxygen towards the surface with subsequent desorption and impedes further incorportion of oxygen deeper in the structure, leading to the saturation of the oxide growth. The point-defect-mediated diffusion of oxygen thus acts as a depth-dependent diffusivity preferentially directed towards the surface. It is interesting to note that a stress-dependent diffusivity would have the same effect. Assuming thermal and enhanced-diffusion coefficients equal to 2×10^{-14} and $10^{-12}\ cm^2\ s^{-1}$ respectively, a remarkably good agreement (Figure 8.45) can be observed between the calculated values and the experimental thicknesses measured in our study at low energy (100 eV–1 keV) and by Augustus *et al.* (1988) for an IBO SiO_2 film formed with 4 keV O_2^+.

8.5.3.1.4 Conclusions

The ISRD model can thus account for all the experimental observations made for IBO of Si: the three stages of IBO growth (Figure 8.49), the dependence of IBO SiO_2 thickness on ion energy (Figure 8.46), and the formation of a sharp interface with the substrate (Figure 8.45). The ISRD model also shows that the saturation regime is not the consequence of a competing process between oxidation and sputtering (IS model), but results from two competing mechanisms of oxygen diffusion, one towards the surface and the other towards the substrate.

Very recently, Todorov *et al.* (1992) have improved their IS model by using TRIM to calculate the oxygen-incorporation profile and included oxygen diffusion. To predict correctly the thickness of an IBO SiO_2 layer formed with 100 eV O_2^+, they assumed a depth-dependent oxygen diffusivity with a cut-off at a certain depth and favoring out-diffusion over further incorporation. This depth-dependent diffusivity has a similar effect to that of our point-defect-mediated oxygen diffusion in ISRD. Although the new model of Todorov and the ISRD model were developed independently and are quite different in their implementation, the conclusions are very similar regarding the role of oxygen diffusion and its depth dependence.

The same ISRD model can be applied to the case of $Si_{1-x}Ge_x$ dioxides prepared by IBO and is expected to lead to the same results. Specifically, it is expected to account for the interface widths observed at 200 eV and 1 keV. However, present lack of knowledge of point-defect diffusivities in $Si_{1-x}Ge_x$, as well as the dependence of diffusivity upon strain, composition and Ge/Si fraction makes this undertaking difficult at the present time.

8.6 CONCLUSIONS

In order to investigate the kinetics of IBS and to account for the observed relationship between ion energy and film properties, three models of increasing complexity were developed. These models progressively take into account the ion incorporation and sputtering, the ion-beam-mixing effects, and the diffusion and chemical reaction leading to the compound formation. This allowed us to highlight the role of each step in the kinetics of IBS. The models can be used for single- (Si, Ge) or multi-element targets ($Si_{1-x}Ge_x$ alloy, GaAs). The simulation results show that a model based only on implantation and sputtering cannot explain the thicknesses observed experimentally, but can give information on the ion-energy dependence of phases formed by IBS (e.g. IBO of Ge). Ion-beam-mixing effects are especially important in the spatial redistribution of the elements in compound targets resulting in a change of stoichiometry with depth (e.g. IBO of $Si_{0.8}Ge_{0.2}$). Thermal and radiation-enhanced diffusions and driving force for reaction must be taken into account to simulate the observed thicknesses and interfaces widths. Good correlation can be made regarding the dependence of the phase formation and stoichiometry upon ion energy as well as the overall kinetics in the three stages of IBO.

Thermal oxidation of $Si_{1-x}Ge_x$ heteroepitaxial alloys has been observed to result in a preferential oxidation of Si, with a corresponding pile-up of Ge-rich alloy near the growing oxide. This is consistent with the lower heat of formation of SiO_2, and suggests that the two-phase region in the Si–Ge–O ternary between [O] = 0 a/o and 66 a/o exhibits equilibrium between SiO_2 and Si_xGe_{1-x}. This separation into two phases has been observed even for ion-beam-synthesized SiGe oxides, wherein the substrate was held at 550 °C under 30 keV $^{16}O^+$ implantation (Chu *et al.*, 1978; Chap. 8; Zhang *et al.*, 1991; Tang *et al.*, 1990) indicating that the kinetics for the separation (presumably by a spinoidal mechanism) will proceed at this temperature. $Si_{1-x}Ge_x$ dioxides prepared at room temperature by IBO at energies of 1 keV and below on $Si_{1-x}Ge_x$ heteroepitaxial alloys do not phase-separate (Vancauwenberghe *et al.*, 1991a, 1992a, 1992b; Vancauwenberghe, 1991). In the same fashion metastable Si_xGe_{1-x} nitrides and GaAs oxides and nitrides can be grown by IBS.

The effect of ion energy upon growth kinetics, which is the additional parameter brought about by the use of IBS as compared to thermal kinetics, can be quite different depending on the material system IBS is applied to. However, the different IBS kinetics observed can be classified in three groups, from a simple atomic collision model to a sophisticated model including reactivity and diffusion. These three models are each found to account quantitatively for different kinetic regimes that have been observed experimentally.

One can thus state that a basic understanding of IBS has been reached which can lead to quantitative predictions in well-characterized systems. However, a large amount of theoretical and experimental work is still required, to establish on a fundamental level the link between IBS kinetics and important thin-film properties, such as strain, electronic and optical properties. Such work can take advantage of the fairly accurate predictions that can already be made about microstructure and phase formation as a function of ion energy, based on the existing experimental data and theoretical developments. The existing body of research already provides a great amount of guidance in the energy and temperature ranges to be explored as well as in the characterization techniques to use. Also, with regard to XPS, extensive characterization by several authors has led to a good supply of reference data for several material systems, which will be useful for phase identification in future experiments.

8.7 ACKNOWLEDGMENT

N. Herbots, P. Ye, and X. Wang acknowledge the support of Grant No. FGIA-AWR-B765.

8.8 NOTES

1 E_d is the bulk displacement energy. Brice *et al.* (1989) suggests that the surface E_d is about half of that value. $E_d \approx 15$ eV is taken here as an estimate for the lower energy boundary for significant ion penetration with the ability to create bulk defects, but depends on the specific ion–target combination.

2 The BCA assumes a pair-like interaction with only two atoms interacting at any time.

3 Alternately, samples can also be rinsed for 120 s in device-grade methanol, etched for 20 s in 10:1:1 ethanol:HF:water, rinsed for 60 s in methanol, blown dry with nitrogen, and immediately placed on the insulated sample holder to be loaded into the vacuum chamber.

4 TEM samples were prepared by cleaving six small pieces of an IBN film substrate (approximately 1.5 mm × 2.5 mm), gluing them together in a stack with Devcon 5-minute epoxy, and polishing one side of this stack by SiC grinding and Al_2O_3 polishing. The polished side of these pieces is then epoxied onto a 3 mm Cu grid, and the opposite side then thinned by grinding and polishing to approximately 500 μm. The sample was then thinned further, to about 25 μm, using a Gatan dimple grinder with a slurry of water and 1 μm diamond paste. Ion milling was used as the last step to thin the samples to approximately 100 nm, thin enough to be transparent to a 100 keV electron beam. Ar ion milling was performed at 14° from the plane of the sample surface, at an ion energy of 6 keV. The standard procedure of rotating the TEM sample in the plane of the sample during ion milling was not used: instead the sample was oriented in the ion mill such that the path of the ion beam projected in the plane of the sample surface was close to the perpendicular to the surfaces of the

individual substrates which composed the sample stack. This was done to minimize decomposition of the epoxy by ion milling, and thus retain sharp interfaces between the epoxy and IBN film surface.

5 The EMOO2B is part of the Arizona State University High Resolution Electron Microscopy Center (HREM), funded by the National Science Foundation under contract DMR-91-15680.

8.9 REFERENCES

Augustus, P. D., G. D. T. Spiller, M. G. Dowsett, P. Kightley, G. R. Thomas, R. Webb, and E. A. Clark (1988) In *Secondary Ion Mass Spectrometry (SIMS VI)*, A. Benninghoven, A. M. Huber, and H. W. Werner (eds), Wiley, Chichester, p. 485.

Alay, J. L. (1993) Ph.D. Thesis, Interuniversitair Mikro-Elektronica Centrum vzw (IMEC), Louvain.

Alay, J. L. and W. Vandervorst (1992a) *J. Vac. Sci. Tech.*, **A 10** (4), 2926.

Alay, J. L. and W. Vandervorst (1992b) *Surface and Interface Analysis*, **19**, 313.

Armour, D. G. (1988) In *Fundamentals of Beam–Solid Interactions and Transient Thermal Processing*, M. J. Aziz, L. E. Rhen, and B. Stritzker (eds), *MRS Proc. Vol. 100*, p. 127.

Belyi, V., L. L. Vasilyeva, A. S. Ginovker, V. A. Gritsenko, S. M. Repinsky, S. P. Sinitsa, T. P. Smirnova, and F. L. Edelman (1988) *Silicon Nitride in Electronics*, Materials and Science Monographs Vol. 34, Elsevier, New York.

Betz, G. and G. K. Wehner (1983) In *Sputtering by Particle Bombardment II*, R. Behrisch (ed.), Springer, Berlin.

Biersack, J. P. and W. Eckstein (1984) *Appl. Phys.*, **A 34**, 73.

Biersack, J. P. and L. G. Haggmark (1980) *Nucl. Inst. and Meth.*, **174**, 257.

Bourguet, P., J. M. Dupart, E. Le Tiran, P. Auvray, A. Guivarc'h, M. Salvi, G. Pelous, and P. Henoc (1980) *J. Appl. Phys.*, **51** (12), 6169.

Breeman, M. and D. O. Boerma (1993) Paper presented at the Fall MRS 1992 meeting, Boston, Nov. 30–Dec. 4 1992. In *Beam–Solid Interactions: Fundamentals and Applications*, M. A. Nastasi, N. Herbots, L. R. Harriott, and R. S. Averback (eds), *MRS Symp. Proceed. Vol. 279*, Pittsburgh, Pa.

Brice, D. K., J. Y. Tsao, and S. T. Picraux (1989) *Nucl. Inst. and Meth.*, **B 44**, 78.

Castle, J. E., H. D. Liu, J. F. Watts, J. P. Zhang, P. L. F. Hemment, S. M. Newstead, A. R. Powell, T. E. Whall, and E. H. C. Parker (1992) *Mater. Sci. and Eng.*, **B 12**, 199.

Chang, P. H., C. Slawinski, B.-Y. Mao, and H. W. Lam (1987) *J. Appl. Phys.*, **61** (1), 166.

Chereckdjian, S. and I. H. Wilson (1986) *Nucl. Inst. and Meth.*, **B 13**, 426.

Chiu, T. Y., H. Bernt, and I. Ruge (1982) *J. Electrochem. Soc.*, **129** (2), 408.

Chu, W. K., J. W. Mayer, and M. A. Nicolet (1978) *Backscattering Spectrometry*. Academic Press, New York.

Colligon, J. S., W. A. Grant, J. S. Williams, and R. P. W. Lawson (1976) In *Applications of Ion Beams to Materials 1975, Proceedings of the Institute of Physics Conference Series*, **28**, 357.

Croydon, W. F. and E. H. C. Parker (1981) *Dielectric Films on GaAs*. Gordon and Breach, New York.

Cuomo, J. J. and S. M. Rossnagel (1987) *Nucl. Inst. and Meth.*, **B 19/20**, 963.

DeLouise, L. A. (1991a) *J. Chem. Phys.*, **94**, 1528.

DeLouise, L. A. (1991b) *J. Appl. Phys.*, **70**, 1718.

DeLouise, L. A. (1992) *J. Vac. Sci. Tech.*, **A 10** (4), 1637.

Dodson, B. W. (1987) *Phys. Rev.*, **B 36**, 1068.

Dodson, B. W. (1988) In *Fundamentals of Beam–Solid Interactions and Transient Thermal Processing*, M. J. Aziz, L. E. Rhen, and B. Strizker (eds), *MRS Proc. Vol. 100*, p. 139.

Dodson, B. W. (1989) In *Processing and Characterization of Materials Using Ion Beams*, L. E. Rehn, J. E. Greene, and F. A. Schmidt (eds), *MRS Proc. Vol. 128*, p. 137.
Doolittle, L. R. (1983) *Nucl. Inst. and Meth.*, **B 9**, 344.
Fenner, D. B., D. K. Biegelsen, and R. D. Bringans (1989) *J. Appl. Phys.*, **66**, 1 July.
Garrison, B. J., M. T. Miller, and D. W. Brenner (1989) In *Atomic Scale Calculations in Materials Science*, J. Tersoff, D. Vanderbilt, and V. Vitek (eds), *MRS Proc. Vol. 141*, p. 419.
Ghose, D., D. Basu, and S. B. Karmohapatro (1983) *Phys. Stat. Solidi (a)*, **77**, 121.
Goleci, I., G. E. Chapman, S. S. Lau, B. Y. Tsaur, and J. W. Mayer (1979) *Phys. Lett.*, **A 71**, 267.
Gösele, U., W. Frank, and A. Seeger (1983) *Solid State Communication*, **45**, 31.
Gossmann, H. J. (1984) Ph.D. Thesis, State University of Stony Brook, Albany, excerpted in L. C. Feldman and J. W. Mayer, *Fundamentals of Surface and Thin Film Analysis*. North-Holland, New York (1986).
Gourrier, S., P. Friedel, and P. K. Larsen (1983) *J. Appl. Phys.*, **54**, 3993.
Grabowski, K. S., A. D. F. Kahn, E. P. Donovan, and C. A. Carosella (1989) *Nucl. Inst. and Meth.*, **B 39**, 190.
Grunthaner, F. J. and P. J. Grunthaner (1986) *Materials Science Reports*, **1**, 65.
Grunthaner, F. J., P. J. Grunthaner, R. P. Vasquez, B. F. Lewis, J. Maserjian, and A. Madhukar (1979) *J. Vac. Sci. Tech.*, **16**, 1442.
Grunthaner, P. J., F. J. Grunthaner, R. W. Fathauer, T. L. Lin, M. H. Hecht, L. D. Bell, W. J. Kaiser, F. D. Schowengerdt, and J. H. Mazur (1989) *Thin Solid Films*, **183**, 197.
Harper, J. M. E., M. Heiblum, J. L. Speidell, and J. J. Cuomo (1981) *J. Appl. Phys.*, **52**, 4118.
Hellman, O. C. (1991) Ph.D. Thesis, MIT, Cambridge, Mass.
Hellman, O. C., N. Herbots, and O. Vancauwenberghe (1992a) *Nucl. Inst. and Meth.*, **B 67**, 301.
Hellman, O. C., N. Herbots, O. Vancauwenberghe, R. J. Culbertson, and W. J. Croft (1992b) *J. Vac. Sci. Tech.*, **A 10** (4), 1631.
Hellman, O. C., O. Vancauwenberghe, N. Herbots, R. J. Culbertson, and W. J. Croft (1992c) *Mater. Sci. and Eng.*, **B 12**, 53.
Herbots, N. and O. C. Hellman (1989) US Patent #4,800,100. (1/24/1989).
Herbots, N., B. R. Appleton, T. S. Noggle, S. J. Pennycook, R. A. Zuhr, and D. M. Zehner (1986a) In *Semiconductor-Based Heterostructures: Interfacial Structure and Stability*. M. L. Green, J. E. E. Baglin, G. Y. Chin, H. W. Deckman, W. Mayo, and D. Narasinham (eds), The Metallurgical Society, p. 335.
Herbots, N., B. R. Appleton, T. S. Noggle, R. A. Zuhr, and S. J. Pennycook (1986b) *Nucl. Inst. and Meth.*, **B 13**, 250.
Herbots, N., B. R. Appleton, S. J. Pennycook, T. S. Noggle, and R. A. Zuhr (1986c) In *Beam–Solid Interactions and Phase Transformations*, H. Kurz, G. L. Olsen, and J. M. Poate (eds) *Mat. Res. Soc. Symp. Proc., Vol. 51*, 369.
Herbots, N., O. C. Hellman, P. A. Cullen, and O. Vancauwenberghe (1988) In *Deposition and Growth: Limits for Microelectronics*, G. W. Rubloff (ed.), *American Vacuum Society Series 4, AIP Vol. 167*, 259.
Herbots, N., O. Vancauwenberghe, O. C. Hellman, and Y. C. Joo (1991) *Nucl. Inst. and Meth.*, **B 59/60**, 326.
Herbots, N., O. C. Hellman, and O. Vancauwenberghe (1992) In *Phase Formation and Modification by Beam–Solid Interactions*, G. S. Was, L. E. Rehn, and D. Follstaedt (eds), *Mat. Res. Soc. Symp. Proc. Vol. 235*, p. 769; *Photons and Low Energy Particles in Surface Processing*, C. Ashy, J. Brandon, and S. Pang (eds), *Mat. Res. Soc. Symp. Proc., Vol. 236*.

Hezel, R. and N. Lieske (1982) *J. Electrochem. Soc.*, **29** (2), 379.
Hill, J. M., D. G. Royce, C. S. Fadley, L. F. Wagner, and F. J. Grunthaner (1976) *Chem. Phys. Lett.*, **44**, 225.
Himpsel, F. J., F. R. McFeely, A. Taleb-Ibrahimi, J. A. Yarmoff, and G. Hollinger (1988) In *The Physics and Chemistry of SiO_2 and the Si–SiO_2 Interface*, C. R. Helms, and B. E. Deal (eds), Plenum Press, New York, p. 219.
Hoffmann, K. R. (1991) In *Silicon Molecular Beam Epitaxy*, J. C. Bean, E. H. C. Parker, S. S. Iyer, Y. Shiraki, E. Kaspar, and K. Wang (eds), *Mater. Res. Soc. Proc.*, *Vol. 220*.
Holland, O. W., C. W. White, and D. Fathy (1987) *Appl. Phys. Lett.*, **51** (7), 52.
Holmen, G. and H. Jacobsson (1988) *Appl. Phys. Lett.*, **53**, 1838, See also: Jacobsson, H., PhD thesis, Cholmers University of Technology, Göteberg, Sweden (1993).
Kijima, K. and S. Shirasaki (1976) *J. Chem. Phys.*, **65** (7), 2668.
Kilner, J. A. (1991) *Vacuum*, **42**, 391.
Kingery, W. D., H. K. Bowen, and D. H. Uhlmann (1984) *Introduction to Ceramics*, New York.
Kitabatake, M., P. Fons, and J. E. Greene (1991) *J. Vac. Sci. Tech.*, **A 9**, 91.
Kreissig, U. (1982) *Thin Solid Films*, **98**, 229.
Landman, U., W. D. Ludetke, and M. W. Ribarsky (1989) *J. Vac. Sci. Tech.*, **A 7**, 2829.
LeGoues, F. K., R. Rosenberg, and B. S. Meyerson (1989) *Appl. Phys. Lett.*, **54** (7), 644.
Littlewood, S. and J. A. Kilner (1988) *J. Appl. Phys.*, **63** (6), 2173.
Liu, W. S., G. Bai, M.-A., Nicolet, C. H. Cern, V. Arbet, and K. L. Wang (1991) *Mat. Res. Soc. Symp. Proc.*, *Vol. 220*, 259.
Liu, W. S., E. W. Lee, M.-A. Nicolet, V. Arbet- Engels, K. L. Wang, N. M. Abuhada, and C. R. Aita (1992) *J. Appl. Phys.*, **71** (8), 4015.
Meekison, C. D., G. R. Booker, K. J. Reeson, P. L. F. Hemment, R. J. Chater, J. A. Kilner, and J. R. Davis (1986) *Vacuum*, **36** (11/12), 925.
Meekison, C. D., G. R. Booker, K. J. Reeson, P. L. F. Hemment, R. F. Peart, R. J. Chater, J. A. Kilner, and J. R. Davis (1991) *J. Appl. Phys.*, **69** (6), 3503.
Namavar, F., J. I. Budnick, F. A. Sanchez, and H. C. Hayden (1985) *Mat. Res. Soc. Symp. Proc.*, *Vol. 45*, 317.
Namavar, F., J. I. Budnick, A. Fasibuddin, F. H. Sanchez, and H. C. Hayden (1986a) *Mat. Res. Soc. Symp. Proc.*, *Vol. 53*, p. 281.
Namavar, F., J. I. Budnick, F. H. Sanchez, and H. C. Hayden (1986b) In *Semiconductor-on-Insulator and Thin Film Transistor Technology*, A. Chiang, M. W. Geis, and L. Pfeiffer (eds), *Mat. Res. Soc. Symp. Proc.*, *Vol. 53*, p. 233.
Netterfield, R. P., P. J. Martin, and W. G. Sainty (1986) *Applied Optics*, **25** (21), 3808.
People, R. (1986) *IEEE J. Quantum Electronics*, **QE- 22**, 1696.
Petruzello, J., T. F. McGee, M. H. Frommer, V. Rumennik, P. A. Walters, and C. J. Chou (1985) *J. Appl. Phys.*, **58** (12), 4605.
Pixraux, S. T., L. C. Feldman, and J. W. Mayer (1982) *Materials Analysis by Ion Channeling*. Academic Press, New York.
Priolo, F., A. LaFerla, and E. Rimini (1988) *J. Mater. Res.*, **3** (6), 1212.
Robinson, M. T. (1993) Plenary Talk, presented at the Fall MRS 1992 meeting, Boston, Nov. 30–Dec. 4 1992. In *Beam–Solid Interactions: Fundamentals and Applications*, M. A. Nastasi, N. Herbots, L. R. Harriott, and R. S. Averback (eds), *MRS Symp. Proceed. Vol. 279*, Pittsburgh.
Roorda, S. (1990) Ph.D. Thesis, FOM Instituut, Amsterdam.
Rutherford, E. (1911) *Phil. Mag.*, **21**, 669.
Schulze, D., J. Finster, E. Hensel, W. Skorupa, and U. Kreissig (1983) *Phys. Stat. Sol.*, **A 76**, K21.
Schwartz, G. P., G. J. Gualtieri, G. W. Kammlott, and B. Schwartz (1979) *J. Electrochem. Soc.*, **126**, 1737.

Skorupa, W., K. Wollschläger, U. Kreissig, and R. Grötzschel (1987) *Nucl. Inst. and Meth.*, **B 19/20**, 285.
Sodini, C. G., S. S. Wong, T. W. Ehstedt, H. R. Grinolds, and W. G. Oldham (1984) *IEEE Trans. Electr. Dev.*, **ED-31** (1), 17.
Spaepen, F. (1974) *Phil. Mag.*, **30**, 417.
Tang, Y. S., J. P. Zhang, P. L. F. Hemment, B. J. Sealy, H. D. Liu, J. E. Castle, S. M. Newstead, A. R. Powell, T. E. Whall, and E. H. C. Parker (1990) *J. Appl. Phys.*, **67** (11), 7151.
Taniguchi, S., T. Shibata, K. Nakamura, L. Xianghuai, Z. Zhihong, H. Wei, and Z. Schichang (1989) *Mat. Sci. and Eng.*, **A 121**, 519.
Taylor, J. A., G. M. Lancaster, and J. W. Rabalais (1978) *J. Electron Spectro. Rel. Pheno.*, **13**, 435.
Tencor Alpha-Step 1000 Users Manual.
Tenner, A. (1986) Ph.D. Thesis, FOM-Instituut, Amsterdam.
Thomas, G. E., L. J. Beckers, F. H. P. M. Habraken, and A. E. T. Kuiper (1982) *Appl. Phys. Lett.*, **41** (1), 56.
Thurmond, C. D., G. P. Schwartz, G. W. Kammlott, and B. Schwartz (1980) *J. Electrochem. Soc.*, **127**, 1366.
Timoshenko, S. (1930) *Strength of Materials*. Van Nostrand, New York.
Todorov, S. S. and E. R. Fossum (1987) In *Photon, Beam and Plasma Stimulated Chemical Processes at Surfaces*, V. M. Donnelly, I. P. Herman, and M. Hirose (eds), *Mat. Res. Soc. Symp. Proc.*, *Vol. 75*, p. 349.
Todorov, S. S., I. R. Chakarov, and D. S. Karpuzov (1994) *Nucl. Instrum. Meth.*, **B**, in press.
Troxell, J. R. (1985) *J. Elect. Mat.*, **14** (6), 707.
Vancauwenberghe, O. (1991) Ph.D. Thesis, MIT, Cambridge, Mass.
Vancauwenberghe, O., N. Herbots, and O. C. Hellman (1990) *Proc. of the SPIE, Growth of Semi-conductor Structures and High T_c Thin Films on Semiconductors*, **1285**, 47.
Vancauwenberghe, O., O. C. Hellman, N. Herbots, and W. J. Tan (1991a) *Appl. Phys. Lett.*, **59** (16), 2031.
Vancauwenberghe, O., N. Herbots, and O. C. Hellman (1991b) *J. Vac. Sci. Tech.*, **B 9**, 2027.
Vancauwenberghe, O., N. Herbots, H. Manoharan, and M. Ahrens (1991c) *J. Vac. Sci. Tech.*, **A 9**, 1035.
Vancauwenberghe, O., O. C. Hellman, and N. Herbots, (1992a) *J. Vac. Sci. Tech.*, **A 10** (4), 713.
Vancauwenberghe, O., O. C. Hellman, and N. Herbots (1992b) *Mat. Sci. and Eng.*, **B 12**, 97.
Vandervorst, W. (1993) In *Beam–Solid Interactions: Fundamentals and Applications*, M. A. Nastasi, N. Herbots, L. H. Harriott, and R. Averbach (eds), *Mater. Res. Soc. Symp. Proc.*, *Vol. 279*, Pittsburgh.
Wang, K. L. and C. Woo (1992) United States Patent: 5,155,571 (10/13/1992).
Wang, Z., Y. Li, and J. Zhang (1986) *Nucl. Inst. and Meth.*, **B 13**, 453.
White, A., E. Short, R. O. C. Dynes, J. P. Grain, and J. M. Gibson (1987) *Appl. Phys. Lett.*, **50** (2), 95.
Wilmsen, C. W. (ed.) (1985) *Physics and Chemistry of III–V Compound Semiconductor Interfaces*. Plenum Press, New York.
Zalm, P. C. and L. J. Beckers (1982) *Appl. Phys. Lett.*, **41**, 168.
Zhang, J. P., P. L. F. Hemment, U. Bussman, A. K. Robinson, J. E. Castle, H. D. Liu, J. F. Watts, S. M. Newstead, A. R. Powell, T. E. Whall, and E. H. C. Parker (1991) *Nucl. Inst. and Meth.*, 691.

9 FILM DEPOSITION FROM LOW-ENERGY ION BEAMS

Denes Marton
Department of Chemistry, University of Houston, Houston, Texas 77204-5641, USA

CONTENTS

Low Energy Ion–Surface Interactions
Edited by J. W. Rabalais

9.1 INTRODUCTION

Film deposition can be carried out by an extremely wide range of technologies; in a broad sense, painting qualifies as well as anodic oxidation. One possible classifying principle is whether particles of the film are deposited below the surface of the substrate or on the substrate surface. Those methods which involve subsurface film-growth processes lead to films with distinguishing qualities such as excellent adhesion, good mechanical and structural properties (e.g. epitaxial films grown at low substrate temperatures), high film density, and formation of metastable phases (e.g. diamondlike films grown at low substrate temperatures).

Several non-equivalent processes may result in subsurface penetration. These include the following:

(a) chemical interaction (e.g. incorporative adsorption),
(b) diffusion,
(c) penetration due to film-particle kinetic energy,
(d) penetration due to assisting kinetic energy.

Obviously, these processes may occur simultaneously, as is the case, for example, with diffusion and chemical reaction during thermal growth of SiO_2 on silicon. The third and fourth processes refer to methods involving energetic particles which themselves become constituents of the film (c), or which are used to assist penetration of other particles by supplying kinetic energy (d). In this chapter we will focus on the third process, i.e. kinetic energy supplied by film particles. It must be realized though that all other processes can play concurrent roles in most practical cases.

Among low energy ion–surface interactions, deposition is one of the fundamental processes. As energetic ions impinge on solid surfaces, they either will stay 'collected' in the surface or will backscatter into vacuum. The collection efficiency, which is sometimes referred to as sticking probability, increases with the ion kinetic energy. As Fontell and Arminen (1969) have shown, at room temperature, for most metals it will approach unity at kinetic energies near or even below 100 eV. However, at kinetic energies above 100 eV, sputtering becomes significant and in most cases the sputtering yield will again decrease only for kinetic energies above 10–20 keV (Behrisch, R. (ed.), 1981). Hence, deposition is efficient with ion beams either at relatively high energies when ion implantation occurs or at very low energies.

The collection efficiencies are significant (0.4–0.8) for most metal ions or atoms even at energies < 10 eV, levels which are attainable with film deposition equipment utilizing evaporation or sputtering (without bias), etc. rather than ion beams (Klabunde, K. J. (ed.), 1985). Such equipment is much simpler than low energy ion beam systems. The difference in the collection efficiency certainly does not justify the more complex equipment. These extra efforts are only justified by the vary different film qualities obtained which apparently are due to the extra kinetic energy of just a few times 10 eV per ion, and are related to subsurface film growth.

A broad variety of deposition methods yield encouraging results by supplying assisting kinetic energy to the growing film in order to facilitate subsurface film growth. Such methods include various evaporation, sputtering and molecular-beam (MBE—molecular-beam epitaxy) techniques combined with ion bombardment either from ion beams (IBAD—ion-beam-assisted deposition) or from a plasma (Itoh, T. (ed.), 1989). In these techniques, inert-gas ions or sometimes ions of one of the constituents of the film are used. Application of these techniques involves processes that are very complex, and the parameters, especially ion species and energies, are not well controlled. It is very difficult, if not impossible, to optimize these techniques in order to obtain desired film properties in a controlled and reliable manner. Use of ion beams as the direct source of the deposited particles enables development of instruments featuring mass- and energy-selection combined with an ultra-high vacuum (UHV) environment. Such instruments have become powerful research tools with excellent parameter control for subsurface deposition. It is likely, however, that once the physical processes are better understood and goals and optimal parameter ranges are well established, much of the technology will turn to less sophisticated and less expensive tools that offer higher productivity.

The first low-energy ion-beam deposition systems (LEIBS) known to the author were reported by Aisenberg and Chabot (1971) and Fair (1971). The former generated considerably more interest by depositing diamondlike carbon (DLC) films than did Fair with his indium film. Nevertheless, Fair correctly pointed out that the future of ion-beam deposition (IBD) lies in the improvement of the ion source. Subsequently, a sophisticated LEIBS with mass- and energy-selection, neutral trap and adequate differential pumping was built by Amano *et al.* (1976) and used for deposition of Pb and Mg films (Amano and Lawson, 1977; Amano, 1982). Many of the important design concepts were first described by this group but more interesting results were achieved by others who pursued applications where the advantages of IBD are more important than for the deposition of simple metal films.

Yagi, Tamura and Tokuyama (1977) of Hitachi reported the first semiconductor (Ge) epitaxial films grown at low substrate temperatures by IBD, thereby opening a second very attractive area of application after DLC.

During the next decade an increasing number of publications were devoted to various aspects of IBD, repeating and refining the results of the 1970s. It is not the goal of this introduction to review the entire literature and we will limit ourselves to those few developments that opened new areas of research. Although there were some early efforts to produce binary films such as BN from a molecular ion beam of borazine (Weissmantel, 1981), the first meaningful result came from a combination of a Si molecular beam and a C^+ ion beam to form β-SiC (3C-SiC) epitaxy on Si (Miyazawa *et al.*, 1984b). However, it was not until 1989 that a group at Oak Ridge National Laboratory reported binary films—GaAs (Haynes *et al.*, 1989) and β-SiC (Withrow *et al.*, 1989)—grown by alternating ion beams. This work was a precursor of the most recent development in the field, i.e. the appearance of multiple-beam LEIBS (Ohnishi *et al.*, 1989; Yoshida *et al.*, 1989; Tamura *et al.*, 1989).

These developments in instrumentation and achievements in film composition and structure were only partially matched in the development of our understanding of the fundamental processes governing IBD. In this respect, it is significant that LEIBS with mass selection and UHV deposition chamber allowing *in situ* surface characterization are used (Kasi *et al.*, 1987; Herbots *et al.*, 1988). Our understanding of the processes in IBD has evolved significantly due to this type of research with well-defined systems. Also, recent molecular-dynamics simulations by several groups (Dodson, 1990; Kitabatake *et al.*, 1990; Kaukonen and Nieminen, 1992; Sprague and Gilmore, 1992) show promise for improving the understanding of fundamental processes and for helping to determine optimal conditions in IBD.

In this chapter we will examine some of the concepts of hardware design and requirements, some of the details of the qualities of the films obtained, and the physical processes in ion-beam deposition.

9.2 LOW-ENERGY ION-BEAM DEPOSITION SYSTEMS

The essential components of a modern LEIBS are schematically shown in Figure 9.1(a). The ions produced in the ion source are accelerated from the source potential, U_s, to U_F, the flight-tube voltage which is usually a few thousand volts. Then, before the ions reach the substrate, they are decelerated to earth potential. The ion kinetic energy is thus determined by the source

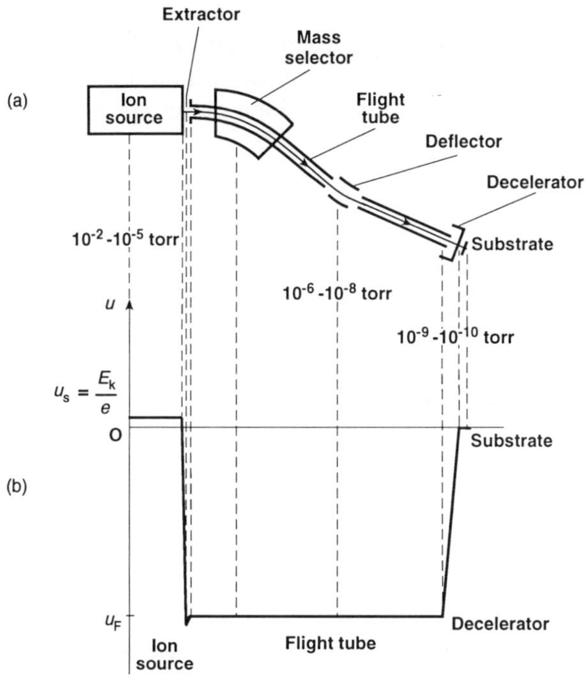

Figure 9.1 (a) Main components of an ion-beam deposition system. (b) Schematic of the electrical potential distribution along the beam line. The source is at a positive potential (for positive ions) and the substrate (target) is at ground

potential (see Figure 9.1(b)). Some laboratories use instruments modified from ion implanters by adding a decelerator lens in front of the target. In this case, the flight tube will be grounded and both the source and the target are at high voltage, a quite significant inconvenience from the points of view of application of *in situ* surface analysis techniques, electron flood sources, etc. Also, in this case the ion energy is regulated as the difference of two high-voltage points and the construction of the source becomes more complicated. On the other hand, in the case of the design shown in Figure 9.1(a), there is some difficulty in establishing a continuous flight-tube high-voltage environment across the mass selector, the deflector, and the gate valves separating different areas of the beam line.

The LEIBS design shown in Figure 9.1 is obviously quite complicated. One could simply position the substrate next to the ion source and utilize all ions produced without worrying about acceleration, deceleration, mass selection and deflection. Indeed, such instruments have been built for ion etching (Kaufman *et al.*, 1982, 1989; Proudfoot *et al.*, 1989), but they are not useful for deposition because they do not produce ion beams that are clean enough. In order to achieve high-quality films, the ion-beam deposition has to be carried out in a clean environment and well-controlled high-purity ion beams are required.

9.2.1 VACUUM CONDITIONS

One of the most important design considerations is the base pressure required in the deposition chamber. As Amano *et al.* (1976) pointed out, at $T = 300$ K the residual-gas molecular flux arriving at the target surface will be described as

$$\mathrm{d}n_\mathrm{a}/\mathrm{d}t = 1.5 \times 10^{23}\,(P/M^{1/2})\mathrm{s}^{-1}\mathrm{m}^{-2}, \tag{1}$$

where P is the pressure in Pa and M is the molecular weight in amu. In a worst-case scenario, when the sticking probability is $S = 1$, equation (1) describes the arrival rate of impurities in the film. The rate of arrival of the ions in the ion beam is

$$\mathrm{d}n_\mathrm{i}/\mathrm{d}t = J/ze, \tag{2}$$

where J is the ion current density, z is the ion charge state, and e is the elementary charge. In order to produce pure films, evidently $\mathrm{d}n_\mathrm{i}/\mathrm{d}t \gg \mathrm{d}n_\mathrm{a}/\mathrm{d}t$ is necessary. For example, with an ion current density of 20 μA cm^{-2} of single-charged Si^+ ions, this means $P \ll 4.4 \times 10^{-5}$ Pa, i.e. UHV is necessary in order to perform deposition without serious contamination from the residual gas.

This vacuum requirement is far from trivial for two reasons. First, high pressures are necessary in the ion sources in order to achieve high ion fluxes. Hence, differential pumping is absolutely necessary, usually in several stages, along the beam line. Second, as high-current ion beams impinge on construction parts of the beam line, on the decelerator, or on the deposition chamber walls, ion-induced desorption and/or sputtering can take place which may significantly contribute to the residual-gas pressure. Hence, powerful pumping is necessary in the deposition chamber itself and precautions have to be taken in the design of the decelerating lenses in order to avoid deposition of lens material on the substrate (Herbots *et al.*, 1988).

9.2.2 ION SOURCES WITH MASS SELECTION

In the past fifteen years, significant developments have occurred in the area of ion sources producing large-area, high-intensity, low-energy beams (Kaufman *et al.*, 1982, 1989; Proudfoot *et al.*, 1989). It is possible now to produce ion beams which are able to process 6-inch semiconductor wafers at ion energies as low as 80 eV and current densities of about 1 mA cm^{-2} (Proudfoot *et al.*,

1989; Kaufman *et al.*, 1989). The ion beams, however, have to be mass-analyzed in order to obtain the pure beams necessary for almost all applications save ion etching, and the transport and mass selection of these broad low-energy beams remains unresolved. The main reason for this is that space-charge repulsion results in ion beams diverging along their path. This repulsion effect varies approximately as $JM^{1/2}E_k^{-3/2}$ (Freeman *et al.*, 1976), where J is the ion current density, M is the ion mass and E_k is the ion kinetic energy. It can be negligible if the beam is transported at high kinetic energy, hence the potential outline of Figure 9.1(b), including the deceleration system.

The space-charge effect is partially compensated by electrons obtained from the residual gas. The ionized residual-gas atoms are slow as compared with those in the ion beam and they are quickly pushed out of the beam by the space charge. This effect is significant in the pressure range of 10^{-6} torr (Kreisler *et al.*, 1984), making this the optimal pressure for the beam line. However, there is a probability for residual-gas atoms to be accelerated to high speed (dragged along the ion beam) and for beam ions to lose charge (recombine with the electrons). Both these effects result in energetic neutrals which have to be later eliminated from the ion beam.

The necessity of mass selection is immediately obvious in the case of ion species which, unlike noble-gas ions, cannot be easily obtained from simple materials. Carbon, silicon, and many metals belong to this group. An excellent compilation on the production of heavy ions has been published by Freeman *et al.* (1977) with information on ion-source feed materials for practically any elemental ion, including the suggested temperature range for the source, and comments regarding toxicity, etc. Mass selection also allows use of the same source to produce several different ion species and deposition of those species sequentially. Some multicomponent films have been produced in this way (Haynes *et al.*, 1989; Withrow *et al.*, 1989) and, at least in principle, deposition of layered systems should also be possible.

Mass resolution and transmission have to be balanced in order to assure feasible deposition rates. This means that high selectivity is not practical and mass interference can be a problem. An interesting example has been recently described for the case of Si deposition (Al-Bayati *et al.*, 1992). Figure 9.2 shows the quality of the Si films deposited on Si as evaluated from medium-energy proton scattering. The quality of the films is described by the dechanneling level, χ, which is the ratio of the intensities of the ion-scattering spectra along a channeling direction and along a random direction. The film quality depends dramatically not only on the pretreatment of the Si substrate and the ion kinetic energy but also on the Si^+ isotope used for the deposition. The explanation for this is that when $^{28}Si^+$ ions are used, CO^+ ions from the residual gas in the source can reach the substrate. The presence of CO is detrimental to the quality of the epitaxial Si films as are the effects of Cl contamination from Cl^+ cleaning of the Si substrate and of insufficient annealing.

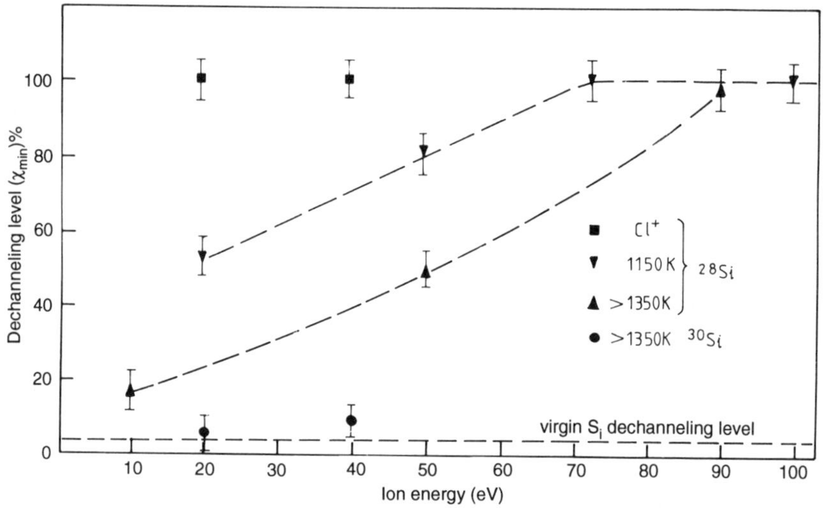

Figure 9.2 Dechanneling level, χ_{min}, as a function of ion energy for $^{28}Si^+$ and $^{30}Si^+$ ions incident on Si(001) surfaces subjected to different pretreatments and maintained at 723 K during deposition. ▲ 1350–1450 K annealed $+^{28}Si^+$ ions. ▼ 1100–1150 K annealed $+^{28}Si^+$ ions. ■ Cl^+ bombarded surface $+^{28}Si^+$ ions. ● 1350–1450 K annealed $+^{30}Si^+$ ions. (Reproduced by permission of Elsevier Science Publishers from Al-Bayati *et al.*, 1992)

Mass selection may not be very important in the case of gas ions like O^+ and N^+. For dual-source systems, it may be reasonable to build equipment designed with a simple beam line for these ions in addition to a mass-selected beam line for the other component (Ohnishi *et al.*, 1989; Yoshida *et al.*, 1989).

The ion energy plays an important role in IBD. The energy spread of the ion beam must be reasonably small for well-defined experimental conditions. This, together with the brightness (intensity), most often decides the choice of the ion source. Many instruments use modified versions of the Freeman ion source which is a magnetically confined plasma source with a hot ribbon cathode. An ion-beam system built by VSW Scientific Instruments for the University of Houston using such sources (Gordon *et al.*, 1991a) has been proven to have an energy spread of $2\sigma < 6$ eV. A negative-ion-beam instrument using a Cs plasma source has been reported to have ≈ 10 eV FWHM (Ishikawa *et al.*, 1986). Kimball Physics, Inc. has announced an electron cyclotron resonance (ECR) microwave plasma source with an energy spread ≈ 7 eV (IMG-31 Microwave Ion Source, private communication, 1992). Colutron Research Corp. offers a Colutron-type plasma source with an energy spread of only 0.5 eV (G-4-D Mod. Ion Beam System, private communication, 1992); this claim is not substantiated with experimental data. All these systems have been designed to produce low-energy ion beams with intensities of the order of 10 μA.

The plasma potential in various ion sources utilizing plasmas is more positive than the cathode potential. Hence, the actual ion energy exceeds the difference between the cathode and target potentials. The plasma potential is a function of the plasma power and other plasma parameters and it is not constant for a given source. The exact value of the ion energy has to be measured for well-characterized experiments.

9.2.3 DECELERATION OF ION BEAMS

Since the ion beams have to be transported at high kinetic energies, it is necessary to decelerate the beams before they reach the substrate. This requires special ion optics designed to avoid significant distortion (focusing or defocusing) of the ion beam. Fast neutrals cannot be decelerated so they must be

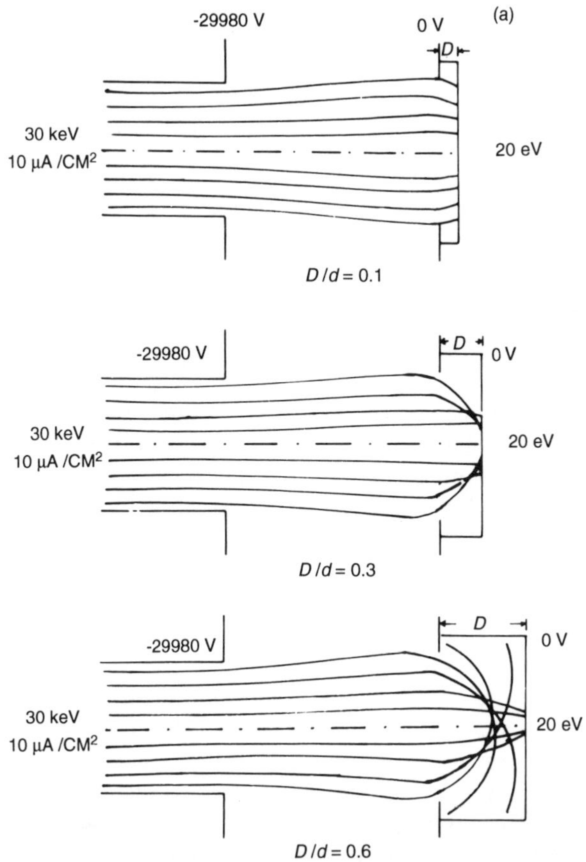

Figure 9.3 (a) Computer simulation of beam trajectories in the retarding field at three collector positions. (Reproduced by permission of Elsevier Science Publishers from Yamada *et al.*, 1985)

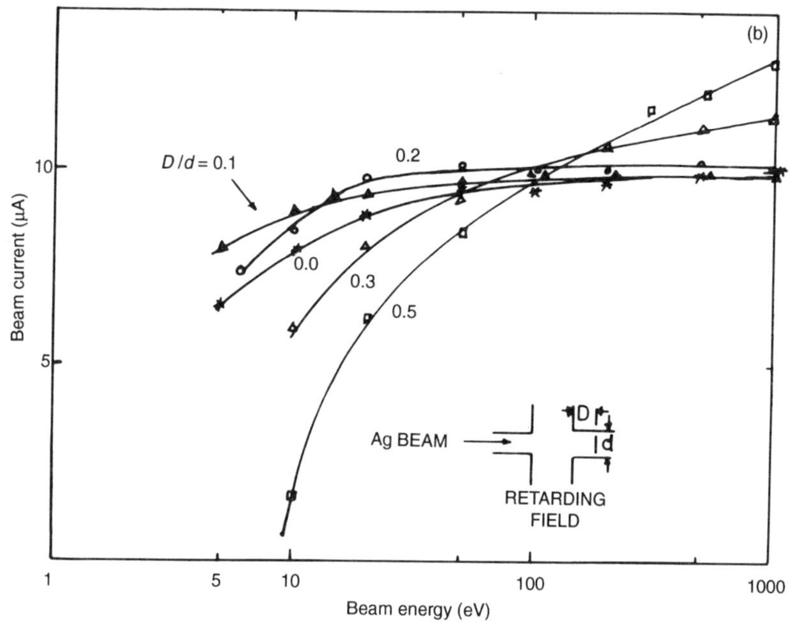

Figure 9.3 (b) Beam current as a function of beam energy at various electrode geometries. (Reproduced by permission of Elsevier Science Publishers from Yamada *et al.*, 1985)

eliminated from the ion beam in order to avoid damage to the substrate and/or to the film. These two functions can be achieved separately or in combination. The usual method for elimination of fast neutrals is deflection of the ion beam.

The separate deflection and deceleration (Figure 9.1(a)) is simpler but it implies that there is a part of the beam line after the deflection where new fast neutrals can form. Hence it is advantageous to combine these two functions. Freeman *et al.* (1976) have shown experimentally as well as using computer simulations that this can be done. They have also shown that strong focusing in the deceleration system can lead to ion trajectories impinging at low (even grazing) angles onto the substrate. Also, trajectories turning back in the beam line can evolve and the ions will be accelerated through the deceleration ion optics. In a more recent study, Yamada *et al.* (1985) show similar results which are presented here in Figure 9.3(a) and (b). These studies indicate that there is no need to apply very complicated deceleration systems (the suggested systems in these studies contain only two or three electrodes), but that the last electrode has to be at a very short distance from the substrate in order to ensure high ion-current transmission.

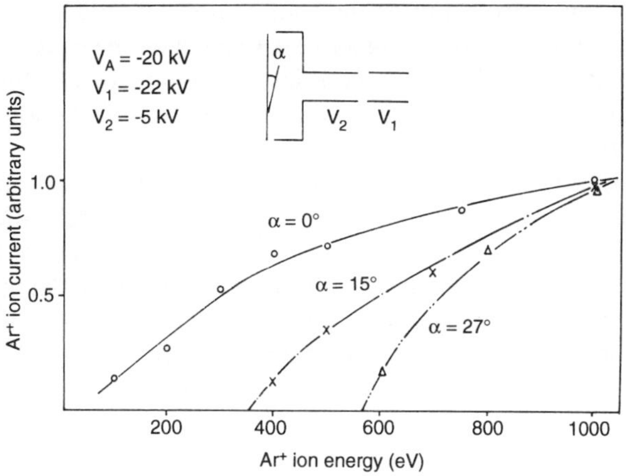

Figure 9.4 Ar^+ ion current obtained at various substrate angles (α). (Reproduced with permission from Shimizu *et al.*, 1985)

An interesting aspect of the deceleration is that the substrate must be perpendicular to the ion beam in order to achieve good results in an instrument with a simple deceleration stage. This is illustrated by the measurements of Shimizu *et al.* (1985) shown in Figure 9.4. These results are significant for the design of multi-ion-beam systems with independent beam lines (see Section 9.2.4). In such instruments, at least one of the ion beams arrives at the substrate at an off-normal angle. Calculations by Freeman *et al.* (1976) suggest that this problem can be solved using specially designed asymmetric deceleration lenses.

Electrons will be accelerated towards the substrate by the decelerating ion optics. In some cases, weak magnetic fields perpendicular to the ion beam are applied in order to deflect electrons from the substrate (Tokuyama *et al.*, 1981). Magnetic fields can also be used instead of electrostatic lenses in order to deflect the ion beam and thus remove fast neutrals in the deceleration stage, but this method leads to more complex geometries and it is not usually preferred.

9.2.4 MULTIPLE-SOURCE SYSTEMS

9.2.4.1 Dual ion-source low-energy ion-beam systems

One of the important requirements for a film-deposition technology is to allow deposition of multicomponent films and films composed of multiple layers of different materials without breaking vacuum. Although it is possible to achieve such films using a single ion source, this ability is limited by the materials that are compatible for use in a single source. More advanced approaches are the use of multiple ion sources and/or the combination of IBD with other deposition methods. Two types of dual-source LEIBS have been reported to date. The principal difference between these is the method of combining the two beams and the closely related problem of deceleration of the ions before they impinge onto the substrate.

The first type of these instruments combines the high-energy ion beams from the two sources in a single beam line and uses common deceleration ion optics as shown in Figure 9.5 (Gordon *et al.*, 1991a). The advantage of this design is that the ion beams arrive at the substrate with perpendicular incidence and well-established deceleration optics can be used. Combining the two beam lines can be accomplished using an electromagnet with switching of the poles (Gordon *et al.*, 1991a; Shimuzu *et al.*, 1991), or using electrostatic deflectors (Gottdang *et al.*, 1991).

The second type of dual-source LEIBS uses two entirely independent ion beam lines as shown in Figure 9.6. The ion beams meet only in the vicinity of the substrate (Ohnishi *et al.*, 1989; Yoshida *et al.*, 1989; Tamura *et al.*, 1989;

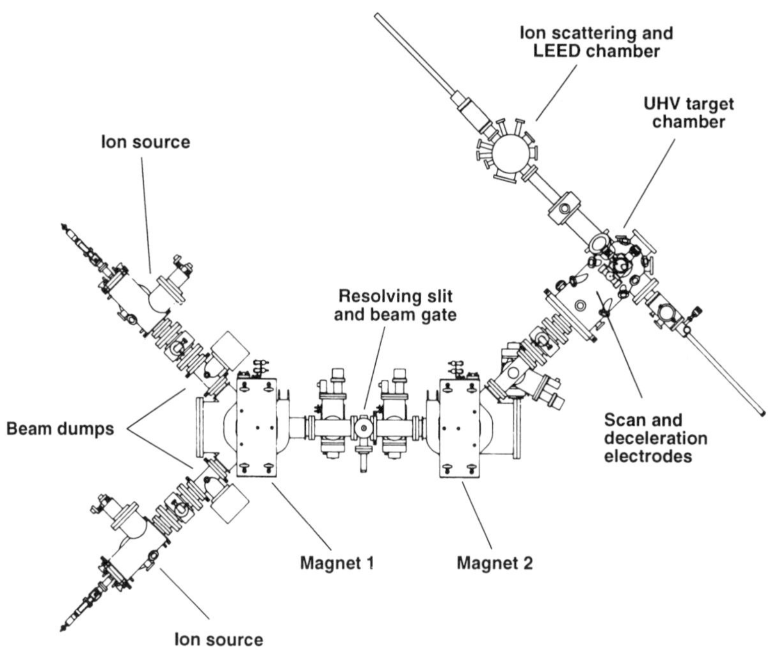

Figure 9.5 Plan view of the dual-source mass-analyzed LEIBS built by VSW Scientific Instruments, England for the University of Houston. The deposition chamber is equipped with AES and RHEED. The films can be transferred from the deposition chamber into the ion-scattering and LEED chamber built in Houston under UHV conditions.

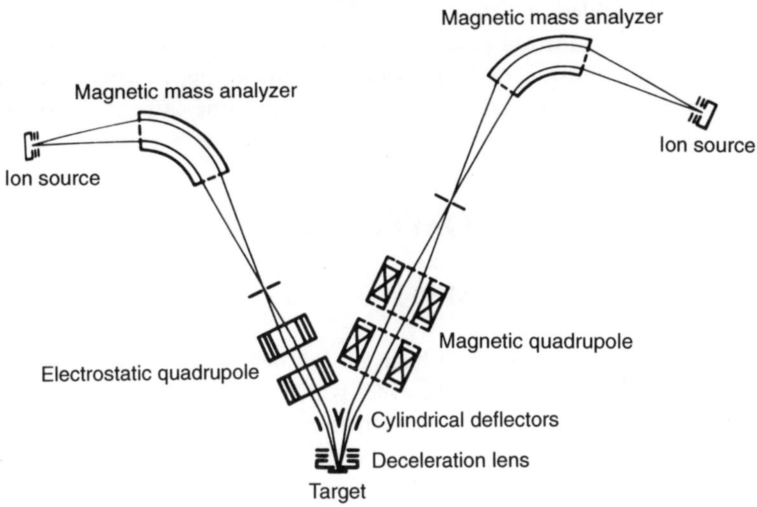

Figure 9.6 Plan view of a dual-source mass-analyzed LEIBS with separate beam lines (Reproduced by permission of the American Physical Society from Qin *et al.*, 1991a)

Qin *et al.*, 1991a). In these instruments, either one (Ohnishi *et al.*, 1989; Yoshida *et al.*, 1989) or both ion beams (Tamura *et al.*, 1989; Qin *et al.*, 1991a) arrive at the substrate surface with an incident angle other than 90°. In the case of the first type of dual-ion-source instruments, fast beam switching is a very important and difficult task; in the case of the second type, the two beam lines can operate continuously. Beam switching may still be necessary to control the arrival ratio of the ion species.

To date, there is very limited information available on the performance of any of these systems and there is not enough experience to argue in favor of certain solutions as compared with others.

9.2.4.2 Combination of IBD with other techniques

Combination of low-energy ion-beam deposition with other film-deposition technologies, especially with MBE, is a very promising and quickly developing area. This is easy to understand since such a combination has two obvious potential advantages. First, each material is best deposited by a particular method and, second, the low-energy ions from a single ion beam can deliver sufficient extra kinetic energy to enhance surface mobility. Hence, the advantages of IBD can be complemented in such instruments with the advantages of other film-deposition methods. A further practical advantage is that existing deposition systems can be equipped with additional facilities. For example, ion beam lines can be added on existing MBE instruments (Miyazawa *et al.*, 1984b; Shimizu *et al.*, 1985; Gordon *et al.*, 1991b). Also, existing ion beam lines could be equipped with effusion cells, provided modifications to the deceleration system are made. As IBD becomes more established, one can expect to see this latter type of instrument combination emerging.

9.3 ION-BEAM-DEPOSITED FILMS

9.3.1 IBD CARBON FILMS

Deposition of diamondlike carbon from an ion beam (Aisenberg and Chabot, 1971), known as i-C, was perhaps the event which initiated serious investigation of the practical application of IBD. Even though the instrument used in this early IBD experiment was not equipped with a mass selector, due to a special design it probably yielded fairly high-purity films. Among other features, the best films obtained were characterized by electrical resistivity of $10^{11}\ \Omega$ cm or more (compared with diamond's approximately $10^{14}\ \Omega$ cm, Samsonov, 1968), dielectric constants of 14 (compared with 16.5 for diamond), high transparency, and an index of refraction in excess of 2.0 (compared with 2.4 for diamond). Although structure could not be firmly established, as is common with almost all i-C films deposited to date, the above characteristics were of sufficient interest to make i-C deposition one of the primary areas of IBD research.

Since this pioneering work in 1971, carbon-film deposition has been pursued very successfully by various technologies that generally involve decomposition of carbon-containing gases (most often methane) and direct deposition on a heated substrate. The substrate temperature must be 850 °C to 1050 °C in order to obtain good-quality films (for reviews see e.g. Bachmann and Meisser, 1989 and Tsai and Bogy, 1987). A number of analytical techniques confirm that films deposited by these methods contain crystalline diamond. The most credible test is probably Raman spectroscopy which normally reveals a weak but clear and sharp peak at 1332 cm^{-1}, which is characteristic for diamond and does not exist in graphite spectra. Scanning electron micrographs of micron-size diamond crystallites and clear electron-diffraction patterns are also convincing. In contrast, films obtained with direct ion-beam deposition always fail these tests. The reason for this may be related to the relatively small film thickness usually achieved and to the small crystallite size. Consequently, ion-beam-deposited carbon films do not meet the definition of diamond films (Bachmann and Meisser, 1989).

However, IBD allows deposition of carbon films that are characterized by excellent mechanical, electrical and optical properties while the substrate is held at room temperature. This alone makes IBD a competitive technology for diamondlike film production on substrates which cannot be exposed to high temperatures. This is so even though the films may not contain crystalline diamond and in spite of such technological shortcomings as the relative complexity, expense and low deposition rate.

Various qualities of i-C films have been measured by a number of laboratories using two different approaches. Structural and chemical characterization are helpful from the point of view of attempts to understand the deposition mechanism. Mechanical, electrical and optical characterization help to assess the potential usefulness of the deposits.

Significant efforts have been made using diffraction techniques to establish the crystallinity, and possibly the microstructure of i-C films. For example, X-ray and/or electron diffraction has been used by many authors (Weissmantel, 1981; Ishikawa *et al.*, 1987; Spencer *et al.*, 1976; Miyazawa *et al.*, 1984a; Chaikovskii *et al.*, 1981, 1986; Qin *et al.*, 1991b). Some of these authors reported polycrystalline films with very small crystallites (Spencer *et al.*, 1976). Chaikovskii *et al.* (1986) claim that their X-ray diffraction (XRD) data show hexagonal diamond crystallites of 5–10 Å size if primary-ion energies of 25–100 eV and deposition temperatures at or below 293 K are used. This result seems to disclaim earlier published data about crystallites with micron dimensions grown as inclusions in an otherwise amorphous film (Chaikovskii *et al.*, 1981). Other authors found their films to be amorphous (Weissmantel, 1981; Ishikawa *et al.*, 1987; Miyazawa *et al.*, 1984a; Qin *et al.*, 1991b).

Chemical examination reveals very little except for the existence of a carbidic interface layer for films deposited on metals (Kasi *et al.*, 1988a) or silicon (Qin *et al.*, 1991b; Kasi *et al.*, 1988a). Also inertness has been noted by several authors (Ishikawa *et al.*, 1987; Kasi *et al.*, 1988a). Contamination content (O, N, S) was below detection limits (Kasi *et al.*, 1988a), and no hydrogen could be detected by low-energy ion scattering or by Rutherford back-scattering (RBS) (D. Marton, M. Shi, J. W. Rabalais, and Y. Lifshitz, unpublished data, 1992). This information is important because it is assumed that the sp^3-type bonding in DLC films deposited from hydrocarbons is related to the stabilizing effect of hydrogen bonds (Badzian *et al.*, 1988).

There is remarkable consensus among many reports concerning the density of i-C films. The most reliable data are probably those obtained by a combination of RBS and surface profilometry which yield 1.7×10^{23} atoms/cm^3 (Anttila *et al.*, 1986); other methods also yield densities of the same order (Ishikawa *et al.*, 1987; Miyazawa *et al.*, 1984a; Chaikovskii *et al.*, 1981, 1986; Kasi *et al.*, 1988a). This is very close to the density of diamond, 1.77×10^{23} atoms/cm^3.

Several authors have noted the mechanical resilience of i-C films and that there is little strain in them (Ishikawa *et al.*, 1987; Spencer *et al.*, 1976; Freeman *et al.*, 1984). More quantitatively, wear measurements have revealed that i-C films can be 10 times more wear-resistant than TiN hard coatings but showed less wear resistance than diamond (Koskinen *et al.*, 1985). The Vickers hardness measured was $18\,000 \pm 2000$ kgf/mm^2 at 100 g load, higher than that of diamond (Koskinen *et al.*, 1985). Similar results were obtained on a film deposited by Lifshitz *et al.* (Y. Lifshitz, J. W. Rabalais, and A. Wu, private communication).

The electrical resistance of i-C films varies from $10^6\ \Omega$ cm to $10^{12}\ \Omega$ cm. Although these values are significantly below the resistivity of diamond, they are quite high. The highest resistivities were achieved at moderate ion energies of approximately 50 eV and at room temperature (Ishikawa *et al.*, 1987). All reports agree that the i-C films are transparent and sometimes colored; the IR transmittance is very close to 100% for films which are several hundred angstroms thick (Ishikawa *et al.*, 1987; Miyazawa *et al.*, 1984a). The refraction index in the visible region varies from 2 to 3.1 as measured by different investigators (Ishikawa *et al.*, 1987; Spencer *et al.*, 1976; Miyazawa *et al.*, 1984a; Kasi *et al.*, 1988a).

In conclusion, there is no reason to believe that significant improvement in the properties of i-C films cannot be achieved if systematic research is continued. For example, there is very little experience with deposition below room temperature although results obtained so far indicate that low substrate temperatures may be helpful in obtaining good i-C films.

9.3.2 IBD METAL FILMS

There is little information on the ion-beam deposition of metal films since most metals can be deposited using simple technologies. Some early IBD studies were directed to deposition of metal films with the aim of clarifying the possibilities and implications (Fair, 1971; Amano and Lawson, 1977; Amano, 1982; Freeman *et al.*, 1976; Tsukizoe *et al.*, 1977). Later research is mainly related to semiconductor research problems such as contacting materials to silicon (Yamada *et al.*, 1985; Shimizu *et al.*, 1991; Zuhr *et al.*, 1988; Bousetta *et al.*, 1992; Thomas *et al.*, 1982), and shallow implantation (Gottdang *et al.*, 1991).

Amano *et al.* (Amano and Lawson, 1977; Amano, 1982) deposited Pb and Mg films on carbon and NaCl(100) single crystals. They found that the films were pure and their density and structure depended on the ion energy. Maximum density, optimal continuous coverage, and maximum thickness were all achieved with deposition energies in the 50–70 eV range. They also observed using transmission electron microscopy (TEM) that the films were polycrystalline on carbon but epitaxial Pb films with fcc(100) orientation were also obtained on NaCl(100) with 120 eV ions. More importantly, scanning electron microscopy (SEM) micrographs show that the Pb deposits on carbon were often discontinuous, granular, with holes and/or channels between the grains.

The density of Mg films (Freeman *et al.*, 1976) deposited at 100–500 eV was also close to the bulk density. Electron micrographs reveal that the Ag film formation (Tsukizoe *et al.*, 1977) on an Al(110) single crystal starts with formation of small Ag particles. Then the Ag particles grow and coalescence takes place, resulting in some island formation. This process continues and the islands grow into a continuous film.

Ag (Thomas *et al.*, 1982), Ca (Shimizu *et al.*, 1991), Pd (Yamada *et al.*, 1985), Ti (Zuhr *et al.*, 1988), and Co (Zuhr *et al.*, 1988; Bousetta *et al.*, 1992) have been deposited on Si single crystals at room temperature, or elevated temperatures. Ag and Ca did not form silicides, and in both cases room-temperature epitaxial film formation was observed. The energy of the Ag^+ ions was above 25 eV and below 100 eV. The Ag on a Si(111) surface formed films with the Ag⟨111⟩ axis parallel to the Si⟨111⟩, and Ag⟨110⟩ parallel to Si⟨110⟩, as detected by reflection high-energy electron diffraction (RHEED) (Thomas *et al.*, 1982). The Ca epitaxy formed at room temperature was unstable, and decomposed at 50 °C. Stable Ca epitaxy was obtained when the substrate was held at 400 °C; according to RHEED data, the Ca⟨110⟩ axis is parallel to the Si⟨110⟩. The optimum ion energy is about 20 eV (Shimizu *et al.*, 1991).

Pd, Ti, Co, Fe, Ni and W all form silicides when deposited on silicon, both at room temperature and at elevated temperatures. When Pd was deposited at energies as low as 10 eV, islands were formed. At 20 eV, RHEED pictures reveal growth of a textured polycrystalline film. In the 50–100 eV range, the film is Pd_2Si with low damage and clean interface. Apparently, the interface remains unclean when ions with lower energy are used. The electrical resistivity of about 10 nm thick films grown at 50 eV ion energy is about $4 \times 10^{-4}\ \Omega$ cm (Yamada *et al.*, 1985).

Similar results were obtained with Co and Ti (Zuhr *et al.*, 1988). Stoichiometric $TiSi_2$ films were formed by Ti^+ IBD on a Si(100) substrate at 100 eV and 550 °C. RBS shows that the films are uniform, and sheet-resistance measurements yield 0.58 Ω/square which corresponds to a bulk resistivity of less than 15 μΩ cm, compared with 22 μΩ cm for a similar thermal film. At a lower temperature (400 °C), IBD with 100 eV ions results in TiSi films. The cobalt silicide films obtained in the same study at 500 °C and 40 eV were not stoichiometric and the Co concentration varied through the film.

In contrast, Bousetta *et al.* (1992) found that, according to medium-energy ion-scattering (MEIS) spectra, stoichiometric $CoSi_2$ was formed even at room temperature with a 50 eV $^{59}Co^+$ ion beam on a Si(111) surface. Evidently, diffusion alone establishes the stoichiometry. However, electrical measurements, which are much more sensitive to fine structural and compositional differences than is MEIS (or other analytical techniques), show significant changes due to deposition at elevated temperatures or annealing. The electrical resistivity of the films deposited at 400–600 °C is 20–40 μΩ cm, which is four to eight times below the resistivity of those films deposited at room temperature. Similar values were obtained by 500–600 °C annealing of films deposited at room temperature, but at higher annealing temperatures instabilities were observed. It is possible that films obtained at room temperature contain defects (e.g. pinholes) that will not anneal perfectly.

A significant difference between the cobalt silicide films of these investigators is the thickness of their films: ≈1000 Å (Zuhr *et al.*, 1988) and 60–120 Å (Bousetta *et al.*, 1992). As Zuhr *et al.* (1989a) pointed out, the difference between the silicide formations of Ti and Co is that in the first case Si is the dominant diffusant, and in the second case it is the metal. As the films grow thicker, the ability to supply silicon diminishes. This effect explains the differences between the results for Ti and Co, and between the results of Zuhr *et al.* (1988) and Bousetta *et al.* (1992). In subsequent experiments, Zuhr *et al.* (1989a) deposited WSi_2 and $FeSi_2$ for which Si is the main diffusant, and Ni silicide films for which the metal is the main diffusant. The WSi_2 film was obtained on a Si(100) substrate at 500 °C with 200 eV ions. The $FeSi_2$ film was grown on a similar substrate with 40 eV ions at 550 °C. Ni^+ ions at 100 eV were used for deposition on a p-Si(100) substrate at 560 °C. As in the case of Co, the films were uneven. Electron micrographs revealed rodlike formations aligned in the ⟨001⟩ direction in the surface plane.

9.3.3 IBD SEMICONDUCTOR FILMS

9.3.3.1 Elemental semiconductors

Low-temperature epitaxial growth of semiconductors is a very important area of research. Although there is consensus among IBD researchers that, at least in principle, epitaxies which are useful for electronic-device applications can be achieved, the research effort made to systematically explore this possibility is far from adequate.

Yagi *et al.* (1977) achieved epitaxial growth of Ge on Ge(111) surfaces using substrate temperatures as low as 300 °C but their attempts with silicon homoepitaxy failed. The crystallinity of the films was checked only by RHEED, and this method does not yield much information on their quality. Although deposition of Ge heteroepitaxy on Si under similar circumstances was successful, secondary-ion mass spectrometry (SIMS) depth profiles showed that the problem was that the silicon surface was contaminated. This is understandable since the vacuum in the deposition chamber was inadequate (1×10^{-5} torr). These results were helpful in raising one of the most important questions related to IBD of epitaxial films: What is the significance of the surface cleaning for this process? The answer is most difficult in the case of silicon which is very sensitive and reactive.

Subsequently, Si epitaxy was achieved on Si in a number of laboratories. Miyake and Tokuyama (1982) obtained films on Si(100) using 200 eV $^{28}Si^+$ ions at 740 °C although both RHEED and TEM showed twin structures in the films and they were rather defective with many dislocation loops. The achievement of Si epitaxy was considered to result from improved vacuum conditions compared to those of Yagi *et al.* (1977). Zalm and Beckers (1982) grew Si epitaxial films on Ge(100), Si(100) and Si(111) surfaces, in all cases with 50 eV ions and the substrate at 400 ± 25 K. In their view, the UHV ($<10^{-7}$ Pa) was essential for this result. No epitaxy could be obtained if the substrate surface was exposed to water vapor.

Zuhr *et al.* (1987) obtained Si epitaxial films from $^{30}Si^+$ ions on Si(100) surfaces with no *in situ* cleaning. The ion energy was varied from 40 eV to 200 eV, and the substrate temperature varied in the 700 K to 900 K range. The crystallinity and film quality were examined by XRD, RBS and TEM; it was concluded that the films had a density of 75% of the bulk silicon. Better films were obtained after *in situ* cleaning with 20 eV Cl^+ ions, and 10 eV deposition ion energy on n-Si(100). For these films, the dechanneling level was $\chi = 3.6\%$ which is a value close to the level for bulk Si. Interestingly, films on p-Si(100) were amorphous when deposited under similar circumstances. The film densities for the better epitaxial films were very close to the bulk Si density and the interface widths to the Si substrate, evaluated from RBS and TEM data, were nearly 1 nm (Herbots *et al.*, 1988). Still better films were produced when the substrate treatment included a short annealing after the Cl cleaning. On such substrates, films with $\chi = 3\%$ were obtained using 30 eV $^{30}Si^+$ ions at a substrate temperature of 400 °C (Zuhr *et al.*, 1989b). TEM showed a clear film–substrate interface and several threading dislocations across the film. It is worthwhile to emphasize that the film growth rate at 400 °C temperature was about 0.6 nm min^{-1} which is nearly 100 times the thermal growth rate for this temperature (Herbots *et al.*, 1988).

Silicon epitaxial films were also obtained on p-Si(001) (Orrman-Rossiter *et al.*, 1990). In this study, the *in situ* cleaning included annealing at 850 °C, which was followed by the deposition, by further reactive ion cleaning, or by self-ion cleaning. Only the immediate deposition led to epitaxial growth, apparently due to damage from 500 eV Si^+ or due to residual Cl. Extended energy-loss fine-structure (EXELFS) measurements showed that the films which in TEM seem to be epitaxial are only $72 \pm 5\%$ crystalline. EXELFS provides information on the localized short-range order. More information on possible CO contamination and methods of obtaining good Si epitaxy were reported by Al-Bayati *et al.* (1992), as mentioned in Section 9.3.2 in relation to mass selection (cf. Figure 9.2). However, the optimization of IBD for Si epitaxial films will require much more research. In particular, there are no data available on the electrical properties of the films.

Germanium epitaxial films obtained by Yagi *et al.* (1977) were analyzed by TEM and RBS (Tokuyama *et al.*, 1981). The films were found to have quite defective structure with many small dislocation loops. The films deposited on Si, somewhat surprisingly, were of better quality than the homoepitaxial ones. RBS showed significant mixing at the interface, which could be related to the better film quality. As in the case of Si, lower beam energies produce better results. Thomas *et al.* (1982) produced Ge epitaxial films on Si(111) at 230 and 350 °C using 50 eV ions, but the film quality was checked only with RHEED. Zuhr *et al.* (1987) deposited $^{74}Ge^+$ on Si(100) and Ge(100). 40 eV ions resulted in very good epitaxy on Ge at 400 °C substrate temperature. The dechanneling level in RBS was $\chi = 4\%$ compared with 3% in the bulk, and TEM also revealed some defects in the film. Epitaxial films could be grown at lower temperatures, even at 200 °C (Zuhr *et al.*, 1989b) with 30 eV ions, although these films were not as good as those grown at 400 °C which were characterized by a dechanneling level of only 3%. Ge films grown on Si(100) substrates under similar conditions showed a much higher dechanneling, 12%, due to defects at the interface (Zuhr *et al.*, 1989b). In the case of a GaAs(100) substrate, where for Ge there is virtually no lattice mismatch, there were still some interface defects but no extended defects could be seen in TEM and the dechanneling was as in the bulk (Zuhr *et al.*, 1989b).

9.3.3.2 COMPOUND SEMICONDUCTORS

9.3.3.2.1 β-SiC (3C-SiC) epitaxial growth

β-SiC may be an important semiconducting material characterized by a wide band gap (2.2 eV) and high electron mobility (1000 cm^2/Vs). It has been produced as a heteroepitaxial film by chemical vapor deposition (CVD), or MBE, however these technologies require a substrate temperature of 1000–1400 °C. Lower deposition temperatures would be very desirable and IBD may be the answer (Yamada, 1989).

β-SiC was deposited by a combination of a C^+ ion beam and Si MBE by Miyazawa *et al.* (1984b). The deposition of β-SiC on Si(100) is difficult because of the big lattice mismatch (20%). They achieved epitaxial growth, as evidenced by RHEED data, using 50–200 eV ions and 820–1000 °C substrate temperatures. Auger electron spectroscopy (AES) depth profiles showed clean films with constant C/Si = 1 stoichiometry. Polycrystalline growth was observed at higher ion kinetic energies and at lower temperatures. It was also observed when the substrate was thermally cleaned rather than using a chemical cleaning (HNO_3–HF); this thermal cleaning probably resulted in an intermediate carbide layer when the sample was annealed in vacuum. On thermally cleaned samples, epitaxial growth was only observed at temperatures of 1000 °C and 1100 °C.

Withrow *et al.* (1989) produced β-SiC epitaxial films on α-SiC, which, being a perfect lattice match, is significantly easier to accomplish than epitaxial growth on Si. $^{13}C^+$ and $^{30}Si^+$ ions at 40 eV were alternately used to produce films on an α-SiC(001) substrate which was held at 700 °C. The films were single crystalline as demonstrated by high-resolution electron microscopy (HREM) with a (111) orientation according to optical diffraction data. The films produced by sequential deposition of equal doses of C and Si produced a 30%

excess C concentration in the film and this excess carbon may be responsible for its poor structure. Defects observed in TEM included amorphous areas, dislocations and twins, and, according to RBS, a large number of point defects were also present.

9.3.3.2.2 Epitaxial growth of III–V compounds

The sticking probabilities of Group V elements with thermal energies are usually small as compared to those of Group III elements. This translates into a difficulty in depositing III–V compounds, including ternary and quaternary compounds, such as GaAsP and InGaAsP, using thermal methods (Shimizu *et al.*, 1985). This problem may be solved using either dual-source IBD, or IBD combined with MBE for the Group III element.

Homoepitaxial growth of GaAs and InP was carried out in a system combining MBE with a low-energy ion beam (Shimizu *et al.*, 1985). The As^+ and P^+ ion energies were 100 eV or 200 eV; at higher kinetic energies sputtering becomes significant. For GaAs, Cr-O doped semi-insulating GaAs(100) was used as the substrate. The deposition was carried out at temperatures between 220 and 550 °C. According to RHEED data, the film growth is two-dimensional; no surface roughness, twinning, or faceting was observed. The ion/atom flux ratio, n/m, varied from 1 to 2.2, leading to excess As in the film for $n/m > 1$. This excess As concentration depended on the deposition temperature but at 500 °C the films were always stoichiometric due to the evaporation of the excess As. Photoluminescence (PL) spectra were characteristic for a nearly perfectly annealed crystal when the substrate was at 550 °C during deposition.

InP was deposited on an Fe-doped semi-insulating InP(100) crystal. Single-crystal films were obtained at substrate temperatures between 210 and 420 °C. The RHEED data indicated that above 390 °C the film growth is two-dimensional and the quality is comparable to what is observed in MBE. PL data indicate that the films grown at 390 °C are better than those obtained at higher temperature, probably because of loss of some phosphorus due to desorption (Shimizu *et al.*, 1985).

In similar experiments, InP and $GaAs_xP_{1-x}$ films were deposited in an MBE system equipped with an IBD source (Maruno *et al.*, 1988a) For InP, the results were in general agreement with Shimizu *et al.* (1985). Unintentionally doped InP films were of n type, with about 1000 cm^2/Vs Hall mobility, and 10^{15} cm^{-3} carrier concentration for films grown below 400 °C. For films deposited at higher temperatures, the Hall mobility was higher, e.g. 3100 cm^2/Vs measured at room temperature for films grown at 480 °C. This mobility value is as high as the mobility of MBE-grown InP films. The PL yield was also optimal at 480 °C deposition temperature. However, at 490 °C, electron micrographs show In precipitations in the films (Maruno *et al.*, 1987). SIMS data show remarkably uniform film compositions except at the surface of the film where Fe from the substrate and, to a lesser extent, P pile up (Morishita *et al.*, 1988).

For $GaAs_xP_{1-x}$ deposition, Cr-doped semi-insulating GaAs(100) substrates were used (Maruno *et al.*, 1988b). As_4 and Ga were deposited from molecular beams and the P^+ ion energy was varied from 50 eV to 250 eV. The As atomic fraction, x, increases from about 0.5 to 0.67 with increasing ion energy, and the growth rate decreases at the same time. In this experiment, the arrival ratios were kept constant, $As_4/Ga \approx 12$. Sharp RHEED patterns were obtained when the deposition temperature was 600 °C. The lattice mismatch between the film and the substrate is estimated to be about 3.6% from XRD data. This value is close to that for GaP on GaAs, which indicates that the stress in the film is relaxed. The morphology of the films grown at low ion energy (< 100 eV) is better than what is observed for MBE films where carbon contamination can cause characteristic oval defects.

Haynes *et al.* (1989) deposited GaAs films on Si(100) and Ge(100) substrates from $^{69}Ga^+$ and $^{75}As^+$ ions using the same ion source and switching the mass selector. Films deposited on Si at 400 °C with 40 eV ion energy were found to be epitaxial and stoichiometric, indicating complete incorporation of both types of ions. The film density, from RBS and TEM, is equal to 4.4×10^{22} cm^{-3}, the same as for bulk GaAs. High dechanneling yields indicated defective epitaxy and TEM pictures revealed a high density of microtwins and stacking faults originating at the interface. These defects may be due to the large lattice mismatch between Si and GaAs (4.1%) and they are common in MBE-grown films (Iyer *et al.*, 1989).

In a similar experiment, using two independent ion sources, GaAs films were grown on Si(100) substrates at 500 °C (Tamura *et al.*, 1989). The energies of both ions were the same and were varied from 40 eV to 240 eV. At energies above 60 eV, there is no film growth, since sputtering exceeds deposition. Films deposited at 40 eV showed crystalline RHEED patterns but SEM images revealed spotty film surfaces. The film lattice parameter was about 5.672 Å from XRD data, as opposed to 5.653 Å for bulk GaAs. RHEED measurements combined with gradual etching of the film reveal an amorphous intermediate layer between the Si substrate and the film. This amorphous layer probably serves as a buffer to relieve lattice-mismatch stress.

There is virtually no lattice mismatch between GaAs and Ge and the films obtained on Ge were indeed of much higher quality. Films were grown using 30 eV ions at 320–520 °C temperatures. The lowest RBS dechanneling was observed in the films grown at 520 °C, which were comparable to horizontal Bridgman-grown films (Haynes *et al.*, 1989).

GaN films were produced in a dual-beam-line IBD machine on Si(111) and sapphire (1102) substrates at 600 °C and 100 eV ion energy (Qin *et al.*, 1991a). Crystalline (according to RHEED) and stoichiometric (according to AES) GaN was formed when the N/Ga arrival ratio exceeded 1.5, indicating a limited incorporation probability for nitrogen.

9.3.4 IBD OF INSULATING FILMS

Very thin insulating films, mostly oxides and nitrides, can be obtained by simply using O^+ or N^+ ion beams and allowing a chemical reaction with the substrate to occur. We will call this process ion-beam oxidation or nitridation, as opposed to ion-beam deposition of oxides and nitrides; this latter process occurs when both compound-forming species are deposited simultaneously.

9.3.4.1 Ion-beam oxidation and nitridation

Most oxides and nitrides of interest are dense materials with significant ionicity and are not characterized by large diffusion rates at low temperatures. Hence, unlike the silicides in Section 9.3.2, the films obtained are very thin. For example, N_2^+ bombardment of boron at 500 eV to 4 keV ion energies at room temperature leads to formation of BN films with maximum thickness of about 40 Å according to AES depth profiles (Mezentzeff *et al.*, 1990). The films are stoichiometric according to X-ray photoelectron spectroscopy (XPS) data but the film thickness is limited by the ion range rather than the dose, indicating that the process involves sputtering of the film.

The quality of ion-beam silicon dioxides produced by broad-beam low-energy machines using a mixture of oxygen and argon is claimed to be suitable for use as the gate dielectric of metal-oxide-semiconductor (MOS) transistors (Todorov *et al.*, 1986). Stoichiometric SiO_2 was obtained at room temperature by 100 eV O_2^+ ion bombardment (Todorov and Fossum, 1988). Data from ellipsometry show film thicknesses which saturate at about 40 Å as the ion dose increases beyond 10^{17} cm^{-2}. In the range between 40 eV and 140 eV, no significant energy dependence was observed. In broad agreement, good-quality 70 Å thick SiO_2 was obtained with 200 eV O^+ ions on Si(100) at 400 °C (Herbots *et al.*, 1988).

Oxides and nitrides were formed in separate experiments using 100 eV and 200 eV $^{18}O_2^+$ and $^{14}N_2^+$ ion beams on $Si_{0.8}Ge_{0.2}$ substrates at room temperature (Vancauwenberghe *et al.*, 1991a). According to XPS data, fully oxidized and fully nitridized films form when sufficient doses are employed. At low kinetic energies the films retain the substrate concentration ratio of Si/Ge but at higher kinetic energies Ge is sputtered preferentially. The dielectric films were stable when annealed up to 500 °C.

9.3.4.2 Ion-beam synthesis of BN

Like carbon, BN can be hexagonal or cubic. c-BN is a material characterized by hardness second only to that of diamond and with chemical inertness which is superior to diamond's. So, the quest for superhard films which started with the deposition of dense carbon films naturally led to attempts to deposit boron nitride. Early results were obtained by Weissmantel (1981) and Shanfield and Wolfson (1983) using ion beams extracted from borazine ($B_3N_3H_6$) plasma. As in the early carbon-deposition experiments, no mass filter was applied, and as in the case of carbon films deposited from plasma, the hydrogen content of the films may have been significant. XRD analysis showed that the films were c-BN and substrates with the BN coating showed increased hardness compared to hard metal substrates such as TiN-coated WC.

More recently, c-BN has been deposited in systems combining vapor deposition of boron with nitrogen ion beams (Sainty *et al.*, 1988; Andoh *et al.*, 1988). As in the case of III–V compounds, an excess in the arrival ratio of the volatile component is required for stoichiometric films; the necessary excess nitrogen increases with the substrate temperature and ion kinetic energy. Higher energies (>200 eV) are less effective for deposition due to sputtering losses. Films grown at room temperature exhibited c-BN XRD patterns (Andoh *et al.*, 1988), while films grown at 300 °C were h-BN, as follows from their density and π-loss features (Sainty *et al.*, 1988). Unfortunately, no experiments have been reported for BN obtained from pure ion beams.

9.3.4.3 Other nitrides and oxides

In an early experiment, silicon and aluminum nitrides were obtained by a combination of Si or Al sputter deposition with 1 keV N_2^+ ion bombardment (Erler *et al.*, 1980). The electrical properties of the films were inferior to films obtained with radio-frequency sputtering, which was probably a result arising from ion-beam damage.

More recently, TaO_x, WO_x, MoO_x and TaN_x films, and also stoichiometric Ta_2O_5 films were obtained in a dual-ion-beam system (Ohnishi *et al.*, 1989; Yoshida *et al.*, 1989). Ta_2O_5 films were deposited on n- and p-Si(100) as the gate insulator in a metal–insulator-silicon (MIS) transistor structure from 100 eV ions at room temperature. Since the oxygen beam line is not mass-separated in this instrument, both O^+ and O_2^+ ions were probably present in the oxygen beam. After a 500 °C annealing and gate metal deposition, I–V and C-V characterization was carried out. These measurements revealed leakage currents in the nA range (for an area of 0.12 mm^2) and a large fixed charge which Ohnishi *et al.* (1989) believe is associated with radiation-induced defects due to fast neutrals and electrons. At ion energies lower (60 eV) and higher (140 eV) than the optimal 100 eV, the oxide films are not stoichiometric, as XPS spectra show. The results obtained show that it will require a large systematic effort to optimize ion-beam-deposition conditions for each oxide film.

9.4 PROCESSES IN ION-BEAM DEPOSITION

9.4.1 PROCESSES IN SOLIDS UPON ION IMPACT

Ion-beam deposition involves the collision of the ions to be deposited with a solid surface. This is a complex process which can take several possible paths depending on the physical and chemical properties of the system which includes the ion and the solid, and on the collisional parameters of the individual ions. Some of these processes are shown schematically in Figure 9.7. The processes of the collision may occur prior to, during, or after the impact, which is the initial close encounter between the ion and a surface atom (primary collision event).

9.4.1.1 Prior to impact: charge exchange

Most deposition experiments utilize positive ions. An exception is deposition of carbon films from C^- ions by the Takagi group at Kyoto University (Ishikawa *et al.*, 1987). At the energies used in IBD it is safe to assume that the ions will be neutralized upon approaching the surface. Reactive ion scattering from a Ni surface has shown that the neutralization efficiency is greater than

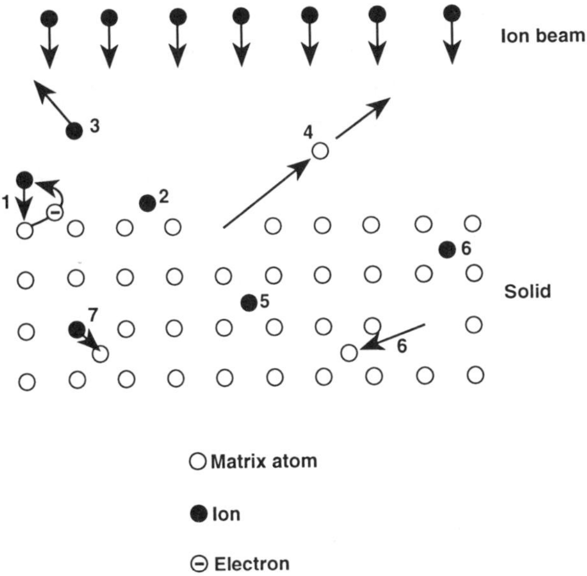

Figure 9.7 Processes in solids upon ion impact: (1) charge exchange, (2) adsorption, (3) back-scattering, (4) sputtering, (5) interstitial without displacement, (6) interstitial with Frenkel pair, (7) replacement

99% for 50–200 eV C^+ and O^+ scattered at a 45° specular angle (Kasi *et al.*, 1988c). This process occurs within a few angstroms of the surface through either resonant or Auger electronic processes (Hagstrum, 1978). The charge-exchange process may contribute to the complexity of the physical processes in IBD when films with good insulating properties (e.g. i-C) are deposited and/or when insulating substrates are used, since electrical charging and breakdown processes may also occur. These effects may be minimized by applying electron-flood sources (Y. Lifshitz, private communication). Even though the particles actually reaching the substrate in IBD are neutralized, we will keep referring to them as ions in order to make a distinction between these particles and those of the substrate.

9.4.1.2 Impact: primary collision event

When a projectile (ion, atom, or molecule) reaches the substrate surface, i.e. at the moment of the impact, the outcome will depend on a number of circumstances. If the projectile has enough kinetic energy, it will *penetrate* below the surface. Or the projectile may be *back-scattered* (reflected back) into the vacuum with a probability determined by the atomic masses of the projectile and the solid, by the solid structure, and by the temperature. If the kinetic energy of the projectile is below a certain threshold, it may not be able to penetrate below the surface and if it will stay on the surface, this process will be *surface deposition*, a process like adsorption. This is the only possible outcome when the kinetic energy of the projectile is very low, i.e. for evaporation and MBE, and perhaps also in sputter deposition.

The kinetic-energy threshold for penetration is not well known, although it has long been shown to exist for noble-gas ion penetration in metals (Kornelsen, 1964). In a recent measurement we found that Kr^+ ions do not penetrate into a graphite substrate below a threshold energy of 45–50 eV (Marton *et al.*, 1993). Although it is likely that in many cases the penetration threshold may be lower (perhaps just a few eV), this example shows that there is a minimum energy below which ion-beam deposition is not truly a subsurface process.

Herbots *et al.* (1988) describe the following experiment: a 10 nm thick Ge film was deposited from 40 eV ions on a Si(100) surface from which the native oxide was not removed. The original amount of oxygen on the surface was 10^{16} atoms/cm^2 according to ion-scattering data. This corresponds to an oxide layer which is 2.5 nm thick. The penetration depth of the Ge^+ ions is less than this thickness and so the film forms originally inside this oxide. The film formed is amorphous and the oxygen detected by AES depth profiling at the Ge–Si interface is only 1.4×10^{14} atoms/cm^2. The plausible explanation is that most of the oxygen remained on the surface during the deposition which occurred subsurface. The authors call this process 'anchoring' the film.

9.4.1.3 After the impact: film growth

9.4.1.3.1 Stopping

Low-energy ions typically will not have enough energy to develop a full-blown collision cascade. For that, an energy transferred in a first collision equal to at least several times the displacement energy (E_d) is necessary. Most metals have displacement energies in the 20–40 eV range; the displacement energies for semiconductors are usually less; for oxides, the displacement energy is typically higher. (See the compilation of available measured E_d data in the Appendix to this chapter.) The stopping of the impinging ion in IBD is therefore a process which is most often determined by the primary collision event. Before we continue a description of the possible outcomes, it will be useful to reconsider the displacement process and the concept of displacement energy for the case of ion impact.

The displacement energy is defined as the minimum energy that must be transferred to a substrate atom in order to create a stable Frenkel pair, i.e. a vacancy and interstitial removed sufficiently far from each other so as to have a negligible recombination probability. When the masses of the projectile (M_1) and the substrate atoms (M_2) are different, the maximum energy transferred, T, in a collision is less than the kinetic energy of the projectile, E_k:

$$T = 4E_k M_1 M_2/(M_1 + M_2)^2. \tag{3}$$

The displacement threshold, E_{th}, is the projectile kinetic energy that can result in the transfer of the displacement energy, E_d, to a substrate atom:

$$E_{th} = E_d(M_1 + M_2)^2/(4M_1 M_2). \tag{4}$$

The displacement energy concept (for a review see Vinetskii and Kholodar, 1986) was developed for high-energy radiation impact and it is fully compatible with the usual method of E_d measurements using electron irradiation. However, in the case of low-energy ion impact, stable defect structures other than Frenkel pairs may be formed at energies below E_{th}.

First, an ion with sufficient energy for penetration but less then E_{th} may simply stop in an interstitial site without displacing a substrate atom and without creating a vacancy. Second, an ion with kinetic energy below E_{th} may knock out a substrate atom from its lattice position and replace it, thereby forming a permanent defect. In fact, two defects will be created if the ion replaces a different type of atom. Because of these possibilities, it should be expected that the threshold for permanent-defect production will be significantly below the displacement energy. Some defects are obviously useful for the deposition process; it is these interstitials which contribute to film growth. It is likely that in most cases ions with energies sufficient to penetrate and perhaps even replace but not create Frenkel pairs, i.e. with energies just below E_{th}, may be optimal for deposition of high-quality films.

In practice, because of technical limitations pertinent to low-energy ion beams, beams with energies that significantly exceed E_{th} are very often used. This is true in particular in the case of semiconductor epitaxy by IBD because of the generally low values of the displacement energies of semiconductors. For example, experiments of Si and Ge epitaxy by Herbots *et al.* (1988; Vancauwenberghe *et al.*, 1991b) were mostly carried out at 40 eV, i.e. several times the displacement energy. Naturally, a very high number of interstitials were created during these experiments, and as calculations by Vancauwenberghe *et al.* (1991b) show, these interstitials are sufficiently mobile even at room temperature in Si to diffuse deep in the substrate and create extensive defects (dislocation loops) which can be readily observed.

9.4.1.3.2 Diffusion and hopping

The excess atoms in a near-surface region of the substrate will diffuse away in order to restore the minimum chemical potential situation. This phenomenon will take place even when a monoatomic substrate and ions of the same material are used such as mentioned above for the case of semiconductor homoepitaxy and also in experiments of carbon IBD on diamond (Freeman *et al.*, 1984) and graphite (Steffen *et al.*, 1991). The less diffusivity the system has, the denser are the films that can be obtained. High diffusivity of the film and/or substrate atoms may preclude obtaining pure films and may reduce the attainable density. For example, no diamondlike films could be produced on a substrate containing lithium (Lifshitz *et al.*, 1990) and silicide films rather than metal films have been obtained when Ti or Co was deposited on Si at elevated temperatures (Zuhr *et al.*, 1988; Bousetta *et al.*, 1992). Also, diamondlike films were not produced on metal substrates in experiments at elevated substrate temperatures (150 °C) in cases where such films were formed at room temperature (Lifshitz *et al.*, 1990). In contrast, Freeman *et al.* (1984) claim that diamond homoepitaxy was grown by 900 eV carbon ions at 700 °C substrate temperature.

The elementary process in diffusion is hopping, i.e. site exchange between atoms with each other or between atoms and vacancies. Hopping is probably also responsible for the ordered growth of certain films deposited by IBD including low-temperature epitaxial films. Diffusion has been known for some time to be stimulated by keV ion beams (Ho, 1978). This process includes the creation of defects which facilitate diffusion, and also the creation of concentration gradients due to preferential sputtering. In IBD, even at very low energies, many defects may be created (Zuhr *et al.*, 1989b; Orrman-Rossiter *et al.*, 1990) together with significant concentration gradients. Still, these defects may not increase diffusion very significantly because they may be confined to a narrow subsurface region. If this is the case, the main process will be a mere rearrangement of the structure due to hopping with no material transfer, i.e. no diffusion.

Herbots *et al.* (1988) argue that the total number of atomic jumps induced by an incoming ion, N_d, can be considered as the sum of replacements, N_r, and displacements, N_v. The ratio of these

$$R_o = N_r/N_v, \tag{5}$$

must be $R_o > 1$ for epitaxial growth rather than defect production to occur. However, they assume that displacement collisions do not take place below the displacement threshold, an assumption which we believe, leads them to overestimate R_o quite significantly.

9.4.1.3.3 Chemical bonding

Chemical reactions (in a broad sense, i.e. including bonding between atoms of the same species) and chemical potential differences which cause diffusion play prominent rôles in all deposition processes. Chemical forces are responsible for surface deposition, and in the case of incorporative adsorption, even ions with very low kinetic energies, insufficient for kinetic penetration, will occupy subsurface positions. These processes, however, do not lead to particularly high-density films and lack all the advantages of IBD.

Chemical forces may play an important role in low-energy ion–solid collision processes; however, very little is known concerning this topic (Dodson, 1990). The deviations from the binary collision picture at these low energies are obvious. It is necessary to acknowledge the effects of attractive forces rather than merely use repulsive ones as in well-known simulation codes like MARLOWE (Robinson and Torrens, 1974; Hou and Robinson, 1979) and TRIM (Biersack and Haggmark, 1980). However, the potentials to be used are not known and the treatment of such situations is not well established. It is likely that the lower the kinetic energy of the impinging ions, the larger will be the influence of chemical factors.

The most significant scenarios of deposition from the point of view of chemical bonding are heterodeposition, i.e. the starting phase of deposition on a substrate different from the film, and deposition of compounds.

In any case, part of the ion kinetic energy may help to overcome reaction barriers. Kinetic energies which may be transferred to substrate atoms in the collisional event of IBD, even if relatively low, significantly exceed thermal energies due to usual substrate heating. This enables chemical reactions otherwise impossible or very unlikely due to energy barriers to occur and compounds which otherwise can only be created under very extreme circumstances may form. On the other hand, too much excess energy may drive a chemical compound (or a structure) from a desired metastable state into a stable one. It is likely that the deposition of some compounds will have to be optimized from this point of view.

Ion-beam nitridation and oxidation can be viewed as examples of utilizing chemical reactions in IBD. Two different approaches exist: (i) reactive ion beams can be used to form thin overlayers with the substrate materials (Todorov and Fossum, 1988; Vancauwenberghe *et al.*, 1991a; Trehan *et al.*, 1990) or (ii) both film constituents are deposited simultaneously (Ohnishi *et al.*, 1989; Yoshida *et al.*, 1989). Interest in both approaches is driven by the possibility of forming films at relatively low substrate temperatures. In experiments of type (i), not only chemical forces but also diffusion plays a great role since reactive material has to be delivered to the reaction area.

9.4.1.4 After the impact: additional effects

9.4.1.4.1 Sputtering

Ions impinging on solid surfaces with transferable kinetic energies exceeding the solid's surface-binding energy will have a non-zero probability of inducing sputtering, i.e. to causing atoms to leave the surface. Although at typical IBD energies this is a small effect, it is mentioned because the early IBD experiments were carried out often at relatively high energies (several hundred eV) and in fact it constitutes a physical limit for IBD, which has been recognized for a long time. Using as high ion energies as possible is attractive because the technological problems related to the handling of the ion beams becomes increasingly difficult as the energy decreases. Therefore, if defects are not a serious concern, e.g. for deposition of hard carbon films for certain purposes, one may choose a kinetic-energy range where sputtering may become significant.

As Lifshitz *et al.* (1990) have pointed out, the subsurface nature of the IBD process leads to persisting surface concentrations of the substrate material even after substantial deposition has taken place and this surface concentration is gradually removed due to the sputtering of the surface.

9.4.1.4.2 Desorption

Some materials have very small surface-binding energies and will leave the surface readily either due to chemical forces (e.g. excess concentration) only or due to chemical forces and energy supplied by the incoming ions. There is no sharp difference between this process and sputtering, but it merits separate treatment because it takes place in a very significant way mostly in the case of materials which are gases at normal conditions. For example, this is the case when oxidation or nitridation is attempted (Todorov and Fossum, 1988; Vancauwenberghe *et al.*, 1991a), or in the case of materials with high vapor pressure at the temperature of the substrate during the deposition process, and particularly in the case when surface segregation of the same component is significant (Noël *et al.*, 1989).

9.4.2 ENERGY THRESHOLDS IN ION-BEAM DEPOSITION

The kinetic energy of the ions used for IBD is probably the most important parameter, the correct choice of which may decide success or failure. In many IBD experiments known to the author this parameter was selected as the highest possible energy which, by some feature of the deposition process (e.g. sputtering or defect production rate), seemed to be still acceptable. This method of energy selection is perfectly understandable in view of the technical difficulty of sustaining sizable ion currents, i.e. reasonable deposition times, at very low ion energies. The results were often disappointing, especially from the point of view of films for the electronics industry where low defect densities are crucial. On the other hand, Kasi *et al.* (1988a) found that diamondlike carbon films are not produced if the C^+ energy is too low, e.g. $E_k < 10$ eV, a result which strongly suggests that there are also low-energy limits. Lifshitz *et al.* (1990) proposed a model which can explain the existence of these high- and low-energy limits, at least in the case when a desired metastable dense phase and a stable phase with lesser density can be formed. In this 'subplantation model', the maximum energy transferred from an ion to a substrate atom, T, is compared to the displacement energies of the different phases. T can be calculated from the ion energy E_k using equation (3).

In the case of carbon, diamond is the dense phase but graphite is the most stable. The displacement energies of such materials are related to the density because the bond strength is higher for the higher-density phase. The diffusivity is higher for the low-density phase. The diffusivity is important because the displacement energy is related to the formation of stable defects. Diffusion is the process which leads to recombination of vacancies and interstitials. In particular, graphite has the lower displacement energy, $E_d(L) \approx 33$ eV, and diamond has the higher displacement energy, $E_d(H) \approx 38$ eV. Figure 9.8 summarizes the schematics of the matrix formation in IBD according to this model. For low kinetic energies, $T < E_d(L)$, the model predicts only the working of a 'mold' effect, i.e. replacement in a phase which already has evolved, and thermal spikes which are electronic and phonon excitations within a short period of time after the ion impact. This time, of the order of 10^{-14} s, is insufficient for the thermalization of the primary-ion energy. Thermal spikes can cause matrix rearrangement, but with the likely end-result of the evolution of a stable phase.

For high kinetic energies when $T > E_d(H)$, the model predicts extensive radiation damage and evolution of both phases. At medium energies when $E_d(L) < T < E_d(H)$, the evolution of the high-density phase is favored because atoms from the low-density phase will be replaced, while atoms in the high-density phase cannot be removed permanently from their positions.

A weakness of the subplantation model is the assumption that the displacement energies in imperfectly evolved phases are the same as in the perfect

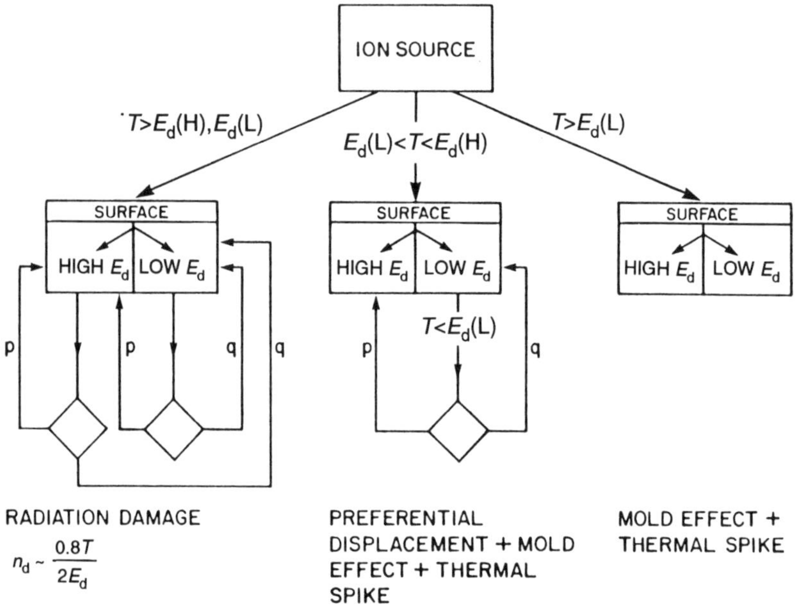

Figure 9.8 Schematic illustration of dense matrix formation for three energy ranges: (i) $T < E_d(L)$, only 'mold effect'; (ii) $E_d(L) < T < E_d(H)$, preferential displacement; (iii) $E_d(H) < T$, radiation damage and amorphization. H and L refer to high- and low-E_d components; N_d is the number of displacements per incident primary particle; p and q are probabilities of high- and low-E_d atomic-site occupancies. (Reproduced by permission of the American Institute of Physics from Lifshitz *et al.*, 1990)

structure. Moreover, the dependence of the displacement energies on the crystallographic orientation and possible surface effects are ignored. More significantly, however, this model, similarly to the replacement/displacement model of Herbots *et al.* (1988), assumes that displacement collisions do not occur below E_d. Nevertheless, the concept of the displacement/replacement collisions coupled with the concept that there are different thresholds for penetration, replacement and displacement in different phases provide the bases on which a model explaining the energy dependence of IBD can be constructed.

For subsurface deposition to occur due to the kinetic energy of the ions, the kinetic energy must exceed the penetration threshold, E_p. If this condition is fulfilled, an excess density of the deposited material will build up in the subsurface region of the substrate. Depending on the temperature and the material parameters, this excess density may facilitate structural changes leading to epitaxial growth and/or growth of metastable phases. It may be necessary though to deliver sufficient kinetic energy to the substrate atoms in collisional events in order to eject them from their original lattice positions so that replacements can occur. For that, the ion kinetic energy must exceed a second threshold which we call replacement threshold, E_r. Both E_p and E_r are normally below E_{th}, even for the phase with the lower displacement energy. High defect densities will occur already if the kinetic energy exceeds E_r for the high-density phase. This model shifts the window for the optimal deposition energies below those energies which have previously been considered in the subplantation model.

For example, let us consider the deposition of 40 eV Ge ions on a Si substrate. According to equation (3), 80% of the ion kinetic energy is transferred in this case. $E_d = 16 \pm 5$ eV for silicon, so one to three permanent defects will be created by each ion. This process can create as many as 10^{15} defects/cm^2 while depositing the first monolayer of the Ge epitaxy, and these defects may diffuse to a significant depth in the Si substrate and combine into extended defects as dislocation loops. After a deposition of only 10^{16} atoms/cm^2 of 40 eV Ge on a Si(100) substrate at 400 °C, the number of displaced atoms observed in the substrate was 5×10^{14} atoms/cm^2 (Herbots *et al.*, 1988). Even in the case of 20 eV Si deposition on Si, significant damage was found in the substrate (Herbots *et al.*, 1988), a result perfectly understandable in view of the present analysis.

Electron-spectroscopic analysis of the sp^3 content of IBD hard carbon films (Steffen *et al.*, 1991) provides us with another example. We found that only about 45% of the bonds were sp^3 in films deposited at 150 eV and even at 50 eV the sp^3 content did not exceed 60%. This result should not be unexpected because 50 eV exceeds the displacement energies of graphite and diamond; thus this result does not contradict the subplantation model (Lifshitz *et al.*, 1990). It is also possible that our data based on Auger electron spectroscopy were influenced by higher sp^2 content at the surface so that we underestimated the sp^3 ratio, although this result was confirmed by recent electron-energy-loss measurements with 100 keV electrons obtained on much thicker films (Kulik *et al.*, 1992). The explanation is that while 50 eV is not much greater than the E_d of diamond, it is well above the threshold for the insertion of interstitial carbon atoms in diamond. Thus, while a diamond phase may be formed, it is simultaneously damaged by the deposition process. This model explains why i-C films are amorphous dense films with little sign of diamond crystal growth.

9.4.3 COMPETING PROCESSES

A number of the processes described in Section 9.4.1 can occur simultaneously. The final result will depend on the kinetics of these competing processes. As previously mentioned, the deposition process will be diminished by back-scattering and sputtering since both of these effectively decrease the number of atoms that constitute the film. Diffusion can increase the deposition rate if substrate atoms become film constituents due to this process, or diffusion can decrease the deposition rate if it allows deposited atoms to escape from the film.

A different type of competition can occur when multiple ion sources are employed in order to deposit composite films. In ion-beam-assisted deposition (IBAD) such situations are common and the concept of arrival ratios has been developed to describe them (Cuomo *et al.*, 1987). The situation for IBD as we treat it here is different from IBAD in that we consider both species arriving in the form of low-energy ion beams.

For a compound film with a composition of A_nB_m the following reaction describes the deposition process:

$$n'A^+ + m'B^+ \rightarrow A_nB_m + (n' - n)A + (m' - m)B, \tag{6}$$

where n' and m' represent the ion fluxes of A^+ and B^+, respectively. The last two terms on the right-hand side of equation (6) describe 'lost' atoms that do not take part in compound formation due to back-scattering, sputtering, desorption, or diffusion. In general $n'/m' \neq n/m$ since the ratio of these lost atoms to those that form compounds may vary for different species. The variation of the ratio n'/m', a basic experimental parameter, is expected to yield varying n/m ratios in those instances where such variation is allowed, i.e. A_nB_m is at least metastable for a range of n and m. In other cases it may lead to the formation of multiphase systems.

There is very limited experience with such processes to date. In the case of GaAs (Haynes *et al.*, 1989; Tamura *et al.*, 1989), stoichiometric epitaxy was formed simply by using equal amounts of Ga and As. In the case of β-SiC (Withrow *et al.*, 1989), again equal amounts of the constituents were deposited but the C-to-Si ratio was later found to be about 1.3. In the case of GaN deposition, stoichiometric crystalline films were formed when the Ga-to-N beam-current ratio was 1.5 to 2 but not for lower ratios (Qin *et al.*, 1991a). These results show that equation (6) or similar relations incorporating diffusion should be considered if good film qualities are to be achieved.

9.4.4 COMPUTER MODELING

Computer simulation of the film-growth processes at typical IBD energies, i.e. at 10–100 eV constitute a major challenge because of the computational complexities and the largely unknown interaction potentials. Attempts to utilize various computational methods to describe IBD processes were made almost as early as the first IBD experiments themselves. A recent review by Dodson (1990) provides insight to many of the problems related to the application of molecular-dynamics simulations to both vapor-phase and IBD film growth.

Earlier calculations experimented with computational methods which had been successful in elucidating sputtering phenomena. For example, Thomas *et al.* (1982) used a modified version of the MARLOWE program (Robinson and Torrens, 1974; Hou and Robinson, 1979) to follow the collection of impinging ions and the creation of damage in the crystal via atomic displacements initiated by the ions. In such a model, the collisions are treated as isolated binary events and the appropriate Molière interaction potential can be used to describe atomic interactions during the collisions. In such calculations, a number of adjustable threshold energies have to be used as input parameters. Such parameters are the displacement energy, binding energy, cutoff energy (the minimum significant energy of a moving atom), and the surface-binding energy. The surface-binding energy replaces the displacement energy of a surface atom in a collision with an atom that has a momentum pointing 'outward' from the crystal. Even though such treatment has obvious problems, it yields some insight into the deposition process. One can conclude that the film growth occurs as a subsurface process and that the depth distribution of both deposited atoms and defects is very shallow, at least at energies $E_k < 2E_{th}$.

Similarly, various calculations based on the TRIM code (Biersack and Haggmark, 1980) can be used to calculate sputtering, back-scattering and damage effects at various deposition parameters. In order to obtain some insight in the actual film-growth process, versions with evolution capability such as TRIDYN (Möller *et al.*, 1988), or versions including replacement/relocation events such as TRIMCSR (Vancauwenberghe *et al.*, 1991b) have to be used. Again, threshold and cutoff energies as input parameters have to be introduced and this can be a serious problem. For example, the displacement energy used by Lifshitz *et al.* (1990) for diamond is 80 eV following earlier data (Clark *et al.*, 1961), rather than 37.5 eV as more recent measurements suggest (Koike *et al.*, 1992). This leads to a significantly lower estimated defect density in the case of diamond than what is probably correct, although the qualitative findings may still be valid. Similar problems may affect the quantitative results in Vancauwenberghe *et al.* (1991b) where the Si displacement energy is taken to be 15 eV. In a more recent report, the same authors use a 13 eV displacement energy (Hellman *et al.*, 1992). The actual merit of these calculations is indeed not in the quantitative estimates of the various defect densities but rather in the qualitative results which help understand previously unclear experimental findings. For example, Figure 9.9 shows a series of carbon concentration depth profiles calculated for the case of C^+ deposition on gold (Lifshitz *et al.*, 1990). These depth profiles clarify the subsurface nature of the deposition process (anchoring) and provide an explanation for the persisting surface gold concentration during IBD which had been an unexplained experimental finding.

A further improvement of understanding of IBD comes from the work of Vancauwenberghe *et al.* (1991b; Hellman *et al.*, 1992) who coupled their collisional calculations with calculations for defect diffusion and recombination. The main result of such calculations for the case of Si IBD are presented in Figure 9.10. These results show convincingly that, even at room temperature,

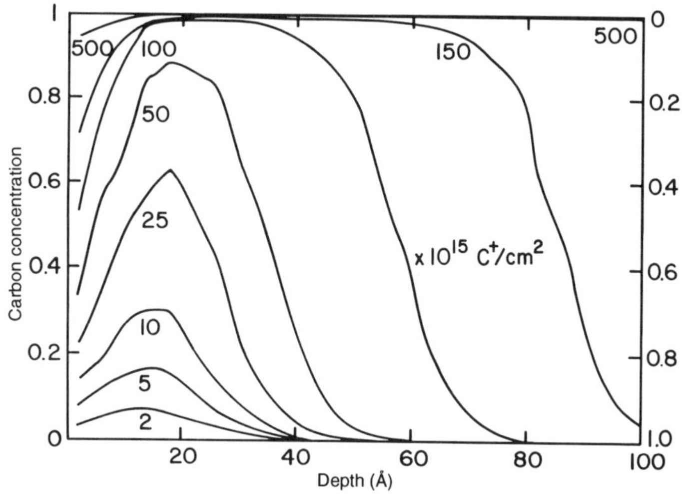

Figure 9.9 Dynamic TRIM (TRIDYN Version 32) calculations of evolution of C concentration with fluence from 150 eV C^+ ion impingement on Au. The fluences are given in $\times 10^{15} C^+/cm^2$. Note the subsurface entrapment of the C^+ and the high fluence needed to remove the Au surface layer. (Reproduced by permission of the American Institute of Physics from Lifshitz *et al.*, 1990)

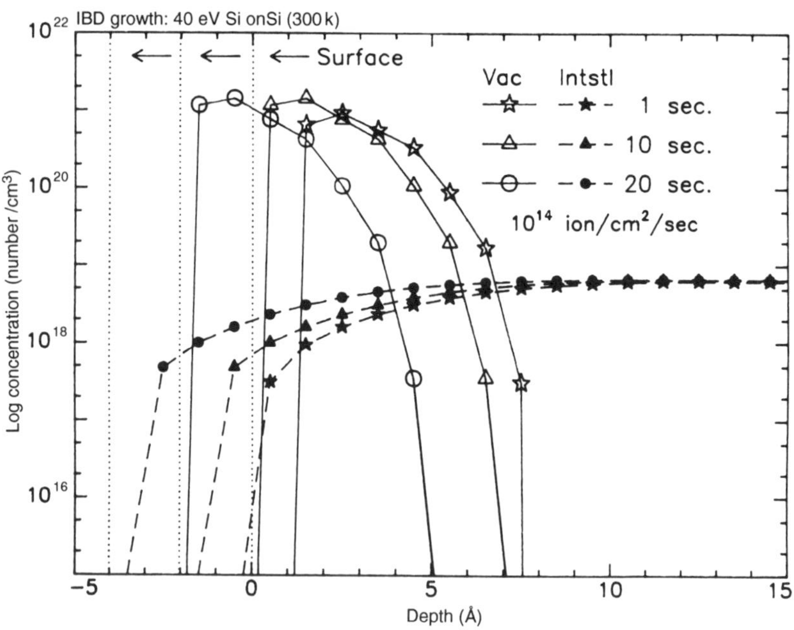

Figure 9.10 Vacancy and interstitial profiles at different times (1, 10, 20 s) of IBD growth of Si on Si at 300 K, 40 eV and 10^{14} ion $cm^{-2} s^{-1}$. The dotted lines indicate the surface position of the growing film. (Reproduced by permission of the American Institute of Physics from Vancauwenberghe *et al.*, 1991b)

the interstitial diffusion in Si is so fast that a large interstitial density (10^{18} cm^{-3}) can extend in the substrate up to several thousand angstroms depth (Vancauwenberghe *et al.*, 1991b). This provides an explanation for the previously observed dislocation loops.

Although the binary-collision calculations mentioned above provide some insight into the gross IBD results of layer growth and damage, they are not very informative for the details of structure evolution. For this, molecular dynamics (MD) calculations are more promising. One has to be aware, however, that such calculations cannot be extended in time much beyond 10^{-12} s, and therefore only collisional events and effects of the thermalization of the ion kinetic energy can be included. Diffusional effects such as mentioned above can significantly modify the ultimate, experimentally observable, picture.

As Dodson (1990) points out, the difficulty in treating IBD in MD calculations stems largely from the fact that, unlike in either the high-energy >1 keV or low-energy <10 eV processes, at medium energies, the energetics of ion–solid interactions is not necessarily fully described by the instantaneous positions of the interacting atoms and requires a velocity- and history-dependent description. The reason for this is that at these medium energies, bond formation becomes significant but there is not enough time to achieve equilibrium bonding configurations. Comparisons of calculations (Dodson, 1990) with and without bonding of low-energy Si beam effects on a Si crystal are shown in Figure 9.11. This figure shows a dramatic decrease of the average penetration depth when bonding is allowed both for 12.5 eV and 50 eV ions. The number of (at least temporarily) displaced atoms, however, changes quite differently; while it decreases for the case of 12.5 eV ions, it increases for 50 eV ions as bonding forces are introduced. Both the extent of penetration and of damage are found to be larger in MD calculations than would be expected from calculations based on the binary-collision approximation (Dodson, 1987b).

Kitabatake *et al.* (1990) have studied the impact of 10 eV Si atoms on a Si(001) 2 × 1 reconstructed surface by utilizing the Tersoff (1988) many-body potential. These calculations show that, depending on the point of impact within the Si surface unit cell, (i) the projectile may end up in an epitaxial position with or without causing damage, (ii) it can stop in several different types of interstitial positions while either forming complex defects or not, (iii) surface defects may be created, or (iv) the projectile may be reflected from the surface. Defects mentioned can be produced in depths down to the fifth layer but no vacancies have been found, in agreement with Vancauwenberghe *et al.* (1991b). Although these calculations do not provide a good estimate for the probabilities of the outcomes, the large number of defects in the results is somewhat surprising. The continuing nature of the deposition was not considered in these calculations. Evidently, some of these defects may anneal during continuing deposition but many of them may be able to diffuse into the Si bulk, as suggested by Vancauwenberghe *et al.* (1991b). More interestingly, in about one-third of the simulations epitaxial growth was observed, providing an explanation for

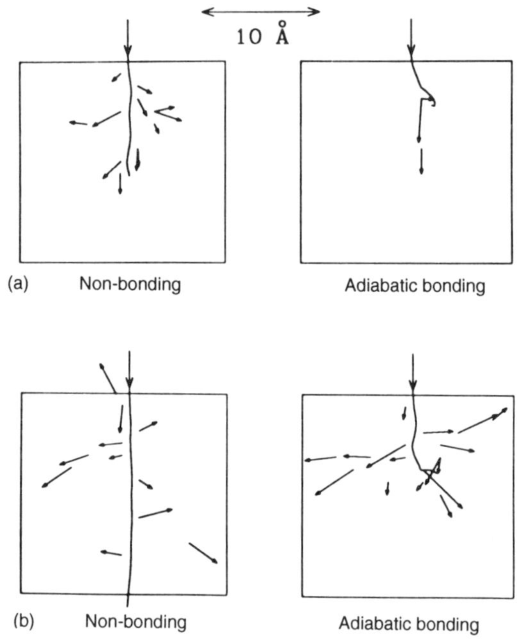

Figure 9.11 Collision cascades produced in a Si(100) substrate by a very low energy neutral Si projectile at normal incidence. The beam energy is 12.5 eV in (a) and 50 eV in (b). The cascades marked 'non-bonding' result from use of the Ziegler–Biersack–Littmark (ZBL) universal nuclear potential (Ziegler *et al.*, 1985), whereas those marked 'adiabatic bonding' use an empirical many-body potential (Dodson, 1987a). In all cases, the substrate atoms interact via this many-body potential to accurately include the effect of lattice vibrations. (Reproduced by permission from Dodson, 1990. Copyright CRC Press, Inc., Boca Raton, FL)

the significant decrease of the substrate temperature required for epitaxial growth in IBD (and in IBAD) as compared with MBE.

The processes in deposition of diamondlike-carbon films have been studied recently also using the Tersoff potential (Kaukonen and Nieminen, 1992). In this simulation, the carbon ions impinged on a (100) diamond surface with 1–70 eV kinetic energy and a continuing deposition process with random impact points was considered. Figure 9.12 shows the densities of the films grown using various kinetic energies. The number of fourfold coordinated atoms (sp^3 bonds), like the atomic density, reaches a maximum (44%) at 40 eV, while at lower and higher energies more sp^2 bonds form. This result is in good agreement with experimental observations of carbon deposition on substrates other than diamond (Steffen *et al.*, 1991; Lifshitz *et al.*, 1990; Kasi *et al.*, 1988b); in fact, the substrate does not seem to play a significant role here. These results conform

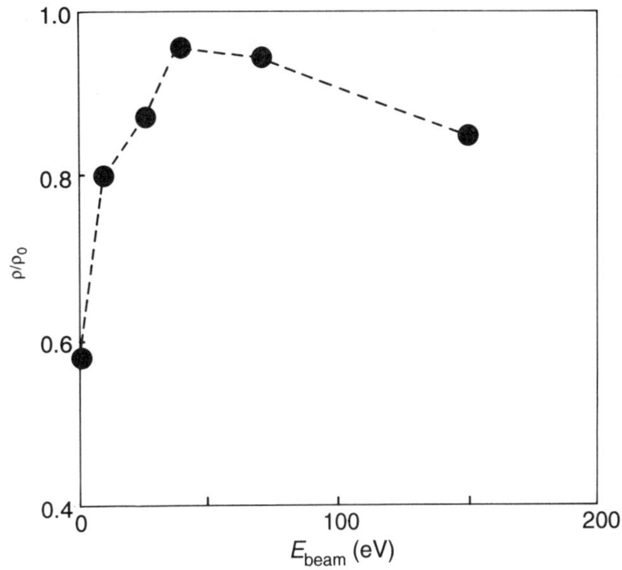

Figure 9.12 Fractional density ρ/ρ_0 of carbon films obtained in molecular dynamics calculations vs beam energy; ρ_0 is the density of diamond. The data at 150 eV were obtained by additional ions impinging on the film produced with 70 eV ions. (Reproduced by permission of the American Physical Society from Kaukonen and Nieminen, 1992)

with the prediction of the subplantation model (Lifshitz *et al.*, 1990), i.e. that the optimal energy for diamondlike-carbon deposition should be between the displacement energies of graphite and diamond, at 33–38 eV. Although no sign of epitaxial growth was observed in these calculations, it is not at all excluded that there exist a narrow energy window within which epitaxial diamond can be grown.

9.5 SUMMARY

Ion-beam deposition is not a mature technology today although the list and the classes of materials which have been explored by this technology are quite impressive. The most important conclusions for future development are, first, that systematic parametric studies are necessary in order to understand the fundamental processes involved in IBD; second, for such studies, systems which allow control of the deposition parameters in a wide range are necessary; third, this wide range must include ion energies as low as 10 eV (and possibly below 10 eV), and deposition temperatures not only above but also below room temperature. UHV and *in situ* analysis techniques are also a must.

The most important *in situ* analysis techniques are ion scattering and RHEED which yield structural information, and AES and XPS which provide chemical information. For *ex situ* analysis, TEM can be very useful, and electron-energy-loss spectra obtained in conjunction with TEM may become one of the most important methods in i-C research. As the IBD technology matures, the importance of electrical and optical measurements will increase dramatically.

For some applications and to elucidate some phenomena, thick films are necessary; so striving for high beam currents is not simply a matter of technological feasibility. Nevertheless, systems with lower currents may be used very effectively in studies of low-energy ion-beam effects, which are vital in order to expand our fundamental understanding of such phenomena.

Dual-beam systems, as well as systems combining IBD and MBE, have great potential but single-beam studies can still be very useful. The potential of the IBD–MBE combination can hardly be overestimated. It is very likely that in such combinations high ion currents are not so important, and that will open the possibility of using focused low-energy ion beams for patterned deposition. Patterned doping during MBE is now being explored (Thompson *et al.*, 1992).

IBD was investigated by semiconductor researchers in the past and some companies have temporarily given up when other technologies showed more promise. Today, there is a certain incompatibility between the thinking of some in semiconductor R&D who would only welcome experiments on 6-inch wafers and the realities of the available IBD systems which for the most part can handle only samples of much smaller size. The requirement to do experiments on large wafers makes IBD more expensive and may not be the fastest road to success.

IBD has the potential to complement other film-deposition technologies in certain special areas which include deposition of hard carbon coatings at room temperature and deposition of various semiconductor epitaxial films at temperatures which are below what is necessary for MBE. These are areas which are more or less feasible today. Beyond these areas, IBD may be considered as a laboratory tool which can help to optimize processes which may then be better carried out on large industrial scale using other approaches, such as sputtering and IBAD.

Calculations, especially molecular dynamics, are a very important part of IBD research. At the same time, one of the likely spin-offs of IBD experiments will be information regarding interatomic potentials in the low-energy range.

9.6 ACKNOWLEDGEMENTS

I am happy to acknowledge the help and encouragement received from Professor J. W. Rabalais in writing this paper. I would like to express my gratitude to all those researchers in the field of IBD who sent their reprints to me. Lastly, I thank Sherry Zwiebel for her generous help in reading and correcting my manuscript. This material is based on work supported by the National Science Foundation under Grant No. DMR-8914608 and the Texas Advanced Research Program under Grant No. 003652-211.

9.7 REFERENCES

Aisenberg, S., and R. Chabot (1971) *J. of Appl. Phys.*, **42**, 2953.

Al-Bayati, A. H., K. G. Orrman-Rossiter, D. G. Armour, J. A. van den Berg, and S. E. Donelly (1992) *Nucl. Instr. and Meth. in Phys. Res.*, **B 63**, 109.

Amano, J. (1982) *Thin Solid Films*, **92**, 115.

Amano, J., and R. P. W. Lawson (1977) *J. Vac. Sci. Technol.*, **14**, 831.

Amano, J., P. Bryce, and R. P. W. Lawson (1976) *J. Vac. Sci. Technol.*, **13**, 591.

Andoh, Y., K. Ogata, and E. Kamijo (1988) *Nucl. Instr. Meth. Phys. Res.*, **B 33**, 678.

Anttila, A., J. Koskinen, M. Bister, and J. Hirvonen (1986) *Thin Solid Films*, **136**, 129.

Bachmann, P. K., and R. Meisser (1989) *Chemical and Engineering News*, **67**, 24.

Badzian, A. R. and R. C. De Vries (1988) *Mat. Res. Bull.*, **23**, 385.

Behrisch, R. (ed.) (1981) *Sputtering by Particle Bombardment, Vol. I.* Springer, Berlin.

Biersack, J. P., and L. G. Haggmark (1980) *Nucl. Instr. and Meth.*, **174**, 257.

Bousetta, A., A. H. Al Bayati, J. A. van den Berg, and D. G. Armour (1992) *Appl. Surface Sci.*, **56–58**, 480.

Chaikovskii, E. F., V. M. Puzikov, and A. V. Semenov (1981) *Sov. Phys. Crystallogr.*, **26**, 122.

Chaikovskii, E. F., V. M. Puzikov, A. V. Semenov, G. H. Rosenberg, and A. B. Kostenko (1986) *Archivum Nauki o Materialach*, **7**, 187.

Clark, C. D., P. J. Kemmey, and E. W. J. Mitchell (1961) *Discuss. Faraday Soc.*, **31**, 96.

Cuomo, J. J., and S. M. Rossnagel (1987) *Nucl. Instr. and Meth. in Phys. Res.*, **B 19/20**, 963.

Dodson, B. W. (1987a) *Phys. Rev.*, **B 35**, 2795.

Dodson, B. W. (1987b) *Phys. Rev.*, **B 36**, 1068.

Dodson, B. W. (1990) *Solid State and Mat. Sci.*, **16**, 115.

Erler, H.-J., G. Reisse, and C. Weissmantel (1980) *Thin Solid Films*, **65**, 233.

Fair, R. B. (1971) *J. of Appl. Phys.*, **42**, 3176.

Fontell, A., and E. Arminen (1969) *Canad. J. Phys.*, **47**, 2405.

Freeman, J. H., W. Temple, D. Beanland, and G. A. Gard (1976) *Nuclear Instr. Meth.*, **135**, 1.

Freeman, J. H., D. J. Chivers, G. A. Gard, and W. Temple (1977) *Ion Beam Studies, Pt. VI: The Production of Heavy Ion Beams*, Chem. Div., AERE Harwell, Oxfordshire.

Freeman, J. H., W. Temple, and G. A. Gard (1984) *Vacuum*, **34**, 305.

Gordon, J. S., D. G. Armour, S. E. Donnelly, J. A. van den Berg, D. Marton, and J. W. Rabalais (1991a) *Nucl. Instr. and Meth. in Phys. Res.*, **B 59/60**, 312.

Gordon, J. S., A. Bousetta, J. A. van den Berg, D. G. Armour, R. Kubiak, and E. H. C. Parker (1991b) *Nucl. Instr. Meth. in Phys. Res.*, **B 55**, 314.

Gottdang, A., K. Eich, A. Hassenbürger, W. H. Schulte, B. Cleff, D. J. W. Mous, R. Koudijs, G. F. A. van de Walle, and J. Politiek (1991) *Nucl. Instr. Meth. in Phys. Res.*, **B 55**, 310.

Hagstrum, H. D. (1978) In *Electron and Ion Spectroscopy of Solids*, L. Fiermanns, J. Vennik, and W. Dekeyser (eds), Plenum Press, New York.

Haynes, T. E., R. A. Zuhr, and S. J. Pennycook (1989) *Mat. Res. Soc. Symp. Proc.*, **144**, 311.

Hellman, O. C., N. Herbots, and O. Vancauwenberghe (1992) *Nuclear Instr. and Meth. in Phys. Res.*, **B 67**, 301.

Herbots, N., O. C. Hellman, P. A. Cullen, and O. Vancauwenberghe (1988) In *Deposition and Growth: Limits for Microelectronics*, G. W. Rubloff (ed.), AVS Series 4, p. 259.

Ho, P. S. (1978) *Surf. Sci.*, **72**, 253.

Hou, M., and M. T. Robinson (1979) *Appl. Phys.*, **18**, 381.

Ishikawa, J., Y. Takeiri, and T. Takagi (1986) *Rev. Sci. Instrum.*, **57**, 6.

Ishikawa, J., Y. Takeiri, K. Ogawa , and T. Takagi (1987) *J. Appl. Phys.*, **61**, 2509.

Itoh, T. (ed.) (1989) *Ion Beam Assisted Film Growth*. Elsevier, Amsterdam.

Iyer, S., H. Morkoc, H. Zabel, and N. Otsuka (1989) *Comments Cond. Mat. Phys.*, **15**, 1.

Kasi, S. R., H. Kang, and J. W. Rabalais (1987) *Phys. Rev. Lett.*, **59**, 75.

Kasi, S. R., H. Kang, and J. W. Rabalais (1988a) *J. Chem. Phys.*, **88**, 5914.

Kasi, S. R., H. Kang, and J. W. Rabalais (1988b) *J. Vac. Sci. Technol.*, **A 6**, 1788.

Kasi, S. R., M. A. Kilburn, H. Kang, J. W. Rabalais, L. Tavernini, and P. Hochmann (1988c) *J. Chem. Phys.*, **88**, 5902.

Kaufman, H. R., J. J. Cuomo, and J. M. E. Harper (1982) *J. Vac. Sci. Technol.*, **21**, 725.

Kaufman, H. R., W. E. Hughes, R. S. Robinson, and G. R. Thompson (1989) *Nucl. Instr. and Meth. in Phys. Res.*, **B 37/38**, 98.

Kaukonen, H.-P., and R. M. Nieminen (1992) *Phys. Rev. Lett.*, **68**, 620.

Kitabatake, M., P. Fons, and J. E. Greene (1990) *J. Vac. Sci. Technol.*, **A 8**, 3726.

Klabunde, K. J. (ed.) (1985) *Thin Films from Free Atoms and Particles*. Academic Press, Orlando, Fla.

Koike, J., D. M. Parkin, and T. E. Mitchell (1992) *Appl. Phys. Lett.*, **60**, 1450.

Kornelsen, E. V. (1964) *Canad. J. of Phys.*, **42**, 364.

Koskinen, J., J.-P. Hirvonen, and A. Anttila (1985) *Appl. Phys. Lett.*, **47**, 941.

Kreisler, P., H. Baumann, and K. Bethge (1984) *Vacuum*, **34**, 215.

Kulik, J., Y. Lifshitz, G. D. Lempert, S. Rotter, J. W. Rabalais, and D. Marton (1992) *Proc. 50th Ann. Meeting of the El. Micr. Soc. of America*, G. W. Bailey, J. Bentley, and J. A. Small (eds), San Francisco Press, p. 1272.

Lifshitz, Y., S. R. Kasi, J. W. Rabalais, and W. Eckstein (1990) *Phys. Rev.*, **B 41**, 10468.

Marton, D., K. J. Boyd, T. Lytle, and J. W. Rabalais (1993) *Surf. Sci.*, **282**, 113

Maruno, S., Y. Morishita, T. Isu, Y. Nomura, and H. Ogata (1987) *J. of Crystal Growth*, **81**, 338.

Maruno, S., Y. Morishita, T. Isu, Y. Nomura, and H. Ogata (1988a) *Surf. Sci.*, **201**, 355.

Maruno, S., Y. Morishita, T. Isu, Y. Nomura, and H. Ogata (1988b) *J. of Electr. Mat.*, **17**, 21.

Mezentzeff, P., Y. Lifshitz, and J. W. Rabalais (1990) *Nucl. Instr. and Meth. in Phys. Res.*, **B 44**, 289.

Miyake, K., and T. Tokuyama (1982) *Thin Solid Films*, **92**, 123.

Miyazawa, T., S. Misawa, S. Yoshida, and S. Gonda (1984a) *J. of Appl. Phys.*, **55**, 188.

Miyazawa, T., S. Yoshida, S. Misawa, S. Gonda, and I. Ohdomari (1984b) *Appl. Phys. Lett.*, **45**, 380.

Möller, W., W. Eckstein, and J. P. Biersack (1988) *Comput. Phys. Commun.*, **51**, 355.

Morishita, Y., S. Maruno, T. Isu, Y. Nomura, H. Ogata, and K. Kuramoto (1988) *J. of Crystal Growth*, **88**, 215.

Noël, J.-P., N. Hirashita, L. C. Markert, Y.-W. Kim, J. E. Greene, J. Knall, W.-X. Ni, M. A. Hasan, and J.-E. Sundgren (1989) *J. Appl. Phys.*, **65**, 1189.

Ohnishi, T., Y. Yoshida, Y. Hirofuji, and H. Iwasaki (1989) *Nucl. Instr. Meth. Phys. Res.*, **B 37/38**, 850.

Orrman-Rossiter, K. G., D. R. G. Mitchell, S. E. Donnelly, C. J. Rossouw, S. R. Glanvill, P. R. Miller, A. H. Al-Bayati, J. A. van den Berg, and D. G. Armour (1990) *Phil. Mag. Lett.*, **61**, 311.

Proudfoot, G., C. M. O. Mahony, and R. Perrin (1989) *Nucl. Instr. and Meth. in Phys. Res.*, **B 37/38**, 103.

Qin, F., X. Wang, Z. Liu, Z. Yao, Z. Ren, L. Lin, S. Su, W. Jiang, and W. M. Lau (1991a) *Rev. Sci. Instr.*, **62**, 2322.

Qin, F., Z. Yao, Z. Ren, S.-T. Lee, I. Bello, X. Feng, L. J. Huang, and W. M. Lau (1991b) MRS *Symp. Proc.*, **223**, 307.

Robinson, M. T., and I. M. Torrens (1974) *Phys. Rev.*, **B 9**, 5008.

Sainty, W. G., P. J. Martin, R. P. Netterfield, D. R. McKenzie, D. J. H. Cockayne, and D. M. Dwarte (1988) *J. Appl. Phys.*, **64**, 3980.

Samsonov, G. V. (1968) *Handbook of the Physicochemical Properties of the Elements*. IFI/Plenum, New York.

Shanfield, S., and R. Wolfson (1983) *J. Vac. Sci. Technol.*, **A 1**, 323.

Shimizu, S., O. Tsukakoshi, S. Komiya, and Y. Makita (1985) *Jap. J. of Appl. Phys.*, **24**, 1130.

Shimizu, S., N. Sasaki, S. Ogata, O. Tsukakoshi, S. Seki, and H. Yamakawa (1991) *Mat. Res. Soc. Symp. Proc.*, **223**, 347.

Spencer, E. G., P. H. Schmidt, D. C. Joy, and F. J. Sansalone (1976) *Appl. Phys. Lett.*, **29**, 118.

Sprague, J. A., and C. M. Gilmore (1992) In *Materials Modification by Energetic Atoms and Ions*, K. S. Grabowski *et al.* (eds), MRS Series, Vol. 268.

Steffen, H. J., C. D. Roux, D. Marton, and J. W. Rabalais (1991) *Phys. Rev.*, **B44**, 3981.

Tamura, S., M. Hyouzho, K. Yokota, and S. Katayama (1989) *Nucl. Instr. and Meth. Phys. Res.*, **B 37/38**, 862.

Tersoff, J. (1988) *Phys. Rev.*, **B 38**, 9902.

Thomas, G. E., L. J. Beckers, J. J. Vracking, and B. R. de Koning (1982) *Crystal Growth*, **56**, 557.

Thompson, J. H., D. A. Ritchie, G. A. C. Jones, E. H. Linfield, J. E. F. Frost, A. C. Churchill, G. W. Smith, D. Lee, M. Houlton, and C. R. Whitehouse (1992) *Surf. Sci.*, **267**, 69.

Todorov, S. S., and E. R. Fossum (1988) *Appl. Phys. Lett.*, **52**, 48.

Todorov, S. S., S. L. Shillinger, and E. R. Fossum (1986) *IEEE Device Lett.*, **EDL-7**, 468.

Tokuyama, T., K. Yagi, K. Miyake, M. Tamaura, N. Natsuaki, and S. Tachi (1981) *Nuclear Instr. and Meth.*, **182/183**, 241.

Trehan, R., Y. Lifshitz, and J. W. Rabalais (1990) *J. Vac. Sci. Technol.*, **A 8**, 4026.

Tsai, H., and D. B. Bogy (1987) *J. Vac. Sci. Technol.*, **A 5**, 3287.

Tsukizoe, T., T. Nakai, and N. Ohmae (1977) *J. Appl. Phys.*, **48**, 4770.

Vancauwenberghe, O., O. C. Hellman, N. Herbots, and W. J. Tan (1991a) *Appl. Phys. Lett.*, **59**, 2031.

Vancauwenberghe, O., N. Herbots, and O. C. Hellman (1991b) *J. Vac. Sci. Technol.*, **B 9**, 2027.

Vinetskii, V. L., and G. A. Kholodar (1986) In *Physics of Radiation Effects in Crystals*, R. A. Johnson and A. N. Orlov (eds), Modern Problems in Condensed Matter Sciences, Vol. 13, North-Holland, Amsterdam, p. 285.

Weissmantel, C. (1981) *J. Vac. Sci. Technol.*, **18**, 179.

Withrow, S. P., K. L. More, R. A. Zuhr, and T. E. Haynes (1989) *Vacuum*, **39**, 1065.
Yagi, K., S. Tamura, and T. Tokyama (1977) *Jap. J. of Appl. Phys.*, **16**, 245.
Yamada, H. (1989) *J. Appl. Phys.*, **65**, 2084.
Yamada, I., H. Inokawa, and T. Takagi (1985) *Nucl. Instr. and Meth. in Phys. Res.*, **B6**, 439.
Yoshida, Y., T. Ohnishi, Y. Hirofuji, H. Iwasaki, and T. Ikeda (1989) *Nucl. Instr. Meth. Phys. Res.*, **B 37/38**, 866.
Zalm, P. C., and L. J. Beckers (1982) *Appl. Phys. Lett.*, **41**, 167.
Ziegler, J. F., J. R. Biersack, and U. Littmark (1985) *The Stopping and Range of Ions in Solids*. Pergamon Press, New York.
Zuhr, R. A., B. R. Appleton, N. Herbots, B. C. Larson, T. S. Noggle, and S. J. Pennycook (1987) *J. Vac. Sci. Technol.*, **A 5**, 2135.
Zuhr, R. A., D. Fathy, and O. W. Holland (1988) In *Dislocations and Interfaces in Semiconductors*, K. Raja, J. Narayan, and D. Ast (eds), Metallurgical Soc., p. 119.
Zuhr, R. A., S. J. Pennycook, T. E. Haynes, and O. W. Holland (1989a) *Mat. Res. Soc. Symp. Proc.*, **128**, 47.
Zuhr, R. A., S. J. Pennycook, T. S. Noggle, N. Herbots, T. E. Haynes, and B. R. Appleton (1989b) *Nucl. Instr. and Meth. in Phys. Res.*, **B 37/38**, 16.

9.A APPENDIX

DISPLACEMENT ENERGIES OF MATERIALS

The entries for each element and compound are arranged in chronological order, the more recent data appearing first. No effort has been made to evaluate the merit of published data. Doctoral theses are cited as referenced by other authors; all other references are made on the basis of the original papers.

9.A.1 Elements

Element	E_d (eV)	Orientation	Temperature (K)	Reference
C (graphite)	35.3 ± 1		300	1
surface	31 ± 1		300	2
	28 ± 2	⟨0001⟩	285	3
	23 ± 2, 31 ± 2	⟨0001⟩	80	3
	23 ± 2, 31 ± 2	⟨0001⟩	6	3
	42 ± 2 in	(0001) plane	285	3
	30 ± 2 in	(0001) plane	80	3
	30 ± 2 in	(0001) plane	6	3
	33 ± 1		77–600	4
	60 ± 10		20	5
	24.7		300	6
C (diamond)	37.5 ± 1.2	⟨100⟩	300	7
	45.0 ± 1.3	⟨111⟩	300	7
	47.6 ± 1.3	⟨110⟩	300	7
	55 or less		≈400	8
	35 ± 5		15	9
	80		288	10
Mg	10 ± 0.5		4.2	11
	10 ± 0.5		4.5	12
Al	17.0	⟨100⟩	393	13
	16.0	⟨110⟩	393	13
	14.0	⟨111⟩	393	13
	16 ± 2			14
	16		10	15
	16 ± 2		6	16
	16		8	17
	32		20	18
Si n-type	11.0 ± 1.6		300	19
	22 ± 1		80, 245	20
n-type	22	⟨111⟩	290	21
n-type	≈13		300	22
n-type	≈20		80	22
	20		80	22
p-type	12	⟨111⟩	≈290	23
p-type	≈20.7		77–300	24
n-type	≈14.4		77–300	24
p-type	≈13		300	25
	12.9 ± 0.6		289	26
Ti	19.2 ± 1.0		293	27
	29		20	18
V	30.0 ± 2.1	⟨100⟩	300	28
	38.7 ± 2.3	⟨110⟩	300	28
	34.2 ± 2.3	⟨111⟩	300	28

(continued)

9.A.1 Elements (*continued*)

Element	E_d (eV)	Orientation	Temperature (K)	Reference
	25 ± 2		4.5	29
	30	⟨100⟩	300	30
	39	⟨110⟩	300	30
	34	⟨111⟩	300	30
	26 ± 2		20	31
	26		6	32
Cr bcc	28 ± 1		8	33
	21	⟨100⟩	300	30
	34	⟨110⟩	300	30
	24	⟨111⟩	300	30
Fe α-Fe	17 ± 1	⟨100⟩	36	34
α-Fe	20 ± 1.5	⟨111⟩	36	34
α-Fe	>30	⟨110⟩	36	34
α-Fe	20	⟨100⟩	25	35
α-Fe	30	⟨111⟩	25	35
	24		20	18
Co	22 ± 1		6	36
α	40 ± 3	⟨0001⟩	4.2	37
α	23 ± 2	⟨1014⟩	4.2	37
α	22 ± 1	⟨1012⟩	4.2	37
α	40 ± 5	⟨1011⟩	4.2	37
h-Co	23 ± 0.5	⟨1120⟩	300	38
h-Co	30 ± 1.0	⟨1010⟩	300	38
h-Co	33 ± 1.0	⟨0001⟩	300	38
c-Co	23 ± 0.5	⟨110⟩	300	38
c-Co	30 ± 1.0	⟨100⟩	300	38
Ni	21	⟨110⟩	300	39
	38	⟨100⟩	300	39
	>60	⟨111⟩	300	39
	24		20	18
Cu	19	⟨100⟩	10	40
	20	⟨110⟩	10	40
	76	⟨111⟩	10	40
	17.5	⟨100⟩	70	41
	10.5	⟨100⟩	300	41
	19	⟨100⟩	≈5	42
	20	⟨110⟩	≈5	42
	59	⟨111⟩	≈5	42
	18 ± 2	⟨100⟩	10	43
	22 ± 6	⟨111⟩	10	43
	20 ± 2	⟨110⟩	10	43
	16.5 ± 0.6	⟨110⟩	10	44
	17.5 ± 0.6	⟨110⟩	70	44
	15	⟨110⟩	160	44
	13.5	⟨110⟩	270	44
	10.5	⟨110⟩	550	44

9.A.1 Elements *(continued)*

Element	E_d (eV)	Orientation	Temperature (K)	Reference
	17		20,77	45
	10–14		220	45
	18.0 ± 1.2	⟨110⟩	300	28
	27.5 ± 1.8	⟨100⟩	300	28
	29.2 ± 1.8	⟨111⟩	300	28
	27	⟨100⟩	300	30
	18	⟨110⟩	300	30
	29	⟨111⟩	300	30
	8.5–13		400	46
	16		78	46
	19.3 ± 0.3	⟨110⟩	4.5	47
	22.3 ± 0.8	⟨100⟩, ⟨111⟩	4.5	47
	19.2	⟨110⟩	300	48
	21.6	⟨100⟩	300	48
	23.6	⟨111⟩	300	48
	19 ± 3		6	16
	17.5 ± 1.5		10	49
	16–19		10	50
	22		20	18
Zn	15	⟨0001⟩	300	30
	19 ± 2	⟨0001⟩	4.2	37
	25 ± 2	⟨1014⟩	4.2	37
	30 ± 5	⟨1012⟩	4.2	37
	14 ± 2	⟨1120⟩	4.2	37
	13.5		4.4	51
	16		7.6	51
Ga	12 ± 1		12	52
Ge	14		78	53
n-type	15	⟨100⟩, ⟨111⟩	78	54
n- and p	15.5		300	55
n-type	16		79	56
	18.0		21	56
	14.5 ± 0.4		289	26
n-type	22.3 ± 1		300	57
	31 ± 6		80	58
Zr	21		7.5	59
	24		8	60
Nb	28.5	⟨100⟩	50	41
	25	⟨100⟩	300	41
	25	⟨100⟩	300	61
	28 ± 2		4.5	29
	36 ± 5		25	62
Mo	34.5	⟨100⟩	70	41
	26.5	⟨100⟩	800	41
	34 ± 2		4.5	29

(continued)

9.A.1 Elements (*continued*)

Element	E_d (eV)	Orientation	Temperature (K)	Reference
	36 ± 1	⟨100⟩	8	63
	33 ± 2		8	64
	37		20	18
Pd	34 ± 2		9	65
	≈ 31		10	66
	26 ± 2		9	67
Ag	26.5		8	68,69
	23	⟨110⟩	300	30
	24 ± 1		7, 75	70
	28		20	18
Cd	≥ 40	⟨0001⟩	4.2	37
	≥ 35	⟨1014⟩	4.2	37
	≥ 35	⟨1012⟩	4.2	37
	19 ± 1	⟨1011⟩	4.2	37
	21 ± 2	⟨1120⟩	4.2	37
	19 ± 2		4.4	71
In	10.5 ± 0.5		6	72
Sn α-Sn	12		10	73
β-Sn	22 ± 2		5	74
Lu	17 ± 1		14	75
Ta	33 ± 1	⟨100⟩	7	76
	38 ± 1	⟨100⟩	7	76
	> 55	other	7	76
	36	⟨111⟩	4.5	77
	53	⟨100⟩	4.5	77
	> 130	other	4.5	77
	32 ± 2	minimum	4.5	77
	32			78
	32 ± 3		23	62
	32			79
W	50.5 ± 1		6.5	68
	40 ± 2		4.2	79
	> 35		20	18
Re	$> \approx 40$		4.5	80
Pt	36	⟨110⟩	5	47
	40	⟨100⟩	5	47
	> 70	⟨111⟩	5	47
	33	⟨110⟩		48
	36	⟨100⟩, ⟨111⟩		48
	36 ± 1		9	81
	37–38			82
Au	17.5	⟨100⟩	473	83
	21	⟨110⟩	473	83
	≥ 23	⟨111⟩	473	83
	25–40		10	84
	≈ 36	⟨100⟩, ⟨110⟩	15	85

9.A.1 Elements (*continued*)

Element	E_d (eV)	Orientation	Temperature (K)	Reference
	35		13	86
	33–36		10	50
	>40		20	18
Hg	19			87
Pb	12.1 ± 0.7		1.5	88
	15 ± 1		6	72
	12.5 ± 1.2		4.7	89
Th	35 ± 3		12	90

9.A.2 Compounds

Compound	E_d (eV)	Element	Orientation	Temperature (K)	Reference
MgO	60 ± 3	Mg	⟨001⟩, ⟨011⟩ ⟨111⟩	125–400	91
	53	O		90	91
	44 ± 3	O	⟨001⟩	700	91
	64 ± 3	O	⟨011⟩	700	91
	46 ± 3	O	⟨111⟩	700	91
	61	Mg		300	92
	91	O		300	92
	34	Mg		1000	92
	51	O		1000	92
	60	O		78	93
$MgAl_2O_4$	20	Mg	⟨001⟩		94
	59	O		77	95
	130	O		300	95
	56	O			96
Al_2O_3	18 ± 3	Al	(random)	300–1070	97
	76 ± 3	O	(random)	300–1070	97
α	53	O	⟨0001⟩	300	98
α	41	O	⟨1120⟩	300	98
α	32	Al	⟨0001⟩	300	98
α	24	Al	⟨1120⟩	300	98
α	18	Al		900–1120	99
α	≈75	O		300–500	99
α	40 ± 5	Al		77	100
α	70 ± 10	O		77	100

(*continued*)

9.A.2 Compounds (*continued*)

Compound	E_d (eV)	Element	Orientation	Temperature (K)	Reference
α	90	O		77	101
α	50	Al		77	101
$Al_{23}O_{27}N_5$	≈50	average			94
AlN	≈50	average			94
SiC	45–90	average			94
Si_3N_4	≈50	average			94
ZnO	57	Zn		313	102,103
	57	O		313	102,103
	≈30	Zn		333	104
ZnS	9.9	Zn		300	105
	15	S		300	105
	9.9	Zn		300	106
ZnSe	10 ± 0.5	Zn		85	107
	8.2 ± 0.4	Se		85	107
	9.9	Zn		85	108
	11.9	Se		85	108
ZnTe	7.35	Zn		10	109
p-type	6.7	Te		77	110
	9.7	Zn		10	111
GaAs	10	average	⟨100⟩	300	112
n- and p	11	average		45	113
	17–18	average		77	114
	8.8	Ga		77–373	108
	10.1	As		77–373	108
	9.0	Ga		300	115
	9.4	As		300	115
GaSb	≈6.2	Ga	⟨111⟩	77	116
	10.3 ± 2.8	Sb	⟨111⟩	77	116
CdS	7.3 ± 0.4	Cd		77	117
	8.7 ± 0.4	S		300	118
CdSe	8.1	Cd		5	119
	8.7	Se		77	120
CdTe	5.6	Cd		16	121
	7.8	Te		16	121
p-type	7.8	Te		77	122
InP	6.6	In		300	107
	8.8	P		300	107
	6.7	In		77	115
	8.7	P		77	115
InAs	6.7	In		77	107
	8.5	As		77	107
	6.7	In		77	115
	8.3	As		77	115
InSb n-type	≈6.4	In	⟨111⟩	77	123
n-type	9.2 ± 0.7	Sb	⟨111⟩	77	123
	6.8	Sb		78	107
	≤5.7	In		77	124

9.A.3 References for Appendix

1 H. J. Steffen, D. Marton and J. W. Rabalais, *Phys. Rev. Lett.*, **68**, 1726 (1992)
2 G. L. Montet and G. E. Myers, *Carbon*, **9**, 179 (1971)
3 T. Iwata and T. Nihara, *J. Phys. Soc. Japan*, **31**, 1761 (1971)
4 G. L. Montet, *Carbon*, **5**, 19 (1967)
5 M. W. Lucas and E. W. J. Mitchell, *Carbon*, **1**, 345 (1964)
6 D. T. Eggen, NAA-SR-69 (1950)
7 J. Koike, D. M. Parkin and T. E. Mitchell, *Appl. Phys. Lett.*, **60**, 1450 (1992)
8 J. F. Prins, T. E. Derry and J. F. P. Sellschop, *Phys. Rev.*, **B 34**, 8870 (1986)
9 J. C. Bourgoin and B. Massarani, *Phys. Rev.*, **B 14**, 3690 (1976)
10 C. D. Clark, P. J. Kemmey and E. W. J. Mitchell, *Discuss. Faraday Soc.*, **31**, 96 (1961)
11 T. N. O'Neal and R. L. Chaplin, *Phys. Rev.*, **B 5**, 3810 (1972)
12 W. E. Faust, T. N. O'Neal and R. L. Chaplin, *Phys. Rev.*, **B 183**, 609 (1969)
13 K. Tanaka and Y. Shimomura, *Jap. J. Appl. Phys.*, **20**, 1393 (1981)
14 L. R. Kirkland Jr, Thesis, Clemson Univ. (1971)
15 H. M. Simpson and R. L. Chaplin, *Phys. Rev.*, **185**, 958 (1969)
16 G. W. Iseler, H. I. Dawson, A. S. Mehner and J. W. Kauffman, *Phys. Rev.*, **146**, 468 (1966)
17 H. H. Neely and W. Bauer, *Phys. Rev.*, **149**, 535 (1966)
18 P. G. Lucasson and R. M. Walker, *Phys. Rev.*, **127**, 485 (1962)
19 N. N. Gerasimenko, A. Z. Dvurechenski, V. I. Panov and L. S. Smirnov, *Soviet Phys. Semicond.*, **5**, 1439 (1971)
20 P. L. F. Hemment and P. R. C. Stevens, in *Atomic Collision Phenomena in Solids*, Edited by D. W. Palmer, M. W. Thompson and P. D. Townsend, North-Holland, Amsterdam, p. 217 (1970)
21 P. C. Banbury, in *Radiation Effects in Semiconductors*, Edited by F. L. Vook, Plenum, New York, p. 280 (1968)
22 T. I. Kolomenskaya, S. M. Razumovskii and S. E. Vitovkin, *Soviet Phys. Semicond.*, **1**, 652 (1967)
23 J. A. Grimshaw, *Phys. Lett.*, **22**, 372 (1966)
24 R. L. Novak, *Bull. Amer. Phys. Soc.*, **8**, 235 (1963)
25 V. S. Vavilov, V. M. Patskevich, B. Ya. Yurkov and P. Ya. Glazunov, *Soviet Phys. Solid State*, **2**, 1301 (1960)
26 J. J. Loferski and P. Rappaport, *Phys. Rev.*, **111**, 432 (1958)
27 C. G. Shirley and R. L. Chaplin, *Phys. Rev.*, **B 5**, 2027 (1972)
28 E. A. Kenik and T. E. Mitchell, *Phil. Mag.*, **32**, 815 (1975)
29 P. Jung and G. Lucki, *Radiat. Eff.*, **26**, 99 (1975)
30 T. E. Mitchell, G. Das and A. E. Kenik, in *Fundamental Aspects of Radiation Damage in Metals*, Edited by M. T. Robinson and F. W. Young, ERDA-CONF-751006, p. 73 (1975)
31 M. G. Miller and R. L. Chaplin, *Radiat. Eff.*, **22**, 107 (1974)
32 R. L. Chaplin, K. Sonnenberg and R. R. Coltman Jr, *Bull. Amer. Phys. Soc.*, **19**, 1105 (1974)
33 M. Biget, P. Vajda, F. Maury, A. Lucasson and P. Lucasson, in *Fundamental Aspects of Radiation Damage in Metals*, Edited by M. T. Robinson and F. W. Young, ERDA-CONF-751006, p. 66 (1975)
34 F. Maury, M. Biget, P. Vajda, A. Lucasson and P. Lucasson, *Phys. Rev.*, **B 14**, 5303 (1976)
35 J. N. Lomer and M. Pepper, *Phil. Mag.*, **16**, 1119 (1967)
36 F. Maury, A. Lucasson and P. Lucasson, *Radiat. Eff.*, **21**, 65 (1974)
37 F. Maury, P. Vajda, A. Lucasson and P. Lucasson, *Phys. Rev.*, **B 8**, 5496, 5506 (1973)

38 L. M. Howe, *Phil. Mag.*, **22**, 965 (1970)
39 A. Bourret, *Phys. Stat. Sol.*, **a4**, 813 (1971)
40 W. E. King, K. L. Merkle and M. Meshii, *J. of Nucl. Mater.*, **117**, 12 (1983)
41 K. Urban, B. Saile, N. Yoshida and W. Zag, in *Point Defects and Defect Interactions in Metals*, Yamada Conf., Edited by J.-I. Takamura, M. Doyama and M. Kiritani, Univ. Tokyo Press, North-Holland, Amsterdam, p. 783 (1982)
42 W. E. King, R. Benedek, K. L. Merkle and M. Meshii, in *Point Defects and Defect Interactions in Metals*, Yamada Conf., Edited by J.-I. Takamura, M. Doyama and M. Kiritani, Univ. Tokyo Press, North-Holland, Amsterdam, p. 789 (1982)
43 W. E. King, K. L. Merkle and M. Meshii, *Phys. Rev.*, **B 23**, 6319 (1981)
44 K. Urban and N. Yoshida, *Phil. Mag.*, **44 A**, 1193 (1981)
45 J. Lauzier, C. Minier and A. Seeger, *Phil. Mag.*, **35**, 1617 (1977)
46 G. Roth, H. Wollenberger, Ch. Zeckau and K. Lücke, *Radiat. Eff.*, **26**, 141 (1975)
47 P. Jung, R. L. Chaplin, H. J. Fenzl, K. Reichelt and P. Wombacher, *Phys. Rev.*, **B 8**, 553 (1973)
48 M. J. Makin, in *Atomic Collision Phenomena in Solids*, Edited by D. W. Palmer, M. W. Thompson and P. D. Townsend, North-Holland, Amsterdam, p. 205 (1970)
49 W. Bauer and A. Sosin, *J. Appl. Phys.*, **37**, 1780 (1966)
50 W. Bauer and A. Sosin, *J. Appl. Phys.*, **35**, 703 (1964)
51 F. Maury, A. Lucasson and P. Lucasson, *Crystal Lattice Defects*, **2**, 47 (1971)
52 S. Myhra and R. B. Gardiner, *Radiat. Eff.*, **27**, 35 (1975)
53 Y. Chen and J. W. MacKay, *Phil. Mag.*, **19**, 357 (1969)
54 Y. Chen and J. W. MacKay, *Phys. Rev.*, **167**, 745 (1968); **173**, 918 (1968)
55 L. S. Smirnov and P. A. Glazunov, *Sov. Phys. Solid State*, **1**, 1262 (1960)
56 W. L. Brown and W. M. Augustiniak, *J. Appl. Phys.*, **30**, 1300 (1959)
57 V. S. Vavilov, L. S. Smirnov, G. N. Galkin, A. V. Spitsyn and V. M. Patskevich, *Sov. Phys. Tech. Phys.*, **1**, 1805 (1956); **3**, 894 (1958)
58 E. E. Klontz and K. Lark-Horovitz, *Phys. Rev.*, **82**, 763 (1951)
59 M. Biget, F. Maury, P. Vajda, A. Lucasson and P. Lucasson, *Radiat. Eff.*, **7**, 223 (1971)
60 H. H. Neely, *Radiat. Eff.*, **3**, 189 (1970)
61 H. Schmid and K. Urban, in *Proc. Fifth Int. Conf. on High Voltage Electron Microscopy*, Kyoto, p. 527 (1977)
62 G. Youngblood, S. Myhra and J. W. DeFord, *Phys. Rev.*, **188**, 1101 (1969)
63 M. Biget, P. Vajda, A. Lucasson and P. Lucasson, *Radiat. Eff.*, **21**, 229 (1974)
64 R. Rizk, P. Vajda, A. Lucasson and P. Lucasson, *Phys. Stat. Sol.*, **a18**, 241 (1973)
65 C. M. Jimenez, L. F. Lowe, E. A. Burke and C. A. S. Sherman, *Phys. Rev.*, **153**, 735 (1967)
66 A. Lucasson and P. Lucasson, *J. Phys. Chem. Solids*, **27**, 1423 (1966)
67 L. F. Lowe, C. M. Jimenez, E. A. Burke and C. H. Sherman, *Bull. Amer. Phys. Soc.*, **10**, 1094 (1965)
68 P. Vajda, M. Biget, A. Lucasson and P. Lucasson, *J. Phys. F: Metal Phys.*, **7**, L123 (1977)
69 R. Rizk, P. Vajda, F. Maury, A. Lucasson and P. Lucasson, *J. Appl. Phys.*, **48**, 481 (1977)
70 C. G. Roberts, W. P. Rickey and P. E. Shearin, *J. Appl. Phys.*, **37**, 4515 (1966)
71 F. Maury, P. Vajda, A. Lucasson and P. Lucasson, *Rad. Effects*, **10**, 239 (1971)
72 D. N. Borton, A. E. Mardignian, H. B. Huntington and G. L. Salinger, *Bull. Amer. Phys. Soc.*, **19**, 257 (1974)
73 J. McIlwain, R. Gardiner, A. Sosin and S. Myhra, *Radiat. Eff.*, **24**, 19 (1975)
74 S. Myhra and R. B. Gardiner, *Phys. Lett.*, **39 A**, 405 (1972)

75 J. N. Daon, J. E. Bonnet, P. Vajda, M. Biget, A. Lucasson and P. Lucasson, *Phys. Stat. Sol.*, **b40**, 101 (1977)
76 M. Biget, F. Maury, P. Vajda, A. Lucasson and P. Lucasson, *Phys. Rev.*, **B 19**, 820 (1979)
77 P. Jung and W. Schilling, *Phys. Rev.*, **B 5**, 2046 (1972)
78 D. Meissner, Thesis, Rep. KFA Jülich, JüL-462-FN (1970)
79 H. H. Neely, D. W. Keefer and A. Sosin, *Phys. Stat. Sol.*, **28**, 675 (1968)
80 H. Vandenborre, L. Stals, J. Cornelis and J. Nihoul, *Radiat. Eff.*, **21**, 137 (1974)
81 W. Bauer and W. F. Goeppinger, *Phys. Rev.*, **154**, 584 (1967)
82 E. A. Burke, C. M. Jimenez and L. F. Lowe, *Bull. Amer. Phys. Soc.*, **9**, 655 (1964)
83 M. Hohenstein, A. Seeger and W. Sigle, *J. of Nucl. Mater.*, **169**, 33 (1989)
84 J. N. S. Hanckock, Ph.D. Thesis, Reading Univ. (1974)
85 W. Bauer, A. I. Anderman and A. Sosin, *Phys. Rev.*, **185**, 870 (1969)
86 W. Bauer and A. Sosin, *Phys. Rev.*, **135**, A521 (1964)
87 F. Ono and H. Maeta, *Nucl. Instr. and Meth. in Phys. Res.*, **B 33**, 784 (1988)
88 R. C. Birtcher and J. S. Koehler, *Phys. Rev.*, **B 13**, 2314 (1976)
89 H. Schroeder, Thesis, Rheinisch-Westfälische Technische Hochschule, Aachen (1974)
90 M. W. Guinan, C. L. Snead, Jr and A. N. Goland, *Radiat. Eff.*, **20**, 33 (1973)
91 G. P. Pells, *Radiat. Eff.*, **64**, 71 (1982)
92 R. A. Youngman, L. W. Hobbs and T. E. Mitchell, *J. Physique (Paris), Colloque*, **41**, C6-227 (1980)
93 Y. Chen, D. L. Trueblood, O. E. Schow and H. T. Tohver, *J. Phys. C*, **3**, 2501 (1970)
94 G. P. Pells, *J. Nucl. Mater.*, **155–157**, 67 (1988)
95 G. P. Summers, G. S. White, K. H. Lee and J. H. Crawford, Jr., *Phys. Rev.*, **B21**, 2578 (1980)
96 G. P. Summers, G. S. White, K. H. Lee and J. H. Crawford, Jr., *J. Physique (Paris)*, **41**, C6-51, 227 (1980)
97 G. P. Pells and A. Y. Stathopoulos, *Radiat. Eff.*, **74**, 181 (1983)
98 G. Das, *J. Mater. Sci. Lett.*, **2**, 453 (1983)
99 G. P. Pells and D. C. Phillips, *J. Nucl. Mater.*, **80**, 207 (1979)
100 W. D. Compton and G. W. Arnold, Jr., *Disc. Farad. Soc.*, **31**, 130 (1961)
101 G. W. Arnold and W. D. Compton, *Phys. Rev. Lett.*, **4**, 66 (1960)
102 D. R. Locker and J. M. Meese, *IEEE Trans.*, **NS-19**, 237 (1972)
103 J. M. Meese and D. R. Locker, *Solid State Commun.*, **11**, 1547 (1972)
104 W. E. Vehse, W. A. Sibley, F. J. Keller and Y. Chen, *Phys. Rev.*, **167**, 828 (1968)
105 F. J. Bryant and S. A. Hamid, *Phys. Rev. Lett.*, **23**, 304 (1969)
106 F. J. Bryant and A. F. J. Cox, *Phys. Lett.*, **20**, 108 (1966)
107 B. A. Kulp and R. M. Detweiler, *Phys. Rev.*, **129**, 2422 (1963)
108 R. Bäuerlein, *Z. für Phys.*, **176**, 498 (1963)
109 J. M. Meese and Y. S. Park, in *Radiation Damage and Defects in Semiconductors*, Edited by J. E. Whitehouse, Inst. of Physics, London, p. 51 (1973)
110 F. J. Bryant and A. T. J. Baker, *J. Phys. C, Solid State Phys.*, **5**, 2283 (1972)
111 F. J. Bryant and A. T. J. Baker, *Phys. Lett.*, **35A**, 457 (1971)
112 D. Pons, P. M. Mooney and J. C. Bourgoin, *J. Appl. Phys.*, **51**, 2038 (1980)
113 G. W. Arnold and G. W. Gobeli, in *Radiation Effects in Semiconductors*, Edited by F. L. Vook, Plenum, New York, p. 436 (1968)
114 J. A. Grimshaw and P. C. Banbury, *Proc. Phys. Soc.*, **84**, 151 (1964)
115 R. Bäuerlein, in *Radiation Damage in Solids*, Edited by D. S. Billington, Academic Press, New York, p. 358 (1962)
116 K. Thommen, *Phys. Rev.*, **174**, 938 (1968); errat. **179**, 920 (1969)

117 B. A. Kulp, *Phys. Rev.*, **125**, 1865 (1962)
118 B. A. Kulp and R. H. Kelley, *J. Appl. Phys.*, **31**, 1057 (1960)
119 H. J. Schulz and B. A. Kulp, *Phys. Rev.*, **159**, 603 (1967)
120 B. A. Kulp, *J. Appl. Phys.*, **37**, 4936 (1966)
121 F. J. Bryant, A. F. J. Cox and E. Webster, *J. Phys. C*, **1**, 1737 (1968)
122 F. J. Bryant and E. Webster, *Phys. Stat. Sol.*, **21**, 315 (1967)
123 F. H. Eisen, *Phys. Rev.*, **135**, A1394 (1964)
124 F. H. Eisen and P. W. Bickel, *Phys. Rev.*, **115**, 345 (1959)

10 THIN-FILM GROWTH: MODELS AND MORPHOLOGIES

Joseph Rudnick and Robijn Bruinsma
Department of Physics, University of California at Los Angeles, 405 Hilgard Avenue, Los Angeles, California, 90024-1547, USA

CONTENTS

10.1 INTRODUCTION

Physicists and materials scientists tend to adopt differing points of view in the study of manufactured thin films. Researchers in materials science, whose focus is the fabrication of materials that meet requirements determined by various applications, concern themselves with properties that are specific to the

Low Energy Ion–Surface Interactions
Edited by J. W. Rabalais

compound being deposited, and that have a demonstrable influence on the efficiency, economy of manufacture, and durability of the finished product. On the other hand, theoretical condensed-matter physicists, by and large, focus their attention on 'generic' properties of films resulting from various model processes, in particular the self-similar, or, alternatively, self-affine structure of the surface of such films. As the authors of this chapter belong to the latter group, we will largely ignore the practical issues that have influenced the development of such production mechanisms as molecular-beam epitaxy (MBE), chemical-vapor deposition (CVD) and laser ablation. We will, instead, concentrate on the predictions of various theoretical models for the evolution of growing fronts. Our focus is, thus, 'pure' rather than 'applied'.

It is important, however, to temper the above dichotomy. The morphological structures of interest here are relevant to the actual physical properties of thin films. The instabilities that lead to rough—ultimately self-similar or self-affine—surfaces in theoretical models are legitimate consequences of real thin-film growth mechanisms. The landscapes that result from these instabilities will almost certainly leave their mark on the interior as the film continues to grow. The primary focus of this chapter is on the physical structures that arise from growth instabilities, and, to the extent that they play a role in the growth of actual films, they will exert an influence on the utility of those films in practical applications, as in the case of the columnar microstructure (for a review see Dirks and Leamy, 1977). In this chapter, we will review predictions of some of the most popular models that have been applied to the problem of thin-film growth, with special emphasis on the consequences of the growth instabilities, both with regard to the roughness and the general morphology of the resulting surface and the structure that develop there.

10.2 MODELS FOR GROWTH-INDUCED ROUGHENING: LOCAL THEORIES

10.2.1 THE KPZ AND WOLF–VILLAIN MODELS

There are, generically, two classes of models for the evolution of a growing surface (and other forms of front propagation). In a purely *local* model, the rate of propagation of an infinitesimal section of surface is controlled entirely by surface properties in the immediate vicinity of that particular section. In a *non-local* model, the propagation rate of a section depends also on the properties from surface regions which are spatially removed. We will first consider local models as exemplified by the Kardar–Parisi–Zhang (KPZ) model (Kardar *et al.*, 1986). This model was an extension of an earlier growth model, called the Eden model (Eden, 1958), which is important in its own right. The Eden model was assumed to provide a description of the evolution of an expanding colony of cells, or a tumor. Physically, one imagines an object that grows by extending its boundaries at a constant rate. The direction of growth of a section of boundary is always perpendicular to that section. There is a helpful analog from the field of geometrical optics: if we think of the initial surface as a wavefront at some point in time then the normal growth described by the Eden model algorithm is equivalent to the evolution of the wavefront according to the Huygens construction, discussed further below. (For a discussion of the Huygens principle in film growth, see Carter, 1986). From this analogy it immediately follows that the Eden model, after a sufficiently long time, always produces a flat surface since this holds for the propagation of wavefronts in free space. There is thus no *intrinsic* instability mechanism (intrinsically unstable models will be discussed in Section 10.2.3). In order to complete the KPZ growth equation, we add two effects to the Eden model: (i) a 'smoothing' term that simulates annealing mechanisms such as evaporation/recondensation or surface diffusion, and reduces the heights of peaks and fills in valleys; and (ii) a stochastic contribution to the rate of growth to model, for instance, the fact that growth by generation of additional material is an inherently discrete process which is accompanied by shot noise. This shot noise is assumed to overwhelm the thermal noise of the substrate so that the 'noise temperature' is not the ambient temperature. The destabilization of the asymptotically flat surfaces of the Eden model is entirely due to this stochastic contribution.

The KPZ equation models the location of the boundary by a single-valued height function $h(\mathbf{r}, t)$ measured along the z-direction perpendicular to $\mathbf{r}$:

$$\frac{\partial h(\mathbf{r}, t)}{\partial t} = \nu \nabla^2 h(\mathbf{r}, t) + \lambda(1 + \tfrac{1}{2}|\vec{\nabla} h|^2) + \eta(\mathbf{r}, t). \tag{1}$$

The first term on the right-hand side is the smoothing term alluded to above. The second term—with λ the growth velocity—represents the constant rate of growth that drives the process. The origin of the square of the gradient in the second term is the fact that the rate at which a front propagates in a given direction is controlled by the angle that the front's unit normal makes with respect to that direction. If the angle is θ, then the rate of growth is enhanced by a factor $1/(\cos\theta)$. If we assume that h is a slowly varying function of $\mathbf{r}$, then this factor can be expanded as a power series in $|\nabla h|$. The terms in brackets are the first two terms in that gradient expansion. Finally, $\eta(\mathbf{r}, t)$ is the shot-noise term with the following 'white-noise' properties:

$$\langle \eta(\mathbf{r}, t) \rangle = 0; \qquad \langle \eta(\mathbf{r}, t)\eta(\mathbf{r}', t') \rangle = \Delta\delta(\mathbf{r} - \mathbf{r}')\delta(t - t'). \tag{2}$$

The version of the KPZ equation that is actually studied is obtained from (1) by replacing $h(\mathbf{r}, t)$ by $h(\mathbf{r}, t) + \lambda$ which removes the first term in the brackets on the right-hand side.

The nonlinearity appearing in the KPZ equation cannot be applied directly to MBE-grown surfaces: during MBE growth we deposit a *fixed* number σ of atoms per unit area. If the density of deposited material ρ is also fixed, then the normal growth velocity is *not* fixed, as it was for bacterial growth, but rather is equal to $(\sigma/\rho)(1 - \frac{1}{2}|\vec{\nabla}h|^2)$. After projection onto the z direction, one finds that the two nonlinear terms cancel, i.e. it is the vertical rather than the normal growth velocity which is fixed. There are two ways around this difficulty: (i) for an amorphous material with many vacancies, the density is not really fixed and the KPZ term may survive; (ii) even for a crystalline material with fixed density, the sticking probability of a particle on a surface depends in general on the angle between the incoming beam and the surface normal. Performing once again an expansion in this angle, a $|\vec{\nabla}h|^2$ term reappears.

The morphology of KPZ-controlled surfaces can be computed analytically for the one-dimensional case (i.e. frontal lines) and numerically for two-dimensional surfaces. This morphology is found to obey *scaling laws.* If we define a height–height correlation function:

$$G(\mathbf{r}, t) = \langle (\delta h(\mathbf{r}, t) - \delta h(0))^2 \rangle \tag{3}$$

with $\delta h(\mathbf{r}, t)$ the difference between $h(\mathbf{r}, t)$ and the average surface height and with $\langle \ldots \rangle$ standing for sample averaging, then for sufficiently large $\mathbf{r}$, and sufficiently long time,

$$G(\mathbf{r}, t) \approx |\mathbf{r}^{2\alpha}| F(t/|\mathbf{r}|^z). \tag{4}$$

Here, α and z are two characteristic exponents. For a one-dimensional KPZ surface, $\alpha = 1/2$ and $z = 3/2$. Numerical simulations for two-dimensional surfaces yield values close to 0.4 for α. The two exponents α and z are related by the 'scaling law' $\alpha + z = 2$ which follows from the rotational invariance of the equations underling the KPZ model.

The scaling function $F(x)$ in equation (4) approaches a constant for large x and goes to $x^{2\alpha/z}$ for small x. Thus, for late times when $t \gg |\mathbf{r}|^z$, the height correlation function approaches a steady-state value $G(\mathbf{r}, t) \approx \mathbf{r}^{2\alpha}$. This predicts that the further two points on the surface are separated, the greater their difference in height will be. For early times when $t \ll |\mathbf{r}|^z$, $G(\mathbf{r}, t) \approx t^{2\alpha/z}$. The surface has a roughness which does not depend on distance (provided $t \ll |\mathbf{r}|^z$), but which increases with time. In earlier work, Family and Vicsek (1985) had already proposed that in general non-equilibrium roughening of growing surfaces should obey this scaling-law description. The scaling hypothesis leads to a remarkable result concerning non-equilibrium roughening: the surface roughness

$$W^2(L, t) = \int \mathrm{d}^2 r \delta h^2(\mathbf{r}, t)/L^2 \tag{5}$$

of a sample of size L should depend on the sample size as $L^\alpha F(t/|L|^z)$, as can be deduced form equation (4). Surfaces whose RMS width $W(L)$ behaves as L^α are termed 'self-affine' provided α does not exceed one (if α were to exceed one, then the surface would 'invade' the sample interior and the assumption of a single-valued height function would no longer make sense). Self-affine surfaces, discussed in Mandelbrot's well-known book on fractals (Mandelbrot, 1982), are examples of scale-invariance; if you magnify a small section of such a surface then, after a simple rescaling, it should look essentially the same as the original section. For sample sizes less than a length $\xi(t)$ proportional to $t^{1/z}$, the surfaces are not scale-invariant and W does not depend on sample size: $W(t) \propto \xi(t)^\beta$ (with $\beta = \alpha/z$).

These results have stimulated some excitement among theoretical condensed-matter physicists. The above discussion has a direct analog in the description of the time evolution of a statistical-mechanical system cooled to its critical point, an example being a magnet in the vicinity of the Curie point. The KPZ equation is indeed very similar to the time-dependent Ginzburg–Landau equations used for dynamical critical phenomena (Ma, 1976), if one interprets the noise term as thermally driven, with a noise 'temperature' proportional to Δ. This logical connection suggests that one should interpret $\xi(t)$ as a time-dependent correlation length: for length scales less than the correlation length one observes critical (power-law) behavior; at larger lengths, the behavior is non-critical. Similarly, one might hope that growth phenomena divide into univesality classes, as in the case of critical phenomena. Each class obeys the scaling hypothesis, equation (4), but with different exponents. An enormous variety of growth models have been investigated (for reviews see Family, 1990; Viesek, 1989; Krug and Spohn, 1991), both numerically and analytically, with and without impurities, and the scaling hypothesis has been theoretically well confirmed, while the models exhibit a range of roughening exponents.

How can one test the scaling hypothesis experimentally? The most tempting procedure would be via standard MBE hetero-epitaxial growth of a thin film, followed by a measurement of the RMS surface roughness, $W(L)$, of sections of the surface of size L. Plotting the results on a log-scale, one extracts the exponent α. The exponent z would follow from a study of the height dependence of the roughness, but this is a lot harder (for a discussion of measured exponents, see Krim, 1992). This is a convenient method for measuring α *provided* the surface actually obeys scaling. In general, it is a rather dangerous procedure to *test* scaling this way for two reasons. (i) Hetero-epitaxial growth is well known to generally exhibit de-wetting phenomena (islanding) induced by substrate-overlayer misfit strains. Unfortunately, roughening of films by islanding, in the presence of annealing mechanisms, minics the scaling hypothesis for roughening at long and intermediate length scales. At rather short length scales (of the order of the film thickness), one discovers that the phenomena observed are associated with the growth of an unstable mode. (ii) From experiments on conventional critical phenomena, we know that measured scaling exponents can only be taken seriously if they are obtained from more than three decades of data, and this gives rise to significant experimental problems. It is possible to guard against some of these problems by a more laborious procedure: a measurement of the *full* height–height correlation function $G(\mathbf{r}, t)$ for a broad range of growth times (e.g. using scanning tunneling microscopy (STM) or grazing-angle X-ray scattering). If the scaling hypothesis is correct, then all the data should collapse on a single plot provided we measure length in units of $\xi(t)^{-1}$ and G in units of $\xi(t)^{-2\alpha}$. The analysis is done most conveniently by performing a Fourier transform on G to get the analog of the time-dependent susceptibility $\chi(q, t) = \langle |h(\mathbf{q}, t)|^2 \rangle$ which should obey a scaling relation

$$\chi(q, t) \propto 1/q^{2+2\alpha}\Gamma(q\xi) \tag{6}$$

with $\Gamma(x) \propto x^{2+2\alpha}$ for small x and $\Gamma(x)$ constant for large x. This function is very sensitive to islanding instabilities and other violations of scaling. If the film roughness through an intrinsic mechanism, such as the islanding instability, then a peak in $\chi(q, t)$ develops at the wave vector of the unstable mode. The growth of this peak violates the scaling hypothesis. If one measures no more than the sample roughness,

$$W(L, t) \propto \int \mathrm{d}^2 q \chi(q, t) \tag{7}$$

one only obtains information on the *average* of $\chi(q, t)$, and the violation of scaling over a limited range of wave vectors may not show up. The 'full' method was applied to the analysis of sputter-eroded surfaces (Eklund *et al.*, 1991) to avoid islanding—and KPZ scaling was found to be reasonably well obeyed (the origin of the KPZ nonlinearity for erosion derives from the dependence of the sputter-erosion rate on the angle between the beam and the surface normal (Bruinsma, 1990). When applied to hetero-epitaxy (Tong *et al.*, 1992), violation of scaling by conventional islanding was found. The surface roughness $W(L, t)$ *still* showed approximate scaling behavior in this case.

Apart from the preceding study on sputter-erosion, experimentally measured roughening exponents did not agree well with the KPZ predictions (Krim, 1992). Although some of the previously mentioned experimental problems may be responsible, there is reason to believe that for non-equilibrium crystal growth a model different from KPZ should be used. To see why the KPZ model may not apply, first recall that for MBE-grown films with perfect sticking (and similarly for other constant-flux procedures), the KPZ non-linearity cancels for films with a fixed density. Another source of problems is related to the nature of annealing in the film. There are two dominant annealing mechanisms for surface roughness: surface diffusion and evaporation–recondensation (Mullins, 1957). For low vapor pressures of the film material, surface diffusion should dominate. The annealing surface-diffusion current is $\mathbf{J} \propto \Delta\mu$ with μ the chemical potential. Above the roughening temperature, the chemical potential of surface atoms is given by

$$\mu = \mu_0 + \gamma \nabla^2 h + (\text{higher-order terms}) \tag{8}$$

with μ_0 independent of the surface geometry and γ the interfacial energy. The rate of change $\partial h(\mathbf{r}, t)/\partial t$ is proportional to $\nabla \cdot \mathbf{J}$ by mass conservation. This gives rise to a $-\nabla^4 h$ annealing term rather than the $\nabla^2 h$ term of the KPZ model. Next, we must determine the leading nonlinearity. If we rule out the KPZ nonlinearity for the reason stated earlier, then it appears reasonable to simply take the next-higher-order term, i.e. $\nabla^2 |\vec{\nabla} h|^2$. The resulting growth equation—proposed by Wolf and Villain (1990; Villain, 1991; Das Sarma and Tamborenea, 1991)—is

$$\frac{\partial h(\mathbf{r}, t)}{\partial t} = -\nu \nabla^4 h(\mathbf{r}, t) + \lambda \nabla^2 (|\vec{\nabla} h|^2) + \eta(\mathbf{r}, t). \tag{9}$$

This model has become quite popular. It obeys scaling and its roughening exponents—which were proposed to be $z = (8 + d)/3$ and $\alpha = (4 - d)/3$ with d the substrate dimensionality—agree better with the reported exponents (Krim, 1992). The main difficulty with the Wolf–Villain model has to do with the origin of the nonlinearity. It can be shown that it does not come from the higher-order terms in equation (8). In fact, as is the case for the KPZ nonlinearity, the Wolf–Villain nonlinearity *must* vanish at zero deposition current and it cannot be derived from any quasi-equilibrium argument. The reason for this is that, if the second term were to survive for zero deposition current, then we must replace the shot-noise term by the (conserved) thermal noise appropriate for the relaxational dynamics of a conserved system in equation (9). The roughening exponents for studied systems will not agree with the known equilibrium thermal roughness ($\alpha = 0$ for $d = 2$) so λ necessarily vanishes for zero deposition current. Neither can the nonlinearity come from interference of the incident atoms with surface diffusion: at deposition rates so high that this indeed occurs, one encounters amorphous rather crystalline deposits and one loses the argument forbidding KPZ nonlinearity. Numerical studies (A. Zangwill, private communication) actually do appear to produce the Wolf–Villain term under non-equilibrium conditions for some, but not all, simulation algorithms. The precise origin of the Wolf–Villain term thus remains unsettled at the moment. Another reason to question the Wolf–Villain model arises when one reflects on the behavior of growing surfaces in the vicinity of the roughening transition. The annealing term in equation (9) is invalid if one is interested in the behavior of a faceted surface below the roughening temperature because it does not encompass the singular dependence of the surface energy on the surface normal of a crystal. If we simply *assume* that growth has driven the surface rough, then we can apply the annealing term, but this approach begs the question of how the surface actually roughens starting from a facet.

10.2.2 COLUMNAR GROWTH, THE HUYGENS CONSTRUCTION AND SUBSTRATE-INDUCED ROUGHENING

How well does the morphology of the KPZ or Wolf–Villain model agree with microscopy studies of thin films grown under normal production procedures such as sputter deposition (Bruinsma *et al.*, 1990)? The most frequently encountered morphology during low-temperature thin-film growth is the 'columnar' structure. The orientation of the columns is controlled by the direction of the incoming beam, and the grooves separating the columns extend from the film surface all the way down to the substrate. The columnar network itself is characterized by a length scale $\xi(h)$ which depends on the film thickness h. In a number of cases of Zone I growth morphologies (e.g. pyrolytic graphite—Messier and Yahoda, 1985; Yehoda and Messier, 1985; Messier, 1986), $\xi(h)$ increases as a power law with h, while for Zone II morphologies—at higher growth temperature—it depends weakly on h. The KPZ and Wolf–Villain surface morphologies bear no visible relation to the observed columnar microstructure. Recall that our self-affine surface structure always has a width $W(L)$ which is small compared to L, while for the surface of a columnar landscape with 'canyons' descending down to the substrate, one would need $W(L) \sim L$.

The formation of the columnar microstructure is not (directly) related to the non-equilibrium roughening theories mentioned above. It is an entirely *deterministic* effect controlled by the *substrate roughness*. The shot noise of the incoming beam does not play a key role. We can get more insight into columnar growth by re-examining the Huygens construction—which is a very useful tool for the analysis of the growth of films under practical conditions. Imagine that we start out with some sort of initial surface configuration, which can have an artibrary spatial structure. If we think of this initial surface as a wavefront at some point in time then the normal growth described by the Eden model is equivalent to the evolution of the wavefront according to the Huygens construction. To effect the Huygens construction, think of the surface as a portion of rough terrain across which one can roll a sphere. The radius of the sphere is equal to the distance that a perfectly flat surface would propagate in the time interval of interest. The new terrain that has resulted from Eden-model-type growth is found by tracing out the location of the center of the sphere. This process is displayed pictorially in Figure 10.1. As can be seen from the figure, rough features are smoothed out, though with one important exception. When the radius of the sphere is large enough, what may have been the smooth contours of a given depression are sharpened into a cusp-like singularity, known as a *caustic*. In Figure 10.1, the locations of two caustics are indicated by cross-hatches on the rolling spheres that give rise to them. As we will see below, more realistic non-local 'shadow' models of amorphous growth produce grooves which appear in the same place as the caustics of the Huygens construction.

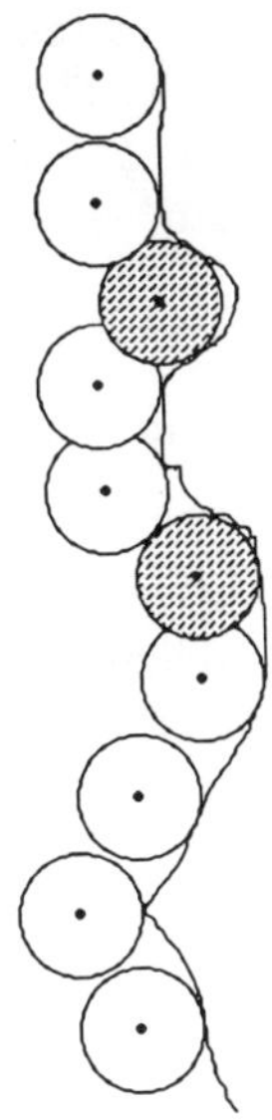

Figure 10.1 Illustration of the Huygens construction in terms of spheres that are rolled over an initially rough surface. Sharp features are blurred, with the important exception that *caustics* are formed when the rolling sphere comes into contact with the surface at more than one point

In other worlds, the caustics are not mathematical oddities but actually trace out the evolution of the groove network. The late-time evolution of the caustic network can be inferred in a simple way. When light hits a modulated refracting surface it produces caustics. The caustics appear as bright lines on a screen parallel to the surface at a distance h. As can be observed from the caustic pattern on the bottom of a swimming pool on a bright day, the pattern rapidly 'coarsens' as we increase h. It can be shown (Tang *et al.*, 1990) that the resulting coarsening length $\xi(h)$ agrees well with the data from pyrolytic graphite. The groove structure is thus essentially a remnant from the substrate roughness which should vanish for thick enough films. In the KPZ model, the caustics of the Huygens model were automatically removed by the annealing term so no trace was left of the initial roughness.

10.2.3 INTRINSIC INSTABILITY MODELS

The KPZ and Wolf–Villain models ascribe growth-induced roughening to the shot noise of the incoming beam. As we mentioned, however, during hetero-epitaxy growing films become unstable without the benefit of any noise terms. What should we expect when this happens? This problem was first analyzed by Grinfeld (1985, 1986) who examined strained overlayers. He found that the linear mode spectrum becomes unstable. The simplest type of mode instability, although not exactly the one obtained by Grinfeld, would be an unstable mode with a q^2 spectrum with a stabilizing q^4 spectrum due to surface diffusion. The equation of motion is then

$$\frac{\partial h(\mathbf{r}, t)}{\partial t} = -\nu\nabla^4 h(\mathbf{r}, t) - \lambda\nabla^2 h(\mathbf{r}, t) + \text{noise}. \tag{10}$$

Note the minus sign of the second term. It is not an annealing term but instead (crudely) describes the destablizing effect of substrate strain. Alternatively, the negative sign of the second term may be considered (Villain, 1991) as the macroscopic expression of the 'Schwoebel effect' (Schwoebel, 1969)—asymmetric diffusion at step edges.

Equation (10) predicts exponential growth for surface-corrugation modes with wave vectors of order $(\lambda/\nu)^{1/2}$. What happens if we include non linear terms in equation (10)? This has been done in two ways. The first approach is to replace the ∇^2 and ∇^4 operators with their correct 'covariant' nonlinear analogs (so $\nabla^2 h(\mathbf{r}, t)$ becomes the true surface curvature and $\nabla^4 h(\mathbf{r}, t)$ the second derivative measured along the surface of the curvature). This has been done by Golubovic and Karunasiri (1991; Siegert and Plischke, 1992) who do not find a self-affine surface at all but, instead, a coarsening mountain landscape with pyramidal structures resembling piles of sand. The dynamics is akin to spinodal decomposition. This should describe the behavior at wave vectors of order $(\lambda/\nu)^{1/2}$. The alternative method is to add our old KPZ nonlinearity to equation (10):

$$\frac{\partial h(\mathbf{r}, t)}{\partial t} = -\nu\nabla^4 h(\mathbf{r}, t) - \lambda\nabla^2 h(\mathbf{r}, t) + \lambda'|\vec{\nabla} h|^2. \tag{11}$$

This is a well-known equation called the Kuramoto–Shivashinski equation. It is encountered in the study of chemical reactions and was introduced in the study of growing films by Sneppen *et al.* (1992). Interestingly, it develops *chaos* and deterministic noise. It has been suggested by Yakhot (1981) that for large length scales (i.e. large compared with $(\lambda/\nu)^{-1/2}$) and large time scales this should produce a self-affine surface described by the KPZ equation, with the deterministic noise playing the role of the thermal noise. Sneppen *et al.*, have confirmed this for one-dimensional surfaces (although it may not be true for two-dimensional surfaces). At any rate, we see that in this 'backhanded' way the islanding instability can still give self-affine roughening without any contribution of shot noise. The study of hetero-epitaxial growth-induced roughening as a form of turbulence may be one of the most fertile areas of future research in this field.

10.3 MODELS FOR GROWTH-INDUCED ROUGHENING: NON-LOCAL THEORIES

Transport of matter and energy in the region *outside* a growing thin film more often than not exerts a profound influence on the evolution of both surface and bulk morphology. Thus, a strict reliance on local growth models entails serious risks. One is in danger of entirely overlooking the key mechanisms responsible for, among other things: growth instabilities in the nearly flat surface; the ultimate self-similar or self-affine structure of the 'fully developed' surface; the shapes of surface features and the resulting bulk structure (i.e. channels, vacancies, patterns of dislocations). Unfortunately, a trade-off is involved. The more nearly complete predictive power of non-local growth models is gained at the cost of considerably more difficult analysis, and the numerical simulation of non-local growth is much more time-consuming than is the case for local growth models. Nevertheless, linear-stability analysis, some renormalization-group arguments and heuristic discussions based upon observation and inspired guesswork have let to progress in our understanding of the complex morphologies that can result from various models that approximate the flow of matter to the evolving surface. Furthermore, there are cases in which approximations that do not surrender all contact with physical reality are amenable to exact analytical treatment. In this section we will briefly review some of the models considered, and catalogue the predictions of these models, both with regard to the exponents characterizing the scaling properties of the surfaces and the shapes of surface features.

Roughly speaking, we can divide the transport of matter to the growing surface into two categories: *ballistic* and *diffusive*. In diffusive transport, molecules scatter several times in the external medium before impinging on the surface. More precisely, the mean free path of an incident molecule is smaller than the characteristic linear dimension of surface features of interest. By contrast, ballistically propagating molecules suffer no collisions as they travel over distances of the order of characteristic surface structures. We will neglect the case of molecules whose mean free path approximates the length scales of surface structures. This undoubtedly important intermediate case has received little attention as yet.

10.3.1 DIFFUSION-LIMITED GROWTH

If the medium whose particles aggregate, condense, or solidify into the resultant material is sufficiently dense and is, additionally, free of convection currents, then transport will be effectively diffusive. As the rate of surface growth is limited by the rate at which material is delivered there, growth with diffusive transport of materials to the surface is said to be *diffusion-limited.* The most widely studied model of such growth is *diffusion-limited aggregation* (DLA) (Witten and Sander, 1981), in which a seed particle grows to a complicated, branched, ultimately fractal, structure as new particles perform, one by one, random walks into the region occupied by the structure, and 'stick' onto the part of the structure with which they first come into contact. Physical examples of diffusion-limited growth processes include electrochemical deposition (Matsushita *et al.*, 1984), solidification from melts, viscous fingering (Langer, 1980), and —because the limiting form of the diffusion equation when transport is asymptotically slow is Laplace's equation—dielectric breakdown (Bensimon *et al.*, 1986).

In contrast to the KPZ and Wolf–Villain models, pattern formation in the form of growth instabilities occurs in DLA *in the absence of noise.* The growth process in its deterministic form amplifies irregularities in the growing surface. In light of the mathematical connection between the diffusion equation and Laplace's equation, this property of DLA is understandable. The impinging particles flow with a current density that can be taken as the gradient of a potential function that—approximately at least—satisfies Laplace's equation. Furthermore, for many applications the growing surface can be treated as an approximate equipotential. The flux of particles onto the surface is thus modeled in terms of the electric field strength at the surface of a conductor, and it is well known that the field strength is much greater at protuberances on a conductor in the presence of an external electric field. In light of this analogy, it is to be expected that hills will grow at the expense of valleys, and that slight irregularities will be considerably amplified. There is, in addition, every reason to believe that growing hills will eventually compete with each other, that the higher ones will win the competition, and that a hierarchy of hills will develop. The addition of a stochastic term to the growth equation, to simulate the atomistic characteristics of the material deposited (i.e. the fact that material is deposited in discrete increments), will generate, on a continual basis, new features, both on the hilltops and in the valleys, which will be subject to the same amplification process. A self-similar outcome is not unsurprising. Indeed, it is to be expected.

Given the success of the renormalization group in describing a wide variety of physical phenomena exhibiting scale invariance (see, e.g. Wilson and Kogut, 1974; Ma, 1976), it is natural to hope that this method will prove adaptable to DLA and other diffusion-limited growth models. To date, however, no entirely suitable formulation of the renormalization group has been discovered. However, heuristic arguments have yielded reasonably accurate results for exponents with diffusion-limited growth processes. Two particularly important ones are, first, an argument based on a mean-field approach and, second, a discussion based on an analysis of the diffusion currents in the vicinity of a cone- or wedge-shaped tip on a cluster.

10.3.1.1 Mean-field approach

Two approaches to the formation of an aggregate via diffusion-limited growth processes are described with the use of the term 'mean field'. The first approach, introduced by Ball, Nauenberg and Witten (1984) utilizes a variant of time-dependent Ginzburg–Landau theory. They obtain an equation controlling the evolution of a 'front' separating the exterior from the interior of a growing spherical cluster. This front decays asymptotically as a power law in space. The power yields the fractal dimension of the cluster. In three and more dimensions they obtain the 'mean-field' exponent $D = d - 1$, where D is the fractal dimensionality of the cluster and d is the dimensionality of the embedding space.

Another approach is also described as a mean-field approximation, although it differs considerably in spirit from the method of Ball *et al.* This involves the introduction of a free-energy-like quantity, which is then minimized (Tokuyama and Kawasaki, 1984). The relevance of a free-energy-minimization principle to a manifestly off-equilibrium process may be rightly called into question, but the argument does lead to a definite prediction for the fractal dimensionality, D, of a DLA cluster in d dimensions. The result is

$$D = \frac{d^2 + 1}{d + 1}. \tag{12}$$

This expression is reminiscent of the famous result of Flory (1971) for the fractal dimensionality of a chain polymer in d dimensions. In a way, this is to be expected, given the essential similarity between the arguments leading to these two results. As it turns out, a more complex discussion, in which one considers the behavior of quantites associated with the invasion of the diffusion cloud into the empty regions that dominate the interior of the cluster, also leads to the above result for D (Muthukumar, 1983).

10.3.1.2 Growth in the vicinity of a tip

There is a natural inclination to think of the fractal dimension of an object as a quantity that has no relationship with the object's gross physical shape. However, Turkevich and Scher (1985), in a remarkable set of arguments, assert an initimate connection between these two features of a DLA cluster (see also Vicsek, 1989 for a discussion of this and the previous approaches to sealing in DLA clusters. They argue that the fractal dimensionality of a growing wedge-shaped cluster is entirely determined by the angle subtended by the wedge. The connection between tip angle and fractal dimensionality follows from the solution of Laplace's equation in the vicinity of the tip. The electrostatic analog for this DLA configuration is a wedge-shaped conductor in the presence of an external electric field. In terms of polar coordinates r and ϕ, the electrostatic potential in the vicinity of the tip of the wedge is given by

$$\psi(r, \phi) = C \cos\left(\frac{\pi\phi}{2\pi - \theta}\right) r^{\pi/(2\pi - \theta)}, \tag{13}$$

where the variables r, ϕ and θ are defined in Figure 10.2. The quantity C is a constant, proportional to the overall field strength. This potential also yields the flux of diffusing particles via $\mathbf{j} = -\nabla\psi$. This means that the flux impinging on the wedge a distance λ from the tip is proportional to $\lambda^{[\pi/(2\pi-\theta)]-1}$. If the length of a side of a growing wedge is R and if the number of particles in the wedge is N then

$$\frac{\mathrm{d}N}{\mathrm{d}t} = K \int_0^R \lambda^{[\pi/(2\pi-\theta)]-1}\,\mathrm{d}\lambda = K\,\frac{2\pi - \theta}{\pi}\,R^{\pi/(2\pi-\theta)}, \tag{14}$$

while the growth rate, $\mathrm{d}R/\mathrm{d}t$, is proportional to the total flux impinging on the wedge very close to the tip, or, given a small distance a,

$$\frac{\mathrm{d}R}{\mathrm{d}t} = K \int_0^a \lambda^{[\pi/(2\pi-\theta)]-1}\,\mathrm{d}\lambda = K\,\frac{2\pi - \theta}{\pi}\,a^{\pi/(2\pi-\theta)}. \tag{15}$$

Thus,

$$\frac{\mathrm{d}N}{\mathrm{d}R} = \left(\frac{R}{a}\right)^{\pi/(2\pi-\theta)}. \tag{16}$$

Integrating:

$$N(R) \propto R^{(3\pi-\theta)/(2\pi-\theta)}. \tag{17}$$

The measured fractal dimensionality of the two-dimensional DLA cluster is $D \cong 1.71$. The angle of a wedge that, according to the argument above, has this fractal dimensionality is $\theta \cong 0.59\pi \cong 106°$.

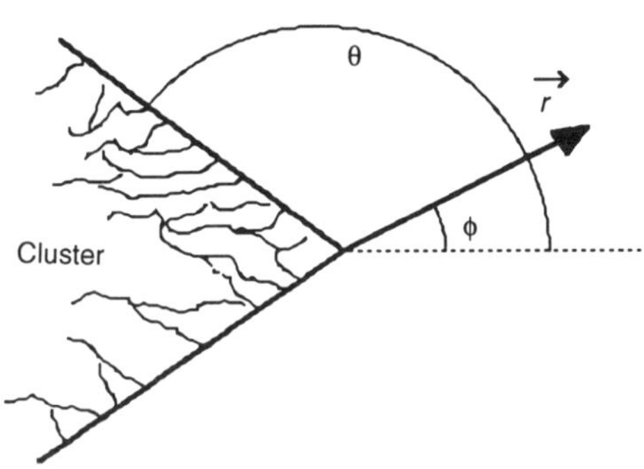

Figure 10.2 A figurative depiction of a wedge-shaped tip on a growing cluster

The Turkevich–Scher argument has been applied to anisotropic on-site DLA clusters—that is, clusters in which the random-walking and aggregating particles occupy the sites of a regular lattice and in which the probability that a particle will stick to the growing cluster depends on the angle from which it impinges on it. In the case of extreme anisotropy of the sticking probability, the cluster shape is found to be well approximated by a highly elongated rhombus, so that there are two sets of tips—two with an angle $\theta \cong 0$ and two with $\theta \cong \pi$. In this case, the agreement between the above analysis and simulations has been found to be excellent (Ball *et al.*, 1985).

The above set of arguments can be extended to higher dimensions, where one assumes cone-shaped tips. Furthermore, the fractal, or more precisely *multi*-fractal character of a DLA cluster that sprouts growing tips having a range of angles has also been discussed (Halsey *et al.*, 1986).

10.3.2 THE SHADOWING MODEL

In films that grow as the result of the deposition of both ballistically and diffusively propagating matter, prominences on the surface will have an effect on the growth of nearby, and not-so-nearby, regions that can be described in terms of the casting of a shadow. The shadowing effect can be visualized without too much effort in the case of ballistic deposition such as MBE. A protuberance intercepts particles that would otherwise land on the portion of the surface that lies in its shadow. In diffusive processes, such as DLA, shadowing is a consequence of Laplace's equation. In this latter context, shadowing can be seen as a manifestation of the 'lightning-rod effect', in which electrical field lines concentrate onto the most elevated features on an equipotential surface. We now turn our attention to a model that *explicitly* incorporates shadowing. This model, which is readily simulated, is also, in a certain limiting case, amenable to an analysis based on renormalization group ideas. Finally—and this may be its most noteworthy feature—the model approximates a growth process that almost certainly occurs in nature.

We begin by imaging a thin film in a two-dimensional world. The surface of the film—the growing front of the system—is a one-dimensional curve. The front evolves, and the film thickens, because a steady stream of ballistically propagating particles arrive at the surface from all directions in an upper semicircle. The configuration is depicted in Figure 10.3. At any given point on the surface, the film grows at a speed proportional to the rate at which matter accumulates there, and this rate of accumulatin is equal to the total flux of particles impinging at that point (Karunasiri *et al.*, 1989; Bales *et al.*, 1990). The rate at which matter is captured depends upon the angle of exposure at the location in question, as illustrated in the figure. It is clear that prominences will have a larger exposure angle than nearby depressions and will, as a result,

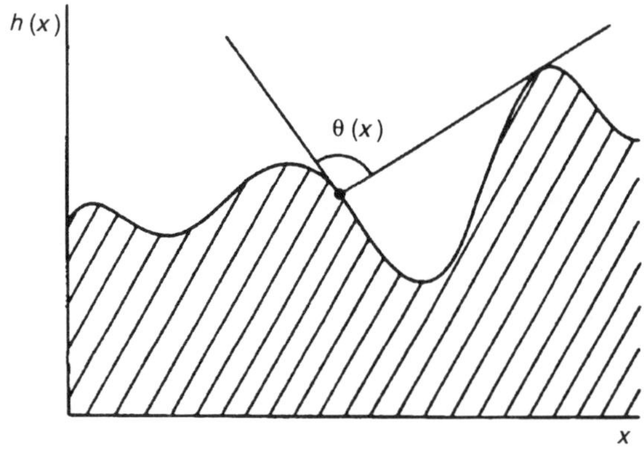

Figure 10.3 Shadowing on a growing front

tend to grow faster. A growth instability like that seen in DLA will occur. Put simply, the growth instability in this model results from the fact that peaks, by intercepting material that would otherwise have landed elsewhere, cast shadows about themselves. As a matter of course, they will grow more rapidly than the valleys that lie in their shadow.

To simplify the model, one can replace the continuous front by an array of vertical columns. A column grows with a speed proportional to the angle of exposure of its tip. One can envision the group of columns as a field of grass, each blade of which grows in accordance with the amount of light received by a photoreceptor at its tip. For this reason it has become common to refer to this crude approximation to thin-film growth as the 'grass model'. To quantify the model, let $h_i(t)$ be the height of the ith column at time t. Then,

$$\frac{\mathrm{d}h_i(t)}{\mathrm{d}t} = R\theta_i(t) + C\,\frac{\mathrm{d}^4 h}{\mathrm{d}i^4} + \eta_i(t). \tag{18}$$

This equation incorporates the shadowing effect, in that the quantity θ_i is the exposure angle at the tip of the ith column. It also contains surface diffusion through the ∇^4 term and stochasticity in the form of the Gaussian white-noise variable η_i (see equation (9)). Because the subscript i is not a continuous variable, the symbol ∇^4 stands for a discrete version of the differential operator. In the 'pure' shadowing model, the second and third terms on the right are absent. Surface diffusion will temper the effects of shadowing, in that it leads to the erosion of hills and the filling-in of valleys. The stochastic term gives rise to irregularities, which may or may not find themselves amplified by the shadowing process.

A very nice feature of equation (18) is that, in the absence of surface diffusion and stochastic contributions, certain key features of the configuration of columns that results from the growth-with-shadowing process can be inferred from an analysis that adheres to the spirit of the renormalization-group method. Simulations of the 'pure' growth-with-shadowing equation reveal a kind of 'Cantor lattice' of columns, as displayed in Figure 10.4. This structure is consistent with the occurrence of a series of competititions between columns that are resolved on time scales that are proportional to the length scales at which the competitions take place. One can envision the process as occurring in the following way. In its first stages, the competition is primarily between immediately neighboring columns. If two such columns are separated by a distance d and their heights are h_1 and h_2, respectively, then, if $|h_1 - h_2| < d$, the equation governing the time evolution of the height difference is

$$\frac{\mathrm{d}(h_1 - h_2)}{\mathrm{d}t} = R\,\frac{h_1 - h_2}{d}. \tag{19}$$

This competition is resolved in favor of the higher column in a time of the order of d/R. Once resolution has taken place, the competition is between the 'winners', the separation between which is essentially equal to $2d$. As a rough approximation, we assume that the 'losers' will have ceased to grow at the end of competition. The equation that describes this new competition is again equation (19). with d replaced by $2d$. The time required to resolve this new competition is clearly $2d/R$. Now, the winners are typically a distance $4d$ apart. According to this reasoning, one can analyze the result of shadowing on

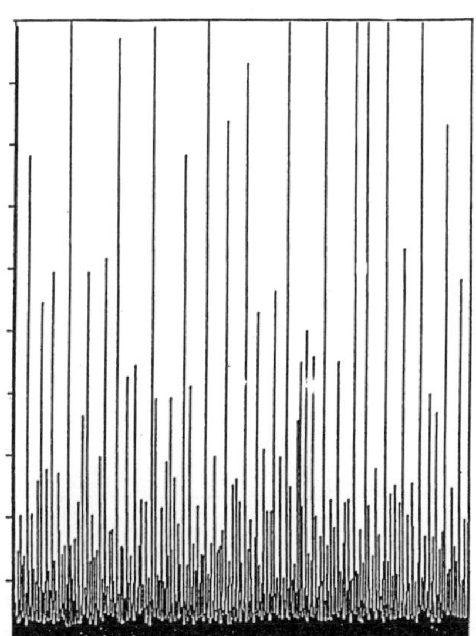

Figure 10.4 The kind of Cantor lattice of columns that result from the 'grass' model

columnar growth by considering competitions on a sequence of geometrically increasing length scales. The corresponding time scales also grow geometrically. The winners of each competition, which accept an incoming flux of matter from the maximum possible angle of exposure, grow linearly in time. Then, if there are N columns on the substrate, $N/2$ of them, the losers of the first set of competitions, will have a height of approximately π/d. Then, $N/4$ of the remaining $N/2$ columns lose the second competition, and will have ceased to grow after having reached a height of approximately $(\pi/d)(1 + 2)$. In general, there will be $N/2^n$ columns with heights of approximately $(\pi/d)(1 + 2 + 4 + \ldots + 2^{n-1})$. This means that the number of columns with a height equal to h, $c(h)$, varies inversely with h, i.e. $c(h) \propto 1/h$. This result is borne out by computer simulations, as displayed in Figure 10.5.

The effects of surfance diffusion on the grass model are substantial and not yet entirely understood. According to simulations, a small amount of surface diffusion leads to a change in the effective exponent for the coverage. That is, when the coefficient D in equation (18) is non-zero, $c(h) \propto h^{-p}$, where $p \geq 1$ is a continuous function of D. This situation is also illustrated in Figure 10.5. Limited simulations indicate that there is a transition from pure power-law decay of the coverage function $c(h)$ to qualitatively more rapid decay at large h when the exponent p is some value less than 2. The task of understanding the above findings is rendered more complicated by the fact that, as long as

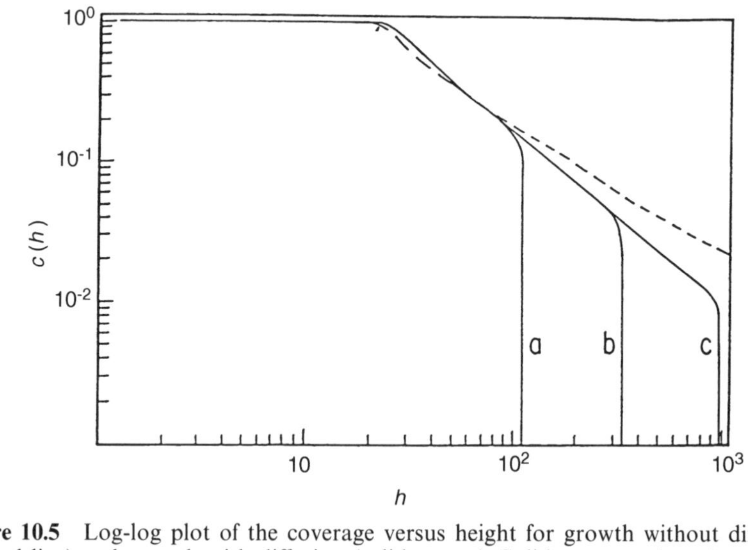

Figure 10.5 Log-log plot of the coverage versus height for growth without diffusion (dashed line) and growth with diffusion (solid curves). Solid curves a, b and c are the coverage versus height at various stages of the growth process, curve a corresponding to the earliest and c to the latest stage

they are non-zero, all coefficients in equation (18) can be set equal to one by a suitable rescaling of the heights, the distances measured along the substrate, and the time. This result implies that the only determinant of asymptotic behavior is the initial distribution of column heights. A dependence of the exponent p on the coefficients alone seems to be explicitly ruled out. In this context, it is interesting to note that growth via the Huygens construction produces a surface whose roughness exponent does, indeed, depend in a crucial way on the initial distribution of column heights. As yet, no reliable large-scale simulation has been performed on the columnar model with a two-dimensional substrate.

Finally, simulations indicate that the introduction of stochasticity into the growth equation does not lead to a significant modification of the resulting distribution of column heights. In this respect, the grass model contrasts markedly with the KPZ model, for which the surface morphology depends decisively on the presence—or the absence—of stochastic contributions to the growth equation.

A more realistic version of growth with shadowing replaces the array of columns by a smooth surface. In two dimensions, the surface is described by a curve. Each point on the curve grows in the direction normal to the curve at a rate v_n, where

$$v_n = R \int_{\theta_1(s)}^{\theta_2(s)} \mathbf{j}(\alpha) \cdot \hat{n}(s)\, \mathrm{d}\alpha + D \frac{\partial^2 \kappa(s)}{\partial s^2}. \tag{20}$$

In this equation, the quantity $\mathbf{j}(\alpha)$ is a flux of uniform amplitude with angle of incidence α. That is, $\mathbf{j}(\alpha)$ is along the direction of a ray that makes an angle α with respect to the vertical. The angles θ_1 and θ_2 are the angles between the vertical and the lines from a point on the surface to the peaks shadowing that point on the right and the left respectively. The quantity κ is the curvature of the surface. This growth equation incorporates shadowing and surface diffusion in a fully 'covariant' form. This equation is a good deal more difficult to simulate than that of the grass model, because of the complications associated with the reparameterization of the evolving curve. The results of a simulation of a limited portion of a smooth-growing front are shown in Figure 10.6. As is evident, variations in the surface are amplified by the growth process. One can see clear evidence for competition between neighboring peaks and for the resolution of this competition in favor of those that are more prominent. Furthermore, the figure indicates that the shadowing process leads to the formation of grooves in the interior, as higher peaks abut and then envelop those that lose the competition. This feature is characteristic of one particular type of microstructure that is known to develop as a result of thin-film deposition.

It is clear that this set of models deserves more attention than has yet been paid to it.

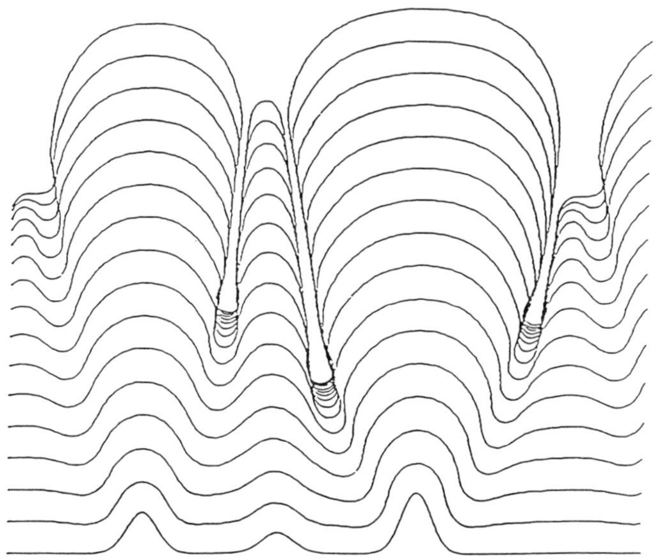

Figure 10.6 Growth of a front with shadowing, illustrating the amplification of small irregularities, the competition between growing protuberances and the grooves that are left in the bulk as a result of the competition

10.3.2.4 COLUMNAR GROWTH WITH A COLLIMATED FLUX

There is another version of columnar growth for which analytical techniques provide essentially exact results for the dynamics. This is columnar growth in which the incoming flux of particles is unidirectional, or *collimated.* This model more closely approximates growth of a thin film via molecular-beam epitaxy, in which the incoming flux is, characteristically unidirectional. The screening mechanism clearly operates here, in that higher columns cast shadows on columns farther from the source of the matter flux than they are. In the limit that the incoming flux is at a right angle to the growing columns, or is at *grazing incidence*, the growth of the columns can be modeled in terms of a set of one-dimensional directed random walkers. Exact results for the distribution of column heights immediately follow (Krug and Meakin, 1989). When the flux is not perpendicular to the columns, the nature of the growth may be different, but the model is again amenable to analytical treatment. Results for various exponents, such as the roughness exponent, $2\alpha/z$, have been extracted, including interesting 'crossover' effects as the angle between the incident flux and the columns is varied from $\pi/2$ (grazing incidence) to 0 (normal incidence). Interestingly, the latter case is found to correspond to KPZ-model-type growth. Two variables of particular importance are, first, the shot-noise amplitude and, second, the angle of incidence of the flux. Among the dynamical transitions that can control the growth of columns in one version of this model are a faceting transition and a roughening transition. Mean-field analyses can also be applied, and the evidence is that predictions derived from them are exact in the limit of infinite substrate dimensionality (Krug and Meakin, 1991).

10.3.4 STRUCTURES RESULTING FROM MOLECULAR-DYNAMICAL SIMULATIONS OF BALLISTIC DEPOSITION

Recent molecular-dynamical simulations of ballistic deposition have given rise to structures on growing clusters that imply intriguing correspondences with the kinds of wedge-like structures that play a central role in the Turkevich–Scher discussion of the fractal structure of anisotropic DLA clusters. Tang has performed simulations in which particles, modeled as hard spheres, propagate ballistically in two dimensions onto a flat one-dimensional substrate (C. Tang, unpublished). The flux of incoming matter has, as in the grass model, a uniform angular distribution. Incoming particles stick to the first portion of the substrate, or growing cluster, that they encounter. The clusters which form appear to be compact. That is, their effective dimensionality is that of the 'embedding space', which is, in this case, two-dimensional. As is evident in Figure 10.6 and 10.7, a noteworthy feature of the 'film' that results from this process is a set of peaks on the surface, whose envelopes are wedge-shaped. The qualitative and quantitative aspects of these features can be understood in terms of the shadowing process. The explanation is as follows.

Consider, first, the pinnacle of a wedge-shaped prominence. This portion of the growing surface will receive incoming matter from all possible directions, and will grow at the maximum possible rate. Furthermore, its rate of growth will be governed by the version of the growth-with-shadowing equation that is appropriate to columns, rather than the version describing the evolution of a smooth front. We then focus our attention on a side of the wedge. This is actually a curve that traces out the positions of the most prominent components of the cluster on that side of the prominence. Each of these local peaks also receives flux in the same way as the tips of columns. The side of a wedge will grow in a direction normal to its orientation with a speed proportional to the angle of exposure of a peak lying in it. If the angle that this side of the wedge makes with respect to the horizontal is ϕ then the exposure angle is $\pi - \phi$. The summit of the wedge has the largest possible exposure angle, namely π. The

Figure 10.7 Clusters grown by ballistic deposition. Note the wedge-shaped tips on local prominences. This figure and Figure 10.8 represents unpublished results of Dr Chao Tang. We gratefully acknowledge his generosity in allowing us to use them in this chapter

summit thus grows perpendicularly with a speed proportional to π, while the sides grow normally with a speed proportional to $\pi - \phi$. Their intersection propagates vertically with a speed proportional to $(\pi - \phi)/\cos\phi$. In order for the summit to grow at the same rate as the intersection, the following equation must clearly be satisfied:

$$\frac{\pi - \phi}{\cos\phi} = \pi.$$

The numerical solution to this equation is $\phi = 0.66$ radians, or 37.83°. This implies a wedge angle of $\theta = 180^\circ - 2\phi = 104.34^\circ$. This is consistent with the angles seen by Tang and is intriguingly close to the wedge angle of approximately 106° associated with the fractal dimensionality of DLA clusters in the Turkevich–Scher picture. At this point there seems to be no indication of a sufficiently intimate connection between the shadowing mechanism as it operates in ballistic deposition and the effective shadowing that one can associate with diffusion-limited growth processes to justify attributing any predictive value with regard to DLA to the result above.

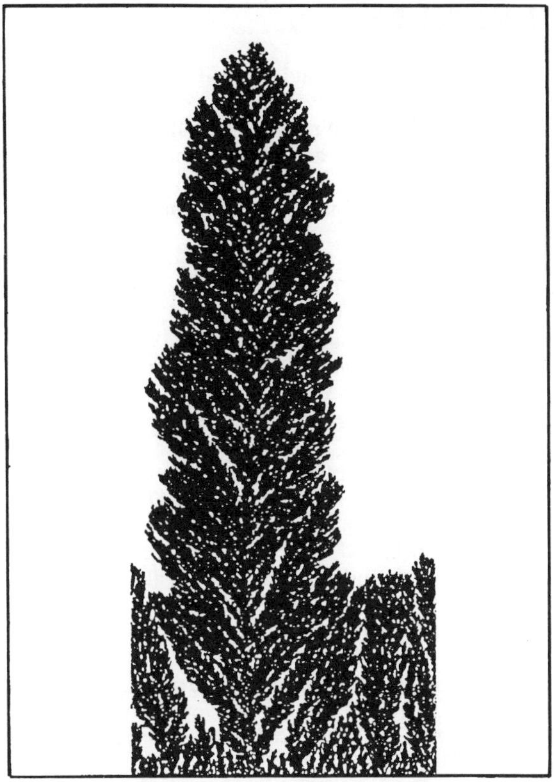

Figure 10.8 An isolated prominence, in which the wedge-shaped tip is more evident. The angle of the wedge is close to 105°

10.4 CONCLUSION

It is clear that much has been left out of this discussion. We have made no attempt to do justice to the full range of models that have been proposed to describe or elucidate one of the many physical processes that are thought to be relevant to thin-film growth. The unfettered imagination of theorists has provided us with an embarrassment of riches. We can only await the experimental results that will anchor these models in physical reality.

10.5 REFERENCES

Bales, G. S., R. Bruinsma, E. A. Eklund, R. P. U. Karunasiri, and J. Rudnick (1990) *Science*, **249**, 264.

Ball, R., M. Nauenberg, and T. A. Witten, Jr. (1984) *Phys. Rev.*, **A 29**, 2017.

Ball, R. C., R. M. Brady, G. Rossi, and B. R. Thompson (1985) *Phys. Rev. Lett.*, **55**, 1406.

Bensimon, D., L. P. Kadanoff, S. Liang, B. Schraiman, and L. Tang (1986) *Rev. Mod. Phys.*, **58**, 977.

Bruinsma, R. (1990) In *Proceedings of the Les Houches Workshop on Surface Disordering: Growth, Roughening and Phase Transitions, March 31–April 9, 1992, Les Houches, France*, R. Jullien, J. Kertesz, P. Meakin, and D. E. Wolf (eds).

Bruinsma, R., R. P. U. Karunasiri, and J. Rudnick (1990) In *Proceedings of the NATO Advanced Research Workshop on Kinetic of Ordering and Growth on Surfaces*, M. Lagally (ed), Plenum Press, New York.

Carter, G. (1986) In *Erosion and Growth of Solids Stimulated by Atom and Ion Beams*, G. Kirikiadis *et al.* (eds). Nijhoff, Dordrecht.

Das Sarma, S., and P. Tamborenea (1991) *Phys. Rev. Lett.*, **66**, 325.

Dirks, A., and H. J. Leamy (1977) *Thin Solid Films*, **47**, 219.

Eden, M. (1958) In *Symposium on Information Theory in Biology*, H. P. Yockey (ed). Pergamon Press, New York.

Eklund, E. A., R. Bruinsma, J. Rudnick, and R. S. Williams (1991) *Phys. Rev. Lett.*, **67**, 1759.

Family, F. (1990) *Physica*, **A 168**, 561.

Family, F. and T. Vicsek (1985) *J. Phys.*, **A 18**, L75.

Flory, P. (1971) *Principles of Polymer Chemistry*. Cornell University Press, Ithaca, NY.

Golubovic, L., and R. P. U. Karunasiri (1991) *Phys. Rev. Lett.*, **66**, 3156.

Grinfeld, M. A., (1985) *Dokl. An SSSR*, **283**, 1139. [*Sov. Phys. Dokl.*, **31**, 831].

Grinfeld, M. A. (1986) *Dokl. An. SSSR*, **290**, 1358.

Halsey, T. C., P. Meakin, and I. Procaccia (1986) *Phys. Rev. Lett.*, **56**, 854.

Kardar, M., G. Parisi, and Yi-Cheng Zhang (1986) *Phys. Rev. Lett.*, **56**, 889.

Karunasiri, R. P. U., R. Bruinsma, and J. Rudnick (1989) *Phys. Rev. Lett.*, **62**, 788.

Krim, J. (1992) In *Proceedings of the Les Houches Workshop on Surface Disordering: Growth, Roughening and Phase Transitions, March 31–April 9, 1992, Les Houches, France*, R. Jullien, J. Kertesz, P. Meakin and D. E. Wolf (eds).

Krug, J., and P. Meakin (1989) *Phys. Rev.*, **A 40**, 2064.

Krug, J., and P. Meakin (1991) *Phys. Rev.*, **A 43**, 900.

Krug, J., and H. Spohn (1991) In *Solids Far From Equilibrium: Growth, Morphology and Defects*, C. Godreche (eds), Cambridge University Press, Cambridge.

Langer, J. S. (1980) *Rev. Mod. Phys.*, **52**, 1.

Ma, S.-K. (1976) *Modern Theory of Critical Phenomena*. Benjamin, Reading, Mass.

Mandelbrot, B. (1982) *The Fractal Geometry of Nature*. Freeman, San Francisco.

Matsushita, M., M. Sano, Y. Hayakawa, H. Hondo, and Y. Sawada (1984) *Phys. Rev. Lett.*, **53**, 286.

Messier, R. (1986) *J. Vac. Sci. Technology*, **A 4**, 490.

Messier, R., and J. E. Yehoda (1985) *J. Appl. Phys.*, **58**, 3739.

Mullins, W. W. (1957) *J. Appl. Phys.*, **28**, 333.

Muthukumar, M. (1983) *Phys. Rev. Lett.*, **50**, 839.

Niemeyer, L., L. Petronero, and H. J. Wiesman (1984) *Phys. Rev. Lett.*, **52**, 1033.

Schwoebel, R. L. (1969) *J. Appl. Phys.*, **40**, 614.

Siegert, M., and M. Plischke (1992) *Phys. Rev. Lett.*, **68**, 2035.

Sneppen, K., J. Krug, M. H. Jensen, C. Jayaprakash, and T. Bohr (1992) *Phys. Rev.*, **A** (Dec.).
Tang, C., S. Alexander, and R. Bruinsma (1990) *Phys. Rev. Lett.*, **64**, 772.
Tokuyama, M., and K. Kawasaki (1984) *Phys. Lett.*, **100 A**, 337.
Tong, W. M., E. J. Snyder, R. S. Williams, A. Yanase, Y. Segawa, and M. S. Anderson (1992) *Surf. Sci. Lett.*, **277**, L63.
Turkevich, L. A., and H. Scher (1985) *Phys. Rev. Lett.*, **55**, 1026.
Vicsek, T. (1989) *Fractal Growth Phenomena.* World Scientific, Singapore.
Villain, J. (1991) *J. Phys. (Paris)*, **I 1**, 19.
Wilson, K. G., and J. Kogut (1974) *Physics Reports*, **12 C**.
Witten, T. A., and L. M. Sander (1981) *Phys. Rev. Lett.*, **47**, 1400.
Wolf, D. E., and J. Villain (1990) *Europhys. Lett.*, **13**, 389.
Yakhot, M. (1981) *Phys. Rev.*, **A 24**, 642.
Yehoda, J. E., and R. Messier (1985) *Appl. Surf. Sci.*, **22/23**, 590.

AUTHOR INDEX

Note: The page on which the reference appears in full is shown in italic type.

SUBJECT INDEX

Indexes compiled by G. Jones

CONTENTS OF PREVIOUS TITLES

CLUSTER IONS

Edited by C. Y. Ng, T. Baer and I. Powis